Integration of Large Scale Wind Energy with Electrical Power Systems in China

Integration of Large Scale Wind Energy with Electrical Power Systems in China

Zongxiang Lu and Shuangxi Zhou

Tsinghua University, China

WILEY

中国电力出版社
CHINA ELECTRIC POWER PRESS

This edition first published 2018 by John Wiley & Sons Singapore Pte. Ltd under exclusive licence granted by China Electric Power Press for all media and languages (excluding simplified and traditional Chinese) throughout the world (excluding Mainland China), and with non-exclusive license for electronic versions in Mainland China.

Registered Offices
John Wiley & Sons, Inc., 111 River Street, Hoboken, NJ 07030, USA
John Wiley & Sons Singapore Pte. Ltd, 1 Fusionopolis Walk, #07-01 Solaris South Tower, Singapore 138628

Editorial Office
1 Fusionopolis Walk, #07-01 Solaris South Tower, Singapore 138628

For details of our global editorial offices, customer services, and more information about Wiley products visit us at www.wiley.com.

Wiley also publishes its books in a variety of electronic formats and by print-on-demand. Some content that appears in standard print versions of this book may not be available in other formats.

Library of Congress Cataloging-in-Publication Data

Names: Lu, Zongxiang, 1974- author. | Zhou, Shuangxi, 1941- author.
Title: Integration of large scale wind energy with electrical power systems in China / by Zongxiang Lu, Shuangxi Zhou.
Description: First edition. | Singapore ; Hoboken, NJ : John Wiley & Sons, 2018. | Includes bibliographical references and index. |
Identifiers: LCCN 2017055294 (print) | LCCN 2018004506 (ebook) | ISBN 9781118910078 (pdf) | ISBN 9781118910085 (epub) | ISBN 9781118910009 (cloth)
Subjects: LCSH: Hybrid power systems–China. | Wind power–China.
Classification: LCC TK1541 (ebook) | LCC TK1541 .L8 2018 (print) | DDC 621.31/21360951–dc23
LC record available at https://lccn.loc.gov/2017055294

Cover design by Wiley
Cover image: © breath10/GettyImages

Set in 10/12pt WarnockPro by SPi Global, Chennai, India

10 9 8 7 6 5 4 3 2 1

Contents

1 Overview

1.1 Wind Energy and Wind Energy Resources

1.1.1 Basic Concepts

Wind energy is the energy of moving air. In a broad sense, wind energy is derived from solar energy. The radiant energy from the sun is constantly transmitted to the earth's surface. Solar radiation does not heat every part of the earth's surface evenly resulting in differences in temperature and pressure and forming the wind.

According to aerodynamic theories, the moving air has energy and the wind energy flowing through the section perpendicular to the wind speed A (m^2) per unit time, namely, the wind power [1], is (Formula (1.1))

$$E = \frac{1}{2}mv^2 = \frac{1}{2}\rho A v^3 \quad (1.1)$$

where E is wind energy with the unit W (kg·m^2·/s^3); m is air mass (kg/m^3); v is air flow speed, namely, wind speed (m/s); A is the area of the section the air passing through that is perpendicular to the direction of the air flow (m^2); ρ is the density of air (kg/m^3).

Wind power is proportional to the square of the wind speed while wind energy (kinetic energy) is proportional to the third power of the wind speed. That is to say, if wind speed doubles, wind power output increases eight times.

Wind energy is a renewable energy source. From a long-term perspective, wind energy is inexhaustible. Meanwhile wind energy is a form of process energy that cannot be directly stored. Instead it has to be converted into other forms of energy in order to be stored.

According to different requirements, wind energy can be converted into a great variety of energy forms, including mechanical, electrical, thermal energy, and so on, in order to achieve pump water irrigation, power generation, sail-assisted navigation, and other functions.

Wind energy resources are kinetic energy resources created by the movement of the air across the surface of the earth. The formation of wind energy resources is affected by multiple natural factors, especially the climate, terrain, and sea and land location. Wind energy is widely distributed in space and meanwhile it is unstable and discontinuous. Since wind is very sensitive to the climate, it is variable and varies from region to region and season to season.

There is an abundance of wind energy resources in nature. According to the World Meteorological Organization (WMO), global wind energy totals 3 × 10^{17} kW of which 2 × 10^{10} kW is exploitable wind energy, 10 times more than the total amount of exploitable hydro energy on the earth [1]. The amount of technologically exploitable wind energy resources totals about 53 TW·h/year (1 TW = 10^{15}W = 10^{12}kW), equivalent to more than two times the world's total electricity demand in 2020 [2]. China is rich in wind energy resources: the total reserves of wind energy resources at an altitude of 10 m above the land are estimated to be 3,226 GW; the reserves of the exploitable onshore wind energy are 253 GW and the reserves of exploitable offshore wind energy are 750 GW, totaling 1,000 GW [3, 4].

The potential for wind energy resources in a certain area of the earth is expressed by the wind energy density and available hours in this area.

Wind energy density is the kinetic energy of the moving air perpendicularly passing through the unit section per unit time, namely, the wind power density. If the area A = 1 in Formula (1.1), then the wind energy density (Formula 1.2) is shown as

$$w = \frac{1}{2}\rho v^3 \quad (\mathrm{W/m^2}) \quad (1.2)$$

Wind energy density is also changing with time and with the change of wind speed. The average value of the wind energy density over a certain period of time (e.g., one year) is called the average wind energy density that is shown in Formula (1.3):

$$\overline{w} = \frac{1}{T}\int_0^T \frac{1}{2}\rho v^3(t)dt \quad (1.3)$$

where $\overline{w}$ is the average wind energy density; T is a certain period of time; $v(t)$ is the wind speed changing

Integration of Large Scale Wind Energy with Electrical Power Systems in China, First Edition. Zongxiang Lu and Shuangxi Zhou.

with time; dt is the duration of a certain wind speed within T. If in the wind speed measurement the wind speeds $v_1, v_2, \ldots, v_n$ and their corresponding durations $t_1, t_2, \ldots, t_n$ within T can be directly obtained (or after data processing), then the average wind energy density can be calculated by Formula (1.4):

$$\overline{w} = \frac{\sum_{i=1}^{n} \frac{1}{2}\rho v_i^3 t_i}{T} \tag{1.4}$$

In the actual use of wind energy, wind turbines only work within a certain range of wind speeds. The wind energy density within a certain range of wind speeds is regarded as effective wind energy density. In China the range of wind speeds corresponding to the effective wind energy density is 3-20 m/s [1, 5].

The air density (ρ) can be calculated by a great variety of formulas that vary in complexity, parameters, and accuracy. Usually the more parameters the formula has, the more accurate it is. It's suggested Formula (1.5) should be used to calculate the values in reference [1].

$$\rho = \frac{1.276}{1 + 0.00366t} \times \frac{p - 0.378e}{1000} \tag{1.5}$$

where ρ is the average air density (kg/m^3); p is the average air pressure (hPA); e is the average water vapor pressure (hPa); t is the average Celsius temperature (°C).

The air density varies with altitude. At an altitude of below 500 m, that is, at a normal temperature and under standard atmospheric pressure, the air density is 1.225 kg/m^3. If the altitude is above 500 m, the relationship between the air density and the altitude can be calculated according to the experience of China's meteorological stations (Formula (1.6)) [5]:

$$\rho_z = 1.225e^{-0.0001z} \tag{1.6}$$

where ρ_z(kg/m^3) is the air density at an altitude of z(m).

Wind speed and wind direction are two important factors in the utilization of wind energy. In order to estimate wind energy resources, we must measure the daily and annual wind speed and wind direction and understand their changing laws. The wind direction in a certain area of the earth is, first of all, related to the atmospheric circulation. Besides, it is also related to the geographical location (its distance from the equator and the south and north poles) and the earth's surface (ocean, land, valley, etc.).

The fundamental basis for the calculation of wind energy resources is the hourly wind speed that can be calculated in three ways: ① average the measured hourly wind speeds; ② average the wind speeds measured in the last 10 minutes of each hour as required in China; ③ average the several selected instantaneous wind speeds in each hour.

Wind speed varies with height. From the surface of the earth to the upper air layer at an altitude of 10,000 m, the moving of the air is affected by factors such as the eddy, viscosity, and surface friction. The higher it is above the earth's surface, the higher the wind speed is. In engineering the index method is usually used to express the change of the wind speed with height (Formula (1.7)):

$$v = v_1\left(\frac{h}{h_1}\right)^n \tag{1.7}$$

where h, h_1 are different heights from the earth's surface; v_1 is the wind speed at a height of h_1 above the earth's surface; v is the wind speed at a height of h above the earth's surface that is to be calculated; index n is related to the surface evenness (roughness), the stability of the atmosphere and other factors, ranging from one-half to one-eighth and being one-seventh in areas with normal stability. China's meteorological departments measure the wind speeds at various heights and calculate the average value of n to be between 0.16 and 0.20 that can be used to estimate the wind speed at different heights. Obviously, the higher the wind turbines are placed, the more wind energy they can capture.

Wind direction is usually expressed using 16 directions. The diagram based on the frequency of winds blowing from different directions is called **wind direction frequency rose diagram**. Shown in Figure 1.1 is the wind rose diagram that displays the average wind direction and corresponding average wind speed at Lvsi Ocean Wind Measurement Station (see Table 1.1 for the corresponding data). It can be seen that according to the average annual frequency of wind direction measured over many years, the prevailing wind directions are N, ESE, and SSE and the corresponding frequency is

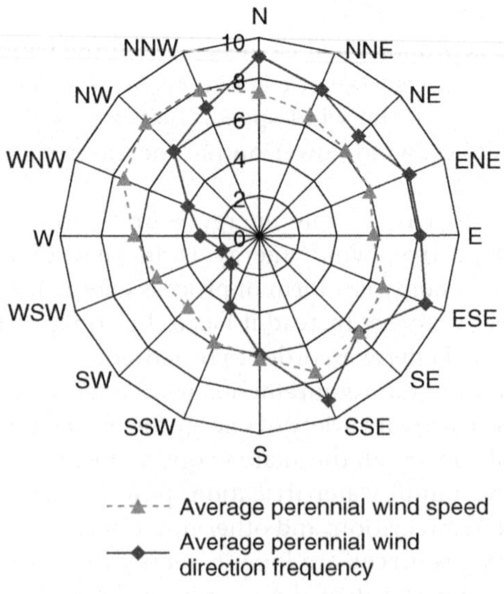

Figure 1.1 Wind direction frequency rose diagram and corresponding average wind speed rose diagram.

Table 1.1 Average perennial wind direction frequency (f_w) and corresponding wind speed (v).

Wind direction	N	NNE	NE	ENE	E	ESE	SE	SSE	S	SSW	SW	WSW	W	WNW	NW	NNW	C
f_w (%)	9	8	7	8	8	9	7	9	6	4	2	2	3	4	6	7	1
v (m/s)	7.3	6.6	6.0	5.9	5.7	6.7	6.9	7.5	6.3	5.8	5.1	5.6	6.3	7.4	8.1	7.9	0.0

9%; the secondary prevailing wind directions are NNE, ENE, and E and the corresponding frequency is 8%; the wind direction NW has the greatest average wind speed, 8.1 m/s, followed by the wind direction NNW, which has the average wind speed of 7.9 m/s. It shows that prevailing wind directions are different from strong wind directions, but generally wind directions from NNW to SSE have higher frequency and wind speed, with the frequency and average wind speed being 72% and 6.7 m/s respectively. The wind rose diagram can accurately display the distribution of wind frequency in a certain area so as to determine the overall arrangement of wind turbines in a wind farm (WF) and facilitate WF micro-siting. It plays an important role in the initial design of WF construction and wind power forecasting.

Generally, a WF consists of a large number of wind turbines. Due to the influence of the wake of the upwind wind turbines, downwind wind turbines capture less wind energy and correspondingly the output of wind turbines is also reduced. In order to have a quantitative analysis of the impact of wind directions on the output power of a WF, the efficiency coefficient of the WF η is defined as (Formula (1.8))

$$\eta = \frac{P_m}{P_f} \tag{1.8}$$

where P_m is the actually measured output power of the WF when the wind blows at a certain speed from a certain direction; P_f is the output power of the WF when the wind blows at a certain speed from a certain direction without the influence of the wake. Shown in Figure 1.2 is the generation efficiency of a WF, with the efficiency at the circumference being 1, namely, 100%. It can be seen from Figure 1.2 that when the wind speed low, for example, 4 m/s, due to the influence of the wake and the surface roughness, the generation efficiency of the WF is low when the wind blows from certain directions. Meanwhile it can be seen that the higher the wind speed is, the higher the efficiency coefficient of the WF is. When the wind speed is higher than the rated wind speed by a certain amount, for example, 18 m/s, the wind speed of the downstream wind turbines will also be higher than the rated wind speed. In this case the wake effect does not affect the output power and the efficiency coefficient of the WF is always 100% when the wind blows from any direction.

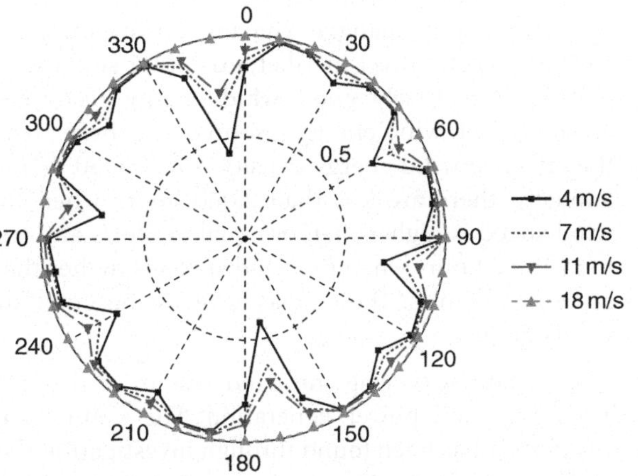

Figure 1.2 Generation efficiency of WF with different wind speeds and wind directions.

1.1.2 Distribution of Wind Energy Resources in China

Wind energy resources in China are distributed intensively in following areas:

1) China's southeast coastal areas and nearby islands abound in wind energy resources. The isoline of the effective wind energy densities higher than or equal to 200 W/m^2 is parallel to the coastline. In offshore islands the wind energy density is above 300 W/m^2; annually there are 7,000 to 8,000 hours when the wind speed is higher than or equal to 3 m/s and 4,000 hours when the wind speed is higher than or equal to 6 m/s.
2) Northern Xinjiang, Inner Mongolia and northern Gansu are also rich in wind energy resources. In these areas the effective wind energy density is higher than or equal to 200 to 300 W/m^2; annually, there are over 5,000 hours when the wind speed is higher than or equal to 3 m/s and more than 3,000 hours when the wind speed is higher than or equal to 6 m/s.
3) Heilongjiang, western Jilin, northern Hebei, and Liaoning Peninsula also have an abundance of wind energy resources. In these areas the effective wind energy density is above 200 W/m^2; annually there are 5,000 hours when the wind speed is higher than or equal to 3 m/s and 3,000 hours when the wind speed is higher than or equal to 6 m/s.

4) In northern Qinghai-Tibet Plateau, the effective wind energy density ranges from 150 to 200 W/m^2. In these areas the effective wind energy density is from 150 W/m^2 to 200 W/m^2; annually, there are 4,000 to 5,000 hours when the wind speed is higher than or equal to 3 m/s and 3,000 hours when the wind speed is higher than or equal to 6 m/s.
5) Yunnan, Guizhou, Sichuan, Gansu, southern Shanxi, Henan, western Hunan, Fujian, Guangdong, mountainous areas in Guangxi, Xinjiang Tarim Basin, and Yarlung Zangbo River in Tibet, except for some areas that have relatively good wind energy resources, generally lack wind energy resources. In these areas, the effective wind energy density is below 50 W/m^2; annually, there are less than 2,000 hours when the wind speed is higher than or equal to 3 m/s and less than 1,500 hours when the wind speed is higher than or equal to 6 m/s. These areas have low potential for wind energy.

China is still surveying the wind resources that can be used for wind power generation. For example, in recent years it has been found through investigation that coastal areas in Guangdong have abundant wind energy resources. In these areas, the average annual wind speed is 6 to 7 m/s or above; the annual effective power generation time is about 7,500 hours; the area where wind turbines can be installed reaches 539 km^2. The available installed wind capacity in eastern and western Guangdong and coastal areas in the Pearl River Delta is over 11 GW. According to the results of the wind resource engineering survey of the nine large wind areas including the Dabancheng Valley in Xinjiang, the area where wind turbines can be installed is estimated to be 54,000 km^2 and the available installed capacity is about 540 GW.

1.2 Characteristics of Wind Power Generation

Wind power generation is different from conventional power generation both technologically and economically, in terms of its power resource, the wind energy conversion system, operating characteristic of its system and its electrical power output. Compared with conventional power generation, wind power has both some outstanding advantages and obvious disadvantages.

1.2.1 Advantages

One of the reasons that wind power has developed very rapidly worldwide is that it has the following advantages.

1) **There are abundant wind energy resources worldwide.** According to statistics, the global wind energy potential is five times the current global electricity consumption. Countries where wind energy resources can meet a part of or most of the electricity demand include Argentina, Canada, Chile, China, Russia, Britain, Egypt, India, Mexico, South Africa, and Tunis, etc. In these countries 20% or more of the electricity is supplied by wind power.
2) **Wind energy is renewable energy.** Conventional energy (coal, oil, etc.) that can be used on the earth is becoming increasingly scanty and will be exhausted sooner or later. However, from a long-term perspective, wind energy is inexhaustible.
3) Wind energy is clean and pollution-free. According to the estimation by the "Greenpeace" and the European Wind Energy Association, by 2020, wind power will be able to meet 10% of the world electricity demand and reduce global carbon dioxide emissions by more than 10 trillion tons.
4) **Construction period of WFs is short.** Wind turbines are produced industrially with simple site treatment and short installation and construction period. The single wind turbine can be delivered and installed in no more than three months. The construction period of a 10 MW WF is less than one year. In addition, after a wind turbine is installed, it can be immediately put into operation.
5) **Wind power development is characterized by less and flexible investment and short payoff period.** WFs can vary greatly in size. Households and villages can invest in building mini- and small-sized WFs while large-sized WFs can be jointly built by the central government, the state-owned enterprises and individual enterprises.
6) **Wind power development takes up less land and has lower requirements for land.** The wind power unit and monitoring, substation equipment, such as building accounts for about 1% of the WF's area. The other area can still be used for farming, grazing and fishing. What's more, WFs can be constructed in areas with various terrain conditions such as hills, seaside, riverbank, and desert.
7) **WFs can be easily operated.** The whole process of production and management is highly automatized and WFs can be operated unattendedly. Compared to thermal power, wind power need less manpower with the same installed capacity.
8) **Wind power generation technologies have become quite mature.** In the past two decades, significant breakthroughs have been made in the commercially operated wind turbines. The availability (for wind turbine generation system, WTGS) has increased from 50% to 90% and the wind energy utilization coefficient has exceeded 40%. The computer monitoring technology has been used to realize the self-diagnostic function of wind turbines, improve safety protection measures and achieve independent control of single wind turbine, cluster control and remote control of

multiple wind turbines. The designed life of wind turbine can reach 20 years and even 30 years. At present multiple series of 100 kW-level wind turbines have been commercialized and 1 MW-level wind turbines have also been massively produced and widely used in large WFs and offshore WFs.

9) **Wind power generation is cost-effective.** At present the cost of wind power generation in European and American countries is lesser than that of oil and gas power generation. With the development of wind power and as time goes on, the price of wind power will continue to fall.

1.2.2 Disadvantages

Meanwhile due to the limitations of wind energy, wind power generation also has some disadvantages:

1) **Small energy density.** In order to obtain the same generating capacity, the size of the wind turbine is dozens of times larger than that of the corresponding hydraulic turbine. For example, the diameter of the 3000 kW wind turbine already reaches 100 m and the limit of the single wind turbine is about 10 MW. As a result, for the electrical power system, wind turbines can only be small generator unit.
2) **Fluctuation and variability.** Wind speed is characterized by fluctuation, variability, and difficult to accurately predict. Accordingly, the output of wind turbines is also mutable and random.
3) **Single wind turbine has small capacity and low efficiency.** In theory the maximum = efficiency of the wind turbine is 59%, but actually it can only reach 40%.
4) **Impact on the ecological environment.** Factors such as shadow flicker, visual effects and harmony with the environment nearby, and so on need to be considered. Wind turbines have electromagnetic noise and should not be installed in residential areas.
5) **The grid integration of the wind power have an adverse impact on the stable operation of the grid and power quality.**
6) **Prime mover is uncontrollable.** Wind power generation uses wind as the prime mover while wind is uncontrollable. Generally, the cut-in wind speed of wind turbines is 3 m/s and the cut-off wind speed is 25 m/s. In other words, the effective speed zone is 3-25 m/s. It is very difficult to adjust and control the wind speed to get stable output. Under the existing technical conditions, adjustment can only be made within a limited range (for example, changing the absorbed wind energy by changing the pitch angle of the wind turbine blade).
7) **So far wind energy cannot be directly stored in large quantities.** Small wind turbines can be equipped with storage battery while large wind turbines must be operated in combination with other power generation modes or operate through integration with large power grid.

1.3 Present Situation and Development of Wind Power Generation

1.3.1 Present

Environmental pollution and energy shortage have become the century problems of the modern civilized society. People have increasingly raised their environmental crisis awareness. Since the mid-1980s more and more attention has been paid to the application of the wind power generation technologies. With the rapid development of modern science and technology, especially the research on aerodynamics, space technology and the application of high-power electronic technology in new types of wind turbines, wind power generation has developed very rapidly in a short span of 10 to 20 years.

With the development of large-scale and industrialized wind power generation, large grid-connected WFs have become the mainstream of wind power generation. As the proportion of wind power in the grid is becoming increasingly large, wind power generation has become the most mature and most realistic clean energy generation mode except the hydro power generation. Many countries have made plans to speed up wind power development and issued various policies and regulations to promote wind power development. Since the 1990s, capacity of wind power generation has grown by 22% annually on average. The average annual growth rate between 1999 and 2005 is 30%, ranked first in a great variety of power generation modes. By the end of 2007 the global installed wind power capacity had reached 94 GW. In 2007 the global newly installed wind power capacity exceeded 20 GW, which means an increase of 27% as compared with 2006. In 2009 the global newly installed wind power capacity reached 27 GW, growing by 35% as compared with 2007. By the end of 2008 the world's installed wind power capacity had reached 120.79 GW. In 2010 the global newly installed wind power capacity was 35.8 GW and China's newly installed wind power capacity was 16.5 GW, accounting for 46% of the world total.

In 2003, European Wind Energy Association proposed the objective of installing 75 GW (75×10^6kW) of wind energy for 2010, accounting for 10.6% of Europe's installed power capacity and installing 180 GW (1.8×10^8kW) of wind energy for 2020, accounting for 21% of Europe's installed power capacity. America has planned to reach 300 GW of installed wind capacity by 2030 so that wind power will supply 20% of America's electricity [6].

China's wind power generation started in the 1980s. Since the first demonstrative WF was established in Rongcheng, Shandong province, in 1986, with the development of more than 20 years, the installed capacity of WFs has constantly increased. Between 2005 and 2009 China's installed grid-connected wind power capacity (excluding that of Taiwan province) had grown by over 90% annually on average. The installed grid-connected wind power capacity was 3,304 MW in 2007 and 6,246 MW in 2008. In 2009 the installed grid-connected wind power capacity was 12.02 GW and the newly installed wind power capacity increased by 92.4% as compared with 2008. By the end of 2009 China's cumulative grid-connected wind power capacity had totaled 22.68 GW and in 2009 wind power generation capacity reached 27 TWh.

The installed wind capacity is mainly distributed in the "Three-North" (Northwest China, North China, and Northeast China) areas and eastern coastal provinces and regions. In these areas, the installed grid-connected wind power capacity in Inner Mongolia, Liaoning, Jilin, Heilongjiang, and Hebei exceeds 1 GW and that in Inner Mongolia is above 7 GW. Based on the results of the surveys conducted in the past few years and the fact that China's wind resources are relatively concentrated, since 2008, China has begun the planning and construction of 10 GW-level wind power bases in Inner Mongolia, Gansu, Xinjiang, Hebei, Qinghai, Jilin, Jiangsu, and other areas where wind resources are concentrated. In these areas matching transmission lines will be constructed according to the wind power development mode of "constructing large bases and integrating wind power into large grids." In 2009, China started the construction of the transmission lines for its first 10 GW-level wind power base, Gansu Jiuquan Wind Power Base.

In the *Medium- and Long-Term Development Plan for Renewable Energy* issued by the National Development and Reform Commission in 2007, China's medium- and long-term objectives for wind power development are: by 2020, the installed wind power capacity and wind power generation capacity will reach 30 GW and 60 TWh respectively, accounting for 3% of China's total installed capacity (1,000 GW) and 1.2% (suppose China's average annual equivalent full-load hours of wind power are 2,000 hours and average annual equivalent full-load hours of power generated using other energy sources are 5,000 hours) of China's total power generation capacity (5,000 TWh) respectively; by 2030, the installed wind power capacity and wind power production will reach 100 GW and 200 TWh respectively; by 2050, the installed wind power capacity and wind power production will reach 400 GW and 1,000 TWh respectively. These objectives will be achieved in advance. According to the estimation by China's wind power industry at the beginning of 2010, by 2020, China's installed grid-connected wind power capacity will reach 102 GW of which 48 GW will be accommodated locally within the province itself and 54 GW will be accommodated in a wider range through trans-regional power grid. By 2030, China's installed grid-connected wind power capacity will reach 160 GW.

Wind power generation has broad development prospects in China. There are two reasons. First of all, China is rich in wind resources and has great potential for wind power development. China's total reserves of wind energy resources are ranked the third in the world, only preceded by America and Russia. Second, the encouragement of the Chinese government opens green channels for China's new energy power generation. Both the National Development and Reform Commission and the Ministry of Science and Technology have issued some policies to encourage power generation using wind energy and other renewable energy sources.

1.3.2 Development Trends

At present the general trends for the development of global grid-connected wind turbines and wind power plants are as follows [7–9]:

1) **Development from small unit capacity to large unit capacity**
   In the early stage, most main wind turbines operating in China's WFs were below the level of 1 MW. However, in recent years their capacity has increased to the level of 1 MW and above. Judging from the wind turbine technology abroad, as the 1 to 2 MW wind turbine has become technologically mature, it has been put into commercial operation and widely used in WFs. The 5 MW large wind turbine was put into operation in Germany in 2005. The rotor diameter of this kind of wind turbine is 126 m and its control panel is 120 m high. The increase in the capacity of the single wind turbine will help to improve the wind energy utilization, reduce the unit cost, expand the scale effect of WFs, and reduce the occupied area of WFs.
2) **Development from the fixed pitch technology to the variable pitch and variable speed and constant frequency technologies**
   Wind energy is a form of energy characterized by low energy density and poor stability. As wind speed and wind direction change randomly, the angle of attack of the blade also changes constantly, which leads to the fluctuation of the efficiency and power of the wind turbine generators, causing the oscillation of the driving torque and affecting the quality of output power and the stability of the power grid. With the development of the wind power technologies,

nowadays a lot of wind turbines use the pitch control technology so that the setting angle of the blade can be changed according to the random variation of the wind speed and the attack angle of the air flow can be kept in a reasonable range when the wind speed changes. As a result, it is possible to maintain good aerodynamic characteristics in a large range of wind speeds and obtain higher efficiency. Particularly when the wind speed is higher than the rated wind speed the output power can still remain stable. The Variable Speed Constant Frequency technology has also been developed based on the pitch control technology so that the rotating speed of the wind turbines can change with the wind speed in order to further improve the efficiency of the wind turbines.

3) **Development from onshore wind power to offshore wind power**
With the wind power development, the size of WFs and the capacity of single wind turbine have kept increasing. In addition, onshore WFs are restricted by environmental factors. As a result, people have begun to develop offshore WFs. Generally, 2.0 MW is thought to be the limit of the onshore WF development. It's because the length of the blade of giant wind turbines will reach 60 to 70 m, posing great difficulties for transportation by land, and the capacity of the cranes used to install the wind turbines will exceed 1,200 to 1,400 tons. Most areas cannot meet these requirements. The noise and huge volume of the wind turbines will make site selection for WFs and transportation of the wind turbines very difficult. However, these problems can be easily solved in building offshore WFs because maritime transport is convenient and floating cranes of over 1,500 tons are very prevalent. More importantly the offshore wind speed is high and stable; the average annual utilization hours can reach 3,000 hours and the annual offshore wind power production can be 50% higher than the annual onshore production.
A lot of countries worldwide are actively developing offshore WFs. Offshore wind power development has advantages such as high wind speed, large power generation capacity, small-scale turbulence, reduced fatigue load, and extended service life of wind turbines. However, the cost of integrating offshore WFs into the power system and foundation of offshore wind turbines is quite high. The construction of offshore WF meets many challenges. For example, the design of the basic structure for the highly reliable offshore wind turbines requires a long service life so that after the first wind turbine is scrapped, the second wind turbine can continue to use the original basic structure, and super large wind turbines with the unit capacity of 2,000 to 5,000 kW need to be developed in order to make full use of the offshore wind energy. Due to its long coastline, China has abundant exploitable wind energy resources. As a result, developing offshore WFs is one of the development directions of China's wind power development.

4) **Development of the compact, flexible, and lightweight wind turbine structure design**
With the growing unit capacity of wind turbines, wind turbines are required to be compact, flexible, and lightweight in terms of structure design, especially the structure design of their top, so that they can be easily transported and installed. Many wind turbine manufacturers have begun and will continue working on these aspects.

5) **WFs will become conventional power plants**
Grid-connected large-scale WF poses great challenges to the operation and control of the grid. From the perspective of the power system operation, the main differences between wind power and conventional power generation are as follows: ① Wind resources used for wind power generation are unstable and variable. However, the supply of resources for conventional power plants, such as coal, oil, gas, water, and nuclear fuel, is stables. ② Due to the small capacity of the single wind turbine, it is difficult to regulate and control the active and reactive power output. Particularly due to the limits of the wind conditions, generally, the power output cannot be increased. However, the active and reactive power output of conventional power plants can be flexibly regulated in a wide range. ③ When the unstable wind power in the power system increases, the grid reserve capacity should also increase. When no wind power is integrated into the grid, the load consumption power is borne by conventional generating sets and only the load disturbance needs to be considered in setting the system reserve capacity. When wind power is integrated into the grid, the load consumption power is jointly borne by conventional generating sets and wind turbines. What's more, the disturbance of load and wind power needs to be considered in setting the system reserve capacity.
With the increase of the proportion of wind power in the power system, a great variety of specifications have been made for the WF integration. WFs are required to take part in frequency regulation, voltage, and reactive power control and support the grid in case of accidents. WFs should satisfy all the grid codes and be able to be controlled and adjusted as easy as conventional power. For these purposes, main technologies that need to be studied and achieved include wind power forecasting technology, control technology of frequency/active power and voltage/reactive power for wind turbine/WF, wind turbine generator fault ride through technology, wind power-included grid optimal dispatch and operation

technology and demand side management and energy storage technology, and so on.

6) **Declining wind power generation cost**
Compared with the technology of solar energy, biomass, and other renewable energy sources, wind power generation is more mature and cost-effective and has less impact on the environment. In the past two decades, breakthroughs have been constantly made in wind power generation technology and its scale-economy has become increasingly obvious. According to the statistics of United State National Renewable Energy Laboratory (NREL), between 1980 and 2005, the cost of wind power generation was declined by more than 90%, whose declining speed is higher than other renewable energy sources. According to the evaluation of the wind turbines installed in Denmark conducted by RIS National Laboratory of Denmark, between 1981 and 2002, the cost of wind power generation was reduced from 15.80 eurocents/(kW·h) to 4.04 eurocents/(kW·h) and it's estimated that the cost would be reduced to 3.00 eurocents/(kW·h) by 2010 and to 2.34 eurocents/(kW·h) by 2020.
With the improvement of the wind power generation technology, wind turbines will become less expensive and more efficient. The increase in the unit capacity of wind turbines will reduce the investment in the infrastructure. Meanwhile, less wind turbines will be required for the same installed capacity, which also saves the cost. As the financing cost decreases and developers accumulate rich experience, the cost of the wind power project is also reduced. The improvement of the reliability of wind turbines also reduces the average operation and maintenance cost.

1.4 Wind Power Conversion System and Technical Route

1.4.1 Wind Power Conversion System

According to the principle of energy conversion, the wind power conversion system (wind power generation system) converts the wind energy (kinetic energy) collected by wind turbines into rotational mechanical energy, which is transferred to the generator shaft through the transmission device and then the generator converts the mechanical energy into electrical energy. The process is shown in Figure 1.3.

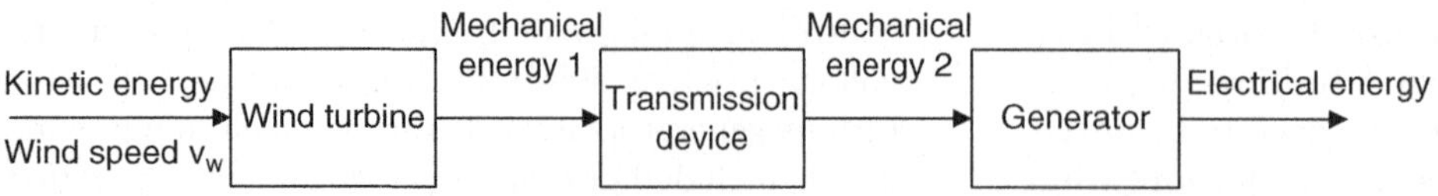

Figure 1.3 Energy conversion process in wind power generation.

As shown in Figure 1.4, the wind power generation system is mainly composed of two core systems, the wind turbine and generator, and the transmission device, control system, energy storage device, standby power supply, and other auxiliary systems.

WTGS (wind turbine generation system) is the key equipment to realize the conversion from wind energy to electrical energy. The wind turbine system includes the blade, hub, principal axis, regulating mechanism by adjusting the pitch of blade (hydraulic or electric servomechanism), yaw mechanism (electrical servomechanism), brake and brake mechanism and wind speed sensor, and so on. The generator system includes the generator, excitation regulator (power electronic converter), grid-connected switch, soft grid-connected device, reactive power compensator, main transformer and speed sensor, and so on.

The transmission device (gearbox) connects the wind turbine and generator (it is not shown in Figure 1.4) and can increase the rotor speed (20-30 r/min) to the generator speed (1,500 r/min), with the speed ratio ranging from 50 to 75. By structure there are mainly two kinds of gear box, two stage helical gear, and helical gears plus planetary gear. The former is more widely used.

As wind resources are variable and the wind speed varies all the time, it's necessary to control the start-up, regulation (of speed, voltage and frequency), outage, fault protection (overspeed, vibration, overload, etc.) of the WTGS and the connecting, regulating, and disconnecting the load based on the change in the wind speed and the electrical energy demand. In the small-capacity wind power generation system, the control device consisting of the relay, contactor, and sensor element is generally used; in the large-capacity wind power generation system, the microcomputer control is widely used.

The energy storage device is designed to ensure users' access to electrical energy in the absence of wind; on the other hand, when the wind energy increases dramatically on account of strong winds, the energy storage device can absorb excess wind energy and smooth the output power. In order to achieve uninterrupted power supply, some wind power generation system is also equipped with the standby power supply, such as diesel generator.

1.4.2 Basic Requirements for Wind Power Generation System

Wind power generation includes the conversion from wind energy to mechanical energy and from mechan-

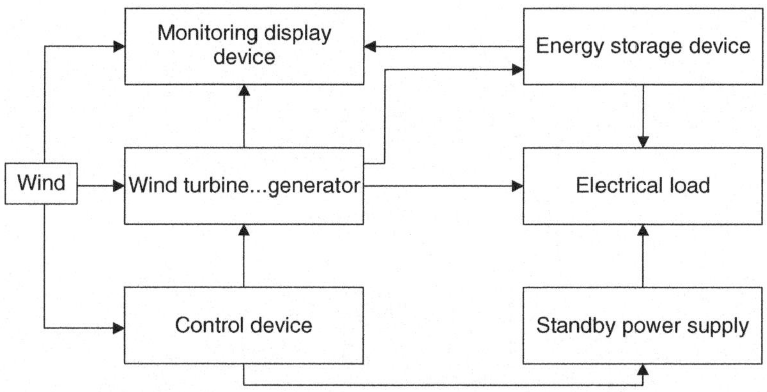

Figure 1.4 Composition of wind power generation system.

ical energy to electrical energy. The latter is achieved through the generator and its control system. The generator system not only directly affects the performance and efficiency of the conversion from mechanical energy to electrical energy and the quality of power supply, but also affects the operation mode, efficiency, and device structure of this conversion. As a result, it is important to develop and select a generator system with reliable operation, high efficiency, and good performance of control and power supply and suitable for the wind power conversion. The main requirements for the generator system are as follows:

1) Converting the mechanical energy of the rotating wind turbine into the electrical energy efficiently;
2) The output electrical energy should meet the requirements of the power system including the frequency, voltage amplitude and waveform, etc.;
3) Operating in combination with the power grid, diesel generator and other power generation devices or the energy storage system in a stable and reliable way to provide users with stable electrical energy supply;
4) Matching with the wind turbine system to maximally utilize the wind power conversion rate of the wind turbine.

1.4.3 Technical Route of Wind Power Generation System

Wind energy can be converted into electrical energy in many ways. Both the constant speed and variable speed wind turbine can be used. The gearbox provides possible solutions for the multi-pole generator system. There are a great variety of wind turbines that have been developed or are under study. Different power electronic converters can be embedded between the power grid and the wind turbine generator system to achieve flexible connection. Power output can be AC or DC. Power converter can also be used for the interface between the WF and the power grid. More detail about the technical route of converting the wind energy (power) into the mechanical power and then into the electrical power can be found in reference [9].

1.5 WF-Included Electrical Power System

The integrated wind power system can be connected to the high-voltage and medium-voltage distribution network to supply power to local load centers. It can also be directly connected to the transmission network to supply power to remote load centers. Shown in Figure 1.5 is the modern WF-included electrical power system.

The integration of wind power into the power system has an impact of varying degrees and different ranges on power generation system, transmission system, and distribution system.

1.5.1 Power Generation System

Modern conventional power plants mainly include thermal power plants, hydraulic power plants, and nuclear power plants. Wind power generation and conventional power generation have different characteristics:

1) Conventional power plants consists of highly concentrated large generator sets with a unit capacity of hundreds of and even thousands of MW. Wind power plants have relatively dispersed small WTGS whose unit capacity ranges from hundreds of kW to several MW;
2) Conventional generator sets are synchronous generators that generate AC that can be directly connected to the AC synchronous power grid. However, WTGS can be synchronous generators or asynchronous generators. Some of the power generated by WTGS can be directly connected to the AC power grid while some must be transformed by the power electronic converter before being connected to the AC synchronous power grid, especially the MW-level wind turbines;
3) Conventional generator sets have stable and reliable power resources while WTGS rely on wind resources that are variable, unreliable, and difficult to control.

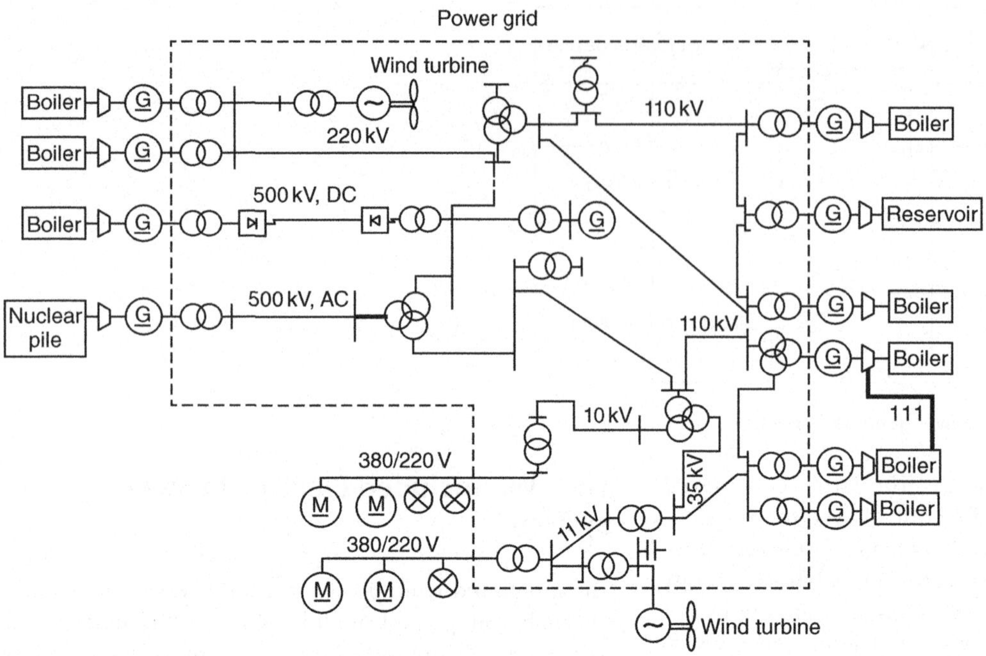

Figure 1.5 WF-included modern electrical power system.

4) Through long-term development and practice, the operation and control technologies of conventional power plants have been quite mature and the power output can be planned and dispatched (increasing or reducing power generation). By contrast, wind power plants lack control technologies and operation experience and most of their power output cannot be planned or dispatched.

These differences pose new challenges to our understanding, operation, and control of the WF-included electrical power system. We need to study the active power and frequency control characteristics and reactive power and voltage control characteristics of WFs (wind power plants) as well as the characteristic interactions between WFs and the electrical power system.

1.5.2 Power Supply and Distribution System

In China the voltage of the power supply network (or high-voltage distribution network) is 35 kV, 66 kV, 110 kV, whereas that of the distribution network is 10 kV and below. When the electrical energy produced by the power generation system is transmitted to the load center through high-voltage transmission lines, the voltage must be reduced to the voltage level of the appliances used by users, which is achieved through the power supply network and distribution network. The power distribution system is required to have high power supply reliability and qualified power quality, namely, to meet the requirements for the voltage level, mutation and flicker of voltage waveform, harmonic content, and other power quality indexes in order to provide users with high-quality services, satisfy their power demands, and boost economic development.

Medium- and small-sized WFs are usually connected to the power supply and distribution system. The impact of the wind power generation system on the power supply and distribution network is also the impact on the local power grid and the impact on the areas electrically adjacent to wind turbines or WFs. The local impact of the wind power on the power supply and distribution system is mainly manifested in the following aspects: ① branch current and node voltage; ② protection scheme, fault current, and rated value of switching equipment; ③ harmonic distortion; ④ flicker, and so on.

For the first aspect, the impact of the wind power on the branch current is mainly determined by the output of the wind turbine generators (or WF) and its impact on the node voltage is determined by choosing to use constant speed or variable speed wind turbines. The rotor speed, active power, reactive power, and the terminal voltage of the constant speed wind turbines-cage asynchronous generator system have a fixed relationship among them and the node voltage cannot be affected by exchanging reactive power with the grid. As a result, it is necessary to add the equipment that generates controlled reactive power. On the other hand, variable speed wind turbines, at least in theory, are capable of changing the reactive power to affect its terminal voltage. However, whether it is feasible depends on the rated value of the power electronic converter and the controller.

The second aspect is related to the fault current. Different types of wind turbines generators make different contributions to the fault current. The constant speed wind turbine is based on the cage induction generator that is directly connected to the power grid. As a result, they make contributions to the fault current and rely on conventional protection schemes (overcurrent, overspeed, over/low voltage) to remove faults.

The double-fed induction generator also makes contributions to the fault current. However, the control system of the converter that controls the rotor current measures different electrical quantities at a very high sampling frequency (several kHz) such as the network side voltage and rotor current so that the fault will be observed soon. Due to the sensitivity of power electronic devices to the overcurrent, at present the double-fed induction generator is cut off rapidly after the fault is detected. It is very difficult for the direct-drive synchronous generator to make contributions to the fault current because the power electronic converter through which the direct-drive synchronous generator is connected to the grid cannot provide the fault current. Generally the direct-drive synchronous generator is also cut off rapidly after the fault occurs.

The third aspect, the harmonic distortion is a problem that occurs in case of variable speed wind turbines because they include power electronic devices that are important harmonic source. However, for modern power electronic converters with high switching frequency and advanced control algorithm and filtering technology, the harmonic distortion should not be a major problem. Well-designed synchronous generators and asynchronous generators that are directly connected to the grid hardly produce harmonic. Therefore, the harmonic distortion is not a problem for the constant speed wind turbines based on asynchronous generators directly connected to the grid.

The last aspect, the flicker, is a unique characteristic of wind turbines generators. Wind is a motive force that changes very quickly. For constant speed wind turbines, the fluctuation of the motive force will directly lead to the fluctuation of the output power because there is no buffer between the mechanical power input and the electric power output. It is related to the strength of connection to the network. The power fluctuation generated may cause the fluctuation of the network voltage, which causes the fluctuation and flicker of the brightness of the light bulb. This problem is called "flicker." Generally, variable speed wind turbines do not have the flicker because the fluctuation of the wind speed does not directly lead to the fluctuation of output power and the rotor inertia serves as the energy buffer.

The impact of diffident types of wind turbines on the local grid is shown in Table 1.2.

1.5.3 Power Transmission System

Wind power is connected to the power distribution network or the power transmission network, but in either way it will make an impact on the power transmission system.

The impact of the wind power on the transmission system is also the impact on the overall behavior of the system. The impact must not be attributed to an individual wind turbine or WF. Instead, it is closely related to the penetration level of the wind power in the system, that is, the contributions of the wind power to the actual load. The main impact on the system includes: dynamic and transient stability of the power system, reactive power and voltage control, frequency control, and load shedding/conventional power generator set dispatching.

Table 1.2 Impact of different types of wind turbines on local grid [10].

Impact on local grid	Fixed speed wind turbine generator	DFIG	Direct-drive wind turbine generator
Voltage control	It is possible only when capacitor bank, SVC and STATCOM are added.	It is possible in theory, but it depends on the rated values of the converter and the controller.	It is possible in theory, but it depends on the rated values of the converter and the controller.
Harmonic	Almost no impact	Having impact in theory, but it should not be a problem.	Having impact in theory, but it should not be a problem.
Flicker	Significant impact, especially in the weak grid.	Due to the role of the rotor as the energy buffer, flicker is not a serious problem.	Due to the role of the rotor as the energy buffer, flicker is not a serious problem.
Contributions to fault current	Yes.	Yes. But at present wind turbines will be rapidly cut off.	No. The converter cannot carry the fault current. At present wind turbines will be rapidly cut off soon.

The cage asynchronous generator uses the fixed speed wind turbine, which may lead to the instability of the voltage and rotor speed. During the fault period, due to the imbalance between the mechanical power absorbed from the wind and the electrical power transmitted to the grid, wind turbines will accelerate. When the voltage is restored, they absorb a lot of reactive power from the system and hinder the recovery of the voltage. When the voltage cannot be recovered very soon, wind turbines will continue to accelerate and consume a lot of reactive power, which will eventually lead to the instability of the voltage and rotor speed. In contrast, for the application of the synchronous generator, as in the low voltage its excitation system will increase the reactive power output of the generator, which can accelerate the recovery of the voltage after the fault occurs.

The sensitivity of power electronic devices of the variable speed wind turbine to the overcurrent caused by the voltage sag may have an adverse effect on the stability of the power system. When the capacity ratio of variable speed wind turbines in the power system (the penetration level is high) is high, at present they will be cut off from the system in case of relatively small voltage sags so that the voltage sags in a wide range may lead to major shortage of power supply. The voltage sags under these circumstances are caused by the faults in the power transmission network. In order to prevent this situation, some power grid companies and power transmission system operators have changed their requirements for the wind power grid integration. They require that wind turbines should be able to withstand a certain amplitude and duration of voltage sags in order to prevent a large number of wind turbines from being cut off from the system in case of faults. In order to meet these requirements, the variable speed wind turbine manufacturers have succeeded in reducing the sensitivity of the variable speed wind turbines to the voltage sag in the network.

First of all, the impact of the wind power on the reactive power generation and voltage originates from the fact that not all the wind turbine can change its reactive power output. Secondly, compared with conventional power generation, wind turbine siting is not very flexible as wind turbines must be installed in places rich in wind resources and WFs must be built in places where they will not affect the landscape. The locations that meet these two conditions are not necessarily good from the perspective of the network voltage control. In selecting the location of the conventional power plant, it is easy to solve the problem of voltage control because the conventional power plant siting is quite flexible. Finally, the coupling between wind turbines and the system is relatively weak because the output voltage of the wind turbines is quite low and these wind turbines are usually installed in remote areas, which further reduces their contributions to the voltage control. When the conventional synchronous generators are replaced by the wind turbines in a large WF in a remote area, the problem of voltage control must be considered.

The impact of the wind power on the frequency control and load tracking is caused by the fact that the wind speed is uncontrollable. So the wind power hardly makes any contributions to the regulation of the primary frequency. In addition, the variability of wind in a long period of time (15 minutes to one hour) makes it complicated to track the load of the conventional power generator sets in the system because the demand curves (each of which is equal to the system load minus the wind power) that these generator sets are required to match are much more unstable than when there is no wind power, which severely affects the dispatching of the conventional power generator sets.

Note that the integrated short-term (<1 min) output power fluctuations of a large number of wind turbine generators are quite stable and generally pose no problem. The power fluctuations of the wind power are caused by the turbulence which is a random quantity and tend to become small when many wind turbines are used to generate power at the same time. As the wind turbine outage resulted from the gusts that exceed the cut-out wind speed is not caused by the random disturbance, it might affect a large number of wind turbine generators at the same time.

When there is a high level of wind power penetration in the system, the impact of wind power on the frequency control and load tracking becomes more severe. The higher the wind power penetration level is, the greater the impact of the wind power on the demand curve to match by conventional generator is. At this point, in order to match the net demand curve and maintain the system frequency fluctuations (caused by the imbalance between the power generation and the load) in the acceptable range, the requirement for the ramp-up capacity of these generator sets must be strict. Due to the differences between different power systems in the composition of the conventional generator, the wind speed, geographical distribution of wind turbines, the demand curve, and the network topology, it is very difficult to quantify the wind power penetration level and the wide range of the impact on the system at this wind power penetration level.

The large WFs directly connected to the power transmission network might also cause problems such as the increase in the change of the power flow of the power transmission system, network congestion, and stability.

1.6 Outline of the Book

This book mainly elaborates on grid-connected WFs (wind power plants) and some issues related to the power system, including the requirements of the power system for WFs and the requirements of WFs for the power system in power system with higher proportion of wind power, the interactions and mutual impact between WFs, and the power system and the analysis, planning, operation, and control of the WF-included power system, etc.

This book altogether has 12 chapters. Chapter 1 briefly introduces the basic concepts and characteristics of wind power generation as well as the rapid wind power development both in China and other foreign countries. Chapter 2 introduces in detail the modern wind power generation system, focusing on the widely used cage asynchronous generator system, doubly-fed induction generator system and direct-drive wind turbine generator system. In addition, it expounds the application of power electronic devices in the wind power conversion system. Chapter 3 explains and compares the plans for WF integration into the power system. Chapter 4 introduces the characteristics of the operation of different grid-connected wind power generation systems, focusing on their power output characteristics and the basic control methods used by the wind power conversion system. Chapter 5 introduces intensively the WF electrical system including the integrated system (main electrical connections scheme), reactive power compensation, grounding system, electrical protection, lightning protection, and energy storage system. Chapter 6 revolves around offshore WFs. It elaborates on the characteristics of offshore WFs that are different from those of onshore WFs, introduces several offshore WFs that have been put into operation, briefly describes the offshore WF electrical equipment and requirements, and discusses the selection of schemas of offshore WF's integration into the power system. Chapter 7 focuses on the analysis of the WF-included electrical power system including the introduction of the model for the WF-included electrical power system research, conventional power flow, and probabilistic power flow analysis, fault analysis, voltage stability analysis, large disturbance, and small disturbance stability analysis and frequency stability analysis, and so on. Chapter 8 discusses the impact of WFs on the power quality, the power quality measurement, and evaluation methods and the measures to improve wind power-related power quality. Chapter 9 revolves around the forecasting of wind speed and output power in WFs. It expounds on the variability and predictability of wind power, characteristics, and basic methods of wind speed and WF power forecasting and the application of wind power forecasting. Chapter 10 discusses WF control and protection issues. Chapter 11 discusses the WF-included electrical power system operation and optimal dispatching strategy and studies the technologies and measures to change WFs into conventional power plants. Chapter 12 describes the assessment techniques for WF-included electrical power system including reliability assessment, confidence level assessment of capacity, and wind power value analysis and finally makes a general analysis of the maximum integration capacity of the electrical power system. These assessment techniques are important to the planning and operation of the wind power's integration into the power system.

References

1 Wang, Chengxu, and Yuan, Zhang. 2003. *Wind Power Generation*. Beijing: China Electric Power Press.
2 Shi, Pengfei. 2003. Forecast of China's wind power development prospects based on the world development trends. *China Electric Power*, **36**(9), 54–62.
3 Bai, Jianhua, *et al.* 2009. Analysis of planning and operation issues on China's large-scale wind power development. *Electric Power Technologic Economics*, **36**(2), 7–11.
4 Hou, Jianzhao, *et al.* 2008. Current situation and policy analysis of the development of world wind energy resources and its enlightenment for China. *China Electric Power*, **41**(9), 65–68.
5 Ni, Anhua. 2005. Introduction to wind power. *Shanghai Medium and Large Electrical Machines*, **2005**(3), 1–8.
6 Smith, J. Charles, and Parsons, Brian. 2007. What does 20% look like? Developments in wind technology and systems. *IEEE Power & Energy Magazine*, November/December, 22–33.
7 Guan, Wei, and Yan, Lu. 2008. General situation and developing direction of wind power generation in China and foreign countries. *Jilin Electric Power*, **5**(6), 22–33.
8 Rohrig, K., Lange, B., Gesino, A., Wolff, M., Mackensen, R., Dobschinski, J., Wessel, A., Braun, M., Quintero, C., Mata, J.-L., Pestana, R. 2009. Wind power plant capabilities: Operate wind farms like conventional power plants. *EWEC*, **36**(1), 50–53.
9 Iov, F. and Blaabjerg, F. 2009. Power electronics and control for wind power systems. Power Electronics and Machines in Wind Applications, *IEEE*, **24-26**, 1–16.
10 Ackermann, Thomas. 2005. *Wind Power in Power Systems*. New York: John Wiley & Sons.

2

Wind Power Generation and Wind Power Generation System

2.1 Wind Power Generation System and WFs

Although the power generated by wind farms (WFs) is the same as that generated by any other power plants, compared with conventional power plants, WFs have quite different characteristics. This chapter will focus on introducing the concepts of a wind turbine generator system (WTGS) and WFs.

2.1.1 Concept of WTGS

In this book, a wind turbine generator system (WTGS) sometimes has the same meaning as a wind turbine (WT). As the wind power generating technology has evolved, the simple fixed speed wind turbine developed into a complete variable speed system that can control the active power output.

The speed of the fixed speed wind turbines is determined by the grid frequency and cannot be adjusted based on the changes of the wind conditions. As a result, the operating efficiency of wind turbines at the maximum wind speed is usually not the maximum. In addition, at the lower end of the wind speed range, the wind turbine spins too fast while at the higher end of the wind speed range the wind turbine spins too slow. However, practice has proved that this kind of wind turbine has a low cost and high quality and its capacity can reach 2 MW and above after optimization.

As for the variable speed wind turbine, due to the application of the power electronic converter, the grid frequency is decoupled from the instantaneous wind speed and the real-time rotation frequency compulsorily controlled by the wind turbine control system, which can improve the performance of the wind turbine and reduce the mechanical load and provide a better means for the WF to become an actively controllable "power plant."

The blade pitch control system is the basic means to realize the variation of the wind turbine speed and to make the wind turbine power completely controllable from an aerodynamics perspective. It is comparable to the fuel throttle for internal combustion engines and gas turbines. The decoupling of the electrical and the rotor frequency absorbs the wind speed fluctuations, which enables the rotor to act as a (accelerating and decelerating) flywheel, and thus smoothing out spikes in power, voltage, and torque.

Until the end of the 1990s, the fixed speed wind turbines still dominated the market and accounted for about 30% of the actually operating wind turbines. However, now newly installed wind turbines are mainly variable speed wind turbines. Due to historical, commercial and technical factors, wind turbines used in different countries and regions vary greatly. As shown in Table 2.1, wind turbines sold in the market can be categorized by characteristic into Type A, B, C, and D. The wind turbine manufacturing industry has been endeavoring to improve the design of wind turbines and increase the controllability of wind turbines in order to meet various requirements for increasing wind power penetration in the power grid. In 1998, the market share of controllable wind turbines (Type C and D) was very limited, but it grew very rapidly. The installed capacity of these two types of wind turbine accounted for about 42% of the newly installed capacity in 1998 and over 80% of the total capacity for sale in 2004.

2.1.2 Concept of WF

Wind power plant, or WF, is the power plant that consists of multiple wind turbines or a power station consisting of a group of wind turbines or wind turbine clusters in a place rich in wind energy resources.

Wind turbine generators are usually allocated in the form of cluster (WF) ranging from several MW to hundreds of MW. Such a cluster is connected to the power grid in the form of a single generation unit. The design of the WF mainly emphasizes effective and economical energy production and it should also follow the regulation of power grid operators. With the improvement of the wind power penetration level, the requirements of the power grid operators are changing. In order to respond to the change of the requirements, in addition to generation of electrical energy, modern wind

Integration of Large Scale Wind Energy with Electrical Power Systems in China, First Edition. Zongxiang Lu and Shuangxi Zhou.

Table 2.1 Wind turbine generator system and characteristic descriptions [1].

<table>
<tr><th>System type</th><th>Characteristic descriptions</th></tr>
<tr><td>A: Fixed-speed (one or two speeds) generator system
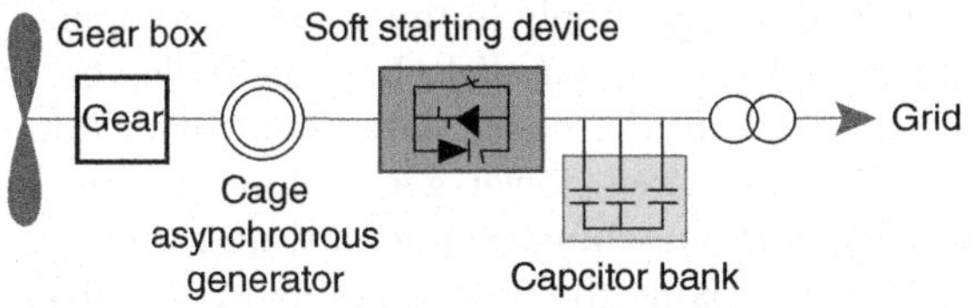
</td><td>It was introduced and widely used in the 1980s. Its concept based on cage asynchronous generator (CAG), its rotor is driven by the wind wheel and the stator is connected directly to the power grid. Its speed can only vary slightly (between 1% and 2%) and is almost “constant” compared with other types of wind turbines. There are two kinds of fixed-speed generator system: single speed and dual speed. Dual speed operation can lead to improved performance and lower noise at a low wind speed. The aerodynamic control associated with Type A WTGS is mainly passive stall. If there is no blade pitch control mechanism, available active control means is very limited, which only contains the connection and disconnection of wind turbines. The active stall structure has been developed and its blade pitch can be switched by the control system to the stall mode.

Manufacturer: Suzlon, Nordex, Siemens Bonus, Ecotècnia

Wind power plant capabilities: voltage control, reactive power control</td></tr>
<tr><td>B: Finite variable speed of wound rotor asynchronous generating system
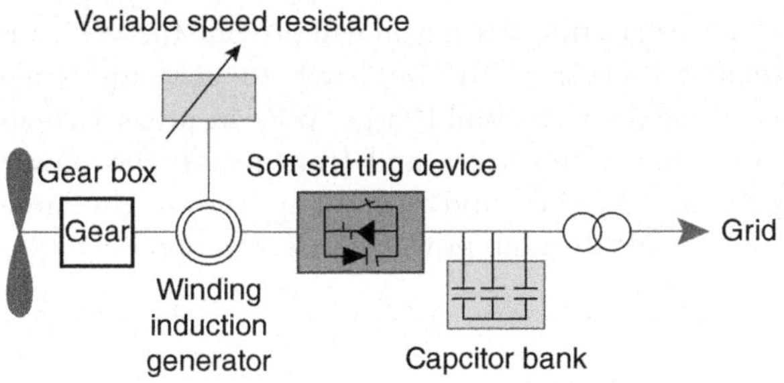
</td><td>Type B wind turbine is the wound rotor induction generator (WRIG) equipped by Vestas in the 1980s and 1990s. Power electronic devices are used to control the rotor resistance and allow the speed of the rotor and generator to change within ± 10% during wind gusts so that the power quality can be optimized and the mechanical stress of the wind turbine components can be reduced (at the cost of slight energy loss). Type B wind turbine is equipped with the active blade pitch control system.

Manufacturer: Vestas (V27, V34, V47)

Wind power plant capabilities: voltage control (power quality)</td></tr>
<tr><td>C: Variable speed generator system with DFIG
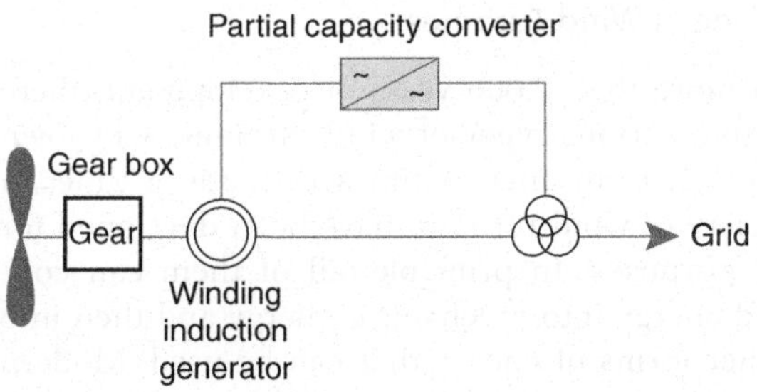
</td><td>Type C is currently the most popular system and combines the advantages of Type A and Type B WTGS. It uses a wound rotor induction generator. Its rotor, connected to the power grid through the back-to-back voltage source converter, controls the rotor excitation system and realizes the decoupling of the mechanical and electrical rotor frequency and the matching of the grid and the rotor frequency. The power electronic devices provide active and reactive power control and can actively control voltage. About 40% of the power output of Type C WTGS is fed into the power grid after being converted and the other power output is fed into the power grid through the stator winding. Its speed change ranges between 40% and −40% of the synchronous speed.

Manufacturer: GE (1.5 and 3.6 MW), Repower, Vestas, Nordex, Gamesa, Ecotècnia, Ingetur, Suzlon

Wind power plant capabilities: reactive power/voltage control, fault ride through (FRT)</td></tr>
<tr><td>D: Direct-drive generator system with full capacity converter
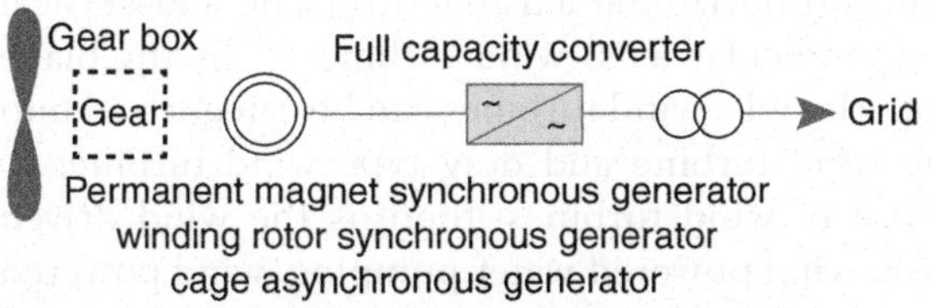
</td><td>Type D wind turbine provides the classical driving chain (geared) and uses the direct-drive concept (using generators that run slowly) and mixed form (low step-up gearbox and medium-speed generator). Generators include wound rotor synchronous generator, permanent magnet synchronous generator and cage induction generator. Type D wind turbine stator is connected to the power grid through the full power converter. Its rotor has excitation winding rotor or permanent magnet. It is completely decoupled with the power grid and can provide a wider range of operating wind speed and reactive power/voltage control capabilities as compared with Type C.

Manufacturer: Enercon, MEG (Multibrid M5000), GE (2.x series), Zephyros, Winwind, Siemens (2.3 MW), Made, Leitner, Mtorres, Jeumont

Wind power plant capabilities: reactive power/voltage control, fault ride through (FRT)</td></tr>
</table>

turbines and WFs are developing toward the so-called wind energy power plant (WEPP) or virtual power plant (VPP), which in essence, has the characteristics of conventional power plants, except that its fuel supply is variable. The wind power plant is designed to provide a certain range of ancillary services for the electrical power system and make positive contributions to the power system stability, fault recovery, and voltage support.

The wind power plant (WF) is equipped with a special WF main controller (WFMC) in order to have control over the whole WF similar to that of conventional power plants, including limiting the generating capacity of the WF when the power output exceeds the maximum power output. The generating capacity of the WF can be regulated up and down to take part in the power balance control and frequency control of the system. In the power grid emergency, the WF output can be regulated down or cut off from the power grid. When the wind speed increases, the output power growth can be limited by WFMC. For the reactive power, the main controller can control each wind turbine through the central control or according to the agreed principle. In case of grid fault or strong winds, the disconnection of wind turbines is individually controlled and then the wind turbines are designed in accordance with the principle that they cannot be cut off at the same time at a high wind speed.

Building WFs can achieve economies of scale. Particularly, it can also save the cost of civil engineering and networking, and lots of purchase discounts can be obtained from wind turbine manufacturers. The economies of scale of building offshore WFs are even better.

2.1.3 Wind Power Generation Operation Mode

The operation modes of wind power generation can be divided into the independent operation mode, complementary operation mode, and grid-connected operation mode.

1) **Independent operation mode**
   The independent operation of a WTGS is a simple operation mode and is used in areas without economical networking power supply, such as remote rural areas, pastoral areas, and islands, to provide residents with electrical power needed for life and production. However, due to the randomness and instability of wind energy and the fluctuation of load, a lot of technical problems need to be solved in the independent operation of wind turbines, including power supply and demand balance and power quality. Independent operation is more difficult than grid-connected operation.
2) **Complementary operation mode**
   The complementary operation mode mainly refers to the combined operation of wind power generation with diesel power generation, photovoltaic power generation, and small hydropower generation. The combined power generation system is aimed to give full play to the advantages of each kind of power generation and achieve complementation of advantages. All kinds of power supply in the system can operate independently or in parallel. Key technologies to be solved include the coordinated control technology of the combined power generation system, the optimal power matching technology, and the technology of improving the power quality in the energy management.
3) **Grid–connected operation mode**
   The grid-connected is the operation mode in which wind turbines are connected to the power grid to transmit power to the power grid. Both large and medium wind turbines can be connected to the electrical power system and the randomness and instability of the wind energy and the fluctuations of the load are mainly compensated by the large power grid, namely, the power grid will provide the WF with ancillary services. The best way to use large-scale wind power is to build large WFs in areas rich in wind energy resources and transmit all the power generated by the wind turbine clusters to the large power grids through power transformation devices.

2.2 Wind Turbine

2.2.1 Type of Wind Turbines

Through more than 2,000 years of development, there have existed various types of wind turbines. As shown in Figure 2.1, from ancient times to modern times, a great variety of wind turbines have been developed for different purposes. In principle, all of them can convert wind energy into mechanical energy and then into some other forms of energy that can be used. Modern wind power generation is the most advanced method for developing and using wind energy with high and new technologies. Wind turbines can be categorized in different ways. For example, ① by the relative position of the rotor shaft and ground, wind turbines can be categorized into horizontal axis wind turbine and vertical axis (perpendicular axis) wind turbine; ② by the blade working principle, wind turbines can be categorized into lift type wind turbine and drag type wind turbine; ③ by the use of wind turbines, there is the wind driven generator, wind powered water pumping, wind powered hay cutter and wind powered threshing machine; ④ by the ratio of the blade tip linear speed to the wind speed, there is high-speed wind turbine (with the ratio greater than 3) and low-speed wind turbine (with the ratio lower than three) as well as medium-speed wind turbine

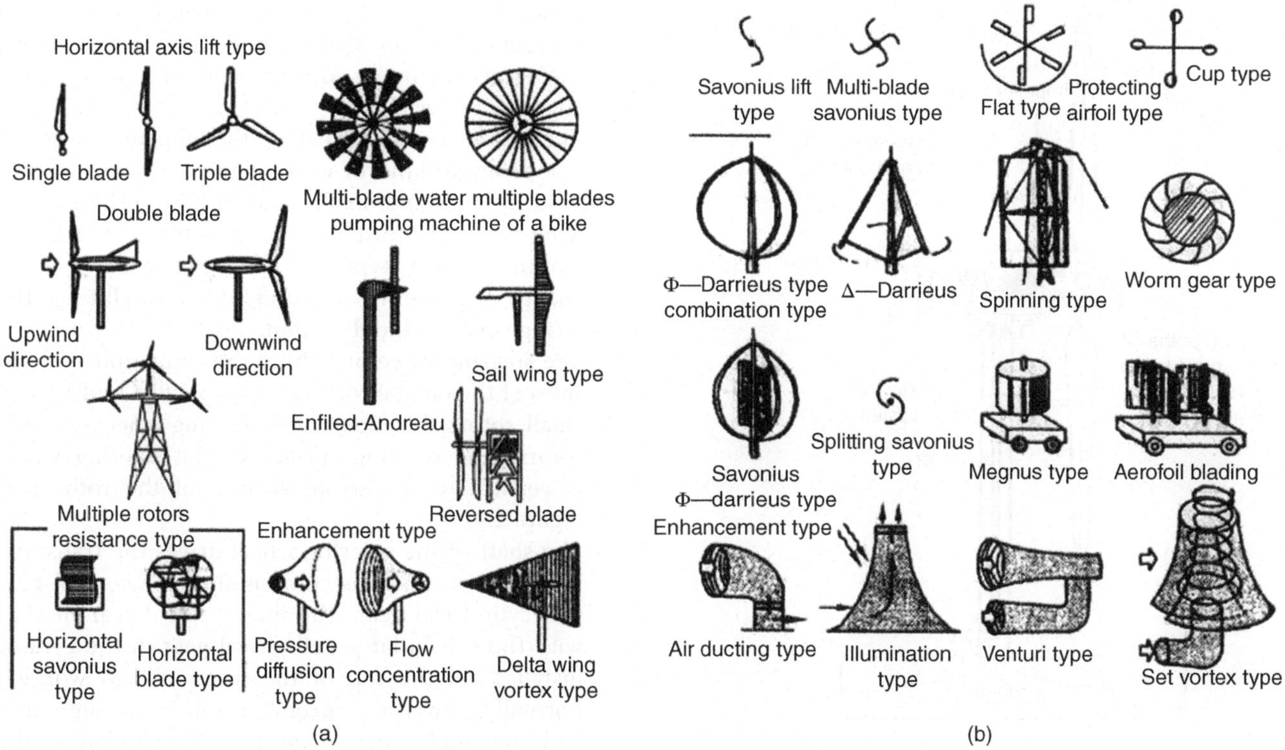

Figure 2.1 Wind turbines: (a) horizontal axis wind turbine (b) vertical axis wind turbine.

(with the ratio ranging between two and five); ⑤ by the capacity of the wind turbine, internationally, wind turbines are generally categorized into small wind turbine (below 100 kW), medium wind turbine (100–1000 kW) and large wind turbine (above 1,000 kW) while in China, wind turbines are categorized into mini wind turbine (below 1 kW), small wind turbine (1–10 kW), medium wind turbine (10–100 kW) and large wind turbine (above 100 kW) and sometimes wind turbines of above 1,000 kW are called giant wind turbines; ⑥ by the relative position of the rotor and tower, there are upwind (front) wind turbine and downwind (rear) wind turbine; ⑦ by the number of rotor blades, there are single-blade, double-blade, three-blade, four-blade and multiple-blade wind turbines. Now the horizontal-axis lift-type wind turbines with less blades (mainly two to three blades) are widely used worldwide; ⑧ by type of blades, there are stall wind turbine (in case of high wind speed due to the shape of the blade or the action of the spoiler at the blade tip, the output torque, and power of the wind turbine is limited) and variable pitch wind turbine (in case of high wind speed the pitch angle is adjusted to limit the output torque and power); ⑨ by rotor speed, there are fixed-speed wind turbine (the rotor maintains a certain speed in operation and the wind energy conversion rate is low) and variable-speed wind turbine (the dual-speed wind turbine can operate at two set speeds to improve the wind energy conversion rate and the speed of the continuous variable-speed wind turbine is continuous and can be adjusted to capture the maximum wind power); ⑩ by transmission gear there are speed-raising wind turbine (with the gear box connecting the low-speed wind turbine and the high-speed power generator) and direct-drive wind turbine (directly connecting the low-speed wind turbine and the high-speed power generator).

2.2.2 Wind Turbine Structure

As shown in Figure 2.2, the section plan of the typical horizontal axis wind turbine is mainly made up of the rotor, tower, and wind aligning device (yawing system).

1) **Rotor**
   The rotor is the most important component of the wind turbine and it is the main sign that differentiates the wind turbine from other power machinery. The rotor is mainly used to capture and absorb wind energy and convert it into mechanical energy, which is transmitted to the transmission devices through the rotor shaft. The rotor is generally composed of the blade, stalk, hub, and rotor shaft (see Figure 2.3).
2) **Headstock and gyrator**
   The upper parts of the wind turbine tower, such as the rotor, transmission devices, wind aligning devices, speed regulating device and generator, and so on, constitute the head, and the parts connecting the head and the tower are called the headstock and gyrator.

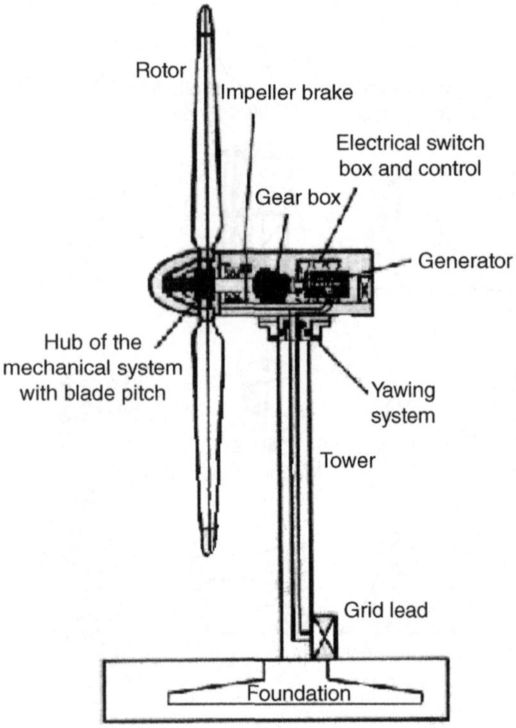

Figure 2.2 Wind turbine structure diagram.

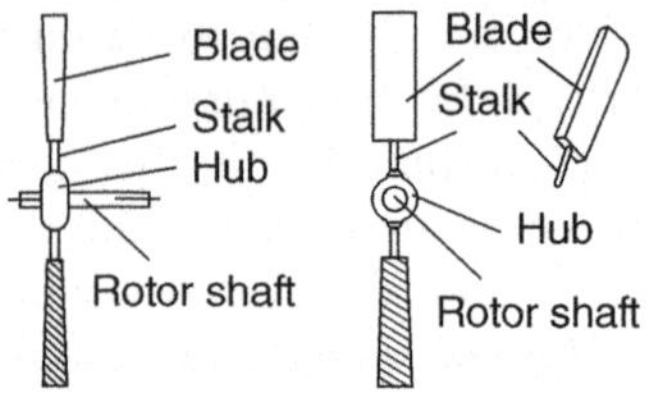

Figure 2.3 Rotor composition diagram.

The headstock is used to support all the devices and auxiliary components above the tower and its firmness is directly related to the safety and service life of the wind turbine.

The gyrator (turntable) connects the tower and the headstock and consists of the fixed sleeve, gyration ring, and the bearing between them. The fixed sleeve is locked in the upper parts of the tower and the gyration ring is connected with the headstock. The fixed sleeve and the gyration ring are connected to the wind alignment devices through the bearing between them. When the wind direction changes, the head can rotate horizontally and then the rotor can work facing the wind.

3) **Wind alignment devices**

As the wind direction and wind speed are always changing, in order to enable the wind turbine to effectively capture the wind energy, the wind alignment devices should be set to track the change of the wind direction and ensure that the rotor can always face the wind. Common wind turbine wind alignment devices include rudder, steering wheel, electrical alignment device, and automatic wind alignment device.

The rudder is also known as the empennage, and it is a common kind of wind alignment device widely used by mini and small wind turbines. The ratio of wingspan and chord length of new types of rudder is two to five. New types of rudder are sensitive to the change of the wind direction and can well track the change of the wind direction.

The steering wheel is behind the rotor, and on both sides of the engine room are two parallel multi-blade small rotors that are called steering wheels or side rotors. The rotating surface of the steering wheel is vertical to the swept surface of the rotor (see Figure 2.4).

The shaft of the steering wheel drives the transmission system, which is composed of the conical gear and cylindrical gear. The intermediate gear meshes with the driven large cylindrical gear in the gyrator installed on the top of the tower. When working normally, the wind turbine rotor is in alignment with the wind direction and the rotating plane of the steering wheel is parallel with the wind direction, and the steering wheel does not rotate. When the wind direction changes, the steering wheel deviates from the wind direction at a certain angle, and under the action of the wind, the steering wheel begins to rotate and the transmission system enables the wind turbine rotor to be in alignment with the wind direction once again. The rotating plane of the steering wheel then once again restores to a position parallel to the wind direction and then the steering wheel stops rotating. The steering wheel wind alignment device works more stably than the rudder version and is often used in medium wind turbines. The transmission device of the steering wheel can also be designed to be the worm and worm gear system.

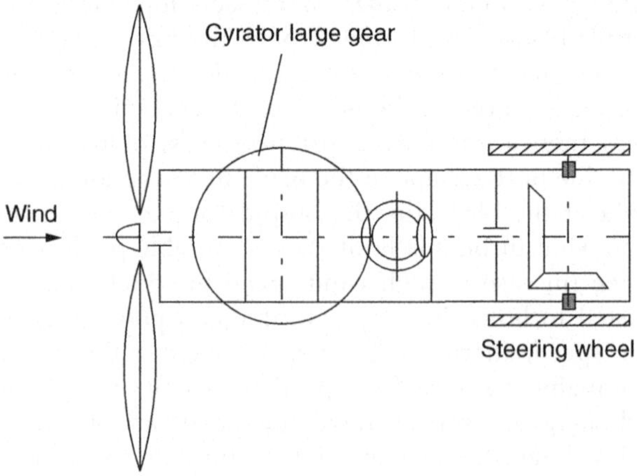

Figure 2.4 Steering wheel wind alignment device.

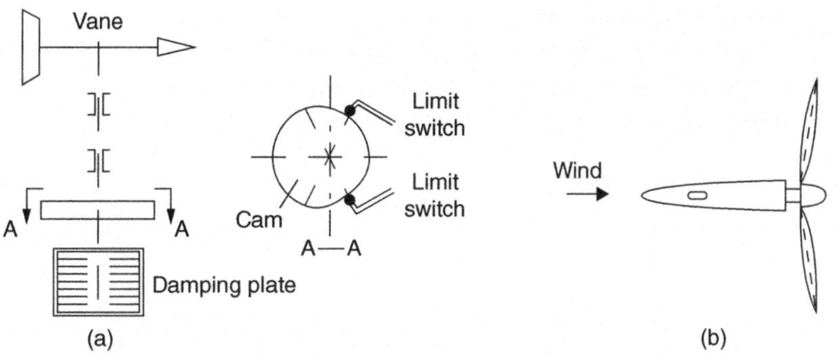

Figure 2.5 Electrical wind alignment device and automatic wind alignment device. (a) electrical wind alignment device (b) automatic wind alignment rotor.

The electrical wind alignment device is often used in large and medium wind turbines. Shown in Figure 2.5a is the schema of the wind aligning device of the domestically produced FD16.2-55 wind turbine. The wind direction signal of this device comes from the vane installed on the top of the engine room. When the wind direction deviates from the rotor axis at an angle of ±15°, the vane drives the cam on its vertical axis to rotate and turn on the left or right limit switch. After 30 s (which can be set arbitrarily) of delay, the alternating current (AC) contactor is closed, starting to turn the wind aligning servo-motor left or right and turning on corresponding indicator lights. The servo-motor drives the turntable in the gyrator, through the reducer, to rotate. After the rotor faces the wind direction once again, the limit switch is turned off; the motor stops working; the indicator lights extinct. Two AC contactors are interlocking so as to ensure that in action only one of them can be closed to avoid a short circuit when both of them are connected.

By the wind, which blows onto the rotor or the engine room first, wind turbines can be categorized into upwind wind turbine and downwind wind turbine and the corresponding rotor is called front rotor and rear rotor. For the downwind wind turbine, its rotor can be designed to be the type shown in Figure 2.5b, which uses the resistance that the wind acts on the rotor to enable the rotor to be automatically in alignment with the wind direction. As a result, it is called the automatic wind alignment rotor and mainly used in the medium and large wind turbines.

4) **Tower**

The tower is used to support the rotor so that the rotor can operate at a high speed on the ground. The tower has to bear the gravity of the wind turbine (downward) and the wind force (resistance borne by the tower). The tower of large wind turbines is basically a tapered cylindrical steel tower frame.

2.2.3 Basic Working Principle of Wind Turbine

The basic function of the wind turbine is to use the rotor to receive the wind energy and convert it into mechanical energy, which is transmitted by the rotor shaft. The working principle of the wind turbine is that the lift force or drag force produced by the air flowing through the rotor blade drives the blade to rotate and converts the wind energy into mechanical energy.

There are a great variety of wind turbines and by principle they can be categorized into lift type wind turbine and drag type wind turbine and by structure, the horizontal axis wind turbine and vertical axis wind turbine. The lift type wind turbine rotates fast while the drag type wind turbine rotates slowly. The horizontal axis lift type wind turbine is widely used both in China and other foreign countries.

1) **Working principle of lift type wind turbine**

Shown in Figure 2.6 are the three different shapes of blade and their stress states. From the knowledge of aerodynamics, we know that when the air flows through a piece of flat type blade, the flat surface and the air flow direction form an angle α, it is known as the angle of attack. Due to the different airspeeds above and below the flat (beacause the airspeed above the flat is higher than that below it), the pressure above the flat is also different from that below it (the pressure below the flat is higher than that above it). The direction of the total force F, which is the aerodynamic force borne by the flat in the moving air is vertical to the flat surface. The force F can be decomposed into two component forces: one component force F_y, which is vertical to the air flow direction and lifts the flat is called lift force; the other component force F_x, which is in the same direction of the air flow

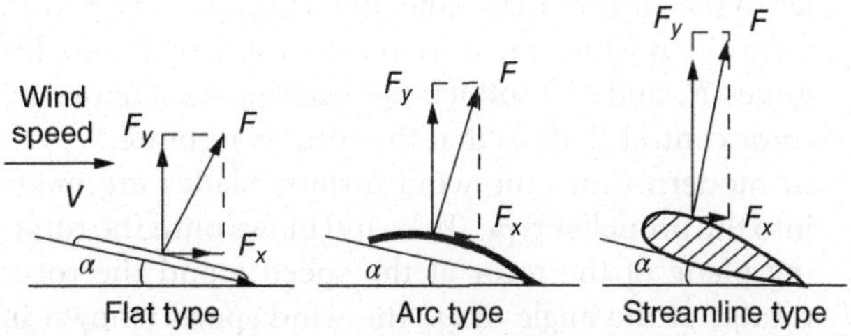

Figure 2.6 Wind force converted into lift force and drag force of the blade.

is called drag force. Both the lift force and drag force are proportional to the projected area S of the blade in the air flow direction, air density ρ, and the square of the airspeed v, which is shown in Formula (2.1):

$$F_y = \frac{1}{2}c_y \rho S v^2$$
$$F_x = \frac{1}{2}c_x \rho S v^2$$
$$F = \frac{1}{2}c_r \rho S v^2 \quad (2.1)$$

where c_y is called the lift coefficient; c_x is the drag coefficient; c_r is the total aerodynamic force coefficient.

The lift is the force that makes the wind turbine work effectively and drives the rotor to rotate. The drag is the positive pressure on the rotor borne by the tower of the wind turbine. In order to make the wind turbine work well, the blade needs to have such an airfoil section so that it can get the maximum lift force and the minimum drag force.

Factors affecting the lift coefficient and drag coefficient are: ① The airfoil. The lift force generated by an airfoil with a streamline cross section is much larger than that generated by a flat airfoil. This is because when the angle of attack is not big, the streamline cross section hardly produces a vortex while the square flat produces a huge vortex at the front edge, thus reducing the lift force and increasing the drag force. ② The angle of attack. For the streamline blade, with the increase of the angle of attack α from zero, the lift coefficient c_y increases from a certain numerical value, generally presenting the linear changes. When the attack angle is increased to a critical angle of attack α_c, the lift coefficient reaches the maximum value $c_{y\max}$. When $\alpha > \alpha_c$, c_y starts to decrease as the angle of attack increases.

Shown in Figure 2.7 is the head of the horizontal axis wind turbine as well as the starting principle of the rotor. Let the central axis of the rotor be consistent with the wind direction, and when the air flows through the rotor at the speed v, it will generate the aerodynamic force F and F' on the blade. Decompose F and F' into the component forces F_x and F'_x (drag forces) along the direction of the air flow and the component forces F_y and F'_y (lift forces) perpendicular to the air flow direction. The drag forces F_x and F'_x form the positive pressure on the rotor while the lift forces F_y and F'_y produce the rotating torque on the rotor central shaft so that the rotor can rotate.

In modern times the wind turbine blades are made into the propeller type. The wind blows onto the rotating plane of the rotor at the speed v and the rotor rotates at the angle of ω. The wind speed (shown in Formula 2.2) corresponding to the airfoil is

$$v_r = \omega r + v \quad (2.2)$$

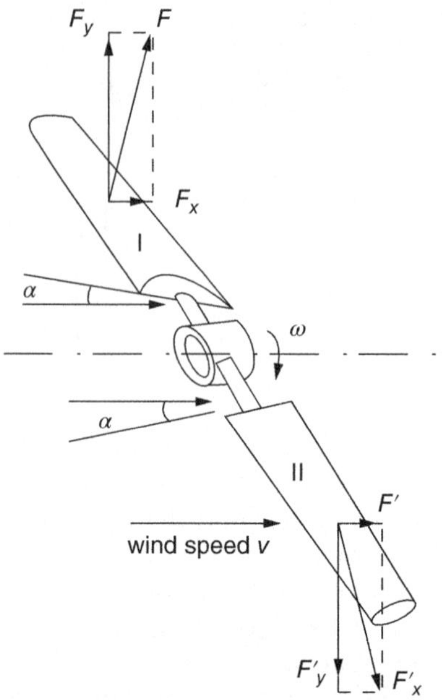

Figure 2.7 Lift force and drag force of horizontal axis wind turbine.

Suppose the angle between the relative wind speed v and the chord of the airfoil is the optimal angle of attack, then the lift coefficient is $C_{y\max}$. However, as the rotation radius of each blade section r is different, the relative wind speed of each section v_r is also different, and even on some sections, the lift coefficient is negative. Therefore, the blade should be made into the form of the twisted spiral along the direction of the blade length so that the chord of each airfoil section of the blade from root to the tip is approximately the same with the corresponding relative wind speed v_r and their angle of attacks are near the optimal angle of attack. In this way, the wind energy can be maximally converted into the lift force of the blade.

2) **Working principle of drag type wind turbine**
Shown in Figure 2.8 is the rotor of the vertical axis drag type wind turbine. It is mainly composed of three curved blades. When the wind blows onto the rotor,

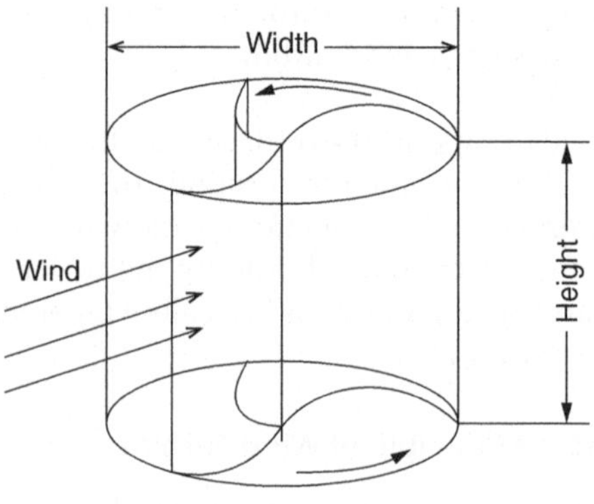

Figure 2.8 Vertical axis S type blade rotor.

the blades generate drag force, which drives the rotor to rotate counterclockwise (top view). The concave blade drives the rotor to rotate while the raised blade hinders the rotor from rotating (Formula 2.3). The value of drag force produced by each blade F_d can be calculated according to the following formula:

$$F_d = \frac{1}{2}\rho(v \pm u)^2 A_V C_d \tag{2.3}$$

where ρ —air density; v —wind speed; u —the linear velocity of the blade, the average linear velocity in the radius direction; A_V —the maximum projected area of the blade (width × height); C_d —the blade drag coefficient (for the rotor consisting of two curved blades, the C_d value of the concave blade is 1.0 while that of the raised blade is 0.12–0.25). For the selection of the sign ± in calculating F_d: for the concave blade (right side), select "−"; for the raised blade (left side), select "+".

For this kind of vertical axis drag type wind turbine, the drag force generated by the concave blade is greater than that generated by the raised blade. Thus, it is only natural that the rotor rotates counterclockwise. Of course, if the wind blowing toward the left side of the rotor is blocked so that the raised blade is not blown by the wind, it will facilitate the rotation of the rotor.

2.2.4 Aerodynamic Characteristics of Wind Turbine

The wind turbine uses the lift or drag of the blade to drive the rotor to rotate, and in this way, converts wind energy into mechanical energy (Formula 2.4). When the blade swept area is A, the wind power captured by using the wind energy converter (rotor) is:

$$P = \frac{1}{2}C_P A \rho v^3 \tag{2.4}$$

where P is the wind turbine output power (kW); C_P —the wind energy utilization coefficient of the wind turbine; $A = \pi R^2$ —the rotor swept area (m^2); R—the rotor radius (m); ρ —the air density (kg/m^3); v —the wind speed (m/s).

The wind energy utilization coefficient C_P is an important parameter that reflects the ability of the wind turbine to convert wind energy into mechanical energy, that is, in the unit time, the ratio of the wind energy absorbed by the rotor and all the wind energy passing through the rotating surface of the rotor. It is related to the specific design of the wind energy converter and the wind direction. At a certain wind speed, the higher the C_P value is, the higher the efficiency of the rotor converting wind energy into mechanical energy is. However, it is impossible for all the wind energy to be completely absorbed by the rotor, so the efficiency of the wind turbine is always less than one. Betz studied a hypothetical plane disc-shaped ideal rotor and found the maximum value of C_P was about 59.3% (Betz limit). To be specific, the possible maximum efficiency of the rotor is approximately 59.3%, but in actual application it ranges between 0.2 and 0.4.

When the wind turbine blade is identified, the wind energy absorbed by the rotor, namely, the wind energy utilization coefficient C_P, is only related to the wind turbine tip speed ratio λ and the pitch angle θ.

The tip speed ratio is the ratio of the tip speed v_t to the wind speed v. It is an important design parameter of the wind turbine. The tip speed ratio λ can directly affect the energy captured by the blade and the wind energy utilization coefficient. It can be calculated according to the formula $\lambda = \omega_r R/v$ where ω_r is the rated mechanical angular velocity of the wind turbine (Rad/s). When the pitch angle is certain, the relationship between C_P and λ is shown in Figure 2.9. C_P reaches its maximum value when λ is a certain value. In the fixed speed wind turbine, the rotation speed of the impeller is constant while the wind speed is variable, so it is impossible for λ to maintain the optimal value. The value of C_P in actual application is usually quite different from its maximum value. As a result, the wind turbine often operates in the low efficiency state.

The pitch angle is defined as the included angle between the blade rotating surface and the chord length of the blade cross section. For the wind turbine with the blade radius R, Formula (2.4) can be rewritten as (Formula (2.5))

$$P = \frac{1}{2}\rho C_P(\lambda, \theta)\pi R^2 v^3 \tag{2.5}$$

For example, The relationship between the wind energy utilization coefficient C_P and the specific tip speed ratio λ and the pitch angle θ can be seen in Figure 2.9. When $\theta = 0^0$ and the tip speed ratio λ is greater than and close to six, the blade can capture the maximum wind energy. In the case of the changing wind speed, in order to maintain the optimal value of C_P, the rotor speed should be proportional to the wind speed. In actual application either the rotor speed remains constant or the tip speed ratio λ remains a fixed value (variable speed).

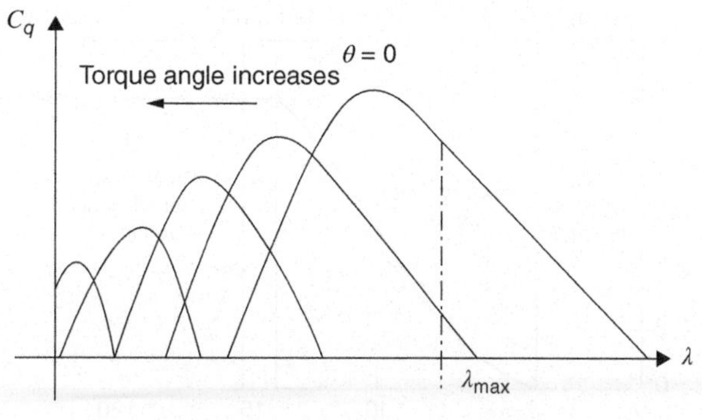

Figure 2.9 Torque coefficient characteristic curve.

The blade of the variable pitch wind turbine can rotate around the central shaft of the blade and the pitch angle θ ranges between 0° and 90°, which can in a certain range regulate the wind energy captured by the rotor. When faults emerge in wind turbines, the blade can manage the feathering of the blade, improve the stress of the wind turbine, and ensure the safety of the system. Under normal operating conditions, the variable pitch wind turbine has better aerodynamic characteristics. It can enable the wind turbine to track the maximum C_P value below the rated wind speed by adjusting the blade pitch angle so as to improve the wind energy utilization efficiency. Through the speed control mechanism, variable speed wind generators can make the angular velocity ω of wind wheel proportional to the wind speed change and always track the maximum C_P value so as to obtain the maximum wind energy utilization rate and wind turbine efficiency when the wind speed changes.

The wind energy captured by the impeller is converted into the mechanical torque on the hub (Formula (2.6)):

$$T_{wt} = \frac{1}{2}C_q \rho v^2 \pi R^3 \quad (2.6)$$

where C_q is the torque coefficient and $C_q = f(\omega_r, v, \theta)$. The curve of the characteristics of the torque coefficient is shown in Figure 2.9. When $\theta = 0^0$, the maximum torque coefficient $C_{q\max}$ is obtained. When the pitch angle changes, both the torque coefficient and the output torque change. At a low wind speed, the pitch angle is adjusted to maintain the optimal C_q while at a high wind speed, the pitch angle is adjusted to limit the rotor output power. The relationship between the power coefficient C_P and torque coefficient C_q is (Formula (2.7)):

$$C_P = \lambda C_q \quad (2.7)$$

The power curve shown in Figure 2.10 presents the relationship between wind speed and the generated energy. Set appropriate values of λ and θ and then formula (2.5) can be used to calculate the power curve. For the fixed speed wind turbine and the variable speed wind turbine, the power curve can be divided into four sections [2]: In section 1, when wind energy is too small, the wind turbine is not starting, and no power is generated. In section 2, when the power generated is less than the rated power. The goal in this section is to optimize the aerodynamic efficiency and the capture of wind energy. The wind speed at the boundary of sections 2 and 3 is called the rated wind speed. The design value of all variables at this wind speed is known as their rated value. In section 3, when wind power is sufficient to generate the rated power, the aerodynamic efficiency is reduced; otherwise the electrical system will be overloaded. In section 4, when no power is generated. Because at the high wind speed, in order to prevent damage to the wind turbine, the wind turbine stops working.

2.2.5 Wind Turbine Power Regulation

Pneumatic power regulation of the wind turbine is one of the key technologies of the wind turbine. When the wind power generation system is operating performance such as the mechanical strength and the wind turbine and converter capacity, the control system is required to stabilize the power output of the wind power generation system near the rated value through the regulation of the pneumatic power of the wind turbine so as to avoid overload operation and meanwhile reduce the load borne by the blade and prevent the wind turbine from being damaged. At present, there are mainly three kinds of pneumatic power regulation: fixed pitch stall regulation, variable pitch regulation, and active stall regulation.

(I) **Fixed Pitch Stall Regulation**
Fixed pitch refers to a rigid connection of the blade and the wind turbine hub. In operation process of the system the blade pitch angle remains unchanged. Namely, when the wind speed changes, the blade windward angle cannot change. With the increase of the wind speed, the angle of attack increases and the surface of the wind turbine blade gradually generates the vortex. The pressure difference between the upper and lower wings decreases, which leads to the increase of the drag force and the decrease of the lift force. The change of drag

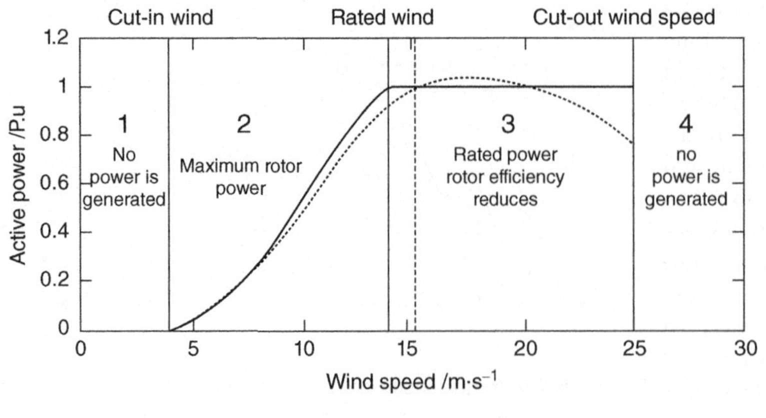

Figure 2.10 Typical power curve of fixed speed stall wind turbine (dotted line); typical power curve of variable speed variable pitch wind turbine (full line).

and lift force leads to the stall of the blade and the decrease of the efficiency of the wind turbine, which then limits the system power output. The power regulation of the fixed pitch wind turbine completely depends on the aerodynamic characteristics of the blade and its output power changes with wind speed. When the wind speed is lower than the rated wind speed, it is difficult to obtain the maximum wind energy utilization coefficient C_P; when the wind speed is higher than the rated wind speed, the constant power can be maintained by reducing C_P through the stall of the blade or the yaw control. Actually, in this case the power will be reduced somehow and it is difficult to make it constant.

Fixed pitch stall control is a traditional control method. The stall regulation method depends on the unique airfoil structure, which is generally used in the fixed speed wind turbine. At a high wind speed, the air that flows through the leeward side of the blade generates turbulence, reducing the blade aerodynamic efficiency so as to limit the maximum output power and keep the generator speed constant. The aerodynamic stall control of the wind turbine is a passive control. As shown in Figure 2.11, if the attack angle exceeds a certain critical value (10°–16°), the boundary layer of air in the upper surface will be separated, which will lead to the wake flow in the upper wing, reducing the flow volumc and the lift force of the wing and increasing the drag force of the wing. So there emerges the stall as the air flows through the wing.

For the fixed speed wind turbine, another characteristic curve can be obtained. If the wind speed is constant, then the dimensionless $CP - \lambda$ curve shows how power changes along with the rotational speed. When the rotor is forced to operate at a constant speed (n), the dimensionless $K_P - 1/\lambda$ curve describes how the power varies with the wind speed. K_P is defined as (Formula (2.8)):

$$K_P = \frac{P}{\frac{1}{2}\rho(nR)^3 A} = \frac{C_P}{\lambda^3} \tag{2.8}$$

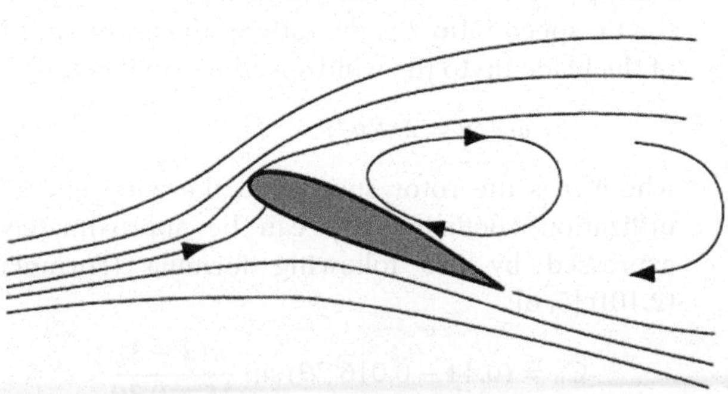

Figure 2.11 Stalled flows around the wing.

For the detail of typical fixed pitch wind turbine, $CP - \lambda$ and $K_P - 1/\lambda$ curves are shown in reference [3]. The above-mentioned $K_P - 1/\lambda$ curve has the same form as the wind turbine power–wind speed characteristics. The disadvantage of constant speed operation is that the efficiency of the wind turbine (presented in $C_P - \lambda$ curve) is largely determined by the change of the wind speed, so the rotor should be designed to achieve the maximum efficiency at the wind speed where the maximum wind energy can be obtained.

An important characteristic of the $K_P - 1/\lambda$ curve is that once stall occurs, the power initially drops and then gradually increases with the wind speed. This characteristic provides the basic principle of the passive power output regulation and prevent overload of generatorwhen the wind speed increases. Ideally, the power should increase to the maximum with the wind speed and then it should be maintained at a constant and has nothing to do with the increase of wind speed. However, the stall-regulated wind turbine cannot present this kind of ideal passive stall performance.

Stall regulation provides the simplest way to control the maximum power of the wind turbine so that the size of the installed generator is matched with that of the gear box. The main advantage of the stall control is its simplicity, but it also has some major shortcomings. Because of the acrodynamic characteristics of the blade, especially the stall characteristics, the power-wind speed curve is fixed. After the stall occurs, the power output of the wind turbine is very unstable, and it is difficult to predict its changes. The stalled blade also presents the low vibration damping and it may cause large vibration amplitude, which will inevitably be accompanied by the large bending moment and stress and lead to fatigue damage.

When the rotor stops in the high-speed turbulent wind with a fixed pitch angle, the blade bears great aerodynamic load, which cannot be reduced through the blade pitch angle adjustment (feathering). Therefore, the stall-regulated wind turbine with the fixed pitch angle bears more severe loads compared with the blade and tower of the pitch-regulated wind turbine.

Since the stall is a very complicated aerodynamic process, it is difficult to accurately calculate the stall effect when the wind speed is variable. When the stall regulation is used, the wind turbine power output curve is shown in reference [4]. In the wind speed section after the maximum wind speed is reached, the wind turbine output power reduces as the wind speed increases.

The fixed-pitch stall-regulated wind turbine generator system is characterized by simple structure, low costs, and a high safety factor. This system usually uses two asynchronous generators varying in power and numbers of pole pairs. The high-power and high-speed generator works in the high-speed section and the low power and low-speed generator works in the low wind speed section, in order to improve the performance of the wind power generation system. However, this kind of pneumatic power adjustment mode depends on the unique airfoil structure of the blade. The structure of the blade is complicated and it is difficult to produce it. With the increase of the system capacity, the blade length increases and meanwhile its rigidity decreases, so that the dynamic stall characteristic is not easy to be controlled. Therefore, the fixed-pitch stall regulation is seldom used in large wind power generation megawatt (MW)-level system and above. But some manufacturers can calculate the stall effect in a very reliable way based on their design experience in small and medium wind power generators. Thus, they still use stall regulation in some MW-level wind turbines.

(II) **Pitch Angle and Pitch Regulation**

In order to improve the wind energy conversion efficiency of the wind turbine and ensure the stability of the output power of the wind power generation system, the wind turbine usually needs pitch regulation. Based on the fixed pitch wind power system, after install pitch regulation devices, the variable pitch wind power generation system is formed. The pneumatic power regulation of this system does not depend entirely on the aerodynamic characteristics of the blade. Instead it is realized by changing the blade pitch angle. When the wind speed is below the rated wind speed, the blade pitch angle is near zero. In this case no regulatory change is needed and it can be considered equal to the fixed-pitch wind turbine. The output power of the system changes with the wind speed according to the blade aerodynamic performance. When the wind speed is higher than the rated wind speed, the variable pitch regulation system adjusts the blade attack angle to ensure that the output power of the generator remains constant within the permitted range. In this process, the variable pitch regulation system will change the pitch angle to make the blade rotate toward the direction with a smaller windward area in order to reduce the angle of attack and the pneumatic power absorbed by the wind turbine blade, the output power, and to limit the system output power in the vicinity of the rated value. The controller must limit the overshoot of the variable pitch regulation system and ensure its stability.

Compared with the fixed-pitch regulation method, the variable pitch regulation method has obvious advantages. First, the wind power generation system using the variable pitch regulation method has a low starting wind speed because the variable pitch regulation system can adjust the blade pitch angle to a proper angle so that the wind turbine has the maximum starting torque. When the wind power generation system needs to be cut out and disconnected from the grid, the variable pitch regulations system can first rotate the blade to reduce the pneumatic power absorbed by the wind turbine and reduce the power output to zero before the system is cut off from the power grid to avoid the impact of the sudden load shedding when the fixed pitch wind power generation system gets disconnected from the power grid and relatively alleviate the mechanical impact force on the transmission shaft. In addition, above the rated power point, the variable pitch wind power generation system has a high wind energy utilization coefficient and stable output power. The fixed pitch wind power generation system usually has a high wind energy utilization coefficient at the low wind speed section. But when the wind speed approaches the rated value, the wind turbine blade begins to stall and the wind energy utilization coefficient of fixed pitch system begins to drop sharply. As the wind speed increases, the rise of the system output power slows down. When the wind speed is higher than the rated value, the blade enters the deep stall mode. At this time if the wind speed continues to increase, the output power of fixed pitch system decreases instead. However, the variable pitch wind turbine has a controllable pitch angle. It has a quite high wind energy utilization coefficient when it operates at a wind speed higher than the rated wind speed and its output power can be stabilized near the rated value.

The output power generated by the rotor absorbing the wind energy is shown in Formula (2.4). For the variable pitch wind turbine, the wind energy utilization coefficient C_P and the tip speed ratio λ and the blade pitch angle θ form the non-linear relationship, the tip speed ratio λ is the ratio of the linear speed of the blade tip to the wind speed (Formula (2.9)):

$$\lambda = \omega R/v = 2\pi Rn/v \tag{2.9}$$

where n is the rotor speed and the wind energy utilization coefficient C_P can be approximately expressed by the following formula (Formula (2.10)) [5, 6]:

$$C_P = (0.44 - 0.016\,7\theta)\sin\frac{\pi(\lambda - 3)}{15 - 0.3\theta} - 0.00184(\theta - 3) \tag{2.10}$$

It can be seen from Formula (2.10): ① At a fixed pitch angle θ, there exists the unique maximum wind energy utilization coefficient C_{Pmax}; ② For a random tip speed ratio λ, when the blade pitch angle $\theta = 0$, the wind energy utilization coefficient C_P is the maximum. With the increase of pitch angle θ, the wind energy utilization coefficient C_P obviously decreases, which is the theoretical foundation of the variable pitch regulation. To be specific, the variable pitch control is to adjust the pitch angle based on the wind speed and the generator speed so as to control the output power of the generator. In this way, the wind turbine can absorb wind energy as much as possible and convert it into electrical energy and maintain the stable power output in the high wind speed section.

At present there are mainly two schemes for the wind turbine variable pitch system to be put into use [6]: the hydraulic control scheme and motor control scheme. The hydraulic control has advantages such as fast response frequency, large torque, sharing the same source of oil with the brake system. So that hydraulic control accounts for a large proportion in the variable pitch control.

The variable pitch regulation is to change the included angle between the blade windward surface and the vertical axis of the blade, thus affecting the lift and drag force of the blade, limiting the increase of the wind turbine output power at the high wind speed and maintaining the constant output power. When the variable pitch regulation method is used, the wind turbine power output curve is smooth and the detail can be seen in reference [4].

The starting wind speed of the variable pitch wind turbine is lower than that of the fixed pitch wind turbine and the impact force of the transmission machinery is relatively small in the cut out. When the wind speed is higher than the rated wind speed, the variable pitch wind turbine mainly uses the power control and the speed of the regulation is determined by the sensitivity of the pitch regulation system. In practical application, as the wind energy is proportional to the third power of the wind speed, small wind speed changes will lead to large wind energy changes. As the wind turbine output power is constantly changing and the pitch regulation system makes adjustments frequently, there is some time delay in their response to the changes of the wind speed. In addition, in case of gusts, the pitch regulation devices may be unable to make immediate responses, thus resulting in the instantaneous overloading of the wind turbine.

The variable pitch control of the wind turbine is shown in reference [7]. The pitch controller is a nonlinear proportional controller, which can compensate for the dead zone and limit of the proportional valve. The actuator of the variable pitch control system is the hydraulic system. The output signal of the pitch controller is changed into the voltage signal through the D/A converter to control proportional valve, drive the hydraulic cylinder piston, promote the variable pitch regulation system, and change the blade pitch angle. The displacement feedback signal of the piston is measured by the displacement sensor and input into the comparator through the A/D converter. The variable pitch electro-hydraulic control system of the large wind turbine usually uses the proportional solenoid valve as the power control element. Based on the size of the value of input voltage signal and through the amplification circuit, the output voltage signal (usually −9 V–+9 V) is converted into corresponding electrical signal (for example, the 1 mV voltage corresponds to the 1 mA current), which is sent to the proportional electromagnet as the input, thus generating the output value proportional to the input signal—the force or displacement. The force or displacement is used as an input to act on the proportional valve and enable it to generate a flow or pressure proportional to the electromagnet output force. In this way, the change of an output voltage signal can not only control the movement direction of the actuator and working parts in the mechanical equipment, but also conduct stepless control over their force and movement speed.

In case of gusts, the impact on the tower drum, blade, and foundation of the variable pitch-regulated wind turbine is much smaller than that for the stall-regulated wind turbine. As a result, we can reduce the material usage and the weight of the entire variable pitch-regulated wind turbine. As the variable pitch-regulated wind turbine can make the blade maintain a good angle of attack at the low wind speed, it has better energy output than the stall-regulated wind turbine. Therefore, it can be installed in areas with a low average wind speed. Another advantage of the variable pitch-regulated wind turbine is that when the wind speed reaches a certain value, the stall-regulated wind turbine must be shut down while the variable pitch-regulated wind turbine can be gradually changed into a full wingspan mode position with no blade load to avoid downtime and increase the generating power. The disadvantage of the variable pitch-regulated wind turbine is the requirement for its quick response to the gust. Due to the wind vibration, the stall-regulated wind turbine has small power fluctuations while the variable pitch-regulated wind turbine, especially the variable pitch-regulated fixed speed wind turbine, has large power fluctuations. As a result, the variable pitch regulation system of the wind turbine is required to respond to the gusts quickly enough to reduce the power fluctuations.

(III) **Active Stall Regulation**

Active stall (i.e., mixed stall) regulation combines the fixed pitch stall regulation and the variable pitch regulation and its regulation system uses the variable pitch regulation and makes full use of the stall characteristic of the blade. Similar to the variable pitch regulation, the active stall regulation is achieved by connecting the wind turbine blade with the hub through the bearing so that the blade can rotate around the axes to adjust the pitch angle. At a low wind speed, the variable pitch regulation method is used. The control system adjusts the blade pitch angle to the position where the maximum power can be obtained. When the output power of the wind power generation system exceeds the rated value, the control system will control the pitch angle through the variable pitch actuator by adjusting it toward the stall direction to limit the maximum power output of the wind turbine. As the wind speed changes, only slight adjustment of the blade pitch angle will be needed to increase the angle of attack and deepen the stall effect of the blade, thus limiting the capture of the wind energy. At time of braking, the blade is adjusted to the deep stall state to gradually reduce the wind turbine speed, which can realize pneumatic brake and significantly reduce the impact force of the mechanical brake on the transmission system. The active stall regulation makes full use of the advantages of both the fixed pitch stall regulation and the variable pitch regulation. The pitch regulation actuator has relatively small power and does not require sensitive regulation speed.

The basic output characteristics of the above-mentioned three wind turbine power control methods can be seen in reference [8] and the main difference is the control and limitation of the conversion of wind energy into mechanical power at a high wind speed. The modern wind turbine generally only uses the active stall control and pitch control.

2.3 Wind Turbine Generator System

The wind turbine generator system is usually composed of the generator, excitation regulator (power electronic converter), grid integration switch, soft grid integration device, reactive power compensator, main transformer, rotating speed sensor, and so on.

WTGS can be categorized in two ways. First, it can be categorized into either an asynchronous generator and a synchronous generator. An asynchronous generator includes a cage single-speed asynchronous generator, a cage dual-speed variable pole asynchronous generator, and a winding asynchronous generator. A synchronous generator includes a electrical excitation synchronous generator and a permanent magnet synchronous generator. Second, the grid connected wind power generator system can be categorized into either a constant speed constant frequency wind power generator system and a variable speed constant frequency wind power generator system. The constant speed constant frequency method means that the WTGS speed does not change with the change of wind speed. Instead the WTGS maintains a constant speed in operation to generate constant frequency AC power. At present this method is widely used as it is simple and reliable, but the wind energy cannot be fully used. The variable speed constant frequency method refers to the WTGS speed changes with the wind speed, but the WTGS still generates constant frequency AC power. This method can improve the wind energy utilization rate and has become a mainstream technology trend. However, a power electronic device is required to be added between the variable speed synchronous generator (AC excitation generator) and the power grid to achieve the constant frequency output, which increases the cost.

A constant speed constant frequency wind power generator system includes: ① a synchronous generator system; ② a cage asynchronous generator system; ③ a winding rotor RCC(rotor current control) asynchronous generator system, and so on. Variable speed constant frequency wind power generator system includes: ① a variable speed constant frequency cage asynchronous generator system (high speed); ② a variable speed constant frequency double-fed asynchronous generator system (high speed); ③ a variable speed constant frequency electrical excitation synchronous generator system (medium and low speed); ④ a variable speed constant frequency permanent magnet synchronous generator system (medium and low speed); ⑤ a variable speed constant frequency transverse flux generator system (medium and low speed), and so on.

From the technical and economic perspectives, the cage asynchronous wind power generator system, characterized by low cost and high reliability, will continue to play an important role in the fixed-speed and variable-speed full power conversion wind power generation system. The double-fed asynchronous wind power generator system with high-cost performance and suitable for variable speed constant frequency wind power generation, will continue to become the mainstream product in the wind power market. The direct-drive (without gear box) variable sped constant frequency permanent magnet synchronous generator and its converting technology have developed very rapidly. The use of new technologies is expected to significantly reduce the volume and weight of low-speed generators.

In China WFs mainly use three kinds of wind turbines: constant speed constant frequency CAG, variable speed constant frequency double-fed asynchronous generator (DFAG), and direct-drive permanent magnet synchronous generator (PMSG).

2.3.1 Constant Speed Constant Frequency Cage Asynchronous Wind Power Generator System

2.3.1.1 Grid-Connected Constant Speed Constant Frequency Cage Asynchronous Wind Power Generation System

The cage asynchronous wind power generation system is shown in Figure 2.12. The natural wind blows and drives the wind turbine. The gear box increases its speed and drives the asynchronous generator to convert wind energy into electrical energy. Widely used both in China and other foreign countries is the horizontal axis fixed pitch (or variable pitch) wind turbine with an effective wind speed range of about 3–30 m/s, designed rated wind speed of 8–15 m/s and rated speed of about 20–30 r/min.

In Figure 2.12 the shunt capacitors (power factor correction capacitors) provide the cage asynchronous generator with excitation and reactive power compensation. The reactive power supplied by the capacitor is about 30% of the capacity of the generator. The soft starting device is composed of thyristors, which limit the impact current through the control of the thyristor conduction angle in grid integration.

2.3.1.2 Cage Asynchronous Generator Structure

As shown in Figure 2.13, the cage asynchronous generator is composed of the rotor iron core and winding, stator iron core and winding, stand, and end cover. Its stator core and winding have the same structure as those of the synchronous generator. The rotor uses the cage structure, and the rotor core, made of folded silicon steel sheets, is cylindrical shaped with aluminium or copper conducting bars embedded in the groove. At both ends of the iron core, aluminium or copper end rings are used to short circuit the conducting bar. The rotor does not require external excitation and has no slip ring or brush, so it is simple in structure, firm, and needs almost no maintenance.

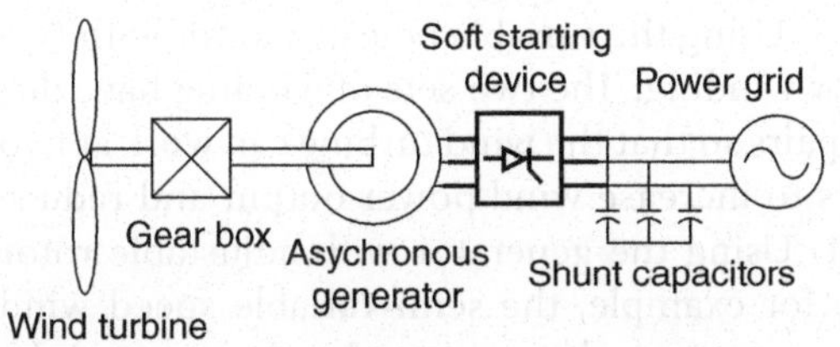

Figure 2.12 Constant speed constant frequency cage asynchronous wind power generation system.

Cage asynchronous generator is designed based on the maximum efficiency achieved at a specific wind speed. In order to increase the generating capacity, some cage asynchronous generators are made into dual-speed generators that have two windings with one used for the low speed (typically eight poles) and the other for the medium and high speed (typically four to six poles).

2.3.1.3 Cage Asynchronous Generator Working Principle

The power grid transmits the symmetrical three-phase AC current into the symmetrical three-phase winding of the stator, forming the rotating magnetic field in the air gap. The speed of the rotating magnetic field n_1 depends on the frequency of the current f and the number of motor pole-pairs p ($n_1 = 60\,f/p$) and n_1 is called the synchronous speed. The rotating magnetic field with the speed n_1 generates in the rotor conducting bar the induction electromotive force e and current i. Bearing the force f in the magnetic field, the current i generates the electromagnetic torque T. If the rotor rotates in the direction of n_1 at the speed $n > n_1$, the T is the brake torque.

When the CAG operates as a motor, its speed n is always lower than the synchronous speed n_1. At this time, the electromagnetic torque generated in the motor T has the same rotating direction as n_1. If the motor is driven by a prime motor, for example, a wind turbine, to a speed higher than the synchronous speed ($n > n_1$), the direction of the electromagnetic torque T is opposite to the rotating direction of n_1 and then this motor operates as a generator whose function is to convert the mechanical energy into the electrical power.

The slip ratio is defined as the ratio of the difference between the synchronous speed n_1 and the rotor speed n to the synchronous speed n_1 and expressed by s (Formula (2.11)):

$$s = (n_1 - n)/n_1, \text{ or } n = (1 - s)n_1 \tag{2.11}$$

When the cage asynchronous generator operates as the motor, s > 0; when the cage asynchronous generator operates as the generator, s < 0, which is shown in Figure 2.14.

The equivalent circuit and power transfer relationship of the asynchronous generator is shown in Figure 2.15. When the asynchronous generator operates as a generator, its power transfer relationship is opposite to that of

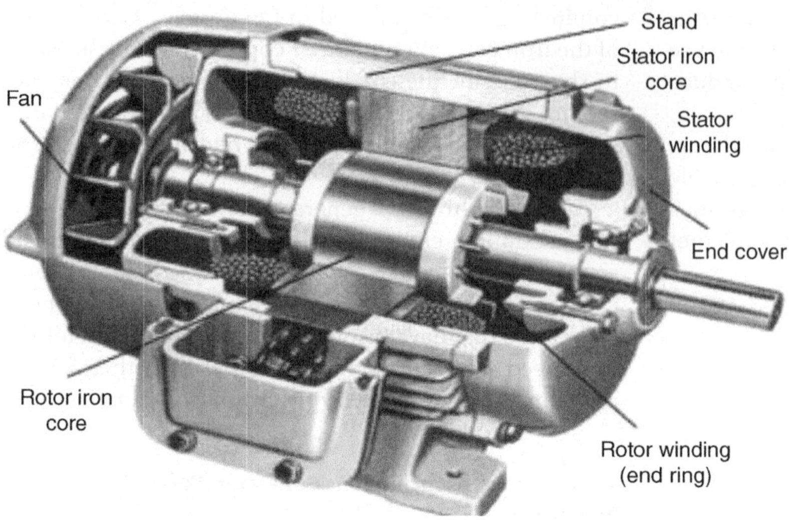

Figure 2.13 Structure of cage asynchronous generator.

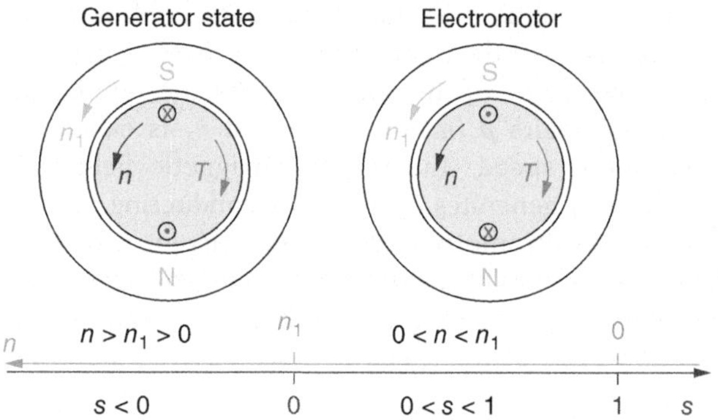

Figure 2.14 Using slip ratio *s* to indicate the operation state of the asynchronous generator.

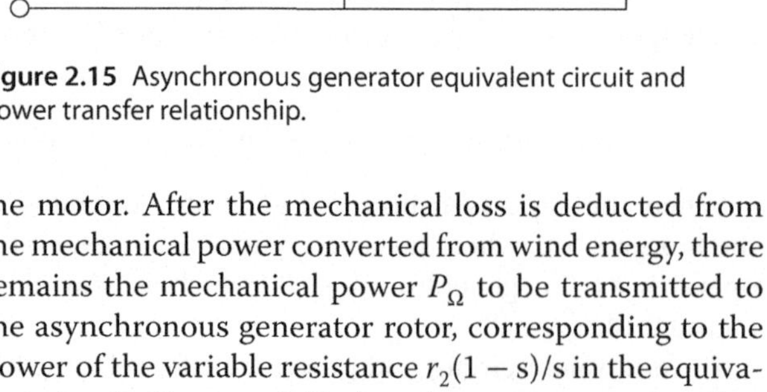

Figure 2.15 Asynchronous generator equivalent circuit and power transfer relationship.

the motor. After the mechanical loss is deducted from the mechanical power converted from wind energy, there remains the mechanical power P_Ω to be transmitted to the asynchronous generator rotor, corresponding to the power of the variable resistance $r_2(1-s)/s$ in the equivalent circuit. Here s < 0. Deduct the rotor copper loss P_{cu2} and iron core loss P_{Fe} to get the electromagnetic power to be input into the stator winding P_m and then deduct the stator copper loss P_{cu1} to get the power P_e to be transmitted into the power grid.

When the generator output power is increased, in order to maintain the constant frequency, the generator speed must be increased accordingly (with *s* increasing within a certain range), which is regulated by the wind turbine and the rising-speed gear box. In order to maintain the constant terminal voltage of generator, the power grid is required to provide more reactive power to increase excitation (increase $\dot{I}_0$).

2.3.1.4 Power Regulation of Cage Asynchronous Wind Power Generation System

The cage asynchronous generator usually works jointly with the fixed speed wind turbine and this wind power generation system regulates power in the following ways: 1) Using the traditional passive stall regulation. As mentioned earlier it will greatly reduce the wind energy utilization rate because when the wind speed increases to the rated wind speed and above, the power coefficient significantly decreases. 2) Using the active stall regulation, for example, using the negative pitch angle to limit the output power in order to keep the rated output power stable at a wind speed above the rated wind speed. Compared with the passive stall regulation, the active stall regulation can increase the wind energy utilization rate by about 20%. 3) Using the variable pole generator with two sets of stator winding. The two sets of winding have different pole pairs so that the wind turbine can work at two fixed speeds to increase wind power output and reduce the noise. 4) Using the generator with adjustable rotor impedance, for example, the semi-variable speed wind turbine. The rotor impedance is regulated to expand the change range of the generator rotor speed, thereby reducing the mechanical power loss.

2.3.1.5 Characteristics of Cage Asynchronous Wind Power Generation System

The cage asynchronous generator is one of the earliest widely used wind power generators and its main characteristics are:

1) Simple structure; good robustness; easy control; no need to be maintained; low cost;
2) Strong overload ability; no problem of desynchronizing;
3) Reactive power consumption is not controllable and changes with the wind speed, so the reactive power compensation device is necessary. The asynchronous generator itself does not have excitation that needs to be provided by the shunt capacitors. As at a high wind speed, the generator generates more active power, the reactive power needs to be increased to provide more excitation for the generator. In addition to meeting the needs of the self excitation, the shunt capacitors at the generator end is required to compensate the power factor requirements for transmitting the wind power into the grid. If the load changes frequently, the fixed capacitor may not be fully compensated, which may lead to over-compensation or under-compensation and the power grid voltage fluctuation. For this purpose, the controllable reactive power compensation device, for example, the static VAr compensator (SVC) is required to be installed in the substation.
4) The asynchronous generator adopts the direct grid connected mode and generates great impact current, which may be seven to eight times the rated current at the moment of being connected to the grid. Therefore, in order to limit the impact current, the asynchronous generator should be connected to the grid step by step when close to the synchronous speed and the soft starting device composed of thyristors should be installed to limit the current.
5) Because of the fixed-speed operation, all the wind speed fluctuations will be further converted to the fluctuations of the mechanical torque and the fluctuations of the electrical power of the power grid. Additionally, in the weak grid, this kind of fluctuation might lead to the voltage fluctuations and significant line loss. The cage asynchronous generator has very limited control ability over the power quality.

2.3.2 Variable Speed Constant Frequency Double-fed Asynchronous Generator System

2.3.2.1 Double-Fed Asynchronous Generator System

As shown in Figure 2.16, the double-fed asynchronous generator system consists of a winding asynchronous generator, a converter and controlling unit. Its stator winding is directly connected to the industrial frequency power grid and the power generated by the generator is directly transmitted to the power grid mainly through the stator winding. The rotor uses the three-phase symmetrical winding and is connected to the grid through the back-to-back bidirectional voltage source converter to provide the generator with AC excitation and output a part of power to the grid.

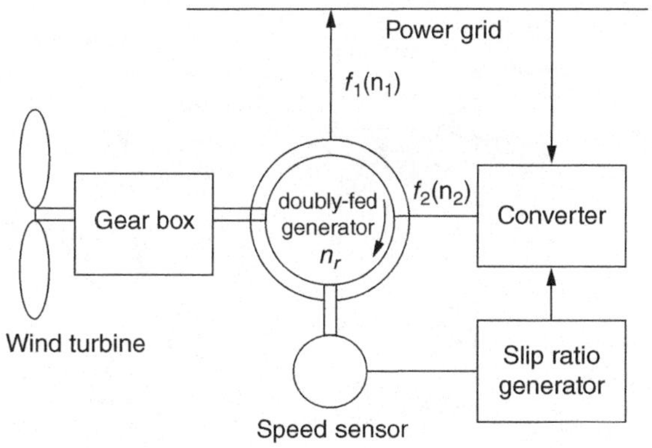

Figure 2.16 Double-fed asynchronous generator system.

2.3.2.2 Double-Fed Asynchronous Generator Structure

The structure of the double-fed asynchronous generator is shown in Figure 2.17. The stator winding is directly connected to the AC power grid. Y connection is usually used for the rotor winding. The wirings of rotor winding terminal are led out by three collecting rings (slip rings) and brush and they are connected to a bidirectional power converter. Connected with the rotor winding is the rotor side converter and connected with the grid is the grid side converter. The controlled variable frequency AC excitation current passes into the rotor winding. When the rotor speed is lower than the synchronous speed, it can also operate in the power generation state. The stator winding port always generates electrical power after grid connection, but the flow direction of the electrical power of the rotor winding port depends on the slip ratio.

2.3.2.3 Double-Fed Asynchronous Generator Working Principle

The stator winding of double-fed asynchronous generator is connected to the industrial frequency power grid and its rotor winding is connected with a three-phase variable frequency power supply of which the frequency, amplitude, and phase can all be regulated. In stable operation, the stator rotating magnetic field and the rotor rotating magnetic field should remain relatively static in space. When the symmetrical three-phase alternating current passes through the three-phase winding of winding asynchronous generator rotor, the rotating magnetic field is generated in the generator air gap. The speed of this rotating magnetic field n_2 is related to the frequency of the incoming AC f_2 and the pole-pair number p of the generator, namely, $n_2 = 60 f_2/\text{p}$. If the rotating magnetic

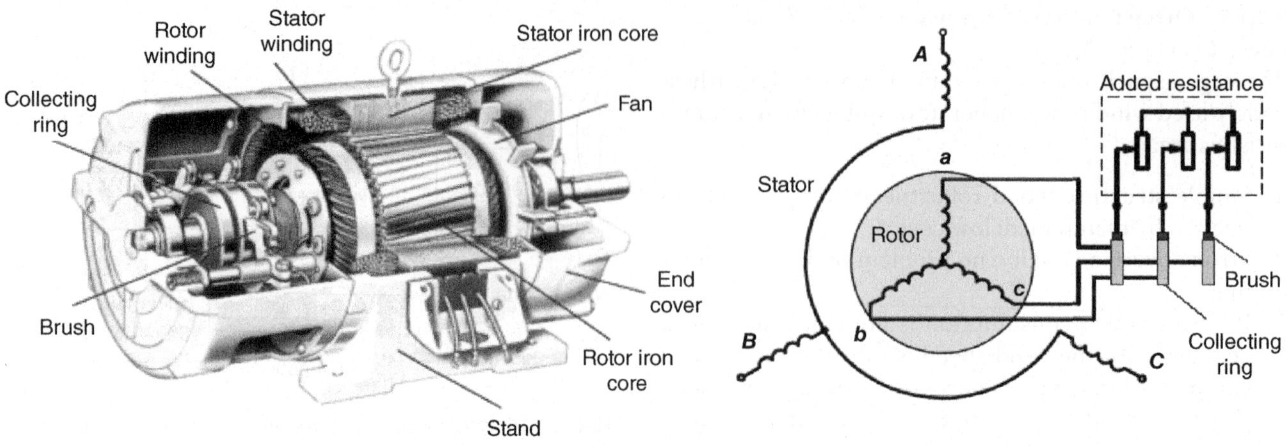

Figure 2.17 Winding rotor three-phase asynchronous generator structure. (a) Double-fed asynchronous generator structure (b) stator and rotor connections.

field of the stator rotates at the speed of n_1, the relationship between the rotating magnetic field of the rotor and the speed n can be expressed as $n_2 = n_1 - n = n_1 - (1 - s)$ $n_1 = n_1 s$ from which $f_2 = f_1 s$ can be further obtained (s being the slip ratio and f_2 is the slip frequency). As long as $n_1 = n \pm n_2$ remains constant, the frequency of the induced electromotive force of the asynchronous generator stator winding will always be f_1. Thus, when the asynchronous generator rotor speed changes, as long as the symmetrical three-phase alternating current of the slip frequency (f_1s) passes through the three-phase winding of rotor, the asynchronous generator stator winding will generate electromotive force with 50 Hz fixed frequency.

According to the changes of the rotor speed n of the double-fed asynchronous generator, the double-fed asynchronous generator can run in three states:

1) Super-synchronous operational state. At this time $n > n_1$; change the phase sequence of the rotor current with the slip frequency f_2, then the direction of the speed of generated rotor magnetic field n_2 is opposite to the rotor rotation direction; $n_1 = n - n_2$; the rotor outputs power to the power grid.
2) Sub-synchronous operational state. At this time $n < n_1$; the direction of rotor magnetic field generated by the rotor current whose slip frequency is f_2 is the same as the rotor rotation direction; $n_1 = n + n_2$; the power grid inputs power into the rotor.
3) Synchronous operational state. At this time $n = n_1$; the slip frequency $f_2 = 0$, namely, the frequency of the incoming rotor current is 0 (direct current). This operation state is the same as that of the conventional synchronous generator.

The operation range of slip ratio for double-fed generator is generally $|s| < 0.3$.

2.3.2.4 Characteristics of Double-Fed Asynchronous Generator System

The main characteristics of the double-fed asynchronous generator system include: 1) The generator speed regulation range is very wide ranging(between 33% and −33% of the synchronous speed). 2) The continuous variable speed operation results in high wind energy conversion rate. 3) Only partial power (about 15%) passes through the converter and the cost of converter is relatively low. 4) Power quality is good, output power is smooth, stator power factor can be regulated, and unit power factor operation is possible. 5) Taking part in the active power regulation of the power grid to improve the stability of power system operation. 6) Simple grid integration without impact current. 7) Reducing the dynamic response requirement of the pitch control. 8) Improving the mechanical stress condition acting on the blade. 9) Complicated bidirectional converter structure and control. 10) There are still slip rings and brushes and mechanical wear exists between brush and slip rings; 11) Poor converter overload capacity.

This type of WTGS has relatively mature technology and ensured reliability. As it can meet the requirements of the grid codes, it is widely used.

2.3.3 Variable Speed Constant Frequency Direct-Drive Synchronous Generator System

Direct-drive wind power generation system does not use the gear box. Instead, the wind turbine and generator are directly connected with the shaft. Developed direct-drive synchronous generator systems include: ① a multi-pole winding synchronous generator type wind power generation system; ② a multi-pole permanent magnet synchronous generator type wind power generation system; ③ a high voltage permanent magnet generator type wind power generation system, and so on. The direct-drive wind power generation system overcomes shortcomings such as big gear box losses and heavy maintenance workload, high cost of large capacity gear boxes and easy to cause overload or premature damage. In this kind of system, due to the slow generator speed, the MW level wind turbine speed range is 10 – 25 r/min. Thus, it cannot use standard generators. Instead, special generators need to be studied and developed. Because to produce a great torque, the generator is large in volume. One converter is used to output the power of the entire wind turbine and this converter (full power converter) converts the variable frequency of the generator into the fixed frequency of the power grid. When the wind speed is higher than the rated wind speed, the power output will also be reduced by variable pitch regulation.

2.3.3.1 Multi-Pole Winding Synchronous Generator Type Wind Power Generation System

The multi-pole winding synchronous wind power generation system is shown in Figure 2.18. The wind turbine and generator of the wind turbine generator system are connected to the shaft. The variable pitch regulation is used to control the wind turbine power. The "flexible connection" between the generator stator side and the power grid is achieved through the power electronic converter. The rotor side regulates the generator excitation current through the excitation controller to control the amplitude of output voltage on stator side, forming the full power conversion variable speed constant frequency wind power generation system.

The multi-pole winding synchronous generator type wind power generation system without the gear box increases the stability of the system and can obtain better power quality. It is connected with the power grid through the power converter and can flexibly control the voltage and frequency of the system to make the generator working frequency and power grid frequency mutually independent of each other. Thus, the grid integration has little effect on the system and it does not have the desynchronizing problem. In addition, the wind turbine of this system uses the variable pitch control, which can enable the wind turbine to achieve the optimal efficiency and increase generating capacity. Compared with other types of generator, the synchronous generator can not only output active power, but can also provide reactive power in operation; it has a relatively mature manufacturing process, thus, its price is relatively cheap. According to the existing technical level of the winding synchronous generator, this structure can be applied to large capacity wind power generation system. The capacity of the "E112" type wind turbine designed in 2002 has reached 4.5–6.0 MW. Like ordinary synchronous generators, the multi-pole winding synchronous generator also needs the excitation controller and contains brushes and slip rings, which increase the system complexity and maintenance workload. In addition, the power converter of the system is connected in serieson stator side. It means that the converter and generator must have the same capacity, which increases the cost and the losses of converter.

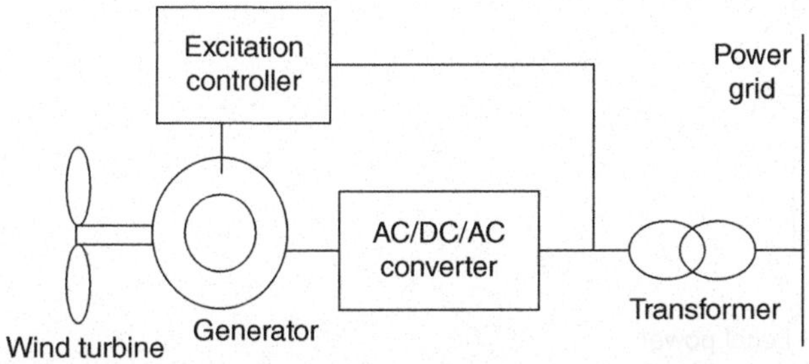

Figure 2.18 Multi-pole winding synchronous generator type wind power generation system.

2.3.3.2 Multi-Pole Permanent Magnet Synchronous Generator Type of WPGS

Mini or small multi-pole permanent magnet synchronous generator type wind power generation system is shown in Figure 2.19. It is commonly used in the independently operated DC wind power generation system. The wind turbine is directly connected with the generator and the AC generated by the generator is rectified into DC output.

The structure of the multi-pole permanent magnet synchronous generator type wind power generation system suitable for grid integration is shown in Figure 2.20.

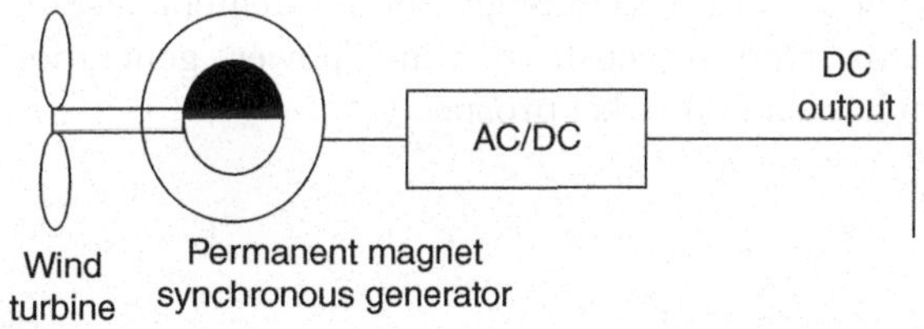

Figure 2.19 Permanent magnet synchronous generator type DC power generation system.

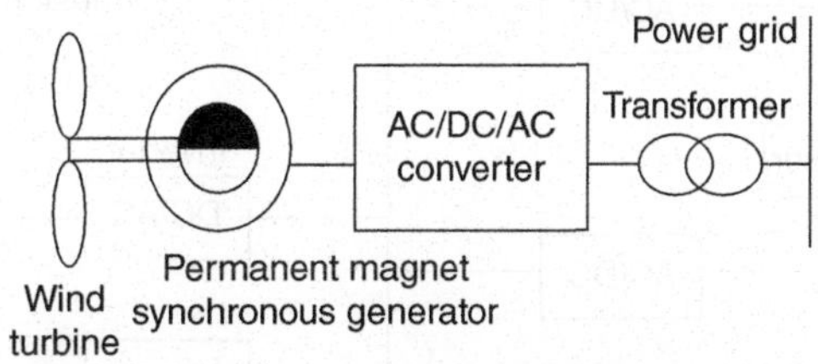

Figure 2.20 Direct-drive permanent magnet synchronous generator type wind power generation system.

The wind turbine is directly connected with the generator and the wind turbine uses the variable pitch power control to achieve the most effective operation. The stator of the permanent magnet generation is the same as that of ordinary AC generator; the rotor is of the permanent magnet structure and no excitation winding is needed. As a result, it does not have excitation winding loss. The rotor does not have slip rings and the operation is safe and reliable.

The direct-drive permanent magnet synchronous generator type wind power generation system is characterized by: ① Continuous variable speed operation and high wind energy conversion ratio. ② Being able to reduce the dynamic response requirements for the pitch control and improve the mechanical stress condition of the blade. ③ Excitation cannot be regulated and the induction electromotive force changes with the speed and load. It can use the controllable pulse width modulated (PWM) rectifier, or it can use the uncontrollable rectifier and then be connected with the direct current AC/(DC) converter, which can keep the DC bus voltage generally constant and meanwhile can control the generator electromagnetic torque to regulate the wind rotor speed. ④ It uses the PWM-controlled converter on the power grid side to invert the power generated by the generator into three-phase AC with the constant frequency and and can adapt well to the grid fluctuations. ⑤ The permanent magnet generator has large volume, heavy weight, and high cost; the full power full control converter control is complicated, which increases the loss and cost.

With the development of the permanent magnetic material technology, modern power electronic technology and control technology, the permanent magnet generator type direct-drive wind power generation system has sound market prospects.

2.3.3.3 High Voltage Permanent Magnet Synchronous Generator Type Wind Power Generation System

In 1998, ABB developed a high voltage permanent magnet synchronous generator type wind power generation system, which is shown in Figure 2.21. This system uses the high voltage permanent magnet generator (Windformer^{TM}) directly connected with the wind turbine and the variable pitch control. The AC output of multiple wind power generators is converted into DC through controllable rectifier respectively and then inverted into the industrial frequency AC centrally to be transmitted to the local high voltage power grid and transmitted to nonlocal power grids through the step-up transformer. It can also be connected to the high voltage direct current (HVDC) or voltage source converter high voltage direct current (VSC-HVDC) to transmit power to other remote areas. The output power of each high voltage permanent magnet generator can reach 3 to 5 MW and its output voltage is no less than 20 kV. This power generation system was first used in a Nassrden III offshore WF in Sweden.

The principle and structure of the generator is shown in Figure 2.22. In Figure 2.22, the material used to wind the armature winding is a kind of traditional circular cross-linking polyethylene cable. The material and processing techniques are identical to those of the same type of cable used in the 500 kV transmission and distribution system. The working electric field strength (RMS) of the insulation system can be as high as 15 kV/mm while that of the insulation system of the rectangular wire rod used in the conventional winding can only reach 2.5 kV/mm. Therefore, the winding made using this material is of high reliability and its nameplate voltage is up to 400 kV. The winding structure is of the concentric type and its end fixing is very simple; the cable in the armature slot is of the double row layout; the winding of each phase

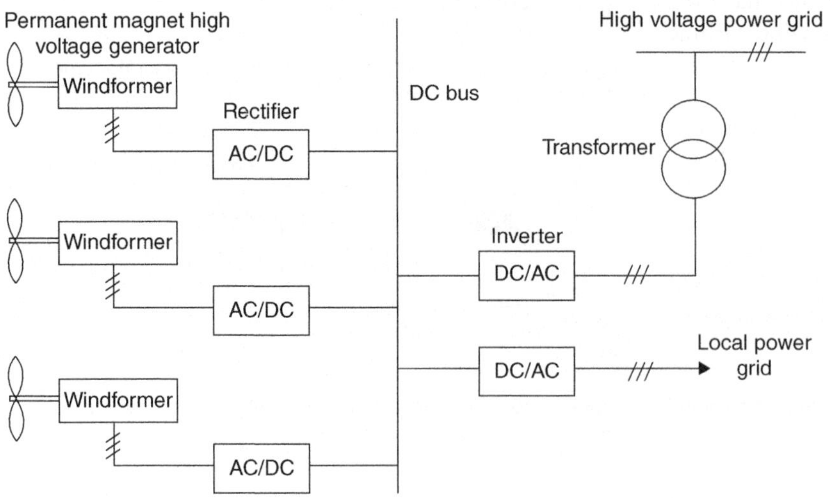

Figure 2.21 High voltage permanent magnet synchronous generator type wind power generation system.

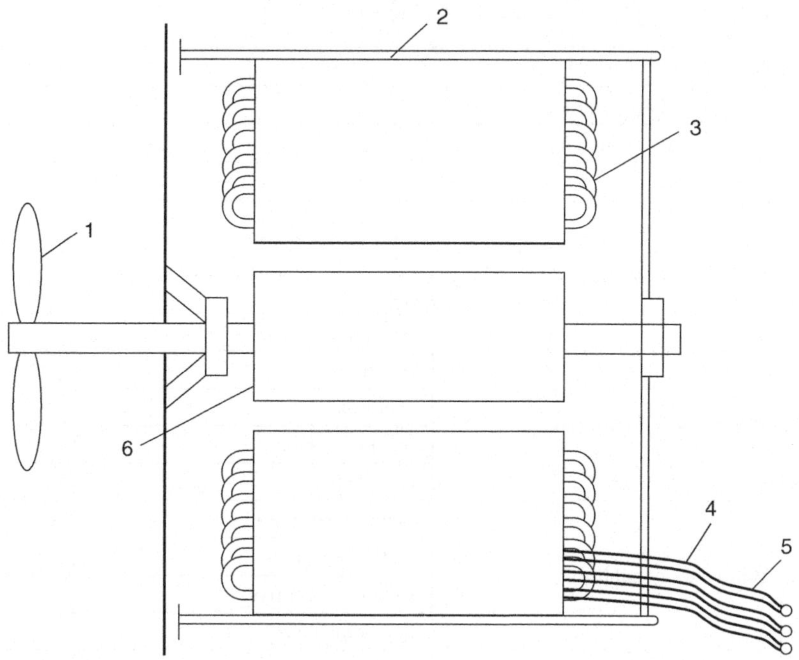

Figure 2.22 Principle and structure of high voltage permanent magnet generator (Windformer™). 1—wind turbine; 2—armature core; 3—high voltage cable armature winding; 4—winding end; 5—cable joint; 6—permanent magnet rotor.

is made by winding a whole cable without any joint in the middle. The winding end is a standard cable end, which is connected with the winding outgoing using the prefabricated vulcanized standard joint. Due to the small working current in the conductor of winding, the rated operating temperature is only 70°C, far lower than the 90°C design standard of the rated operating temperature of XLPE cable, which makes the winding have quite high overload capacity. Therefore, in the design of the generator, the cooling system, and the mechanical stress system can be designed independently and optimally.

The generator rotor is made of new permanent magnet materials NdFeB and SmCo and it is multi-pole. Its structure is similar to that of the above-mentioned multi-pole permanent synchronous generator type wind power generation system. The main difference between them is that this system uses the high voltage permanent magnet generator stator, which is the cable armature winding made of a circular cross-linked XLPE cable and characterized by simple structure, high reliability, light weight, small size, high efficiency, and easy maintenance. However, it has the disadvantage that the field in the air gap cannot be controlled or regulated. But this disadvantage can be completely avoided as all wind turbines use the parallel operation of DC in which it is uncontrolled rectify output mode.

Windformer™ has significant characteristics: ① The wind turbine directly drives the generator rotor, which makes the speed-up gear box unnecessary. ② The rotor is made up of multi-pole permanent magnet steel without the excitation device and slip ring. The entire stator winding is made by winding a high voltage cable (so named Windformer™) without any cable joint whose rated voltage reaches 20 kV and frequency is 5 to 10 Hz, saving the step-up transformer on the generator end. ③ Each generator end is equipped with a set of rectifier to convert the AC into DC and connect it to the DC bus. ④ The WF, through a large common inverter, converts the DC of the DC bus into the 50 Hz high voltage AC, which is then directly connected to the local distribution network or transmitted to nonlocal remote areas after being boosted to a higher voltage through a step-up transformer.

The Powerformer wind power generation system is still in the experimental stage. The long-term operation performance of the whole system is still to be further studied. As the high voltage generator rotor used in the system needs a lot of permanent magnet materials, the system has higher requirements for the stability of the material performance. Meanwhile, it has higher requirements for other aspects of the whole system, which increases the cost of the WTGS. The development of permanent magnet materials has greatly promoted the development and application of permanent magnet synchronous generators. At present, most small- and medium-power synchronous generators in the world use the permanent magnet structure. Compared with conventional rotor, both the weight and size of the permanent magnet generator rotor decrease by about one third while its efficiency increases by 2.2%. At present the wind power generator capable of variable speed operation, which is made of permanent magnet materials, is quite technologically mature (Table 2.2).

Table 2.2 Comparison of Three Kinds of Wind Turbines [2].

Comparison item		CAG	DFAG	PMSG
Cost, size, and weight		+	+/−	−
Grid frequency match		−	−	+
Blade noise		−	+	+
Wind energy output	variable speed	−	+	+
	gear box	−	−	+
	generator	+	+	−
	converter	+	+/−	−
Reliability and maintenance	brush	+	−	−(PM:+)
	gear box	−	−	+
	mechanical load	−	+	+
	complexity	+	−	−
Wind energy quality	power fluctuation	−	+	+
	voltage control	−	+	+
	frequency control	−	+	+
	harmonic level	+	−	−
Impact on power grid fault	short−circuit current contribution	+	+	+/−
	voltage recovery capability	−	+	+

Notes: + indicates relative advantages while – indicates relative disadvantages.

2.3.4 Comparison of Three Kinds of Wind Power Generation Systems

2.3.4.1 Cost, Size, and Weight

In terms of price, the cage synchronous generator (CAG) is about 25% cheaper than the double-fed asynchronous generator (with the winding type rotor), which is cheaper than the direct-drive synchronous generator. In terms of size, the converter of the double-fed synchronous generator is smaller than that of the direct-drive synchronous generator. The direct-drive generator is more expensive because it is large and heavy and must be specifically developed. However, the direct-drive wind turbine generator does not need the heavy gear box.

2.3.4.2 Match with 50–60 Hz Power Grid Frequency

The wind turbine with the generator directly connected to the power grid (CAG and DFAG) needs different gear boxes to match different power grid frequencies, which is different from the use of the converter that can well adapt to the two different power grid frequencies.

2.3.4.3 Blade Noise

In a well-designed wind turbine, the blade is the main source of noise. In the variable speed wind turbine, the rotor rotates very slowly at a low wind speed, generating audible noise, which, however, does not happen in case of the fixed speed wind turbine. At a high wind speed, the noise from the blade tip is covered by the noise caused by the wind blowing past obstacles beside the observer. However, for the wind turbine that is not designed so well, the mechanical resonance can produce other noise.

2.3.4.4 Energy Acquisition

In order to obtain the maximum wind energy, the rotor speed should be proportional to the wind speed (section 2 in Figure 2.12). Thus, the wind energy captured by the variable speed wind turbine is larger than that captured by the fixed speed wind turbine. The components with the energy loss include the gear box and the power electronic converter, which affect the efficiency of the system. The low-speed direct-drive generator is less efficient than 1,000–1,500 r/min standard asynchronous generator because the direct-drive generator needs to produce a greater torque at a very low speed.

2.3.4.5 Reliability and Maintenance

The synchronous generator with brushes and the double-fed asynchronous generator need to be checked regularly and brushes need to be replaced. However, it is not the case with the permanent magnet synchronous generator and the cage asynchronous generator.

As is well known, the gear box is widely used in many fields. But the application record of the reliability of the wind turbine gear box has been quite negative.

In the fixed speed wind turbine, the gusts can directly lead to the changes of the torque while in the variable speed wind turbine, the gusts will lead to the changes of the speed, but only slight changes of the torque. Therefore, the fixed speed wind turbine suffers from heavy mechanical load, which reduces its reliability and increases the maintenance workload. Usually the complicated system suffers from more faults than the simple system.

2.3.4.6 Wind Energy Quality

The wind speed sequences of the three kinds of generator systems measured around the rated speed, their rotating speed, pitch angles, and output power indicate that the power output of the variable speed wind turbine is much more stable than the fixed speed wind turbine (with less flicker). It is because rapid changes of captured wind power are buffed by the rotor inertia while the power output of the fixed speed wind turbine will change rapidly with the wind speed and the tower shield.

If the rated power of the converter is large enough, the variable speed wind turbine can also be used for the power grid voltage and frequency regulation (which is limited by the actual wind speed) while the fixed speed wind turbine can not do this.

The power electronic converter generates harmonics and as a result the filter device may be needed to filter out the harmonics.

2.3.4.7 Power Grid Faults

In the case of power grid fault due to voltage sag, the three kinds of wind power generation system make different responses. In case of fault, the constant speed wind turbine might produce large fault current and the protection system must be started. However, when the voltage is recovered, the wind turbine needs a lot of reactive power and this will impede the voltage recovery following the voltage sag. Meanwhile, the power grid fault and re-connection to the grid will lead to large torque offset, which might damage the gear box.

In case of the power grid fault, the rotor current of the double-fed asynchronous generator increases rapidly. If the special protection system is not available, the wind turbine should be cut out from the power grid within several milliseconds to protect the converter. In the power grid with a large proportion of wind power, when all the wind turbines used in the power grid are disconnected from the grid, there will emerge the problem of power imbalance in the power grid (the generated power at any time must be equal to the consumed power) and power balance is the basic requirements for the self-adjustment of the electrical power system. After the fault is eliminated and the conventional power plant voltage is recovered, the wind turbine can be re-connected to the power grid.

In Table 2.2, it can be assumed that the DFAG wind turbine is equipped with the auxiliary hardware such as CROBAR protection so that the wind turbine does not get disconnected from the power grid and produce the reactive power to help restore the grid voltage in case of the grid fault.

When the fault occurs, the wind turbine in which all the power passes through the converter can maintain its connection with the power grid. When the grid voltage drops, the converter can limit the current to the rated value and continue to transmit active and reactive power to the grid under the reduced voltage. Therefore, in the case of the power grid fault, these wind turbines can help the conventional power station reconstruct voltage.

2.3.5 Other Types of Wind Turbines

In addition to the above-mentioned three kinds of mainstream wind turbines, in order to make full use of wind energy, a great variety of wind turbines have been developed. Here we are going to introduce some variable speed constant frequency wind turbine generator systems with good development prospects [9, 10].

2.3.5.1 AC/DC/AC Wind Power Generation System

The AC/DC/AC wind power generation system is shown in Figure 2.23. In this system, the variable speed constant frequency control strategy is realized in the stator circuit. As the wind speed is constantly changing, the wind turbine and generator also rotate at varying speeds and produce electrical power with variable frequency. The variable frequency AC generated by the generator is first rectified into the DC through the three-phase bridge rectifier and then through the inverter converted into the constant frequency AC which is transmitted into the power grid. As a result, the capacity of the converter is the same as that of the wind power generation system.

In the grid integration, this system suffers no impact current and hardly affects the system. Meanwhile as the frequency conversion device uses the static self-commutated inverter, although the reactive power can be regulated, there is still high frequency current flowing into the power grid.

In this system, generators that can be used include the synchronous generator, cage generator, winding generator, and permanent magnet generator. A comparison of these generators is shown in Table 2.3. If the permanent magnet generator is used in this system, then it uses the

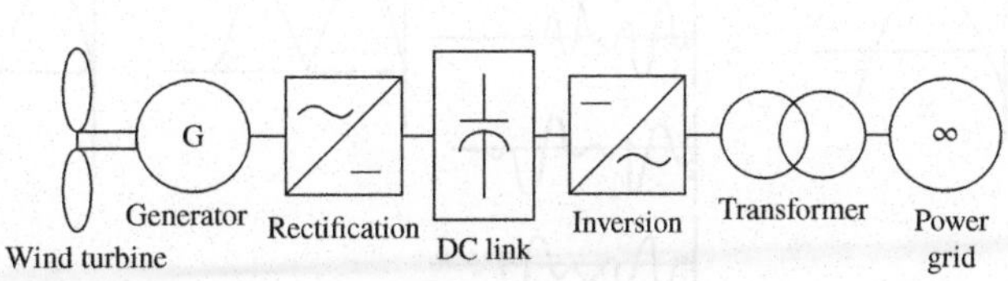

Figure 2.23 AC/DC/AC wind power generation system.

Table 2.3 Comparison of generators in the AC/DC/AC system [10].

Generator used	Synchronous generator	Cage generator	Winding generator	Permanent magnet generator
Rotor structure	As the rotor has the excitation winding and the damper winding; its structure is complicated.	The rotor is of the cage shape. It is simple in structure, easy to manufacture, and reliable in operation.	The rotor is made of winding cable and it is complicated in structure.	The rotor is of the permanent magnet structure and is easy to maintain.
With or without brush and slip ring	with	without	with	without
Excitation mode	The excitation device and the excitation regulation device are required.	Exitation current is acquired from the power grid. The excitation device and the excitation regulation device are not required.	Exitation current is acquired from the power grid. The excitation device and the excitation regulation device are not required.	The external excitation is not required.

direct coupling of the wind turbine and the generator and makes the gear box unnecessary, which can improve its reliability and reduce the system noise as well as the maintenance cost.

2.3.5.2 Magnetic Field-Modulated Generator System

Figure 2.24a shows the block diagram of the single-phase output system of the magnetic field-modulated generator as well as the output voltage waveform of each part. The generator itself has a high rotation frequency f_r and different from the ordinary synchronous generator, it uses the low frequency AC excitation with the frequency f_m (f_m is the required output frequency and generally is 50 Hz) instead of the DC excitation. When the frequency f_m is much lower than the frequency f_r, the output voltage waveform of the generator's three-phase winding will be the amplitude modulated wave composed of two components with the frequency of $(f_r + f_m)$ and $(f_r - f_m)$ respectively. As shown in Figure 2.24b, the frequency of the envelope of the amplitude modulated wave is f_m and the higher frequency of wave contained in the envelope is f_r. Connect the three-phase winding into a group of parallel bridge rectifiers and obtain the full wave rectified sinusoidal pulse wave with the basic frequency of f_m (several ripples with the frequency of $6f_r$), which is shown in Figure 2.24c. The reverse half of the sine pulse wave through the thyristor switching circuit is shown in Figure 2.24d. Finally, filtering out the ripples with the filter to obtain the constant frequency sine wave output is shown in Figure 2.24e. The frequency and phase of the output voltage are determined by the frequency and phase of the excitation current, which makes the magnetic field-modulated generator very suitable for the grid-connected wind power generation system.

The variable speed constant frequency wind power generation system is composed of a specially designed high frequency AC generator and a set of power electronic conversion circuit.

Compared with the AC/DC/AC system, the magnetic field-modulated generator system has the following advantages:

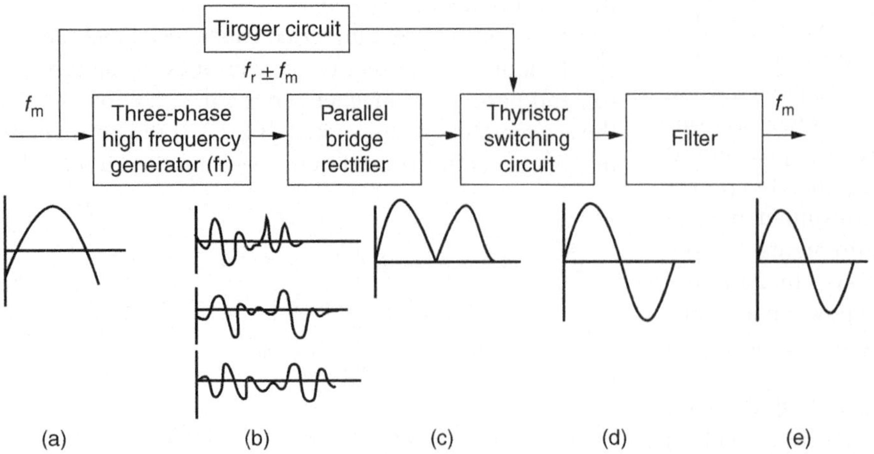

Figure 2.24 Magnetic field-modulated generator system.

1) As the sinusoidal pulse wave is obtained through the parallel bridge rectifier, after the thyristor switching circuit, generally the switch commutates when the waveform crosses zero. As a result, the commutating is simple and easy. Due the low commutating cost, the system efficiency is high.
2) The harmonics account for a very small proportion in the output waveform of the thyristor switching circuit. In addition, the harmonic frequency is very high and easy to be filtered out to get quite good sinusoidal output waveform.
3) As the output frequency of the magnetic field-modulated generator system is the same in principle as that of the excitation current frequency, the parallel operation of this kind of variable speed constant frequency wind turbine with the power grid or the diesel generator set is very simple and reliable.

The main disadvantages of this kind of generator system are similar to those of the AC/DC/AC system. To be specific, the power electronic conversion device is installed in the main circuit and therefore has large capacity. It is suitable for small and medium wind power generation systems with a capacity from tens of kilowatts to hundreds of kilowatts.

2.3.5.3 Brushless Double-Fed Generator System

As shown in Figure 2.25, the brushless double-fed generator is engaged. Its stator has two sets of winding with different number of pole-pairs. One is called power winding that is directly connected to the power grid; the other is called the control winding that is connected to the power grid through the bidirectional converter, which is shown in Figure 2.25. Its rotor is in the cage or the magnetic resistance structure with no brush or collecting ring. The number of the rotor poles should be the sum of the pole-pairs of the two sets of stator winding.

In terms of function, the power winding and control winding of the stator of this kind of brushless double-fed generator are respectively equivalent to the stator winding and the rotor winding of the AC excitation double-fed generator. Therefore, although these two generators have essentially different operating mechanisms, they can realize the variable speed constant frequency control through the same control strategy. For the brushless double-fed generator,

$$f_p \pm f_c = (p_p + p_c)f_m \tag{2.11}$$

where f_p—frequency of stator's power winding current, which is the same as the grid frequency as it is connected to the grid; f_c—frequency of stator's control winding current; p_p—number of the pole-pairs of the stator's power winding; p_c—the number of the pole-pairs of the stator's control winding. In case of super-synchronization, $f_p + f_c = (p_p + p_c)f_m$, while in case of sub-synchronization, $f_p - f_c = (p_p + p_c)f_m$. It can be seen from Formula (2.11) that when the generator speed n changes, namely, when f_m changes, if f_c changes accordingly, f_p will remain constant, namely, remain consistent with the grid frequency, thus realizing the variable speed constant frequency control.

Although this kind of variable speed constant frequency control scheme is realized in the stator circuit, the power passing through the stator control winding is only a small part of the total power of the brushless double-fed generator. It is because the propotion of control winding power and power winding power is $p_c/(p_p + p_c)$. As a result, the capacity of the bidirectional converter shown in Figure 2.25 is also only a small part of the generator capacity.

This control scheme using the brushless double-fed generator can not only realize variable speed constant frequency control and reduce the capacity of the converter, but also achieve the flexible control of the active and reactive power, compensating the reactive power for the power grid. Meanwhile as the generator has no brush or collecting ring, it not only decreases the cost of the generator, but also improves the reliability of the system operation.

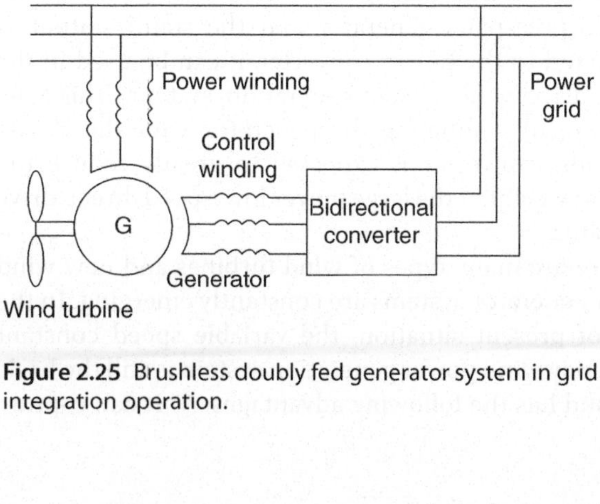

Figure 2.25 Brushless doubly fed generator system in grid integration operation.

2.3.5.4 Claw-Pole Generator System

The only difference between the claw-pole brushless self-excited generator and ordinary synchronous generator lies in its excitation system. Its stator iron core and armature winding are basically the same as those of the ordinary synchronous generator. As the magnetic circuit system of the claw-pole generator is a parallelstructure-and the magnetomotive force of all pole-pairs is from a common set of excitation winding, compared with the ordinary synchronous generator, the excitation winding therfore uses very few materials and needs smaller excitation power. For an eight-pole generator, when each pole has identical magnetic flux and magnetic intensity, the copper wire needed for the claw-pole generator excitation winding and its consumed excitation power will be less than half of those of the ordinary synchronous generator. Thus, it has higher efficiency. In addition, like the permanent magnet generator, the brushless claw-pole generator has no brush and so basically does not need maintenance. Compared with the permanent magnet generator, the brushless claw-pole generator hardly has starting resistance torque except the mechanical friction torque. Another advantage is that it has good regulation performance. Its output characteristics can be easily controlled through the regulation of the excitation and it has the potential to make wind turbine achieve the optimal tip speed ratio operation and obtain the best operation efficiency. This generator is very suitable for the kilowatt-level wind power generation device.

2.3.5.5 Switched Reluctance Generator System

The switched reluctance generator type wind power generation system is shown in Figure 2.26. This system takes the switched reluctance generator as the core of electromechanical energy conversion and is generally applied to the small wind power generation system with the output power below 30 kW.

The switched reluctance generator is the doubly salient motor and relies on the reluctance torque for operation. Both its stator and rotor are of the salient-pole slot structure. Its stator is equipped with concentrated winding. Its rotor has no independent excitation winding or permanent magnet. Instead, it is made of laminated silicon steel sheets. Therefore, the mechanical structure is simple and solid. It achieves excitation and power generation through the time-sharing control of the controller and it has high reliability. In addition, it is characterized by large starting torque, low speed, and good performance. If it is designed to be of multiple pole-pairs, it can be directly connected to the wind turbine, forming the direct-drive wind power generation system without gear box. If the switched reluctance generator matches well with the wind turbine, through the control over the wind power generation system, it can enable the wind turbine to work at the optimal power load curve. The switched reluctance generator has the characteristics of numerous controllable parameters, nonlinearity, and lack of clear mathematical models. Compared with the conventional brushless DC generator and the rotating rectified brushless synchronous generator, the switched reluctance brushless DC generator has advantages such as

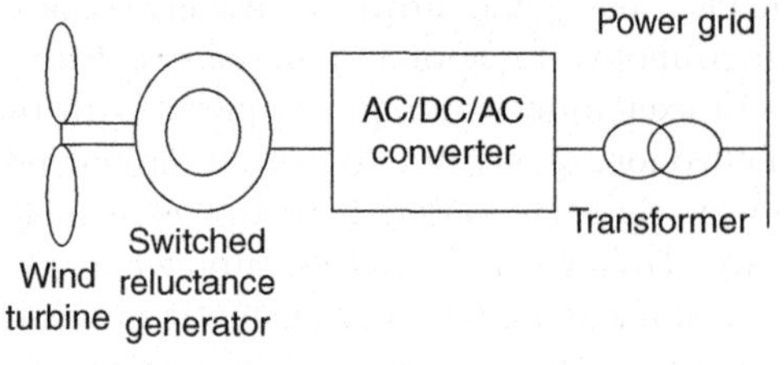

Figure 2.26 Switched reluctance generator type wind power generation system.

strong fault tolerance capability, easy combined starting and power generation, suitability for high temperature (>250°C) and high-speed (30000 r/min) operation, large capacity, high efficiency, and high power density operation. The switched reluctance generator has no independent excitation winding. Instead it combines the excitation winding with centralized embedded stator armature and achieves the excitation and power generation through the time-sharing control of the controller. As a result, the simplified structure improves its reliability. Meanwhile, there is no electrical coupling among the generator phase windings, which greatly increases its fault tolerance capability.

The converter in the switched reluctance generator system is directly connected with the generator stator and its capacity is also required to be the same as the rated capacity of the generator. The power density of switched reluctance generator is not large, so the performance of the wind power generation system with this kind of structure is not as good as that of the permanent magnet generator type wind power generation system. In addition, it has higher requirements for the performance of the power electronic conversion device and the control of its system is complicated.

2.3.5.6 Transverse Flux Generator Type Wind Power Generation System

The transverse flux generator type wind power generation system is the same as that shown in Figure 2.27. The stator cog-slot and armature winding of the transverse flux generator are perpendicular to each other in space; the direction of magnetic circuit is along the axial direction of the rotor; the stator and coil dimensions are independent of each other. This kind of system realizes the decoupling of the circuit and magnetic circuit. To be specific, it can also achieve high electrical load and high magnetic load. Moreover, the magnetic circuit of the transverse flux generator is three-dimensional and can be categorized into several types according to the way to place the rotor permanent magnet poles. The transverse flux generator is a kind of synchronous generator and its operation mechanism has the characteristics of the permanent magnet generator. If it is designed to be a multi-pole-pairs generator and the stator output is connected to the power converter, it can be used in the direct-drive wind power generation system. This new structure of combining the generator with the power electronic converter is expected to greatly reduce the size and weight of the large power low-speed direct-drive generator.

There are many types of wind turbines and new wind power generator systems are constantly emerging. In the light of present situation, the variable speed constant frequency wind power generation system is still the basic type and has the following advantages: ① It can capture

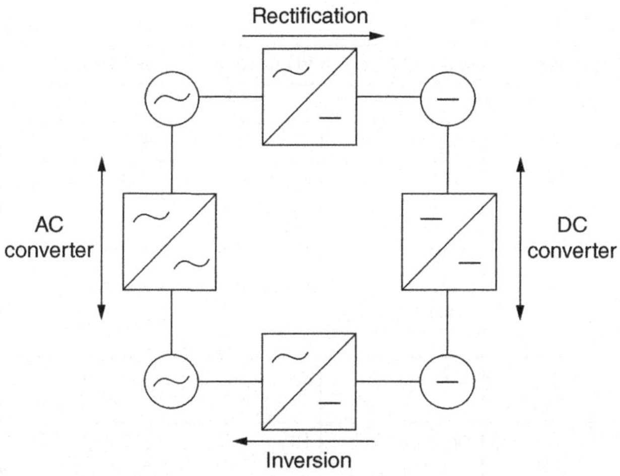

Figure 2.27 Power conversion through power converter.

the maximum wind energy. ② It has a wide speed range to adapt to the wind turbine speed changes caused by the changes of the wind speed. ③ It adopts a certain control strategy (such as the vector PWM) so as to be able to flexibly regulate the active and reactive power of the system and regulate the power factor of the grid. ④ It uses the advanced PWM control technology to suppress harmonics, reduce switching loss, improve efficiency and reduce the cost.

2.3.6 Technical Parameters of Wind Turbine Generator

The main technical parameters of several types of large wind turbine generators are listed in Table 2.4.

2.3.7 Development Trends of Wind Power Generator System

2.3.7.1 Some New Types of Wind Turbines

1) **Variable speed cage asynchronous generator** It is the variable speed wind turbine based on the cage asynchronous generator and the generator is connected with the full power converter. Compared with the double-fed asynchronous generator, the generator has the following advantages: 1) The generator is cheaper. 2) The generator has no brush. 3) It can be used as the standard industrial transmission. 4) It can work in the 50 Hz and 60 Hz power grid. Its disadvantages include: 1) The converter relatively large and expensive (its capacity is the full rated power of the asynchronous generator). 2) The converter loss is high because the converter needs to transmit all the power. The scheme is rarely used because of its obvious disadvantages.
2) **Variable speed wind turbine system based on the gear box and synchronous generator** It is the wind turbine based on the gear box and the brushless synchronous generator system using the full power converter. Compared with the double-fed asynchronous generator system, it has the following advantages: 1) The generator has higher efficiency. 2) The generator is cheaper. 3) The generator has no brush. 4) It can work in both the 50 Hz and 60 Hz power grid. Its disadvantages include: 1) The converter loss is greater because the converter needs to transmit all the power. 2) The cost of the power electronic system can be reduced gradually (in the past 10 years it has been reduced to one tenth and in the near future this system will become more attractive).
3) **Double-fed asynchronous generator system based on the gear box** The double-fed asynchronous generator system is suitable to operate in case of the grid fault because it has good fault ride through capability, which has been proven feasible in theory. The future goal is to manufacture the wind turbine which can auxiliary control the grid voltage and frequency and be used in practice.
4) **Direct-drive generator system** Existing direct-driven generators mainly include the electrical excitation synchronous generator and permanent magnet synchronous generator. In the early stage, due to the high price of the permanent magnet motor, the direct-drive permanent magnet synchronous generator did not develop as well as the electrical excitation synchronous generator. But in the past 10 years, the price of permanent magnet has been reduced a lot. Compared with the electrical excitation generator, the permanent magnet generator has the advantages such as lower loss (without excitation loss), lighter weight (its weight is one half of the electrical excitation generator) and lower cost. Its disadvantage is that the excitation cannot be controlled, but the permanent magnet generator has become more and more competitive due to the constantly decreasing price of the permanent magnet.

The direct-drive generator is not standard ready-made equipment, so it is very valuable to study the topology structure of the AC generator, which will be able to reduce its weight and cost in the future.

2.3.7.2 Voltage Level Trends

A few years ago, manufacturer mainly produced 400 V and 700 V wind turbines. However, in the 1990s, ABB became a pioneer in the field of wind power generation and produced a medium-voltage-level generator. In the past few years, several wind turbines of higher voltage level have been produced: Vestas and Made uses 1 kV; Zephyors uses 3 kV; NEG-Micon even uses 6 kV for its rotor winding. In principle, there is no connection between the voltage class and the electromagnetic design of the generator.

Table 2.4 Comparison of some technical parameters of wind turbine generators.

Product model	WTG600C	WTG750A	WTG850B	WTG1200A	WTG1250E	WTG1500H	WTG2000D
Rotor							
Diameter/m	43	50	52	62	66	77	80
Wind-swept area/ m^2	1466	1963.5	2124	3019	3421	4657	5027
Rotational speed/r·min $^{-1}$	17.9–26.8	22	14.6–30.8	11–20	13.5–20.3	10.1–20.4	9.0–19.0
Power regulation	Fixed pitch		Variable pitch				
Cut-in wind speed/m·s $^{-1}$	3.2	3.5	4	3	3	3.5	4
Rated wind speed/m·s $^{-1}$	14	15	16	12	14	15	16
Cut-out wind speed/m·s $^{-1}$	25	25	25	25	22	25	25
Survival wind speed/m·s $^{-1}$	70	70	70	59.5	70	55	59.5
Generator							
Type	Cage asynchronous generator		Double-fed asynchronous generator	AC permanent magnet synchronous generator	Asynchronous generator with 4/6 poles	Double-fed asynchronous generator	
Capacity/kW	600	750	800	1200	1250	1500	2000
Voltage/V	690			620	690		
Power factor	0.99 (with reactive power compensation)		0.99	1.0	0.99	Adjustable between ±0.95	1.0
Frequency/Hz	50						
Blade							
Length/m	19.1	23.5	25.3	30	31	34	39
Gear box							
Transmission ratio	1:56.56	1:70.02	1:61.74		1:74.917	1:90	1:100.5
Brake system							
Main brake system	Tip pneumatic brake		Blade full feathering	Tip pneumatic brake	Blade independent variable pitch brake		Disc brake
Second brake system	Hydraulic mechanical brake		Mechanical disc brake	Generator brake	Hydraulic mechanical brake		
Control system							
Control system	Open-loop control			Closed-loop control			
Tower							
Form	Steel cone cylinder						
Height/m	48.5	60/65	55/65	70	74.5	65	67/78
Weight							
wheel/t	13	14	15.9	30	32	32.9	37.2
Engine room/t	22	22	23	50	47	51	61.2
tower/t	52	53/76.2	57/73	93	106	82	135/190

By doubling the number of the coils in the slot and then dividing the cross section of the coils equally, we can double the voltage without increasing the amount of copper used in the slot. However, in case of the low voltage class, a large cable is required to connect the generator terminal and the transformer. When the voltage is higher than several kV, the wire insulation takes up more space and the amount of copper used in the slot is reduced, which requires a larger generator to convert the same amount of energy. Generally, in the power plant, even the voltage of the turbine generator with large power (500 MVA) does not exceed 20 kV. In addition, power electronic converter whose voltage is up to several kV has already been produced. As a result, it is possible to increase the voltage to several kV, but by estimation it is impossible to exceed 5 kV [2].

2.3.7.3 Offshore Wind Turbine Generator

Offshore wind turbine generators are required to be firmer and need no maintenance. It is because the maintenance and repair of wind turbine generators on the sea is very expensive and difficult and even impossible in some weather conditions. In addition, in the marine environment the insulation materials can be extremely easily damaged and the iron or permanent magnet might rust. Therefore, special anti-rust measures as well as the methods for regulating the air temperature and humidity are required.

Given the special conditions at sea, the direct-drive permanent magnet synchronous generator is more suitable for offshore WFs because they need no maintenance for the brush and the gear box.

2.4 Power Electronic Technology in Wind Power Conversion System

2.4.1 Application of Power Electronic Technology in Wind Power Conversion System

In the development of the modern wind power conversion system, the application of the power electronic converter is an important milestone. Thanks to the application of the power electronic converter in 50 kW wind turbine generators in the late 1980s, great breakthroughs were made in the wind power generation system technology. Later on, the direct-drive permanent magnet synchronous generator and double-fed asynchronous generator that developed rapidly were combined with the power electronic converter so that wind turbine generators could operate at variable speed and the output active and reactive power of wind turbines could be regulated, greatly improving the operation and control performance of the wind power generation system. In a certain sense, it can be said that the modern wind power generation technology is based on the development of the power electronic technology and that without the power electronic technology, it is impossible for the wind power development to reach the present level. In the future, power electronic technology will continue to promote the development of wind power in a better and more economical way. Power electronic technology has a wide range of applications in the wind power generation system. Listed below are the main areas of the application of power electronic technology in the wind power conversion system.

2.4.1.1 Application in the Wind Power Generation System

First, in the cage asynchronous generator system, the soft starting grid integration device made up of power electronic devices is needed to achieve the smooth starting and grid integration of wind turbines. Second, in the wound rotor asynchronous generator system, mainly DFAG, the back-to-back voltage source converter is needed to provide excitation for the rotor winding and regulate the voltage and reactive power output. And third, in the direct-drive wind power generation system (e.g., the permanent magnet synchronous generator system), the back-to-back current or voltage source converter is needed to be connected to the power grid to provide the AC system with voltage and reactive power support.

2.4.1.2 Application in the Wind Power Output System

WFs are usually built in areas rich in wind energy resources and far away from cities and towns. As a result, the transmission capacity of the transmission lines is a restrictive factor for the wind power development. Now the AC transmission mode is mainly used, but it also has some shortcomings. High voltage direct current (HVDC) and VSC-HVDC, also known as HVDC Light as well as the FACTS device can be potentially applied to wind power transmission system,

2.4.1.3 Application in the Wind Turbine and WF Control System

In order to meet the power grid operation specifications, the power electronic devices are needed in low/high voltage ride through (LVRT/HVRT) or fault ride through (FRT) of the wind turbine. In WFs, especially those mainly composed of conventional cage asynchronous generators, the SVC or STATCOM is usually required to improve the voltage and reactive power control of the AC system. With the increase of propotion of wind power in electrical power system (e.g., wind power accounts for more than 20% of the total capacity of the power system), in order to stabilize the WF power output, effective large power capacity storage technologies are required to regulate the output power. Technologies that can hopefully store large quantities of electrical energy such as the battery energy storage, flywheel energy storage, super capacitor energy storage, and compressed air energy storage technologies generally require the power electronic device and AC system interfaces. At present in the most cost-effective pumped hydro energy storage system that can store large quantities of energy, proper power electronic technologies are also required in the starting and operation conversion of variable speed generators/motor pumped storage units.

2.4.1.4 Improving WF Power Quality

The power electronic converter connects the generator and the power grid and can be used in the separate control of the active and reactive power to ensure that high-quality electrical energy can be transmitted to the power grid at variable wind speeds. The WF dynamic reactive power compensation device can maintain the PCC point voltage and power factor.

Through the introduction of the power electronic devices, many wind turbine generator systems acquire performance similar to that of conventional power plants. Modern wind turbines have the ability to quickly respond to the requirements of the power grid operation, but the actual power generated by the wind turbine generator is determined by the available wind speed. In some solutions, for example, the wind turbine based on the full rated power converter can transmit reactive power without generating any active power. This kind of wind turbine generator can be used to reconstruct the grid voltage in case of grid fault. Even if more power can be obtained from the wind, it is possible for the wind turbine to operate at a low power (part load operation mode). As a result, it can be used as the spinning reserve capacity of the electrical power system. Finally, some systems can manage islanded operations in case the power grid collapses.

2.4.2 Power Electronic Converter

The power electronic converter, the so-called power converter, is the most common solution for the conversion and control of electrical power. Figure 2.27 shows the low loss energy conversion through the use of the power converter. Power converters include the AC converter and the DC converter. In the AC converter, the rectifier converts a single-phase or three-phase AC system into the DC system and the power flows into the DC side. The inverter converts the DC into the single-phase or three-phase AC and the power flows into the AC side. The AC conversion is to convert the AC with a given voltage, frequency, and number of phases in order to be used by the AC system with different voltages, frequencies, and number of phases. The DC conversion is to convert the DC with a given voltage and polarity in order to be used by the DC system with different voltages and possibly opposite polarities.

The main components of the power conversion system include the power part (with the so-called power converter valve, which can transmit power) and electronic signal processing unit (for digital control, protection, and regulation) part.

In the wind power conversion system, the application of the power electronic converter can increase the degree of wind energy conversion and regulate the generator frequency and voltage to meet the grid requirements, especially in the variable speed wind turbine system. The application of connecting the wind turbine to the power grid through the converter (completely through the stator or partly through the rotor circuit) has become increasingly common, especially in the MW-level wind turbine system.

As the wind turbine is always equipped with the three-phase generator, so the three-phase converter is used for power and frequency conversion. The converter in the wind power conversion system can be categorized into direct converter (AC-AC converter or cycloconverter) and indirect converter (AC-DC-AC converter).

2.4.2.1 Direct Converter

The direct converter converts the AC of a certain frequency into the AC of another frequency without the DC link, so it is also called frequency converter, which is shown in Figure 2.28. The whole system is composed of two sets of converters. The control system alternately controls the two sets of anti-parallel converters based on the polarity of the load current to make them in the rectification and inversion state in turn in order to obtain the AC voltage with adjustable frequency. The output voltage of the AC-AC converter is determined by the control angle of the rectifier, which can produce the square wave AC voltage or sinusoidal AC voltage.

Direct conversion is particularly used to reduce the frequency. When the power supply is from the 50 Hz grid, the output frequency usually operates within the frequency range of 0–25 Hz and also allows the frequency conversion of the same magnitude. Without intermediate links, it improves the conversion efficiency and can operate in four quadrants.

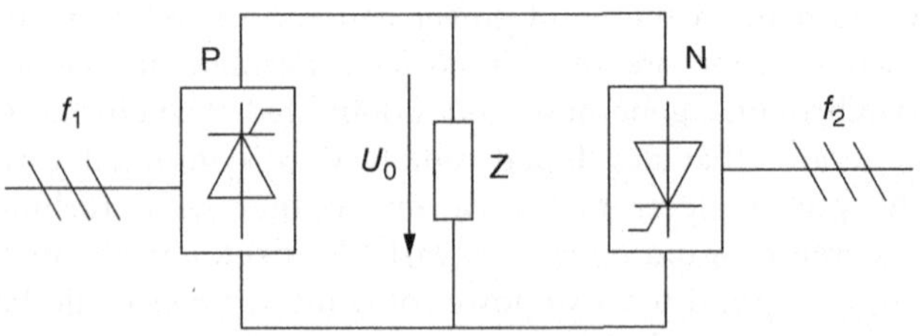

Figure 2.28 Schematic diagram of AC-AC converter basic circuit principle.

2.4.2.2 Indirect Converter

The indirect converter, namely, the AC-DC-AC converter, is composed of the rectifier, the constant voltage, or constant current DC link and the inverter. The indirect converter first converts the constant frequency (e.g., 50 Hz) AC into DC through the rectifier. Then after the filter, it inverts the smooth DC into the AC with adjustable frequency and amplitude. The frequency variation of the AC-DC-AC converter is controlled by the thyristor and the output current frequency can be higher or lower than the industrial frequency so as to achieve a wide conversion range of frequency.

According to the characteristics of the DC power supplied to the inverter or the different components of the filter, the AC-DC-AC frequency conversion mode is categorized into current source type and voltage source type.

2.4.2.3 Current Type Converter

The main circuit structure of both the current type converter and the voltage type converter is the three-phase bridge structure and the difference between them is that the DC part of the current type converter is connected with the large capacity inductor L instead of the capacitor C, which is shown in Figure 2.29a. The output impedance of the converter is very large and the amplitude of current output to the load is generally not affected by the load and remains the same. This characteristic is similar to that of the current source, so it is called current type converter. The AC output by the current type converter is the square wave current while its voltage is close to a sine wave. For the rectifier, the current type converter generally uses the phase control rectifier or the DC chopper, which can be used to regulate the DC power supply andcontrol DC current by controlling DC voltage so as to contro the ouput.

Due to its controllability, the current type inverter can limit the overcurrent caused by the commutation failure of the inversion device or the load short circuit. Due to its high protection reliability, it is mainly used when frequent acceleration and deceleration or four-quadrant operation is required.

2.4.2.4 Voltage Type Converter

For the DC part, the large capacity capacitor is used for filtering. The DC circuit voltage waveform is relatively flat and has small output impedance so that the voltage is less susceptible to mutations. Judging from the inverter, the DC power supply is the voltage source with small internal resistance, equivalent to the constant DC voltage source, so it is called the voltage type converter. The typical form of the main circuit of the voltage type converter is shown in Figure 2.36b.

The AC output by the inverter of the voltage type converter is square wave voltage or square wave voltage series while the current is close to a sine wave after the motor load filtering. As the voltage type converter provides the AC power for the AC load as a voltage source, its main advantage is that its operation is almost not affected by commutation or the power factor of load. Its disadvantage is that when the load is short-circuited or the load is input in the running status of the converter, it is prone to overcurrent and protective measures must be taken in a very short period of time.

The advantage of the voltage source converter is that it has good adaptability to the load, but there exist fifth and seventh harmonics in the system, which can make

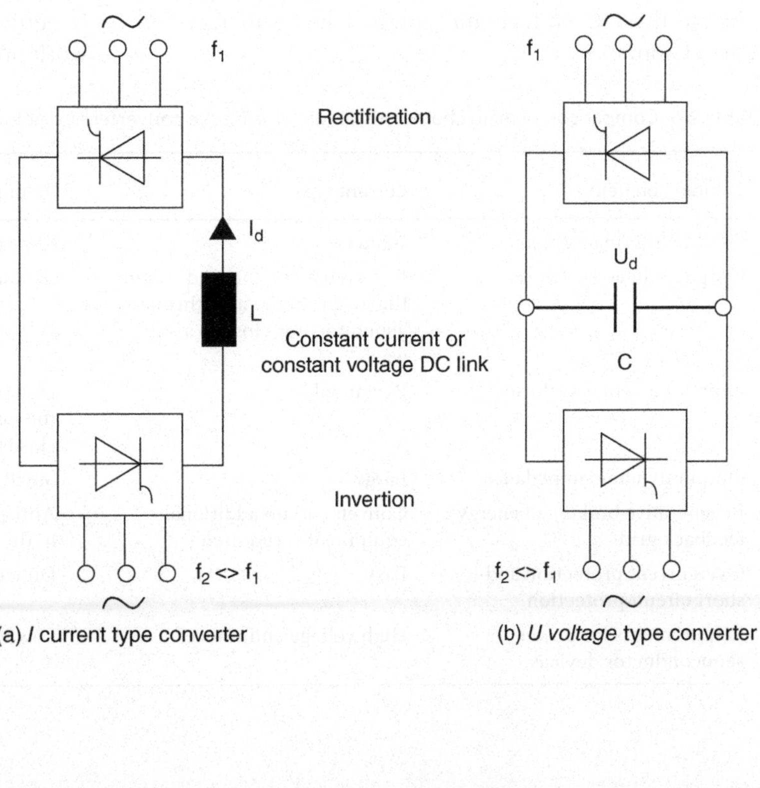

Figure 2.29 Indirect (AC-DC-AC) converter.

the generator get heated and generate the torque ripple. If the capacitor is used for filtering, it is difficult to control the current. In case of output short circuit, it will definitely lead to inversion failure and damage the thyristors. The current source converter has the advantages such as simple major loop, low cost, wide speed range, and high efficiency up to 95%. In addition, as it uses the reactor for filtering, the current is easy to control. the current inner loop is set in the control system to regulate the current, which increases its current limiting capability and leads to low system harmonic pollution. Even in case of the inversion failure of the inverter or load short circuit, it can limit the fault current to avoid accidents. As a result, it has high operation reliability. Table 2.5 shows the comparison of the main characteristics of the current type converter and the voltage type converter.

2.4.2.5 Comparison of AC-DC-AC Converter and AC-AC Converter

Based on the above introduction of the thyristor AC-DC-AC converter and AC-AC converter, it is easy to compare their main characteristics, which are shown in Table 2.6 [11].

The indirect converter has obvious advantages in energy conversion and connecting the variable speed wind turbine to the power grid. The direct converter is only used in very few cases such as power supply of the double-fed asynchronous generator rotor circuit.

2.4.2.6 Chopper DC-DC Converter

The AC voltage can be changed through the transformer. How to change the DC voltage? The DC voltage can be changed through the resistance voltage division, but the efficiency is very low. The chopper is a device that can change the DC voltage and conduct high-efficient DC conversion.

The power transistor or IGBT ca be used as a switch to control the on-off time so as to control the output DC voltage.

Mainly there are two ways to convert the DC into DC. One is through the chopper that can directly convert DC into DC of varying voltages through the on-off operation of the switching device. The other is through the indirect conversion of the transformer.

For the DC chopper, the average voltage on the load is changed through the on-off operation of the switch. In order to eliminate the voltage ripples on the load, the filter can be inserted into the circuit. For the same average DC voltage, with the increase of the switching frequency, the filter can be used to easily achieve the filtering of the DC voltage waveform.

Choppers can be categorized into buck chopper, boost chopper, buck-boost chopper, and so on.

2.4.2.7 Buck Chopper

Shown in Figure 2.30 is the schematic diagram of of the buck chopper. When the switch S is turned on, the voltage at both ends of the load (including the filter inductor L) is $E_{d,}$ and the inductor current increases accordingly. When the switch S is turned off, the current flows through the free-wheeling diode D and then decreases. Turn on the switch s once again before or after the current reduced to zero, with the former being the current continuous conduction mode and the latter the current discontinuous conduction mode. Both conduction modes have an important impact on the performance of the system. Generally, in order to reduce the current pulsation, the circuit is connected in series with the filter inductance to maintain the current continuity. The effect of the buck chopper on the DC voltage is equivalent to reducing the power supply voltage to work on the load.

Table 2.5 Comparison of main characteristics of current type converter and voltage type converter [11].

Comparison item	Current type	Voltage type
DC link filtering elements	Inductor	Capacitor
Output voltage waveform	Determined by the load. When the load is in the asynchronous generator, it is close to a sine wave.	Rectangular
Output current waveform	Rectangular	Determined by the load. When the load is the inductive load, it is close to a sine wave.
Output dynamic impedance	Large	Small
Regenerative braking of energy feedback grid	Convenient; no additional equipment is required	Anti-parallel converter is needed in the rectification side.
Overcurrent protection and short circuit protection	Easy	Difficult
Requirements for power semiconductor devices	High voltage endurance	Low voltage endurance

Table 2.6 Comparison of main characteristics of thyristor AC-DC-AC converter and AC-AC converter.

Category / Comparison item	AC-DC-AC converter	AC-AC converter
Energy conversion mode	Energy is converted twice; low efficiency	Energy is converted once; high efficiency
Commutation mode	Forced commutation or load resonant commutation	Power supply voltage commutation
Number of device elements	Small	Large
Frequency variation range	Wide frequency regulation range	Generally, the highest output frequency is one third to one half of the grid frequency.
Grid power factor	When the controlled rectifier is used to regulate the voltage, the power factor is low at a low voltage; when the chopper or the PWM is used to regulate the voltage, the power factor is higher.	Low
Use occasion	It can be used in various electrical power drive devices such as constant frequency constant voltage power sources and uninterruption power sources.	Especially suitable for low-speed and high-power drive [12].

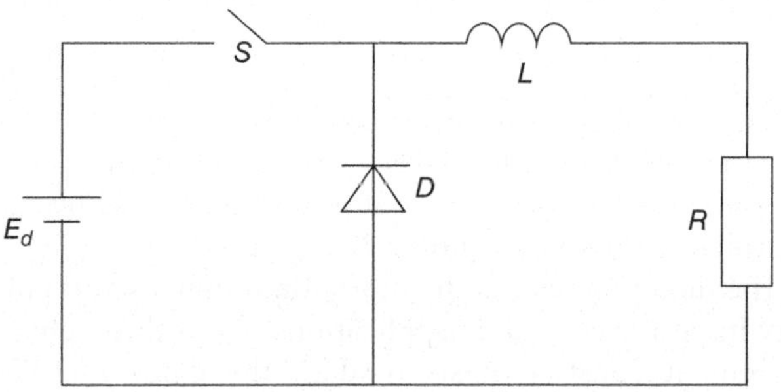

Figure 2.30 Buck chopper.

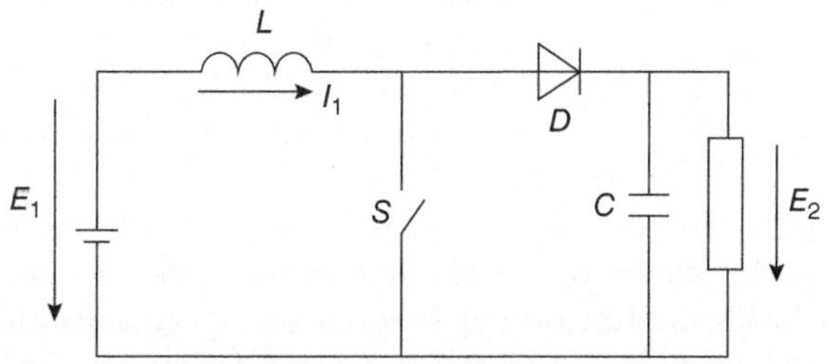

Figure 2.31 Equivalent circuit of boost chopper.

2.4.2.8 Boost Chopper

Shown in Figure 2.31 presents the principle of the boost chopper. When the switch S is turned on, the power source stores energy in the inductor L; when the switch S is turned off, both the inductor and the power source provide the energy for load at the same time. If the inductor L is large enough, the current flowing through the inductor is a certain value I_1 and the energy stored in the inductor L during the period of time t_{on} when the switch S is turned on is $E_1I_1t_{on}$. If the filter capacitor is large enough, its output voltage can be regarded as a certain value and the energy released by the inductor to the load during the period of time t_{off} when the switch S is turned off is $(E_2-E_1)I_1t_{off}$. In steady state, the energy absorbed by the inductor should be equal to the energy released by it. It can be deduced that the output voltage is higher than the input voltage, that is, the circuit has the ability to increase the input voltage.

2.4.3 Topology of Converter in Wind Power Generator System

The task of the power converter is to regulate the voltage and frequency of the wind turbine to the voltage and frequency of grid. Its impact on the grid should be as small as possible. The connection between the wind turbine and the grid is basically determined by the characteristics of the generator and the permitted influence on the grid. In the application of the converter, the conversion loss and the additional technical cost must be compensated through the acquisition of the corresponding higher energy. Therefore, low loss and low cost must be the target of feed.

The electrical power supplied to the grid should be approximately sinusoidal and contain few harmonics. It is necessary to control and stabilize the oscillation tendency (especially in the permanent magnet synchronous generator) with the help of appropriate generator current in order to ensure the reliable operation of the device.

In addition, in the design of the frequency converter, the system criteria made based on the generator and power grid requirements must be observed. Here, the electromagnetic compatibility, system loss, normal operation, and protection measures taken in fault conditions as well as the control of the energy flowing into the grid have special significance.

At present, two types of power converter topology with full control of power generation are used in the wind power generation system: partial rated power converter and full rated power converter [8]. The former refers to only part of the power generated by the wind turbine is converted by the converter; the latter refers to all the power generated by the wind turbine is converted by the converter. Now we are going to introduce several commonly used power converter topologies.

2.4.3.1 Bidirectional Back-to-Back Two-Level Power Converter

The back-to-back pulse width modulation-voltage source converter (PWM-VSC) is a bidirectional power converter composed of two conventional PWM-VSCs. It is a kind of indirect converter and its topology is shown in Figure 2.32.

As shown in Figure 2.32, the system can be divided into three subsystems: the generator side rectifier converts the generator voltage or current into DC and controls the operation of the generator so as to control the operation of the wind turbine. At this time, the decisive variable is the driving torque, rotating speed, and rotating angle, as well as the generator voltage and current. The DC link (DCL) with the function of energy storage can separate the grid frequency and the generator frequency. The performance of the DC link is influenced by its voltage and current level. The grid supplies power through the inverter. The operating state of the inverter is characterized by the voltage, current, frequency, and the generated active and reactive power.

PWM-VSC is the most commonly used three-phase frequency converter and its technical advantage is the decoupling between the power grid side converter and the generator side converter by using capacitors. In addition to providing some protection, this kind of decoupling also provides the separate control of the two converters, allowing the asymmetric compensation of the generator-side converter and the grid side converter. The boost inductor circuit in the DC circuit increases the number of components, but the positive effect is to reduce the performance requirements of the grid-side harmonic filter and provide some protection for abnormal states of the power grid side.

However, the back-to-back PWM-VSC also has its disadvantages. One of the disadvantages is that the DC capacitor is large and heavy, which increases the cost and reduces the life of the whole system. Another important disadvantage is the switching loss. Every commutation in the grid-side converter and the generator-side converter is related to the hardware switch and the natural commutation. As the back-to-back PWM-VSC is composed of two converters, its switch loss is more significant. For the power grid, high switching speed may require additional electromagnetic interference (EMI) filters. In order to avoid the high stress to the insulation components in the generator and shaft current question, the output filter must be used to limit the voltage gradient.

The back-to-back PWM-VSC topology is widely used in the wind turbine based on large DFAG. However, now some wind turbine manufacturers use this kind of topology in the full rated power converter wind turbine with the cage asynchronous generator. This kind of topology can also be used in the permanent magnet synchronous generator.

2.4.3.2 Unidirectional Power Converter

For the generator-side converter, the winding rotor synchronous generator only requires a simple diode bridge rectifier, as shown in Figure 2.33.

The diode rectifier is the most commonly used topology in power electronic applications. For a three-phase system, it consists of six diodes. The diode rectifier can only be used in one quadrant, which is simple and uncontrollable. It can be used in some applications with DCL. The variable speed operation of the wind turbine can be achieved by an additional power converter which feed to the excitation winding. The grid-side converter will provide the decoupling control of active and reactive power transmitted to the grid and plays the important role of providing all kinds of power grid support. These wind turbines can have a gear box or be directly driven. In order to achieve the variable speed operation, for the wind turbine equipped with the permanent magnet synchronous generator, the boost DC/DC converter

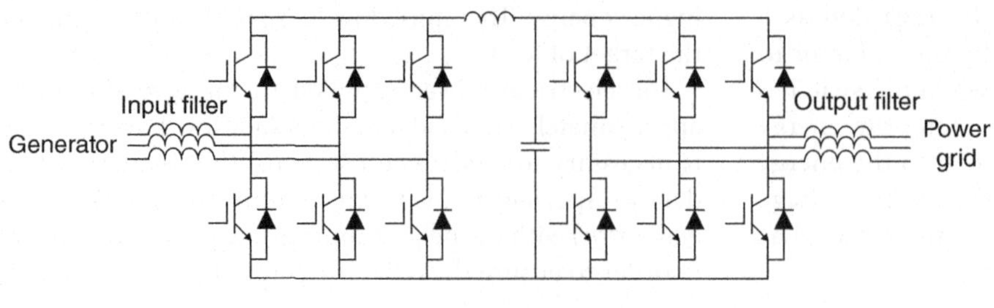

Figure 2.32 Structure of back-to-back voltage source power converter.

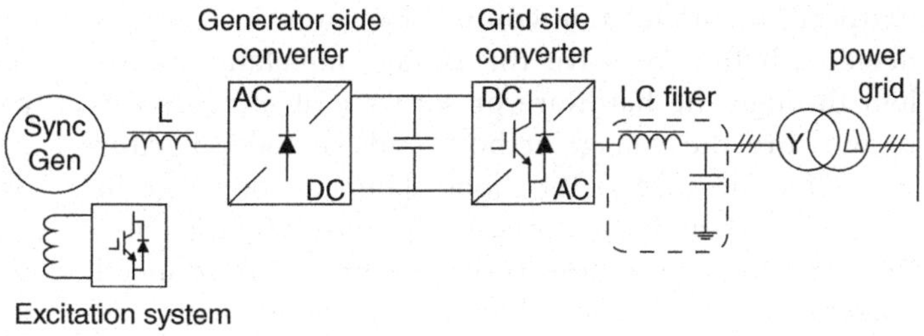

Figure 2.33 Variable speed wind turbine with synchronous generator and full rated power converter.

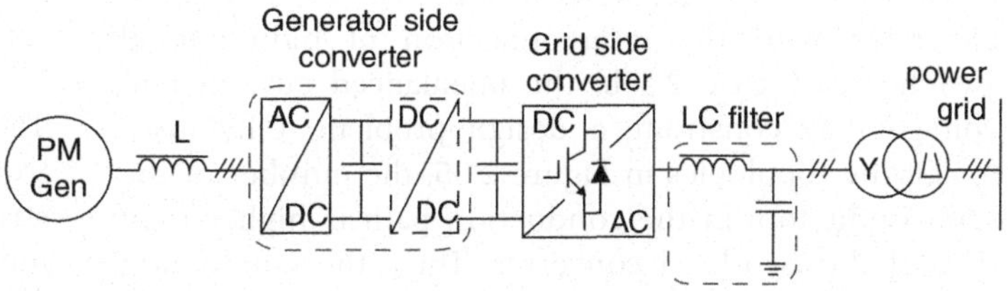

Figure 2.34 Wind turbine with permanent magnet synchronous generator and full rated power converter.

(chopper) is required to be embedded in the DCL circuit, as shown in Figure 2.34.

2.4.3.3 Multilevel Power Converter

At present, interest in the multilevel power converter is increasing, especially its application in the medium-large capacity and high voltage wind turbine. The general idea of the multilevel converter is to create a sinusoidal voltage from several voltage levels, typically obtained from the capacitor voltage source. The proposed different multilevel converter topologies can be categorized into the following five categories: diode-clamped multilevel structure, multilevel structure with interconnection of the bi-directional switches, multilevel structure with flying capacitors, multilevel structure with multiple three-phase converters, and multilevel structure with cascaded single-phase H-bridge converters. These topologies are shown in Figure 2.35.

However, the multilevel converter also has some disadvantages. The most common one is that for three-level converter with the function of isolating DC faults, the voltage between the upper and lower capacitors in DC link is not balanced. Still, for the three-level converter this problem is not very serious. The problem in the three-level converter is mainly due to the difference of the actual capacitance of each capacitor, the error in the execution within dead time, and the unbalanced load. The imbalance problem can be solved through the proper modulation control of the switch. The voltage imbalance problem can be solved through the

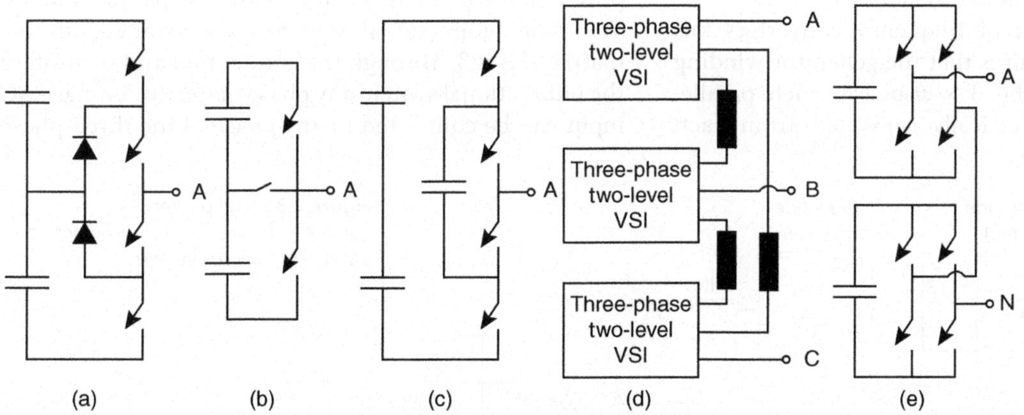

Figure 2.35 Multilevel topology. a) One arm of the three-level diode-clamped converter; b) One arm of the three-level converter with interconnection of the bidirectional switches; c) One arm of the three-level converter with flying capacitors; d) Three-level converter using three two-level converters; e) One arm of the cascaded three-level H-bridge converters. Initially, the main purpose of the multilevel converter is to achieve higher performance of voltage. With the increase of rated value of the components and the improving of the performance of the switch and the breakover, the second effect of using multilevel converter becomes more and more favorable. It reduces the harmonic content and the electromagnetic interference (EMI) in the input and output voltage. Under the same harmonic conditions, the switching frequency of the multilevel converter is reduced; the switch loss is small; the conduction loss is large; the total loss is determined by the proportion of the two.

hardware or modulation control. However, in order to decide whether to solve the voltage imbalance problem through the hardware or software, it is necessary to measure the voltage at both ends of the capacitor in DCL. The three-level diode-clamped converter in Figure 2.35a and the three-level converter with flying capacitors in Figure 2.35c present an unbalanced current in the semiconductor. Compared with the intermediate switch, the upper and lower switches in a converter branch may be non-rated. For the multilevel converter with the interconnection of bidirectional switches in Figure 2.35b, the unbalanced current and voltage may constitute a design problem. Obviously, for all the topologies in Figure 2.35, the number of the semiconductors in the conduction path is higher than that of the two-level converter. Thus, the conduction loss of converters is increased. On the other hand, each semiconductor only needs to block half of the total DC voltage, and because of the low rated value of voltage the conduction loss of each switch is reduced, which to a certain extent might justify more semiconductors in the conducting path.

2.4.3.4 Modular Power Converter

At the low wind speed, the generated power is low. When the full rated power converter is used, the utilization rate of the (power) main switch is low, which increases the power consumption. The maximum output of the conversion and transmission system might be limited by the rated values of the available components. For the high-power wind energy converter, its output must be divided into several groups. Therefore, the parallel operation of several power converters is used, which is shown in Figure 2.36. In this scheme, the power converter may be one of the above-mentioned structures.

The parallel operation of frequency converters has different options. It requires that the generator winding should be designed to be 3 × n-phase. Such parallel converter units are electrically separated from each other, thus providing favorable conditions for the high reliability and low power grid impact. If part of the parallel operating converter units fails, the wind energy converter can continue to run at the partial rated output. The harmonic of the frequency converter in parallel operation may be widely neutralized by the superposition, which reduces the influence on the network. The use of seveal modules with the same output types makes the development, manufacturing, storage, and services much easier, thus reducing the cost.

2.4.3.5 Matrix Converter

The AC-DC-AC converter usually uses the uncontrollable rectifier with capacitor filter circuit as input structure. Due to the combination of the nonlinear elements and the energy storage elements, its input power factor is low and the input current waveform is seriously distorted. Although the double PWM converter can adjust the power factor and its input current is continuous, it needs also large capacitors for filtering. The use of these passive elements leads to the increasing cost, low integration level, reduced reliability, and so on.

The matrix converter is a new type of "green" converter. Except for the small capacity high frequency filter, which is required to eliminate the switching ripple, it contains no passive element. If the good control algorithm implemented, basically the buffer circuit do not need to switch. The schematic diagram of the three-phase—three-phase matrix converter, which is used for the brushless double-fed generator system is shown in Figure 2.37. It is composed of nine bidirectional switches (S_{ij},i = A, B, C; j = a, b, c). Each bidirectional switch has the bidirectional conduction and turn-off functions and can be made up of two connected turn off power semiconductor devices with anti-parallel diodes. The nine bidirectional switches are arranged in the matrix of 3 × 3. Through the conduction and turn-off of the bidirectional switch, any phase of the three-phase AC input can be connected to any phase of the three-phase

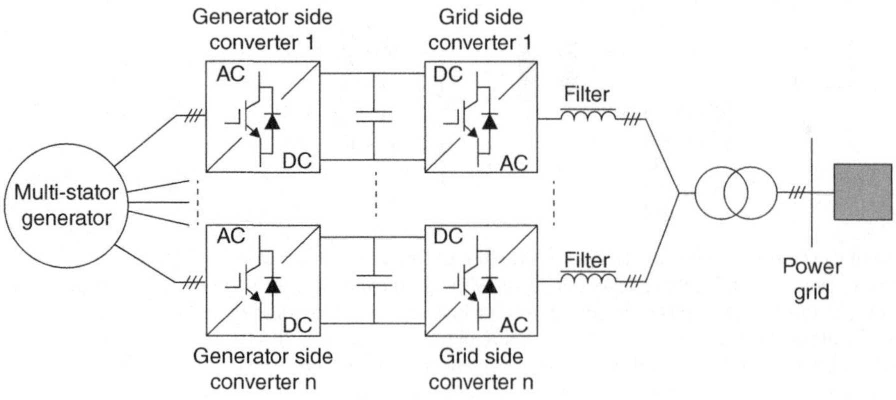

Figure 2.36 Full power converter-based wind turbine with n parallel power converters.

Figure 2.37 Matric converter brushless double-fed generator power generation system.

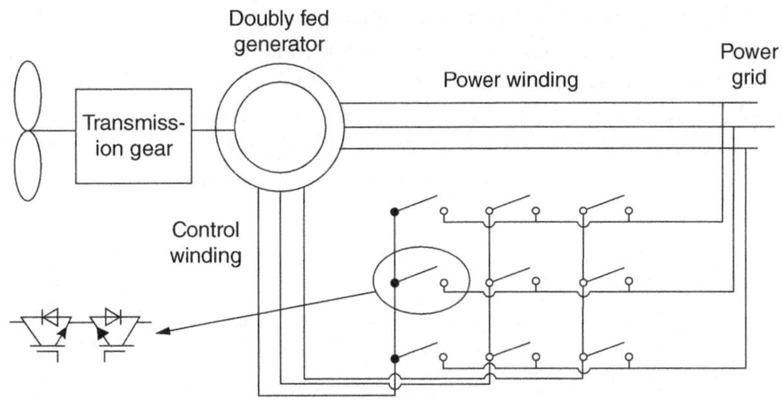

AC output. The input side of the matrix converter also needs the three-phase inductance capacitance filter to filter out the high frequency harmonics in the input current caused by the switching operation.

The matrix converter is actually a kind of AC-AC converter. Compared with the traditional AC-AC converter and the AC-DC-AC converter, the matrix converter has significant advantages [13, 14]: its degree of freedom of control is great; the output voltage is adjustable; its output frequency can be higher or lower than the input frequency; its adjustable input power factor can be lead or lag and can even be regulated to be close to the unit power factor; as it uses the four-quadrant switch, the energy can flow bidirectionally. With no intermediate DC link, it is characterized by compact structure, small size, and high efficiency. However, in the actual application, the matrix converter also has some problems. For example, in its commutation process, it does not allow two switches to be turned on or off at the same time, which is difficult to be realized. What's more, the output side of the matrix converter must be connected with filter capacitors which, like the AC capacitor at the output side of the AC-DC-AC current source converter, needs to bear the AC of the switching frequency caused by the switching devices.

References

1 European Wind Energy Association (EWEA). 2005. Largescale integration of wind energy in the European power supply: Analysis, issues, and recommendations. http://www.ewea.org/fileadmin/ewea_documents/documents/publications/grid/051215_Grid_report.pdf

2 Polinder, H., de Haan, Sjoerd, W.H., and Maxime R. Dubois, Maxime R., *et al.* 2005. Basic operation principles and electrical conversion systems of wind turbine. *EPE Journal,* **15**(4), 43–50.

3 Burton, Tony, Sharpe, David, Jenkins, Nick, and Bossanyi, Ervin. 2001. *Wind Energy Handbook.* New York: John Wiley & Sons.

4 Chen, Lei, Zuoxia, Xing, Jian, Pan, and Mingfang, Zhong. 2003. Large wind turbine generator system technology development trends. *Renewable Energy Resources,* **2003**(1), 27–30.

5 Abdin Ezzeldin, S. and Winlson, Xu. 2000. Control design and dynamic performance analysis of a wind turbine-induction generator unit[J]. *IEEE Trans. On EC,* **3**(15), 91–96.

6 Tian, Huifang, Zhilin, Gao, and Zude, Zhou. 2007. Wind turbine variable pitch system design and simulation. *Machinery & Electronics,* **2007**(3), 20–23.

7 Dai, Yun, Zhixin, Wang, and Yifan, Zhu. 2006. Study on variable speed constant frequency wind turbine blade electrohydraulic proportion control system. *Mechatronics,* **2006**(1), 16–20.

8 Iov, F. and Blaabjerg, F. 2009. Power electronics and control for wind power systems: Power electronics and machines in wind applications, *IEEE,* **24-26**, 1–16.

9 Ma, Hongfei, Dianguo, Xu, and Lijie, Miao. 2000. Comparative analysis of several VSCF wind power generation system control plans. *Electrotechnical Journal,* **2000**(10), 1–4.

10 Huan, Yi and Zhizhong, Wang. 2002. Comparative analysis of wind turbine and its control system. *S&M Electric Machines,* **29**(4), 41–45.

11 Chen, Shousun, Lipei, Huang, and Kexiong, Tan. 2001. *Wheel of Modern Civilization.* Jinan, China: Shandong People's Press.

12 Dettmer, R. 1998. The heart of a new machine [Powerformer, cable-wound stator machine]. *IEE Review*, **44**(6), 255–258.

13 Wang, Xiaohong, Jinming, Yang, and Wujie. 2009. Power electronic technology in variable upped constant frequency wind power generation system. *Electric Machines & Control Application*, **36**(11), 12–17.

14 Li, Shanshan, Fengyou, He, Xianzhao, Lv, and Na, Li. 2008. Variable speed constant frequency AC motor wind power generation technology. *Electric Machines & Control Application*, **35**(4), 13–16.

3

Operation of Grid-Connected WTGS

The wind energy is an unstable energy source. Without the complementary operation of the energy storage device or other power generating devices, it's difficult for wind power generating devices alone to provide stable power output. In order to achieve stable wind power supply, China and other foreign countrieshave taken the following measures: large wind turbine (above 1,000 kW) grid connection; grid connection of medium wind turbines (from dozens of kW to hundreds of kW) or parallel complementary operation of medium wind turbines with diesel generators or other power generating devises; small wind turbines (below 10 kW) mainly adopts DC power generating system operation independently and cooperate with battery energy storage devices. The main problems of the wind turbine grid-connected operation to be solved are grid connection control and power regulation. As the wind power system uses different types of wind turbines, the grid connection modes and problems arising from it also vary. This chapter focuses on describing the grid connection of various wind turbines, the power regulation of the grid-connected wind turbines, reactive voltage control of wind turbines and response characteristics of wind turbines in grid disturbance.

3.1 Wind Turbine Grid Connection

Like conventional generators, wind turbine grid connection is required to meet the power grid voltage and frequency requirements and have a small impact on the power grid. Different wind turbines have different conditions and methods of grid connection.

3.1.1 Synchronous Generator Grid Connection

Shown in Figure 3.1 is the circuit of the grid-connected synchronous generator driven by wind turbine [1]. In addition to wind turbine and speed-up gear, the electrical system also includes the synchronous generator, excitation regulator, and circuit breaker. The generator is connected to the power grid through the circuit breaker.

Like conventional synchronous generators, the synchronous generator driven by wind turbine is connected to the power grid in two ways: quasi-synchronization grid connection and self-synchronization grid connection. For the quasi-synchronization grid connection, due to the deviation of voltage, frequency, and phase in actual operation, it will generate some impact current and the State Grid Corporation of China requires the deviation of voltage should be no more than 5% to 10% and the deviation of frequency no more than 0.1% to 0.3%. The self-synchronization grid connection has advantages such as simple and rapid operation, but one of its disadvantages is that switching on will generate large impact current and the grid voltage will be reduced in a short period of time.

Before the synchronous generator is connected to grid, in order to avoid the current shock and the sudden torque on the rotation shaft, the following grid connection conditions should be met: ① The terminal voltage of the wind generator is equal to the voltage of the grid. ② The frequency of wind generator is the same as the grid frequency. ② The frequency of the wind generator is equal to the frequency of the power grid. ③ At the moment of getting connected to the power grid, the electric potential of the circuit composed of the wind generator and the power grid is zero. ④ The wind generator has the same phase sequence as the power grid. ⑤ The waveform of the wind generator voltage is the same as the waveform of the grid voltage.

As the wind generator has a fixed rotation direction, as long as the three phases at the output end of the wind generator correspond with the three phases of the power grid, the condition ④ is met. The condition ⑤ can be met in the equipment selection and manufacturing. Therefore, in the process of the grid connection the first three conditions should mainly be checked and met.

The starting and quasi-synchronous grid connection process of the wind generator is as follows: (1) Use the wind direction sensor to detect the wind direction and make the yaw controller act in order to align the wind turbine to the wind direction. (2) When the wind speed exceeds the cut-in wind speed, the pitch controller will

Integration of Large Scale Wind Energy with Electrical Power Systems in China, First Edition. Zongxiang Lu and Shuangxi Zhou.

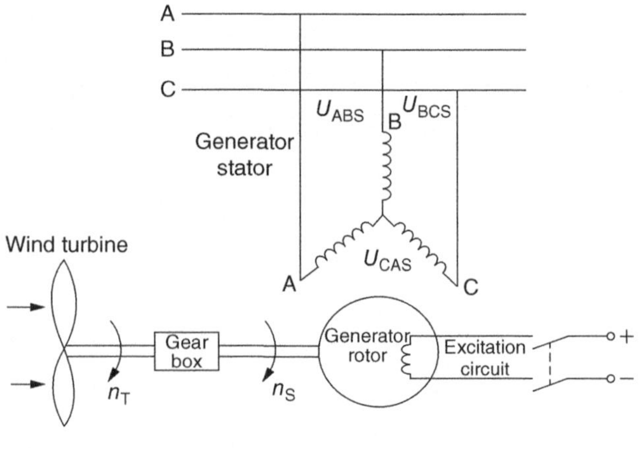

Figure 3.1 Grid-connected wind turbine-driven synchronous generator.

adjust the blade pitch angle to start the wind turbine. (3) When the speed of the generator driven by the wind turbine is close to the synchronous speed, use the excitation regulator to provide the generator with excitation and regulate the excitation current so that the terminal voltage of the generator is close to the voltage of the power grid. (4) When the speed of the wind generator is accelerated to almost reach the synchronous speed, the electric potential of the generator or the amplitude of the terminal voltage will be approximately the same as the grid voltage. Due to the slight difference in frequency, the phase difference between the terminal voltage of the generator and the grid voltage changes slowly between 0°and 360°. (5) When the phase difference of voltages on both sides of the circuit breaker is zero or very small, the circuit breaker will be closed for grid connection. After the closing, due to the self-sychronization, as long as the rotor speed is close to the synchronous speed, the generator can be pulled into synchronization mode. In other words, the frequency of the generator will remain the same as that of the power grid.

The above steps can be automatically operated through the microcomputer. This synchronous grid connection method can minimize the transient current of the grid as well as the impact on both the wind generator and the power grid. However, this method requires that the wind generator speed governor should regulate the speed so that the wind generator can be connected to the power grid when the difference between the generator frequency and the grid frequency reaches the allowable value. So the requirements for the regulator are quite high. If the grid connection time is not controlled properly, it may produce large impact current and even lead to the failure of grid connection. In addition, the cost of the control system needed in order to achieve the above-mentioned synchronous grid connection function is quite high. It will account for quite a large part of the total cost of the small wind generator. Therefore, the synchronous generator grid connection mode is usually used in large wind generators.

3.1.2 Cage Asynchronous Wind Turbine Grid Connection

The cage asynchronous wind turbine can be directly integrated to the power grid or connected to it through the thyristor voltage regulator. The asynchronous wind turbine grid connection conditions include: ① The rotating direction of the rotor should be consistent with that of the rotating magnetic field in the stator. In other words, the phase sequence of the asynchronous generator should be the same as the phase sequence of the grid. ② The speed of the generator should be as close as possible to the synchronous speed.

The first grid connection condition must be satisfied; otherwise after being connected to the grid, the generator will be in the state of the electromagnetic brake. The second condition is not required to be strictly satisfied, but the closer the speed of the generator is close to the synchronous speed, the faster the impact current attenuates.

The asynchronous wind generator can be directly connected to the grid or in other ways such as quasi-synchronous grid connection, step-down grid connection and thyristor soft grid connection [1, 2].

1) **Direct grid connection**
   When the wind speed reaches the starting conditions, the wind turbine starts. When the asynchronous wind generator is driven and its speed is close to the synchronous speed (usually 90% to 100% of the synchronous speed), it is connected to the power grid. This kind of grid connection is simpler than the quasi-synchronization grid connection of the asynchronous wind turbine. However, at the moment of the grid connection there exists the problem of three-phase short circuit and the impact current is five to six times the rated current, which will lead to the instantaneous voltage drop of the power system. As the generator is connected to the grid without voltage, it is bound to experience a transition process, which can be transferred to the steady state after a couple of tenths of a second. Although the asynchronous generator speed in grid connection has a certain impact on the transition time, it is generally not a big problem. So the requirements for the wind generator speed when it is closed for grid connection are not very strict and its grid connection is very simple. When wind turbines are connected in parallel with the grid, the impact current at the closing moment will not have a great impact on the generators and the safe operation of the large power grid system. However, for the power grid system with small capacity, at the grid connection moment,

the voltage of the power grid will drop significantly and affect the normal operation of other electrical equipment on the same grid and even the stability and safety of the small power grid system. As a result, the direct grid connection is only applicable when the asynchronous generator capacity is less than 100 kW and the power capacity of the grid is quite large. For example, both the 55 kW wind turbine China imported in the early stage and the 50 kW wind turbine that was domestically produced later on are both directly connected to the grid.

2) **Quasi-synchronous grid connection**
Like the quasi-synchronous grid connection of the synchronous generator, when the speed is close to the synchronous speed, the capacitor excitation is first used to establish the rated voltage, and then the generator voltage and frequency established through excitation and regulated and rectified to make them synchronize with the system. When the voltage, frequency, and phase of the generator are consistent with those of the system, the generator is then connected into the power grid for operation, which is shown in Figure 3.2. Although the impact current at the moment of connection is very small, it must be controlled within the maximum allowed torque range in order to avoid runaway accidents. As this kind of grid connection has an extremely small impact on the voltage of the system, it is suitable for the small capacity power grid.

3) **Step-down grid connection**
In this grid connection mode, resistances or reactors are connected in series between the asynchronous generator and the power grid or the auto transformer is accessed, in order to limit the amplitude of impact current and the voltage reduction of power grid at the moment of connection. As the resistance and reactance are energy-consuming components, they must be short-circuited after the grid connection is stable. This grid connection mode is suitable for wind turbines of hundreds of kilowatts and above. It is used in the 200 kW asynchronous wind turbines introduced in China. Obviously, this kind of grid connection is less economical.

4) **Thyristor soft grid connection**
For the large asynchronous wind turbine, the advanced grid connection mode is the soft starting grid connection based on the bidirectional thyristor control, which is shown in Figure 3.3. When the wind turbine drives the generator to make its speed close to the synchronous speed, the circuit breaker D at the output end of the generator is closed so that the generator is connected to the grid through a set of bi-directional thyristors whose trigger angle is gradually opened from 180° to 0 ° and the conduction angle of the bidirectional thyristor is gradually increased from 0° to 180°. Through the control of conduction angle of bidirectional thyristors by current feedback, the impact current at the grid connection moment can be limited within 1.5 to 2 times of the rated current to obtain a relatively smooth grid connection process. After the transient process is over, the controller sends out a signal and a set of switch K is used to short-circuit bidirectional thyristors so as to end the wind generator grid connection process and enter the normal power generation state. In China, both the introduced and domestically produced 250 W, 300 W, and 600 kW wind turbines are connected to the power grid in this way. This method requires that the performance of the three-phase thyristors should be consistent with each other, trigger voltage, and trigger current of the control electrode should be consistent and after the full opening their voltage drop remains the same, which can ensure the synchronous increase of the thyristor conduction angle from 0° to 180° and ensure the balance of the three-phase current; otherwise, it will have an adverse impact on the generator. In addition, in this grid connection mode, in the grid connection process each phase current is a non-sine wave with symmetric positive and negative half wave, which contains more odd

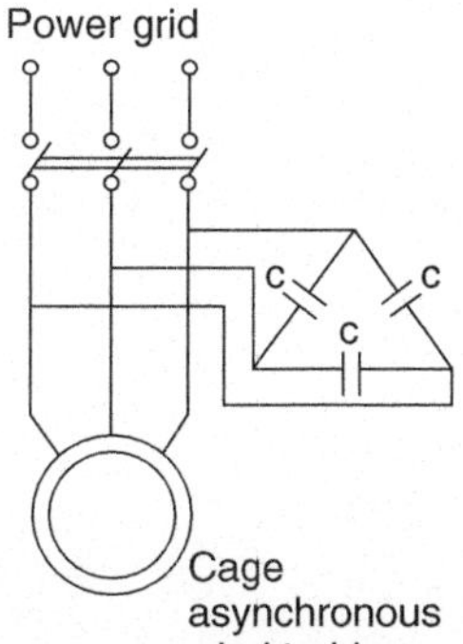

Figure 3.2 Quasi-synchronous grid connection of cage asynchronous wind turbine.

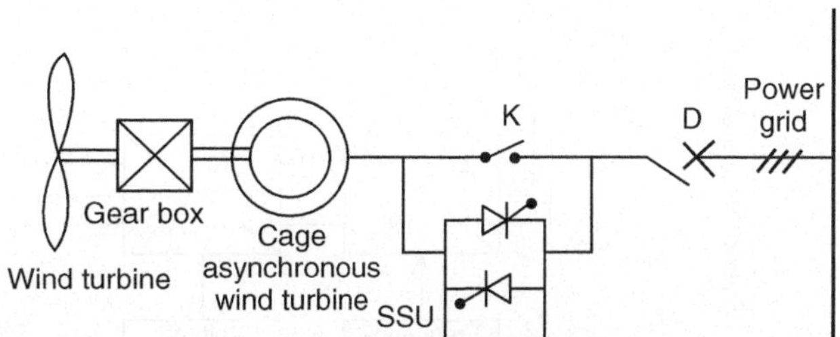

Figure 3.3 Soft grid connection of cage asynchronous wind turbine.

harmonics that should be restrained and eliminated through some measures.

3.1.3 Direct-Drive Permanent Magnet Synchronous Generator Grid Connection

Shown in Figure 3.4 is the schematic diagram of the direct-drive permanent magnet synchronous generator connected to the power grid. In order to ensure that at the grid connection moment, the voltage, frequency, and phase of the generator are consistent with those of the power grid, parameters such as the power grid voltage, frequency, and phase are collected through the controller and then compared with the output voltage and other parameters of inverter. When the grid connection conditions are met, the direct-drive permanent magnet synchronous generator is connected to the power grid. As in this grid connection mode no impact current will be produced, it will not lead to the power grid voltage drop or damage the generator stator windings and other mechanical components.

3.1.4 Double-Fed Asynchronous Generator Grid Connection

The double-fed asynchronous generator includes the general type and the brushless double-fed generator. Shown in Figure 3.5 is the schematic diagram of the general type double-fed generator, which consists of the body of the wound-rotor asynchronous generator, the converter, and the controlling components. Its stator winding is directly connected to the grid and its rotor adopts the three-phase symmetrical windings. It is connected through the back-to-back pulse width modulated (PWM) bi-directional voltage source converter to the power grid. In other words, it supplies the low-frequency excitation current through the three-phase power supply with adjustable frequency, phase, and amplitude to form the low-frequency rotating magnetic field in the rotor, which plus the rotor mechanical angular velocity is equal to the rotating angular velocity of the stator magnetic field so as to obtain the power frequency voltage corresponding to the synchronous speed in the stator magnetic field through induction. The operation of the

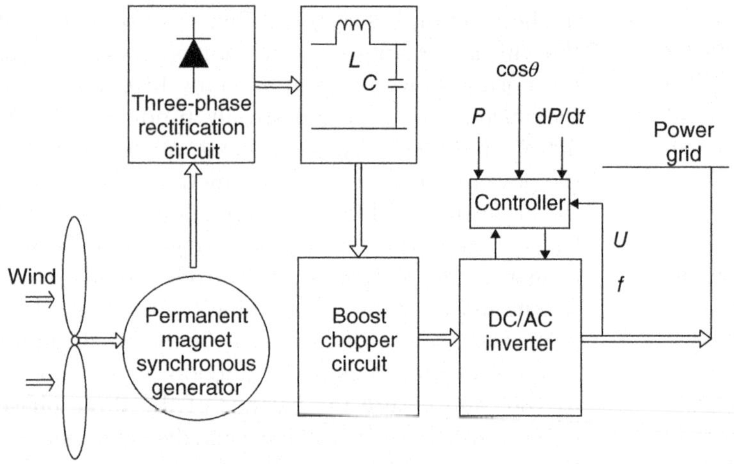

Figure 3.4 Diagram of permanent magnet synchronous generator connected to power grid.

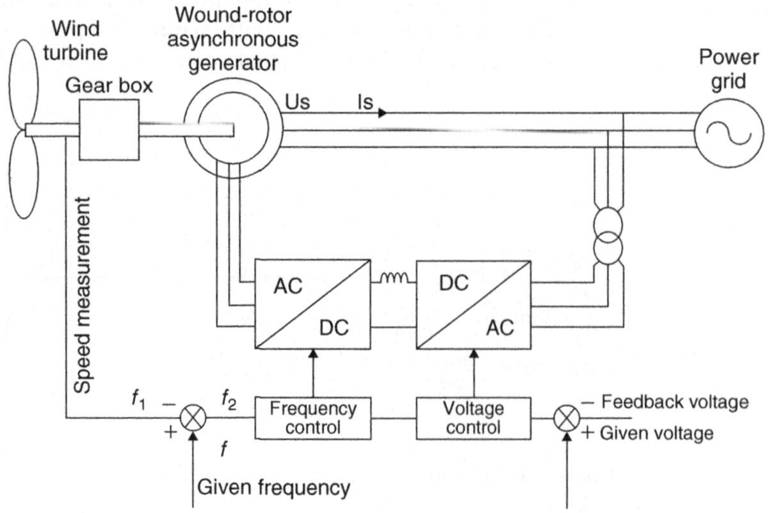

Figure 3.5 Diagram of working principle of double-fed asynchronous generator.

grid-connected double-fed asynchronous generator is characterized by:

1) After the wind turbine gets started, it drives the generator to make its speed close to the synchronous speed. The voltage matching, synchronization, and phase control are controlled by the cycloconverter so that the wind generator can be rapidly connected to the power gird. At the grid connection moment there is generally no impact current.
2) The speed of the wind turbine can be adjusted based on the change of the wind speed and the load so that the wind turbine operates at the optimal tip speed ratio to generate the maximum electrical energy output.
3) The double-fed asynchronous generator has three adjustable excitation variables: frequency, amplitude, and phase of the excitation current. The excitation current can be adjusted to ensure that the wind turbine generates constant frequency electrical power in case of variable speed operation. The amplitude and phase of the excitation current can be changed to regulate the output active power and reactive power. When the phase of the rotor current is changed, the rotor magnetic field generated by the rotor current produces a displacement in the air gap space of the generator, which changes the relative position of the electric potential of double-fed asynchronous generator stator and the grid voltage vector, namely, the power angle of the generator. So regulating the excitation can regulate both the reactive power and active power.

At present, grid connection modes suitable for the AC excitation double-fed wind generator mainly use the quasi-synchronous grid connection control technology based on the stator-flux -oriented vector control including no-load grid connection, independent load grid connection, islanded grid connection and grid connection by series resistance starting in rotor [2–4].

1) **No-load grid connection**
In the no-load grid connection, before grid connection, the double-fed wind generator operates with no load and its stator current is zero. The voltage information of the power grid (amplitude, frequency, and phase) is extracted as the basis to be provided to the double-fed wind generator control system. The stator-flux-oriented technology is introduced to regulate the generator output voltage so that the frequency, phase, and amplitude of the no-load voltage of double-fed wind generator stator are consistent with those of the power grid voltage. When the grid connection conditions are met, the double-fed wind generator is connected to the power grid. After the grid connection is achieved, the control strategy shifts from the grid connection control to the power generation control. In this grid connection mode, before the grid connection the generator has no load and does not participate in the energy and speed control. In order to prevent the speed out of control resulted from the energy imbalance of the generator before the grid connection, the prime mover should be used to control the speed of wind turbine. Therefore, in the no-load grid connection, the prime mover is required to have sufficient speed regulation capacity.
2) **Independent load grid connection**
Shown in Figure 3.6 is the independent load grid connection, which is based on the idea: before grid connection the double-fed wind generator operates with load (e.g., resistive load) and the double-fed wind generator and the load are controlled based on the power grid information and the voltage and current of stator. When the grid connection conditions are met, the double-fed wind generator is connected to the power grid. The independent load grid connection is characterized the double-fed wind generator having independent load and the stator having current before grid connection. As a result, the information needed

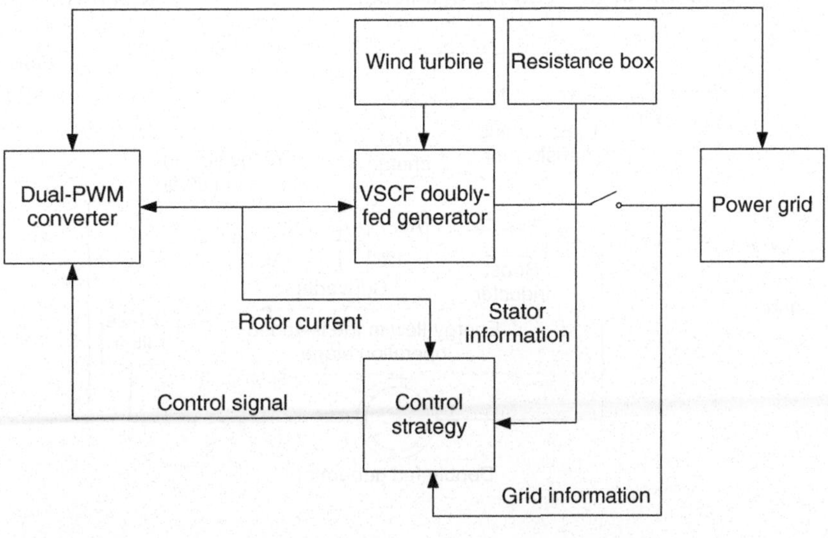

Figure 3.6 Schematic diagram of independent load grid connection control.

for grid connection control is not only obtained from the grid side, but also from the stator side of the double-fed wind generator. The participation of the generator in the energy control of the prime mover is manifested, on the one hand, in changing the generator load and regulating the energy output of the generator, which helps achieve the rough energy regulation of the generator. On the other hand, under a certain load, it is manifested in changing the speed of generator as well as the distribution of energy inside the generator, which helps achieve the fine energy regulation. Independent load grid connection enables the generator to regulate the energy to a certain degree. Moreover, this method realizes the control of rotor speed in corporation with prime mover and the reduction the requirements of the prime mover speed regulating ability. However, the regulation is very complex so voltage compensation and measuring more voltage or current are necessary.

3) **Islanded grid connection**
Shown in Figure 3.7 is the "islanded" grid connection that can be divided into three stages: the excitation stage, the islanded operation stage, and the grid connection stage.
   a) Excitation stage. A pre-charging circuit is introduced from the power grid side and connected to the DC side of the AC-DC-AC converter. The pre-charging circuit is composed of the switch K1, pre-charging transformer, and DC charger. When the wind turbine speed reaches a certain required speed, K1 is closed and the DC charger is used to charge the DC side of the AC-DC-AC converter through the pre-charging transformer. After the charging is ended, the converter on the generator side begins to work to supply the double-fed generator with the rotor side excitation current. At this point, the stator side voltage of double-fed generator gradually increases until the output voltage reaches the rated value when the excitation stage is coming to an end.
   b) Islanded operation stage. First, the K1 is disconnected, and then the grid side converter is started for boost operation to increase the DC side voltage to the required value. At this point, the energy flows between the grid side converter, the generator side converter, and the double-fed generator to form an islanded operation mode.
   c) Grid connection stage. As in the islanded operation stage, the amplitude, frequency, and phase of the stator side voltage are the same as those of the grid side voltage. So at this point the switch K2 is closed and the no-impact grid connection can be achieved between the generator and the power grid. After the grid connection, the pitch angle of the wind turbine can be regulated to increase the energy input by the wind turbine so as to achieve the purpose of power generation.

4) Grid connection by rotor resistance starter

For the vertical shaft type of double-fed wind turbine (also known as the Darrieus wind turbine), as it does not have the self-starting capability, the starting of the wind turbine in the stationary state can be achieved by the double-fed wind turbine running in the motor working conditions (shown in Figure 3.8). In order to achieve the system starting, a single pole double throw switch K3 is installed between the rotor winding and the rotor side converter. In conducting the grid connection operations, K3 is first operated to short-circuit the rotor of the double-fed wind turbine through resistance and then K1 is closed to connect the grid with the stator winding. Under the action of the grid voltage, the double-fed generator will gradually get started by series resistance in asynchronous motor rotor. When the speed of the double-fed asynchronous generator gradually increases and approaches the synchronous speed, the current of the rotor will drop to zero. Under this condition, the K3 is operated to disconnect the resistance and then the rotor winding is connected with the rotor side converter. Meanwhile the rotor side converter is triggered to input

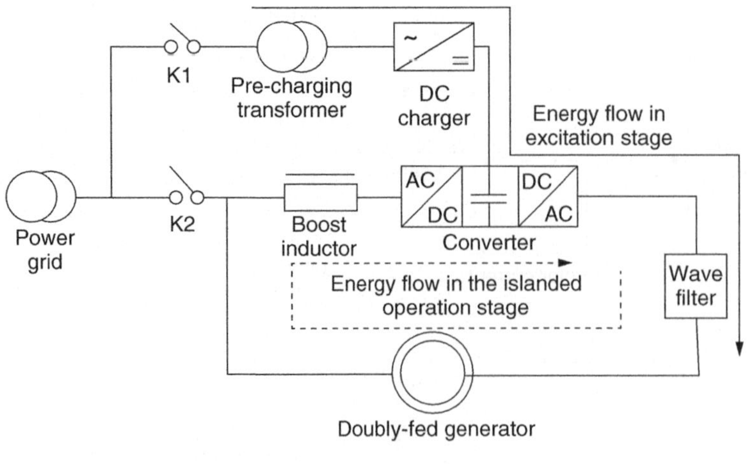

Figure 3.7 Schematic diagram of "islanded" gird connection.

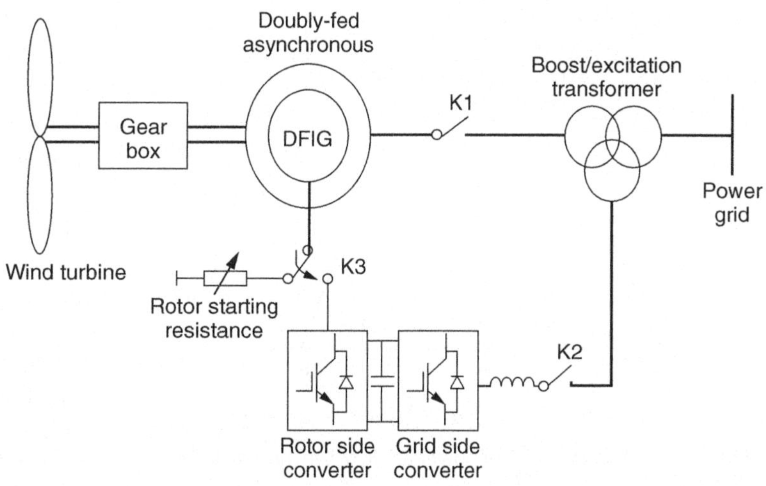

Figure 3.8 Schematic diagram of grid connection by connection resistance in series with rotor.

excitation. Finally, after the successful input of the excitation, the excitation is regulated so that the double-fed generator can rapidly enter the stator power control or speed control state. By then the starting process of the wind turbine is completed.

This grid connection method is simple and can achieve the starting and grid connection of the double-fed wind turbine without self-starting ability through proper sequence control. If the CrowBar protection device is installed on the generator rotor side, the CrowBar can be switched on and off through the controller to achieve the starting and quasi-synchronous grid connection of the system.

It can be known from the above-mentioned generator grid connection methods that the grid connection of the AC-excited double-fed wind generator is different from that of the traditional DC-excited synchronous generator and ordinary asynchronous generator. Due to the "rigid connection" between the synchronous generator and the power system, the output frequency of the generator is completely dependent on the speed of the prime mover and has nothing to do with excitation. The generator must be strictly synchronized and (quasi-)synchronized before grid connection and must strictly keep the constant speed after grid connection. The grid connection of the asynchronous generator has low requirements for the speed regulating accuracy of the wind turbine and after grid connection the out-of-step oscillation will not exist. However, it is required that the speed should be close to the synchronous speed (90% to 100%) when the synchronous generator and the asynchronous generator are connected to the grid, which still has certain restrictions on the speed. After the double-fed asynchronous generator uses the AC excitation, there forms the "flexible connection" between the double-fed generator and the power system. The excitation current can be regulated based on the grid voltage and the generator speed and then the output voltage of the generator is regulated to meet the grid connection conditions. The whole grid connection regulation process is implemented through the rotor converter, requiring no additional hardware devices. It is characterized by high-regulation accuracy and small-grid connection impact.

3.2 Power Regulation of Wind Turbines

3.2.1 Power Regulation of Synchronous Generator

1) **Active power regulation**

After the wind turbine is connected to the power grid, most of the mechanical power P_m transmitted from the wind turbine into the generator (except a small part of it that is used to compensate the mechanical loss of the generator q_m, the iron loss q_{Fe}, and the additional loss q_{ad}) is converted into the electromagnetic power P_{em}, namely (Formula (3.1)),

$$P_{em} = P_m - (q_m + q_{Fe} + q_{ad}) \tag{3.1}$$

Subtract the copper loss of the stator winding q_{cu} from the electromagnetic power to obtain the active power output by the generator (Formula (3.2)):

$$P = P_{em} - q_{cu} \tag{3.2}$$

For the synchronous wind turbine connected in parallel to an infinite power grid, in order to increase its output power, the mechanical power input from the wind turbine should be increased. With the increase of the output power, when the excitation is not regulated, the generator's power angle δ will definitely increase. The characteristics of the power angle of the synchronous generator are shown in Figure 3.9. It can be seen from it that when δ=90°, the output power reaches the maximum value. The maximum power that occurs when sinδ=1 is called out-of-step power. When the out-of-step power is

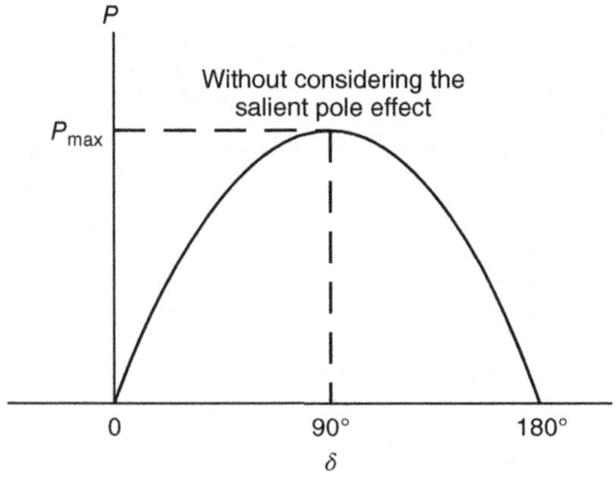

Figure 3.9 Characteristics of power angle of synchronous generator.

reached, if the mechanical power input by the wind turbine continues to increase, δ will exceed 90°, the output power of the generator will drop. In this case a new balance cannot be established and the generator speed will continue to rise and desynchronize. As the synchronous generator can no longer run stably, this maximum power is also known as the limit power of the generator. If a wind generator is operating at the rated power, due to a sudden gust of wind, the output power might exceed the limit power of the generator and desynchronize. In order to avoid this situation, a kind of method is to design the wind turbine rotor and control system so that it has the fast pitch adjustment function and can respond to the sharp changes in the wind speed. Another method is to increase the excitation current in a short time to increase the power limit and improve the steady state stability and the third method is to select a generator with large overload multiples. In other words, the maximum power of the generator has a larger margin as compared with its rated power.

Another point that can be seen from the power angle characteristic curve is that when the power angle δ becomes negative, the output power of the generator also becomes negative, which means when the generator operates as motor and absorbs power from the power grid, the wind turbine will become a huge fan. Of course, we do not want this to happen. So when the wind speed drops below a critical value, the generator should be disconnected with the power grid to prevent operating as motor.

2) **Reactive power regulation**

When the generator is connected with the power grid in parallel, if the wind turbine power remains unchanged, the reactive power of the generator can be changed by adjusting the excitation current of the generator. The synchronous wind power generator can send out both active power and reactive power to the power grid, which is a big advantage of using synchronous generator.

3.2.2 Power Regulation of Cage Asynchronous Generator

3.2.2.1 Power Output in the Operation of Grid-Connected Asynchronous Generator

In grid connection operation of asynchronous generator, the current and the power factor depend on the slip ratios and the generator parameters. The slip ratio is related to the load size of the asynchronous generator. As the parameters are given at the design stage, they cannot be controlled or regulated. After grid connection, the asynchronous generator operates in the stable region of the torque-speed curve, which is shown in Figure 3.10. When both the mechanical power transmitted from the wind turbine to the generator and the torque increase with the wind speed, the output power and the counter torque of the generator will also increase. The original torque balance point A_1 shifts along its operating characteristic curve to a new balance point A_2 whose speed is slightly higher than the previous one (A1) and the stable operation continues. However, when the output power of the generator exceeds the power that its maximum torque corresponds to, its counter torque decreases, leading to the rapid increase in speed and runaway accidents, which is very dangerous. As a result, we must have a reasonable and reliable stalled blade or speed limiting mechanism to ensure that when the wind speed exceeds the rated wind speed or in case of gust of wind, the mechanical power input by the wind turbine is limited to a maximum range to ensure that the output power of the generator does not exceed the power that its maximum torque corresponds to.

It should be pointed out that the maximum torque of the asynchronous generator is proportional to the square of the power grid voltage. The voltage drop of power

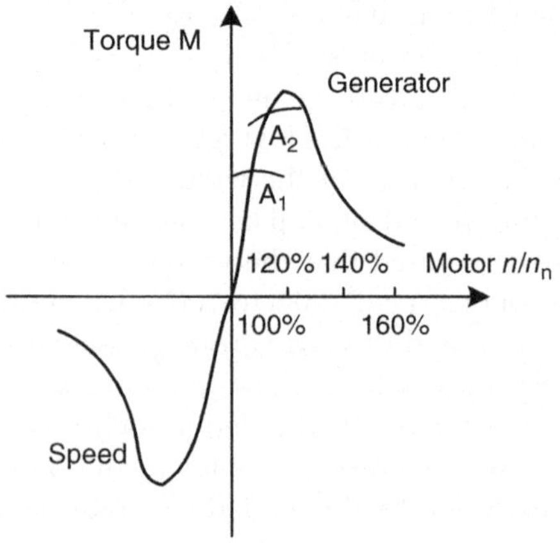

Figure 3.10 Torque-speed characteristics of asynchronous generator.

grid will cause the maximum torque of the generator to decrease by its square, and therefore, a serious decline in power grid voltage will cause the rotor runaway. On the contrary, if the power grid voltage increases to an excessively high value, it will cause the excitation current of the generator to increase and the power factor to decrease and is likely to result in the overload operation of the generator. So on the one hand, the small capacity power grid should be equipped with reliable overvoltage and under-voltage protection devices. On the other hand, the generator with strong overload capacity (its maximum torque should be at least 1.8 times the rated torque) should be used.

3.2.2.2 Reactive Power and Its Compensation

The asynchronous generator needs lagging reactive power (inductive reactive power) mainly due to the need for excitation, as well as the need for supplying the reactive power of leakage flux of the stator and rotor. In terms of the need for excitation, generally, for medium and large asynchronous generators, the excitation current is about 20% to 25% of the rated current. As a result, the reactive power required for excitation reaches 20% to 25% of the generator capacity. With the reactive power of leakage flux of the stator and rotor taken into consideration, the reactive power needed by the asynchronous generator is about 25% to 30% of the total capacity of the generator.

For the load connected to the power grid, generally, its power factor is lagging, which means it requires lagging reactive power. As the asynchronous generator connected to the power grid also needs to absorb reactive power from the power grid, which will undoubtedly increase the burden on other synchronous generators connected to the power grid by supplying reactive power, resulting in adverse effects. Therefore, for the wind turbine equipped with the cage asynchronous generator, generally the power factor correction capacitor (PFC) is used for proper reactive power compensation. The PFC can optimize the grouping switching based on the wind turbine output and the power grid voltage level. Of course, if the wind farm (WF) is equipped with dynamic reactive power compensation devices (such as SVC, D-STATCOM, SMES, etc.), these devices will be very effective in improving both the voltage level of the WF and the voltage stability of the power system.

3.2.3 Power Regulation of Direct-Drive Permanent Magnet Synchronous Generator

3.2.3.1 Power Regulation

The permanent magnet synchronous wind power generator uses the permanent magnet to replace the rotor excitation winding. The output power of the wind power generator is connected to the power grid through the

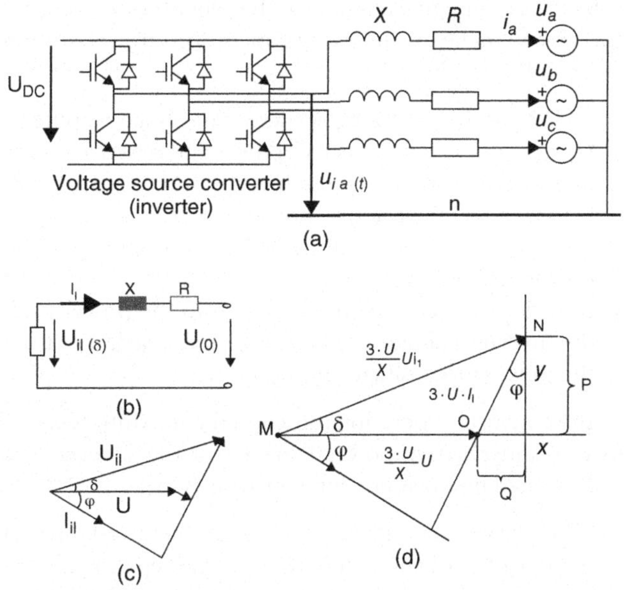

Figure 3.11 Voltage source converter connected to power grid. (a) Voltage source converter (inverter) connected to power grid; (b) VSC equivalent circuit connected to power grid; (c) phase diagram; (d) P-Q axis phase diagram.

power electronic converter. The converter on the generator side is the rectifier, and the converter on the power grid side (line side) is the inverter. These two sides are connected with DC. In controlling the power factor, the active and reactive current components input into the power grid are added to the inverter controller.

Shown in Figure 3.11a is voltage source converter (inverter) on the power grid side connected to the power grid. Shown in Figure 3.11b is the equivalent circuit. The pulse width modulated voltage source inverter (PWM-VSI) is regarded as an ideal power source, which generates the fundamental frequency voltage and the instantaneous voltage harmonics are negligible. The power grid is represented by the Thevenin equivalent circuit in which X represents the reactance of the power grid at the point of common coupling (PCC) with filter reactance included. As generally the filter reactance is larger than the reactance of the power grid, the reactance of the power grid is negligible. Shown in Figure 3.11c is the phase diagram in which each phasor is multiplied by 3U/X to get the phase diagram with P-Q coordinates system. In Figure 3.11d, the point O is the end of the phasor $3U^2/X$; the x-axis represents the reactive power injected into the power grid at the PCC point and the y-axis represents the active power.

The active power and reactive power at the PCC are (Formulas (3.3) and (3.4))

$$P = \frac{3UU_{i1}}{X}\sin\delta \tag{3.3}$$

$$Q = \frac{3U}{X}(U_{i1}\cos\delta - U) \tag{3.4}$$

Based on the same analysis of the synchronous generator, we can make the following conclusions from Formula (3.3) and Formula (3.4):

- As long as the power angle δ and the voltage amplitude U_{i1} are controlled, the active power and reactive power injected or absorbed by the inverter can be controlled.
- In order to inject active power into the power grid, the voltage of inverter must be lead one angle δ of the power grid voltage.
- In order to inject reactive power into the power grid, the inverter voltage amplitude U_{i1} must be larger than the power grid voltage amplitude U.

The converter operation limit and the maximum active power and reactive power capacity of the inverter are determined by the following constraints [5]:

1) The maximum effective value of the fundamental component of the converter voltage depends on the voltage of DC link, the modulation method used, and the maximum amplitude of modulation coefficient allowed under steady-state conditions. After the modulation method and the modulation coefficient are determined, the reactive power capacity is determined by the voltage of the DC link.
2) The rated current of the power switching element or any other limiting element in the power grid interface: The limited value of the maximum apparent power that the inverter can be injected into the power grid is S=3UI, or can be expressed as the circle with the origin of coordinates as the center and the radius being 3UI, namely, $P^2 + Q^2 = (3UI)^2$.

3.2.3.2 Reactive Power Regulation

The reactive power control of the direct-drive permanent magnet synchronous generator system includes the constant power factor control and the constant voltage control:

a) **Constant voltage control**
   In this operation mode the permanent magnet synchronous generator can absorb or send out reactive power to keep the voltage on the generator side constant. Within the regulation range of wind turbine reactive power, the WF can be regarded as PV node. The regulation range of permanent magnet synchronous generator reactive power is mainly restricted by the maximum current of the converter.
b) **Constant power factor control**
   As the generator is excited by permanent magnet and permanent magnet supplies constant excitation, there is no reactive power exchange between the generator and the rectifier. The active power P_g and reactive power Q_g exchanged between the inverter and the power grid is controlled through the control of the components in the d axis and q axis of the current on the power grid side inverter in order to meet the power factor regulation requirements. Generally, the maximum power tracking characteristic of the wind turbine is used to determine reference value of the active power to ensure that the wind turbine operates at the optimal power point. When the constant power factor control is used, if the power factor is set to be $cos\varphi$, then (Formula (3.5))

$$Q_g = P_g tan\varphi \tag{3.5}$$

In this control mode, the WF can be regarded as PQ node.

3.2.4 Power Regulation of Double-Fed Asynchronous Generator

The characteristic of the double-fed asynchronous generator lies in that it can be used to maximally utilize wind energy or improve the power factor of the grid and support the power grid. The double-fed asynchronous generator has two control modes: (1) Speed control mode: when the output active power and reactive power of the generator are variable, with the generator speed as the controlled object, the amplitude and phase of the rotor excitation voltage is regulated to make the speed of the generator equal to the given value. In this mode the purpose is to maximally use wind energy. (2) Power control mode: when the generator speed is variable, with the output active power and reactive power of the generator as the controlled objects, the amplitude and phase of the rotor excitation voltage is regulated to make the output active and reactive power of the generator equal to the given values. In this mode, the purpose is to improve the power factor of the grid and stabilize the voltage of the power grid.

3.2.4.1 Maximally Use Wind Energy

An important advantage of variable speed constant frequency (VSCF) wind power system is that it can enable the wind turbine to operate with the best efficiency in a large wind speed range. It can be seen from the operation principle of the wind turbine that it requires that the speed of the wind turbine should changes with the change of wind speed and keeps being proportional to the wind speed. Additionally, a constant optimal tip speed ratio should be maintained so that the maintain maximum value of the wind turbine wind energy utilization coefficient C_P can remain unchanged and the wind turbine can output the maximum power. Therefore, in terms of the requirements for the VSCF wind power system, in addition to stable and reliable parallel operation, the most important point is to realize the maximum power output control.

In wind power generation, the wind energy utilization coefficient C_P is an important parameter that reflects the ability of the wind turbine to convert wind energy into mechanical energy. At a certain wind speed, the higher the C_P value is, the higher the efficiency of the wind turbine converting wind energy into mechanical energy is. When the tip speed ratio is a certain value λ_m, the wind energy utilization coefficient reaches the maximum value. For the constant speed wind turbine, the angular speed ω of the rotor remains almost unchanged while the wind speed is changing. As a result, the λ value will constantly change instead of remaining at λ_m and the wind energy cannot be fully utilized. Only when the wind turbine angular speed ω changes with the wind speed and keeps being proportional to the wind speed change, can the constant maximum C_P value be kept. The speed of the double-fed asynchronous generator can be adjusted promptly based on the change of the wind speed and its load to make the wind turbine always operate at the optimal tip speed in order to generate the maximum electrical power output and maximally utilize wind energy.

3.2.4.2 Improve Power Factor of the Grid

The stator structure of the AC-excited double-fed generator is the same as that of the cage asynchronous generator, but different from the cage asynchronous generator, the rotor winding of the AC-excited double-fed generator has slip rings and resistance so that the electrical energy can also be input and output on the rotor side. When the AC-excited double-fed generator is used, the rotating speed of the rotor magnetic field is related to the excitation frequency. The internal electromagnetic relationship of the double-fed generator is different from that in the cage asynchronous generator or the synchronous generator, but it has some characteristics of the two kinds of generators.

When the stator winding of the double-fed generator is connected to the power grid with industrial frequency, the rotor winding is connected with a three-phase variable frequency power supply with adjustable frequency, amplitude and phase. When it is operating stably, the stator rotating magnetic field and the rotor rotating magnetic field should remain relatively static in space. When the stator's rotating magnetic field is operating at the speed of ω_0 in space, the angle speed of the rotating magnetic field of the rotor relative to the rotating angle speed of the rotor is (Formula (3.6)):

$$\omega_s = \omega_0 - \omega = \omega_0 - \omega_0(1-s) = \omega_0 s \tag{3.6}$$

in which ω_0 is the magnetic field angle frequency of the stator; ω is the rotating angle frequency of the rotor; ω_s is the angle frequency of rotor's rotating magnetic field relative to the angle frequency of the rotor; s is the slip ratio(negative).

The above formula shows that the rotor magnetic field relative to the rotor rotating angle frequency is proportional to the slip. If the rotor rotating speed of the AC-excited generator is lower than the synchronous speed, the rotation direction of the rotor magnetic field is the same as the rotation direction of the rotor. If the rotor speed is higher than the synchronous speed, their rotating direction are opposite to each other.

As the angle speed of the rotor's rotating magnetic field relative to the rotor's rotating angle speed is $\omega=2\pi f$, the frequency of the excitation current input into the rotor winding should be the slip frequency whose relationship with the stator frequency is (Formula (3.7)):

$$f_s = f_0 s \tag{3.7}$$

in which f_0 is the stator current frequency (e.g. 50Hz); f_s is the frequency of the excitation current of the rotor.

From the above analysis, it can be known that in the rotor winding of the double-fed generator there are always two power supplies with the frequency of $f_0 s$. One is the rotor induction electromotive force sE_{20} (E_{20} is the induction electromotive force when the rotor is open-circuited) and the other one is an external voltage U_2 on the rotor winding. The equivalent circuit of the double fed asynchronous generator is shown in Figure 3.12.

Assume the phase angle of the additional electromotive force on the rotor side U_2 is δ. The included angle between U_2 and the rotor induction electromotive force phasor sE_{20} approximates π-δ and the rotor current can be determined using the rotor circuit voltage equation (Formula (3.8)):

$$E_2 + \dot{U}_2' e^{j(\pi-\delta)} = \dot{I}_2' Z_2 \tag{3.8}$$

So the rotor current (Formula (3.9))

$$\begin{aligned} I_2' &= [E_2 + U_2' e^{-j(\pi-\delta)}]/Z_2 \\ &= sE_{20} e^{-j\varphi_2}/|Z_2| + U_2 e^{j(\pi-\delta-\varphi_2)}/|Z_2| \end{aligned} \tag{3.9}$$

or (Formula (3.10))

$$I_2' = E_{20}[se^{-j\varphi_2} - U_{2*} e^{-j(\delta+\varphi_2)}]/Z_2 \tag{3.10}$$

in which: $\varphi_2 = \arccos(r_2/|Z_2|)$

$$U_{2*} = U_2'/E_{20}$$
$$z_2 = r_2' + jsx_2'$$

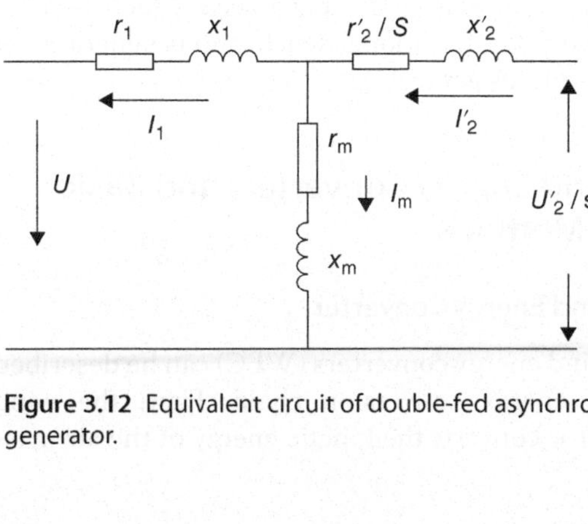

Figure 3.12 Equivalent circuit of double-fed asynchronous generator.

$$|z_2| = \sqrt{r_2'^2 + s^2 x_2'^2}$$

$$E_2 = E_{20} S$$

Expand the formula to get the active component I_{2a} and the reactive component I_{2r} of the rotor current (Formula (3.11)):

$$\begin{cases} I_{2a} = E_{20}[s\cos\varphi_2 - U_{2*}\cos(\delta+\varphi_2)]/|Z_2| \\ I_{2r} = -jE_{20}[s\sin\varphi_2 - U_{2*}\sin(\delta+\varphi_2)]/|Z_2| \end{cases} \tag{3.11}$$

The expression of the electromagnetic torque can be obtained from the active component of the rotor current (Formula (3.12))

$$T = C_T E_{20} I_{2a} \tag{3.12}$$

in which C_T is a constant that is determined by the structure of the generator.

From Formula (3.12), it can be known that the electromagnetic torque of the double-fed asynchronous generator is proportional to the active current through which the active power of the double-fed asynchronous generator can be determined. The reactive component of the rotor current determines the reactive power input into the rotor circuit and affects the reactive power at the stator end.

The active and reactive current component Formula (3.11) indicates that the active and reactive power of the generator can be controlled by independently regulating the amplitude of the additional electromotive force on the rotor side U_{2*} and the phase angle δ. In other words when the slip ratio remains unchanged, if the active current changes, the reactive current can remain unchanged or a certain relationship between the active and reactive power can be maintained through appropriate regulation of U_{2*} and δ. In addition, Formula (3.11) shows that if U_{2*} and δ take proper values, when the slip ratio is positive ($0 \leq s \leq 1$), the active component of the rotor current might be negative; when the slip is negative ($s \leq 0$), the active component of the rotor current might be positive. It indicates the AC-excited double-fed asynchronous generator can operate under power generating working conditions in both the synchronous speed area and the super synchronous speed area, which is quite different from the traditional synchronous generator or the asynchronous generator.

3.3 Wind Energy Converters and Basic Control Methods

3.3.1 Wind Energy Converter

Various wind energy converters (WEC) can be described using the energy conversion chain. First, the wind turbine rotor converts the kinetic energy of the air mass

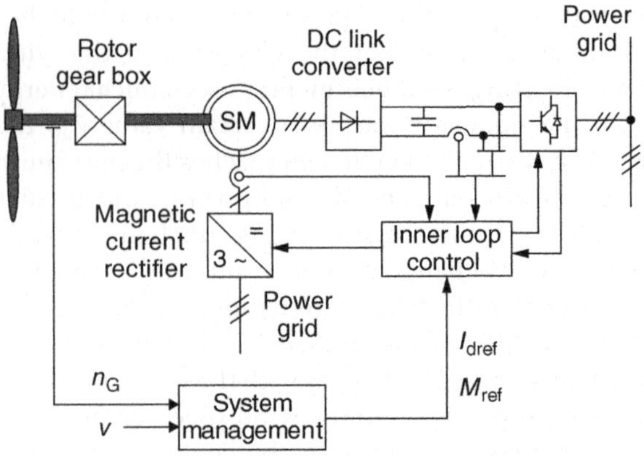

Figure 3.13 Typical structure of variable speed WEC.

into mechanical energy. The driving shaft transmits the rotating mechanical energy to the generator. Most WECs use the gear box to make the rotor speed of wind turbine well matched with the speed of generator. Then the electrical energy is fed into the connected power grid. The generator can be directly connected to the main power grid or supplied to the power converter coupling to the generator and power grid (Figure 3.13). From the point of view on the system side, the WEC is required to have high system efficiency, long service life, and high equipment availability. In addition, in terms of application, the special requirements are the maximum energy output, low impact on the power grid and low strain/stress in the power driving chain. In order to obtain the maximum power output, the WEC must be able to operate at variable speeds, which requires the decoupling of the generator speed with the frequency of power grid voltage. As shown in Figure 3.13, in the variable-speed WEC, the DC link connects the generator to the power grid to achieve the decoupling of the generator speed with the frequency of power grid voltage [6].

The C_P—λ curve is usually used to describe the wind turbine rotor characteristics. In order to facilitate the understanding, the C_P—λ curve can be converted to the curve of the rotor power to the rotor speed (P—n_R) which is shown in Figure 3.14 [7]. The rotor power at a given rotating speed can be read out based on the wind speed. The operating point of the rotor changes with the wind speed (for example, from Point 1 to Point 2 in Figure 3.14). Therefore, in order to achieve the maximum power output (Curve P_{opt}), the controller must make the rotor speed change from Point 2 to Point 3. Depending on the dynamic characteristics of the control method used, this process may lead to the additional torque and power change that fail to meet the requirements.

3.3.2 Basic Control Methods of WEC

The control of wind energy conversion system can be divided into inner loop control and outer loop control

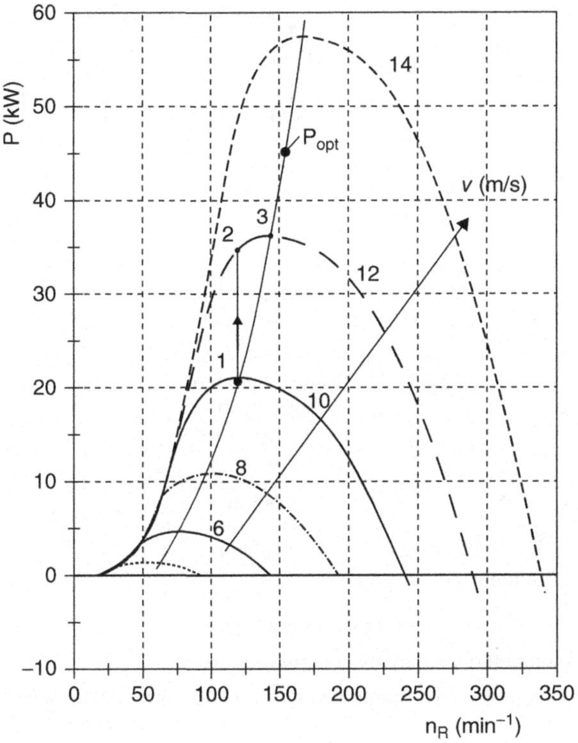

Figure 3.14 Wind turbine P—n_R curve.

(system management). The inner loop control sets the wind turbine to the desired operating point (which is given by the outer loop, i.e., the system management). The variable that affects the inner loop control is the generator electromagnetic torque who counteracts the rotor mechanical torque. Due to the torque difference, the rotating speed of the driving chain is changed. If the mechanical torque is larger than the electromagnetic torque, it will lead to the acceleration of the driving chain while if the mechanical torque is smaller than the torque of the generator, it will cause the driving chain to decelerate.

The controllable variable of the system management depends on the implemented control method. Generally, the single variable control method pursues different objectives, such as maximum power, low torque oscillation, or low power output fluctuation. The system management is also responsible for the operation of the entire wind turbine, for example, the starting process, emergency shutdown, and so on.

In this section we are going to introduce four different control methods [6, 7].

3.3.2.1 Optimal Rotor Speed Control

The optimal rotor speed control is aimed to make the wind turbine rotor always operate at the rotor speed with the maximum capture of power. The correlation between the wind speed and the optimal rotor speed

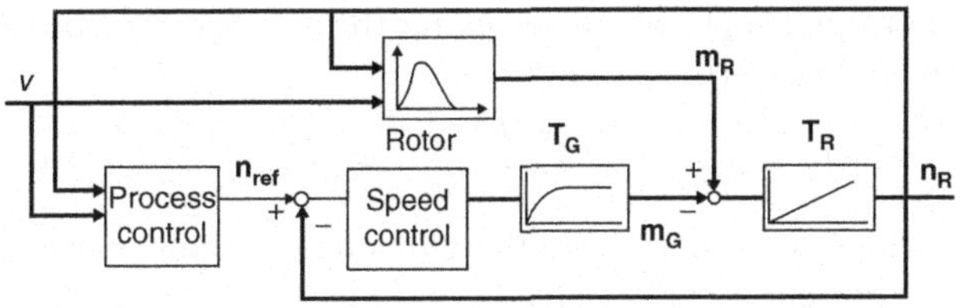

Figure 3.15 Optimal rotor speed control block diagram.

can be obtained from the wind turbine characteristic power curve (P—n_R) (Figure 3.14). Figure 3.15 shows the structure of the rotor speed control loop [6].

The disadvantage of this kind of single speed control is in case of the rapid increase of the wind speed v, due to the large inertia of the system, the increase in the rotor speed n_R will be slower than the reference speed n_{ref}. Entering a large input value to the speed controller may cause the sudden drop of the output power as the speed controller attempts to rapidly speed up the rotor by reducing the counter torque of the generator. When the wind speed decreases rapidly, the controller must act in the opposite direction. To be specific, by setting the generator to the state of maximum power output, it produces the large opposite torque so as to make the wind turbine rotor to decelerate, which will result in the violent fluctuation of the output power of the wind turbine under unstable wind conditions.

One measure to reduce the undesired power fluctuation is to limit the control deviation so that the reaction of the maximum speed controller is limited, resulting in a fast control characteristic for small control deviation (small time constant) and a greater time constant with small power fluctuation caused by large deviation. Therefore, due to the slow adaptation, the small control deviation limit will also result in an increase in the deviation of maximum power factor (C_P).

3.3.2.2 Torque Control

The torque control is designed to avoid the load peak in the driving chain of the WEC. This load peak might occur during the optimal rotor speed control. It is similar with the optimal speed control with a wind turbine rotor model. The actual torque generated by the rotor (M_R) is calculated based on the measured wind speed and the rotor speed control and is sent to the generator as the reference torque. Therefore, for the instantaneous wind speed, the generator torque follows the rotor torque. In this way, the peak of the output power can be avoided. The rotor torque can be calculated by the following formula (Formula (3.13)).

$$M_R = \frac{P_\omega}{\omega_R} = \frac{\frac{1}{2}c_P\rho\pi R^2 v^3}{2\pi n_R} = \frac{2c_P\rho\pi^3 R^5}{\lambda^3} n_R^3 \tag{3.13}$$

According to the requirements, in order to obtain the maximum power, λ= constant $=\lambda_{opt}$. The torque of the generator (M_G) can be set using the following formula (Formula (3.14)):

$$M_G = \eta_G \eta_{gear} \frac{2c_P \rho \pi^3 R^5}{\lambda^3} n_R^2 \tag{3.14}$$

in which η_G and η_{gear} are the efficiency of the generator and the gear box respectively.

3.3.2.3 Average Power Control

Another way to control WEC is to determine the reference value of output power. Based on Figure 3.14 and the measured values of v and n_R, the controller determines whether the wind turbine is operating at the maximum power factor; otherwise, the controller will generate a given power time series to converge to the desired state. The reference power output is calculated using the average wind speed while the optimal average wind speed value is calculated using the average regression value. Therefore, each latest measured value v_{actual} is weighted (for example, by using a weighted factor of 7%), which is added to the old average value $\overline{v}_{old}$ (i.e., the old average value is weighted with the factor of 93%) (Formula (3.15)).

$$\overline{v}_{new} = (1-\varepsilon)\cdot\overline{v}_{old} + \varepsilon\cdot v_{actual} \tag{3.15}$$

in which for a given example, ε=0.07, but generally $0 \leq \varepsilon \leq 1$. So add the latest measured value v_{actual} to the old average value $\overline{v}_{old}$ every one second. The old average value constantly loosens its impact on the calculation of the average value, but never completely abandons it. Increasing the value of ε will make the rotor speed more quickly adapt to the changed wind speed, resulting in sound wind power utilization. This trade-off (alternate use), however, will lead to more intense power output fluctuations, resulting in greater stress on the electrical and mechanical subsystems.

3.3.2.4 Random Dynamic Optimization

A new WEC control method is stochastic dynamic optimization, repeated adaptation to the probability distribution of wind speed, which is also known as an iterative self-adaptive system management (ISSM). Here WEC is understood as a stochastic process in which the wind speed is a stochastic disturbance z. The objective of the control is to find the operation state with the minimum fluctuation of output power and the maximum utilization of wind power.

Figure 3.16 shows the schematic representation of the control method for the system management with identified WEC set points [6]. The state values of the process are the wind speed v and the rotating speed n_R among which only the rotating speed is affected by the controller. The current in the DC-link $I_{d,ref}$ is a controllable variable determined by the control algorithm based on the measured state values. This algorithm is applicable to the distribution of changed wind speed (this distribution is in a specific location) and the frequency distribution of the current in the DC link (due to the WEC operation).

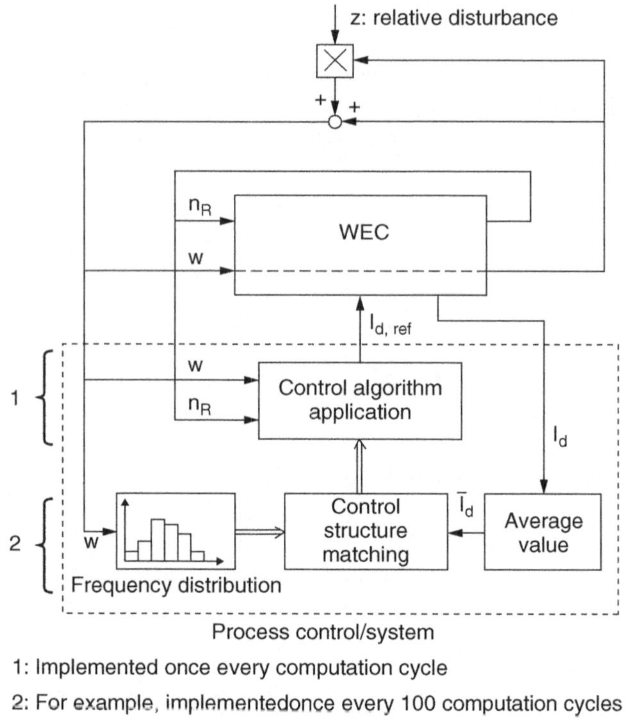

Figure 3.16 Flow chart of random dynamic optimal control.

The control structure is not static, but keeping updated much more slowly than the updating of the power set point. Therefore, this control method can adapt to the change of the wind conditions in the location where WEC is installed.

The optimal control algorithm is implemented in the first execution layer of the device including looking the table lookup of the reference value of the current in the DC link. The input variables of the table lookup are v and n_R. During the process, the statistical data of the wind speed and the DC link are collected during the operation and stored as the wind speed frequency distribution and the average generator current.

For each operation state, the control calculates the following two values:

1) The deviation of the average power output (P-index) (Formula (3.16))

$$P_{index} = \frac{\sum_{i=1}^{\infty} |P_i - \overline{P}|}{\overline{P}} \tag{3.16}$$

It is also the index of the output power fluctuation rate.

2) The deviation of the rotor optimal power coefficient (C_P - index) (Formula (3.17))

$$C_{P,index} = \frac{\sum_{i=1}^{\infty} |C_{P,i} - C_{P,opt}|}{C_{P,opt}} \tag{3.17}$$

It shows the aerodynamic efficiency of the operation of the wind turbine.

The control objective is to keep the power output fluctuation as low as possible and meanwhile the wind turbine operates at the highest efficiency, so the two indexes must be as small as possible. In order to express the two indexes in one formula, both the P- index and C_P- index are put in a formula called the cost function φ (Formula (3.18)):

$$\varphi = g\cdot\frac{|P_i - \overline{P}|}{\overline{P}} + (1-g)\cdot\frac{|C_{P,i} - C_{P,opt}|}{C_{P,opt}} \tag{3.18}$$

In which $0\leq g \leq 1$ is the weighting coefficient. The stochastic dynamic optimization controller is to minimize the cost function in the whole operation process.

For each possible operating state at the next time step (t + 1), the actual cost can be calculated based on Formula (3.18). For any given v_t-to-v_{t+1} transition, the next operating state can be calculated using the rotor torque, the actual rotating speed and the assumed generator torque for each possible value of $I_{d,ref}$. Based on the torque difference and the system inertial, the next operating state (related to the applied value of $I_{d,ref}$) can be calculated. For WEC, the transition of the wind speed v is stochastic. So the expected cost is calculated by applying the frequency distribution of the transitions of wind speed in the calculation. The objective of the stochastic dynamic optimization control is not only to minimize the cost of the next operating state, but also to minimize the cost of all future operating states. So the principle of the optimization control is implemented by calculating the residual cost of each operating state. Then the stochastic dynamic optimization control selects the value of $I_{d,ref}$ to make the WEC approximate the state with minimum transition and residual costs. These $I_{d,ref}$ value-related operating states are listed into the lookup table and passed on to the control algorithm in the matching period of each control structure. With the change of the frequency distribution of wind speed and the average value of the Id, the residual cost also changes and therefore has to be recalculated. Therefore, the wind energy converter can adapt to the change by updating P (z|v), the frequency distribution of wind disturbance.

3.4 Voltage/Reactive Power Control Characteristics of Three Mainstream Wind Turbines

3.4.1 Three Mainstream Wind Turbines

As the three mainstream wind turbines mentioned here, namely, the cage asynchronous generator, double-fed asynchronous generator and direct-drive permanent magnet synchronous generator were already introduced in Chapter 2, now we are going to briefly describe the three kinds of wind turbines.

The cage asynchronous generator converts mechanical energy into electrical energy. As the speed of the impeller is different from the speed of the generator, the gear box is required to match these two speeds. Instead of being a constant, the slip ratio of the generator changes slightly with the output electrical power. As the change of the generator speed is about 1%, this type of generator is usually referred to as "constant speed" wind turbine or "fixed-speed wind turbine." Although the constant speed wind turbine with variable pitch control has been manufactured, the design of this kind of wind turbine is almost always combined with the stall control.

Like the cage asynchronous generator, the double-fed asynchronous generator also needs the gear box. The stator winding of this kind of generator is connected to the power grid while its rotor winding is connected to the converter. Now the back-to-back voltage source converter with current closed-loop control is usually used. In this way the electrical and mechanical frequency of the rotor do not affect each other. As the converter injects current with variable frequency into the rotor to compensate the difference between electrical and mechanical frequency, the variable speed operation is possible. In this case the mechanical speed of the rotor can be controlled based on certain objective functions (such as maximum generation power or minimum noise). The mechanical speed of the rotor is controlled through the control of the electrical power of the generator. This type of wind turbine usually uses the variable pitch control to control its wind power.

As the third kind of wind turbine does not need the gear box, it is known as the direct-drive wind power generator. The multi-pole synchronous generator with same rotational speed as the rotor blade converts mechanical energy into electrical power. The generator has a wound rotor or a rotor equipped with permanent magnet. The stator is not directly connected to the power grid. Instead, it is connected to the power grid through the power electronic converter that can be the back-to-back voltage source converter or a diode rectifier with a single voltage source converter. The converter enables the wind turbine to operate at variable speeds. Like the double-fed asynchronous generator, the direct-drive wind power generator can limit the input of the mechanical power by controlling the pitch angle.

3.4.2 Voltage/Reactive Power Control Capability of Three Mainstream Wind Turbines

In the power transmission and distribution network, the node voltage depends on the network impedance characteristics and the branch current. As the node voltage is related to the reactive power, the node voltage can be controlled by changing the reactive power sent out or absorbed by the generator. Now we are going to discuss the voltage control capability of the above-mentioned three kinds of wind turbines one by one [8].

The cage asynchronous generator in the constant speed wind turbine always absorbs reactive power and the amount of the absorbed reactive power is determined by the terminal voltage, the output active power and the rotating speed. Figure 3.17 shows the relationship between the terminal voltage, the rotor speed, the output active power, and the absorbed reactive power, which shows that the cage asynchronous generator cannot be used to control the voltage because it can absorb, but cannot output the reactive power. Moreover, the reactive power that cage asynchronous generator exchanges with the power grid cannot be controlled and can only be determined by the rotor speed, the active power, and the terminal voltage.

Absorbing reactive power is one of the disadvantages of the cage asynchronous generator, especially when the large wind turbine or the large WF is connected to the weak power grid. In this case the reactive power demand might cause the serious drop in the voltage of the node (including PCC). Therefore, in most cases, the reactive power absorbed by the wind power generator should be compensated by capacitors. In this way, the reactive power is exchanged not only between the generator and capacitors, but also between the generator and the power grid. The reduction of the reactive power exchanged with the power grid can generally improve the power factor that the wind turbine inputs into the power grid.

The conventional capacitor is uncontrollable reactive power. Through the additional compensation capacitors, the impact of the wind turbine on the node voltage will be reduced. However, it only improves the quality without improving the voltage control capability because there is a unique relationship between the speed, terminal voltage, active power, and reactive power. The voltage control capability of the constant speed wind turbine can only be improved by using more advanced controllable reactive power supply, for example, the switched capacitor or capacitor group, the static var compensator (SVC) and the static compensator or the static synchronous var compensator (STATCOM). For example, American Superconductor used the dynamic voltage regulator (D-VAR) system in WFs equipped with

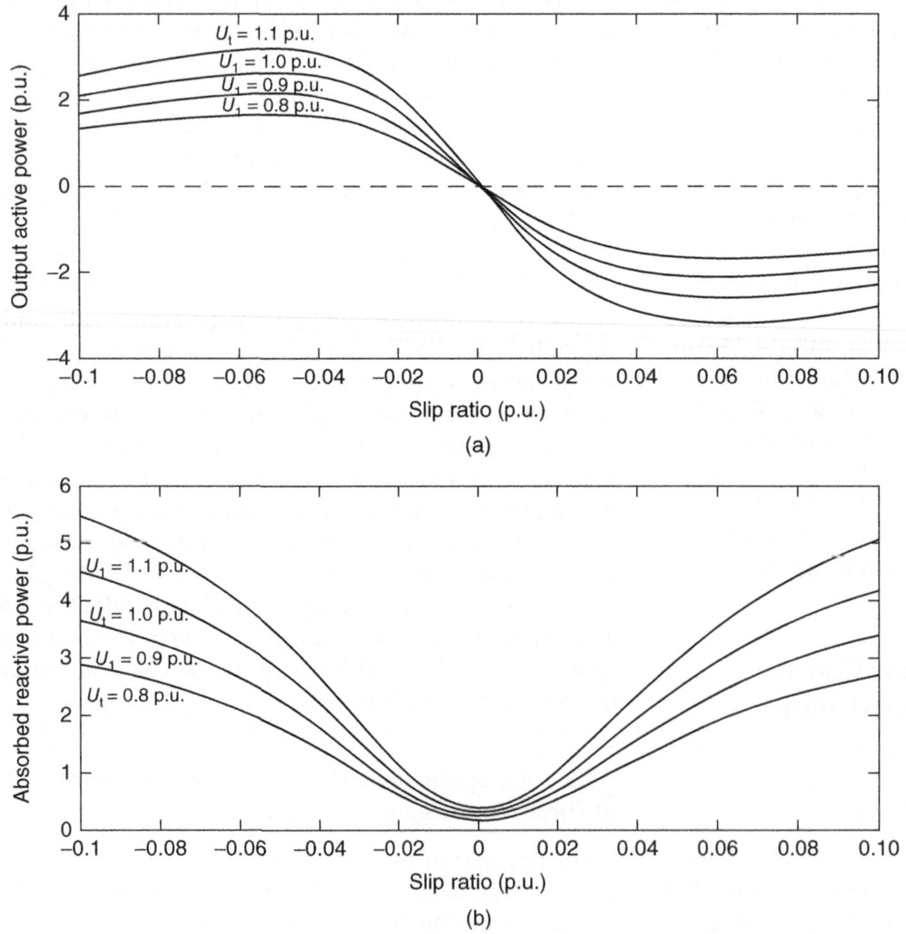

Figure 3.17 Power (P, Q)-slip characteristics of cage asynchronous generator. (a) Relationship between active power and rotor speed of cage asynchronous generator with different generator terminal voltage U_1 (b) Relationship between reactive power and rotor speed of cage asynchronous generator with different generator terminal voltage U_1 s.

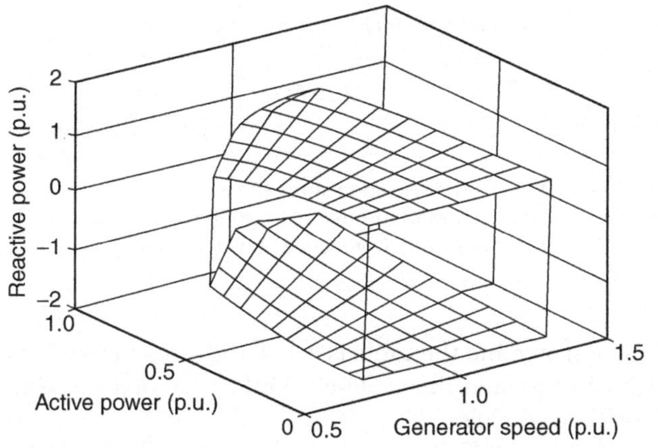

Figure 3.18 Operating range of double-fed asynchronous generator.

constant wind turbines in Minot, North Dakota of the United States.

The reactive power produced by the double-fed asynchronous generator is controlled by the rotor current. In this case, there is no unique relation between variables such as the rotor speed and active power. The output or absorbed reactive power can change in a large range.

Figure 3.18 shows the operating range of the double-fed asynchronous generator at the rated terminal voltage. It shows that the reactive power component is affected by the rotor speed and the output active power to a certain extent, which is just like the cage asynchronous generator, although it is not directly determined by these variables. The reason is that both the electromagnetic torque and the output or absorbed reactive power of the generator are directly determined by the current supplied by the converter to the rotor. The current generating the electromagnetic torque is determined by the setting value of the electromagnetic torque, which is obtained through the speed controller based on the actual rotating speed of the rotor. Conversely, the current required for generating the desired electromagnetic torque also determines the capacity of the converter so that the converter allows the current outputting or absorbing the reactive power to flow.

For the direct-drive variable speed wind turbine, the reactive power that the wind turbine exchanges with the power grid is not determined by the characteristics of the generator, but by the characteristics of the converter in the power grid side. The generator is completely isolated from the power grid. As a result, the reactive power exchanged between the generator and the generator side of the converter and the reactive power exchanged between the power grid side of the converter and the power grid are decoupled, which means that the power factor of the generator and of the power grid side of the converter can be controlled separately.

Figure 3.19 shows the operating range of the variable speed wind turbine equipped with the direct-drive synchronous generator with the terminal voltage U1 as the variable parameter without considering the change of the generator speed. As the generator is isolated from the power grid, the change of the generator speed can hardly affect the power grid. Assume that the wind turbine can operate with the leading/lagging power factor $\cos\varphi = \pm 0.9$ at the rated voltage and rated power. Figure 3.19 indicates that the variable speed wind turbine equipped with the direct-drive synchronous generator allows the reactive power or terminal voltage control because the multiple values of reactive power correspond to the single value of active power. What needs to be explained is that the curve shown in Figure 3.19 represents the boundary of the operating range while the curves shown in Figure 3.18 represents its unique relationship.

3.4.3 Factors Affecting Voltage Control

Based on what has been discussed above, the wind turbine with the voltage control capability can control the node voltage by changing the amount of the reactive power generated or absorbed. In order to control the voltage, the voltage at a certain node is required to

Figure 3.19 Operating range of wind turbine equipped with direct-drive synchronous generator with terminal voltage U_1 as variable; the curves show the boundary of operating range when the terminal voltage U_1 changes.

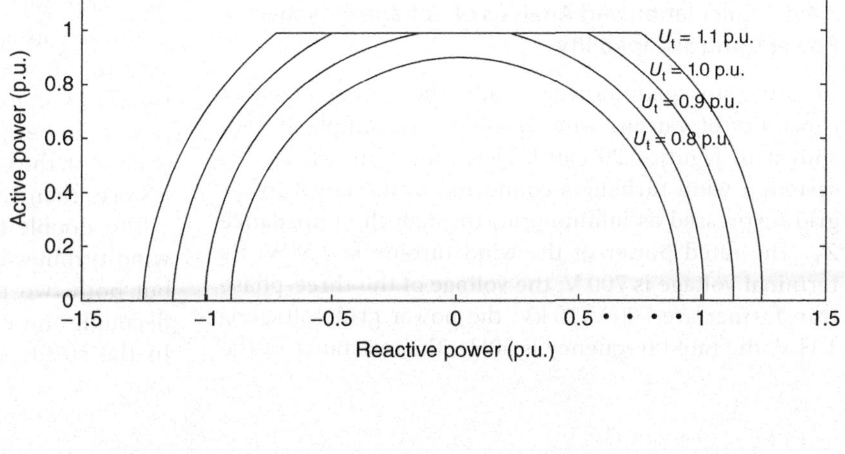

be measured and regarded as the input of the voltage controller. The voltage controller determines the size of the reactive power needed to be generated or be absorbed based on its transfer function. It is easiest for the wind turbine to control its terminal voltage, but sometimes we have to select a certain node in the power grid to control its voltage. Because of the local characteristics of the voltage of power grid, this node must be in the vicinity of the wind turbine. When the voltage is measured to be lower than the set value, the output reactive power is increased. When the voltage is measured to be higher than the set value, the output reactive power is then reduced.

The following factors can affect relationship between the amount of reactive power output or absorbed and the converter current that controls the reactive power:

1) The parameters of the generator (only the wind turbine equipped with the double-fed asynchronous generator);
2) Setting point of the voltage controller;
3) The values of the resistance (*R*) and reactance (*X*) of the connection line used to connect the wind turbine to the power grid;
4) The amount of the active power flowing through the connection line connecting the wind turbine to the power grid.

Among all the above factors, the generator parameters are closely related to the size of the generator. Double-fed asynchronous wind turbines of the same rated power have similar parameters of generator. As a result, the impact of these parameters is very small. As the setting point of the terminal voltage is almost always equal to the rated voltage value and has a very small change range, the setting point of the terminal voltage controller cannot play the major role, either. See references [9, 10] for the discussion about the impact of the resistance *R* and reactance *X* of the power grid connection line on the amount of the reactive power needed to maintain the reference value of the voltage. Now we are going to calculate and analyze the impact of the last factor (the amount of the active power flowing through the connection line).

3.4.4 Calculation and Analysis of Voltage/Reactive Power Control Capability

In order to quantitatively study the voltage control capability of various wind turbines, the simple system shown in Figure 3.20 can be used for analysis. In this system a wind turbine is connected to the large power grid (expressed as infinite grid) through the impedance Z_L. The rated power of the wind turbine is 2 MW; its terminal voltage is 700 V; the voltage of the three-phase transformer are 700 V/10 kV; the power grid voltage is 10 kV; the rated frequency is 50Hz; the resistance of the transformer and the cable is 2.5 Ω; the inductance of the transformer and the cable is 16 mH and corresponding reactance is j5 Ω.

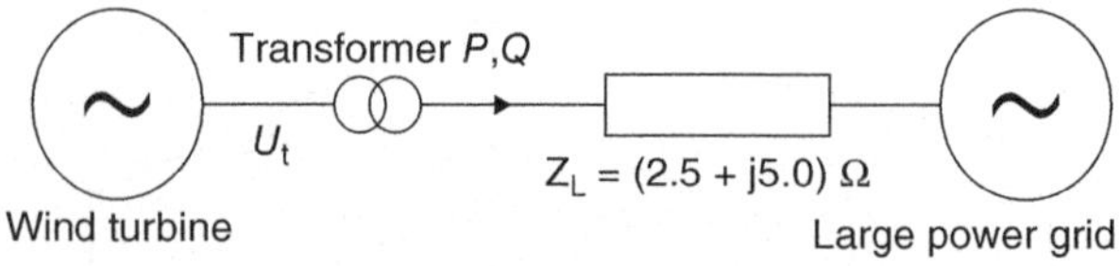

Figure 3.20 Test system used to study the voltage control capability of various wind turbines. U_1-terminal voltage of generator; *P*-active power; *Q*–reactive power; Z_L–line impedance; j-$\sqrt{-1}$.

3.4.4.1 Steady-State Analysis

For the constant speed wind turbine, the wind turbine itself does not have the ability to control the voltage/reactive power. The products produced by the wind turbine manufacturer might be equipped with the compensation capacitor with fixed capacity, which, however, cannot control the voltage/reactive power, either. The external voltage/reactive power control device must be used to control the voltage/reactive power.

The 2 MW cage asynchronous generator can absorb 1.05Mvar reactive power at the rated voltage and rated output power. If the reactive power is to be compensated completely, a 6.8 mF capacitor is required to be installed for each phase.

Figure 3.21 shows the results of the constant speed wind turbine steady-state analysis. In the diagram, the horizontal axis represents the active power generated by the wind turbine and the vertical axis represents the terminal voltage and the reactive power supplied to the power grid which is equal to the reactive power generated by the capacitor minus the reactive power absorbed by the cage asynchronous generator. Figure 3.21 shows that with the increase of the active power generated by the generator, the reactive power absorbed by the cage asynchronous generator will also increase. As a result, only a small part of the reactive power generated by the capacitor is supplied to the power grid. In the diagram, the generator terminal voltage increases because the reactive power is completely compensated when the generator generates the rated power, which belongs to the over-compensation mode. Actually in order to prevent the overvoltage, the under-compensation mode is usually used, for example, taking 30% to 35% of the rated power. To be specific, for the 2 MW cage asynchronous generator, the reactive power compensated in each phase is lower than 1.05 Mvar.

The double-fed and the direct-drive variable speed wind turbines have different power generation systems, but both two types of generators have full capacity to decouple the control of the active and reactive power. In the steady-state analysis, only the terminal voltage

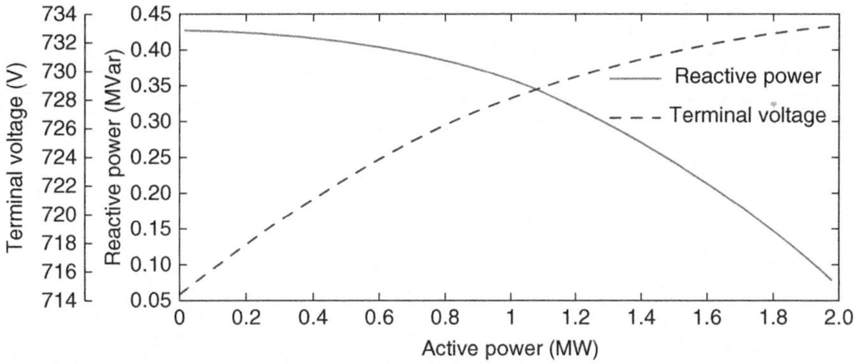

Figure 3.21 Results of the steady state analysis of the constant speed wind turbine with the test system shown in Figure 3.20.

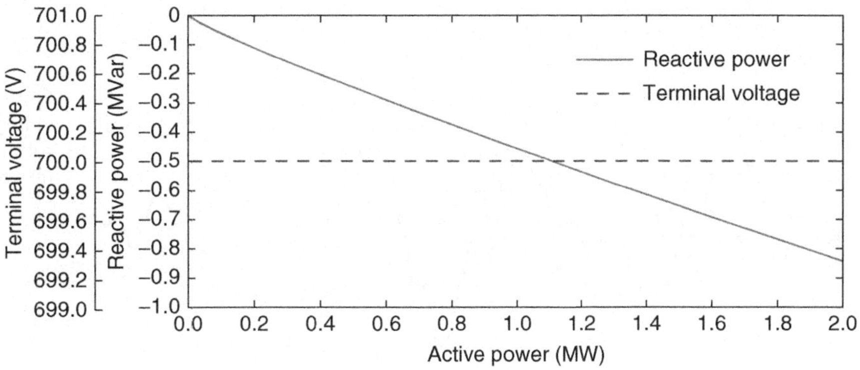

Figure 3.22 Results of the steady state analysis of the variable speed wind Turbine equipped with voltage controller with the test system shown in Figure 3.20.

of the wind turbine and the active and reactive power exchange between the wind turbine and the power grid are required to be studied.

Shown in Figure 3.22 are the results of the variable speed wind turbine steady-state analysis. In two cases, the wind turbine controller controls the voltage on the low-voltage side of the transformer (700 V). In the diagram, the horizontal axis represents the active power output by the wind turbine while the vertical axis represents the terminal voltage of the wind turbine and the reactive power output. Different from the terminal voltage shown in Figure 3.21, the terminal voltage in Figure 3.22 is constant.

At present the converter of the variable speed wind turbine is controlled so that it cannot exchange the reactive power with the power grid, which indicates that the current injected into the power grid and the power grid voltage have the same phase (the power factor $cos\ \varphi = 1$). This kind of operation mode is called unit power factor operation. Figure 3.23 shows the simulation results of this operation mode. As in the test system shown in Figure 3.20, the impedance connected to the power grid is quite large, the voltage control capability of the variable speed wind turbine operating in the unit power factor mode is not very good. Therefore, in order to prevent the voltage from rising, it is unnecessary to achieve completely the unity power factor operation. Instead, the power grid can supply some reactive power to the wind turbine, which is due to the same consideration as the constant speed wind turbine does not require full compensation or overcompensation.

3.4.4.2 Dynamic Analysis

The wind turbine dynamic voltage/reactive power control capability can be simulated through the large disturbance of the power grid voltage. It is assumed here that the power grid voltage drops from 10 kV to 9.75 kV in 30 seconds, which is possible in case of a large load put into operation or the tripping of a nearby synchronous generator.

In the analysis of the dynamic voltage/reactive power control capability, the two different variable speed wind turbines must be discussed respectively because they use different controllers to control the reactive power. The double-fed asynchronous generator controls the output reactive power by controlling the rotor current while the direct-drive synchronous generator controls the reactive power through the converter on the power grid side. They have different transfer functions and the performance of their control systems may also varies. Therefore, these two types of variable speed wind turbines must be simulated respectively.

Shown in Figure 3.24 are the simulation results under a certain (measured) wind speed sequence [8]. Figure 3.24 a represents the wave form of this wind speed series.

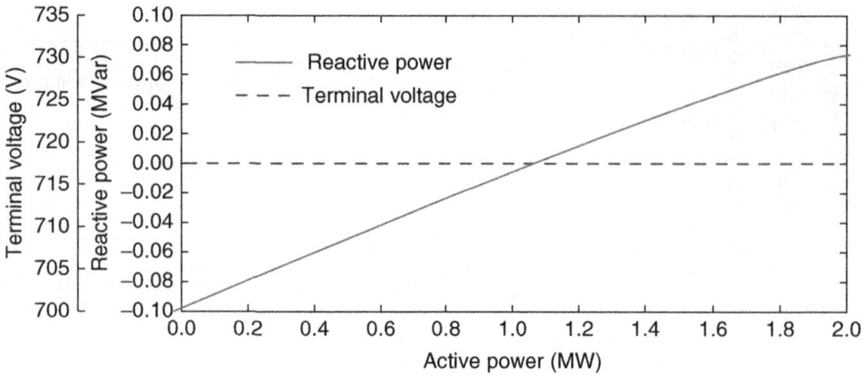

Figure 3.23 Results of the steady state analysis of the variable speed wind Turbine in the unit power factor operation mode with the test system shown in Figure 3.20.

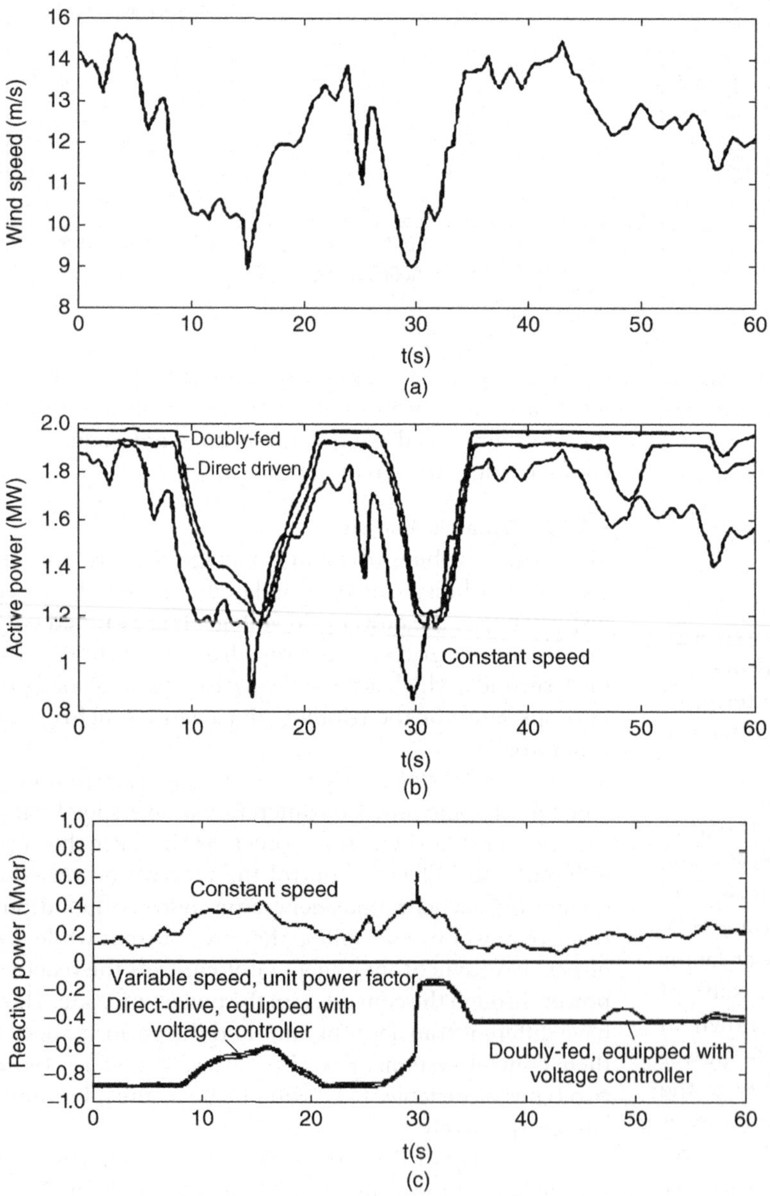

Figure 3.24 Simulation results of dynamic voltage/reactive control capability of various wind turbines. (a) wind speed series; (b) active power; (c) reactive power; (d) terminal voltage.

Figure 3.24 (*Continued*)

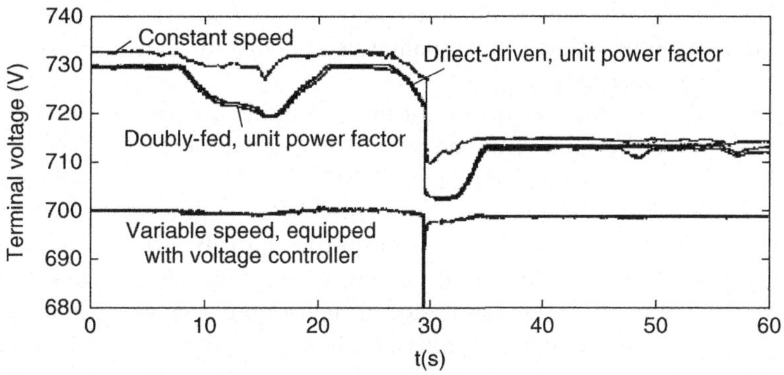

(d)

Figure 3.24b, Figure 3.24c, and Figure 3.24d represent respectively the change of the output active power, reactive power, and terminal voltage. From Figure 3.24 it can be seen that:

a) The voltage fluctuation of the constant speed wind turbine is the fastest because the generator is directly connected to the power grid and its rotor speed is constant. Therefore, like the variable speed wind turbine, the rotating mass of the constant speed wind turbine cannot be used for energy buffering. The change of wind speed and the tower shadow effect directly affect the change of active power. The reactive power needed by the generator depends on the active power and in turn the active power affects the reactive power generated and the terminal voltage. If the tower shadow effect is taken into consideration, the impact will be more obvious;
b) The amplitude of the terminal voltage change of the constant speed wind turbine and the variable speed wind turbine in the unit power factor operation mode is quite large due to the high impedance of grid connection line;
c) The fluctuation of the terminal voltage of the variable speed wind turbine equipped with the voltage controller is the flattest. As expected, the wind turbine equipped with the terminal voltage controller has better control performance than the wind turbine withoutterminal voltage controller;
d) Although the voltage controller works in different ways, but the control performance of two kinds of variable speed wind turbine is similar. If the controller is well designed, the terminal voltage control performance of two kinds of variable speed wind turbine could be equally excellent;
e) Finally, it is obvious that the wind turbine equipped with the voltage controller can compensate the drop of power grid voltage and keep the voltage at the set value. Only the variable speed wind turbine equipped with a voltage controller has the ability to control the terminal voltage and has nothing to do with the voltage of power grid unless beyond its operating limits.

3.4.4.3 Voltage Control Capability and Rated Value of Converter

The above analysis shows that the variable speed wind turbine equipped with the voltage controller has the best voltage control performance, but it comes at a cost. For the two types of variable speed wind turbine, the rated capacity of the converter required for the voltage control is higher than the rated capacity of the converter required for the unit power factor. How to determine the rated values of the converter? We need to analyze the current that might flow through the converter in different working conditions. For example, we have to study the current that might flow through the converter when the wind turbine generates the rated active power and ① does not generate the reactive power, ② generates the rated reactive power and ③ absorbs the rated reactive power.

Here we assume that the power grid voltage and frequency are kept at the rated value and the rated reactive power exchanged with the power grid is equal to 1Mvar, which is equivalent to the power factor $\cos\varphi = 0.9$. Compare it with the unit power factor operation and calculate the relative change of the reactive power generated and absorbed flowing through the converter current respectively. Table 3.1 lists the results. Although

Table 3.1 Requirements for rated current of converter of double-fed asynchronous generator and direct-drive synchronous generator.

Q	Rated current of converter (A)	
	Double-fed asynchronous generator	Direct-drive synchronous generator
0Mvar (in the unit power factor mode)	537	1650
Generating 1 Mvar (increase in percentage as compared with generating 0Mvar)	845 (+57.4)	1845 (+11.8)
Absorbing 1 Mvar (increase in percentage as compared with absorbing 0Mvar)	618 (+15.0)	1845 (+11.8)

the quantitative values obtained here are just for the wind turbines studied, but qualitative conclusions can used in a wider range.

In Table 3.1 the parameters of the two kinds of wind turbine are the same: frequency f=50 H_Z, active power P=2 MW, and terminal voltage U=700 V.

Table 3.1 shows that compared with the unit power factor operation mode, in case of outputting and absorbing reactive power, the rated current of the converter must be larger. For the double-fed asynchronous generator, the rated current of converter when the generator absorbs reactive power is smaller than when the generator outputs reactive power because as this kind of generator is connected to the power grid, its excitation current can be supplied by the power grid instead of the converter. In this mode of operation, the current of converter when the generator absorbs reactive power is smaller than the current when the generator outputs the same amount of reactive power. The full voltage control ability requires the wind turbine should be able to both output and absorb reactive power. Therefore, for the double-fed asynchronous generator equipped with the voltage controller, the size of the output reactive power is the decisive factor for determining the rated capacity of the converter. Table 3.1 also indicates that in order to meet the requirements for the ability of voltage control, in terms of the required increase in the rated value of the converter's current, the required increase in percentage of the capacity of the converter for the double-fed asynchronous generator system is the highest. However, we cannot make the conclusion from it that the voltage controller equipped by the double-fed asynchronous generator is more expensive than that equipped by the direct-drive synchronous generator. Table 3.1 shows that the increase in percentage of the converter current of the direct-drive synchronous generator in order to meet the requirements for the ability of voltage control is lower, but the total capacity of the converter is larger as it is required to realize the full power conversion, which means that, although the cost increase percentage of the converter of the direct-drive synchronous generator is lower than that of the double-fed asynchronous generator, the absolute value is higher. Therefore, the cost needed to equip the wind turbine with the voltage controller must be studied in terms of both the relative and absolute value of its increase.

References

1 Wang, Chengxu and Yuan, Zhang. 2003. *Wind Power Generation*. Beijing: China Electric Power Press.

2 Lin, Li Jian, Dongli, Zhao, Yaxi, Li, and Honghua, Xu. 2006. Comparison and analysis on network synchronization modes of several kinds of variable speed constant frequency wind power generator. *Electric Power Construction*, **May**, 8–17.

3 Qi-hui, Liu, Yi-kang, He, and Song-Jiang, Bian. 2004. Study on the no-load cutting-in control of the variable-speed constant-frequency (vscf) wind-power generator. *Proceedings of the CSEE*, **24**(3), 5–11.

4 Li Yaxi, Wang Zhihua, Zhao Bin, and Xu Honghua. 2006. Study on 'isolated island' cut-in strategy of high-power doubly-fed induction generator. *Acta Energiae Solaris*, **27**(1), 1–6.

5 Chinchilla, M., Arnalte, S., Burgos, J.C., and Rodrıguez, J.L. 2005. Power limits of grid-connected modern wind energy systems. *Renewable Energy*, **31**, 1455–1470.

6 Sourkounis, C., and Bingchang, Ni. 2006. Drive train control for wind energy converters based on stochastic dynamic optimization. IEEE Industrial Electronics, IECON, 32nd Annual Conference on November 6–10, pp. 4701–4706.

7 Bingchang, Ni and Constantinos Sourkounis, Constantinos. 2008. Power output characteristics analysis of wind energy converter control methods. Power Electronics and Motion Control Conference, EPE-PEMC, September 1–3, pp. 2331–2336.

8 Ackermann, Thomas. 2005. *Wind Power in Power Systems*. New York: John Wiley Sons, Ltd.

9 Shukla, Meera and Sekar, Arun. 2003. Study of the effect of X to R ratio of lines on voltage stability. Proceedings of the 35th Southeastern Symposium, March 16–18, pp. 93–97.

10 Holdsworth, L., Jenkins, N., and Strbac, G. 2001. Electrical stability of large offshore wind farms. *Seventh International Conference on AC-DC Power Transmission*, November 28–30, pp. 156–161.

4

Connection of WFs to Power Systems

Wind farms (WFs) are usually on the verge of power grids, small-scale WFs are connected to distribution grids, and large WFs are connected to transmission grids. With the development of wind power technology, and the constant increase of installed capacity of the WFs, the configurations of the relay protection device and power flow direction at the connection point between WFs and power grids have changed, and even the operation state of local power grid system has radically changed. Therefore, the connection modes between WFs and power grids are very important when determining the connection schemes. Selection of correct connection schemes can reduce risks and avoid adverse effects on the stability of the system. In addition, connection scheme of the system will directly affect the level of investment, and even the ultimate economic and social benefits of the entire project.

4.1 Requirements of the Power System for Grid Connection of WFs

The main goal of the grid is to economically and reliably transmit the power generated by power plants to individual users. Uniform dispatching and graded management shall be adopted to the operation of power grids, in order to meet the needs of social development requirements on modern power systems, ensure power quality, and improve the economic efficiency of power grid operation. Grid connection operation of conventional power plants shall meet relevant rules on the power plants and grid connection operation, including that power plants of different types and sizes are connected to grids of different voltage classes; power plants are required to be equipped with relevant equipment (stable control devices, compensation devices, and automatic generation control devices, etc.), relay protection and automatic safety devices are ready for putting into operation; the technical grade of the electricity metering devices are in line with relevant rules of the state and are accepted as qualified; communication systems between power plants and the power grid dispatcher have been established, and so on.

Grid-connected WFs only accounts for a very small part of the total electricity demand in most parts of the world. In power grids of some areas, such as Germany, Denmark, Spain, and other places, wind power supply has accounted for a large proportion of the total electricity demand. The annual average penetration levels of some power systems are higher than 50%. During strong wind weather, the wind energy is available for meeting the load demand of the entire power grids, and exporting to neighboring grids [1].

If the existing interconnected electric power systems integrate a high penetration level of wind power (20% to 30%), then the existing grids and their mode of operation may need to be redesigned. But this may be mainly an economic issue rather than a technical problem. For most of the power system, the current challenge is not the high penetration level of wind power that requires urgent resolution, but the gradual increase of wind power to be addressed.

Problems on integration of wind power into power systems can be divided into two aspects: one is how to maintain an acceptable voltage level for all users, that is, users are able to use the equipment they used before; the other is how to maintain the power balance of power systems, that is, how the wind power generator and other generator meet the needs of users.

General systems always face these fundamental problems. When inflexible nuclear power is introduced to the power systems, based on security considerations, frequent adjustments on the nuclear power is not allowed, but load is always in change, therefore, other power sources that can be flexibly controlled, such as hydro power plants, shall be introduced to power systems. In Japan, a set of very flexible pumped storage power stations have been built to increase the penetration level of nuclear power. When constructing the Dayawan nuclear power plant in China, there was the supporting Guangzhou pumped storage power station under construction and the Tianhuangping pumped storage

Integration of Large Scale Wind Energy with Electrical Power Systems in China, First Edition. Zongxiang Lu and Shuangxi Zhou.

Table 4.1 Basic requirements of the grid code on wind turbines.

Active power control	In order to ensure stability of power system and avoid line overload, the grid code requires positive control of the WF active power, but the power regulation degrees of different systems required by the grid code are quite different.
Frequency control	Frequency of the power system shall be maintained within acceptable limits, to ensure safe power supply and avoid electrical equipment overload, and meet power quality standards.
Frequency and voltage range	Even when voltage or frequency of the power system deviate from normal values, power system is able to continue operation.
Voltage control	Support network voltage by adjusting reactive power, which means reactive power compensation is required.
Voltage quality (rapid change, flicker, and harmonics)	This is a set of different requirements included in the national code.
Transformer tap changing	WFs are required by some grid codes to be equipped with interconnected tap-changing transformers, so that the voltage ratio between the WF and power grid can be changed when necessary.
WF protection	Such requirements are specifically prepared for potential failures or disturbances of the power grid. The relay protection system shall be able to operate under the circumstances of large short-circuit current, low-voltage/over-voltage during or after the fault period. It shall ensure that the WF will comply with the requirements for normal operation of power grid, and be able to support the power grid during and after the fault period. The WF shall be equally safe against the hazards caused by the power grid fault. The so-called FRT request falls into this category.
WF models and conformity	Some codes require WF owners/developers to provide model and system data, to enable operators to understand the interaction between WF and power system through simulation research. They also require the installation of monitoring systems to verify the actual performance of the WF during the fault period, in order to validate the model.
Communications and external control	Different from the above requirements, this requirement is in full conformity with that of the national code. WF operators should provide important signals about the parameters for the system operators, to enable proper operation of power system (representative parameters include voltages, active and reactive power, running state, wind speed and direction, etc.). In addition, some codes require that the wind turbine can be connected and disconnected from outside.

power station, which was the supporting project of the Qinshan nuclear power plant.

Many problems related to the grid connection of wind power depends on the specific conditions of existing power systems. General methods that have long been used by power engineers can also be used in grid connection of wind power, of course, some of these methods may need to be modified, allowing the design and operation of power systems to meet the needs of users.

The target of grid connection specification requirements is to ensure that the WFs will not have negative impacts on power supply security, reliability, and power quality, and so on, which are related to the power system operation. With the continuous development of wind power technology, grid connection specifications of WFs have been revised in parallel with the increasing of wind power levels and development of wind technology. Changes in existing wind turbine technology also reflects the development history of grid connection specifications.

Grid connection specifications and other technical requirements shall reflect the actual technical needs, and shall be developed by power grid companies, power generation companies, and administrative agencies in a cooperative manner.

Basic requirements of the grid code reported by European Wind Energy Association on wind turbines are listed in Table 4.1. The basic requirements of the grid code are related to frequency, voltage, and wind turbine behavior under the power grid fault. Depending on the wind power penetration levels, the robustness and flexibility of the power system, and the experience, knowledge, and policy of system operators, grid codes of wind power connection are very different in many countries. But for wind turbine manufacturers, although they have the ability to comply with different codes, compliance with grid codes for wind power connection of different regions requires a large controlling cost, so they always expect less diversity between codes.

The following requirements and specifications shall be added on in the grid connection of regular wind turbines or WFs

- Fault ride through (FRT) capability: maintain the grid connection operation of wind turbines during the power grid fault period;
- Grid connection control: the wind turbine can operate within the range of 47 to 52 Hz (power grid of 50 Hz);

- Active power control: be able to control the active power during frequency variation (fluctuation) period;
- Ramp rate constraints: constrain the power of the wind turbine/WF to increase/decrease to a certain rate;
- Reactive power control: generate or absorb reactive power according to the request of power system;
- Voltage control: support network voltage by adjusting reactive power.

In general, the wind power industry has the ability to follow additional requirements of the new grid codes. However, in order to meet the requirements of the new codes, it is necessary to significantly increase the costs of wind turbines and WFs. Therefore, new requirements of grid codes shall be added to the extent of meeting safe operation requirements of power system. In other words, countries and regions with little wind power do not have to adopt all these new grid codes mentioned above.

4.2 Connection of WFs to Power Distribution Grids or Transmission Grids

With the development of wind power, existing WFs have been connected to different voltage classes of the distribution grid and transmission grid. Related requirements on WF grid-connection shall be considered when WFs are connected to different voltage classes, which shall be determined after comprehensive considerations on the scale and control capability of WFs and characteristics (robustness, etc.) of the existing power systems. Small-capacity WFs shall be connected to distribution voltage classes, and large-scale WFs shall be connected to transmission voltage classes. WFs connecting to distribution grid and transmission grid have different connection requirements.

WFs connecting to distribution grid can benefit the operation and economic efficiency of power system to some extent. As a power supply close to load, in many cases, wind power has positive effects on reducing distribution power loss, maintaining qualified quality of power supply, and reducing network construction investment. However, due to their radial structure, the robustness and reliability of distribution grid is lower than that of transmission grid, and the so-called "active" distribution grid management is very little. Wind power connection makes distribution grid an active two-way network, changes the power flow direction, and affects power grid control and the operation of protection devices, so it is necessary to modify or add the design and operation codes of distribution grids.

Grid codes for wind power to connect into distribution grids (low-voltage and medium-voltage networks lower than 110 KV) are related to the distribution companies, wind turbine manufacturers, and network operators. The codes aim to establish guidelines, recommendations, and requirements for wind turbines and power grids under the premise that appropriate voltage quality and power-supply reliability are satisfied. The guidelines and requirements shall be irrelevant to the design of wind turbines, and shall be fully applied to synchronous and asynchronous generators, and generators with or without power electronic converters. The codes mainly add contents about regulating and dynamic characteristics of wind generators for maintaining power-supply reliability and voltage quality (long-term and short-term), and contents about grid stability may not be included.

When connecting large scale WFs to high-voltage distribution grids of 35 kV to 110 kV, if other users are electrically close to the WFs, the network voltage control requirements are more stringent. In practice, the operating voltage range is about 0.95 to 1.05p.u.

When connecting to 35 KV systems, the main concern is the fierce fluctuations of voltage. The maximum and minimum loads on 35 KV bus and voltage changes produced by maximum and minimum wind speeds shall be considered on analysis of voltage fluctuations. In the United Kingdom, the allowed voltage fluctuations and the corresponding time are shown in Table 4.2 [2].

Grid codes for wind power to connect into transmission grids (higher than 110 KV) require that the safe operation of the system shall be ensured during or after fault, including fault-ride through, WF model, communication requirements, and the possibility of external control of WFs. Issues to be concerned include long-term and short-term power-supply security and reliability, power quality, and impacts on the stability of power system. For this reason, specific requirements have been put forward on active power control, frequency control, voltage (reactive power) control, change of transformer taps, fault-ride through, WF protection, models and communications, and so on.

In terms of active power control, power variations allowed within different time ranges (one hour, ten minutes, one minute, etc.) is specified, as well as power control in the process of startup and shutdown, to avoid starting and stopping of all wind turbines of the WF at the same time.

Table 4.2 Voltage fluctuation allowed by the 35 KV system and the corresponding time.

Voltage fluctuation (%)	Interval time between two changes (s)
0.4	1
0.8	10
1.6	100
3.0	1000

In terms of frequency control, in addition to frequency range of the normal operation, requirements and range on participation of wind power in primary and secondary frequency regulation, as well as the upper and lower limits for the allowed wind power instantaneous trip-off rate. Under normal operation, the WF power output changes within 15 minutes can reach 15% of its installed capacity. Wind power output changes in extreme wind conditions may affect the system frequency. The codes require that WFs shall have the capacity of participation in frequency control, such as to reduce WFs output by stopping part of the wind turbines or pitch control in case of system over frequency. Some codes require WFs to intentionally maintain wind turbines partial load operated, in order to participate in secondary frequency regulation in case of system low frequency, so as to increase the power output of the WFs. But in general, capacity of WFs for participating primary frequency regulation is very limited, and reducing regulation will result in wind curtailment losses.

In terms of voltage control, the codes require WFs to help maintain the voltage of point of common coupling (or grid connection points) within the set limits, and be able to make a contribution to voltage regulation of power system. Voltage quality requirements related to voltage control include providing limits for rapid change of voltage, voltage flicker, and harmonics at point of common coupling.

In terms of change of transformer taps, the codes require that tap changing transformer shall be equipped if the WF capacity is larger (such as larger than 100 MW), and specify the voltage ratio limit adjusted by one gear shift of the tap of transformers connected to different voltage classes, such as 2.5% for 110 kV, and 1.6% for 400 kV.

In term of fault-ride through (FRT), the codes require that new technology shall be introduced for ensuring that the wind turbines will not trip off during fault period, and even in tripping circumstances, wind turbines shall still be able reconnect into the grid and continue to generate electricity in the shortest time.

In term of WF protection, two targets shall be achieved, one is compliance with requirements for normal power grid operation, and to support the power grid during and after fault period; the other is ensuring the safety of WFs, and avoiding hazards on WFs caused by power grid fault. During grid fault, small WFs can be immediately disconnected from power grids, but larger WFs shall not be disconnected in order to avoid increase of disturbance of power system. The codes provide operating values for over-current protection, over-frequency protection, low-voltage protection, and over-voltage protection. Intelligent system protection should be established, in order to minimize the loss of wind power during power grid fault period, and ensure recovery of the normal operation in the shortest time.

Table 4.3 Connecting capacities of WFs (United Kingdom).

connection position	WF maximum capacity (MW)
11 kV grid	1–2
11 kV bus	8–10
33 kV grid	12–15
33 kV bus	25–30
132 kV grid	30–60

Table 4.4 Connecting capacities of WFs (France).

low voltage	LV (1F)	(230 V)	P≤18 kVA
	LV (3F)	(400 V)	P≤250 kVA
medium voltage	1 kV < U≤50 kV	(15 kV, 20 kV)	P≤12 MW
high voltage	50 kV < U≤130 kV	(63 kV, 90 kV)	P≤50 MW
	130 kV < U≤350 kV	(150 kV, 225 kV)	P≤250 MW
	350 kV < U≤500 kV	(400 kV)	P > 250 MW

Specifications on connecting capacity of WFs to different voltage classes are slightly different from country to country. Specifications on installed capacity of WFs in Spanish are the technical conditions shall meet the requirements of installed power <5% Psc (short-circuit power), power oscillation (fluctuation) <2%, installed power <50% of the line capacity, and installed power <50% of the transformer capacity. Therefore, WFs of 250 to 450 MW shall be connected to voltage class of 400 kV, WFs of 100 to 250 MW shall be connected to voltage class of 220 kV; WFs of 40 MW shall be connected to voltage class of 132 kV; and WFs of 15 MW shall be connected to voltage class of 66 kV. Capacity specifications of WFs connecting to different voltage classes in United Kingdom and France are listed in Table 4.3 and Table 4.4 [3, 4].

4.3 Direct Connection to AC grids

4.3.1 General AC Connection Schemes

Usually we have two schemes for WFs to connect into AC girds [2, 5]. Scheme 1: Transmit the power generated by the wind turbines to the nearby power line through WF step-up substation transmission line by T-type or π-type connection, as shown in Figure 4.1. Scheme 2: Transmit the power generated by the wind turbines to the regional load-center substation through WF step-up substation transmission line, to supply power for the regional load center, as shown in Figure 4.2. Making specific analysis on the wiring scheme used for connecting WFs into the system shall be based on the installed capacity of WFs and the distance from WFs to load center.

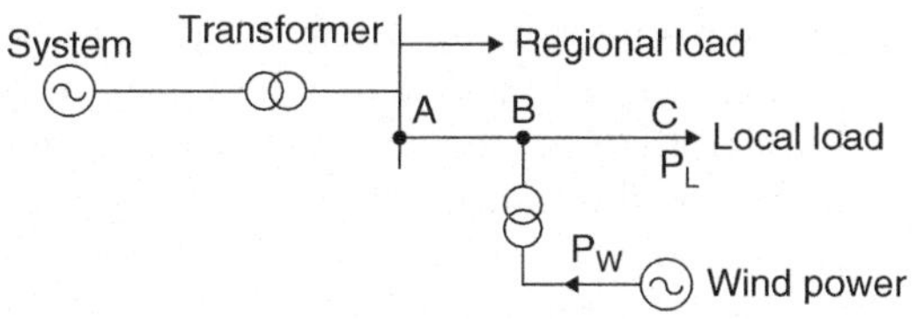

Figure 4.1 WF grid connection; Scheme 1.

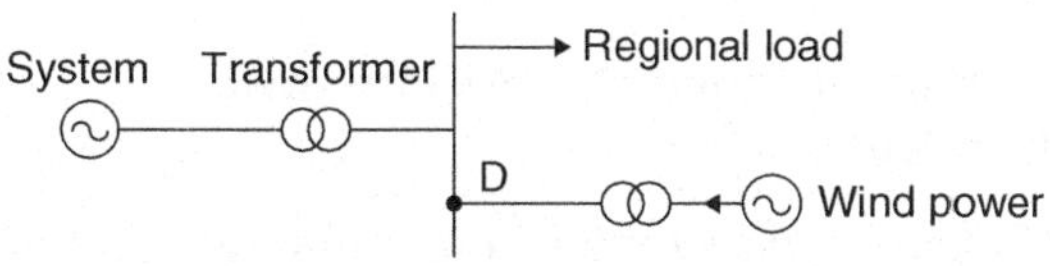

Figure 4.2 WF grid connection; Scheme 2.

1) **Wiring Scheme 1**
Analyzing from the prospective of WFs construction gross investment, the proportion of occupied funds shall be relatively low. But from the perspective of the installed capacity of WFs, the installed capacity of WFs will be constrained by the transmission capacity (i.e., line diameter) of the T-connected or π-connected original line, and the voltage parameter variation range shall be considered in order to ensure that the wind turbines operational requirements are met. However, due to the geographical location of WFs are usually far from the *regional load* center, load around WFs is relatively low, so designed diameter of original transmission lines are relatively small, thus the transmission capacity is small. In addition, the size of WFs installed capacity will have different impacts on relay protection configurations and power flow direction at point A of power system. Assume that the wind power capacity is P_W, local load is P_L, then when P_W is less than P_L, if the power flow direction of the line is not changed, power flow from point A to point B will combined with the wind power and flow to point C together, so the power flow direction is not affected. If P_W is slightly larger than P_L, part of the wind power will flow from point B to point C to meet local load needs, and the other part will flow from point B to point A and flow into the system. Due to random changes in wind speed and wind direction, the power flow direction of power source will change with the wind speed and wind direction, thus bring difficulties to power flow direction control of the line and correct setting of relay protection configuration at point A, and therefore random adjustment shall be carried out on the relay protection configurations at point A. If P_W is much greater than P_L, flow direction of the line, that is, the wind power flow direction, is relatively stable; a small portion of the power will flow from point B to point C to meet local load needs, and the majority of the power will flow from point B to point A to flow into the system. Relay protection configurations at point A does not need to be adjusted.
2) **Wiring Scheme 2**
Analyzing from the prospective of WFs construction gross investment, the proportion of occupied funds is relatively higher than that of Scheme 1, and the amount of one-time investment is large. However, since the WF step-up substation and the power transmission lines connected to the system are constructed based on the ultimate installed capacity and that the line voltage parameters can meet the wind turbines operation requirements, therefore, regardless of the operation mode of wind turbines, the installed capacity will not subject to the constraints in Scheme 1. Power flow direction at connection point D between the system and the transmission line of the wind farm is essentially does not change, but the power size will change with random changes of the wind speed and wind direction, so it is not necessary to adjust the relay protection configurations at point D. Therefore, wiring Scheme 2 is suitable for large-capacity WF step-up substation.

We can clarify from the above analysis and comparison that selection of system connecting schemes mainly depends on the ultimate installed capacity and the location of the WF in the power grid, that is, the distance between the WF and the regional substation. Specific options are as follows: for Scheme 1, to plan the ultimate installed capacity of a WF, we first determine the transmission power of the transmission line according to the diameter parameter of the existing transmission line, transmission line terminal voltage parameter variation range, and other factors, and then determine the total installed capacity of the WF based on the transmission power we have obtained. This scheme is suitable for connection of WFs of medium and small capacity into the system. Scheme 2 is suitable for connection of large-capacity WFs into the system. Actually, Scheme 1 is basically a scheme for WFs connecting into distribution grids, and Scheme 2 can be used for connection to distribution grids or transmission grids, that is, high voltage alternating current (HVAC) grid.

4.3.2 Transmission Power, Line Impedance, and Voltage Distribution of Transmission Lines

When WFs are connected into AC power grids, voltage is the most important issue. Integration of WFs have changed power flow distribution in power grid, thus changing voltage distribution of power grid. Main network elements in the impedance path between WFs and power grid are transformers, underground cables, and overhead lines, of which the impedance will affect the voltage distribution of the grid.

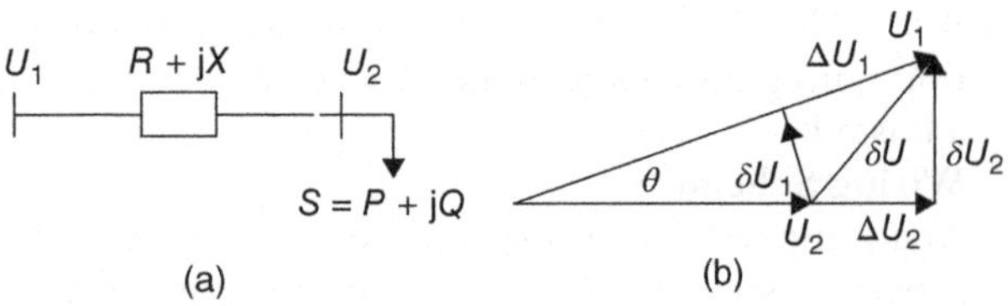

Figure 4.3 Basic model for transmission of active power and reactive power.

The transformer winding resistance is much smaller than the reactance, so its effect on the voltage can be ignored. Effect of cables in the WF on voltage can also be ignored. Of course, if the cable of the WF is very long, it may cause problems to voltage and resonance. The main factor that affects the voltage distribution of the grid is the overhead line of the nearby distribution or transmission systems.

For transmission line shown in Figure 4.3a, the starting node is one, and the terminal node is two. This system can be seen as a generator transmits power $S = P + jQ$ to the load via series impedance $Z_L = R + jX$. When the voltage difference between the two terminals of the line is not too large, and the power transmission angle θ is relatively small (θ represents phase difference between voltages at two ends of the line), voltage drop along transmission line is as follows (Formula (4.1)):

$$\Delta U = \frac{PR + QX}{U_2} + j\frac{PX - QR}{U_2} = \Delta U_2 + j\delta U_2 \quad (4.1)$$

Voltage drop can be divided into two components, as shown in Figure 4.3b, the longitudinal component $\Delta U_2 = (PR + QX)/U_2$ is in phase with U_2, and the transverse component $\delta U_2 = (PX - QR)/U_2$ is orthogonal to U_2.

RMS voltage difference between two ends of the line is called the voltage loss. When the voltage phase difference between two ends θ is small, the impact of transverse component of the voltage drop on voltage loss can be ignored, and longitudinal component of voltage drop can be approximately considered as voltage loss, $\Delta U \approx (PR + QX)/U_2$.

For high-voltage transmission lines, if R ≪ X, the longitudinal component of the voltage drop mainly depends on the reactive power transmitted, and the transverse component of the voltage drop mainly depends on the active power transmitted. The longitudinal component mainly affects the voltage amplitude, and the transverse component mainly affects the phase angle θ of voltages between two ends. In other words, voltage amplitude difference between two ends of the high-voltage overhead transmission line mainly depends on the reactive power transmission, and the voltages phase difference between two ends mainly depends on the active power transmission.

For medium- and low-voltage distribution overhead lines or cable lines, since R and X may be equal, or R can even be larger than X, the main factor affecting voltage distribution is the transmission of active power. For transformers, since the transformer resistance is much smaller than the reactance, the reactive power transmission is the main factor causing the transformer voltage drop in all cases.

4.3.3 Analysis on Connecting WF into Power Distribution Grid

Small WFs can be directly integrated into distribution grids, or connected to the power distribution substation busbars or T connected to the power distribution line. An important feature of the distribution line is that load demand is connected directly to the distribution line or via a transformer with a fixed tap. This circuit is mainly designed according to the standard voltage, the sending end users can reach the maximum allowable voltage, and the receiving end users can be supplied at the lowest allowable voltage. Voltage drop between the sending and receiving ends is the result of changes in load demand, as well as a function of distance and line impedance. When the wind power is not integrated, the voltages at sending and receiving ends are within a load-acceptable range. After integrated, the injected wind power is opposite in direction to the original load flow. In occurrence of minimum load and maximum wind power on the distribution line, voltages throughout the distribution line will increase, which may exceed the load-acceptable range, and threaten the operation of the wind turbine units. In this case, voltage increase or decrease at the sending end depends on the percentage of load line drop compensation (LDC), and voltage increase or decrease at the receiving end depends on the load level and the wind power injection.

How to avoid wide voltage fluctuations at the receiving end? Reduce line impedance can play a little role, which is very limited. One solution is to install a voltage regulator on the distribution line. Installation of a voltage regulator on the distribution line allows voltages at the sending and receiving end to be controlled within the allowable range. Control of the voltage regulator requires acquisition of information on voltages at the sending and receiving ends, wind power injection, and load. The purpose of voltage control is to maintain voltages of the WF within the scope prescribed via regulating the voltage at the sending end and the available WF reactive power.

In order to maximize the use of wind power, the regulator shall first control the reactive power, only when the voltage cannot be maintained at a satisfactory level via reactive power control shall we reduce the active power.

If there are several embedded wind power on the distribution line, a wide-area controller is required.

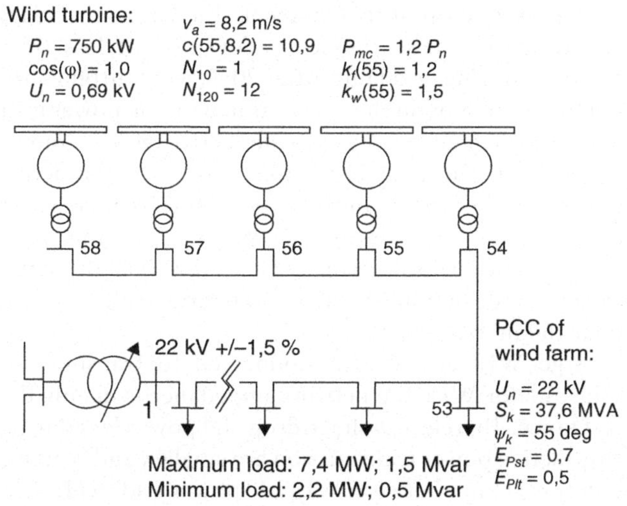

Figure 4.4 Small WFs connected to the distribution line.

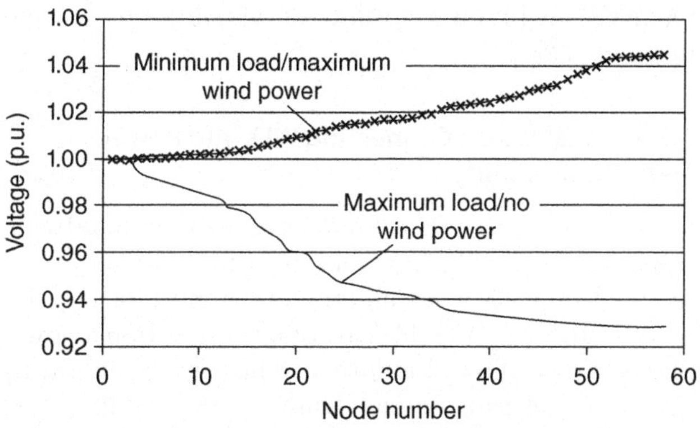

Figure 4.5 Voltage distribution of the distribution line connected with WFs.

If there are several embedded wind power connected to multi-circuit lines on the same bus, a controller of broader range might be required, and limits of the transformer shall be considered.

An example of WFs [6] connected to medium voltage distribution line is shown in Figure 4.4. A WF consists of five 750 kW wind turbines is connected to nodes 54 to 58, node 53 is the point of common coupling (PCC) of the WF and the medium distribution line, node 1 is the medium-voltage node of the high-voltage transformer, of which the voltage is set to the constant value of 1.0 (p.u.). Technical parameters of the grid and wind turbines have been indicated in the figure. The wind turbines are typically designed with constant speed and stall control to limit the power during high wind speed. Each wind turbine is equipped with a soft-start device to limit the surge current when starting asynchronous generator, and a switched capacitor to sustain running state at cosφ = 1.0.

Voltage distributions on the medium-voltage (MV) distribution line in the system shown in Figure 4.4 at the maximum voltage level (corresponding to minimum load and maximum wind power) and minimum voltage level (maximum load and no wind power) can be obtained through power flow calculation, as shown in Figure 4.5.

The load is supplied via a low-voltage transformer and a low-voltage line, so the voltage of users at the low-voltage end can be expressed as (Formula (4.2)):

$$U_{LV} = U_{MV} - \Delta N_T - \Delta U_T - \Delta U_{LF} \tag{4.2}$$

In the equation, ΔN_{T} represents deviation from rated voltage ratio of the low-voltage transformer; U_{LF} represents the voltage drop between the low-voltage transformer and the user, and ΔU_{T} represents the voltage drop caused by transformer loss. In the system, when at peak load, $\Delta U_T = 2\%$, $\Delta U_{LV} = 5\%$, whereas the voltage drop at minimum load can be ignored. This means that setting the low-voltage transformer tap at node 53 to −5% can make the minimum voltage on the low-voltage side higher than 0.90p.u. If the tap is fixed at −5%, maximal voltage on low-voltage side will be 1.09 p.u. If more wind turbines of the same type are installed in the WF, the maximal voltage at load end will continue to increase. If the low-voltage system requires that the slow voltage change deviation shall be set within the range of −15%/+10% of the rated voltage U_{n}, so for the system in Figure 4.4, slow voltage changes will put constraints on further expansion of WFs.

In this situation, ways to increase wind power installed capacity are: ① adjust cos(φ) of the wind generators. If adjusting cos(φ) from 1 to 0.98 (inductive) can reduce the maximal voltage by 1.5%, more wind turbines can be installed. In fact, if cos(φ) = 0.98, before the slow voltage changes puts constraints on further expansion, another eight sets of 750 kW wind turbines can be added. Of course, to alleviate the constraints of slow voltage changes by adjusting the cos(φ) will increase network loss. ② Increase the voltage regulating range of the HV transformer may change the minimum and maximum voltage. ③ Enhance the grid structure and reduce the voltage drop by building new lines. ④ Reduce the WF's power according to the voltage, but this will result in a waste of wind energy.

In the system shown in Figure 4.4, the factors that constrain the WF capacity also include power quality, according to the analytical results of paper [6], the flicker value has overshot, and harmonic voltage and voltage drop do not create constraints on the WF expansion and operation. The most direct way to overcome the flicker constraint is to select different wind turbine types of smaller flicker coefficient values. For constant-speed and stall-controlled wind turbines, flicker coefficient value c = 10.9 is too high; for constant-speed and pitch-controlled wind turbines, the c value is also relatively high; but c value of variable-speed wind turbines is low.

4.3.4 Analysis on Connecting WFs into Power Transmission Grid

Grid codes requirements for WFs integration into transmission grid are different from those for integration into distribution grid, wherein, the most important requirement is that, the WFs, like any other conventional power plants, is not allowed to lose stability or trip off during the period of short-circuit fault or other faults. More generally, during grid fault, wind turbines must be able to survive a short "dead" time (about 100 ms), once the fault is cleared, the voltage begins to recover to resume production, and return to the original operating point. The codes also call for increased control ability of WFs, requiring WFs to reduce its output from full load to a level between 0% ~ 20% within several seconds. These codes requirements involve power and power control (including start up); frequency and voltage (e.g., WFs must be able to operate at deviating frequencies and voltages); effects of WFs on the voltage quality; interactions between WFs and power grid under fault in the power system; and protection and communications, and so on.

The major concerns for WFs connection to transmission grid are also different from those of integration into distribution grid. When connecting to transmission grid, thermal overload, voltage and frequency stability, and other problems have been highlighted.

Thermal overload is likely to occur. During high wind speed and low load period, the local wind turbine power output feeding in the medium voltage (MV) grid is many times more than the local load demand, resulting in wind power feeding in the high voltage (HV) transmission grid, and the power flow redistributed in the high-voltage grid, thus causing overload of some lines and voltage overlimit of high voltage power grid.

In the development of modern electric power systems, synchronous interconnected power systems can extend to hundreds or thousands of kilometers, so stable operation must be considered. Due to various reasons, grid operators often force power systems to operate close to their stability limits. When large-scale wind power is injected into high-voltage power grids, its frequent changes will affect the system stability, frequency, and inertial oscillation damping.

In case of bus-bar fault, conventional power plants only require an individual plant to quit operation, but many wind turbines in a WF are close to each other, which may experience continuous mass active power quit in a large scale, of which the total amount could exceed that of a conventional power plant, just like the result of transmission line fault in an ultra-high voltage power grid. Faults in transmission networks (disturbances) can cause a static frequency deviation which will be exacerbated if wind power is lost at this moment. The Germany USTE system allows static frequency deviation of less than 200 mHz during grid fault. After 2003, wind turbines are required to maintain the connection with the power grid during grid fault events, provide active power immediately after the fault is cleared, and reach the active power level like before in 10 s to avoid further increase of tripped wind power during fault period. After this measure was adopted by Germany USTE, by 2005, the static frequency deviation during serious grid fault was less than 150 mHz.

Large WFs are directly connected to transmission grids, so, for WFs, the network impedance is mainly the reactance. Therefore, voltage drop and power loss during large active power transmission are smaller, and reactive power exchange between WFs and power grids will affect the voltage. Transmission components can generate and absorb reactive power, reactive power in the network is surplus under a light load, but insufficient under heavy load, which may cause the network voltage out of the acceptable range. In this respect, WFs connected to transmission grids shall be able to adjust the network voltage. This requires WFs to be equipped with equipment that generates and absorbs reactive power, such as reactors, capacitors, and power electronic devices, or power electronic converters, which can be controlled by triggering. Large-scale WFs also require tap-changing transformers, slow response reactive power controllers, as well as fast response reactive power controllers. The following analysis shall be made while determining how much static (slow response) and dynamic (fast response) reactive power is required by WFs [5]:

a) power flow calculation of reactive power demand under normal network structure and maximum load;
b) power flow calculation of reactive power demand under normal network structure and minimum load;
c) power flow calculation of reactive power demand under the greatest expected maintenance and minimum load;
d) power flow calculation of reactive power demand under the greatest expected maintenance, least safe generation, outage of transmission lines, and minimum load;
e) increased reactive power demand under least safe power and quit operation of lines.

Studies also take into account neighboring power systems, because the fast wind power development affects the entire power system. In most cases, the damping is adequate, but in some cases the damping is poor. To maintain reliable operation of the system, these two requirements shall be met: on the one hand, such low frequency oscillations must not excite wind turbine shafting subsynchronous resonance; on the other hand, wind turbine unit must not excite power oscillation in the power system, especially in low frequency range.

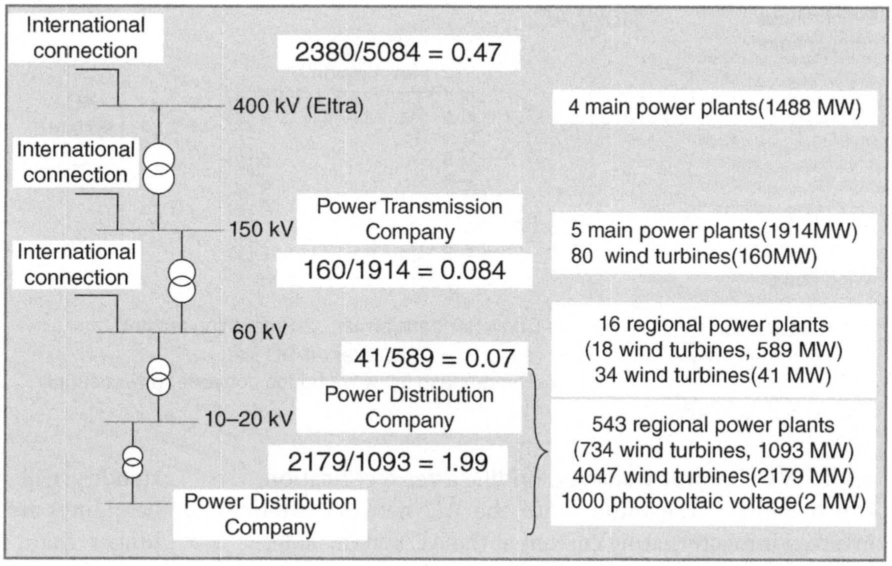

Figure 4.6 Generation capacity at various voltage classes in Denmark ELTRA.

Different from synchronous generators directly coupled to the AC grid, all types of wind turbines can only provide limited short circuit current. Studies have shown that, with the increase of wind power connected to power grid, wind turbines must maintain the grid connection during system fault and quickly return to its former operating point after the fault. However, due to converter constraints and other restrictions, wind turbines maintaining grid connection during the fault cannot provide reactive power required by the system, resulting in lower voltage levels after the fault is cleared, especially for 110 kV voltage level grid.

Depending on the types of wind turbines of the WF, the robustness and flexibility of the grid, integration of large WFs into transmission systems can bring both positive and negative effects on the system stability. Stability analysis of large WFs integration into transmission grids will be discussed in Chapter 7. Issues about maximum installed capacity of WFs connected into transmission grids will be discussed in section 4.9.

4.3.5 Decentralized and Centralized Interconnection

WFs can be connected to the grid in a decentralized or centralized mode. Because of the geographical distribution of wind energy, wind turbines are scattered over a broad geographical range, which are naturally suitable for decentralized connection to the grid. Actually, the installed capacity of a single WF is generally not as large as a conventional power plant.

The benefit of centralized connection is that the utilization ratio of some equipment such as transformers can be enhanced, thus facilitating operation and management. But the main problem is that, the installed capacity will be affected by short circuit capacity and other limitations at the connection point.

Although decentralized connection is also constrained by short circuit capacity and other limitations, the connection capacity can reach a very high level from the prospective of system. In Germany, Denmark, and Ireland, quite a lot of wind power capacity is or will be connected to distribution grids (of which the voltage is lower than 110 kV). Figure 4.6 illustrates generation capacity of various voltage levels in ELTRA region of Denmark: total conventional power generation capacity of the system is 5084 MW, and the wind power generation capacity is 2380 MW (data of 2006.6) [7]. The proportions of wind power installed capacity at various voltage levels in conventional power generation capacity are also listed in the figure. As can be seen in the figure, wind turbine capacity ratio of decentralized connection is very high, up to 1.99.

4.4 WFs Interconnection via Conventional HVDC (PCC—HVDC)

4.4.1 Grid Connection of Conventional HVDC and Its Characteristics

Integration of WFs into AC power grid may have some effects on the electric parameters, such as voltage fluctuations, harmonic interference, reactive power exchange limit, and so on. In order to improve transmission capacity, reliability, and stability of the system, HVDC technology is an ideal choice. Especially for offshore WFs, because there is no nearby large AC grid available.

Figure 4.7 illustrates a scheme of WFs interconnection via conventional HVDC. Power output generated by the WFs collected at an 110 kV AC bus, and is transmitted

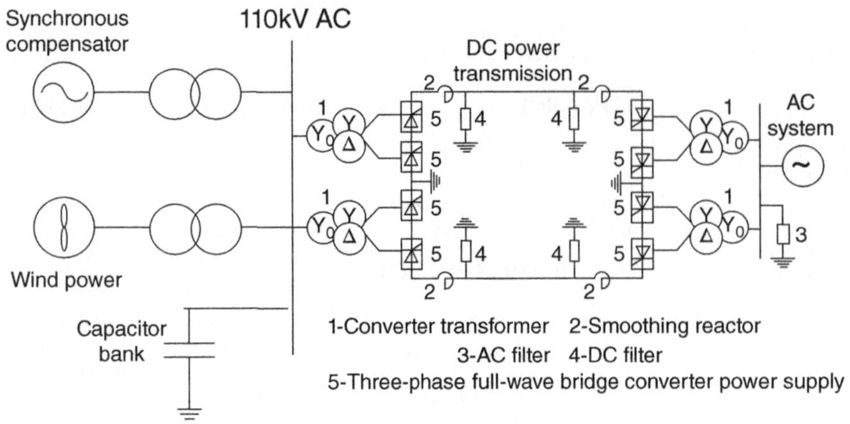

Figure 4.7 HVDC scheme for WFs.

via a HVDC power transmission line after rectification, which will be transmitted into the AC network after inverted into alternating current at the AC system side.

The core of conventional HVDC transmission is the phase commutated converter (PCC), which is also called line commutated converter (LCC). Its principle is based on the AC bus line voltage zero-crossing point, the corresponding valves are turned on by a delay trigger, so the current can flow via two conducting valves at the same time on the same half-bridge to form a two-phase short-circuit with the AC system for a short time. If the short-circuit current causes the current flowing via the first conducting valve smaller than the valve maintain current, the valve will be turned off, and the DC current will continue to flow via the second valve. The valves are turned on and off in sequence by sending trigger pulses in order, so as to achieve power conversion between AC and DC grid.

As a semi-controlled power electronic device—the thyristor is used, compared to AC transmission, the HVDC transmission technology based on PCC has unparalleled advantages in the following areas [8–10]:

1) Enables asynchronous communication between AC systems of different or the same rated frequency, improves the ability of the AC systems at both sides to back each other up and the ability of emergency accident support, and thus improves the system's stability and power supply economy. Back to back HVDC transmission system is used in the Panjiakou pumped storage power station to interconnect the 50 Hz AC system with three reversible pump-storage units within the station, of which the frequency varies within the range of 36 to 50 Hz, enabling the units to operate at peak efficiency.
2) Is especially suitable for high voltage, long distance large-capacity power transmission. HVDC transmission does not change the power-angle relationship, so the transmission capacity will not be reduced because the static stability or transient stability will not be affected. In addition, if overhead lines are longer than 700 km or cable lines are longer than 40 km, HVDC transmission is more economical than AC transmission.
3) Is particularly suitable for interconnection of large regional power grids. The purpose of interconnection of large regional power grids is to achieve resource sharing and emergency support after the fault, so fast-changing power flow and bi-direction transmission are required on the line, which can only be realized by HVDC transmission using power electronic devices.
4) Frequency at sending end is variable. Decoupling between generators and no need for synchronizing with large AC grids are the major advantages of HVDC, each end can operate according to their own control strategies, and have virtually no interaction.
5) The transmission distance is no longer a technical limit. Cable charge current has great influence on interconnection of AC cables, or even becomes the main influencing factor, but the influence on DC system is negligible.
6) The WF equipment is isolated from interfere of the main power grid. Fault of any end will have a very small impact on the other end. In some cases, the DC connection also helps fault recovery.
7) The power flow is completely determined and controllable. Under either normal or fault conditions, the power flow of the transmission line can be controlled by some control strategies.
8) The reliability of the devices has been confirmed. Having been developed and applied for about half a century, high-voltage direct current transmission (HVDC) technology has made considerable progress, and has had very good operating experience in various parts of the world. Therefore, PCC-HVDC can be readily applied to WFs.
9) The transmission line power losses is small, and do small harm to the environment. When transmitting the same power at the same voltage level, the losses

of DC transmission line is smaller than that of the AC line. The space charge effect of HVDC allows DC overhead lines to produce smaller corona loss and radio interference, which makes it easier to meet environmental requirements.

10) Suitable for cables power supply, and has high power transmission capability. Transmission capacity of a HVDC transmission line is 1.7 times as large as that of a three-phase AC transmission line with the same specifications. Ability to withstand voltage of cables with the same specifications under DC is almost three times as large as under AC. When at the same voltage, transmission power density of two-core DC cables is about three times as large as that of the three-core AC cable.
11) Faster and more accurate adjustment. The convertor station employs the thyristor power devices to conveniently, quickly, flexibly, and accurately achieve increase or decrease of active power flow and bi-direction transmission.

Since the thyristor is semi-controlled, the PCC technology can only control it to turn on, and it can only be turned off with the help of the voltage zero crossing of the AC bus bar: by reducing the valve current below the valve maintain current, the valve can naturally turn off. Therefore, HVDC transmission technology based on PCC has the following disadvantages:

1) Cannot supply power to small capacity AC systems or load without rotating motor. If short circuit capacity at the receiving end is insufficient, enough commutation current cannot be provided, and reliable commutation cannot be guaranteed, so the inverter is prone to commutation failure. If the receiving end system is a load without rotating motor, the inverter will not be able to supply power to the AC system due to commutation failure.
2) Harmonics produced by the converter are of low frequency and large capacity, which increase the burden on filtering.
3) Converter of the converter station has absorbed more reactive power. When in normal steady-state operation, the reactive power absorbed by the rectifier and inverter station respectively account for 30% ~ 50% and 40% ~ 60% of the transmitted DC power, and the converter will absorb more reactive power in transient operation.
4) Converter station requires large investment and large space occupation. To meet harmonic standards and converter reactive power needs, the converter station is equipped with a large number of reactive compensation devices and filtering equipment, which increases the investment and area occupation of the converter station.

Since conventional HVDC transmission based on PCC technology is a mature technology, its large transmitted power and variable frequency at sending end are very suitable for large-scale WF power output.

4.4.2 Selection of Wind Turbines When Using Direct-Current Transmission

DC power transmission provides more possibilities for selection of wind turbines of WFs interconnecting via HVDC transmission, the synchronous generation system and asynchronous generation system illustrated in Figure 4.8 can be considered. Figure 4.8a shows how the power generated by the synchronous generator is transmitted to the DC transmission line after diode rectifier and chopper; Figure 4.8b shows how the power generated by the conventional asynchronous generator

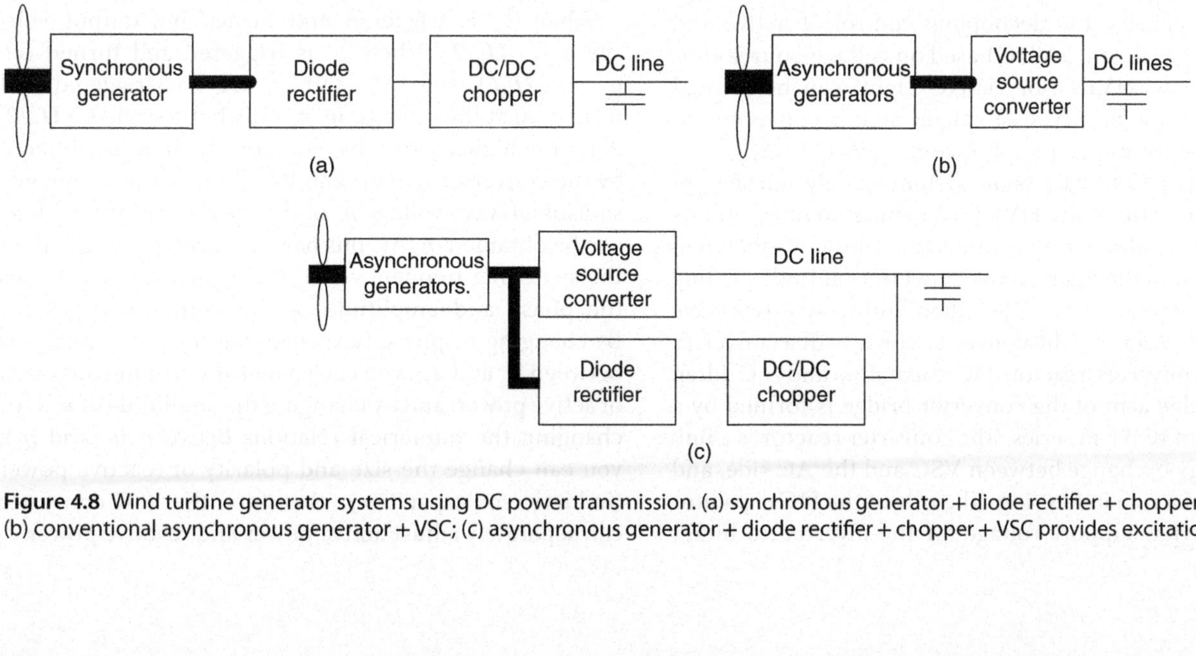

Figure 4.8 Wind turbine generator systems using DC power transmission. (a) synchronous generator + diode rectifier + chopper; (b) conventional asynchronous generator + VSC; (c) asynchronous generator + diode rectifier + chopper + VSC provides excitation.

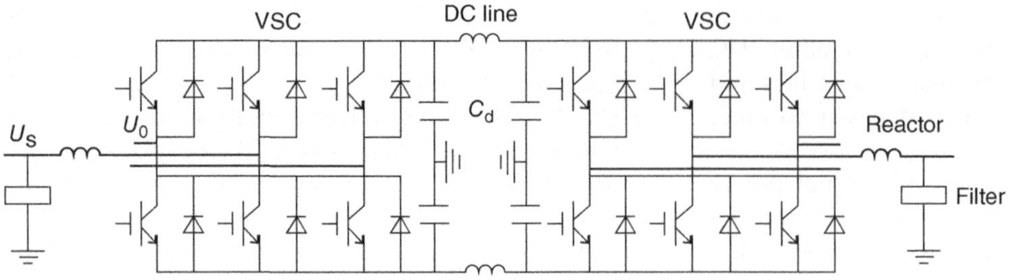

Figure 4.9 Schematic diagram of VSC-HVDC transmission.

is converted into direct current by the voltage-source converter (VSC) and transmitted to DC transmission line, and how the voltage-source converter provides excitation to the asynchronous generator; Figure 4.8c shows improvements made based on Figure 4.8b, wherein the asynchronous generator output is transmitted after diode rectifier and chopper; the voltage-source converter provides excitation current to the asynchronous generator, and its rated capacity can be greatly reduced.

Design of wind turbine generator sets for DC power transmission has two main trends in development, one is making modification on design and analysis of the original wind turbine sets, such as using a larger generator, increasing the parameters, improving component redundancy, and so on, which is a widely used technology at present; and the other is to consider the wind turbine generator sets and the HVDC transmission as a whole for development of entirely new forms of wind turbine structure.

4.5 WFs Interconnection via VSC-HVDC

4.5.1 Work Principle of VSC-HVDC

Voltage-source converter consisting of IGBT and combined with pulse-width modulator (PWM) will be able to easily achieve the decoupling control of active and reactive power, and HVDC based on voltage-source converters (SVC-HVDC, or HVDC Light) can be formed by using these new type of voltage source converters to replace conventional thyristor converters [11, 12].

VSC-HVDC transmission system mainly consists of converter stations and HVDC transmission lines, and its main electrical scheme of connections design is shown in Figure 4.9. In the figure, if the converters at both sending and receiving ends are VSC, then both converters have the same structure. The converter consists of a converter bridge, converter reactor, DC capacitor, and AC filter. Each bridge arm of the converter bridge is formed by a number of IGBT in series. The converter reactor is a link for energy exchange between VSC and the AC side, and also plays a role in filtering. The role of the DC capacitor is to provide support voltage for the inverter, to buffer inrush current when the arm is turned off, and to reduce harmonics at the DC side. The role of AC filter is to filter the AC side harmonics.

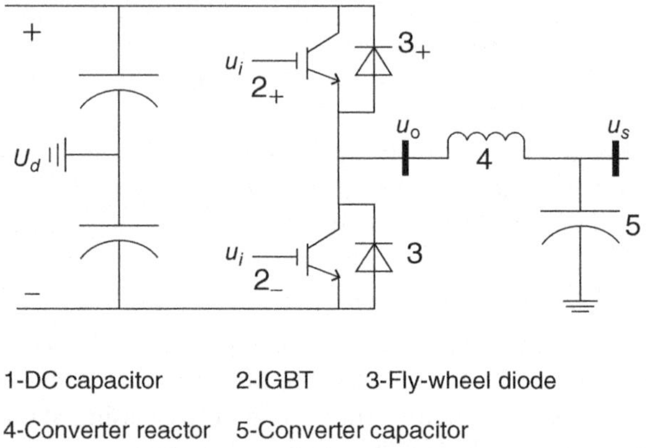

Figure 4.10 Single-phase diagram of the VSC converter consisting of IGBT.

For VSC that consists of high-frequency switching devices IGBT and adopts sinusoidal pulse width modulation, the single-phase circuit of the converter is shown in Figure 4.10. Its working principle is [12]: a trigger signal $\boldsymbol{u}_\mathrm{i}$ is produced by comparing the operating frequency sinusoidal wave signal $\boldsymbol{u}_\mathrm{c}$ with the triangular wave carrier signal $\boldsymbol{u}_\mathrm{tri}$, as shown in Figure 4.11.

When 2_+ is triggered and turned on, output voltage $\boldsymbol{u}_\mathrm{o} = \boldsymbol{U}_\mathrm{d}/2$; When 2_- is triggered and turned on, $\boldsymbol{u}_\mathrm{o} = -\boldsymbol{U}_\mathrm{d}/2$, since 2_+ and 2_- are not triggered and turned on at the same time, $\boldsymbol{u}_\mathrm{o}$ only has to values $\pm\boldsymbol{U}_\mathrm{d}/2$. After the higher harmonic components in $\boldsymbol{u}_\mathrm{o}$ are filtered by the converter reactor and the filter, power-frequency sinusoidal wave voltage $\boldsymbol{u}_\mathrm{s}$ of the same waveform with $\boldsymbol{u}_\mathrm{c}$ can be obtained on AC bus bar. Wherein, $\boldsymbol{u}_\mathrm{tri}$ determines the switching frequency of the switches, $\boldsymbol{u}_\mathrm{c}$ determines the phase and amplitude of the output voltages $\boldsymbol{u}_\mathrm{o}$. By changing $\boldsymbol{u}_\mathrm{c}$ phase (i.e., changing the phase relations between $\boldsymbol{u}_\mathrm{o}$ and $\boldsymbol{u}_\mathrm{s}$), you can change the size and direction of active power; and by changing the amplitude of $\boldsymbol{u}_\mathrm{c}$ (i.e., changing the numerical relations between $\boldsymbol{u}_\mathrm{o}$ and $\boldsymbol{u}_\mathrm{s}$), you can change the size and polarity of reactive power (inductive or capacitive). Therefore, the VSC converter can separately adjust active power and reactive power.

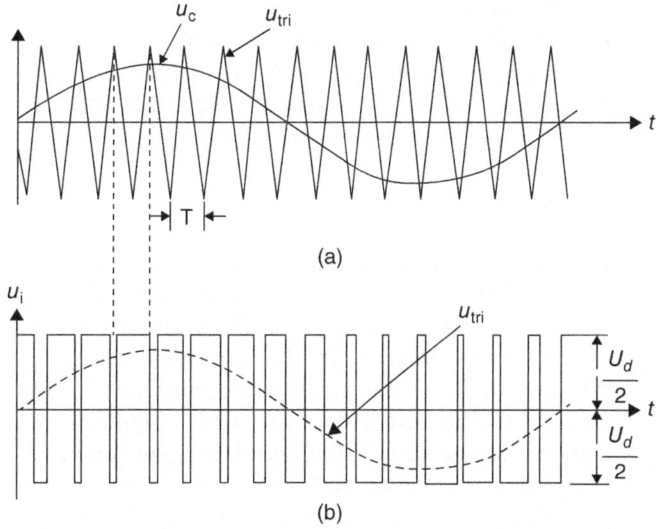

Figure 4.11 VSC waveforms.

Assume that the converter reactor is lossless, when ignoring the harmonic components, the active power P and reactive power Q transmitted between the converter and the AC grid are as follows:

The active power is calculated as (Formula (4.3)):

$$P = \frac{U_S U_0}{X} \sin \delta \tag{4.3}$$

The reactive power is calculated as (Formula (4.4)):

$$Q = \frac{U_S(U_S - U_0 \cos \delta)}{X} \tag{4.4}$$

In the equation, U_0 and U_s respectively represent the effective values of VSC output voltage and the bus voltage of fundamental frequency component, δ represents the phase difference between $\boldsymbol{u}_0$ and $\boldsymbol{u}_s$, X represents the reactance of the converter reactor.

The AC grid side voltage U_s is generally considered a constant. From Formulas (4.3) and (4.4) we can draw that the sizes and flow directions of active power P and reactive power Q can be easily and quickly controlled by controlling the VSC output voltage U_0 and phase angle δ; ① When $\delta > 0$, $P > 0$, the VSC conducts inversion operations, and transmits active power to the power grid. ② When $\delta < 0$, $P < 0$, the VSC conducts rectifier operations and absorbs active power from the power grid. ③ When $(U_s - U_0 \cos \delta) > 0$, $Q > 0$, the VSC supplies reactive power, that is, transmits reactive power to power grid. ④ When $(U_s - U_0 \cos \delta) < 0$, $Q < 0$, the VSC absorbs reactive power, that is, absorbs reactive power from power grid.

Thus, transmission of active power mainly depends on δ, and transmission of reactive power mainly depends on U_0. And U_0 is controlled by the pulse width of the pulse width modulation (PWM) voltage output by the converter. Therefore, by controlling δ we can control the direction of DC current and active power size transmitted, and by controlling U_0 we can control VSC to absorb or generate reactive power and its size being absorbed and generated. VSC can improve the power factor, and play a role of static synchronous compensators (STATCOM), to dynamically compensate reactive power for AC bus, and stabilize AC bus voltage.

4.5.2 Contrast of VSC—HVDC and PCC—HVDC

1) **PQ control**

   Different from the transmission technology based on PCC—HVDC, VSC—HVDC employs novel full controlled devices, such as GTO and IGBT to form converters, and achieve independent control of active power and reactive power by combining pulse width modulation (PWM), thus overcomes the needs of absorption large number of reactive power and the problems of commutation failures and other problems in PCC—HVDC transmission. As shown in Figure 4.12, VSC converter can both absorb and supply inductive reactive power, and conducts four-quadrant operation in the PQ plane [10]. While PCC—HVDC can achieve power control by turning off the filter and the series inductance, and controlling the trigger angle to some extent. It can only absorb inductive reactive power, and this control requires additional equipment and investment.

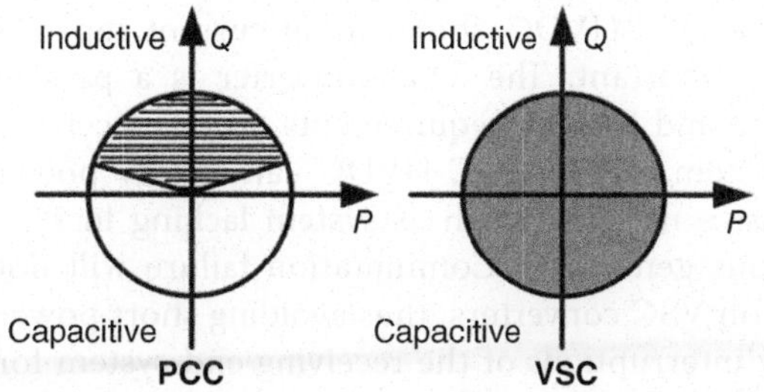

Figure 4.12 Contrast of PQ diagrams of PCC and VSC.

2) **Power range**
PCC-HVDC mainly operates in a high-power range, approximately above 250 MW, but the power transmitted by VSC-HVDC ranges from a few MW to 300 MW, of which the DC voltage can reach ±150 kV at present.
3) **Model components**
VSC-HVDC modular design significantly reduces the design, production, installation, and commissioning period of the VSC-HVDC, and main equipment of converter station can be assembled and tested in the factory in advance. The heaviest module weighs 20 tons and can be easily transported by truck directly to the installation site, which significantly reduces on-site installation and debugging time and the intensity of labor, and significantly reduces the occupied area of the converter station, of which the installation occupies far less area than conventional PCC converter stations. A VSC-HVDC converter station of 20 MW, 30 kV covers an area of less than 250 m^2, and the delivery time is no more than 12 months. But conventional HVDC equipment is often custom-designed according to the demand of system operation and some special usage.
4) **Converter circuit**
The VSC-HVDC converter circuit is natural dual polarity, and the DC current is ungrounded, so they need two wires, but because the PCC-HVDC current is grounded, we can use a single wire (cable). So the VSC-HVDC losses are greater than the PCC-HVDC losses.
5) **Circuit structure of the converter station**
The VSC-HVDC controls the switching of IGBT via a voltage-source converter (VSC), so the circuit structure is very different from that of PCC—HVDC. VSC-HVDC has not stability problems, and the VSC converter station is passive inverter, the AC voltage of receiving-end can be controlled, also can adjust the DC line voltage, the receiving-end systems do not need large short-circuit capacity and rotating motor load, thus can be directly connected to weak AC power grid or directly power supply to remote load.
6) **Operational independence**
The VSC-HVDC does not rely on the AC system to maintain voltage and frequency stability. Compared with the PCC-HVDC, short circuit current capacity is not important. The VSC converter is a passive inverter, and has not requirements on the receiving end system, so the VSC-HVDC can supply power to passive network (such as system lacking in synchronous generator). Commutation failure will not occur on VSC converters, thus avoiding short power supply interruptions of the receiving end system for several sustained periods, and improving power quality of the receiving end system. But the PCC-HVDC must have synchronous generators at the receiving end grid.
7) **Reactive power and harmonics**
The VSC-HVDC can obtain excellent sinusoidal wave by employing high frequency pulse width modulation (PWM) technique, which only requires reactive power compensation of a small capacity and simple high-pass filter devices, and is not required to be installed with converter transformer, so the switch can be simplified. The PCC-HVDC converter station requires a large amount of reactive power compensation, and produce large amounts of harmonics.

Contrast of main technical indicators of the VSC-HVDC technology and the PCC-HVDC technology is shown in Table 4.5.

4.5.3 Application of VSC-HVDC Transmission

4.5.3.1 Application Areas

The VSC—HVDC or HVDC Light transmission technology is still in development, and their voltage classes and transmission capacity continue to increase. Currently, their applications are mainly concentrated in the following areas:

1) **Supplying power to remote areas**
Generally, remote areas are far from the grid, so the load is light and daily load fluctuation is high. Economic reasons and low line transmission capacity is the major factors that constrain the development of AC transmission lines, which restrains economic development and growth in the living standard of people in remote regions. Using the VSC-HVDC supply power, there is an increase of unit power transmission of cable lines, reduce line maintenance, and increase supply reliability.
2) **Offshore power supply**
Offshore power loads that are far from land, such as island or offshore oil drilling platforms and other loads, usually generate electricity by using expensive diesel or natural gases, of which the generation costs are high, the power-supply reliability is difficult to be assured, and can cause damage to the environment. Using HVDC light can effectively address these problems, and send electricity generated by superfluous gases (such as natural gas from oil drilling) back to the system.
3) **Capacity expansion and renovation of urban power distribution grid**
There is no room for development of overhead line transmission corridor in cities, especially large cities, and the existing overhead distribution network cannot meet the requirements of power capacity increasing, so the reasonable approach is to use cable transmission (overhead cables or underground cables). The DC cable can transmit more power than the AC cable does while occupying smaller area, so the stability problems at the receiving grid can get

Table 4.5 Contrast of main technical indicators of VSC-HVDC and PCC-HVDC.

Indicator comparison	VSC-HVDC technology	PCC-HVDC technology
Converter devices	GTO and IGBT, etc. (Fully controlled, can be arbitrarily on or off)	Ordinary thyristor (semi-controlled, controllable turn-on, uncontrollable turn-off)
Conversion and control Technology	Voltage source converter technology (VSC); pulse-width modulation (PWM) control; will not occur commutation failure; do not require inter-station communication	Phase control commutation technology (PCC); cannot supply power to small capacity AC systems and loads without rotating motor; may occur commutation failure; inter-station communication is required
Reactive power and voltage	Can instantly change the phase and amplitude of the AC output voltage to achieve independent regulation of active and reactive power, can omit a converter transformer, and simplify the switch	Rectifiers and inverters absorb a large amount of reactive power from AC grid in operation, which accounts for about 40% to 60% of the DC transmission power
Harmonics and power quality	Small harmonics. Can be controlled in system fault to eliminate the short-circuit current, power, voltage drop it outputs to the AC system, and control power quality	Harmonic problems are serious, several smaller converters of different phase shifts connected by a more complex transformer are used in order to reduce harmonics
Economy	The system requires less equipment, smaller size, smaller occupied area, the cable is grounded, and wide transmission corridor is not required	Converter stations require a lot of reactive compensation devices and filtering equipment, and transmission lines need to occupy a large line corridor, so the investment is large
Voltage and transmission power	10–150 kV, 1–200 MW (at the present)	500 kV or above, can be more than 1,000 MW
Transmission form	In order to give full play of its potential advantages, new DC cables are used	Overhead lines with point-to-point transmission are commonly used
Applications	Can connected to any suitable points in the entire AC system, is especially adapted to independent loads with difficult problems and severe pollution that cannot be solved by traditional transmission techniques	Applicable for large power grids, usually long-distance power transmission

a better solution. Therefore, using VSC-HVDC to supply power to urban center has the potential to become an attractive approach for urban capacity expansion in the future.

4) **Clean energy generation**
Affected by environmental conditions, clean energy generation capacity is generally small, power quality is not high, and the network is far away from the main network, such as small and medium hydropower plants, wind power stations (including offshore wind power stations), tidal power stations, solar power stations, and so on. Due to the high operating costs, low AC transmission capacity, and other reasons, employing of AC interconnection solution is both economically and technically difficult to meet the requirements, and the use of HVDC light interconnected with the main network is the best way to make full use of renewable energy generation and other generation ways.
5) **Asynchronous operation between AC systems with the same or different rated frequency**
6) **DC looped network power supply**
The power-supply reliability of looped network is higher than that of radiation network and chain network. Multiple VSC converters can easily constitute multi-terminal parallel looped DC power supply network, so as to increase the reliability and flexibility of HVDC transmission.
7) **Improve power quality of the distribution network**
Nonlinear load and impact load can cause power quality problems to the distribution network, such as harmonic pollution, voltage interruption, voltage sag/swell, waveform flicker, and other problems, and cause malfunctions to some sensitive equipment such as industrial process control equipment, modern office equipment, electronic security systems, and so on, resulting in great economic losses. HVDC light can quickly control the active/reactive power respectively, and maintain a substantially constant voltage, so that the voltage, current, power quality can meet the standard requirements.

4.5.3.2 Application of VSC-HVDC in Wind Power Interconnection

Since the successful trial operation of the world's first HVDC light project—Sweden Hellsjon engineering in 1997, so far, many HVDC transmission lines have been completed and put into operation over the world, in which some projects are listed in Table 4.6 [13, 14].

The Hellsjon engineering in Sweden is the world's first HVDC light test engineering based on VSC, which is used for interconnecting different parts of the 10 kV

Table 4.6 VSC-HVDC project profile.

Item Name	Country	Length (km)	Voltage (kV)	Capacity (MVA)	Putting into operation Time	The main purpose
Hellsjon Test engineering	Sweden	10	10	3	1997.3	HVDC light transmission test, interconnection of two AC power systems
Gotland HVDC Light	Sweden	70	80	60	1999.12	Wind power generation
Direct-Link	Australia	65	80	180	2000.6	Asynchronous AC grid interconnection and electricity transaction
Tjaereborg	Denmark	4	9	8	2000.9	Small scale wind power generation
Eagle Pass	United States	Back-to-back	15.5	36	2000.12	Asynchronous interconnection between AC power grids
Murray Link	Australia	180	250	200	2002.10	Interconnection of South Australia grid and Victoria power grid
Cross-Sound Cable	United States	40	150	330	2002.7	Achieve power systems interconnection via undersea cables

AC network. The project connected the Hellsjon engineering in Central Sweden and Grangesberg together with a 10 km long 50 kV AC transmission line. Since the project put into operation, a complex pilot scheme has been completed, which shows that the transmission is stable, and the converter can meet the requirements of level of noise, harmonic distortion, phone interference, electromagnetic fields, and other technical requirements.

In June 1999, Sweden Gotland Island HVDC light project was put into operation. This is the world's first HVDC light system for commercial operation. Gotland Island is in the Baltic Sea, approximately 90 km to the east of Sweden mainland. Its total annual power consumption is about 900 GWh, most of the power is supplied via undersea PCC-HVDC network from Sweden mainland. In the 1990s, the total installed capacity of wind power generation has reached 40 MW. The rapid development of wind power generation has affected the Gotland Island power grid, and additional transmission capacity is required. In addition, better ways need to be found for maintaining good quality. After the study, selection of HVDC light scheme can minimize the network impact, rapidly control active and reactive power, and provide high quality electric power to users.

HVDC light is suitable for WFs due to its great technical advantages, the following features make it have good prospects in the development of wind power.

1) Due to the flexibility of HVDC light, its capacity can be easily expanded by adding new units, and WFs are usually expanded after a few years, or expanded by connecting neighboring new WFs. Therefore, HVDC light is practical in WFs.
2) Many places suitable for WFs are generally in more remote areas, where the grid structure is weak; in this case, sizes of the WFs are often limited by short-circuit ratio. In order to have a better performance, direct interconnection of WFs with AC power grids requires that the short-circuit ratio shall be around 10%, and application of HVDC light technology in transmission connection makes short-circuit ratio no longer a constraint.
3) Because HVDC light can independently provide wind generators with the reactive power they need to absorb, the requirements of wind generators to absorb reactive power from AC grid for excitation can be met.
4) Employing voltage source converter (VSC) by the HVDC light has the characteristics of independent control of active and reactive power, which can interconnect the wind power (a power source of unstable operating conditions) to power system without affecting the power quality level of the grid. Besides, it overcomes the problem of great loss of small capacity AC transmission, and is conducive to the rational development and utilization of renewable energies. In particular, VSC technology allows variable-speed operation of wind turbines and water turbines, so that they are able to run at a rotational speed which can output maximum power. The variable-speed operation can increase the output power of the wind turbine by 3% to 10%, therefore it has obvious economic benefits, which is suited to China's national conditions.

4.6 Contrast of WF Integration Schemes

HVAC, PCC-HVDC, and VSC-HVDC have been applied for WFs integration into systems; their characteristics and adaptability are listed in Table 4.7. Applications of HVAC, PCC-HVDC and VSC-HVDC have been shown

Table 4.7 Contrast of schemes of WFs integration into power grids.

Function	HVAC	PCC-HVDC	VSC-HVDC
Maximum allowable transmission capacity	800 MW(400 kV) 380 MW(220 kV) 220 MW(132 kV)	Can reach 600 MW by using undersea cables	The maximum capacity of completed projects has reached 350 MW, projects with the capacity of 500 MW has been presented, and projects with the capacity of 1080 MW has been designed.
Voltage levels	Projects with the voltage level of 132 kV has been completed, the voltage levels of 220 kV and 500 kV (400 kV), etc. have been in development	Can reach ±500 kV	Can reach ±150 kV, has presented the design of ±300 kV
Engineering construction	Many small projects	No	Sweden Gotland
Black startup ability	Yes	No	Yes
Ability to support grid	Limited, SVC is needed to provide inductive reactive power	Limited, capacitor and STATCOM are needed to provide valves with inductive reactive power	Yes, within a range; VSC can absorb and generate inductive reactive power
Offshore substations	Yes	No	In an oil field platform
Decoupling connected grid	No	Yes	Yes
Cable model	Resistors, capacitors and inductors	Resistance	Resistance
Require auxiliary equipment	Not required	Needed in low wind speeds	Needed in low wind speeds
Scales of offshore substations	Smallest	Largest	Medium
Construction costs	The cost of the station are small, only a small amount on the transformer, but the cost of cables is high.	The costs of the station are relatively high (transformers, thyristor valves, filters, capacitors); and the cost of cables is low.	The costs is 30% to 40% higher than that of power stations of PCC technology (IGBT is more expensive than thyristor), and the cables are also more expensive.

in Figure 4.13 from the prospective of WF capacity and transmission distances [15, 16].

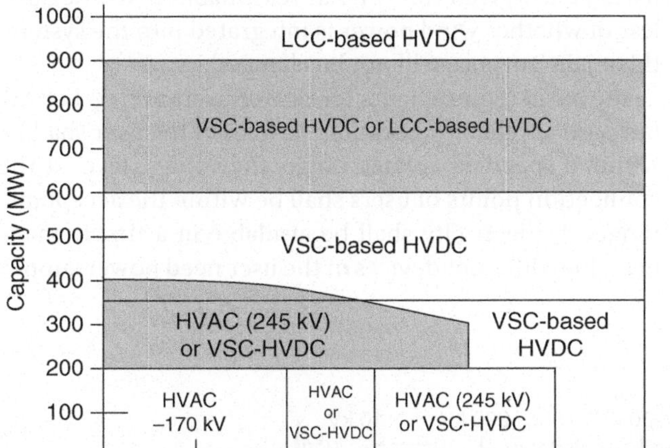

Figure 4.13 Applications of HVAC, PCC-HVDC, and VSC-HVDC.

4.7 Integration of Large-Scale WFs into the Grid

Due to different distributions and characteristics of the wind resources of different countries, WF scales are very different. At present, apart from offshore WFs, the WFs constructed in Europe are mostly of medium and small capacity. Many WFs of 50 MW level have been built in China, large and extra-large WFs under planning and construction are of 1 million kW level and 10 million kW level. Like the construction of large thermal power bases and large hydropower stations, power transmission must be planned and designed in advance. For integration of such large and extra-large WFs into the grid, the power output practices of the TGP hydropower station can be used for reference, but WFs also have characteristics of variability and limited predictability regarding power output, which shall be given special consideration when integrating into the grid.

For large-scale WFs that are closer to the load center, such as the coastal area of Jiangsu Province, construction of multiple 220 kV substations in the WF area can be taken into consideration, for supplying part of the wind power to regional and township grids, and the other part to the main grid of 500 kV.

For areas that are far away from the load center, such as Inner Mongolia, Gansu and other areas rich in wind power resources, since the local loads are small, the wind power is mainly sent outside, choosing of integration schemes shall be according to the load conditions. If the wind power is transmitted to the load center by centralized approaches, it shall be transmitted directly from the power supply point to the receiving-end grid, so as to avoid the adverse effects of wind power on the grid. Choosing of high voltage AC or HVDC, and voltage levels, shall be determined by technical and economic comparison after considering the transmission capacity and distance.

For DC transmission, there is no other local power source available, other power sources are difficult to integrate into the middle of the transmission line (not a multi-terminal DC), and utilization rate of the DC circuit is low. In the selection of DC transmission system, PCC-HVDC shall be employed, because it can use higher voltage levels, greater transmission capacity, and the technology is more mature.

High voltage AC transmission of wind power can be used in combination with other conventional power to improve line efficiency. For long-distance transmission of wind power via AC high-voltage, line voltage control is a prominent problem. Because the power will change significantly and frequently, the line is on no load or light loads for most of the time, adopting controllable shunt reactor for smooth line voltage shall be considered.

The impact of wind power transmission on the receiving end grid shall also be considered, wind power transmission is different from transmission of conventional large power to the receiving end grid, and the wind power is of large variability and predictability error. Ability of the receiving end grid for absorption of wind power is related to its size, flexibility, and robustness.

Construction of power grids in areas of large-scale WFs shall be strengthen. Construction of large substations shall be dispersed to collect wind power together. Collecting voltage of general WFs is of 35 kV level, WF collection system shall employ ring wiring (see offshore WFs) to improve reliability.

4.8 Determination of Maximum Wind Power Capacity Allowed to be Integrated into the Grid

Any kind of power generation equipment integrated into the gird for operation will bring certain influences on the operation, stability, and security of the grids, so we need to study the feasibility of the grid connection. The maximum wind power capacity allowed to be connected into the grid discussed here refers to the maximum wind power capacity integrated into the nearest substations under existing network conditions, rather than the wind power capacity allowed to be integrated into the entire system, or the wind power capacity allowed to be integrated after the grid has been strengthened.

4.8.1 Basic Analysis on WFs Integration Into Power Systems

WF integration scenarios can be represented by Figure 4.14 (the transformer is not drawn in the figure, and its impedance is merged into the line impedance). Figure 4.14a shows a system with local loads, Figure 4.14b shows the system without local load. In the figure, 220 kV (some may be 110/35 kV substation) bus is regarded as point of common coupling (PCC).

Determination of maximum wind power capacity allowed to integrate into the system shown in Figure 4.14, is to determine the maximum wind power injection that can meet constraints of the load, the WF and the grid.

The main purpose of power systems (enterprises) is to provide a sufficient quantity of qualified electricity to users at any given time and at reasonable costs. Regardless of whether wind power is integrated into the system, this main purpose will not be changed.

The basic requirements for power users are: ① For the purpose of reliable operation, devices of the user shall be within a specified voltage range, the voltage level at the connection points of users shall be within the acceptable range. ② Electricity shall be available in a timely manner when different devices of the user need power supply.

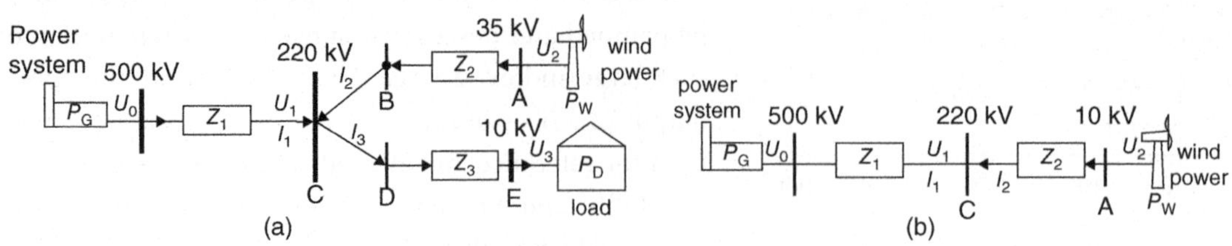

Figure 4.14 Diagram of WFs integration into the power system. (a) with local load; (b) without local load.

③ The power price shall be relatively reasonable. The first two requirements are regarding reliability of the power supply, and the last one is talking about the economy. These two aspects are contradictory, the higher the reliability of power supply is, the higher the electricity costs are.

Similar to the users, WF owners or operators also have certain requirements on the power system, including: ① The WF requires that voltage at the point of common coupling shall be maintained at a certain level, because the wind turbines are usually designed to operate in a predetermined voltage range (e.g., rated voltage ± 10%). Of course, the WF's requirements on voltage are not so strict compared to user's requirements. ② WF owners require to sell all the electricity generated to the grid, otherwise the wind energy will be wasted, and WF owners may lose potential revenue. ③ WF owners also require power system at point of common coupling to be reliable. Again, this is a compromise between cost and reliability, the higher the required reliability is, the higher the cost is.

Challenges caused by wind power integration into power grid is to meet the requirements of the user and the WF in a cost-effective way, even when wind power penetration level in the system is high.

How to determine the maximum power of the WF allowed to be integrated? For the scheme of wind power integration into power system shown in Figure 4.14a, the maximum capacity of wind power allowed to be integrated can be determined by the following calculation:

1) Through analyzing of power balance, check whether reserve capacity of the existing power system can meet system power balance compensation requirements, that is, meet the frequency quality requirements.
   In power systems with wind power, the power balance requires that the result of generated power of conventional power plants minus the wind power shall equal load plus loss.
   That is (Formula (4.5))

$$P_G = P_D + P_L - P_W \tag{4.5}$$

   In the equation, P_G is the generated power of conventional power plants; P_D is the load power consumption; P_L is network power loss (mainly loss of transmission lines and transformers); P_W is the wind power generated power.
   If no wind power is connected to the power grid, the user is directly supplied by the conventional power plants. Formula (4.5) is changed to (Formula (4.6))

$$P_G = P_D + P_L \tag{4.6}$$

   When load P_D increases, output power P_G will directly increase. If increase of the generated power (including standby power) is less than (or greater than) increase of the load, the steam turbine, the shaft, and the generator rotating system will decelerate (accelerate). Decrease (increase) of rotor speed will cause decrease (increase) of the electrical frequency. Therefore, increases (decreases) of the load will result in decrease (increase) of the electrical frequency (increasing).
   The power system shall always maintain the power balance, since a large amount of electrical energy cannot be economically stored in power system, any changes in load demand or wind power, must be balanced by conventional generators of other power sources within the power system. Wind power integration into the system has brought additional variation sources to the power system. As the level of wind power penetration increases, demand of power system balance has become higher.
2) Determine the WF injection power limit for maintaining the normal operation of the system by calculation and analysis of the steady-state power flow. Basic problems about normal operation of power systems include: to keep the voltage of power system at an acceptable level for all users, namely, voltage problems; to maintain a balance of the power system, namely how can the wind power generators and other power generators continuously meet users' power requirements, that is, system frequency problem. For this reason, the following calculations shall be made: ① calculate the voltage levels on the load bus bar and at point of common coupling under maximum or minimum load when there is no wind power (equivalent to N-1 of conventional power system); ② calculate the voltage levels on the load bus bar and at point of common coupling under maximum wind power and minimum load; ③ calculate the voltage levels on the load bus bar and at point of common coupling under minimum wind power and maximum load. The wind power at the voltage in excess of the prescribed limits obtained by calculation can be considered as the maximum wind power allowed to be injected.
   For the power system shown in Figure 4.14, when wind power P_W changes, the current I_2 and current I_1 will also change, thus changing the voltage U_1, and also likely to change the voltage U_3 at the connection point close to users. Impact of wind power fluctuations on voltage U_3 mainly depends on the size of impedance Z_1. On one hand, if Z_1 is large, there is a strong link between wind power and fluctuations of voltage U_3. On the other hand, if Z_1 is small, voltage U_3 will more or less be related to fluctuations of wind power. In fact, only users who are very close to the WF may be under the influence of wind fluctuations.
3) Identify measures to maintain the voltage levels required at user connection points. With or without

wind power, voltages at connection points will change as the load changes. Voltage changes are mainly affected by sizes of impedances and currents between users and supply points. For integration of wind power into power system as shown in Figure 4.14a, measures to maintain or control voltage level U_3 at the load connection point include: ① Use strong grid, that is, the impedance between the user and the power supply point is smaller. This can also be achieved by using high line voltage and large transformers. ② Use controllable transformers to control voltages at user connection points. ③ Employ controllable transformer to adjust voltages at point of common coupling (PCC) or equip voltage control equipment such as shunt capacitors and/or shunt reactors at point of common coupling. ④ Use the local control of WF voltage U_2. Some wind turbine technologies are capable of controlling terminal voltages and supporting system voltages.

4.8.2 Penetration Limit and Short-Circuit Capacity Ratio

1) **Wind power penetration limit**
Wind power penetration refers to: the ratio of WF installed capacity in total load of the system. Wind power penetration limit is defined as the percentage of maximum WF installed capacity connected to the system in the maximum load of the system on conditions that certain technical standards are met. That is (Formula (4.7))

$$\text{Wind power penetration limit} = \frac{\text{The maximum wind power capacity can be withstood by the system}}{\text{maximum load } S_{\text{L.max}}} \tag{4.7}$$

The concept of wind power penetration limit is from the perspective of the entire network, characterizing the maximum wind power capacity can be withstood by the power grid of a given size, which aims to consider the impact of WFs on the system frequency. To determine this indicator, first the randomicity and limited controllability of wind power shall be taken into consideration. In the two cases of wind power is put into/out of operation, the adjustable capacity of power system shall be able to ensure that changes in frequency are within the allowed range.

In most power systems of China, the grids are of large scales, and cover a wide range. When a WF is connected to the system, its capacity will take up relatively low percentage of the total capacity of the grid, so the influence of wind power on the grid frequency is very small.

2) WF short-circuit capacity ratio: short-circuit capacity is calculated from short-circuit current and rated voltage $S_{SC} = \sqrt{3}I''U_n$. WF short-circuit capacity ratio is defined as the ratio of WF rated capacity P_{wind} to the short-circuit capacity S_{sc} at point of common connection (PCC) between the wind power and the power system. That is (Formula (4.8))
WF short-circuit capacity ratio

$$K = \frac{P_W}{P_{sc}} \times 100\% \tag{4.8}$$

where P_w denotes the WF rated capacity, P_{sc} denotes the short-circuit capacity at connection point between the WF and the power system

Short-circuit capacity ratio is a concept often used in voltage stability studies. It usually refers to the ratio of short-circuit capacity at electrical equipment installation point to its installed capacity, and electrical equipment may be large capacity motor load, HVDC converter station, static reactive power compensator, and so on. The concept of WF short-circuit capacity ratio is actually the reciprocal of short circuit ratio under normal circumstances, and the meanings expressed here are the same. PCC generally refers to the high voltage side outlet point of a WF substation. Short-circuit capacity represents the strength of the network structure, and large short-circuit capacity indicates that the electrical distance between the node and the system power source point is short, and the connection is close. Short-circuit capacity at the WF connection point reflects the sensitivity of voltage at this node to changes in injection power of WF. Low WF short-circuit capacity ratio shows that the system has good ability to withstand wind disturbance. Using this indicator to indicate the scale of the WF integrated, is to focus on the study of impacts of wind power integration into local grid on power quality and voltage stability from the prospective of the local power grid that the wind power is connected to.

For the indicator of WF short-circuit capacity ratio, the empirical data given by European countries is 3.3% to 5%, and Japan scholars believe that a short-circuit ratio of about 10% is also allowable. In China, affected by distribution of wind resources, grid structure in regions suitable for construction of WFs is usually relatively weak, short-circuit capacity of the WF at connection point is small. In many cases, in order to meet the required short-circuit ratio of 3.3% to 5%, the size of the WF shall be severely limited. Determination of the wind power installed capacity according to the short-circuit ratio shall be based on calculation and analysis on specified WFs, and the short-circuit ratio is possible greater than 10%.

4.8.3 The Major Factors that Affect the WF Installed Capacity

The size of the WF installed capacity not only depends on the WF operating characteristics and regulation abilities of other power generation equipment in the systems, but also closely related to the network structure of system to be connected and other factors. To sum up, the main factors are:

1) **The load capability at WF connection point PCC**
It can be characterized by the PCC short-circuit capability, which determines the network capability to withstand wind disturbances. The smaller the short-circuit capacity is, the greater the sensitivity of node voltage on power changes is, and the weaker the tolerance capacity on withstanding wind power disturbances is.
2) **Modes for connecting the WF to the power grid**
Including the voltage levels, length, and impedance parameters of the link circuit. For example, change in values of the line x/r will have an effect on the distribution of nodal voltages of WFs and local grids, and will also have an impact on installed capacity of WFs.
3) **The regulation ability of conventional generator units in the system**
Mainly including the spinning reserve capacity and the abilities of the units to regulate the voltage and frequency. This factor mainly affects the wind power penetration levels. Due to the instability and variability of wind power, the system is required to have certain spinning reserve capacity in order to prevent a decline of the system frequency caused by the loss of wind power capacity. While increasing the response capacity of other units in the system on frequency and voltage can improve the system frequency and voltage fluctuations caused by wind power variation. Therefore, increasing the spinning reserve of conventional units in the system, and improving their ability to regulate voltage and frequency can increase wind power penetration limit of the system. If the effect is not significant, then the constraints of the network structure play a decisive role.
4) **Types of wind turbines and reactive power compensation capacity.**
When the WF is equipped with constant-speed constant-frequency wind turbines, since the cage asynchronous generator itself has not excitation, it needs to absorb reactive power from the grid in order to establish a magnetic field, so the wind turbine/WF reactive power compensation status has a significant impact on output characteristics of the WF, thus affecting the maximum integration capacity of the WF. Installation of dynamic reactive power compensation devices (such as SVC, SMES, etc.), can improve power quality and stability of wind power systems, as well as effectively improve the maximum integration capacity of the wind power.
VSCF wind turbines have a high utilizing rate for the wind energy and the ability to regulate voltages, and their interactions with the power grid is not obvious. By employing appropriate control strategies, the pitch angle adjustment device of the variable-speed wind turbine can effectively improve the efficiency of the wind turbine, reduce the impact of fluctuations in wind speed on the network to a certain extent. By connecting to the grid via a converter, it can adjust the power factor of wind generator, and improve operational characteristics of the wind power system, which has great potential for improving the maximum integration capacity of the WF.
5) **Local load characteristic**
Mainly presented in the following aspects: voltage and frequency regulation characteristics of local load near the WFs, and requirements of the load on voltage and frequency quality. If requirements of the load on power quality is too high, maximum wind power that can be withstood by the power system will be limited.

To sum up, determination of the maximum integration capacity of the WFs is constrained by many factors. It is unscientific to simply provide a unified indicator (such as, the WF short circuit capacity ratio K) without connecting to the specific conditions. To determine the reasonable construction scale and maximum integration capacity of the WFs, analysis shall be made on the system by combining with practice.

4.8.4 Calculation and Analysis on Examples

Methods to analyze the maximum integration capacity mainly include: time domain simulation method, steady-state power flow simulation method, static security constraints and optimization method, steady-state frequency constraints method, and other methods. The focuses of different methods are different, so their scopes of application are different. Studies at home and abroad have shown that, restrictions on WF integration capacity are mainly about voltage problems. So, based on impacts of the WF on the local system voltage level and stability, here we use the WF short-circuit capacity ratio K at the WF connection point as an index, to determine the WF installed capacity allowed to be integrated into the grids by combining of steady-state power flow analysis and time domain simulation analysis [17].

1) **System Overview**
Calculation is based on the actual power grid shown in Figure 4.15. The system is a typical example of WFs connected to the distribution network. The WF is connected at the end of the distribution network, and connected to the power grid through double circuit transmission lines. Simplify a network of 220 kV voltage level or above to as an equivalent

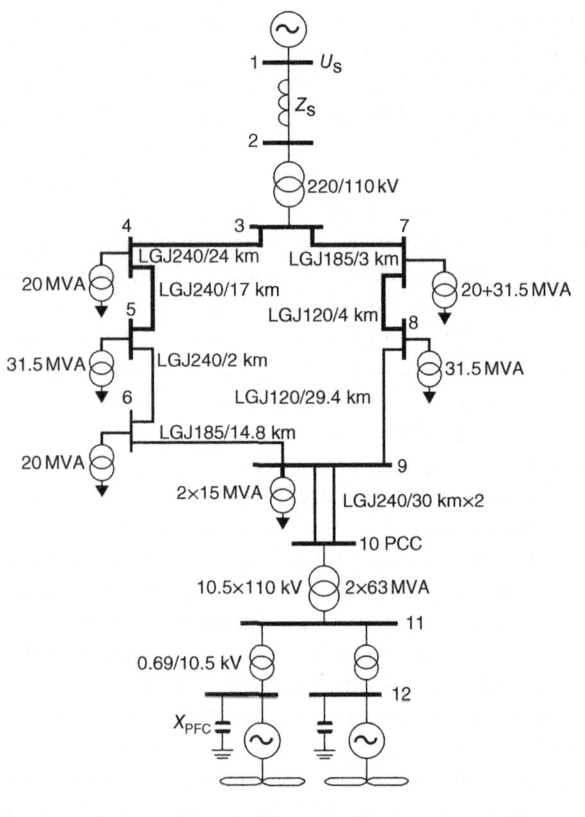

Figure 4.15 System wiring diagram.

power source, the voltage of the power source (i.e., the voltage at node 1) is denoted as *U*s, impedance between node 1 and node 2 represents the equivalent impedance of the network, which is denoted as Zs. Node 10 is the connection point of the WF, between node 2 and node 10 is the regional grids and loads near the WF. The wind generator used by the WF is an asynchronous wind turbine generator set with the rated capacity of 600 kW, and the parameters of a single generator is shown in Table 4.8, in which the capacity base value used for calculation is 100 MVA.

2) **Steady-state analysis**
   Voltage limit violation at local system nodes caused by integration of wind power can be studied by means of power flow calculation. According to the national standard on allowed voltage deviation of China's power supply, the voltage upper and lower limits used for calculation are: 0.69 kV, +5% to −10%; and 110 kV, +4% to −6%. Wind power maximum integration capacity is the result of the combined action various factors, and variable factors considered here include: ① Changes of the system operation mode; ② Size of reactive power compensation of the wind turbine generator system; ③ Impedance parameters of the link line for wind power integration into the system.

Table 4.8 Single asynchronous wind generator parameters (with the rated capacity as the base value).

Rated capacity	2 MW	600 kW
Rated voltage	0.69 kV	0.69 kV
Stator reactance x_1	0.09241	0.09985
Stator resistance r_1	0.00488	0.00833
Rotor reactance x_2	0.09955	0.10906
Rotor resistance r_2	0.00549	0.00373
Excitation reactance x_m	4.95279	4.54708
Rated slip s_N	−0.004	−0.004
Step-up transformer x_T	0.05	0.033
Rated power factor $cos\phi_N$	0.89	0.89

4.8.4.1 Effect of System Operation Mode

In the network shown in Figure 4.11, load capacity under normal operation of network is calculated as 80% of the transformer capacity. Changes in power supply voltage U_s and the load represent changes in the system operation mode.

Under constant load, the voltage levels of equivalent power source will affect the voltage levels of all nodes of the system, thus significantly affecting the maximum integration capacity of the WF. When power supply voltage U_s is 1.05, 1.02, and 1.0, the WF voltage with wind power output change curve are shown in Figure 4.16. Table 4.9 shows WF maximum integration capacities under different power supply voltages, ratio of WF capacity to short-circuit capacity at the connection point K and voltage limit violations. It is thus clear that, in this case, the wind power installed capacity is mainly constrained by the voltage drop at load nodes and WF terminals.

When the load level is reduced by a half, the status is shown in Table 4.10. Where U_s = 1.02 p.u., WF terminal voltage changes with the changes of WF output under two load conditions are shown in Figure 4.17. Since there

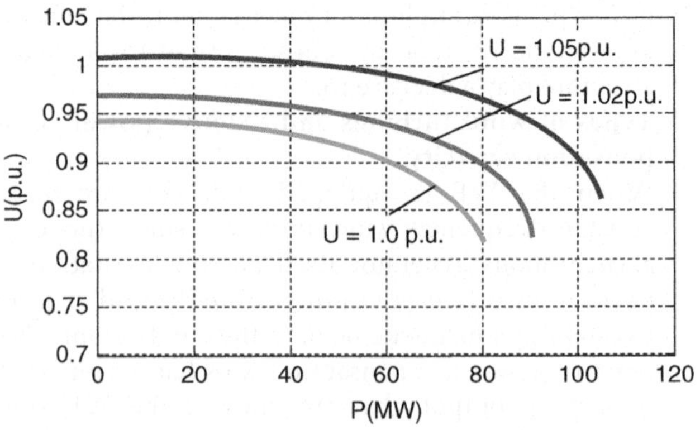

Figure 4.16 WF terminal voltage changes with the WF output under different U_s values.

Table 4.9 Calculation results with different U_s values.

U_s(p.u.)	1.05	1.02	1.0
Maximum capacity MW	100.8	74.2	36
Short-circuit capacity ratio K	24.55%	18.35%	9.21%
Voltage limit violations	$U_{12} < 0.9$	$U_6 < 0.94$	$U_6 < 0.94$

Table 4.10 Calculation results when the load is reduced by a half.

U_s(p.u.)	1.05	1.02	1.0
Maximum capacity MW	108	89.4	72
Short-circuit capacity ratio K	27.30%	22.41%	18.41%
Voltage limit violations	$U_{12} < 0.9$	$U_6 < 0.94$	$U_6 < 0.94$

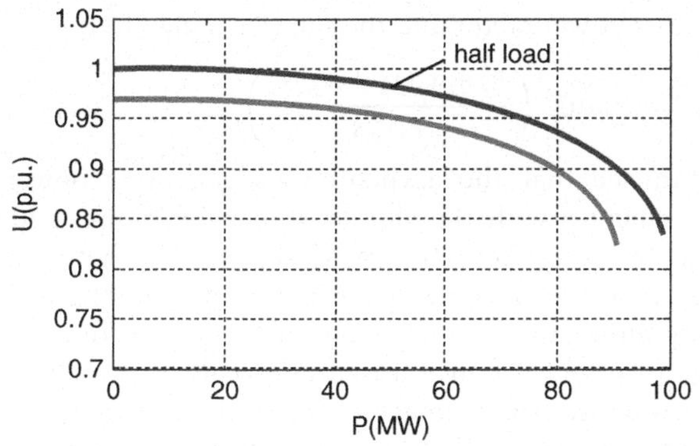

Figure 4.17 ($U_s = 1.02$) WF voltage changes with WF power output under different loads ($U_s = 1.02$).

are no other central buses in the power grid, the heavier the load is, the lower the voltage level is, thus the wind power maximum integration capacity is constrained. If taking into account the voltage regulation at power source nodes, the system voltage conditions will be improved and the maximum integration capacity of wind power will increase.

4.8.4.2 Effects of Wind Turbine Generator System Reactive Power Compensation Size

The asynchronous generator absorbs reactive power from system while generating active power, so usually, a certain number of compensating capacitors are parallel connected to the generator terminal, which is called power factor corrector (PFC). Generally, under the rated condition (wind turbine generator rated terminal voltage is 1.0p.u.), reactive power compensation is made for 30% of active power output of the wind turbine generator system (denoted as PFC = 30%). The above calculation results are all based on the compensation capacity of 30%. The amount of reactive power compensation has certain influence on the output characteristics, as well

Table 4.11 Calculation results under different compensation capacities.

PFC	30%	15%	40%
Maximum capacity MW	74.2	49.8	100.8
Short-circuit capacity ratio K	18.35%	12.49%	25.27%
Voltage limit violations	$U_6 < 0.94$	$U_6 < 0.94$	$U_{12} < 0.9$

as the maximum installed capacity of the WF. Where power supply voltage $U_s = 1.02$, calculation results of the compensation capacity of 15%, 30%, and 40% are respectively shown in Table 4.11. It can clearly be seen that the maximum installed capacity of the WF can be increased by improving power factor of the WF. And it can be seen from the voltage change curve in Figure 4.18, under the condition that the reactive compensation capacity is relatively high and the wind power output is relatively small, if the output increase, the WF terminal voltage will tend to increase, which means appropriate injection of wind power under this condition will improve the end voltage of distribution system, but due to constraints of the asynchronous generator's self-excitation and other factors, the compensation capacity cannot be too large, and using lots of fixed capacitors will reduce the system voltage stability under heavy loads and low voltages.

4.8.4.3 Effects of Link Line x/r

Wind farm terminal voltage is U_{10}, voltage drop on link line (Line 9-10) is approximately $\Delta U \approx \frac{P\,r+Qx}{U_{10}}$, where P and Q respectively represent active and reactive power generated by WF, where P > 0, Q < 0, r, and x represent the impedance parameters of the link line. Thus, when the WF is transmitting active power P to the system, a voltage component produced on the transmission line resistance r will causes the WF terminal voltage to increase, and the reactive power Q absorbed by the WF from the system will produce a component on the transmission line reactance x, which causes the WF terminal

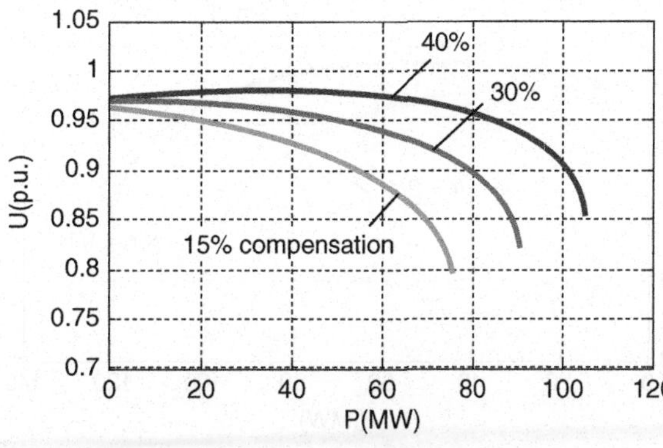

Figure 4.18 WF voltage changes with the wind power output under different compensation capacities.

voltage to decrease. Therefore, impedance parameters of link line have great influence on power and voltage characteristics of WF.

If power supply voltage $U_s = 1.02$, when the compensation capacity is 30%, keep short-circuit capacity of the WF at the connection point unchanged, and change the ratio of the link line x/r to make it 3/1, 1/1, and 1/3, which respectively correspond to the typical parameters of 110 kV overhead line, 35 kV overhead line, and cable. Results of such calculations are shown in Table 4.12.

Under the three cases, WF terminal voltage changes trends with the change of active power output are shown in Figure 4.19. From the results we can see that, when x/r is low, the voltage increases with the increase of wind power, and when x/r is high, the voltage decreases with the increase of wind power. When short-circuit capacities at the WF connection points are the same, employing of the cable lines with smaller x/r for connection will increase the maximum installed capacity of wind power.

From the above steady-state analysis we can see that, the main reason that constrains the integration capacity of the fixed speed wind turbine-asynchronous generator system is the node voltage limit violations caused by wind power injection. Wind power maximum integration capacity is the result of interaction between voltage level at local central node where the WF is connected, system load amount, WF reactive power compensation capacity, as well as the link line x/r for connecting the WF to the system, and other factors. Improving voltage regulation ability, appropriately increasing WF reactive power compensation and using link line with lower x/r will help increase the maximum integration capacity of WF. For a real wind power system, to determine the appropriate wind power integration capacity, analysis and calculation shall be made on the system.

Table 4.12 Calculation results when x/r is in change.

x/r	3/1	1/1	1/3
Maximum capacity MW	74.2	78.6	90
Short-circuit capacity ratio K	18.35%	19.71%	22.57%
Node of voltage limit violations	$U_6 < 0.94$	$U_6 < 0.94$	$U_6 < 0.94$

4.8.4.4 Transient Analysis

Wind turbine generator system integration into power grids will affect the system steady-state characteristics, as well as the transient stability after a major disturbance. In order to ensure the safe operation of wind power systems, transient stability after a network fault must be verified. When short-circuit fault occurs in the network, the wind power system is prone to voltage instability and rotor stall, and the two phenomena are intertwined. Based on the relationship between the asynchronous generator power factor and the slip (Formula (4.9)):

$$\varphi = \tan^{-1}\left(\frac{r_2^2 + x_k(x_k + x_m)s^2}{r_2 x_m s}\right) \tag{4.9}$$

We can see that the asynchronous generator power factor changes with the slip, as shown in Figure 4.20. When deviating from the rated slip, power factor of the asynchronous generator will decline rapidly, and corresponding to some active power output, the reactive power absorbed also increases dramatically.

When short-circuit fault occurs in the network, the generator voltage will decrease, only a small amount of electrical power can be transmitted to power grid, but the output mechanical power of the wind turbine is essentially the same, due to the unbalance between mechanical power and electric power, the generator will begin to pick up speed, and the slip will increase, thus the asynchronous generator will absorb more reactive power from power grid, resulting in voltage drop at the terminal of generator. When the fault is cleared, the

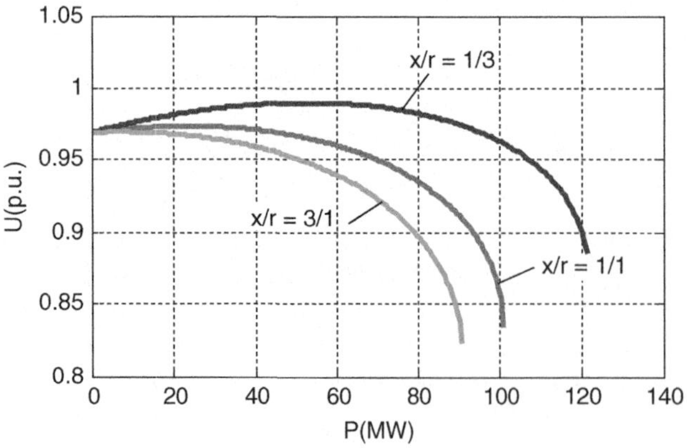

Figure 4.19 WF voltage changes with wind power output when the link line x/r is in change.

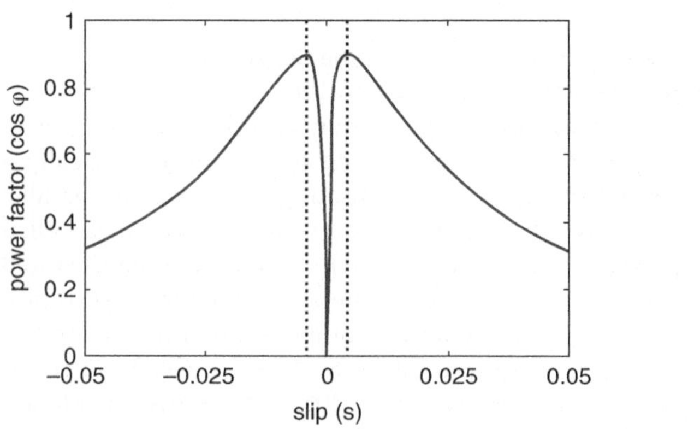

Figure 4.20 Asynchronous generator power factor changes with slip.

rotor speed and the terminal voltage will enter a recovery process. If the voltage recovery process is shorter than speed recovery process, the generator will get into a virtuous circle, and recover to the steady state as before, or establish a new steady state. If the recovery of voltage is too slow, the generator power output shall not increase rapidly, otherwise the mechanical power and electrical power will fail to balance in time, and unable to suppress acceleration of the rotor, thus the slip will further increase, and the reactive power absorbed by the generator will continue to increase, resulting in a vicious circle of the system. The process of positive feedback like this will cause voltage of the wind power system to collapse and rotor to stall.

Static stability of asynchronous generator is usually described in terms of torque-slip curve, increasing the maximum torque and critical slip will help improve the stability. In order to observe the effects of various factors on the wind power system transient stability, dynamic simulated calculation is made on the real system shown in Figure 4.19, of which the variable factors taken into account include: ① short-circuit capacity size at wind power connection point; ② the network impedance parameters; ③ capacitance compensation capacity. In the simulation, the installed capacity of the WF is assumed as 75 MW, three-phase short circuit fault occurs at the WF outlet (node 10), and a single-circuit transmission line is cut off after the fault has lasted for 100 ms. Simulation results under different conditions are shown in Figures 4.21 to 4.24.

Under the conditions that the supply voltage U_s = 1.05 p.u., capacitance compensation capacity equals 30% of the active power output, the network impedance parameter $x/r = 3{:}1$, and short-circuit capacity $S_{SC} = 3.91$, change curves of WF outlet voltage and equivalent wind turbine generator system slip are shown Figure 4.21, from which we can see that, in this case, the system is able to maintain stability after a fault happens.

Figure 4.22 shows the simulation results after the capacitance compensation capacity is reduced to 15%, and other conditions remaining unchanged. We can see the loss of stability of the system after a fault occurs, the voltage drops to an unacceptable level, and the slip increases continuously.

Figure 4.23 shows the simulation results under the conditions that the compensation capacity is maintained at 15%, the short-circuit capability remains unchanged, and the link line impedance parameter is changed to x/r = 1:1. We can see from the simulation results that the system is able to restore to steady state.

Figure 4.24 shows that, under the above unstable situation, the network parameters are changed to double the short-circuit capacity, in occurrence of the same fault, the system can restore to steady state.

These results indicate that if the wind power system stability after short-circuit fault has close relations with the short-circuit capacity size at the connection point of the WF, the impedance parameters of the link line that connects the WF and the system, and the reactive power compensation capacity of the wind turbine generator system (i.e., wind generator power factor). An appropriate increase in wind power short-circuit capacity at the connection point and wind generator power factor, and the use of a line with smaller x/r to connect the WF and power system will help improve the stability of wind power system, of which the result is consistent with the steady-state analysis.

From the results of the steady-state analysis we can see that, in the network shown in Figure 4.19, when the compensation capacity of wind turbines is 30%, and the

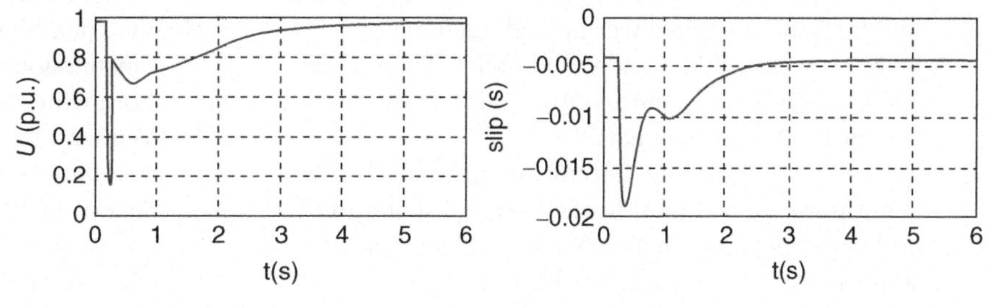

Figure 4.21 Changes in WF voltage and equivalent wind turbine generator set slip.

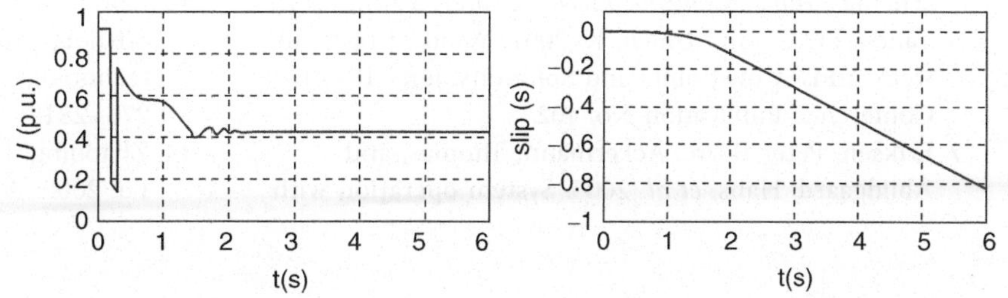

Figure 4.22 Changes in WF voltage and equivalent wind turbine generator slip.

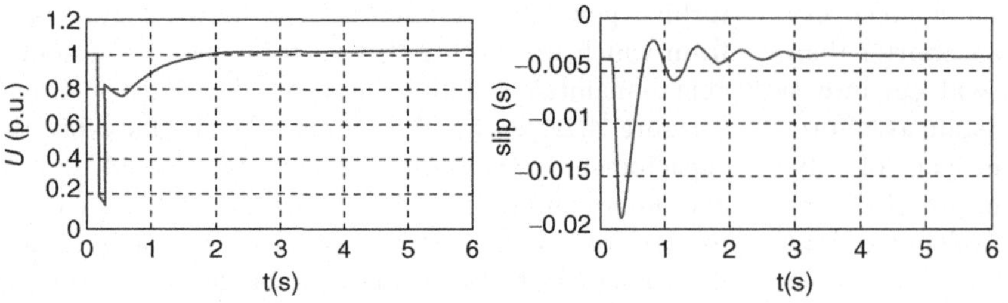

Figure 4.23 Changes in WF voltage and equivalent wind turbine generator slip.

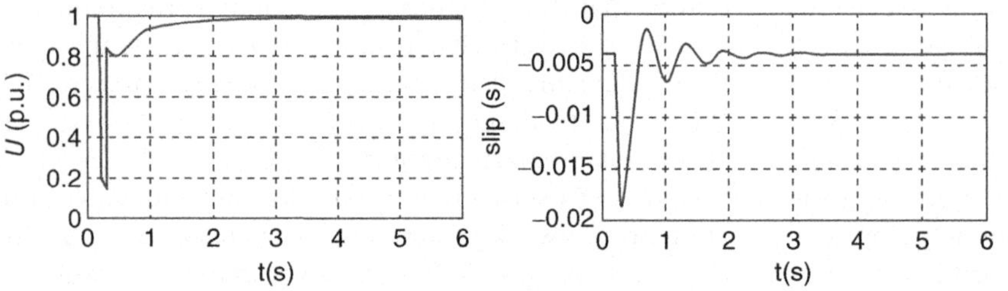

Figure 4.24 Changes in WF voltage and equivalent wind turbine generator set slip.

WF is connected to the system via a 110 kV line, if the system can maintain a high voltage level at the power source node (U_s = 1.05), the maximum wind power integration capacity will be 108 MW. Combining with the calculation results under other operating conditions and taking into account a certain margin, and after validating through transient simulation, we can determine that the integration of 75 MW WF into system is feasible. Short-circuit capacity of the WF is 18.27% at this situation.

References

1 European Wind Energy Association. 2005. Largescale integration of wind energy in the European power supply: Analysis, issues, and recommendations. http://www.ewea.org/fileadmin/ewea_documents/documents/publications/grid/051215_Grid_report.pdf

2 Fox, Brendan, Flynn, Domian, and Bryans, Leslie, *et al.* 2007. Wind Power Integration. Beijing, China: The Institution of Engineering and Technology.

3 Jenkins, Nick, Allan, Ron, and Crossley, Peter, *et al.* 2000. *Embedded Generation(M).* London: The Institution of Electrical Engineers.

4 Belhomme, Régine and Corenwinder, Chantal. 2004. Wind power integration in the french distribution grid regulations and network requirements. Nordic Wind Power Conference, March 1–2.

5 Jiashu, Xu. 1999. Analysis on wind farm integration schemes. *Inner Mongolia Electric Power*, **17**(6), 6,14.

6 Tande, J.O.G., and Uhlen, K. 2001. Wind turbines in weak grids–Constraints and Solutions. June 18–21, Conference Publication No. 402.

7 Eriksen, Peter Børre, Ackermann, Thomas, and Abildgaard, Hans, *et al.* 2005. System operation with high wind penetration. IEEE Power and Energy, **November/December**, 65–74.

8 Kirby, N.M., Luckett, M.J., Xu, L., and Siepmann, W. 2001. HVDC transmission for large offshore wind farms. Seventh International Conference on AC-DC Power Transmission, November 28–30, pp. 162–168.

9 Xingyuan, Li. 2010 HVDC Transmission Systems. Bejing: Science Press.

10 Wei, Yao, Shijie, Cheng, and Jinyu, Wen. 2007. DC transmission technique and its applications in offshore wind farms. *China Electric Power*, 40(10): 10.

11 Weimers, Lars. 1998. HVDC light–New technology for a better environment. *IEEE Power Engineering Review*, **18**(8), 19–20.

12 Jun, Wen and Gongyi, Zhang, et al. 2003. HVDC light—A new generation HVDC technology. *Power System Technology*, **27**(1), 47–51.

13 Hörle, N, and Eriksson, K. 2002 Electrical supply for offshore installation made possible by use of VSC technology. CIGRE Conference, Paris, France, pp. 276–281.

14 Zhaoqing, Hu, Chengxiong, Mao, and Jiming, Lu. 2005. A new type of HVDC–HVDC light.

Transactions of China Electrotechnical Society, **20**(7), 12–16.
15 Ackermann, Thomas. 2005. *Wind Power in Power Systems*. New York: John Wiley Sons.
16 Martínez de Alegría a, Iñigo, Martín, Jose Luis, Kortabarria, Iñigo, *et al.* 2009. Transmission alternatives for offshore electrical power. *Renewable and Sustainable Energy Reviews*, **13**, 1027–1038.
17 Junling, Wu. 2004. Study on some technical problems of large grid-connected wind farm operation. Master's Degree Thesis in Engineering, Tsinghua University.

5

WF Electrical Systems

Electrical systems in a wind farm (WF) include the electric energy collection system (power collection system), grounding system, lightning protection system, electrical protection system, reactive power compensation system, energy storage system, and so on.

5.1 Power Collection Systems

5.1.1 The General Structure of the Power Collection System

Figure 5.1 shows the electrical system diagram of a typical fixed speed wind turbine generator system. The main circuit starts from the cage asynchronous generator, and is connected to the molded case circuit breaker (MCCB) via three soft pendulous cables. The MCCB is provided with a default instantaneous trip for overcurrent protection and a delayed (thermal) function overcurrent protection, to detect generator over-current. The bi-directional thyristor between the generator and the circuit breaker constitutes a soft starter unit (usually this also has a bypass contactor), for reducing the surge (impulse) current of the generator when connected to the grid. There are many additional circuits, including power factor capacitor (PFC) circuits and auxiliary power supplies. PFC is switched in stages to provide reactivity power and also to limit capacitive switching inrush currents. Small inductors (not shown) can be connected in series with the capacitor to reduce the inrush current.

A lot of auxiliary equipment use alternating current (AC) power, but some requires direct current (wind turbine controller). The auxiliary circuit is protected with a smaller MCCB, of which the fuse rated current is smaller. Surge diverter 1 is used to protect the internal electrical system from the overvoltage transferred from the substation electrical network, surge diverter 2 is connected in parallel at the generator terminal.

Generator voltage levels are generally less than 1,000 V. The standard voltages adopted internationally are a line voltage of 690 V (50 Hz system) and 575 V (60 Hz). The voltage is very low for a large generator, and this results in a large operating current. For example, 600 kW wind turbine generator system requires a current higher than 500 A when the voltage is 690 V. However, choosing to limit all voltages of the wind turbines below 1,000 V is convenient and cost-efficient. As in many countries, when voltage is higher than 1,000 V, security rules become more and more severe, and requires special preventive measures, including dedicated grounding devices. Another more important reason is the wide selection of switchgears and pendulous soft cables for 1,000 V, which can effectively reduce the cost of generator system.

Generator's low voltage needs to step-up by the transformer located in or adjacent to the tower foundation. In some of the early constructed WFs, a few small wind turbines were connected to a transformer (Figure 5.2a). But since the increase of wind turbines rated value is accompanied by greater current, this connection will result in additional changes in voltage and electrical losses in low-voltage cables, thus it will only be more economical when several small wind turbines locate very close to each other and are connected to the same transformer. And another way is connecting one wind turbine to one transformer (Figure 5.2b), which is the typically and generally used connecting type.

The generator step-up transformer is a three-phase model, either oil cooling or a dry-type. The primary winding is a Y-shaped grounding or delta shape, and secondary side voltage is 575 V or 690 V (winding Y-shaped grounding). Impedance of the transformer can be used to limit the short circuit current of generator system. But impedance of the generator step-up transformers must be designed carefully in order to avoid possible resonance caused by harmonic currents generated by the induction generators. Impedance values of typical generator step-up transformers are 5% (based on 750 kVA wind turbine design) and 6% (based on 1,500 kVA wind turbine and other designs).

Medium voltage (MV) levels of WF power collection systems are usually determined by the experience of the local distribution companies. So that cables and switching equipment are relatively easy to obtain. Voltage levels

Integration of Large Scale Wind Energy with Electrical Power Systems in China, First Edition. Zongxiang Lu and Shuangxi Zhou.

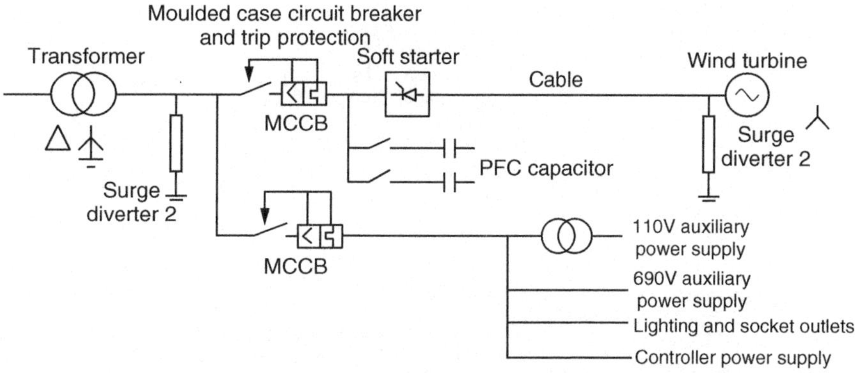

Figure 5.1 Fixed speed wind turbine electrical system schematic.

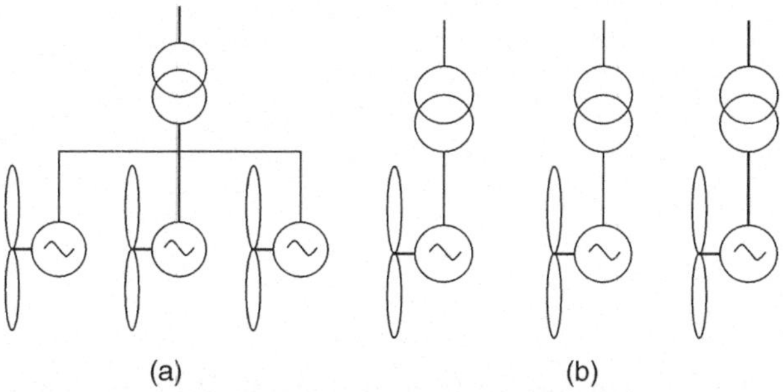

Figure 5.2 Connection of generator and transformer.
(a) connection of multiple wind turbines to one transformer
(b) connection of one wind turbine to one transformer.

are generally chosen between 10 kV to 35 kV, which can be 10 kV, 15 kV, or 20 kV. Large-capacity wind turbines and WFs mainly choose 35 kV voltage level.

The local distribution company's experience also influences the selection of a neutral grounding method of the MV circuit. In some countries, neutral point of the 10 kV circuit is solidly grounded or grounded via a low value resistor (typically 6.35 Ω, the single-phase grounding current allowed to flow through is 1,000 A). Neutral-point solid ground requires no additional equipment, which is more economical, but may lead to large ground fault current, and large currents can cause hazard and high-step voltage or contact voltage. Medium voltage systems are not grounded in some countries, allowing WFs to continue to operate in case of the single-phase ground fault, but the non-fault-phase voltage will be $\sqrt{3}$ times that of original.

Power collection systems are designed very similarly to the conventional MV grids, but generally, redundancy facilities are not required to be provided in the circuit for MV equipment fault considerations. Operation experience and reliability calculations have shown that WF power collection systems are much more reliable than wind turbines. If the WF power collection circuit is in partial failure, the loss is simply less energy in the output of the WFs. This loss is easily assessed and usually limited. In contrast, if a public distribution company is in circuit failure, user power supply may be interrupted, and the corresponding economic losses can never be quantified, usually, several degrees of magnitude higher than the actual power price. Public distribution circuits often require a duplicate supply, WF power collection networks often consist of a single radiation circuit, which is equipped with a finite number of switching equipment for isolating and converting. A wind turbine transformer is directly connected to the radiation circuit.

In Europe, almost all WF power collection circuits adopted underground cable networks. This is mainly for landscape amenity and safety (erected wind turbines require a large crane), but construction costs for these networks are high. In the United States and India, in order to save costs, MV networks of WFs also use overhead lines, as in China. If a cable is used, then the lower series inductance of the cable will make less voltage fluctuates due to the current changes. Taking into account the scattered layout of wind turbines in WF, and the length of their power collection lines, which are longer, electrical losses need to be considered when designing the power collection circuits.

5.1.2 Power Collection Systems of Large WF

Figure 5.3 shows the design scheme of a large WF power collection system. Wind turbine voltage is stepped up by the transformer from 690 V to 30 KV, forming a 30 kV radial power collection bus network in the WF. And the voltage is stepped up to 220 kV in the WF main substation and transmitted to the main grid. Generally, an on-load tap-changing transformer is chosen as the main transformer in order to meet the needs of large voltage fluctuations of the WF. A bus connected between the 220 kV substation high-voltage side outlet and the main network is the point of common coupling (PCC) between the WF and system. In the operating process of WF, the PCC is the most important point for power (active and reactive) and voltage monitoring.

Apart from the wind turbine (cage asynchronous generator) reactive power compensation, there is centralized reactive power compensation in the WF substation (capacitive or inductive), to ensure the voltage and power factor at PCC within the required range of power grid.

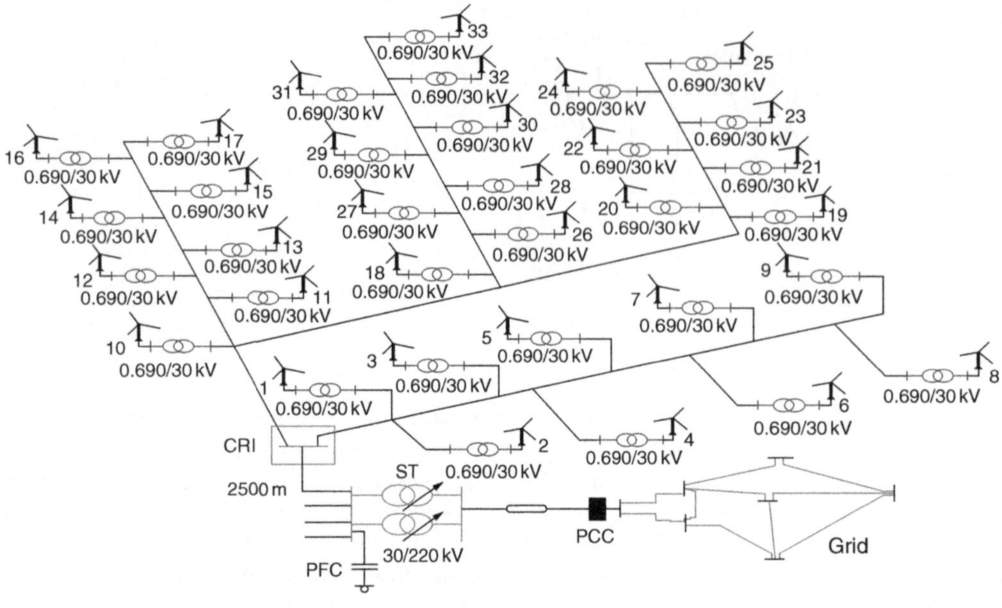

Figure 5.3 Large WF power collection system.

Numbers of wind turbines of different radial circuits are different, and the section areas of cables are different. Power flow at different points of the 30 kV power collection bus are different, so the bus cross-section areas are different.

5.1.3 Wind Turbine Transformer

A step-up transformer associated with the wind turbine generator systems is called a wind turbine transformer. Its role is to rise the voltage of the electric energy generated by wind turbines from 690 V to 10 kV or 35 kV, which is transmitted to the WF substation via buried cables or overhead lines. For high voltage wind turbines, the rated terminal voltage is high, such as the high-voltage generator without transformer that the ABB Company has been working on since 1988, called Powerformer technology. Generators can directly generate the required high voltage (up to a maximum of 400 kV), rather than using transformers to step up. This high-voltage generator can be used for hydro-electric and thermal power generation, as well as for wind power generation. When used for wind power generation, the generator can be directly connected to the step-up substation of WF without a transformer.

Combining with the characteristics of the wind power system, special technical requirements for wind turbine transformers are as follows [1]:

1) Transformer no-load time is long. Wind power generation has a significant seasonal nature and the annual average load rate of transformer is only about 30%. However, since main equipment of wind power generation is expensive, the equipment losses and the no-load loss of transformers shall be as low as possible.
2) Overload time is short. Since the transformer capacity is generally larger than the capacity of wind turbines, micro computer monitoring technology is used to realize self-diagnosis function of wind power generators with perfect safety protection measures. When the wind power generator is overloaded, it will automatically limit the speed or even stop the operation, to avoid overload of the transformer. The service life of the transformer is longer than those of ordinary distribution transformers; regardless of transformer overload protection, at the high voltage side current-limiting fuse protection is directly used that is combined over-current and quick break which can effectively reduce internal contacts, increase reliability, and reduce the cost of combined transformers.
3) Adapt to harsh environments. China's abundant wind resources are generally concentrated in the coastal areas, northeast and northwest regions, so transformers are operating in the wild. Therefore, we must consider the equipment weathering problems. Salt-proof, mold-proof, and moisture-proof shall be considered for equipment used in coastal areas; and effects of ultra-low temperature and sandy wind shall be considered in the northeast and northwest regions.
4) Combined transformer's high voltage side must be configured with arrester, so as to form an over-voltage absorbing circuit with the over-voltage protection device of wind power generators. Effects of arrester residual voltage on the transformer shall be taken into account when designing the insulation property of the transformer.

Combined transformer in the box-type substation is usually used as a generator transformer and placed near the base of the wind turbines. Currently, there are three types of box-type substations: pre-installed cabinet type transformer substation (European box-type transformer substation), combined transformer substation (American box-type substation), and floor-transformer.

A pre-installed cabinet type transformer substation (i.e., European box-type transformer substation) is composed of high voltage switchgear, power transformers, low voltage switchgears, electricity metering equipments, reactive power compensation equipments, auxiliary equipments, and connected components, which are arranged in one or several box shells in the factory, that is, the transformer and the high and low voltage equipment are closely connected as a whole, and can be separated from each other. High-voltage and low-voltage loop circuits are separated with a steel plate, high and low-voltage equipments and the transformer are independent from one another, and the integrated box-type substation as well, which has compact structure, small size, and flexible configuration.

A combined transformer (i.e., American box-type substation) is different from European box-type transformer substation, the American type combines the body of transformer, load switches, fuses, tap switches, and the corresponding auxiliary equipment together, and high-voltage switches and fuses are all connected to the oil tank. Transformer and oil tank are both of close structure, and small external dimensions.

Because all the devices of the floor-transformer are open layout, a dropout fuse is used for the high-voltage switch, and is installed on the concrete pole with the arrester; a low voltage circuit breaker is installed in the outdoor rain-proof distribution box, and is connected to the transformer with a copper bus.

Through technical and economic comparative analysis, we can know that a European box-type transformer substation has flexible configuration, and is fast to install and more reliable than an American box-type substation, but its weaknesses are higher price, greater volume than the American type; for the American box-type substation, the configuration is flexible, the wiring construction is difficult, and is not so reliable compared to the European type, but its advantages are small volume, and easy to install. For the floor transformer, the price is lower, but it occupies more area (about 42 m^2 for each transformer, which is over five times that of the box type), so the land expropriation cost is higher, and all devices shall be installed on-site, which will take about three days to install each transformer, but only a few hours to install box-type transformers. Because all devices are installed in open space, due to poor natural conditions of WFs, the failure rate of floor-transformer is higher than that of box transformers. And since the floor-transformer is not automated, its failure conditions cannot be detected in a timely manner, and aerial work is required for maintenance, which increases the work difficulties, working time, and outage time, thus the parts and labor costs are higher than that of the box type transformers, and the WF output is affected. Therefore, the first two types of substations (in particular, American type) are mostly selected at present.

5.1.4 WF Secondary System

Step-up substation employs automatic computerized monitoring system, namely automatic control technology, micro computer and network communication technology, to reorganize the functions and optimize the design of the step-up substation secondary equipment, so as to carry out monitoring, protection, measurement, operation management, information transfer and their coordination of the step-up substation [3].

An independent direct current (DC) system is established in the step-up substation, to supply power for control, measurement, protection, signal, and a uninterruptible power supply (UPS) device of the electrical equipment. A set of alternating current (AC) UPS is set up in the substation to supply power for host machine/operator station/engineer station, fire-prevention station, fire alarm controller in the main control room, and the remote master station, network communication cabinet, data network connecting to device screen, remote electricity charge screen in the relay protection room, and other loads.

Protections for the entire substation are set up according to the GB/t 14285-2006 "Technical code for relaying protection and security automatic equipment" whereas micro-computer protection is used for the protection of main transformers, 35 kV lines, 35 kV capacitors. A double set of configuration is adopted for the main transformer protection, and single set of configuration is adopted for non-electricity protection of the main transformer and protection of other equipment. The substation is equipped with a microcomputer-based anti-misoperation blocking device for operation blocking of the breakers and isolating switches.

The wind turbine generator system (WTGS) electrical control system uses a programmable controller as its core, of which the control circuit is composed of a PLC principal controller and its extended function modules, to mainly achieve the normal operation control of wind turbines, safety protection of the generator system, fault detection and treatment, setting of operating parameters, as well as data records and manual operation, and the system is equipped with various communication interfaces, to realize local communications and telecommunications. The electrical control system consists of distribution cabinets, control cabinets, nacelle control cabinet, sensors and connecting cables, and so on, and its functions include normal operation, operating status monitoring, security and protection.

WTGS monitoring system is divided into central monitoring system and remote monitoring system. While the central monitoring system is composed of the local communication network, monitoring computer, protection devices, and central monitoring software, and so on. These functions are mainly to the facilitate centralized management and the control of wind turbines. And the remote monitoring system consists of central supervisory control computers, modems, data transmission medium (telephone lines, Wi-Fi, Internet), protection systems, remote monitoring computers, and remote monitoring software. These functions are mainly used to enable remote users to view operating status and historical information of the wind turbines in real time. Local communication networks physically connect the wind turbines via cables, optical cables, and other mediums, of which the network structure can be a chain, star, or tree structure. The central monitoring system monitors host communicates with the connected local communication network via optical cables, to realize centralized management and control of each wind turbine in the WF.

WTGS are equipped with overload, locked rotor, instantaneous trip, over current, open-phase, three-phase unbalance, overvoltage, voltage loss, and other protection, after the operation of protection devices, the generator outlet air circuit breaker will trip off and send operation signals of the protection devices.

5.2 WF Grounding Systems

Grounding systems are very important to conventional power plants, and there are special requirements for grounding systems in WFs. Because a WF often extends to a range of several kilometers, the height of modern wind turbines makes them vulnerable to lightning strikes. Some wind turbines are located in high resistance bases at the mountaintop, so grounding practices used for conventional power plants and substations cannot be simply used to WFs, for which special considerations are required.

In order to ensure personal safety, the electrical equipment enclosure should be grounded, the joint of the ground strip and ground connector should be welded, and welding should comply with the construction requirements. Backfill of the ground grid should be clay or original ground surface soil and backfill should be rammed to reduce the contact potential.

WF grounding systems are required to work effectively in power frequency (50 Hz) current and lightning surge current (of which the rise time is typically less than 10 μs). Although it is common to use physical grounding systems for power frequency currents and lightning surge currents, a response of the grounding system to lightning current high frequency component is completely different from that of 50 Hz system.

The performance of WF grounding systems is qualitatively illustrated in Figure 5.4. Each wind turbine is configured with a local grounding device, of which the horizontal grounding electrode is a ring conductor, surrounding the base in the depth of 1 m (sometimes it is called counterpoise ground); the vertical grounding electrode is rods, buried. Usually the basic steel reinforcing members of the wind turbine base are combined to the local grounding grid. The purpose of this local grounding is to provide equipotential connection for curbing the lightning current and power frequency fault current, and to be used as a part of the entire WF grounding system. An effective grounding system shall ensure that the lightning current is conducted into ground, in order to provide maximum safety for both people and animals, as well as protection of wind turbine parts from damage.

In the figure, R_t is the wind turbine grounding resistance, because the wind turbine grounding network only consists of a circle, for example, a circle 15 m in diameter and a rod inserted into the underground, it can be considered as a pure resistance.

A horizontal electrodes connected to wind turbines one by one have some complex characteristics (similar to transmission lines), as the equivalent π equivalent

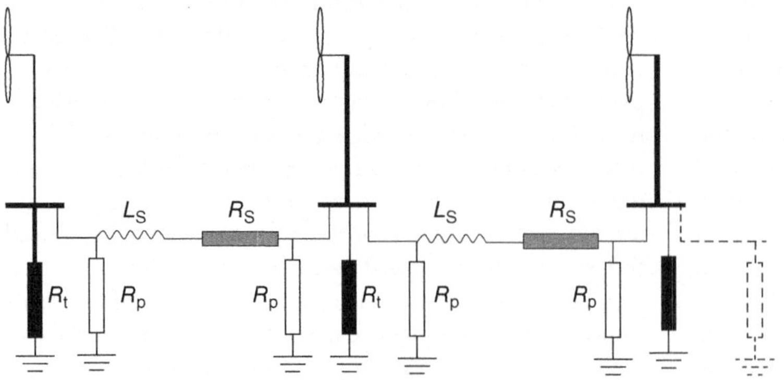

Figure 5.4 Schemes of WF grounding systems.

Table 5.1 Grounding impedance values measured from two WFs.

WF capacity (MW)	Number of wind turbines	Horizontal grounding conductor Length (km)	Substation grounding Impedance (Ω/50 Hz)
7.2	24	6.7	0.89 + j0.92
33.6	56	17.7	0.46 + j0.51

circuit shown in Figure 5.4, where grounding resistance is R_p, series impedance is the combination of R_s and L_s. R_s is the resistance of the grounding wire, and L_s is the self-inductance of the ground circuit. A WF grounding network is very long, and its series resistance cannot be ignored. We can directly see that, for the high frequency components of lightning strike on the wind turbine, series inductance (large impedance) can effectively simplify the grounding network to a single local wind turbine grounding system. Even for a fault current of 50 Hz, series resistance can also lead to grounding impedance that is much greater than that of a small geographic grounding system (of which the series resistance can be ignored). Although it becomes very obvious at a very high frequency, shunt admittance is ignored here.

Table 5.1 shows the resistance values of two WF grounding networks obtained from site measurements (in the United Kingdom).

We can see from Table 5.1 that at 50 Hz the resistance and reactance of the grounding impedance is almost equal (X = R), which is important for design and testing of WF grounding systems. It is impossible to use conventional calculation methods (which are used for small, pure resistance grounding network) here, and it is necessary to consider the effects of long grounding conductors.

Current injection test is an efficient way to determine the grounding impedance of 50 Hz WFs. In this test, a typical current of 10 to 20 A is injected to the WF ground electrode, and then the potential rise of the "positive" electrode is measured. But the injection current return path must be kept away from WF (usually 5 to 10 km) to ensure that the measurement in WF is not affected. Thus, the easy way is to use the de-energized connection circuit between the WF and the main power network as the test current pathway. The potential rise of the WF grounding system is measured against a remote ground (the traditional way is to transfer to somewhere on the metallic phone circuit) ground.

5.3 WF Lightning Protection

5.3.1 Lightning and Its Harm

Lightning is a larger natural disaster that threatens the safe operation of WFs, because the huge energy released by the lightning will cause burst of wind turbine blades, and burning out of electrical equipment insulation, a control components breakdown, and a series of accidents. Lightning protection of wind farms is the important link to guarantee safety in production.

Wind turbines are valuable equipment used by WFs, of which the price accounts for 60% of the wind power project (onshore) investment. If hit by lightning (in particular, valuable components such as blades and generator were struck by lightning), in addition to losses of electricity generation income during the repair period, the huge costs for disassembly, assembly, and update of damaged components shall be covered. According to the information of the Denmark LM company: in 1994 when lightning activities are few in Denmark, more than 6% of the registered operating wind turbines have suffered from lightning damage, of which the estimated repair costs were 15 million korunas (the wind turbine installed capacity of the year in Denmark was 540 MW, average 280,000 kroner/MW). The LM Company estimated that, 1% to 2% of the world's wind turbine blades are hit by lightning each year. For blades hit by lightning, most of the damaged parts are in the tip, which are relatively easy to be repaired, but in a few cases, the entire blade shall be replaced. Lightning striking on wind turbines often leads to electrical system over-voltage, resulting in the burnout of wind turbine automatic control and communication components, wind generator breakdown, electrical equipment damage, and other accidents. Coastal areas in China with complex topography have more thunderstorm days, so we shall pay full attention to the huge threats caused by lightning strikes on WTGS and operators. For example, the Red Bay WF has suffered from several lightning accidents since it was put in operation. According to the statistics, the blade hit rate is 4%, and the hit rate of other communication components is as high as 20%. Therefore, lightning is a serious problem threatening safety and economic operation of wind turbine [4].

Lightning is a very complex natural phenomenon, which contains a series of discharging currents. The term "lightning flash" is used to describe the discharge sequences, which uses the same path and may last up to 1 s. The individual components of a flash are called "strokes."

Lightning strike rate indicates the relationship between the frequency of lightning strikes and the building height. In areas with annual thunderstorm days of 10, relationship between the building height h and the general lightning strike rate n is shown in Table 5.2. The higher the building is, the more lightning strikes it will subject to.

Table 5.3 shows parameters commonly used to characterize lightning and some aspects of their potential harm to wind turbines.

Peak current of a single lightning strike is higher than 200 kA, but median value is only about 30 kA.

Table 5.2 Relationship between the building height *h* and the general lightning strike rate *n*.

Height *h*(m)	**20**	**50**	**100**	**200**
Lightning strike rate *n* (time.a^{-1})	0. 02	0. 1	0. 3	1. 0

Table 5.3 Impact of lightning on different parts of the wind turbine.

Parameters	**Impacts on wind turbines**
Peak current (A)	Conductor overheating, shocks, electromagnetic force
Specific energy (J/Ω)	Conductor overheating, shocks
Current rate of rise (A/s)	Induced voltage on the wires, shocks, spark over (flashover)
Long lasting charge transfer (c)	Ground arc attachment point or other arc sites. Harm (such as bearings)

Corresponding transferred charge peak is 400 C, and the median value is 5 C. The specific energy peak is 20 MJ/Ω, and the median value is 55 kJ/Ω. A wide range of these parameters means that the lightning protection of any WF or any wind turbine, only the effects of wind turbine locations on lightning frequencies and intensities were considered in the early stage, which can change significantly with the geography and topography.

Table 5.4 Hazard frequency of lightning on wind turbines.

Country	**Cycle**	**Wind turbine/year**	**Lightning Failure**	**Failure times/100 wind turbines/ear**
Germany	1991-1998	9204	738	8.0
Denmark	1991-1998	22000	851	3.9
Sweden	1992-1998	1487	86	5.8

Table 5.4 shows historical data of lightning hazards in some countries. Although wind turbine design and lightning protection systems have developed rapidly in this period, these data do suggest the annual number and scope of lightning strikes. The large number of accidents shown in Table 5.4 are caused by indirect effects of lightning on wind turbines and WF control systems. Many failures are generally subject to failures of the control and electrical systems, and blade damage causes greatest losses to the repair costs and wind turbine availability, thus reducing WF income.

Lightning protection for wind turbines is carried out in partition. Lightning protection zone was proposed in order to better protect the wind turbine system components. The wind turbine system can be divided into several different areas by utilizing the Rolling-Sphere Method of 30 m radius. The lightning protection system is zoned in accordance with standards, to reduce the electromagnetic interference and foreseeable coupling interference. The International Electrotechnical Commission divides the lightning-proof overvoltage protection area into LPZ0 area (including LPZ0A and LPZ0B), LPZ1 area, and LPZ2 area [5, 6] as shown in Figure 5.5.

The LPZ0A area is subject to the risk of direct lightning strikes (twist lightning), and it is completely in the electromagnetic environment, with possible lightning surge damage. This area includes blades, nacelle cover lightning rod system, tower, overhead power lines, WF and communication cables. The LPZ0B area has no risk of direct lightning strikes, but the electromagnetic environment and lightning surges do not reduce. This area includes the heating part of the blade, environmental measurement sensor, nacelle interior space unshielded by navigation lights, generators, gearboxes, cooling

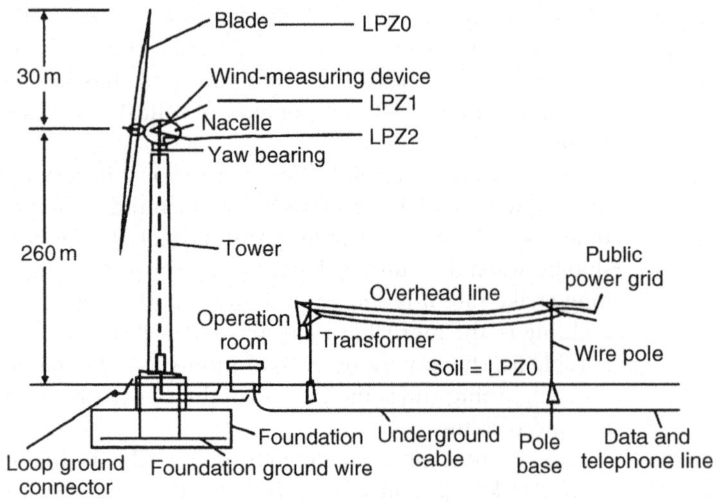

Figure 5.5 Wind turbine lightning protection zoning.

system, transmission system, electrical control cabinet, sensors, and cables. The above areas are those areas subject to direct lightning strikes (twist lightning) or areas not subject to direct lightning strikes but with no attenuation of electromagnetic field.

Surge protection devices (SPD) can be chosen to protect equipment in LPZ1 area, where a risk of a surge damage exists, and electromagnetic has been weakened due to the shielding effect. Such area includes the nacelle interior, equipment inside the tower, such as cables, generators, gear boxes, and so on.

Surge damage is further weakened in LPZ2, and the effects of electromagnetic damage becomes smaller. These areas include equipment in the electrical cabinet inside the tower, especially the well shielded weak current part.

Lightning protection configuration shall have electrical coupling and transition between different lightning protection areas, and install a SPD. Add shielding measures in the protected space, to reduce electromagnetic interference induction effect. Equipotential connection shall be installed at the junction of lightning protection for large metal parts and the cable shielding layer, in order to improve the electromagnetic environment.

5.3.2 Blade Lightning Protection

The mechanism of blade damage caused by lightning is when the lightning current forms an arc in the air side of the blade, enormous energy is released, so the temperature of the blade rises rapidly, and the decomposed air gas expands due to high temperature, causing rise in pressure and burst-out damage, or the pressure less dramatically causes cracks in the blade structure. The American Academy of Transient Characteristics used artificial corona generators to create a lightning test on all-composite blades. High-voltage, long arc impact (35 kV, 20 kA) strikes were applied to the set of blades without lightning protection, so the conclusion is that the blade must be equipped with lightning protection devices. There is a glass steel lightning-proof blade, with a stainless steel blade tip riveted in the top of the blade, and a copper wire mesh attached on both sides, to connect the tip and the blade root as a conductor. The copper mesh on the one hand can conduct the thunderstorm electricity on the tip into the ground, and on the other hand can prevent the lightning from striking the blade body.

When an object is struck by lightning, the lightning current will always choose the best conductive path, that is, the path of least impedance. The blade protection method now widely used is to install a lightning receptor at the blade tip to capture lightning, and a lightning guide wire is placed in the blade cavity, to conduct the lightning to the hub, and then into ground. If the blade is equipped with tip brake mechanism, the steel wire rope can control

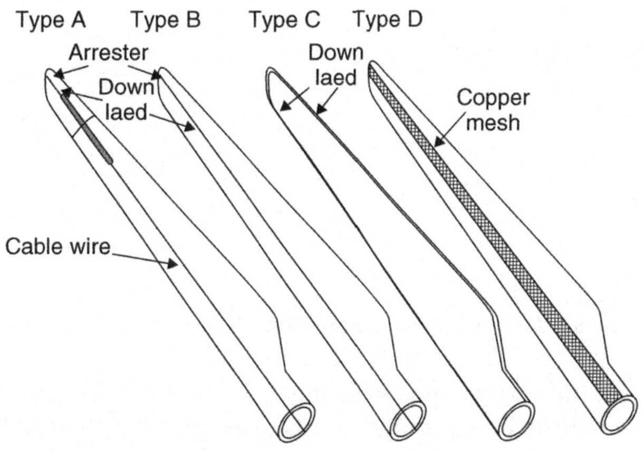

Figure 5.6 Blade lightning protection methods.

the braking function of the blade tip, and conduct the lightning current to the hub as a down conductor.

Figure 5.6 shows the common blade lightning protection methods. The main difference between these methods is whether there are a minority of lightning receptors used to capture lightning (type A and B), or whether it is trying to protect the entire blade (type C and D). Type A (with one or two receptors at the top) illustrates how the control steel wire is used for braking and as a down conductor. There are no movable tips in the blade (type B), so an additional conductor shall be installed. Down conductor of type C is located on the leading and trailing edges, but it is practically difficult to fix a suitable conducting path to the leading edge. Type D illustrates the utilization of conducting mesh on both sides of the blade.

Operating experience shows that the use of a tip lightning receptors is effective for blades of 20 m long. However, for longer blades, it seems to be a problem for expecting good protection just from the installation of a single lightning receptor in the tip.

Once the lightning current has been conducted to the root of the blade, to ensure safe passing of the current through the outside of the tower and grounding will be a problem. The current must be guided through the pitch, shaft, yaw bearing without damaging the generator and sensitive control equipments in the nacelle. Wind turbine bearings may be damaged due to lightning current passing through, usually large, heavy-loaded bearings are unlikely to be suddenly damaged, but their working life may be reduced. So far, there is no effective method to shunt the lightning current around large bearings, since the bearing itself is the lowest inductance path, it is the preferred route for high frequency lightning current. Protection of the control and electrical systems against lightning current is to divide the wind turbine into several sections according to the direct lightning stroke they are likely to suffer and its current amplitude size, and estimate the electromagnetic field in each section. Within each section, each element of lightning

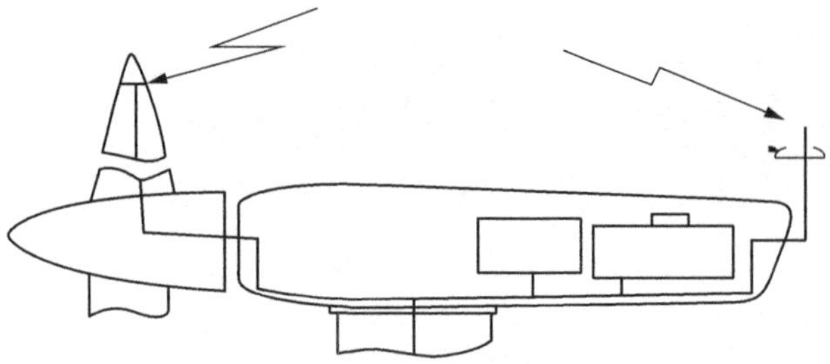

Figure 5.7 Nacelle lightning protection design.

protection withstands the expected influence. The main mechanism of lightning hazards to electrical and control systems is direct conduction and magnetic coupling, and the main protection measures are good welding, effective shielding, and adoption of suitable overvoltage (surge) protection at partition boundary.

5.3.3 External Direct Lightning Protection Design

Blade lightning protection as described above, contains a lightning receptor and a guide wire laid inside the blade cavity, which is connected to the blade root. The aluminum root of the blade is connected to the hub, and is led to the main frame of the nacelle, then to the ground. The main objective of the blade lightning protection system is to prevent direct lightning striking on the blade body, resulting in blade overheating and expansion, thus bursting out and becoming damaged.

The system is connected to the main frame of the nacelle, and the lightning rod atop the nacelle, seen as Figure 5.7. The lightning rod is used to protect the anemometer and weather vane from lightning strokes. Main frame is then connected to the tower and foundation grounding grid.

A dedicated down lead is connected between the nacelle and the tower, to reduce voltage drop, and cross the yaw ring; the nacelle and the yaw brake disc is connected by a grounding wire, and therefore, it will not be damaged when struck by lightning, which will be introduced smoothly through the down lead into ground.

5.3.4 Internal Lightning (Over-Voltage) Protection

For internal overvoltage caused by lightning, the following measures can be taken [7]:

1) Equipotential bonding. Anemometer and weather vane, together with the lightning rod are equipotential grounded; all the components in the nacelle, such as the main bearings, generators, gear boxes, and hydraulic station are grounded by an appropriately sized grounding wire that is connected to the nacelle main cabinet as equipotential; the ground switch plate of tank consists of a closed metal box, which is connected to the ground equipotential.
2) Isolation. Communications between the processors in the nacelle and the ground controller, are connected with fiber optic cables, and separate DC power supply is used for processors and sensors.
3) Installing overvoltage protection devices. Arrester or varistor overvoltage protection are used for the generator, the switch plate, the controller module electronic components, signal cable terminals, and so on.

5.3.5 Lightning Protection Standards and Grounding Resistance Requirements

Modern lightning protection can be divided into external and internal lightning protection. According to IEC1024-1 standards, the protection is divided into level I to IV based on five important parameters of the lightning (Table 5.5).

Today, lightning protection of wind turbine blades (such as LM blades) is designed in accordance with level I protection of the IEC1024-1, and has passed the relevant modeling tests, so the ability of the blade to avoid damage from direct lightning strokes has improved significantly. When the external direct lightning striking on the blade, it is not difficult to induct it into ground. However, the wind turbine is installed in the nacelle at a height of 40 to 50 m (or even higher) above the ground, and has some connection with the ground control cabinet equipment and lightning down lead system, so overvoltage caused by lightning current, resulting in damage to these devices has become a problem.

Overvoltage caused by lightning current depends on the down lead system and the grounding grid. Currently, requirements of international wind turbine manufacturers on the grounding resistance (Table 5.6) are very different: Denmark (Vestas, Micon) allows greater grounding resistance, followed by the United States (Zond) and Spain (Made), while Germany (Nordex, Jacobs) requires the minimum grounding resistance value.

The DL/T621-1997 "Ground Connection of AC Electrical Installations" of China provides that wind turbine foundation grounding parts shall be connected to ground

Table 5.5 Lightning protection level parameters.

Lightning parameters	Protection level		
	I	II	III-IV
Current peak I_{max} (KA)	200	150	100
Total charge Q_{tol} (C)	300	225	150
Charge impulse Q_{imp} (C)	100	75	50
Specific energy $W \cdot R^{-1}$ ($kJ \cdot \Omega^{-1}$)	10000	5600	2500
Average steepness ($kA \cdot \mu s^{-1}$)	200	150	100

Table 5.6 Requirements of manufacturers on wind turbine resistance.

Manufacturer	Design standard	Grounding resistance value/Ω
Vestas	IEC-1024-1/2	10
Micon	IEC-1024-1/2	6
BTK	IEC-1074-1/2	4
Zond	IEC-1074-1/2	6
	IEC-61024-1	
Made	IEC-60364-54	4
Nordex	IEC-1074-1/2	2
Jacobs	Germany Guide	2

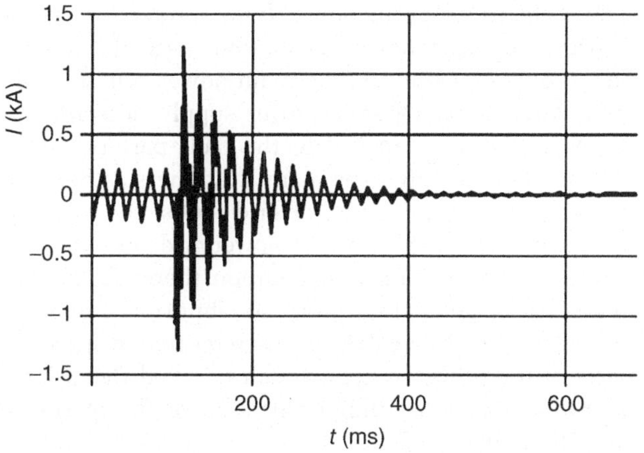

Figure 5.8 Induction generator terminal three-phase fault current.

or the neutral. Specific requirements are: ① arrangement of grounding grid within the effective radius of the impact grounding, power-frequency grounding resistance of a single wind turbine generator foundation grounding grid shall not be larger than 10 Ω. ② Based on the layout of wind turbines, after the wind turbine grounding grids being reliably connected to form a local grounding network, its power frequency grounding resistance must not be larger than 4 Ω.

5.4 WF Electrical Protection

All parts of high-voltage power system are equipped with relay protection, thus, the relay is responsible for detecting abnormal conditions, and the circuit breaker will open and isolate fault circuits. Some low voltage circuits are protected by fuses, but this approach cannot identify control levels provided by the relay and the circuit breaker. However, fuses are cheap, with very fast action, which can effectively limit the energy transfer to the point of fault. In the distribution network, the protection system is mainly designed to detect over-currents caused by insulation faults in the circuit, such as interphase overcurrent or phase-to-ground overcurrent caused by air or solid insulation breakdown. This large current is allowed to only last about 1 s to limit the harm [8].

The basic principles of wind turbine and WF electrical protection are the same as those applied to any electrical equipment protection, but there are two significant differences: ① Because WFs are often connected to the edge of the power grid, usually fault current obtained is quite small. While the small current is pleasing from the standpoint of harm reduction, it can cause considerable difficulties for rapid and reliable detection of faults. In particular, some high-voltage fuses are dependent on the arc energy to correctly operate, so when the arc energy is low, they cannot rely on it to cut off the small fault current. ② Fixed speed wind turbines interface with the grid via asynchronous generators, or variable speed wind turbines interface the grid via voltage source converter, but neither the asynchronous generator nor the voltage source converter use a reliable fault current source, so it is necessary to use voltage or frequency relays to detect that the wind generator is fed in its non-normal state.

Figure 5.8 shows the current when a three-phase fault occurs at the asynchronous generator terminal, and it can be seen that the fault current decreases rapidly with attenuation of electromagnetic energy stored in the motor, and there is no sustained fault current. Because the asynchronous generator absorbs excitation current from network or a local capacitor, a three-phase short circuit at the generator terminal is impossible to cause voltage collapse. Some asymmetric faults (e.g., two-phase fault) may lead to a sustained fault current, and its size can reach to two to three times of the full output, but this current is often lower than the operating threshold value of relay protection. In WFs, it is necessary to ensure that the fault current supplied by the wind turbine will not lead to incorrect tripping of the distribution network protection, and this is achieved by changing the fault current flows in the network, which, in fact, is achieved by providing local network voltage support.

Protection of the WF is similar to the protection of large industrial load devices in many aspects, and large industrial load equipment may be drive motors that may back feed power into the grid. For protection, the WF can be considered as a collection of large motor-drive networks, but the WF driving torque is applied to the motor drive shaft rather than taken to drive mechanical load. Since in industrial motor-drive cases, the terminal voltage and frequency of motor are determined by the network conditions. Distribution network provides reliable source of fault current (the rotating motors may also provide feedback current), which can be used to detect insulation failures.

WF electrical protection follows protection specifications of the general distribution network, and the public power grid is still the main power source of the fault current. In a public power supply network, the important thing is to isolate the fault part or circuit and maintain supply to as many customers as possible. In the early stage, in case of grid faults, the WFs are usually disconnected from the power grid, and this does produce some power loss. In addition, there are still risks of electricity generation under an abnormal frequency or voltage when the WF is disconnected from other parts of the power system, which is called the "island" operation that is an important issue requiring special considerations.

Generally speaking, for the WF, a simple and low-cost protection may be more appropriate.

5.4.1 WF and Generator Protection Configuration

Figure 5.9 shows a typical protection configuration with a constant speed wind turbine, where the generator voltage is 690 V, a connection voltage is 11 kV (which is fed from the 35/11 kV Delta/Star winding connection transformer), the transformer 11 kV neutral point is directly grounded or grounded via the resistor, the 11/0.69 kV transformer is also Delta/Star winding connected, therefore, and the 690 V neutral points of each circuit can be directly grounded. The generator neutral point is not grounded.

WF protection can be divided into a number of protected zones [8, 9]:

Zone D: the base of the wind turbine tower, 690 V circuit breaker (usually armored circuit breaker) is used to protect the pendant cables and generator.

Zone C: the 690 V connecting cable from the transformer to the wind turbine tower-base cabinet. A fuse or circuit breaker is used as cable protection at the 690 V side of the wind turbine transformer. In some wind turbine design, the main bus bar introduced into the bottom of the distribution cabinet at the tower base is composed of bare conductors. Electrical protection of such area requires careful consideration as there is the possibility of dropping tools or other equipment falling on these bus.

Zone B: the 11 kV/690 V transformer and the region around its 690 V terminal. This is a zone particularly difficult to be protected, because the 11 kV fuse must avoid the transformer excitation inrush current. And when the fault current is limited by the impedance of transformer, the detection of 690 V terminal fault requires enough sensitivity. This problem is common for all 10 kV/400 V transformers used in the public distribution network, and one solution is to use a 11 kV load switch (combined fuse-disconnect). A fault in the transformer 11 kV winding will result in a large fault current, which will be cleared by the 11 kV fuse. For fault on the low voltage-terminals, the fuse cannot effectively clear the fault current, so it is cleared by the disconnect switch. A solution of a more expensive industrial installations is the use of so-called restricted earth fault protection (REFP), which is for the detection of the leakage current from the low-voltage winding and the terminal region to ground, but this requires a 11 kV circuit breaker. For the sake of security, all credible faults can be detected and cleared, but this requires increase of investment.

Zone A: the 11 kV cable circuit, which is protected by a conventional manner, with overcurrent protection and ground fault protection acting on the 11 kV circuit breaker.

The protection of 35/11 kV transformer is similar to the protection of grid transformer with the same capacity (differential protection, constant time-lag instantaneous trip, constant time-lag overcurrent protection, overload protection and zero-sequence current protection, gas relay, etc.).

11 kV collection bus bar adopts high or low impedance differential protection, to minimize the clearing time.

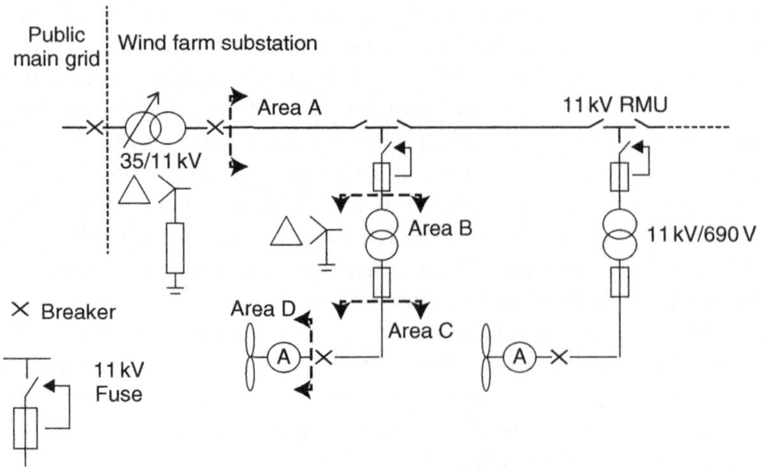

Figure 5.9 A typical configuration of the WF protection.

If a low-impedance differential scheme is adopted, the feeder line TA shall not be connected in parallel. Otherwise, in case of TA saturation during external fault, wind turbine generator (WTG) fault contribution may cause a mis-operation. A blocking scheme can be used as an alternative to the bus differential protection. In case of downstream faults, the overcurrent protector of the intelligent electronic device (IED) in each feeder will send a blocking signal to the overcurrent protector of the IED in the transformer breaker. In case of a bus bar fault, no blocking signal is sent.

With the increase of wind turbine rated capacity and larger WFs are constructed, the cost efficiency of the 11 kV system is getting worse, thus many large WFs use wind turbines of 35 kV voltage class for electrical collection. Figure 5.10 shows a typical configuration of 35 kV WF, which is directly connected to the 35 kV public power grid, and is not equipped with a main transformer. For electrical protection, this configuration brings many difficulties. Fuses of 35 kV are not readily available, so for the entire range of the expected fault current, it is difficult to provide comprehensive protection for a 35 kV/690 V transformer. For a single-phase ground fault at the low voltage side, it is particularly difficult to achieve effective protection. An attempt to overcome these difficulties is the use of comprehensive computer protection to detect the all fault of the 35 kV circuit, transformer, and low-voltage terminals.

In the configuration shown in Figure 5.9, when considering electrical protection, a 35/11 kV transformer provides two useful features—one is that its impedance makes the protection level of the overcurrent relay easy to be distinguished, because it blocks the zero-sequence current path. The other is that the fast operating ground fault relay can be applied to WF circuits. But the scheme shown in Figure 5.10 does not have these two useful features since it is directly connected to the public power grid. Therefore, some main substation is equipped with

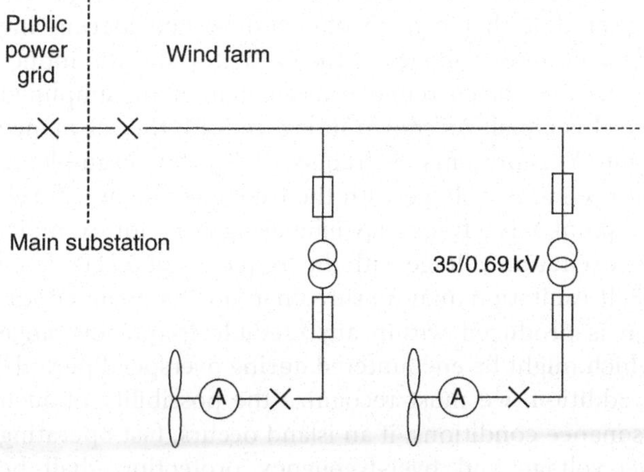

Figure 5.10 Protection of WFs with 35 kV circuits.

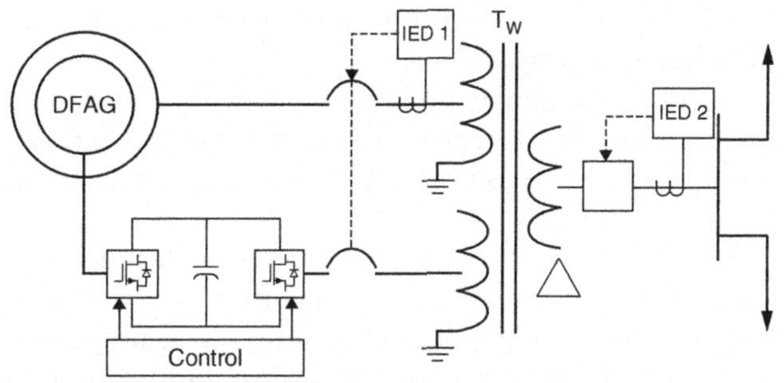

Figure 5.11 DFAG single-line diagram.

an interface transformer, that is, a main transformer with the voltage ratio of 1: 1. This is not economic, but it tells that it is very important to equip a main transformer for the WF.

Different types of wind turbines have different requirements for protection. Single line diagram of a double-fed asynchronous generator (DFAG) is shown in Figure 5.11, which includes the IED configuration for protection. The DFAG provides variable speed control via the back-to-back voltage source converter that is connected to the generator rotor circuit. Wind turbine transformers T_W has three windings, where the high voltage windings is Δ connected, both low-voltage windings have grounded-Y connections. One low-voltage winding is connected to the stator winding circuit, with the other connected to the rotor winding circuit, and the high-voltage winding is connected to the grid through a circuit breaker or fuse. Due to Y/Δ connection of the transformer, the ground fault at the low voltage side of the wind turbine transformer is not detected by the upstream protection. The IED uses instantaneous overcurrent element to provide protection for this fault, because the element may respond to zero-sequence, residual current, or transformer neutral current, but is not required to coordinate with other protection devices, and is allowed to operate with minimal time delay. If the element measures zero-sequence current based on the relationships between the phase current and the residual current, then possible current transformer (CT) saturation shall be considered when determining the set value during external faults.

IED also provides low-voltage phase fault (phase-to-phase and phase-to-ground) protection. Instantaneous action elements will interrupt severe faults with minimal time delay. Note that the DFAG will provide short-circuit current to external faults. An instantaneous action element shall be set according to lower than the minimum current expected for a phase fault at the generator terminal. For network faults, the instantaneous action element shall be set higher than the expected maximum generator terminal fault current above. Overcurrent delay element will detect internal generator phase faults, and the upstream overcurrent delay protection shall be coordinated with this element.

An IED with similar protection elements can also be applied to the converter circuit. The IED can detect faults at both ends of the converter, but cannot detect a fault in the rotor winding.

The DFAG also needs auxiliary protection functions, which can be embedded into a wind turbine controller, or may be achieved within IED, including voltage unbalance protection, overheat protection, reverse phasing protection, out-of-step protection, as well as voltage and frequency out of limits protection.

The WTG must also be able to make itself isolate from feeder faults. Ideally, it shall operate at minimum delay. Meanwhile, for the faults of neighboring feeders or WTG, external fault protection shall never operate. In fact, it is impossible to achieve this level of performance solely through measurement of local current and voltage. General network fault detection relies on the low-voltage and over-voltage elements, which need to be delayed to allow upstream protection to an open feeder breaker, thereby preventing a trip for fault on another feeder. Finally, a wind turbine IED should have the ability to record the voltage and current waveforms during faults or disturbance, as well as the time sequence data, which is very valuable for fault analysis that determines the protection system performance.

5.4.2 Isolated Operation and Self-Excitation of Asynchronous Generator

When a fixed speed wind turbine employs an asynchronous generator to generate power, there is no excitation system directly available. An asynchronous generator excitation current comes from the stator of the generator, which requires it to absorb reactive power from system. To reduce the reactive power absorbed from the grid, usually the PFC is locally configured at the generator terminal. As long as the asynchronous generator is connected to the distribution network, its terminal voltage is fixed, and the PFC serves only to reduce the reactive power absorbed from the grid. Once the asynchronous generator is in isolated operation after being disconnected from system, it is possible to cause an asynchronous generator self-excitation, resulting in a high overvoltage. In addition, after the asynchronous generator is disconnected from the grid, it will lose load and accelerate. The increase of the generator rotary speed will increase the frequency and the possibility of self-excitation. There are many reports about wind turbines that are connected to the distribution grid becoming islands and about over-voltages causing harm to the user equipment.

Figure 5.12 illustrates the equivalent circuit between the asynchronous generator and the PFC, which is not

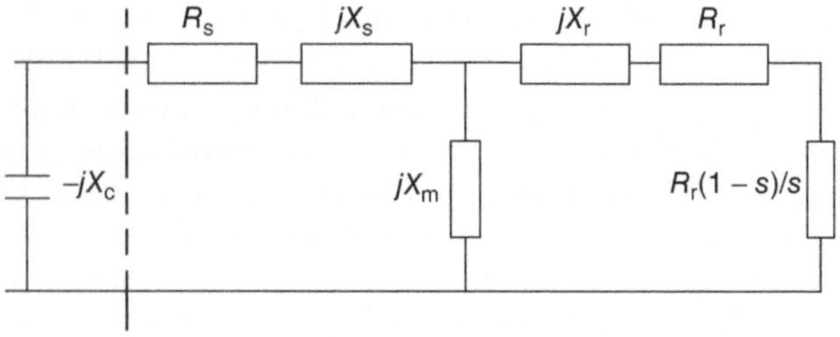

Figure 5.12 Equivalent circuit of asynchronous generators and power-factor correction capacitor.

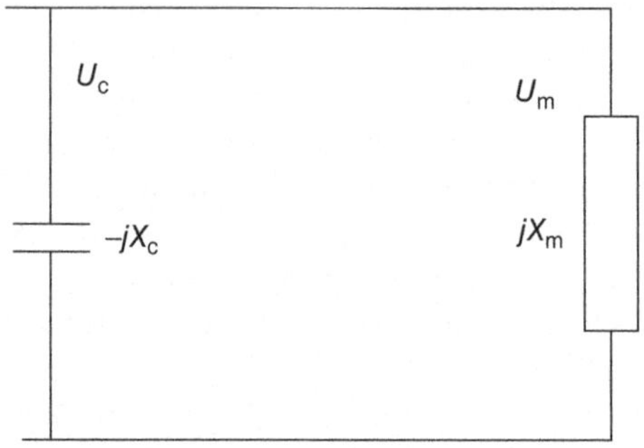

Figure 5.13 Simplified equivalent circuit of the PFC and asynchronous generators.

connected to the network (is disconnected from the network and became an island). In the normal operating speed and frequency, asynchronous generator slip (s) is very small, so the rotor branch resistance is large, and it is approximately an open circuit. Because the stator impedance ($R_s + jX_s$) is much smaller than excitation reactance jX_m, the equivalent circuit can be simplified as parallel branch of the PFC and excitation reactance, and the equivalent circuit becomes the diagram shown in Figure 5.13. This is a LC parallel circuit, and the resonance condition is reactance $X_m = X_c$. However, in this case, due to excitation saturation, X_m value is a nonlinear function of voltage.

Energy is added to this circuit from the wind turbine, a voltage value that may occur due to wind turbine rotor acceleration that can be obtained by considering the intersection of voltages of the capacitor and the inductor, because the current of each element in this simplified model is equal. As shown in Figure 5.14, the resonance point X represents a frequency f_1, for example, a self-excitation voltage with the frequency below 50 Hz, and point Y is a frequency increasing to f_2, for example, self-excitation voltage with the frequency of 55 Hz.

Self-excitation may ensure that no resonant condition is produced within any credible frequency range (which might be encountered during overspeed period). In addition, we must recognize the possibility of such resonance conditions, if an island occurs, fast operating over-voltage and over-frequency protection shall be arranged to stop the wind turbine.

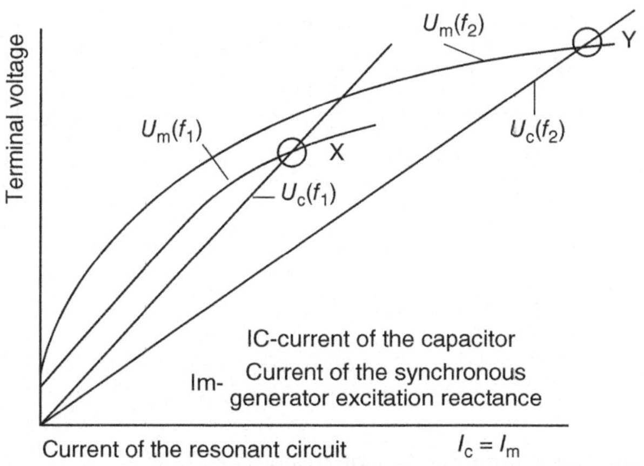

Figure 5.14 Description of self-excitation of two frequencies.

5.4.3 Interface Protection

The connection part of the WF and the grid requires a protection to ensure that the WF will not provide fault current to the distribution network or supply power to the isolated portion of the network. This protection can be called "Interface Protection" between the WF and the grid. The problem is illustrated in Figure 5.15. For a fault on the network (point k in the figure), because the wind turbine is not a reliable fault current source, circuit breaker B cannot be tripped by overcurrent protection, and circuit breaker A is opened by network current operated protection, so the wind turbine is isolated, which will start to accelerate since it still has the wind power input, but can no longer output power to the network. In fact, the acceleration might occur simultaneously with the fault, because before the circuit breaker trips, the fault has reduced the network voltage, thus restricting the power output. Then circuit breaker B will be tripped by the overfrequency relay, which detects the rotating speed increases in the wind turbine (increase of wind turbine output frequency) or by overvoltage/low-voltage relay, which detects the voltage changes at the network connection point.

Usually over/low frequency relay and over/low voltage relay will have a time delay (e.g., 500 ms), in order to reduce mis-operation. Therefore, at the situation of strong wind, wind turbines will shut down owing to the mechanical overspeed protection setting value is reached.

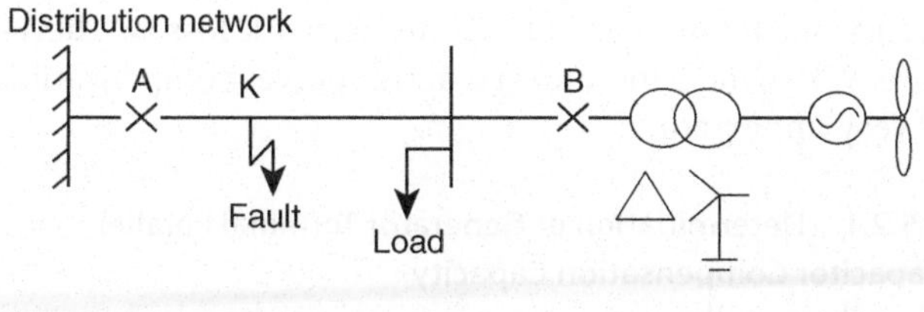

Figure 5.15 Analysis of distribution protection action caused by wind turbines.

Once the wind turbine is isolated, utilizing relays of voltage and frequency sensitive to operate sequential tripping is not considered as good practice in conventional power system protection. However, due to the application of asynchronous generators or voltage source converters, the options are few, so this arrangement is generally recognized for WFs.

In isolated operation, wind turbine output (active and reactive power) might precisely match with the local load. In this case, even if the circuit breaker A is open, the wind turbine voltage and frequency will not change, therefore circuit breaker B will not be opened.

From the WF operator's point of view, the main risk is in terms of out-of-phase reclosing. Many distribution circuits have automatic reclosing, which allows transient faults, especially the faults in overhead lines, cleared without extending power outage time of customers. Thus, circuit breaker A can be arranged to reclose after being opened for a few seconds. If circuit breaker B is still in the closing position, the network voltage adding to out-of-phase wind turbine will result in large current and torque. This is the technical reason for fast islanding protection, especially for wind generators, because it is impossible to require the wind generator output (active and reactive power) continuously match with loads in any length of time.

Great effort has been made to design a robust protection system for electric island, but according to the reclosure principle, if no current flows in circuit breaker A, we cannot know whether the network status of the wind turbine is closed or not. Commonly used relays are the so-called frequency changing rate devices or vector displacement relay (for measuring voltage vector jump when islanding occurs). According to a current flowing in circuit breaker A, both relays can adjust its sensitivity level. But if they are set too sensitive, due to the power system external disturbances (e.g., a large conventional power plant trips in the distance), they may be prone to mis-trip.

If the fault shown in Figure 5.14 does not occur in inter-phase but rather in single phase to ground, the situation will be more complicated. This type of fault is very common on the overhead lines. It can be seen that wind turbine transformer (e.g., 35/0.69 kV) has a Delta winding connection at the high-voltage side, but has no accessible neutral point and easy-to-get no ground connection point. Thus, there is no path for ground current to flow, thus in principle, single phase to ground fault is still uncertain. Actually, some of the stray capacitance currents will flow through, but it is not enough to make the conventional grounding fault protection act, and may result in intermittent arcing. Conventional solution is to utilize neutral-point voltage displacement relays to detect the grounding circuits phase voltage, because the neutral point has phase displacement. The disadvantage of this solution is its high cost, because the high voltage circuit requires a complex five legged voltage transformer, though this applies to large WFs, but the cost is relatively high for a single wind turbine.

Interface protection is very different in different countries. Some countries like to use transfer tripping, when the upstream public circuit breaker is opened, this action is communicated to the WF circuit breaker, then it will immediately be opened. Although this provides a guarantee against isolated operation, it will be very expensive to achieve, as many channels will be needed for communication of remote circuit breakers.

If all of the distribution systems use underground cables, automatic recloser for reclosing the circuit after overhead line transient fault is not needed, there is no requirement for loss-of-mains protection. Application of positive-sequence low-voltage relay, appears to be effective in detecting islanded operation.

5.5 WF Reactive Power Compensation

The voltage/reactive characteristics (see section 3.4.2) of wind generators are different, double-fed asynchronous generator (DFAG) and the direct-drive permanent-magnet synchronous generator (DDPSG) has good voltage/reactive control ability. Reactive power compensation of fixed-speed asynchronous generator (FSAG) WFs is the main focus here.

5.5.1 Voltage Characteristics of FSAG WF

For FSAG WFs, the system shall provide reactive power since the asynchronous generator has no excitation, so each wind turbine is usually equipped with a parallel capacitive compensation of 30% of the rated power.

During normal operation, over big wind power penetration will lead to voltage collapse. Because general type reactive power compensation is capacitor compensation, of which the compensation amount is in direct proportion to the square of the voltage at the connection point, when the system voltage level is not high, reactive power compensation amount will decrease a lot while WF reactive power net demand is rising, which further deteriorate the voltage level, resulting in the voltage collapse and wind turbines are forced to shut down.

In the case of no power-angle instability after fault or operation, part of the wind generators are shut down due to their own low-voltage protection, the system loses part of the reactive power load, resulting in higher voltage levels, and even WF bus voltage overlimit. If fault is not cleared in time, transient voltage instability will occur.

Figure 5.16 is the calculated result of a WF. The terminal voltage of the wind generator is 690 V, it is connected to the regional power grid through the transformer, and the regional power grid voltage is 35 kV. As can be seen from the figure, when the wind generator output is zero, the voltage is at rated value, and regional grid is in light load. As wind generator output increases, wind turbine terminal voltage and voltage at high-voltage side of the transformers increases; when it reaches the rated output, reactive power demand increases, and the voltage dropped drastically. When the WF output reaches 140% of rated output, the power flow will be unable to converge, indicating that the voltage collapse may occur.

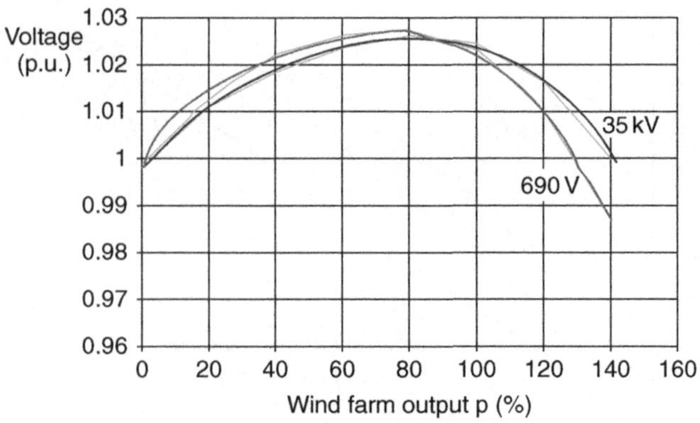

Figure 5.16 Output and voltage change of FSAG WFs.

5.5.2 Reactive Power Compensation of Cage Asynchronous Wind Generator WF

Wind power generation system used by the WF is different, so the reactive power compensation and control schemes are also different. For cage asynchronous wind generators, lagging reactive power is mainly needed for excitation, and the supply of required reactive power for magnetic flux leakage of stator and rotor. In general, for large asynchronous generators, excitation current is approximately 20% to 25% of rated current, and thus the reactive power required for excitation will reach 20% to 25% of the generator capacity, plus the second item, the total reactive power required is about 25% to 30% of the generator capacity, which will increase to about 60% at rated power.

Reactive power flow in the grid is the main reason causing voltage instability in the grid due to power transmission line voltage drop. When squirrel cage asynchronous wind generators are connected to weak grid, gusts will cause voltage flicker, voltage drop, and even tripping of the wind generators. Reactive current also results in the increase of system network losses. Therefore, WF (wind generators) reactive power compensation is very important.

5.5.2.1 Determination of Generator Terminal Parallel Capacitor Compensation Capacity

Usually parallel capacitors are used at asynchronous generator terminals to enhance the wind power output power factor, so as to maintain the voltages at the connection point and the WF high-voltage and low-voltage sides in a reasonable operating range. Compensation capacitor equipped at each wind generator terminal is commonly in three to five groups, of which the maximum compensation capacity is designed based on the power factor of the asynchronous generator at rated power, the usual reactive power compensation amount is 30% of its rated power. After capacitor compensation, generator output power factor at rated load is about 0.98.

The power factor of the Danish N43/600 kW asynchronous generator is 0.62 to 0.87, with two capacities of reactive power compensation at rated power and at no load. Reactive power compensation provided by a single generator is 200 to 300 kvar, and grouping on/off optimization can be made according to wind turbine output and the grid voltage class, and so on. Power factor after compensation can usually be up to 0.98 (see Table 5.7).

For the actual wind power systems, for the WF equivalent system as shown in Figure 5.17 [10], the connection point between the WF and the large-scale system is considered as the balance node, and its per-unit value of the constant voltage $U_0\angle 0^\circ = 1\angle 0^\circ$, the WF step-up transformer reactance jX_b, impedance of the transmission line connected to the large-scale system is $Z_1 = R_1 + jX_1$, and the WF load is $P_{L1} + jQ_{L1,}$ WF voltage at the high-voltage side is $U_1\angle\theta_1$, and low-voltage side is $U_2\angle\theta_2$. Then the wind turbine generator ends reactive power compensation capacity and switching strategy can be determined by simple calculation of power flow.

To calculate the different operating conditions with and without generator terminal PFC, as well as the operating conditions of the PFC automatic grouping switch with the changes of wind power, check the reactive power produced or absorbed by the wind generator

Table 5.7 Wind generator power factor compensation.

Generator load (%)	0	25	50	75	100	125
Power factor after compensation ($\cos\varphi$)	1.0	1.0	1.0	0.99	0.99	0.98

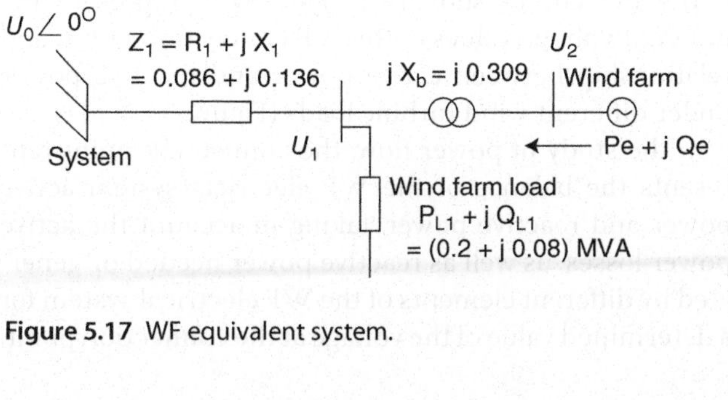

Figure 5.17 WF equivalent system.

from the system, and the terminal voltage (low-voltage and high-voltage sides) changes. It is better that wind generators are not sending reactive power (zero) to the system that the power factor at full load is close to 1.0, as voltages are within the scope prescribed by the grid code. Usually, the guideline is that power factor corrective capacity at no load can just meet the needs of wind generators for rated voltage, and there is no reactive power (current) exchanging with the grid.

The general rule is that if there is no PFC capacitor, with the increase of wind speed, the wind generator will absorb more reactive power from the system, voltages at the WF, both high-voltage and low-voltage sides raise first and then drop, and the average power factor is low. With fixed capacitance compensation, the wind generator absorbs less reactive power from the system, and may transmit reactive power to the system at low wind speeds. If capacitance compensation automatically switches in the group, as wind speed increases, the active power transmitted to the system increases; accordingly, the reactive power absorbed from the system also increases, which approximately keeps a linear relationship between reactive and active power, that is, the power factor is essentially the same, voltage levels at high-voltage and low-voltage sides of the WF are also good. It may reduce the network losses when there is compensation, but power factor at WF end shall not be 1.0, because that might make the WF terminal voltage out of limit and unacceptable.

5.5.2.2 Determination of WF Reactive Power Compensation

WF reactive power compensation is to meet the reactive power needs of the WF collection network and the transformer. Reactive power compensation can be centralized compensation or decentralized compensation.

Centralized compensation is to concentrically install reactive power compensator in the step-up substation for reactive power compensation, which is usually installed at the low-voltage side. The main purpose is to improve the power factor of the entire WF, maintain the voltage level at the point of common coupling between the WF and the system, and compensate for reactive power demand. The problem of centralized compensation are: ① Centralized capacitor switching operations has greater impact on the WF, the equipment is easy to be damaged due to the impact of large inrush current in switching process, and equipment faults will affect the power factor of the entire WF and voltage stability at the system connection point. ② Centralized compensation can compensate for the overall reactive power throughout the WF, but cannot resolve the internal network reactive voltage optimization balance.

Decentralized compensation uses mathematical or intelligent algorithms to select several reactive power compensation points with the optimal compensation effect within a reasonable investment range for local compensation, thus reducing WF internal network losses and improving voltage quality. Compared with centralized compensation, decentralized local compensation has the following advantages: ① The reactive power required by the WF is locally supplied by the reactive power compensators dispersedly installed, which reduces the WF collection system losses. ② Decentralized local automatic compensation can realize monitoring of reactive voltage level within a certain area of the WF in real-time, rapid response to reactive voltage changes in the monitoring area, and give quick compensation. ③ The decentralized compensation points can coordinate with each other by a certain communication mechanism, to greatly reduce the chance of under-compensation or over-compensation, and optimize the compensating effect of the whole WF.

To choose centralized compensation or decentralized compensation is more an economic issue. Basically, we are using centralized compensation at present. Only large-scale WFs with wide distribution will consider the need for decentralized compensation.

Centralized compensation capacitance should be determined by power flow calculation and analysis. For simplified wiring of a WF shown in Figure 5.18, the WF PQ curve obtained from power flow calculation is shown in Figure 5.19, which represents the reactive power values correspond to each of the active power value generated by the WF. Calculation steps to determine the WF PQ curve are: ① Setting different voltage values at the connection point PCC (e.g., take three values of high, medium, and low voltage, the minimum voltage is 0.93 p.u., normal voltage is 1.0 p.u., and the maximum voltage is 1.07 p.u.) and different WF load levels, ② for each step, calculate the reactive power absorbed or generated by the WF (assuming the load of each wind turbine is the same) at the connection point by increasing 10% of

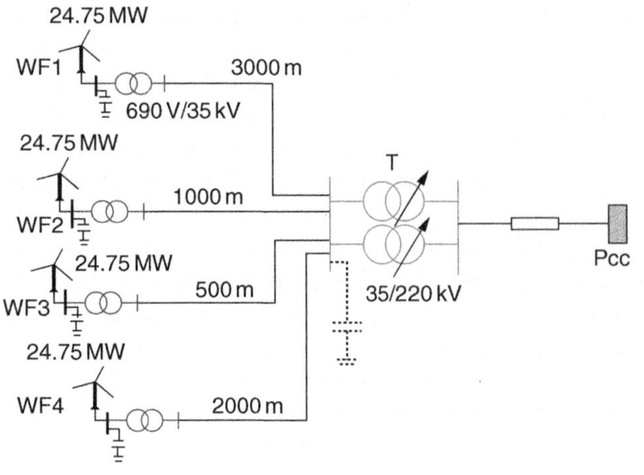

Figure 5.18 WF simplified wiring.

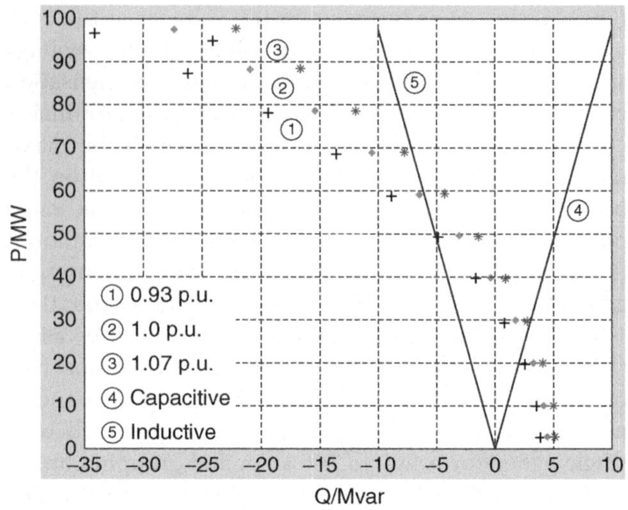

Figure 5.19 P-Q relation curve of different voltages at the connection point.

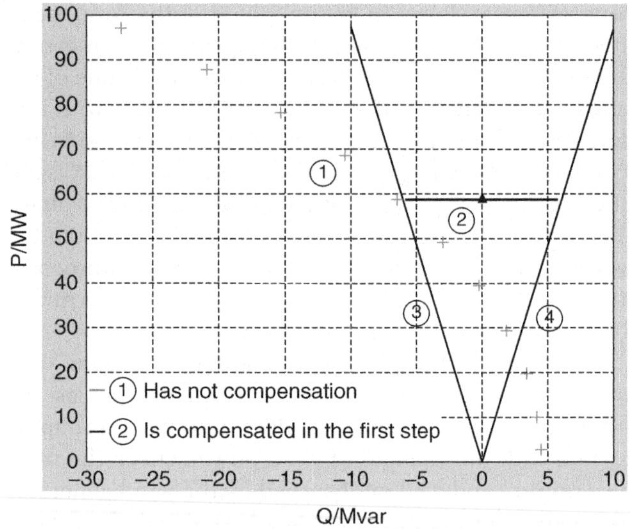

Figure 5.20 PQ curve without medium-voltage compensation.

the normal power (from technically the minimum to the maximum active power); ③ draw out the corresponding PQ curve ①, ②, and ③ of different connection point voltages, and WF operation range curve ④ (the WF is capacitive), and ⑤ (the WF is inductive) on the PQ plane.

The PQ curves shown in Figure 5.19 represent the different voltage values at the WF connection point, and relationship between active power and reactive power under different wind turbine loads (Figure 5.20) [11].

In the study of power flow, the connection point represents the balance of the WF electrical system active power and reactive power, taking in account the active power losses, as well as reactive power needed or generated by different elements of the WF electrical system for a determined value of the voltage at the connection point.

WFs must operate in a power factor close to 1.0. For this reason, when designing the capacitance compensation, the WF operation range shall be selected within 0.995 inductive to 0.995 capacitive. Figure 5.20 shows the PQ curve ① in normal operation voltage (no centralized compensation at medium voltage) and capacitor bank designed operating range (between straight line ③ and ④). Some observations from Figure 5.20 include that for electricity generation levels of 10% and 20%, the operation is beyond the predetermined range, and the WF has a capacitive characteristic, which means there is reactive power injecting in the high-voltage grid at the point of common coupling. For generation levels of 30%, 40%, and 50%, the operation is within the allowable operating range, and therefore, the compensation is not required. In other generation range, when the generation level is larger than 60%, operation is also beyond the predetermined range, and the WF also has inductive characteristic, therefore, it is needed to be compensated with capacitor banks. The thick straight ② in the figure represents the amount of centralized reactive power compensation required for ensuring the WF to operate within the predetermined range. In the WF being studied, the compensation amount in this step is 12.334 Mvar.

Figure 5.21 represents the role of the capacitor banks connected to the substation medium-voltage side in the first step. When the generating level is at 60%, the reactive value Q has been working within the setting range. In addition, when the generating levels are 70%, 80%, and 90%, the WF is also operating within the inductive range of 0.995. Despite the compensation in the first step, the WF still operates beyond the setting range at 100% of the power generating level, so it is necessary to compensate in the second step, in order to maintain the operating point within the setting range. Of course, from an economic perspective, investment of the most optimized capacitor banks step by step is too expensive, because every step of the compensation requires a separate medium-voltage cell. On the other hand, from a technical point of view, it is unnecessary to compensate the reactive power at a generating level that only operates for a few hours every month.

Installation of inductors, capacitor banks, or static var compensators, and so on, in the WF can only partially solve the problem of reactive power control. Traditional wind power systems must switch these devices from time to time, and mechanical switches of the capacitors and reactive power compensators cannot keep up with the speed of rapid frequent changes in reactive power, but these devices themselves will generate large transient currents, which leads to poor power quality, and serious damage to the sensitive computer equipment connected to power grid. So, in the case of vulnerable power grids and gusts, as the reactive power compensation equipment connected to wind generator does not react to

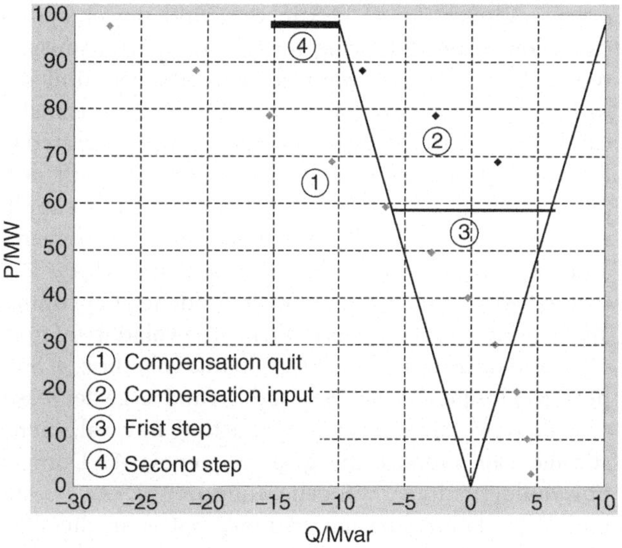

Figure 5.21 Effects of centralized reactive power compensation at substation medium-voltage side.

voltage sags, the grid will lose its stability. Moreover, a large number of low-voltage PFC capacitors will be required in order to satisfy the requirements, so the total cost is higher. If the network structure is strong, the high-voltage capacitors can be used.

For large-scale WFs, centralized reactive power compensation equipment, such as SVC and STATCOM can be used to provide a smooth dynamic reactive power adjustment.

5.6 WF Energy Storage Systems

Large-scale electricity storage can solve the problem of peak-valley difference in power production, improve the power-supply reliability of the power system, and avoid troubles and losses brought about by sudden blackout; energy storage equipment can improve system stability, and avoid system instability in large-scale disturbances of the power system; energy storage equipment is an essential equipment in the power generation systems of renewable and instable energy resources, such as wind power, and solar PV power or solar thermal power, and so on. Energy storage equipment enables unstable power generation systems to supply stable powers to users. Operation of micro-grids must be coordinated with energy storage devices.

With the rapid development of wind power, the costs continue to drop. However, rapid development of wind power is faced with the basic problem of wind energy variability. Wind turbines require flexible backup power supply and sufficient energy storage. Energy storage system (ESS) has significance for the strengthening of wind power. Application of ESS in the wind power system can achieve the following functions: ① Transmission reduced: mitigation power transmission constraints due to insufficient transmission capacity; ② Time transfer: can stabilize and rectify the electric energy generated by the wind turbine by storage of wind power at off-peak period (purchasing electricity from the grid as supplementary when the wind power generation is insufficient), and discharging during the peak load period; ③ Predicted hedge: reduce quotation error (deficiency) on day ahead power market, thus reducing the volatility of node offer, and mitigating users' risks on this volatility; ④ Support grid frequency: during a sudden and large decreases of wind power over short discharge time interval, energy storage can support the grid frequency; ⑤ Dampen power fluctuations: WF generating power can be stabilized by dampening fluctuations, that is, in the short term, during the wind power output fluctuation period, the storage device can absorb and discharge energy.

5.6.1 Modern Energy Storage Technology

Storage technology follows the energy conversion principle, and its main energy conversion modes are: ① chemical energy storage-batteries, synthesis fuels, and chemical heat storage; ② electromagnetic energy storage – superconducting magnetic energy storage and capacitance electric field energy storage; ③ mechanical energy storage – flywheel, pumped storage power station, spring, and compressed air; ④ thermal energy storage – sensible heat storage, and latent heat storage.

Energy storage systems can be divided into two categories by function: power-type energy storage device (PES) and energy-type energy storage device (EES). The former includes super capacitors, superconducting energy storage, flywheel energy storage, batteries, and so on, shown in Figure 5.22. They have high power density, quick power regulation, characteristics of high power rating values, and smaller energy capacity, which are applied to improving power quality or reliability (uninterrupted power source), but not suitable for large-capacity energy storage. The latter includes pumped storage, compressed air energy storage, fuel cells, solar cells, large batteries, thermal energy storage, and so on, shown in Figure 5.22. They have a high energy density, but response speed is slow or not suitable for frequent charge and discharge conversion, so they are applied to energy management systems.

Table 5.8 lists the different types of modern energy storage systems. Technically, they can be used for wind power generation systems. Appropriate storage mode shall be chosen according to different purposes of the WF storage systems.

5.6.2 Configuration of WF Energy Storage System

When selecting WF energy storage system (ESS) type and capacity, specific application requirements of the WF shall be considered.

WF energy storage configuration can be centralized or decentralized. As shown in Figure 5.24, centralized configuration is a large energy storage device serve the entire WF, as shown in Figure 5.23a; decentralized configuration are small energy storage devices installed in the wind generator end, exclusively used for a wind generator, as shown in Figure 5.23b.

Figure 5.24 shows the different interface modes of centralized and decentralized energy storage systems with the grid. Figure 5.24a shows the centralized storage structure, which is usually connected to the AC bus of the WF. Batteries and other energy storage devices that output DC voltage must be connected to the AC bus via DC/AC converters. For energy storage devices outputs lower voltage, a intermediate stage DC/DC converter (chopper) shall be added to raise voltage. Flywheel energy storage and other storage modes that output AC voltage shall go through a voltage source AC/DC/AC converter to be connected to the AC bus. Figure 5.24b shows a decentralized energy storage mode, batteries, and other energy storage devices that output DC voltage are connected to the wind turbine DC bus via bidirectional DC/DC converters. Flywheel energy storage that outputs AC voltage is connected to the wind turbine DC bus via a voltage source AC/DC converter.

Both configurations can effectively dampen power fluctuations and voltage fluctuations, significantly reduce frequency offset, and improve local power quality. Compared with decentralized energy storage system, centralized storage system requires less power capacity, because the WF has a smoothing effect on power

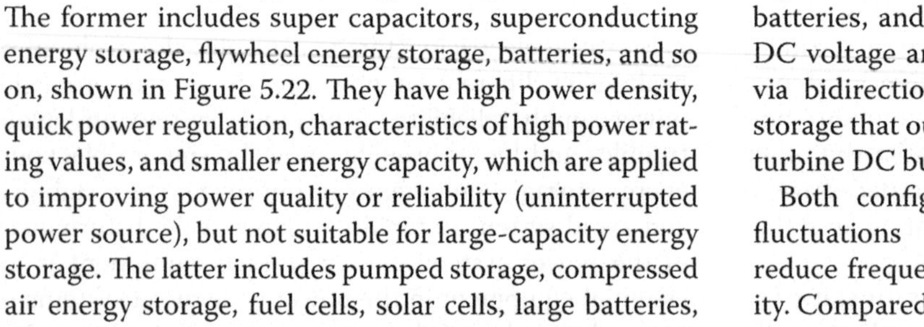

Figure 5.22 Feature-related storage system classification.

Table 5.8 List of modern energy storage systems.

No.	Storage system type and subclasses	
1	Pumped Hydroelectric Storage (PHS)	
2	Compressed air energy storage (CAES)	
3	Battery	Lead acid batteries
		Nickel-cadmium batteries (NiCd)
		Sodium sulphur batteries (NaS)
		Sodium-nickel-chloride batteries (ZEBRA)
		Lithium ion batteries(Li-ion)
4	Fuel cell	Hydrogen fuel cell
		Direct-methanol fuel cells (DMFC)
		Molten carbonate fuel cells (MCFCs)
		Solid oxide fuel cells (SOFCs)
		Metal-air battery
5	Flow battery	Vanadium redox battery (VRB)
		Zinc-bromine flow battery (ZnBr battery)
		Polysulphide bromide battery (PSD)
6	Solar fuels	
7	Superconducting magnetic energy storage	
8	Flywheel energy storage	
9	Capacitance electric field energy storage	Capacitor
		Super capacitor
10	Thermal energy storage (TES)	Low-temperature thermal energy storage (LT-TES)
		Aquiferous low-temperature thermal energy storage (AL-TES)
		Cryogenic energy storage (CES)
		High-temperature TES (HT-TES)
		Molten salt storage and room temperature ionic liquids(RTILs)
		Concrete storage
		Phase change materials (PCMs)

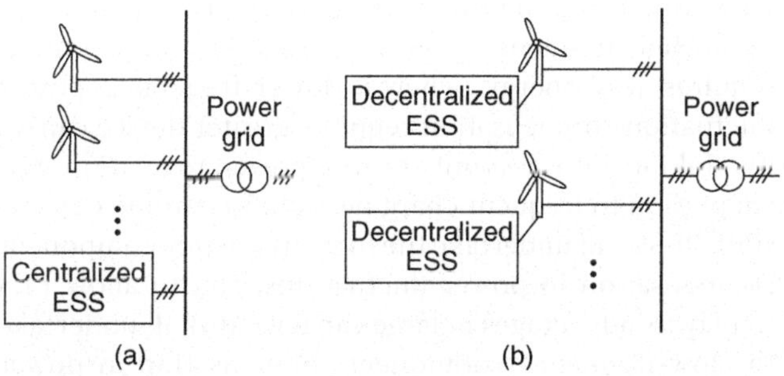

Figure 5.23 WF energy storage system structures. (a) Centralized; (b) Decentralized.

fluctuations. But when connected to the AC system, large capacity transformers are required.

In the WF, in order to stabilize WF power output, and improve power quality and grid-connection stability of the WF, two types of energy storage equipment shall be configured: PES and energy-type energy storage EES equipment. They constitute a hybrid energy storage system (HESS), which has both the advantages of the two types of energy storage equipment, and can greatly improve compensation performance while reducing the costs of the energy storage equipment.

Typical structure of WF systems with hybrid energy storage systems is shown in Figure 5.25 [12]. In the figure, EES and PES form a HESS, and the local load and the reactive power compensation device (RPC) are directly or indirectly connected to the WF collection bus bar B_w, then connected to the power grid via a step-up transformer (T). Bus bar B_{PCC} is a point of common coupling (PCC).

In view of active power balance of the system, the power injected by the WF to the collection bus bar is P_w, the local load power is P_{Load}, the compensation power of power-type energy storage PES and energy-type energy storage EES is P_{PES} and P_{EES}, respectively, the power of hybrid energy storage is $P_{HESS} = P_{PES} + P_{EES}$, the active power injected into the grid via the step-up transformer is P_{Grid}. Thus, the collection bus bar B_w has the following active power balance relationship:

$$P_w - P_{Load} + P_{HESS} = P_{W,L} + P_{HESS} = P_{Grid}$$

Since wind power P_w has significant volatility, whereas local load P_{Load} also has a certain degree of volatility, that means $P_w - P_{Load} = P_{W,L}$ is fluctuating. Thus, the power injected into the grid is not stable, which may not be able to meet the requirements of the grid code.

Hybrid energy storage system compensates the power fluctuations of wind power and load by real-time control of the power generated by HESS P_{HESS}, so that the power

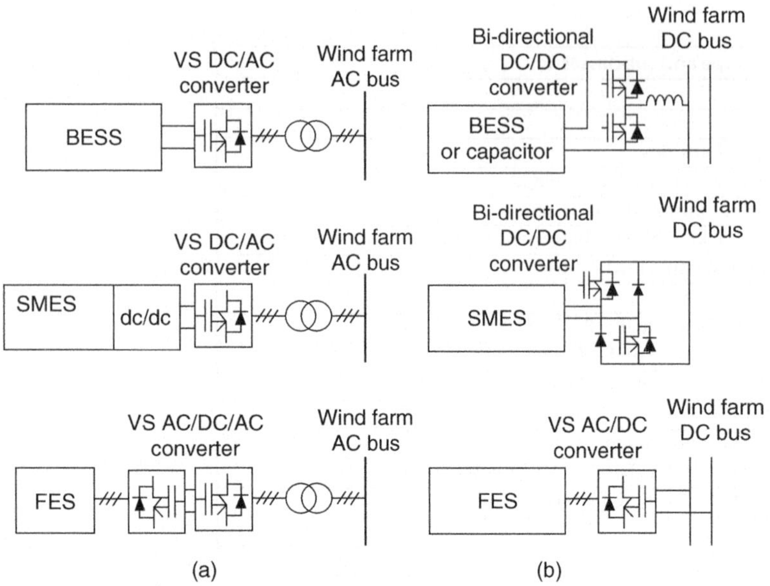

Figure 5.24 Energy storage system units and interfaces. (a) centralized configuration; (b) decentralized configuration.

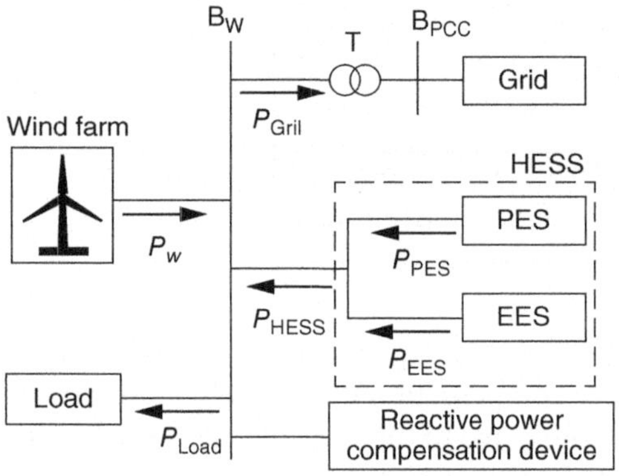

Figure 5.25 Application of hybrid energy storage systems in WFs.

P_{Grid} injected into the grid is as close to its target value P^*_{Grid} (the ideal value P_{Grid} in line with the requirements of the grid) as possible. The HESS compensation task may be shared by PES and EES.

Because during normal operation of the grid, the role of active compensation is to make P_{Grid} remain stable, here its target value P^*_{Grid} is taken as the low-frequency component in $P_{W,L}$. When the grid faults happen, to make wind power generation system ensure its own security, and maintain a certain degree of active support on the grid, the purpose of active compensation is converted into compensation of active unbalance due to fault, which is a high-frequency component compensation. High-frequency power fluctuation compensation requires less energy, whereas lower-frequency power fluctuation compensation requires greater device capacity. Taking into account its fast power regulation and adaptation to frequent charging and discharging conversion, PES shall undertake the high-frequency component compensation in power fluctuations, and to allow EES to play its advantages in large capacity, it shall undertake the low-frequency component compensation in power fluctuations.

References

1 Fang, Su. 2008. Technical characteristics and design of 35 kV combined transformer in wind farm. *Electrical Manufacturing*, **10**, 36–38.
2 Yuxiang, Zhao. 2008. Selection scheme of cabinet transformer using wind turbine group. *Forestry Science and Technology Information*, **40**(4), 52–54.
3 Dawei, Zhang and Fang, Wu. 2008. Huadian Huitengxile 2# (Kulun) 200 MW electrical design features of wind farm. *Inner Mongolia Petrochemical Industry*, **6**, 69–70.
4 Jian-xue, Yuan and Hua-yi, Tian. 2008. Lightning protection of large-scale wind turbine. *Journal of Yanan University (Natural Science Edition)*, **27**(4), 51–54.
5 Xing Zuoxia, Xing, Chen Lei, Chen, and Yao Xingjia, Yao. 2004. Lightning Protection of on-grid scale wind turbine. *Renewable Energy*, **115**, 55–56.
6 IEC/TR 61400-24. 2002. Wind turbine generator systems[s]. IEC.

7 Zhi-yuan, Lin and Cong, Huang. 2001. Lightning protection of aero generator sets. *Guangdong Electric Power*, **14**(5), 15–18, 81.

8 Burton, Tony, Sharpe, David, Jenkins, Nick, and Bossanyi, Ervin. 2001. *Wind Energy Handbook*. New York: John Wiley & Sons.

9 Yuling, Wen and Qin, Chao. 2008. Tursun – yibulayin: Study of windfarm protection. *Electric Power Equipment*, **9**(9), 33–35.

10 Min, Guo. 2006. Research on reactive compensation for generator end of wind power sit. *Inner Mongolia Electric Power*, **24**(3), 1–17, 31.

11 Villacorta, A. Ríos, González, M.V. Gascó, Gómez, Arnaltes, and Amenedo, J.L. Rodríguez. 2004. Implementation of the wind park PQ curve in the dimensioning of the reactive compensation system. Nordic Wind Power Conference, March, 1–2.

12 Hong fu, Wang, Jun, Cao, Jiaju, Qiu, and Xinying, Wei. 2009. An active power compensation model for grid-connected distributed generation system. *Automation of Electric Power Systems*, **33**(8), 94–98.

6

OWFs

The offshore wind farm (OWF) is a new development area of wind power. Compared with onshore WFs, it has many different characteristics. This chapter briefly describes the characteristics of OWFs; the OWF collection systems and offshore substations; and the connection of OWFs with onshore power grids, that is, different transmission systems of OWFs. This chapter also introduces a new OWF wind energy conversion system and several instances of OWFs.

6.1 OWF and Its Characteristics

In recent years, international attention has been given to construction of OWFs, for example, the European Union (EU) has put forward the strategic objective of development of 18,000 megawatts of offshore wind power by 2020. At present, development of offshore wind power is mainly concentrated in Europe. According to statistics, the capability of OWFs that has been constructed in Europe are 2.75 to 165.6 MW, and the WF capacity in planning is up to 4.5 MW to 1 GW. In Denmark, Germany, the Netherlands, and other countries, offshore wind power generation started early and has been developing quickly, and its technical and policy support are leading at the international level. In 2002, the world's first truly large-scale OWF was completed in Denmark's Horns Rev. In 2006, Germany's offshore wind turbine capacity had exceeded 500 MW. In the long-term goal by 2030, the installed capacity in Germany coast areas, exclusive economic zones, and an area of 12 miles in the periphery of its territory will reach 25000 MW, and the electricity generated will reach 70 to 85 TWh. In 2004, the United States' first OWF was completed and put to use in Cape Cod. Asia is bounded to the east by the Pacific Ocean, and to the west by the Indian Ocean with abundant offshore wind energy resources. China, Korea, Japan, India, and Asia's other leading countries with wind power experience have turned their attention to the sea, but are currently at the beginning stages.

Offshore wind power is the extension of onshore wind power utilization, "offshore" here refers to intertidal zone and nearshore zone a little away from the shore, thus, maritime wind power is also called offshore wind power or inshore wind power. At present, offshore wind technology can only be achieved in inshore sea surface, not ocean surface.

Compared to onshore WFs, the main characteristics of offshore wind power are [1–4]:

1) Offshore wind resources are abundant, and the wind speed is more stable. A large contiguous area is available at sea, and the development is subject to less space limits.
   For wind power generation, the annual full load hour is used to measure the size of wind resources in the region. Full load hour refers to the maximum electricity a wind turbine can produce in one hour. The annual generated energy of a wind turbine is divided by its full load hour, that is, the annual full-load hours of the wind turbine. According to observations, the number of onshore annual full load hours is 1,500 to 3,000, with an average of 2,000 hours. And the number of offshore full load hours of a wind turbine is 3,500 to 4,000, almost twice as that of onshore wind power. The farther away from coastline, the greater the wind speed is, since wind power is proportional to the third power of wind speed, when the wind speed increases by 10%, the wind power will increase by 33%, so the use of OWFs can get more wind power. When the wind speed is 20% higher than that on land or onshore, the generated energy will increase by 70%.
   Offshore wind power generation allows the use of more high-capacity wind turbines, which can increase the total installed capacity per unit area, generally the capacity of offshore wind turbines are above 2 MW, or as high as 5 MW to 10 MW. Because of the low noise requirements, higher electricity output can be achieved through higher speed and voltage.
   There are very few calm periods at offshore, so the utilization of the wind turbine capacity can be more efficient; the sea surface is relatively flat, and the friction is small, so wind shear generated is small, and wind speed changes very little at different heights,

Integration of Large Scale Wind Energy with Electrical Power Systems in China, First Edition. Zongxiang Lu and Shuangxi Zhou.

it is not required to build high towers, thus the capital construction cost can be reduced;

Offshore wind turbulence intensity is low, the air temperature difference between the water surface and the above is smaller than that between the land surface and the above, and there is no complex topography to affect the airflow on the sea, so the effect of fatigue loads on the blades is reduced, thus the service life of wind power equipment is extended, which is usually 20 years on land, and even 25 to 30 years or longer at offshore.

2) Little impact on environment. As a renewable energy source, wind power can achieve zero emission of pollutant. Whether the establishment of OWFs will affect other elements of the marine environment, is also one of great concern of the public. Study of the environmental impact of OWFs, especially long-term observations of offshore wind power demonstration projects that have been constructed since the 1990s shows that the environmental impact of offshore wind power is very small. Some preliminary conclusions are that during the construction of OWFs, large marine animals, such as seals and dolphins, will leave the area due to the noise other disturbances, but will eventually return to the waters of OWFs; migratory birds will change the flight path, passing by the WFs; the wind turbine foundation has a good influence on the lives of fish, and the foundation can become a new fish habitat and help increase fish food; WTG is far from the coast, so the visual disturbance is small, and the warning lamps on the wind turbines form a beautiful sigh line in the night. Noise standards at sea can be improved, the turning speed of wind turbines can be increased by 10%, and the effectiveness can be increased by 5% to 6%. In a word, it can save land resources, reduce the noise and visual impact to the public, and has less negative impact on the environment.
3) The technical difficulty in power transmission and grid-connection is great. Due to the wide range of changes in wind power generation and the difficulty in prediction, traditional onshore wind power is transmitted by 110 kV low-voltage cables. But because the capacity of OWFs is large, which is difficult to be transmitted by low-voltage cables, therefore high-voltage transmission lines of over 110 kV have been used in OWFs since 1998, of which the technical difficulty is great. In addition, OWFs are situated far from land and in sparsely populated areas, therefore, power must be transmitted over a long distance in order to be connected to the main power grid, which is also a challenge for the power transmission network.
4) The technical difficulty in construction and maintenance are great and costs are high. Offshore wind power projects' initial investment is higher than that of onshore wind power projects, because the support structure and grid connection costs are high. Wind power projects are constructed in the sea, so they need to bear more load and impact of the waves, be able to resist salt spray corrosion of the marine environment, and their lifting and maintenance work, as well as the infrastructure construction are difficult than those of on land. According to the data of OWFs completed in European, grid connection costs take up about 15% of the initial capital investment, which is much higher than that of onshore wind projects, and the support structure costs take up 24% of the total initial capital investment, and the wind turbine costs take up about 35% of the initial capital investment. In addition, the offshore wind turbine operation and maintenance costs are also very high. For the same capacity, total costs of offshore wind turbines are 60% higher than onshore wind turbines, and the quantity of electricity is 50% higher.

Offshore wind power business has raised new technical requirements on wind turbine, and the construction, operation, and maintenance of the power grid. Fundamental research on seabed, ocean currents, and marine environment are needed. In order to make up construction and transmission costs of offshore wind power, large wind turbines of 1 to 2 MW or above must be used; the concrete or steel structure of wind turbine foundations used at presents only apply to shallow sea, so new breakthrough in the research and development of wind turbine foundations is in urgent need for deep area; and measures must be taken to protect electrical equipment against sea water corrosion. Anti-corrosion and anti-salt spray treatment shall be carried out on blades; and anti-scour and cold-proof shall be considered when selecting of tower frame coating design and material. The protection grade of control cabinet design shall be improved; ventilation and anti-moisture shall be carried out on the nacelle; the key parts with high maintenance rate shall be of military-grade; and the airtightness and anti-moisture treatment shall be added to the transformer, and so on. These needs will promote the further development of wind power technology and industry.

6.2 OWF Electrical System

OWF electrical system and its connection to the main power system on land is a new topic. On land, the standard solution is to step up the voltage level of the electricity generated by wind turbines to the voltage level of the power grid via the transformer near to the wind turbine foundations, and then collect the electricity and transmit it via the alternating current (AC) network of the WF into the power grid. Most wind turbines run at 690 V voltage level, the voltage of the WF and the voltage of the distribution network are usually at the same level, within the range of 10 to 110 kV.

The capacity of OWFs constructed in the early stage is not very large, and the electrical connections are relatively simple, which usually follows the practice of electrical systems on land. But for large-scale OWFs, the OWF efficiency, cost, operation performance, and reliability depend in large measure on its electrical system, so the system shall be specifically designed.

The electrical system connects the generators, transformers, electrical switchgear appliances, transmission cables, power electronic converters, and other electrical equipment together according to a certain rule, in order to converge the electric energy and transmit it via the transmission system to the main power grid.

Electrical systems for OWFs consist of four parts: OWF collection systems, offshore substations, undersea transmission systems, and onshore substations, as shown in Figure 6.1. The simplest collection system is of radial serial structure, and the capacity of each radial series is up to 30 to 40 MW, while the optimal voltage is below 35 kV. Whether to establish offshore substation depends on the scale of the WF, its distance from the coast, and other conditions. The offshore substation is equipped with one transformer (can be two transformers if necessary). AC or DC cable can be used for submarine transmission, which is connected to the high-voltage side of the offshore substation and the low-voltage side of the onshore substation. Onshore substation is similar to other substations on land, of which the low-voltage side shall match with the submarine transmission voltage.

6.2.1 OWF Voltage Levels

At present, for OWFs using wind turbines with the rated voltage of 690 V (50 Hz), the voltage level of its collection system can reach 35 kV. If larger wind turbines are used, the WF can be scaled up to 250 to 1,000 MW, so a higher voltage level may be required in order to reduce the loss. If the wind turbine rated voltage is 4,000 V or higher, the collection system voltage shall be 110 kV or higher. A corresponding high-voltage transformer and switch station shall be configured for the high-voltage system with a well-trained maintenance personnel when a higher safety is needed, which will increase the costs.

OWFs are usually of a larger scale, the distance between the offshore wind turbines is typically further than the distance between the wind turbines on land, because the big tail wave (of ships) will affect the sea wind. Usually the distance from the WF to the next (offshore) substation is also much further than the distance from the onshore WF to the substation on land. For large OWFs with AC networks, in order to minimize the power loss, higher voltage is required, which may lead to larger transformers with higher costs. Wind turbine transformer is placed in the nacelle, tower, or a cabinet near the wind turbine, so the transformer shall be small in size. In addition, the switch station cost and size will increase as the voltage level increases. Using of 35 kV voltage level for the WF collection system is considered the standard solution for OWFs at present.

WF size and the type of wind turbine shall also be considered for determination of voltage level. Small OWFs are close to land, so lower voltage levels shall be selected, because in this case, the savings in power loss is not enough to offset the increase in equipment costs of higher voltage.

The voltage levels of completed OWF collection system are different. Completed small-scale OWFs are all close to land, and they all select lower voltage levels. The collection system voltage of Sweden Utgrunden OWF (10 MW) is 20 kV, the cable length to land is 8 km, and the operating voltage is 20 kV. The voltage

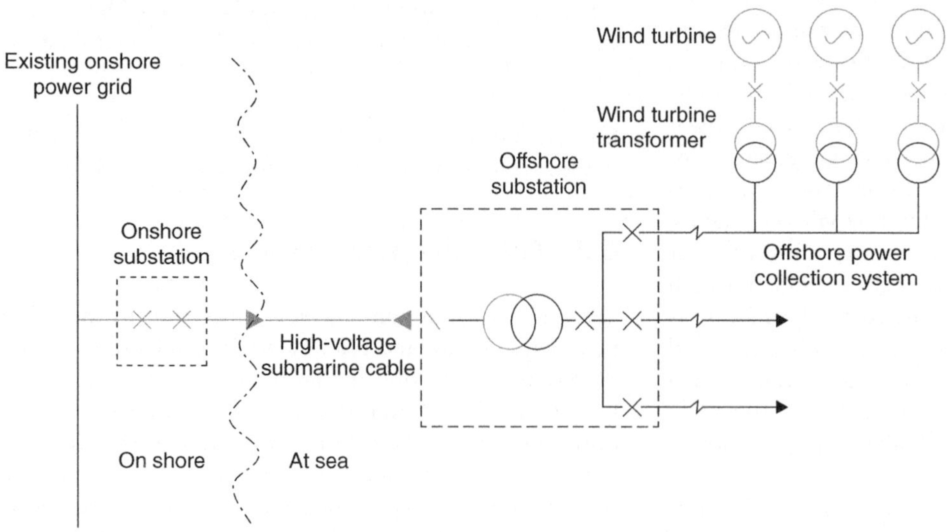

Figure 6.1 Single line diagram of typical electrical system for OWFs.

is stepped up to 50 kV on land. The collection system voltage of Denmark Middelgrunden OWF (40 MW) is 30 kV, the cable length to land is 3 km. The collection system voltage of Denmark Horns Rev OWF (160 MW) is 36 kV. Horns Rev OWF is equipped with a transformer platform, by which the voltage is raised to 150 kV and transmitted via a 15-km length of the AC cable to the shore. The collection voltage of Denmark N ysted Rødsand WF (158 MW) is 33 kV, which is raised by the offshore transformer to 132 kV, and then transmitted via a 10-km length cable to the shore.

6.2.2 Offshore Substation

For an offshore substation, its functions are the collection and distribution of electric power, control of the power flow, as well as the transformation and adjustment of the voltage. Since the substation needs to converge and transmit the power generated by the offshore wind turbines to the shore, its support structure, transformer, and switchgear appliances are very complex, and the operating environment conditions are much worse than those on land. The operation and maintenance are much more difficult, and the investment and operation costs are relatively high, so we should use the simplest method, that is, the most economic method, rather than the most energy coefficient method [5]. Whenever possible, try to avoid installing an offshore transformer platform. Such as the NoordZee wind power project, its capacity is 100 MW, and is 8 km away from the shore, the voltage level of the WF and the cable connected to the shore is 33 to 36 kV.

Large-scale OWFs far away from the shore might require more substations. Each sub substation is connected to the largest main substations, then stepped up or converted to high-voltage direct current (DC) for transmission to onshore substation, or the sub substations may be connected to a onshore substation.

The main electric connection of offshore substation (power collection system) can have a different structural form [6], as shown in Figure 6.2. Schemes are different in configurations of the main transformers (number and type), the serial number and set number, and the consideration of redundant of wind turbines. Decisive factors include the size of OWFs, and the connection mode to onshore main grid. The reliability and operating performance of different wiring schemes are different, so the costs (investment and operation) are also different.

Figure 6.2a shows a shared transformer and separate transmission system, so the wind turbines share the transformer and the transmission system consists of separate cables. It has two transformers and tow transmission lines. The general number of transformers and cables can be from one to four.

Figure 6.2b shows a two-winding transformer set, without redundancy and utility transmission lines, which has L series, and the set number can be from one to the series number.

Figure 6.2c shows a pair of independent transmission lines with redundancy that share a two-winding transformer. The wind turbine clusters are divided into several sets in series, and each set is integrated and connected to a two-winding transformer. Every two sets on the transformer high-voltage side is connected by a bus coupler to achieve redundancy. Each transformer cable is connected to each transformer.

Figure 6.2d is formed after the redundancy couplers of 6.2c are removed.

Figure 6.2e shows the wiring system using a three-winding transformer with a split winding, each split winding is connected to the wind turbine set. Every two sets of wind turbines are connected to a transformer and a transmission line. There is a redundant cable between two transmission lines.

Figure 6.2f shows the system using a three-winding transformer and a transmission line. Each three-winding transformer is used to connect two sets of wind turbines. Multiple transformers share a transmission cable. The redundancy is not considered.

Figure 6.2g shows a structure using the HVDC transmission system, the offshore substation is not equipped with transformer (nor the onshore substation), and only equipped with an AC/CD converter station.

Substations have different layouts. When designing the substation, prevailing wind direction shall be taken into consideration, as well as the transmission method, investment size, operation costs, overall reliability, and location-related issues, such as the depth of water throughout the region.

Equipment configuration of offshore and onshore substations is also different. The main electrical equipment of the offshore substation includes [6, 7]:

1) 110/35 kV main transformer, with adjustable tap. For substation using a DC transmission, an offshore converter station, including AC/DC converter, reactive compensation, and filtering devices are equipped.
2) A 35 kV switch station for the feed lines and local power supply;
3) A 35/0.4 kV transformer for local power supply;
4) 35 kV system grounding resistors;
5) 0.4 kV switch station system for local power supply;
6) 220 V-AC and 110 V-DC standby system;
7) Emergency standby diesel generator;
8) Control/monitoring system;
9) Mechanical vibration protection device (which trips all electrical equipment in case of ship collision)
10) Lighting, including emergency lighting;
11) Grounding system;
12) Lightning protection system;
13) Automatic fire alarm and fire suppression systems;
14) Cathodic protection;

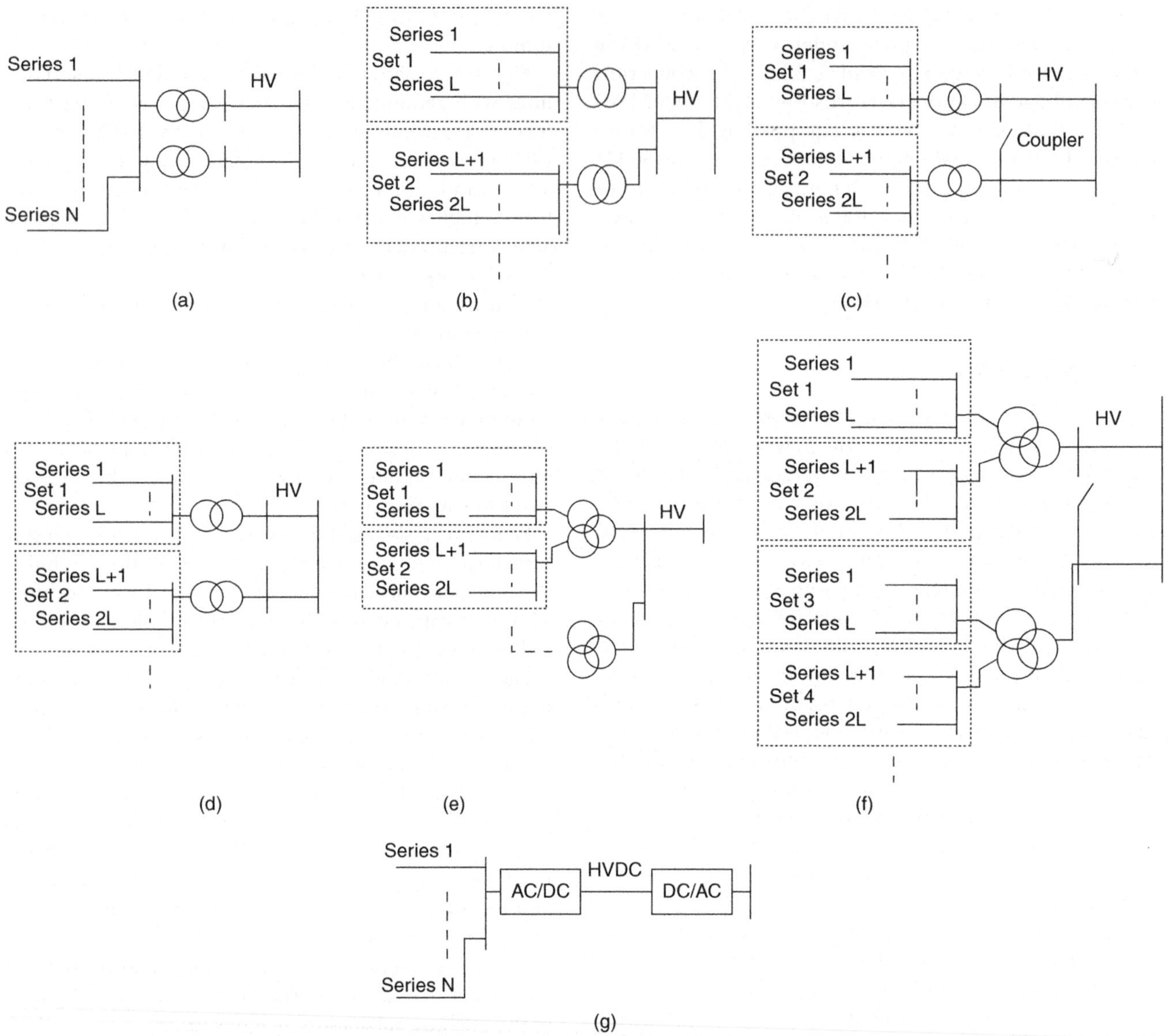

Figure 6.2 OWF main electrical wiring system.

15) Signal lights;
16) Telephone system, including emergency telephone system;
17) Crane;
18) Lifeboats;
19) Oil catcher (a tank of about 40 m^3, used for main transformer oil leakage).

If the offshore substation only has a main transformer, the operational risk is high. Transformer fault may cause outage of several months. To reduce risk, apart from the standard protection, monitoring, and control equipment, the transformer is also installed with a gas sensor in transformer oil, an intelligent fault monitor that can read out the composition of the gases. Gases generated by a flashover caused by transformer internal fault and the composition of the gas will indicate the types of the faults. So that early warning can be realized, informing operators the development conditions of the fault and the transformer fault and unplanned outage it may cause.

6.2.3 OWF Power Collection System

The task of OWF power collection systems is to converge the electricity generated by each wind generator to substation bus according to a certain rule, which mainly includes cable lines, switchgears, and a transformer platform. Now different structures for OWF power collection systems have been put forward, of which the costs, reliability, and system efficiency are different. Five basic structures of the OWF power collection system are listed below [4–8]:

1) **Radial structure**
Figure 6.3a shows a radial structure, which is also called a chain structure, wherein many wind generators are connected to a single cable in series. The maximum number of wind turbines on each feeder is determined by the generating capacity and the maximum rated capacity of the cable (the cable cross areas are different). This structure is economic, and simple to be controlled, of which the main cable is relatively short, and the cross section from the collection bus bar to the feeder tip cable can taper off. The main shortcoming of this structure is in its poor reliability, in case of cable or switch fault at the radial series collection terminal, all downstream wind turbines cannot output power. Denmark Horns Rev's OWF also adopts this structure, and many other OWFs are also recommended to use this structure.
2) **Star structure**
As shown in Figure 6.3b, the structure can reduce the rated capacity of the cable, and improve the overall safety of the WF, because the fault outage of a cable only affects a wind turbine. Voltage regulation along the cable between wind turbines may be better. However, the long diagonal cables and some short cables of large rated capacity may cause extra costs. In this structure, wind turbines in the center of the star structure require more complex switches. This star-shaped arrangement is suitable for WFs of which the wind direction changes frequently, but capture of wind power is not ideal, so the structure is rarely used.

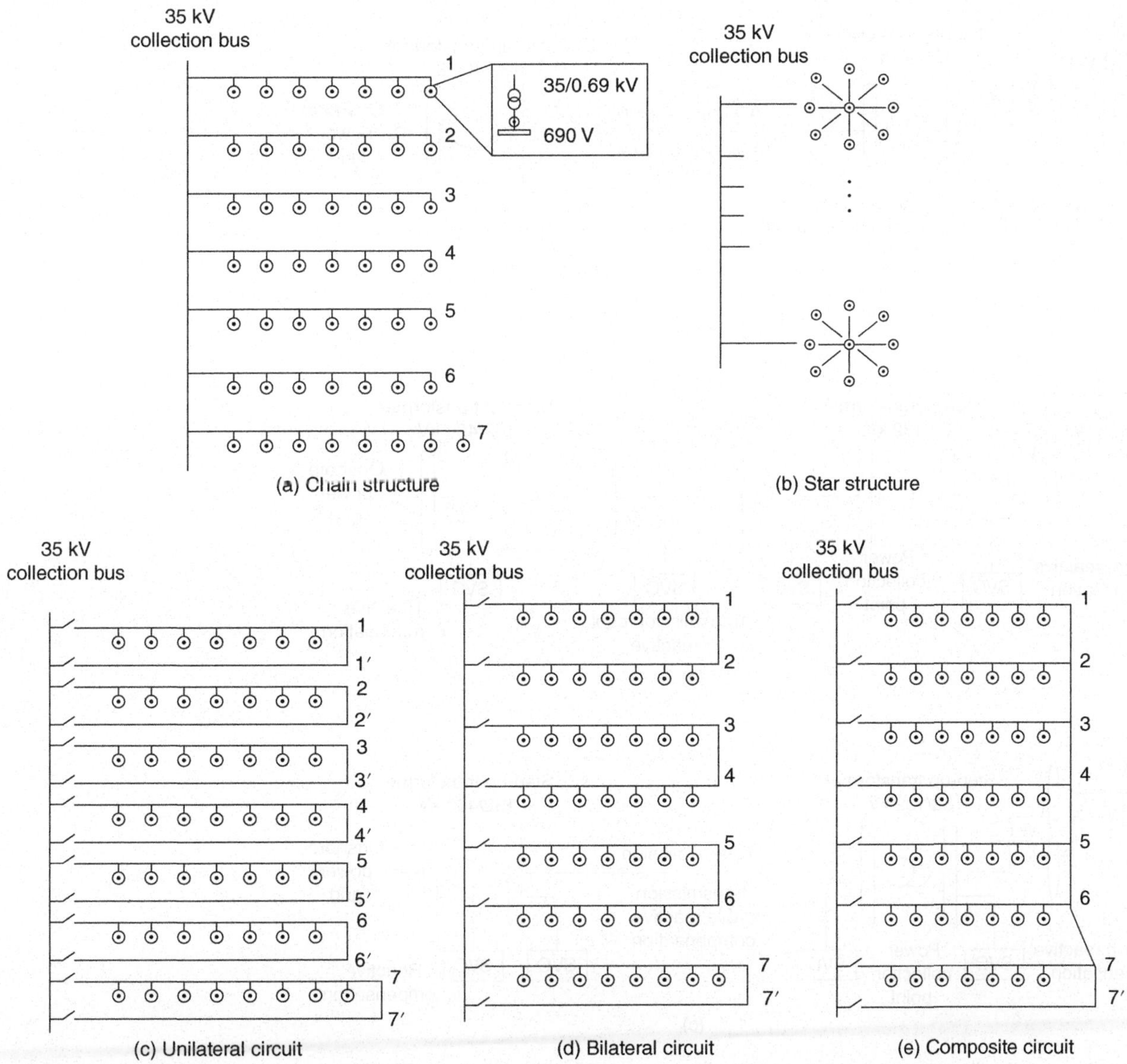

Figure 6.3 Structures of OWF power collection system.

3) **Unilateral circuit**
As shown in Figure 6.3c (take the first set as example), during normal operation, the switches of line 1 and 1′ are in closed state, and the active output generated by the wind turbines can flows from the two paths into the collection bus; if line 1 is in fault, the switch on the collection bus will trip off, and all power on line 1 is transmitted through 1′, that is, the unilateral circuit operation model is changed into chain model, at this moment, the maximum power flows through 1′ may be the sum of the power of all wind turbines. Compared to the radial structure, unilateral circuit improves the operation security through redundant power flow path (line 1′), but increases cable lengths and rated capacity of the cable while the number of wind turbines remain unchanged. Line 1′ has to enable the power of all the wind turbines on line 1 to flow through, so the costs are increased.
In unilateral circuit structure, redundant circuits are available during normal operation and fault period.
4) **Bilateral circuit**
As shown in Figure 6.3d, the tail wind turbines of series 1 and 2, series 3 and 4, series 5 and 6 are respectively connected, series 7 is connected in bilateral circuit structure. Take the first set as example, during normal operation, the switches of line 1 and 2 are in closed state, the active output generated by the wind turbines can flow from the two paths into the

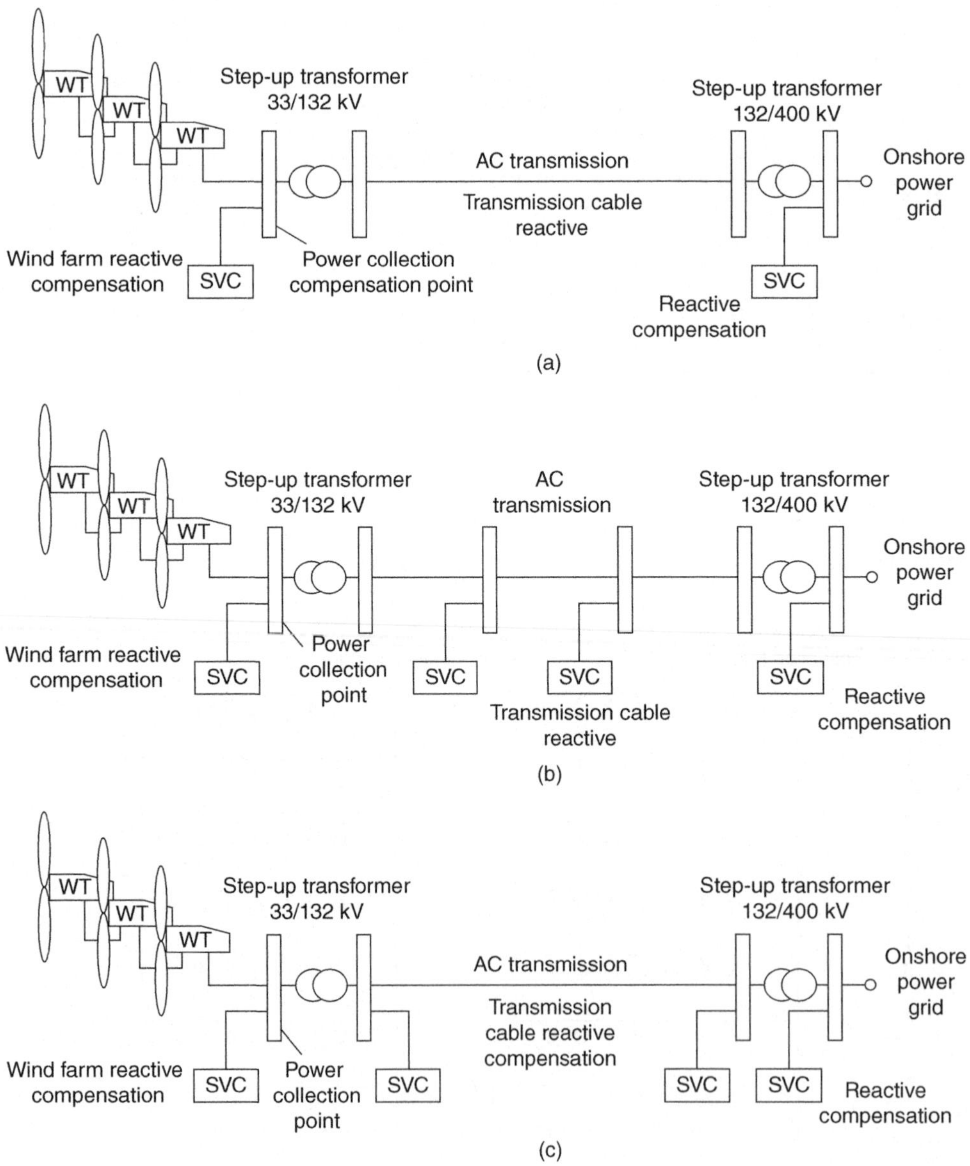

Figure 6.4 Transmission line reactive compensation. (a) The transmission cable has no reactive compensation; (b) Reactive compensation is provided along the transmission cable; (c) Reactive compensation is provided at both ends of the transmission cable.

collection bus; if line 1 is in fault, the switch on the collection bus of serial 1 will trip off, and all power is transmitted through serial 2, that is, the unilateral circuit operation model is changed into chain model, at this moment, the maximum power flows through series 2 may be the sum of the power of all wind turbines. In this structure, two series of cables are interconnected to back each other up ("mutual standby") and achieve redundancy. The rated capacity (cross-sectional area) of each series of cables has increased.

5) **Composite circuit**
As shown in Figure 6.4, starting from series 1, the tail wind turbines of two adjacent series are connected by cable, and the tail of series 7 is returned to the collection bus through a redundant cable. Redundant cable can transmit all output power of the fault series of the four-series structure, this structure is of high reliability. As long as we can ensure any of the four cables in normal operation, all active power generated by the wind turbines can be output. The possibility of two series in fault at the same time is very small, so we don't need to consider increasing the cable capacity and costs for this reason.

6.3 Redundancy Design of Collection System

The redundancy design of WF collection system aims to keep as many wind turbines connected to power grid during fault period as possible. Onshore WF collection systems generally do not take into account redundancies. But different from onshore WF collection systems, OWFs require much longer maintenance hours. For example, the estimated repair time for submarine cable fault is 720 h (in summer) and 2,160 h (in winter). Wind turbines may be connected into a radial network, which means a lot of wind turbines are connected on one feeder. The maximum number of wind turbines that can be connected to each feeder cable is determined by its maximum rated capacity. If the cable is damaged, the entire feeder will be outage, until the damage is repaired. To avoid this loss of maintenance and outage, OWF collection systems need to consider redundancy measure [4, 9].

Wind turbine feeder redundancy can be achieved by allowing the interconnection between feeders. For example, the last wind turbine of a feeder is connected to the last wind turbine of an adjacent feeder. Thus, the two feeders can back each other up. During normal operation, the connection switch between the two feeders is opened. If a fault occurs in one of the feeder, the connection switch will quickly close. For this reason, more equipment shall be added and the rated capacity of the equipment shall be increased, because the maximum rated capacity of each feeder (cable) shall be able to withstand twice the total capacity of all wind turbines on one feeder, or reducing the number of wind turbines connected to a feeder shall be considered.

At this stage, OWFs rarely consider or don't consider the redundancy of the transmission cable line, because the assumed cable fault probability and costs are lower than additional equipment costs. In order to reduce the possibility of cable damage, usually the cable is buried 1 to 2 m deep into the seabed to avoid becoming damaged by anchor and strong ocean currents. However, a fault of a cable between the OWF and onshore network will result in the loss of the entire WF. The economic consequences caused by this fault could be huge. It may take several months to fix the fault, depending on the availability of cable repair ship and the current location. It is difficult to avoid cables subjected to harm, particularly if the cable passes through large shipping channels. According to the report, when at anchor, a large ship can dig up to a depth of 13 m into the ground, but cables cannot be buried as deep as this. To this end, in order to achieve redundancy between the OWFs and the shore, standby cables are used. The standby cable passes through a different path to the shore. Apart from the high cost of connecting the second cable to the shore, the subsequent cable path may also be difficult, because of environmental conditions.

In short, for a small OWF (e.g., <100 MW), in view of its fault probability and that relevant costs might be lower than the costs of additional equipment, redundancy is not considered. But for large OWFs (e.g., >100 MW), where expected equipment maintenance periods are longer than for those on the shore, and the loss of system outage is expected to be heavy, it is necessary to consider redundant structure. At first, the operating experience of some large OWFs also shows that redundancy is necessary for transmission cables connected from the OWFs to the shore.

6.4 OWF Transmission System

Various schemes for WFs integrated into the power system described in Chapter 3 have their own applicability, and the grid-connection mode to be applied depends on the distance of the OWF to the shore, the scale of the WF, the existing onshore power grid, the grid development, and other specific conditions. For OWF transmission systems (connected to on-shore power grids), the main problems to be solved are network layout, appropriate transmission voltage levels, using cables or overhead lines for transmission, high-voltage AC or DC for power transmission, reactive power compensation schemes, and the number and location of onshore and offshore substations.

6.4.1 AC Transmission System

AC power transmission is the preferred transmission way for OWFs, because the main onshore power grid is high-voltage AC power grid. When the OWF is relatively small (around 10 MW) and the WF is near the coast (10 to 36 km), OWFs are generally connected to onshore grids through AC cables and static var compensators (SVC). The transmission systems of OWFs using AC transmission for grid-connection include AC high-voltage submarine cables, WF side transformers, and thyristor-controlled reactor (TCR) type static reactive power compensation units at both ends of the AC cable, and onshore transformers. Usually the voltage level of the OWF is 35 to 110 kV.

Cross-linked polyethylene (XLPE) is used for modern submarine cable insulation. Cables at the voltage level of 150 kV are used for the connection of WFs. At this voltage class, the easily installed triple core cable can be used. The maximum voltage of XLPE submarine cables available in the current market is 245 kV. Application constraints of submarine cable are mainly rooted from the capacitance characteristics of the cable itself. The amount of reactive power generated by the cable depends on the length of cable and square of the system voltage, and the typical value range of XLPE cable are 100 to 150 kvar/km (33 kV), 1,000 kvar/km (132 kV), 6 to 8 Mvar/km (400 kV). The reactive current it generates will limit the current capacity available when transmitting active power. Technically, as long as the reactive power generated along the line is compensated, the transmission distance can be increased, but the effect of extending transmission distance only by installing compensation reactor at the line terminal is limited.

The main features of using AC transmission for grid-connection are: ① the transmission line using three-phase cable is of high cost and large loss, but power electronic devices are not required at both ends, so the cost of the substation based on the power frequency is relatively low; ② since AC high voltage power transmission cable has a lot of charging reactive power, the inductive reactive compensation equipment of large capacity shall be installed for the substations at both ends. Since wind power fluctuations will cause great fluctuations in voltage, generally dynamic reactive compensation equipment such as SVC are used; ③ in order to maximize the use of wind power, usually VSCF wind turbines are used, so it is necessary to equip sophisticated power electronic devices at the rotor side or stator side; ④ there are stability problems to connect with onshore AC grids. The AC power grid where the wind turbine is connected is synchronously connected with the main grid, so any fault in the power grid will spread to other power grid, which also puts constraints on the transmission capacity and distance; ⑤ due to cable capacitance, resonance can occur between the onshore and offshore power grid, leading to voltage waveform distortion.

When using AC transmission lines (cables) to connect OWFs and onshore power grids, the problem of reactive power compensation shall be considered. Transmission lines with and without reactive compensation are shown in Figure 6.4a, b, and c. Figure 6.4a shows the transmission line without reactive compensation. Figure 6.4b shows the reactive compensation method along the transmission line, which is difficult to realize. Figure 6.4c shows the compensation provided on the substation bus bar at both ends. The reactive compensation plan shall be determined through power flow analysis. The allowable deviation range of transmission line voltage can be ±0.05 p.u.

Figure 6.5 shows the constraints on AC power transmission through power flow analysis. For three

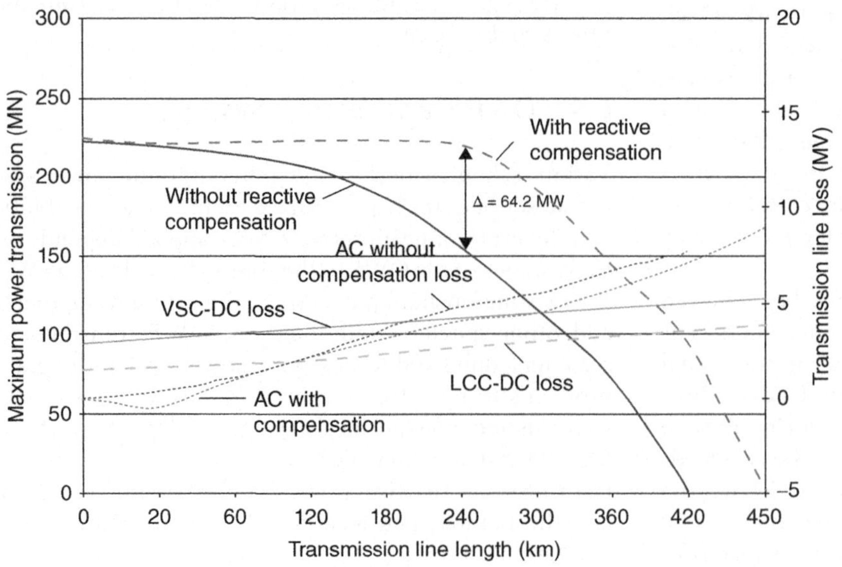

Figure 6.5 Transmission limit of AC transmission cable (132 kV).

single-core (XLPE) AC cables, the maximum transmitted power at 132 kV is 231 MVA. As can be seen from the figure, since the line reactor absorbs excessive reactive power generated by the cable, it is more likely to transmit more power on long distance, without having to increase cable size or MVA ratings.

In the case of AC power transmission without reactive power compensation, as the transmission distance increases, the transmission power is reduced at the beginning; in the case of AC power transmission with compensation, the maximum active power transmission distance can be up to 240 km.

From the perspective of transmission power loss, there is no large difference between AC transmission with and without reactive power compensation. Within the distance of 120 km, AC transmission with compensation has no obvious advantage of compensation. As the distance becomes longer, the transmission line with compensation has certain advantages.

The losses of the two DC power transmission ways are also shown in the figure 6.5. By comparison of the losses of AC and DC power transmission, it can be seen that, the loss of AC transmission increases more quickly with the increase of the length of the transmission line (steep slope), whereas DC transmission increases slower with the increase of the length of the transmission line (slope flat). This is because AC transmission loss mainly depends on resistance, whereas the loss of converter station takes up a large proportion for DC transmission loss, and the switching loss of the VSC-HVDC converter station is more serious. In this example, the distance between the intersection points of losses curves of AC and DC transmission is roughly equal to the maximum distance of AC transmission with compensation.

Figure 6.6 shows the reactive compensation capacity at both ends without reactance compensation on the transmission cable. It can be seen that within the distances of 20 to 60 km, the voltage levels of the WFs and onshore power collection point can be maintained within the qualified range without compensation.

Figure 6.7 shows that there are two reactors on the cable line (respectively, near the offshore and onshore power collection points). Because of the effect of line reactor, the maximum power transmission distance can be up to 240 km. Onshore compensation only requires the reactor to absorb reactive power after the length exceeding 240 km.

The above analysis is made from a technical perspective, and economic evaluation shall be conducted for the determination of the final compensation scheme. Taking into account investment costs, operating costs (loss) and other costs, the average unit transmission costs within the lifetime of OWFs can be obtained.

One way to increase AC transmission distance is the low frequency transmission, such as 16⅔ Hz. This requires special variable frequency equipment, as well as a larger transformer and reactor. Another way is the multi-phase transmission.

6.4.2 Transmission System Based on PCC-HVDC (or LCC-HVDC)

With the increase of transmission distance and scale of OWFs, the use of HVDC technology to connect WFs and onshore grids for transmission becomes necessary, especially for systems with the rated capacity of higher than 500 MW. The advantages of HVDC using thyristor-based phase commutated converters (PCC), also known

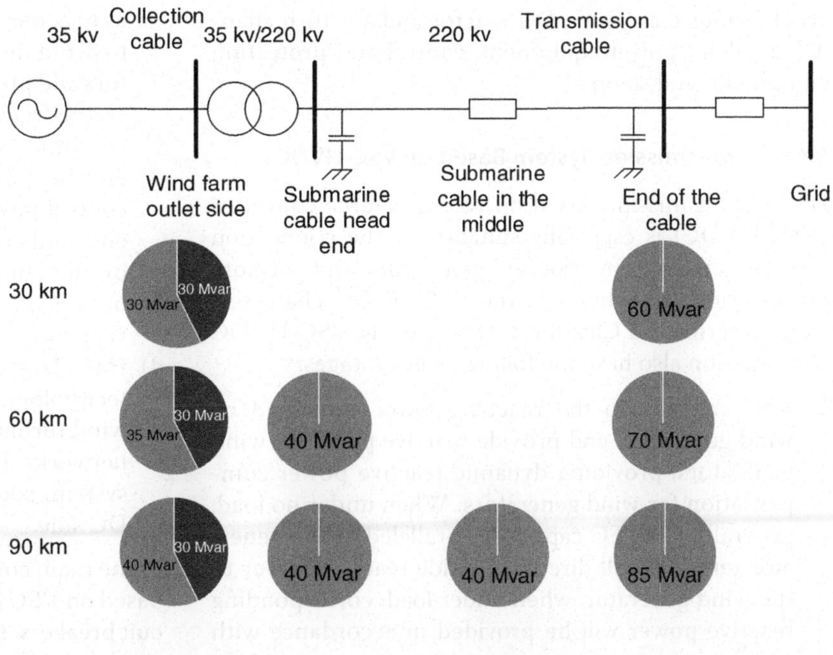

Figure 6.6 The transmission line has no reactor, and requires offshore and onshore compensation.

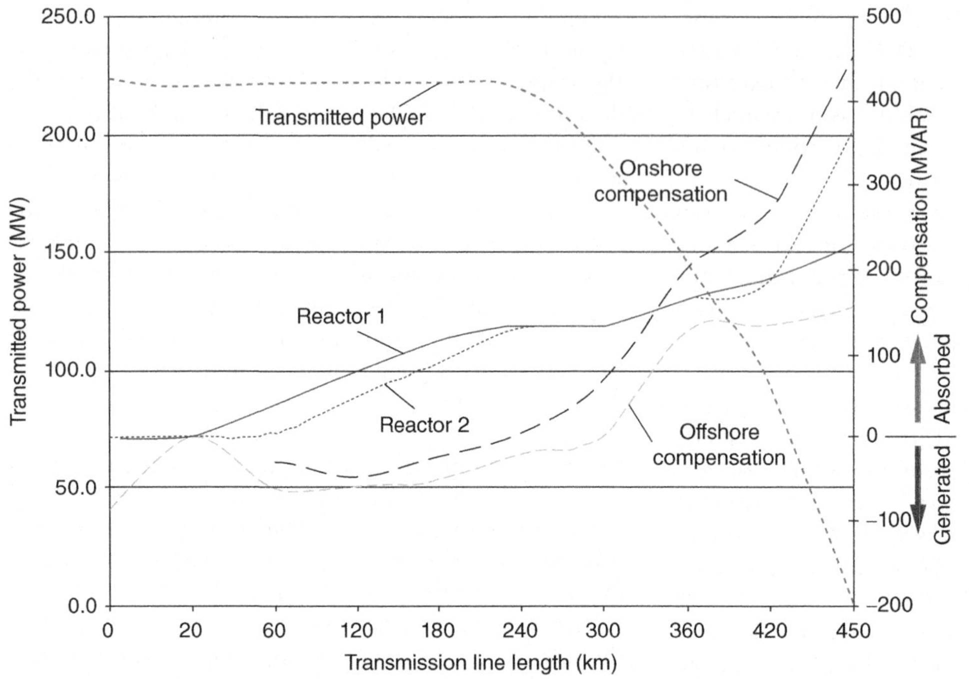

Figure 6.7 The transmission line has reactor, and requires offshore and onshore compensation.

as HVDC based on line commutated converters (LCC-HVDC) or thyristor-based HVDC (SCR-HVDC), are obvious. Although the construction costs of converter stations at both ends of the HVDC transmission line is more expensive, from the point of view of costs of the whole project, including cable and wind turbine costs, the increased cost of using this transmission method, compared with AC transmission, is not so prominent.

An OWF HVDC transmission system based on PCC technology consists of the following parts: DC and AC filter, converter transformer, thyristor valve-based converter, smoothing reactor, capacitors and STATCOM or synchronous condenser, DC reactor and a return channel, auxiliary power equipment, control and protection equipment, and so on.

6.4.3 Transmission System Based on VSC-HVDC

HVDC transmission based on voltage-source converter (VSC-HVDC) is especially suitable for the connection of OWFs with asynchronous generators and onshore power grids. In addition to the VSC-HVDC characteristics described in Chapter 3, OWFs using VSC-HVDC transmission also have the following advantages:

1) VSC can absorb the reactive power generated by wind generators and provide reactive power to wind generators, providing dynamic reactive power compensation for wind generators. When under no load, generally, the PFC capacitor paralleled at the generator terminal will directly provide reactive power to the wind generator; when under load, corresponding reactive power will be provided in accordance with the size of the wind generator power output, and this dynamic reactive power can be provided by controlling the VSC on bus side. Even under the condition that power generation and load are changing fast, it can also add great stability margin to the AC power grid, as well as eliminate voltage flicker caused by turbulence and tower shelter.
2) Can achieve maximum wind energy capture through frequency control. When wind turbines are capable of variable speed control, the rotating speed of wind generators can be adjusted based on the wind speed by controlling the VSC at wind generator bus side, so as to achieve maximum wind energy capture. In order to avoid the magnetic flux invariance of wind generators and prevent saturation, when changing wind generator frequency, the wind generator output voltage amplitude shall also be changed at the same time.
3) Flexible power control characteristics. Can precisely control power exchange at the sending and receiving end, and any changes in wind energy can be detected by the control system, thus enabling the input and output power of the generator to achieve a balance very soon.
4) Has black start capability. A HVDC line using VSC technology can make the asynchronous motor of wind turbine reverse start and connected to a "black" network. This capacity requires an energy storage system, such as installing batteries or SMES on the DC side.

The main components of offshore transmission system based on VSC include [10, 11]: VSC converter station circuit breakers, harmonic filter on system side, connection transformer, harmonic filter on converter station side, VSC unit, VSC DC capacitor, DC harmonic filter, DC reactor, DC cables or over-head transmission lines, and auxiliary power equipment.

VSC-HVDC transmission is very easy to implement parallel extensions, and the grid-connection has many forms. Figure 6.8 illustrates three topological structures.

The grid-connection topological structure of the WF shown in Figure 6.8a is suitable for the connection of existing or new WFs to a specific power level, which depends on the DC system topology, in this case, the entire WF will operate at the same speed and frequency. Since OWFs generally cover a larger area, the wind speed of the same WF will change greatly, thus, the topological structure using the radial series for connection shown in Figure 6.8b can increase overall energy capture. In this case, the generators of each series can operate on their own speeds and frequencies, and are connected to the power grid through a separate DC transmission system. This topology can be chosen as the main connection mode of large WFs. The last operating mode is, all series are converged by a DC bus, and connected to the topology of the power grid through single converters, as shown in Figure 6.8c, and the rated power of the converters is the rated power of the whole WF. This kind of connection is suitable for the increase of WF capacity.

6.4.4 Parallel Transmission Systems Based on PCC-HVDC and VSC-HVDC

Thyristor-based (SCR) PCC-HVDC and IGBT-based VSC-HVDC are two DC transmission concepts, and they have different advantages and disadvantages. The way to get an idea transmission system with large transmission capacity, good transmission quality, which suitable for OWFs is to combine the two HVDC transmission systems together, as shown in Figure 6.9.

The two different concepts of HVDC are connected through a circuit breaker to the HVDC transmission line on DC side, and on AC side, the two HVDC are connected to the public grid or OWF through a transformer. During the maximum power transmission period, the circuit breaker has to disconnect the VSC-HVDC from the transmission line, because the DC voltage level of

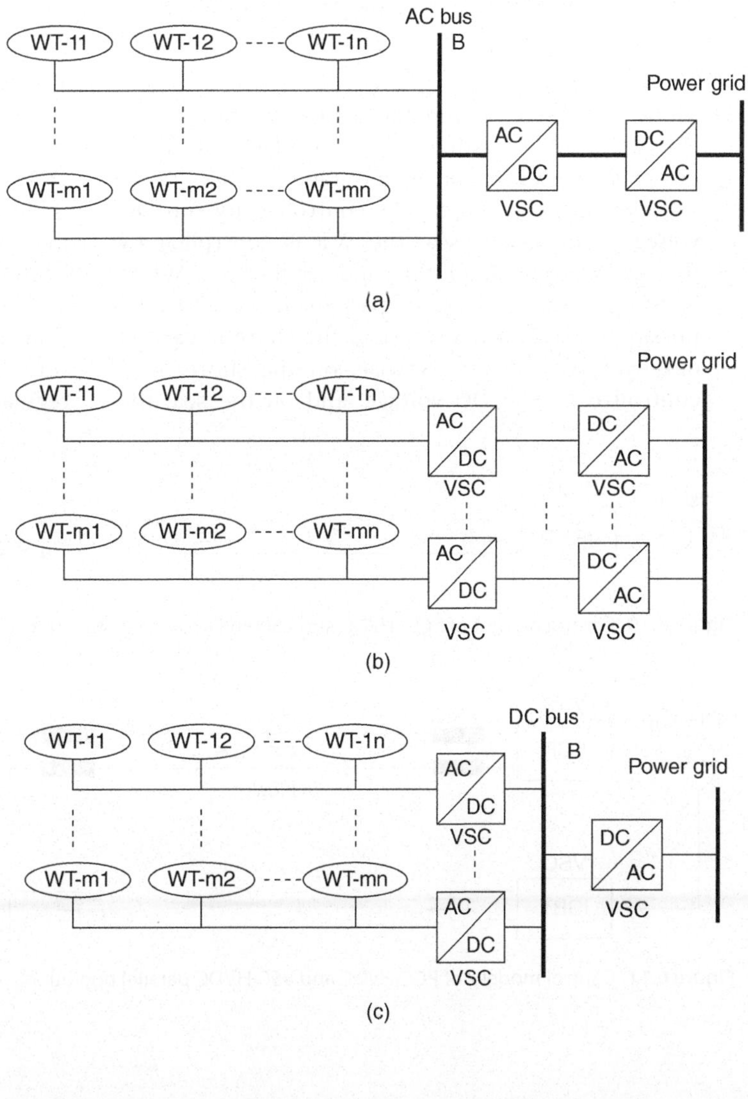

Figure 6.8 Topological structure of OWFs using VSC-HVDC for grid connection.

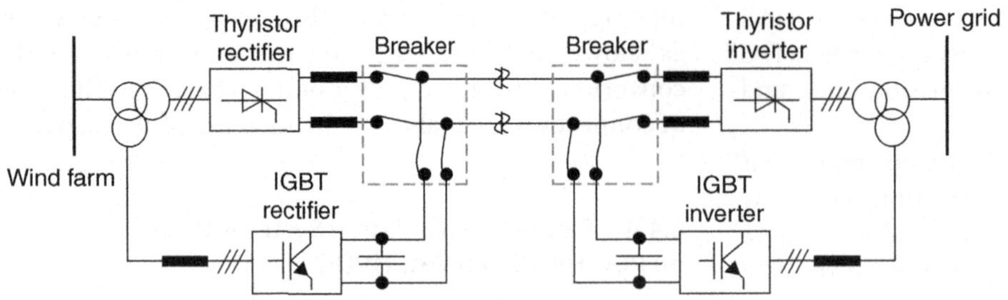

Figure 6.9 Thyristor-based PCC-HVDC and IGBT-based VSC-HVDC.

PCC-HVDC is usually much higher than that of the VSC-HVDC. In this case, the IGBT converters are operating as an active filter and a reactive power compensator for the thyristor unit.

There is a very interesting point that, when starting up large number of wind turbines, starting energy is required to activate the auxiliary equipment, such as cooling system, oil system, control system, and so on, where VSC-HVDC can supply power to OWF during black-start of the power grid. When there is no wind, it can supply power to the heating, cooling, and control systems of WF.

To optimize the parallel operation based on PCC-HVDC and VSC-HVDC, it is necessary to determine the different operating modes. Here are two modes of operation.

1) Mode 1: WF starting and part load generation mode. When the WF is starting up, VSC-HVDC will supply power to the WF and create a voltage for the offshore AC grid. On the shore, it is controlled by the DC voltage, and on the sea, the WF is controlled by the AC voltage (amplitude and frequency). When in the generation mode of part load, power transmission between the OWF and the shore is carried out by VSC-HVDC. Likewise, on the shore, it is controlled by the DC voltage, and on the sea, the WF is controlled by the AC voltage (amplitude and frequency). In mode 1, PCC-HVDC stays open, and only VSC-HVDC is in operation, as shown in Figure 6.10.
2) Mode 2: Full load generation mode. All power is transmitted through PCC-HVDC from OWFs to the shore, and voltage of the DC side is set to maximum. On shore, it is controlled by the DC voltage, and at sea, the WF is controlled by the DC current according to the wind speed. VSC-HVDC is disconnect from the DC line and operating as a SVC (reactive compensator), as shown in Figure 6.11.

6.4.5 Multi-Terminal HVDC Applications

Two multi-terminal DC suitable for OWF applications are introduced here, that is, multi-terminal DC based on the voltage source (MVSC-HVDC) and multi-terminal DC based on current sources (MCSI-HVDC).

Electricity generated from OWF must be sent to the shore through cables. When the WF is far away from the shore, the DC transmission is the best choice. VSC-HVDC concept is ideal for large OWFs, because the sinusoidal pulse width modulated voltage source converters (SPWM-VSCs) can provide power and voltage regulation and control for the WFs.

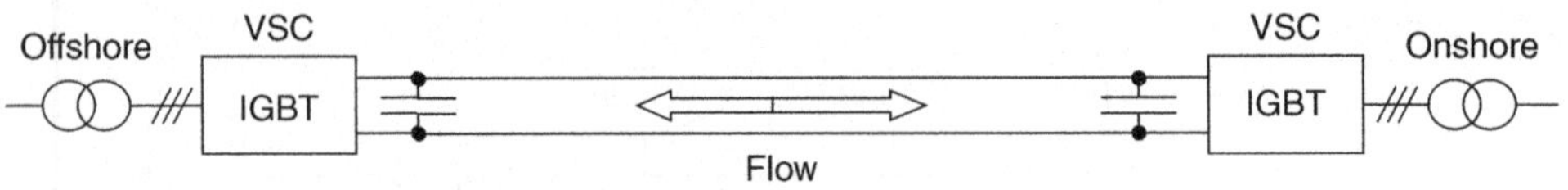

Figure 6.10 Control mode 1 of PCC -HVDC and VSC-HVDC parallel operation.

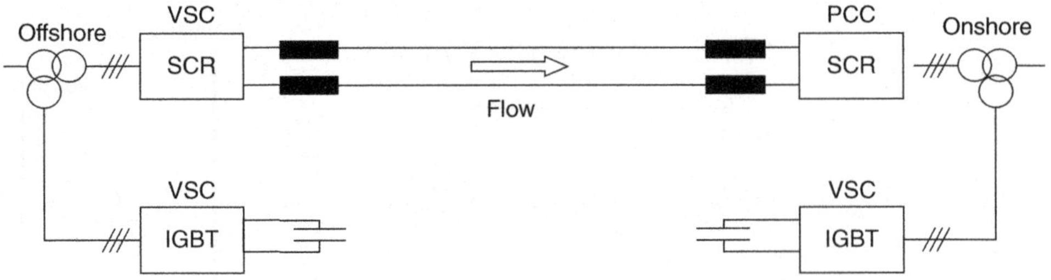

Figure 6.11 Control model 2 of PCC-HVDC and VSC-HVDC parallel operation.

6.4.5.1 Multi-Terminal Voltage Source High-Voltage DC (MVSC-HVDC)

The schemes of multi-terminal voltage source high voltage direct current transmission system for OWFs are shown in Figure 6.12, and the wind turbine-synchronous generator is connected to the pulse-width modulated voltage source converters (PWM-VSC). Four VSCs are connected to DC bus through the DC network. On the AC side of the VSC (WF collection network), each VSC is connected with a harmonic filter at the high voltage side of the transformer. A six-IGBT bridge circuit is applied to each voltage source converter. Phase-locked loops (PLL) enable PWM-VSC to synchronize with its wind turbines.

Due to the variability of wind speed at different locations, each wind turbine will run at different frequencies and different voltage amplitude values. However, PWM-VSCs is an ideal current source at the DC side, as its DC terminal can be directly paralleled to the DC bus. Each PWM-VSC injects direct current into a DC power system with the voltage of U_{dc}.

The main functions achieved by each PWM-VSC are: 1) controlling the speed of wind turbine, so that it can automatically operate at the optimal wind speed to obtain optimal power; 2) making voltage amplitude of the synchronous generator that is related to the speed match with the DC bus voltage.

PWM-VSC responses by tracing the active power reference set value P_{ref} and AC voltage amplitude reference set value U_{ref}. Reference set values can be obtained by looking up the preset data sheet P_{ref}—ω_s and U_{ref}—ω_s. The output frequency of the phase-locked loop (PLL) ω_s serves as a pointer for lookup table.

The structure and working principles of each part of the system are outlined below [12]:

6.4.5.2 The PWM-VSCs Used for HVDC

1) **Voltage source inverter**
   Figure 6.13a and b show the circuit diagram and the simplified block diagram of a pulse-width modulated voltage source converter circuit. Assume that the DC voltage of the DC bus U_{dc} is adjustable, that is regulated by dedicating one PWM-VSC to operate as a DC voltage regulator. U_{dc} can be obtained by the conversion of three-phase AC voltage $u_{SA}(t)$, $u_{SB}(t)$ and $u_{SC}(t)$ by the ON-OFF switching of the IGBTs in each arm of the bridge converter. The typical voltage waveform, such as u_{SA} (t), is shown in Figure 6.14a.
2) **PWM**
   The voltage waveform $u_{SA}(t)$ shown in Figure 6.14a is the pulse-width modulation form of the input signal $m_A(t)$ as shown in Figure 6.14b. At a high sampling frequency, after filtering the switching noise, the output fundamental frequency component of $u_{SA}(t)$ is the linear amplification form of input $m(t)$. Thus, the grounded voltage of phase a is (Formula (6.1)):

$$u_{SA}(t) = G\, m_A(t) \tag{6.1}$$

Where, G is a linear gain. Next, consider only the fundamental frequency components of three-phase AC voltage $u_{SA}(t)$, $u_{SB}(t)$, and $u_{SC}(t)$.

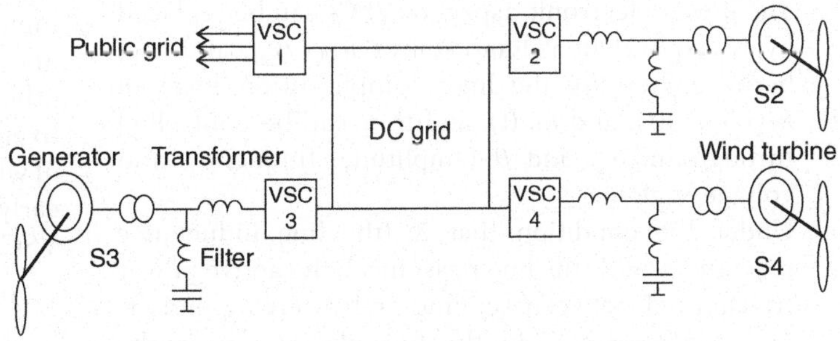

Figure 6.12 Multi-terminal HVDC system based on PWM-VSC.

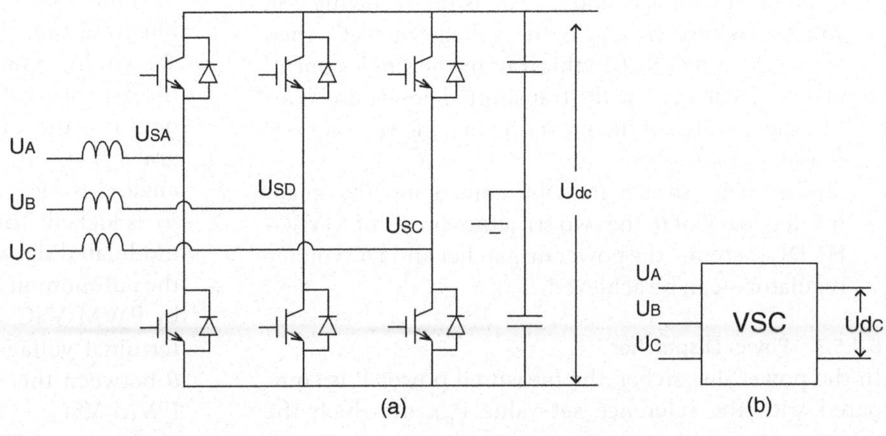

Figure 6.13 Pulse-width modulated voltage source converter circuit. (a) PWM-VSC; (b) Simplified block diagram of PWM-VSC.

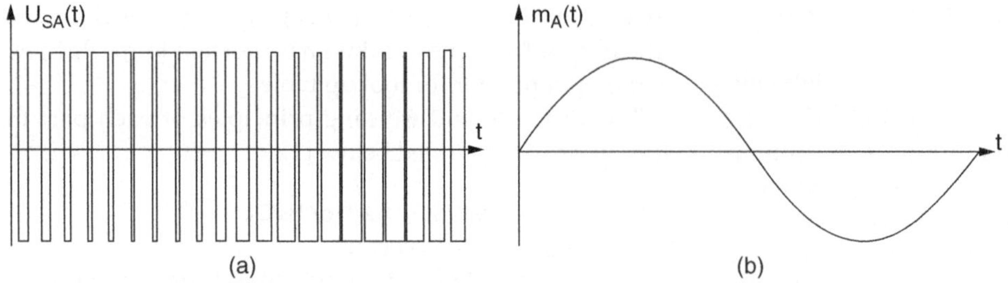

Figure 6.14 PWM-VSC voltage waveform. (a) Typical voltage waveform; (b) Input modulation signal m (t).

3) **The amplitude, frequency, and phase angle control of PWM-VSC**
Here the PWM-VSC is considered as a black box with terminal voltages u_A, u_B, and u_C, as shown in Figure 6.13b. What behind the filter inductance of each phase are voltage u_{SA}(t), u_{SB}(t), and u_{SC}(t), which are actually three linear amplifier outputs (with constant gain), of which the inputs are the modulating signals m_A(t), m_B(t), and m_C(t).
If the modulated signal generated on the electricity has the following forms (Formula (6.2)):

$$m_A(t) = M\sin(\omega_s(t) + \theta) \tag{6.2}$$

Then, the phase a voltage of PWM-VSC is (Formula (6.3)):

$$u_{SA}(t) = GM\sin(\omega_s(t) + \theta) \tag{6.3}$$

Similarly, we can obtain $m_B(t) = M\sin(\omega_s(t) + \theta - 120°)$ and $m_C(t) = M\sin(\omega_s(t) + \theta - 240°)$, thus balanced three-phase voltages u_{SA}(t), u_{SB}(t), and u_{SC}(t) are generated from the PWM-VSC.
4) **Control of three-phase voltage source**
The power electronic aspect of VSCs can be replaced by its equivalent voltage source u_{SA}, u_{SB}, and u_{SC}, because they are the linear amplification form of m_A(t), m_B(t), and m_C(t), and they can be controlled through m, ω_s, and θ (amplitude, frequency, and voltage angle).
Under the condition that X (the line inductance reactance) ≫ R (the line resistance), the active power transmitted between sending and receiving ends is $U_S U_R \sin(\delta_S - \delta_R)/X$, in the formula, $U_S\angle\delta_S$ is the sending end voltage, and $U_R\angle\delta_R$ is the receiving end voltage. Where $U_R\angle\delta_R$ is the voltage of SVC, then U_R = GM and $\delta_R = \theta$. In this way, by the single control of θ, SVC can reverse the transmitted power and control the amount of the power by making $(\delta_S - \delta_R) < 0$ or $(\delta_S - \delta_R) > 0$.
Studies have shown that, by controlling the negative feedback of θ, the two structure boxes of MVSC-HVDC system—the power dispatcher and DC voltage regulator—can be achieved.

6.4.5.3 Power Dispatcher

In the power dispatcher, the measured power P is compared with the reference set value P_{ref}, of which the error is applied to the negative feedback of θ after going through the proportional-integral box (P-I), so the error is zero. P_{ref} can be either positive (rectifier) or negative (inverter).

6.4.5.4 DC Voltage Regulator

At least one VSC of the MVSC-HVDC system has to be DC voltage regulator. The measured DC bus voltage is compared with the reference set value U_{DCref}, of which the error is applied to θ after going through the proportional integral box (P-I), in order to maintain the charge on the DC capacitor, adequate AC power is rectified or inverted to make the error zero. There is no doubt that DC voltage regulator is a power slack to ensure the balance of the power in DC network.

6.4.5.5 AC Voltage Amplitude Control

The AC voltage amplitudes u_{SA}, u_{SB}, and u_{SC} are controlled by M.

6.4.5.6 Feedforward and Feedback Control

Figure 6.15 shows the block diagram of which the AC terminal voltage, current, and frequency are measured at one end, and M, ω_S, and θ are the inputs of the VSC at the other end. The main purpose of control between them is to ensure that: ① VSC frequency matches the frequency of wind turbines; ② AC terminal voltage amplitude is linearly proportional to ω_S; ③ the power absorbed by the VSC corresponds to the optimal power.

1) **Phase-locked loop**
The angular frequency of a wind turbine is measured by phase-locked loop (PLL). Some manufacturers integrate the PLL into VSC system. The PLL input shown in Figure 6.15 is PWM-VSC AC terminal three-phase voltage (u_A, u_B, and u_C), and the PLL output is the angular frequency ω_s and its time integral $\xi = \omega_S t$. As its name implies that the PLL output angle ξ is locked. For example, the phase of voltage u_A is locked. This way, when $\Xi = \omega_s t$ is applied to the modulated signal, such as in $m_A(t) = M\sin(\omega_s(t) + \theta)$, the autonomous voltage (u_{SA}, u_{SB}, and u_{SC}) generated by PWM-VSC synchronize with the wind generator terminal voltage (u_A, u_B, and u_C). The phase angle θ between them and M are the control elements of PWM-VSC.

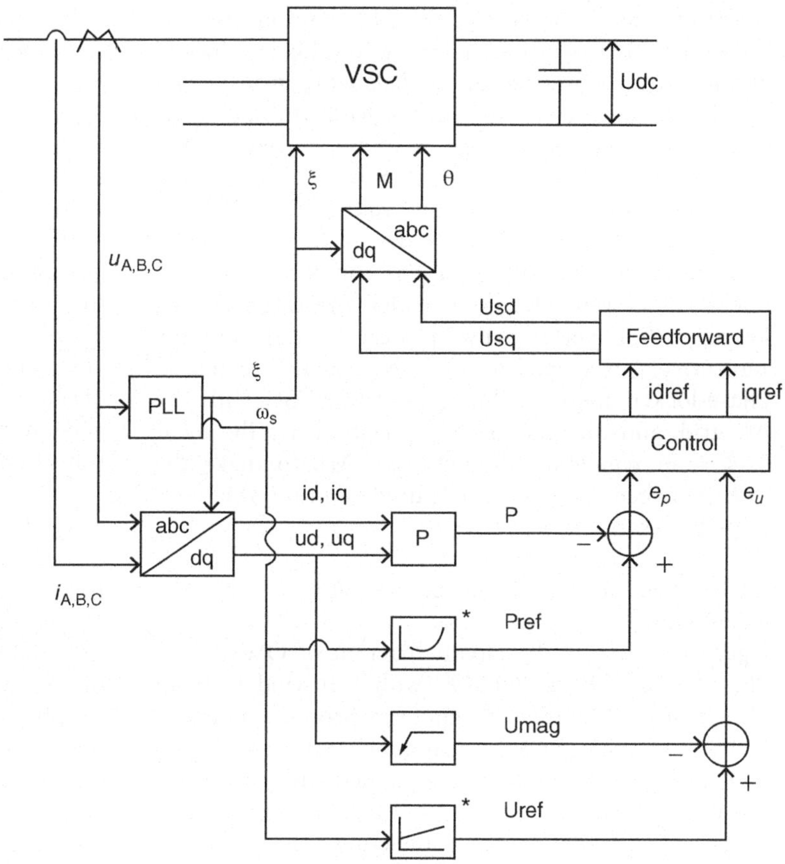

Figure 6.15 Feedforward and feedback control (* refer to obtain the corresponding variable values by table lookup).

2) **Coordinate transformation from a-b-c to d-q**
When PLL provides $\xi = \omega_S t$, $\sin\omega_s(t)$ and $\cos\omega_s(t)$ (based on table lookup), digital signal processor (DSP) will calculate to achieve coordinate transformation from the coordinate a-b-c to the reference coordinate d-q for voltage and current measurement. Thus, (u_A, u_B, u_C) and (i_A, i_B, i_C) is respectively converted into (u_d, u_q) and (i_d, i_q). In the d-q coordinate, current (i_d, i_q) tracks the reference current (i_{dref}, i_{qref}). The rough rule in general is, the reference active power p_{ref} or i_{dref} is tracked by using θ, and the reference reactive power Q_{ref} or i_{qref} is tracked by using M.
3) **Current source controller**
PWM-VSC first is a three-phase voltage source, which will generate a large devastating current in short-circuit faults. Many feedforward schemes enable the VSC to track current references (i_{dref}, i_{qref}), so that the short-circuit fault does not exceed the reference value. Basically, they consist of the known AC terminal voltages (u_d, u_q) and the voltages calculated by PWM-VSC (u_{sd}, u_{sq}), and the voltage drop generated by currents (i_{dref}, i_{qref}) on Z_F, wherein Z_F is the impedance between voltages (u_{sd}, u_{sq}) and the voltages at the measurement point (u_d, u_q).
4) **AC voltage amplitude feedback control**
The measured AC voltage amplitude value is $U_{mag} = \sqrt{u_d^2 + u_q^2}$. When the reference voltage set value is U_{ref}, the error signal is $e_u = U_{ref} - U_{mag}$. Through i_{qref} in the negative feedback (proportional and integral), the error can be made zero.
5) **AC active power feedback control**
The measured active power is $P = u_d\, i_{d+}\, u_q\, i_q$, where the reference power is P_{ref}, and the error signal is $\varepsilon_p = P_{ref} - P$. Through i_{dref} in the negative feedback (proportional and integral), the error can be made zeor.
6) **Inverse transformation from d-q to a–b–c**
If the impedance Z_F, terminal voltage (u_d, u_q) and current references (i_{dref}, i_{qref}) are known, we can calculate the voltage of PWM-VSC (u_{sd}, u_{sq}). The d-q axial voltages are inversed to (u_{SA}, u_{SB}, and u_{SC}) in an a-b-c coordinate axis. In order to produce this voltage output, the modulating signal of PWM-VSC (m, θ) shall also be brought under control.

6.4.5.7 DC Voltage Regulator

Another one is PWM-VSC, operating as an inverter, which is connected to the DC bus, and converge all direct currents injected by the wind generators. The AC side of VSC is connected to the public AC grid. Because wind power is not the main part of the system, it is considered lowering load demand of the system. Handling the random characteristics of wind power is slightly different from handling the random characteristics of load.

The inverter's task is to enable the wind power collected directly used to AC power grid. This is achieved by the current absorbed by the balancing inverter and the direct current injected by the wind turbines. Any imbalances will become the charging current of the DC capacitor between collection buses, which will increase or decrease the DC bus voltage. Zero balance is maintained by the inverter DC voltage regulator. The measured DC voltage U_{dc} is compared with the reference set value U_{dcref}, of which the error is used for i_{dref} (θ) (inverter active power control), and the zero error can be realized through negative feedback. Phase-locked loop synchronizes PWM-VSC with the AC grid, and the AC voltage regulation use the feedback loop regulation through i_{qref} (M), to make the error between AC voltage amplitude and the reference value be zero.

6.4.5.8 Multi-Terminal Current Source HVDC (MCSI-HVDC)

Figure 6.16 shows the circuit diagram of OWFs [13]. The WF capacity is 200 MW, with 100 wind turbines (direct-drive permanent-magnet synchronous generator (PMSG) of 2 MW, and the generator terminal voltage is 4 kV. The generator system is equipped with converters of full-rated capacity, so the wind turbines can operate at variable speeds. In the connecting scheme shown in Figure 6.16, the generator converter is not required, because converters at DC transmission line side can implement similar functionality.

The speed of all wind turbines in the same set is the same. When in variable speed mode, wind turbines of each set can operate at the optimal speed. The speed of wind turbines can be regulated by the controller of the current source inverters (CSI), and the reference speed and frequency can be exported according to the optimal average speed of each set.

The application of MCSI-HVDC allows only a few generators connected to a single converter, which means the speed of each wind turbine can be better regulated. Number of wind turbines in a set is determined by the efficiency of the wind turbine set and the rated power of the transmission system. Since CSIs are used, the generator AC Ig is directly related to the DC (power transmission) current, that is (Formula (6.4)):

$$I_g = \frac{M_{mg} I_{dcg}}{\sqrt{2} N_g} \tag{6.4}$$

In the equation, I_g is the rated generator current (RMS); I_{dcg} is the rated DC current(RMS); $M_{mg} \approx 0.95$ is the nominal value of modulation factor; N_g is the number of generators in one set. In formula (6.4), in order to simplify calculations, it is assumed that all the generators in the set are operate under the same power.

Table 6.1 lists parameters of topologies using different electrical systems shown in Figure 6.16. For Figure 6.16, Formula (6.4) can be used to study the size of wind turbine sets.

If a set of smaller size is selected (smaller N_g), more converters and higher generator insulation level will be

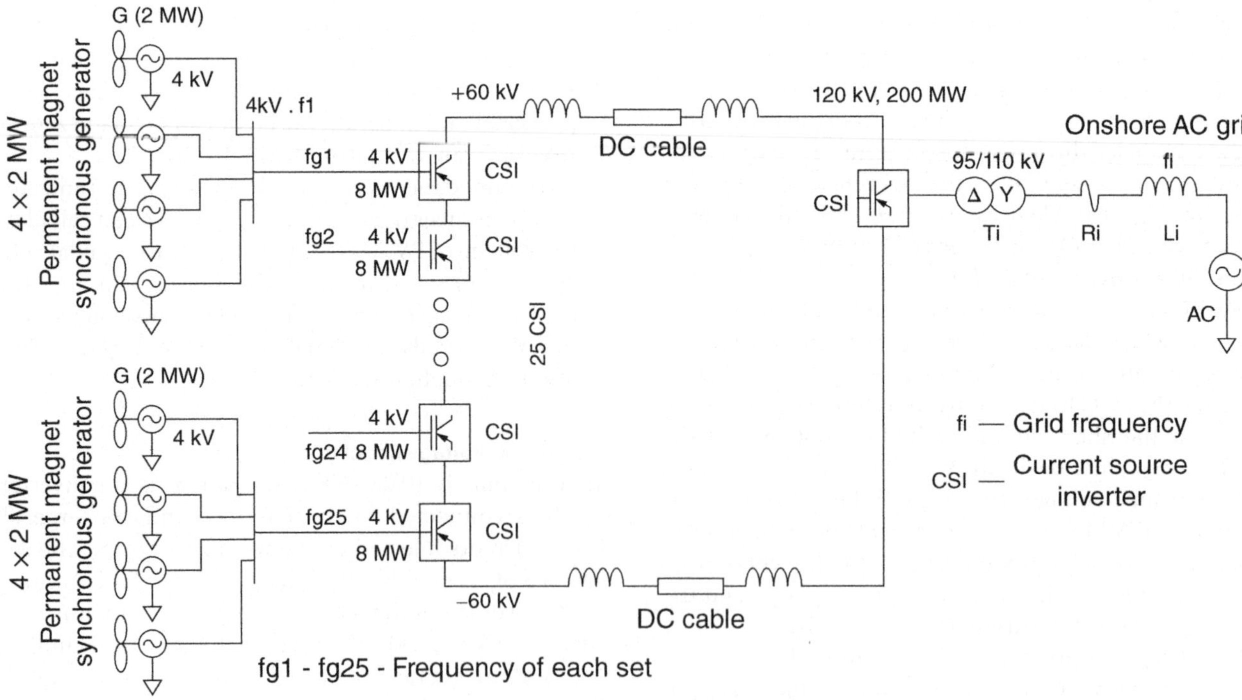

Figure 6.16 A 200 MW WF of MCSI-HVDC transmission with series connection.

Table 6.1 System ratings related to OWF topologies.

Ng	*Nc*	I_{dc} [kA]	U_{dc} [kV]	U_{nom} [kV]
2	50	0.86	245	123
3	34	1.3	163	82
4	**25**	**1.72**	**122**	**61**
5	20	2.25	98	49

symbols: Nc—number of offshore converters; U_{dc}—DC voltage; U_{nom}—voltage to ground of the largest neutral point on the generator, for insulation

required. And if a set of larger size (larger N_g) is selected, large DC rated current (high loss) and poor wind speed tracing of the wind turbine will be generated (poor characterization factor). In order to get a good overall performance, it is determined that four generators are used in one set (bold data in Table 6.1). Offshore transformer is not required, because the required voltage of DC transmission can be archived by converter in series. Using 25 self-commutated converters can make DC voltages reach 120 kV or above, which is enough for long distance transmission.

Since the generator converters and transformers are omitted, the cost of HVDC converter can be offset, thus the scheme has the possibility of saving cost.

6.4.5.9 Comparison of MVSC-HVDC and Two-Terminal VSC-HVDC

Figure 6.17 shows the comparison of two connection schemes of OWFs. The WF shown in Figure 6.17a is connected through MVSC-HVDC in parallel, in which four generators form a set, and each set uses 25 converters. Such a connection requires an offshore transformer, if the low frequency and the generator are directly coupled, the transformer price may be surprisingly high. The typical 22-pole 2 MW permanent-magnet synchronous generator operating with the optimal tip speed ratio of around 7, and transmits power within the frequency range of 1 to 3 Hz (common wind speed). In order to increase the frequency, this connection way may need a gear box. For reliability reasons, generators of OWFs do not want gear boxes, in this case, cage asynchronous generator is probably the best choice.

The 200 MW WF shown in Figure 6.17b uses a two-terminal VSC-HVDC operating at a constant frequency for connection. Assume they are standard wind turbines (based on permanent magnet synchronous generators), local converters can enable them to operate at variable speed. Compared with Figure 6.17a, the rated capacity of the converters has increased by 100%. Table 6.2 gives the WF equipments required by the three topology schemes, wherein multi-terminal series connection seems to be the most attractive one.

It should be noted that, from a reliability standpoint, Figure 6.16 and Figure 6.17a also have some disadvantages, because the fault of a converter means four generators out of operation. Issues need to be further considered include the insulation issues of generator and the system, and the loss issues of transmission cables. Because converters are connected in series and connected directly to the DC cable (away from the grounding point), the line has high potential. All converters have the same voltage and current ratings, but the insulation levels are different. Line voltage is lower, while the voltage of neutral-point to ground is higher, so the insulation problem needs special consideration. And transmission cable loss issues are related to the converter control mode (constant current, constant voltage, and current reduction control).

6.4.6 Other Transmission Systems

The transmission schemes for the transmission of OWF power to the shore are different from the solutions for onshore WFs. Onshore WFs are generally connected to the transmission or distribution network, which already has some load in some part of network. But usually there is no user connected to OWFs or offshore transmission systems, only points of common coupling are required to satisfy the grid specifications established for the shore. Moreover, the HVDC transmission solutions can decouple OWFs and onshore power grids, which makes the application of different wind turbine design concepts possible (such as different generator technologies or different control methods), and eventually achieve the most economical goal for the whole wind turbine system and transmission system. The following are some of the most meaningful system solutions under discussion.

6.4.6.1 Schemes for Low-Frequency Power Grids

Currently, an AC power system of lower than 50 Hz is mainly used in electrified railways. If an OWF uses HVDC transmission, a low-frequency AC transmission (such as 20 Hz) can only be applied to the OWF collection system. If the HVAC scheme is applied, a low-frequency AC transmission can be used to the OWF collection system and connected to the onshore transmission system. The onshore frequency conversion station needs to change low frequency current of offshore grid to the onshore grid frequency.

A low-frequency transmission scheme can simplify offshore wind turbine structures. Because the wind turbine speed of large capacity WFs is rather slow, the speed of 3 to 5 MW wind turbines is approximately 15 to 20 rpm. Lower AC frequency allows the use of lower speed rising ratio for wind turbines applying gearboxes or the reduction of the number of pole-pairs of direct-drive wind generators, making wind generators

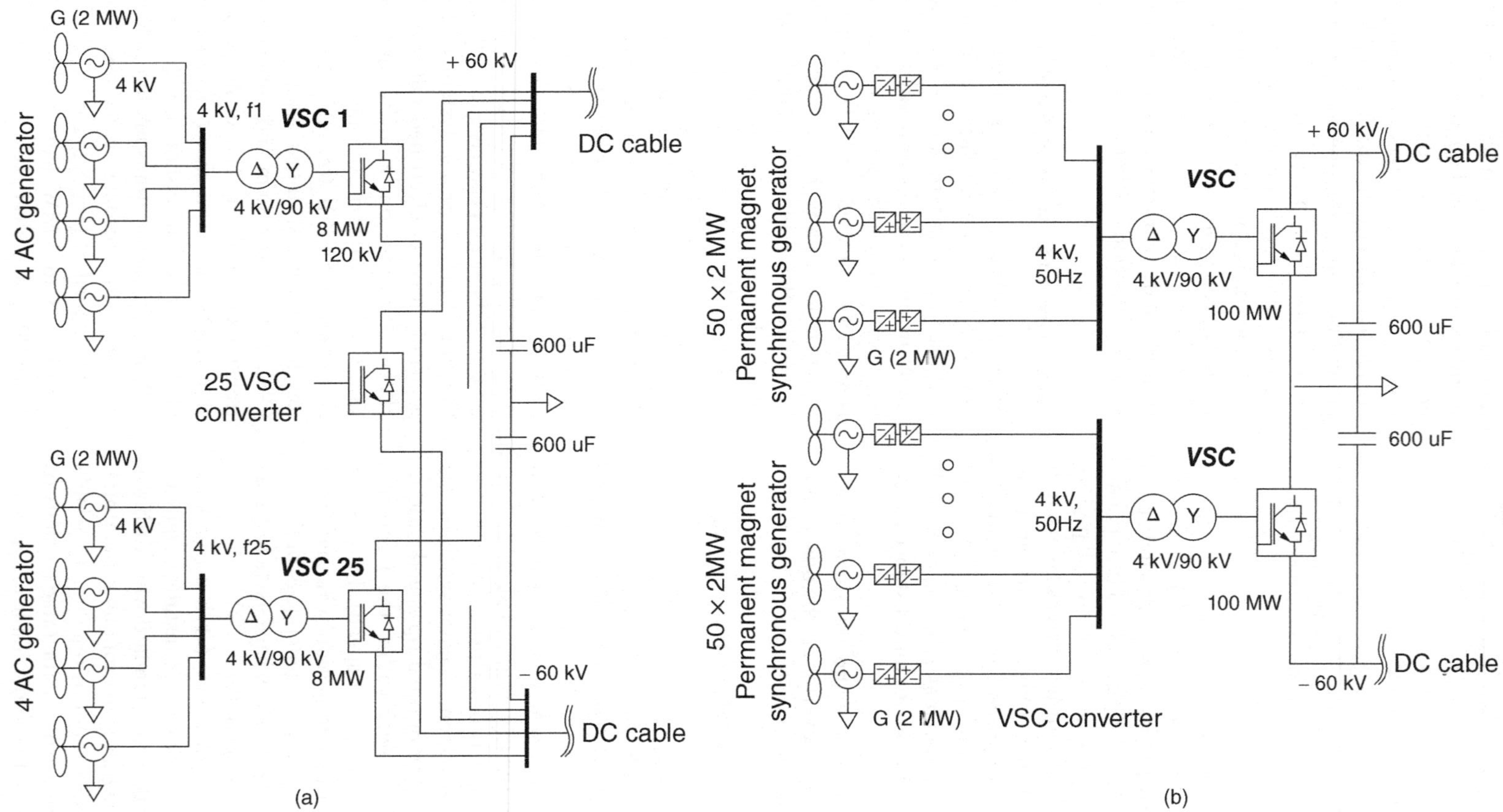

Figure 6.17 Two connection modes of OWFs. (a) Connection in parallel, MVSC-HVDC; (b) Two-terminal VSC-HVDC.

Table 6.2 Comparison of multi-terminal HVDC topologies.

HVDC connections	Generator converter	Transformer requirements	High insulation level	Gear box requirements
Connection in series	No	No	Yes	No
Connection in parallel	No	Yes	No	Yes
Two terminals	Yes	Yes	No	No

lighter and cheaper. On the other hand, low-frequency AC transmission may increase HVAC transmission capacity or increase transmission distance, because low frequencies can significantly reduce the charging capacity of the cable. The disadvantages of a low-frequency transmission scheme is a large increase in the size of the transformer, which will increase the cost.

6.4.6.2 Scheme for DC Power Grids Based on Wind Turbines with AC Generators

When using wind turbines with back-to-back (AC/DC/AC) converters, theoretically it is possible to separate the converter into AC/DC converters installed at the wind turbines (and then connected to the shore through DC power transmission line) and DC/AC converters close to the point of common coupling. In other words, the converter DC bridge is replaced by a VSC-HVDC transmission system. Because the AC generator usually operates at 690 V, and VSC-HVDC transmits power at around 150 kV, so the DC/DC converter must be added after the wind generator AC/DC converter (DC/DC switching mode converter/boost chopper), to increase the voltage to transmission voltage level.

The disadvantage of this method is that if all the wind turbines are connected to the same DC/DC converter, they will all work at the same speed, which may change in a very wide range. However, a large OWF will cover a large area, so only a few turbines are operating at the same wind speed at any given time. The running speed of most wind turbines cannot reach optimal aerodynamic efficiency. Therefore, usually about 5 generators are connected to a DC/DC transformer in group. The 5 generators in the same set will operate at a same speed, which is changeable throughout the time, and wind speed among the 5 generators is also changeable. For this solution, the overall aerodynamic efficiency is still lower than that of each wind turbine operating at variable speeds. However, the benefit gained from the practices of using wind turbine grouping is greater than the loss of decreasing aerodynamic efficiency.

6.4.6.3 The Comparison of OWF Transmission Schemes

Technological development will affect OWF power transmission schemes. Advocates of PCC-HVDC and VSC-HVDC look forward to the decrease of power electronic devices prices in the near future; advocates of HVAC hope to use higher voltage levels. Table 6.3 compares the main characteristics of the three available transmission modes for OWFs.

6.4.7 OWF Connected to Weak AC Power Grid

This section is devoted to the discussion of schemes for the connection of large OWFs that are formed of asynchronous wind generators across long distance to weak AC grids onshore. In this case, when applying AC transmission, the voltage drop issues are outstanding; while conventional DC scheme requires the AC network at the DC connection point to have a high short-circuit ratio. This section provides a detailed description of the connection scheme based on VSC-HVDC and its control strategy [14].

6.4.7.1 System Structure

The HVDC connection schemes for OWFs based on VSC are shown in Figure 6.18. The sending ends are OWFs, and the receiving ends are the AC power grids. The connection between the offshore voltage source converter station SVC_1 and the onshore converter station SVC_2 is of high voltage using a submarine HVDC cable. Each converter station consists of a VSC, coupling transformer, a DC capacitor, and a AC filter. Figure 6.19 shows its VSC bridge and valve structures, VSC is a two-level six-pulse type, and connected by IGBT in the series. Each IGBT has an anti-paralleled diode.

6.4.7.2 Control Principle of VSC-Based HVDC

Active power P between each converter and the adjacent AC network is controlled by changing the phase angle between the fundamental frequency voltage U_g produced by the converter and AC bus voltage U_n (Figure 6.18). Ignoring the transformer losses, active power can be calculated by the Formula (6.5):

$$P = \frac{U_g U_n}{X} \sin\delta \quad (6.5)$$

X—transformer leakage reactance; the control of power angle δ is achieved by modulating the angular frequency of AC voltage U_g created by the VSC. For example, when the wind power increases, angular frequency of point K tends to increase, which is determined by VSC_1, and VSC_1 gives instructions to increase the phase angle δ_1 of PWM sinusoidal reference signal, therefore, the phase

Table 6.3 Comparison of OWF transmission schemes.

Function	HVAC	PCC-HVDC	VSC-HVDC
Maximum allowable transmission capacity	800 MW(400 kV) 380 MW(220 kV) 220 MW(132 kV)	Can reach 600 MW by using undersea cables	The maximum capacity of 350 MW has been constructed; The maximum capacity is 500 MW; the design of 1080 MW has been presented
Whether transmission capacity is affected by the distance	Yes	No	No
Voltage levels	132 kV has been constructed, 220 kV and 400 kV are in development	Can reach ±500 kV	Can reach ±150 kV, has presented ±300 kV
Engineering construction	Many small projects	No	Has a test project in Norway
Black startup ability	Yes	N/A	Yes
Ability to support grid	Limited, SVC is needed to provide inductive reactive power	No, capacitor bank and STATCOM are needed to provide valves with inductive reactive power	Yes, VSC can absorb and generate inductive reactive power
Offshore substations	Yes	No	On an oil petroleum platform
Decoupling connected network	No	Yes	Yes
Cable model	Resistors, capacitors and inductors	Resistance	Resistance
Require auxiliary equipment	No	Needed in low wind speed	Needed in low wind speed
Total system loss	Depends on distance	2–3% (plus maritime auxiliary services)	4–6%
Fault level	Higher than HVDC	Lower than HVAC	Lower than HVAC
Offshore substation space requirements	Smallest	Largest	Medium
Construction costs	The costs of the power station are small, only a small amount on the transformer, but the costs of cables are high.	The costs of the station are relatively high (transformers, thyristor valves, filters, capacitor banks), and the cable cost is low	The costs have a 30%–40% percent higher than that of the power stations of PCC technology (IGBT is more expensive than thyristor), and the cables are also more expensive

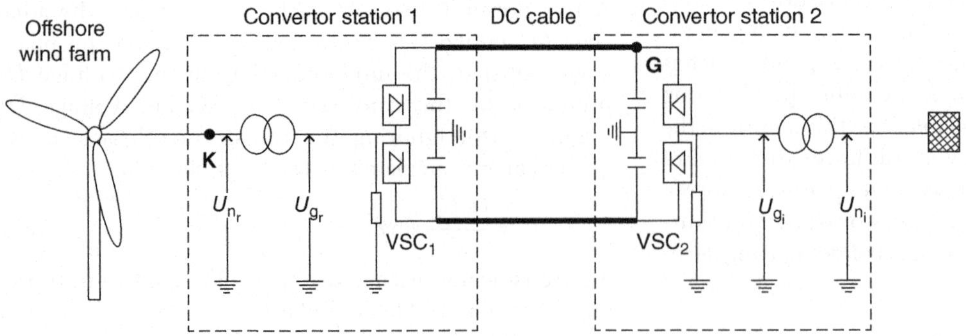

Figure 6.18 VSC-based HVDC connection.

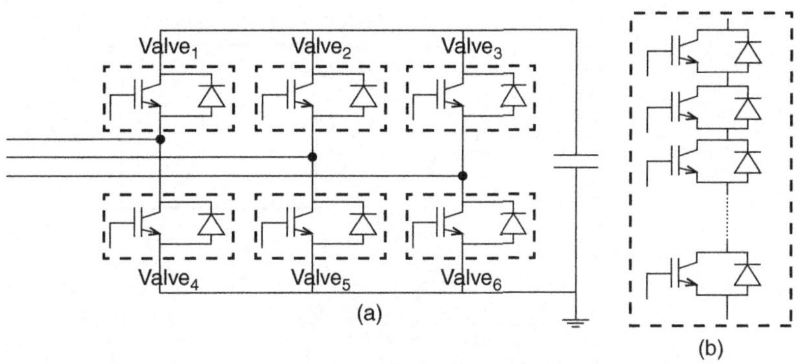

Figure 6.19 VSC bridge and valve structure. (a) single-level VSC bridge; (b) VSC valve structure.

angle of the fundamental frequency AC voltage U_{gr} of VSC$_1$ affects WF bus AC voltage U_{nr}. In accordance with the formula (6.5), when increasing the phase angle δ_1, more power will be transmitted to the DC system, and the angular frequency will return to its reference value.

In normal operation, an active power provided by HVDC must be equal to the active power absorbed by the power grid on the other side of the HVDC adding the losses. Power balance can be achieved by maintaining constant voltage at point G (Figure 6.18). Therefore, VSC$_2$ plays the role of voltage regulator at point G, for example, when WFs transfer more power to the DC connection, then the DC voltage at point G will increase. The control system of VSC$_2$ detects that the DC voltage is deviating from its reference value and gives instructions to increase the phase angle δ_2 of pulse-width modulation (PWM) sinusoidal reference signal, thus, the fundamental frequency AC voltage U_{gi} output by VSC$_2$ affects the AC voltage U_{gn} on the other side of the transformer. Thus, in accordance with Formula (6.5), when more power is absorbed by the AC network, the DC capacitor will discharge and DC voltage will return to the reference value. When DC voltage value at point G remains the same, the active power flows between two converters will automatically keep balance. Communication line is not needed between two converter stations.

In addition, by adjusting the angular frequency at point K (Figure 6.18), the amount of power absorbed by the WF is changed, and maximum wind power can be absorbed based on the wind speed.

Both VSCs control the AC voltages on their own sides by adjusting the modulation index of PMW sinusoidal reference signal "m." If the voltage produced by the converter is lower than the reference value, the modulation index will increase, in order to increase the AC voltage generated. In contrast, when the AC voltage exceeds the reference value, the index "m" will decrease.

6.4.7.3 The Description of the Control System

The control system is based on fuzzy logic (FL). This type of control uses tolerances, uncertainty, fuzziness, and impreciseness in the decision-making process, seeking to provide satisfactory performance without the need for detailed mathematical model of the system, as long as expert knowledge is combined into the rule system. In addition, it has a processing capacity of imprecise or noisy data, and the control ability can be extended to the operating conditions of which linear control technology is powerless (i.e., large parameter fluctuations).

The main drawback of the FL-based control system is that the tuning of its membership function (MF) needs a lot of trial and error. To reduce the time consumption in MF tuning process and improve performance when it cannot meet the specifications, online tuned adaptive fuzzy control system (AFCS) can be used. Its basic principle is to use the system strategy of the master-slave mode, using a fuzzy controller (FC) to online adjust parameter of another FC, that is, the MFs of master FC monitors slave FC to follow expert reasoning by monitoring the slave FC on-line regulation. This expert reasoning knowledge base can manually tune the MFs. In addition to the automatic MFs adjustment, the controller may provide adaptive tuning of MFs in accordance with the system behavior. Thus, by learning to adapt to the environment and acquiring new knowledge, AFCS can lend itself well to the system control of large asynchronous generator WFs with great fluctuations.

The control system has two AFCSs and two simple fuzzy control systems (FCS), as shown in Figure 6.20. Of which the functions are discussed in detail below.

1) **Angular frequency regulator**

   The regulation of angular frequency at point K (Figure 6.20) through VSC$_1$ is managed by AFCS$_1$ of the adaptive FCS. This system consists of FC$_1$, which is the main controller of the adaptive FCS. The main role of FC$_3$ is to fine-tune FC$_1$ and FC$_2$, to dynamically detect the angular frequency of point K online, and correspond it to the wind turbine maximum aerodynamic efficiency of the specific wind speed ω_{ref}. Its control principle is shown in Figure 6.21.

   Meanings of graphical symbols: P—the active power of point K; ω_{el}—the angular frequency of point K; ω_{ref}—the angular frequency of point K at the maximum aerodynamic efficiency; δ_1—the phase angle of PWM sinusoidal reference signal; $\Delta\delta_1$—the phase

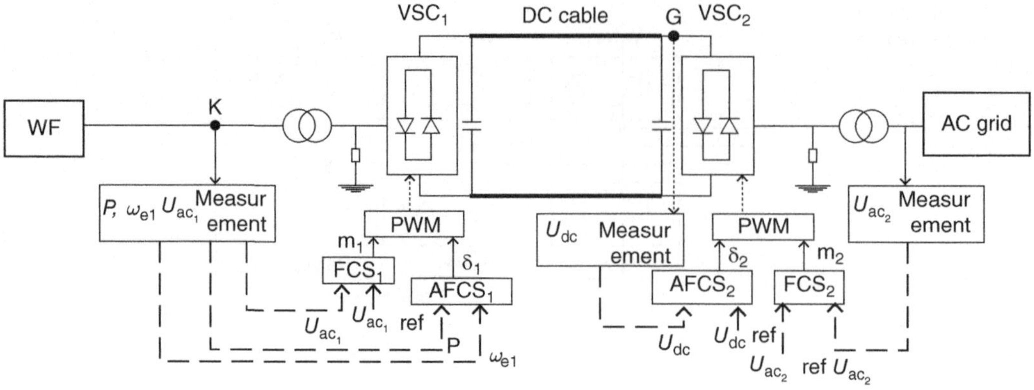

Figure 6.20 VSC-HVDC-based connection and its control circuit diagram.

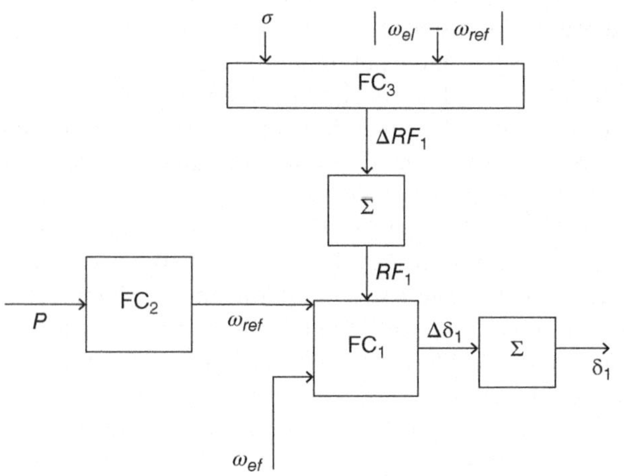

Figure 6.21 Angular frequency modulator controls system FCS_1.

displacement of phase angle δ_1 (increase of δ_1 is positive, and decrease of δ_1 is negative); RF_1—the regulatory factor of the main fuzzy controller MF (FC_1 MF); ΔRF_1—step change of RF_1; σ—the variance of ω_{el}; $|\omega_{el} - \omega_{ref}|$—the absolute mean deviation of ω_{el} against ω_{ref}.

The frequency of point K ω_{el} is compared with ω_{ref}, of which the error is passed through FC_1, which produces a signal $\Delta\delta_1$ (Figure 6.21), by accumulating the successive values of $\Delta\delta_1$, δ_1 is generated according to Formula (6.6).

$$\delta_1^{new} = \delta_1^{ord} + \Delta\delta_1 \tag{6.6}$$

In the formula: δ_1^{new} is the new value of phase angle δ_1; δ_1^{ord} is the previous value of phase angle δ_1; ω_{ref} is the angular frequency of point K (Figure 6.20), which corresponds to wind turbine maximum aerodynamic efficiency for a specific wind speed.

By using maximum power point tracking (MPPT) technology, the values of ω_{ref} dynamically approximates to the values from FC_2 in real time (Figure 6.21). This is achieved by changing the frequency reference value of point K $\Delta\omega_{ref}$ and monitoring of the corresponding changes of wind farm power output P. If the increment of reference frequency causes the increase of power P, the search will continue in the same direction; otherwise it will search the opposite direction. This searching method makes the ultimate ω_{ref} to oscillate near the optimal value.

The detailed FCs structure is as follows:

a) **FC_1 structure**: FC_1 is the main FC of angular frequency control system. Its input is the deviation of ω_{el} from its reference value $\omega_{el} - \omega_{ref}$ and differential. The output is the phase displacement $\Delta\delta_1$ of δ_1. The initial (MFs) of input $\omega_{el} - \omega_{ref}$ is shown in Figure 6.22. This MFs are the first MFs before their tuning, and the general form of the final MFs will remain the same.

   As shown in Figure 6.22, input $\omega_{el} - \omega_{ref}$ needs 5 fuzzy subsets: negative big (NB), negative small (NS), zero (OK), positive small (PS), and positive big (PB).

   For the differential of $\omega_{el} - \omega_{ref}$, the fuzzy sets needed are NB, NS, PS, and PB, of which the MFs are shown in Figure 6.23.

   For phase angle $\Delta\delta_1$, the fuzzy set needed are decrease very fast (DVF), decrease fast (DF), decrease slow (DS), decrease very slow (DVS), OK (OK), increase very slow (IVS), increase slow (IS), increase fast (IF), and increase very fast (IVF), of which the MFs are shown in Figure 6.24.

   In order to form fuzzy control rules, the following points must be considered:

   - If the angular frequency is much higher than the normal value, regardless of acceleration or deceleration, the correction action must be very strong. In order to decelerate, the generator must rapidly increasing electrical power flow into the DC link, which is managed by quickly increasing the phase angle δ_1.

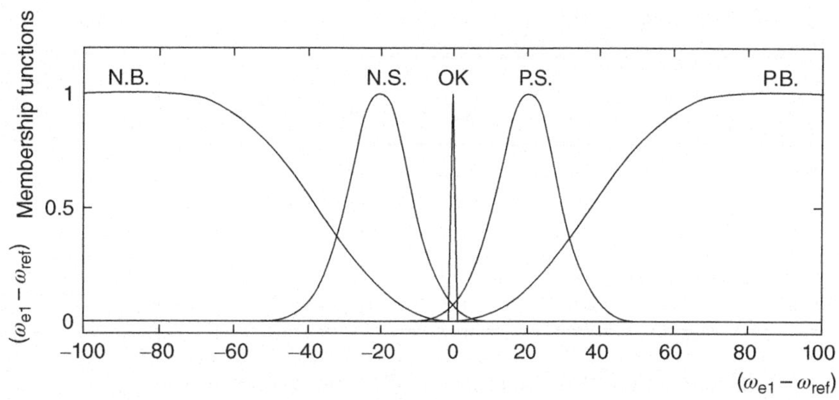

Figure 6.22 MFs of fuzzy set $\omega_{el} - \omega_{ref}$.

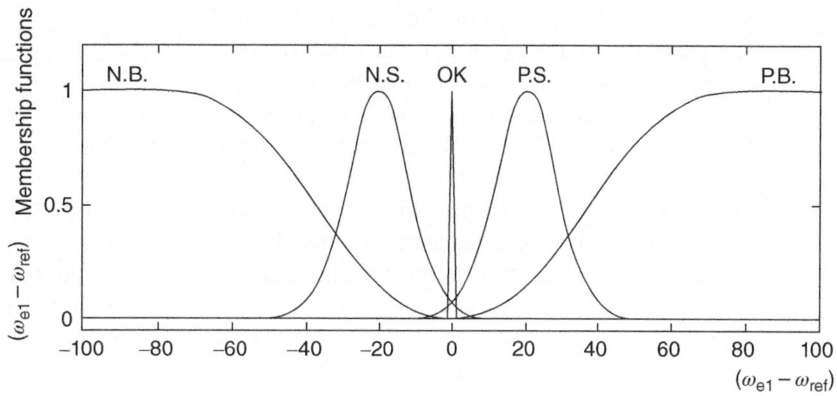

Figure 6.23 MFs of $d(\omega_{el} - \omega_{ref})/dt$.

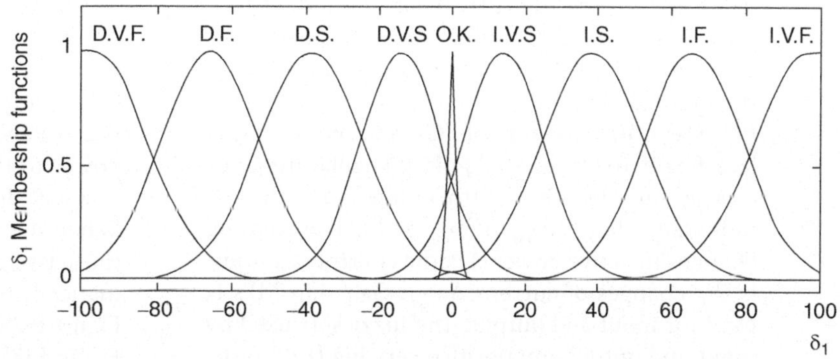

Figure 6.24 MFs of δ_1.

- If the angular frequency at the point K is higher than its normal value and is accelerating, then deceleration must be achieved by increasing phase angle δ_1.
- If the angular frequency is almost equal to its normal value and accelerating quickly, then the increase of the phase angle must be very slow, in order to avoid oscillations near the set value.

These conclusions are extensions of all possible combinations of all possible angular frequencies and their change rates. The corresponding output choices are presented in Table 6.4.

b) **FC_3 structure:** The role of FC_3 is to online adjust the MFs of FC_1. Obviously, no parameter combinations can ensure the optimal performance under all operating conditions. By regulating FC_3, MFs of FC_1 can be regulated online, making its parameters most optimized under any environment. For this reason, continuous monitoring of the system and adjustment of the MFs are required.
Inputs of FC_3 include the absolute deviation value $|\omega_{el} - \omega_{ref}|$ between ω_{el} and its reference value, and the variance σ of ω_{el}. The controller output is signal RF_1 (Figure 6.22).
Continuously accumulate the values of ΔRF_1 according to Formula (6.7), to obtain the regulatory factor RF_1, which makes the MFs of FC_1 be most optimized. In this way, the final output of FC_3 controller will be:

$$RF_1^{new} = RF_1^{old} + \Delta RF_1 \tag{6.7}$$

Table 6.4 Fuzzy rules for FC_1.

$\omega_{el} - \omega_{ref}$, \d($\omega_{el} - \omega_{ref}$)/dt	PB	NB	PS	NS
PB	IVF	IVF	IVF	IVF
PS	IF	OK	IS	OK
OK	IVS	DVS	OK	OK
NB	DVF	DVF	DVF	DVF
NS	OK	DF	OK	DS

In particular, the increase of RF_1 or decrease of MFs range is corresponding to output δ_1, so that FC_1 can become "more severe" or "looser." For example, in the case that the wind speed is changing greatly, the system must become more "severe." To this end, the increase of δ_1 range will cause greater changes in δ_1. The priority of the controller is to adjust the MFs of FC1, in order to drive the ω_{el} as close as possible to its reference value. When this goal is reached, FC_1 may be further adjusted in order to minimize the fluctuations of ω_{el} around its reference value.

Gaussian MFs is used for input and output. Each input needs three fuzzy subsets: big (B), middle (M), and small (S). The fuzzy sets required by ΔRF_1 are positive big (PB), positive middle (PM), positive small (PS), OK (OK), negative small (NS), and negative big (NB). FC_3 fuzzy control rules are shown in Table 6.5.

c) **FC_2 structure:** the role of FC_2 is to online calculate the reference value ω_{ref} of the angular frequencies used by FC_1. The controller inputs are: the last modified value $L\Delta\omega_{ref}$ of ω_{ref} and corresponding changes in active power ΔP; the controller output is the changes of current $\Delta\omega_{ref}$. Gaussian MFs is used for input and output. The fuzzy sets used by input and output are: positive very big (PB), positive middle (PM), positive small (PS) positive (P), zero (ZE), negative very big (NVB), negative big (NB), negative middle (NM), negative small (NS), and negative (N). Fuzzy control rules used by FC_2 are listed in Table 6.6.

2) **DC voltage regulation**

Active power imbalance between sending end and receiving end of the DC link causes corresponding

Table 6.5 Fuzzy control rules of FC_3.

σ / $\lvert\omega_{el} - \omega_{ref}\rvert$	S	M	B
S	OK	NS	NB
M	PS	PM	PM
B	PB	PB	PB

Table 6.6 Fuzzy rules of FC_2.

$L\Delta\omega_{ref}$ / ΔP	P	ZE	N
PVB	PVB	PVB	NVB
PB	PB	PVB	NB
PM	PM	PB	NM
PS	PS	PM	NS
ZE	ZE	ZE	ZE
NS	NS	NM	PS
NM	NM	NBV	PM
NB	NB	NVB	PB
NVB	NVB	NVB	PVB

changes in DC capacitor voltage, so VSC_2 (as a DC voltage regulator) must maintain constant DC voltage across its capacitor, which is achieved by adjusting the power flow by the DC link to the AC system. In order to achieve this goal, the DC link voltage (G point in Figure 6.20) is measured and compared with its reference value, of which the error is sent to $AFCS_2$ of the adaptive fuzzy controller, to produce signal $\Delta\delta_2$ (Figure 6.25). According to Formula (6.8), we can obtain

$$\delta_2^{new} = \delta_2^{ord} + \Delta\delta_2 \tag{6.8}$$

In the formula, δ_2^{new} is the new value of phase angle δ_2; δ_2^{ord} is the previous value of phase angle δ_2. When the DC voltage is too high, δ_2 is adjusted to send more power to the AC system. Thus, the cable and capacitor will discharge, and the DC voltage will reduce. When the DC voltage is adjusted at a constant value, the active power flow between the two VSCs is automatically balanced.

Figure 6.25 shows that $AFCS_2$ consists of FC4 of main FC and FC5 of auxiliary FC, FC_5 of FC is responsible for the online regulation of MF_s of FC_4. The structures of FC_4 and FC_5 are very similar to that of FC_1, which regulates FC_3. Their input and output variables are listed in Table 6.7.

3) **AC Voltage Regulator**

The AC voltage amplitude ($U_{ac\,1,2}$) generated by VSC_S is compared with a reference voltage amplitude

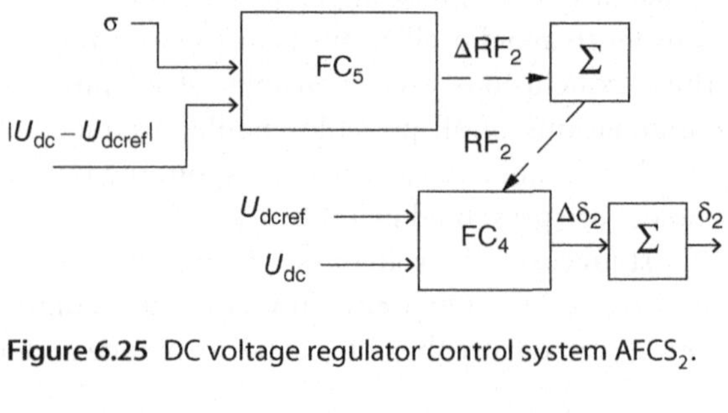

Figure 6.25 DC voltage regulator control system $AFCS_2$.

Table 6.7 Fuzzy rules of FC4 and FC_5.

	Input 1	Input 2	Output
FC4	$U_{dc} - U_{dcref}$	$d(U_{dc} - U_{dcref})/dt$	$\Delta\delta_2$
FC5	$\lvert U_{dc} - U_{dcref}\rvert$	σ	ΔRF_2

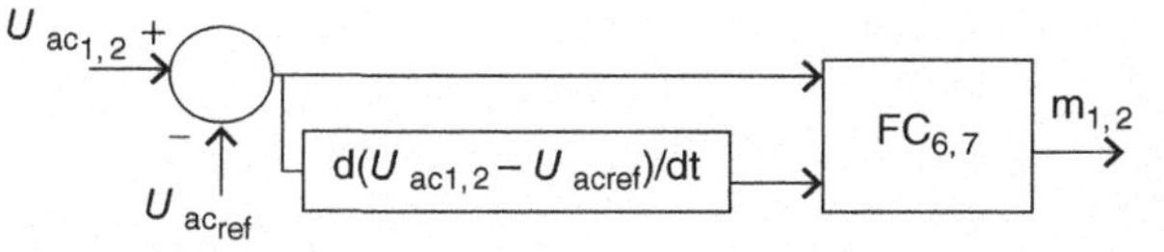

Figure 6.26 AC voltage regulator ($FCS_{1,2}$).

($U_{ac\ ref}$) by the AC voltage regulator in each converter, of which the error is added to an FC (Figure 6.26). The regulator output is the modulation index "m" of PWM sinusoidal reference signal, therefore, it is the modulation signal of the AC voltage amplitude generated by the VSCS. Wherein, the structure of FCs is similar to that of FC_1.

Literature [14] carries out simulation analysis on the system of a 90 MW OWF consisting of cage asynchronous generators connected through a 20 kV VSC-HVDC to a weak power grid (of which the short circuit capacity is 45 MVA). Due to changes in wind speed (wind speed decreases from the average speed of close to 9 m/s, to an average speed of close to 8 m/s; after a few seconds, the wind speed suddenly rises to the average speed of 11 m/s), through the above control, wind turbines can capture the maximum power according to the wind speed changes, and the power at sending and receiving ends can smoothly follow the changes. DC voltage can quickly return to its reference value. During the disturbances, AC voltages (effective values) at the power grid are within the acceptable limits.

6.5 A New Type of Offshore Wind Energy Conversion System

The existing HVDC offshore wind systems are imitating the architecture of onshore WFs. Each wind turbine uses a double-fed asynchronous generator coupled to the 50 (60)Hz transformer to provide 50 (60)Hz output, then the power of each wind turbine is feed to an AC-DC conversion platform, where the power is converted into a dc voltage through another transformer. Such a system is a huge, complex installation with lots of required maintenance, and also very expensive. Literature [15] proposes an alternative structure for OWFs, using a multi-pole permanent magnet machines as the main generator (its rated capacity is from 0.75 MW to 8 MW), and using intermediate frequency transformer and simple power converter, to obtain a compact and lightweight offshore wind power generation system. The system diagram is shown in Figure 6.27, each turbine converts the power it obtained to DC current, and is connected in series or parallel to a 60 to 130 kV DC system. The target capacity of such an OWF is within 100 to 200 MW, of which the system can be achieved by 50 to 70 wind turbines. An important feature of this system it that it does not require centralized converter platform, and the converter in each wind turbine is lighter and smaller.

The rated capacity of the recommended generator is 750 kW at the speed of 4,000 rpm, and the gear ratio required for the gear box is 1:200. The generator can be stacked together on a common shaft, and its power grade can reach 3 MW or higher. Each generator weighs 340 kg. The gross weight of 3 MW generators is 1,360 kg. It is much lighter than conventional DFIG or PM machine structure without gear boxes. And the structure of 3 MW generator is compact, its length is less than 1.8 m, and its diameter is 0.79 m. The technical specifications of the proposed 3 MW generator are listed in Table 6.8. The generator, wind turbines, and transmission system are represented by two masses, as shown in Figure 6.28.

The generator is cooled with liquid, the output frequency at maximum power is 1,200 Hz. When the generator slows down, the output frequency will linearly decrease at a constant ratio, and will reduce to 300 Hz when the corresponding cut-in wind speed is 4 m/s. Thus, intermediate frequency transformer can be used according the generator output, to provide the insulation isolation. Coaxial winding transformer promise robust structure with low and controlled leakage inductance, which has sufficient margin for achieving the required level of insulation, can be extended to higher power levels, and may provide a suitable way to achieve high-voltage insulation. The weight of 3 MW, 1,200 Hz transformer is 8% less than that of a comparable 50 Hz transformer. The use of liquid cooling can further reduce the weight and size.

Electromechanical model of the generator can be described by Formulas (6.9) to (6.12):

$$\frac{d}{dt}i_d = \frac{1}{L_d}u_d - \frac{R}{L_d}i_d + \frac{L_q}{L_d} \times PP \times \omega_{rm} \times i_q \quad (6.9)$$

$$\frac{d}{dt}i_q = \frac{1}{L_q}u_q - \frac{R}{L_q}i_q - \frac{L_d}{L_q} \times PP \times \omega_{rm} \times i_d - \frac{\lambda \times PP \times \omega_{rm}}{L_q} \quad (6.10)$$

$$T_e = 1.5 \times PP \times \lambda \times i_q \quad (6.11)$$

$$\frac{d}{dt}\omega_{rm} = \frac{1}{J}(T_e - B \times \omega_{rm} - T_m) \quad (6.12)$$

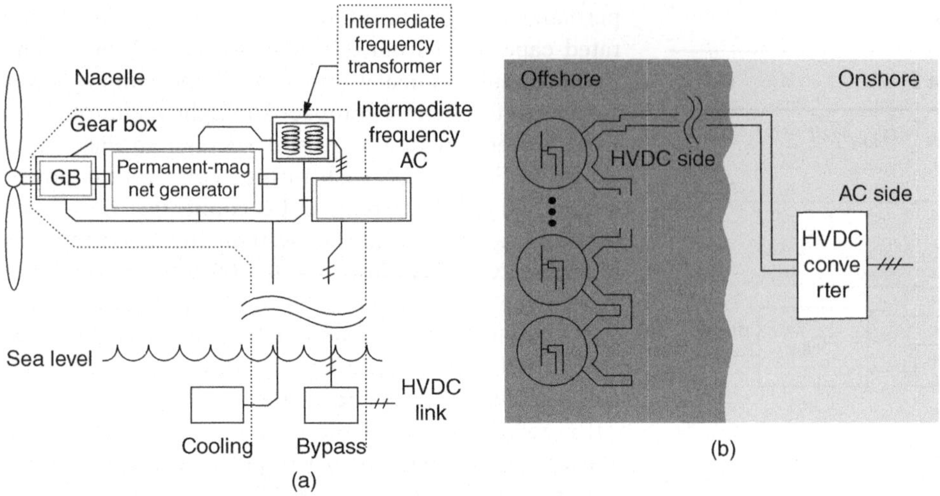

Figure 6.27 Topology of proposed WF. (a) Wind turbine unit with a permanent magnetic generator and intermediate frequency transformer; (b) OWF with wind turbines of series or parallel connection.

Table 6.8 Wind turbine performance specifications.

Rated power	746 kW
Maximum rotational speed	4000 rpm
Effective rated power	96%
Number of pole-pairs	18
Maximum torque	2712 N-m
Maximum current	335 A rms
Single-phase inductance	100 μH
Single-phase resistance	30 mΩ
Rated voltage	700 V peak
Frequency constant	0.30 Hz/rpm
Voltage constant	0.20 V/rpm
Torque constant	8.1 Nm/A rms
Weight	340 kg

Where i_d, i_q, u_d, u_q—the currents and voltages of axis d and axis q; L_d, L_q—the inductances of axis d and axis q; R—stator winding resistance; ω_{rm}—rotor angular velocity; λ—the flux amplitude induced by the permanent magnet; PP—the number of pole-pairs; T_e—electromagnetic torque; J—the combined inertia of the generator and the wind turbine; B—the combined viscous friction of the generator and the wind turbine; T_m—shaft mechanical torque.

In order to provide controllable power to stabilize the DC voltage or current (depending on connected in parallel or in series), a power conversion system is required. Converter must also provide active torque control to smooth the impact of output on the shaft and the gear box, and provide speed control for the blades. A simple single switch converter is shown in Figure 6.29, of which each generator can provide all control requirements. The switches embedded in the diode bridge can switch at a constant frequency, and its duty cycle, which controls the power flowing to the DC system, is controllable. This switch can be realized by an IGBT, which can be placed on the primary side of the intermediate frequency transformer, as shown in Figure 6.29a, or on the secondary side (before single-phase rectification takes place). It is more advantageous to place the switch on the primary side, because the switch counts is fewer, the equipment is subject to less stress, and the complexity of high-voltage design on the DC side will be reduced in the future. The IGBT is rated at 1,200 V and 800 to 1,000 A.

The electrically superimposed structure of four permanent magnet generators is shown in Figure 6.29b, as a particular phase of each generator is superimposed onto the same phases of the other three generators through the intermediate frequency transformer, of which the primary side is connected in parallel, and the secondary side is connected in series. Final output of the wind

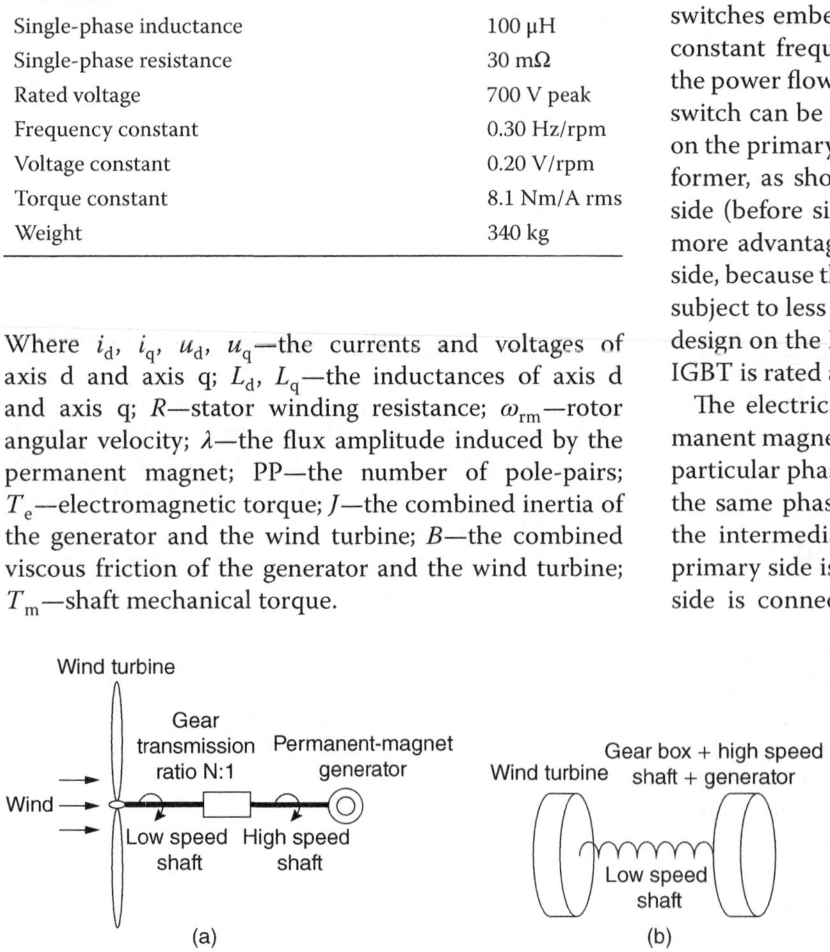

Figure 6.28 Wind energy conversion systems and two-mass model. (a) Mechanical diagram of the wind energy conversion system; (b) Two-mass model of the wind turbine generator system.

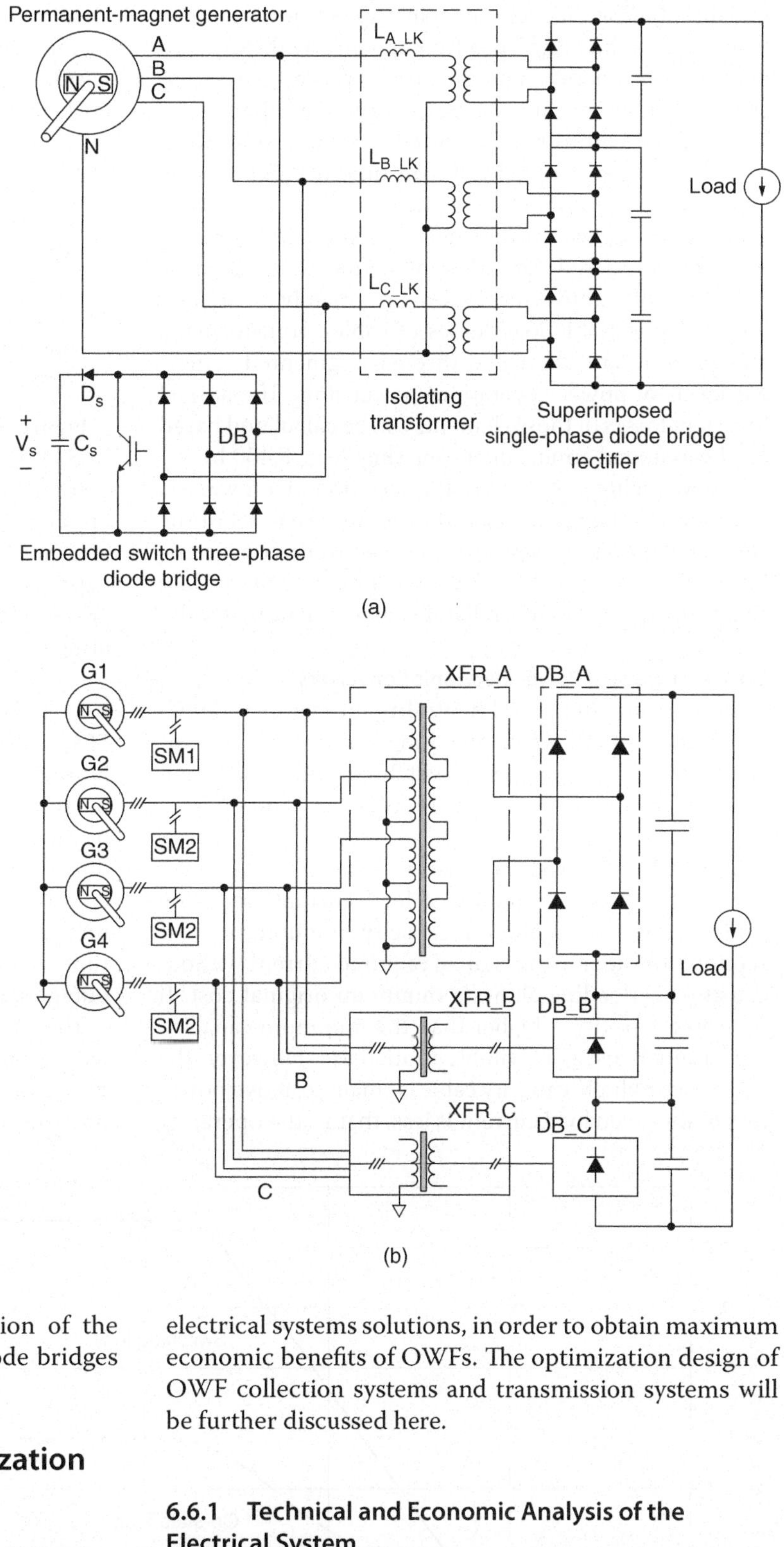

Figure 6.29 Circuit structure of ac/dc converter driven by PM motor. (a) Detailed structure of a single generator; (b) Converter structure of four generators using multi-phase superposition, SMx in the figure represents the switching mode.

turbine is obtained through the superposition of the output of each phase by the single-phase diode bridges and the capacitors.

6.6 OWF Electrical System Optimization Design

OWF electrical systems include WF power collection system and the transmission system for power transmission from offshore to onshore grid. In the offshore wind power project, the costs of the electrical system account for about 10% to 20% of the total costs, so we need to economically assess the technical feasibility of the electrical systems solutions, in order to obtain maximum economic benefits of OWFs. The optimization design of OWF collection systems and transmission systems will be further discussed here.

6.6.1 Technical and Economic Analysis of the Electrical System

Technical analysis of electrical system solutions primarily evaluates the annual system losses of different technical solutions.

In order to assess the losses, the annual generating capacity of WFs shall be considered. Starting from the distribution function of wind speed, for example, assume that the wind speed compliance with Rayleigh distribution function, and the annual power generation curve is simplified to a staircase curve that operates in N segments with different power levels and durations, so the equivalent annual full load time (h) and average capacity factor (%) can be obtained.

Losses of cables, transformers, reactors, and converters are considered for different connection schemes. Wherein the converter losses are mainly switching losses. The variable load losses of cables, transformers, and reactors are calculated through segmented (different levels of power generation) power flow. The annual invariant losses of these elements can be calculated based on the assumed annual operating time (e.g., 8,400 h).

The following part is a brief description of a few basic questions in the technical and economic analysis of the electrical system: losses of power electronic converters, life cycle cost of cable, determination of energy loss caused by unavailability, reliability assessment methods.

6.6.1.1 Losses of Power Electronic Converters

Assume that a three-level neutral-point-clamped (NPC) converter is used, when operating at full load, its efficiency is higher than 98%, switching and conduction losses of the converter valves are shown in Figure 6.30.

6.6.1.2 Life-Cycle Cost of the Cable

Under the given conditions of load current, selecting cables with the minimum capacity (minimum cross section) for the transmission of required current without causing overloading allows a minimum original cost. If the cable capacity is higher than the minimum capacity, the original cost is inevitably to increase. However, the resistance of high-capacity cable is small, so active power loss of the given load current is less, that is, the operating costs will be reduced. The total life-cycle cost is the sum of the original installation costs plus the running costs within the selected analysis period. And the operating costs include three parts [16]:

1) Fixed losses cost, which does not change with the output power of the WF. This is mainly transformer excitation losses.
2) Variable load losses cost, which changes with the square of the output current.
3) Corresponding energy not supplied cost (mandatory constraint) due to unavailability of the electrical system (such as cable fault).

Figure 6.31 shows a comparison of the lifetime costs of 33 kV cable types used by the OWF power collection system. Obviously: ① the installation costs of large cables are high, but losses are low; ② when current of the cable is larger (right curve intersection), the greater the cable capacity is, the smaller the life cycle cost of the cable is; ③ the calculation of losses above must consider the operating regime and load factors.

6.6.1.3 Determination of Corresponding Energy Losses of System Unavailability

No-load and load losses of the electrical system can be directly determined by experiment or calculation, however, the calculation of corresponding energy losses due to system or component unavailability is much more complex. The reason is that the losses caused by system unavailability are affected by many factors, such as the time of unavailability, the frequency, and the duration of outage state.

The method of analysis is to select a reliability evaluation method for determination of the frequency and duration of various system operating conditions depends on the establishment of appropriate mathematical

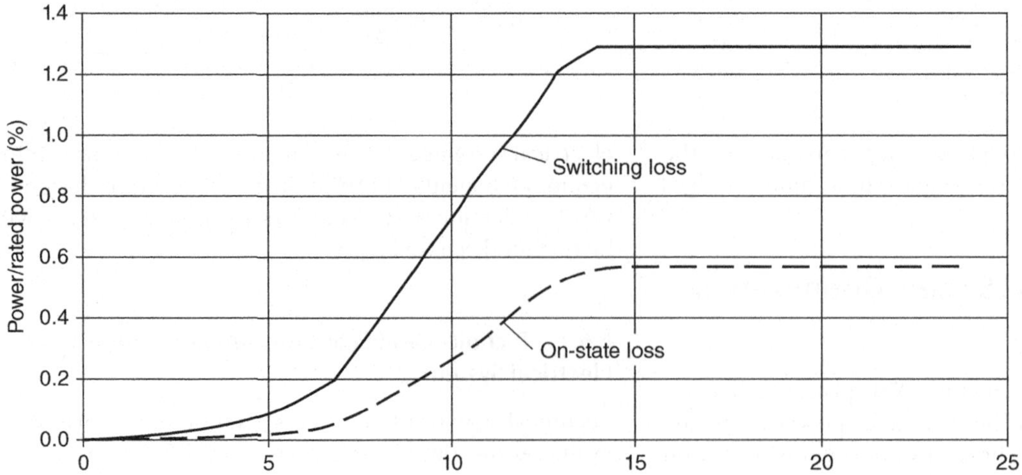

Figure 6.30 Losses in the converter.

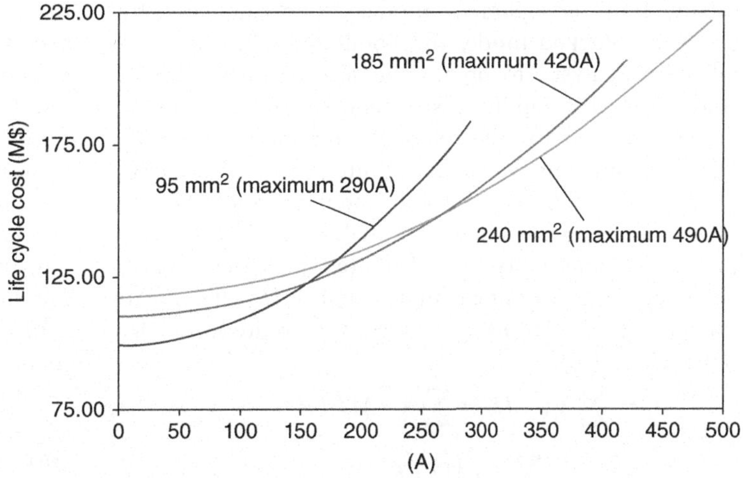

Figure 6.31 33 kV cable life-cycle cost.

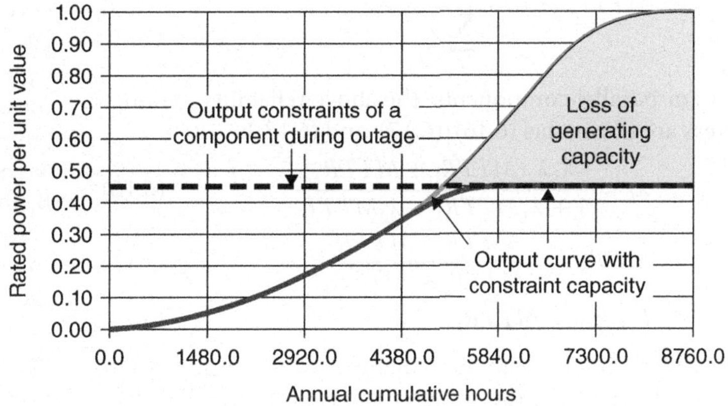

Figure 6.32 Determination of electrical energy loss caused by transmission capacity constraints.

models for the selected electrical system structure and the utilization of some typical approximation.

The calculation period chosen is one year, which is represented by one step per hour (hourly steps). All components are using two (state) state models, namely, operating or outage. The definition of models is described by using fault rate (λ) of each element and its mean time to repair (MTTR). OWFs are largely different from onshore WFs, wherein MTTR is highly unpredictable. This factor is expected to have a significant impact on the assessment of the losses.

For a relatively simple structure, such as wind turbines series, a submarine cable failure is the main fault source, and the outage duration can be easily obtained by multiplying MTTR by the fault rate.

For fault occurrence moment correlation, the WF capacity factor may be roughly used as an acceptable approximation for wind production variability.

For example, if the transmission capacity of the entire electrical system is damaged in the fault duration, the power output loss is equal to the WF installed capacity multiplying by the capacity factor. But in the case of a part of the transmission capacity that is damaged, we can introduce a similar capacity constraint factor to represent the equivalent capacity factor of the damaged capacity of the working condition (constraint) accounting for the total transmission capacity (see Figure 6.32), then the loss of generating capacity as shown in the shaded area in Figure 6.32 can be calculated as follows (Formula (6.13)):

$$P_{lost} = P_{rated}[CF_e - CCF_e(P_{cap})] \tag{6.13}$$

Where P_{rated} is the rated power of the WF; P_{cap} is the constraint power caused by partial unavailability of electrical power system transmission capacity; CF_e and CCF_e respectively represents the capacity factor and constraint capacity factor. When the power system partially or completely fails to deliver the power output of a WF, the energy loss E_{lost} caused by system unavailability can be determined by the lost generating capacity multiplying by the duration (Out_{dur}) (Formula (6.14)).

$$E_{lost} = Out_{dur} \times P_{lost} = \lambda \times MMTR \times P_{lost} \tag{6.14}$$

6.6.1.4 WF Electrical System Reliability Assessment [17]

The optimization of the entire OWF electrical system is a heavy work, which requires a comprehensive and integrated reliability analysis model. In the simplified optimal choice, each subsystem, such as the WF collection system or the transmission system connected to the shore, is separately modeled and calculated by using the blocked mode, and combined with the optimization analysis, to obtain the sub-optimal solution. And classical reliability assessment techniques can be directly applied for the reliability analysis of the subsystem. Annual unavailability of integrated components of the subsystem can be calculated by applying the hybrid failure mode and effects analysis method. In case that the components are integrated in series, equivalent fault rate λ, annual unavailability U_s, and mean time to repair $MTTR$ is respectively represented by (Formula (6.15)):

$$\lambda_s = \sum_i \lambda_i; \quad U_s = \sum_i \lambda_i \cdot MTTR_i;$$
$$MTTR_s = \frac{\sum_i \lambda_i \cdot MTTR_i}{\sum_i \lambda_i} \tag{6.15}$$

For parallel components, the above reliability parameters are (Formulas (6.16), (6.17), and (6.18)):

$$\lambda_P = \frac{\lambda_1 \lambda_2 (MTTR_1 + MTTR_2)}{1 + \lambda_1 MTTR_1 + \lambda_2 MTTR_2} \tag{6.16}$$
$$MTTR_P = \frac{MTTR_1 \cdot MTTR_2}{MTTR_1 + MTTR_2} \tag{6.17}$$
$$U_P \cong \lambda_P \cdot MTTR_P \tag{6.18}$$

Failure mode analysis is mapped to the minimal cut set, which causes the system outage. From the WF electrical system to the grid-connection point, any component or subsystem faults have led to the outage of the entire system, forming a series system from the reliability point of view. Therefore, the unavailability of part or the entire of system can be determined by applying the Formula (6.15).

6.6.2 Economic Analysis of the Electrical System

To compare the different electrical systems solutions, the investment costs, operation and maintenance (O&M) costs, loss value, and energy not supplied (ENS) shall also be considered [17].

Investment costs include the costs of WFs connected to the HV cable on shore and its installation (AC or DC) costs, the costs of transformers (converters), the costs of support structure of reactors, and offshore platform. Cable installation costs depend on the laying technology, the conditions of existing infrastructure and machinery protection requirements, and so on.

The assessments of O&M costs are limited to the high voltage part of the transmission system. For transmission systems (including submarine cables and substations), costs are primarily cable maintenance costs, so it can be considered that the annual O&M costs, as well as the availability ratio of AC and DC are the same (e.g., the cable will fail once in the lifetime, and the mean time to repair is 30 days). Annual maintenance costs of substation can be estimated as a percentage (e.g., 0.4%) of total investment in transmission lines.

ENS costs are assessed by reliability analysis of the transmission system. Some reliability statistic data may refer to that of onshore grids. Economic analysis should also consider the financial parameters, such as the nominal interest rate (%), inflation rate (%), and economic life time.

In the assessment of life-cycle cost, there are three main types of power/energy loss, and they have the greatest impact on the designed life-cycle cost. These three loss types are fixed loss costs, variable load loss costs, and corresponding energy loss caused by system unavailability. By introducing economic assessment factors corresponding to these three loss types, we can calculate the related assessment factors equivalent to initial capital cost (ICCEq). Thus, ICCEq provides a consistent measure to compare the configuration design of different electrical system. These three quantitative economic assessment factors are fixed loss factor A, load loss factor B, and energy loss factor caused by unavailability C.

Factor A, measured in [currency/unit power], indicates the increase of capital investment needed to reduce the unit no-load loss, which is calculated by using the expected rate of return. For the assumed life of the project, the equation that the annual cost NPV of capital investment for reduction unit no-load loss equals the impact on net income, as well as the present value of electricity purchase costs, can be used to derive value A. Due to the relatively low dielectric loss of modern submarine cable structure, the economic equivalent impact of no-load loss is negligible.

Factor B is defined as the increase of capital investment for reducing per unit of the load related losses, it can also be calculated by using the expected rate of return, and measured by [currency/unit power]. Wind varies at random, so the load loss factor is defined as the ratio of the average loss of wind power random changes in the evaluation cycle time to the loss at rated power. The equation between NPV of increased capital investment and the increased NPV of annual net income caused by avoiding 1 kW load loss is the equation for calculating factor B.

Factor C calculates the losses caused by system unavailability. It is measured by [currency/unit of energy/year], and is represented as the increased capital investment required for per unit of energy loss caused by system unavailability. It is also calculated by using the expected rate of return while considering the capacity factor of WFs. For a given WF location, the capacity factor is the measurement of random variation of the wind.

Capacity factor is the ratio of the power output within 1 year (presupposed wind speed distribution curve is

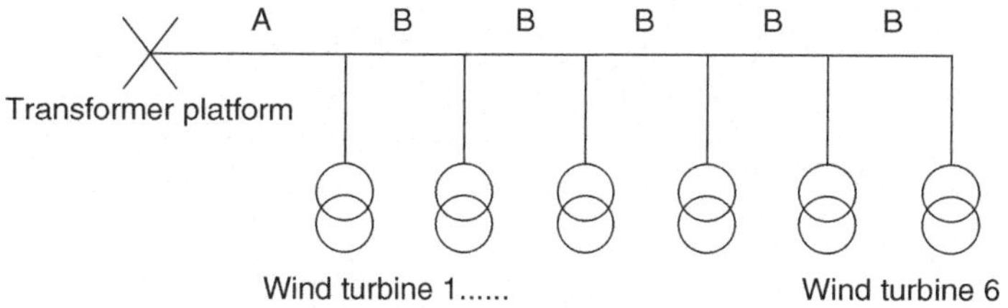

Figure 6.33 Radial series feed lines.

given) to the 1-year power output at peak power output all the time. Large WFs as well as all wind turbines can be measured by the capacity factor.

Once the estimated loss value of each electrical system is determined, we can calculate the estimated life-cycle cost by summing up the products of different types of losses multiplying their corresponding assessment factors. Because the loss ICCEq and the investment costs required to achieve a given design is similar in nature. Finally, calculate the total estimated costs (TEC) as the basis for selection, including both the loss of ICCEq and equipment installation costs.

On the basis of economic analysis of different scheme, the sensitivity analysis on the main factors affecting the economic evaluation results shall also be carried out. The factors can be considered for sensitivity analysis of power transmission schemes include transmission distance, cable installation and protection costs, converter station costs, converter station loss levels, availability of the converter station, and so on. Parameters of each factor is selected within the range of "low-medium-high."

6.6.3 Optimization of Transformer Platform Location and Cable Capacity

6.6.3.1 Optimization of Transformer Platform Location

The main objective of the transformer platform location optimization is to minimize the total length of the 35 kV cable that connects the wind turbine and the transformer platform. Transformer platform optimal location is based on the minimum sum of all distances between the transformer platform and each wind turbine. To simplify the analysis, we assume that the transformer platform is located near the middle of adjacent wind turbines. In fact, the transformer platform located in the middle place among the wind turbines, is also for minimizing the interference of the transformer platform with the airflow.

When calculating the distances between transformer platform and the associated wind turbines, wind turbine arrangement is assumed to follow certain rules of geometry, and the distances between all wind turbines are the same. At the same time, we should take into account the possibility of several wind turbines on the same cable line, so the calculation shall be made according to the number of feeders connected to the transformer platform. For example, for the radial series feeders shown in Figure 6.33, their connection distances can be calculated as follows:

The distance of the first wind turbine is A, the distance of the second wind turbine is A + B, and the distance of the third wind turbine is A + 2B, and so on, the distance of the sixth wind turbine is A + 5B. The total connection distance is 6A + 15B. Assume that A = B, the ratio of the total connection distance and the actual circuit length is 21/6 = 3.5. However, if A = 3B (corresponds to the large area WFs, which require expansion of the "end connection"), the ratio increases to 33/8 = 4.1.

As described above, for a WF with 40 wind turbines in specific arrangement, transformer platform optimal location S is shown in Figure 6.34.

6.6.3.2 Cable Capacity Optimization

Suppose a 150 MW OWF needs a 20 km, 132 kV cable for the connection to onshore grid. Technical assessment has determined that the minimum cable cross-sectional area for WF safe output is 630 mm^2. Developers seeking economically optimal design within the life of project prefer the use of larger cables. Taking 6 p/kWh as the cost of power loss, considering the WF output expected load duration characteristics, the annual loss costs for each cable size is shown in Table 6.9.

The life-cycle cost of cables with considered size can be calculated according to Table 6.9, as shown in Figure 6.35 (assuming that the discount rate is 6%, and the loss cost is calculated as £ 60/MWh).

Analysis shows that for projects of short lifetime, the life-cycle costs are heavily dominated by the initial costs. In this case, the smallest cable is cheap. However, as the project life time is prolonged, the affect of loss becomes more apparent, for lifetime longer than 13 years, the smallest cable is no longer the most economical one. As the lifetime is further extended, cables of a larger cross section gradually become the most economical cables.

The increase of the loss costs or lifetime will benefit cables of large capacity. Conversely, minimum initial cost option tends to be used for shorter lifetime and lower energy loss.

6.6.4 Transmission System Optimization

There are different schemes for OWF transmission systems: HVAC, HVDC, and so on, we should select the feasible schemes in accordance with the specific

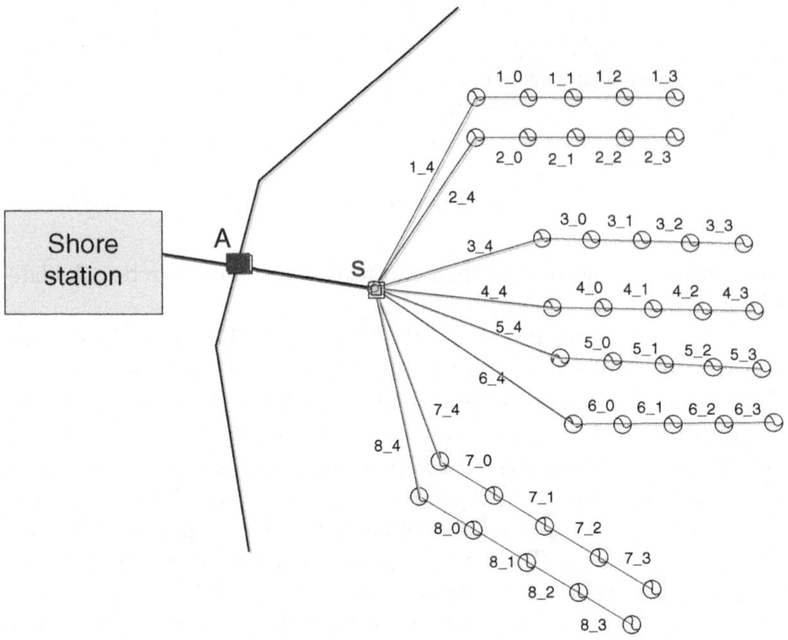

Figure 6.34 Optimization of the offshore transformer platforms.

Table 6.9 132 kV cable costs and losses.

Cable size	Annual loss value	Installation costs
630 mm^2	£176 k	£5.6 M
800 mm^2	£137 k	£5.9 M
1000 mm^2	£109 k	£6.3 M
1200 mm^2	£94 k	£6.8 M

conditions of WF, the distance from shore, the chosen offshore wind power generation system technology, and so on, and then carry out technical and economic comparison of different schemes to determine the optimal scheme. Literature [18] calculates and evaluates the transmission systems designed for OWFs. Evaluate system loss and determine the compensation device ratings required for connecting AC cables to the shore. When assessing loss, assume that the wind speed compliance with Rayleigh distribution function, the annual power generation curve is simplified to 18 operating segments, and the equivalent annual full load time is 3,820 h (average capacity factor of 43.6%). The losses of cables, transformers, reactors, and converters are considered for different connection schemes. Assume that all AC and DC cables are using XLPE insulation. The loss in transformer winding and iron core is considered as a constant. The loss of parallel compensation reactor is assumed have nothing to do with the actual reactive power. Assume that a three-level neutral-point-clamped (NPC) converter is used, when operating at full load, its efficiency is higher than 98%, the switching and on-state losses of the converter valve are shown in Figure 6.30.

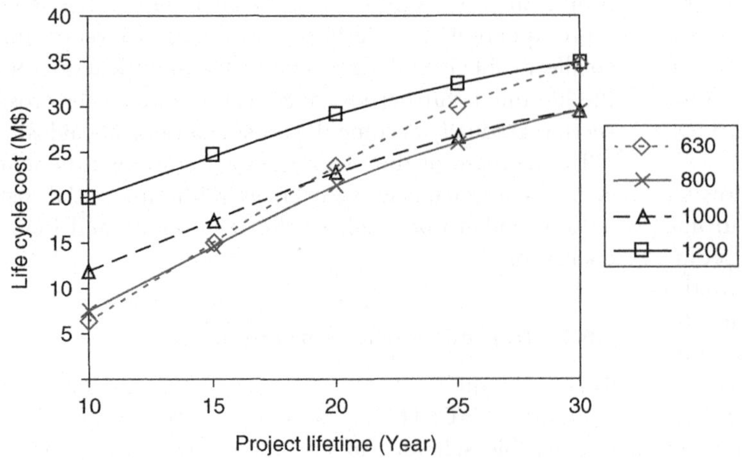

Figure 6.35 Comparison of 132 kV cable cost and life-cycle cost.

6.6.4.1 Transmission System Schemes

Carry out technical and economic analysis on WFs of three different capacities: small WF (100 MW), medium WF (100 + 100 MW), and large WF (250 + 250 MW). In the basic case, it is assumed that the WF is 60 km away from the shore.

1) **100 MW WF**

   In the case of 100 MW, in AC connection scheme, 150 kV submarine cable is used for the connection to the onshore main grid (Figure 6.36), and the VSC-HVDC connection scheme is shown in Figure 6.37.

   For VSC-HVDC connection, converter station (transformer) allows the required reactive power to be transmitted at the rated power of 120 MVA, and the transmission capacity of DC cable is 140 MW. Voltage level of the AC section of the converter has no significant effect on power losses, so the optimization is not required. Table 6.10 summarizes the performance of DC and AC cables.

2) **200 MW WF**

   For 200 MW case, two schemes are considered: AC, 150 kV cable connection; and VSC-HVDC, 150 kV cable connection. AC connection scheme is shown in Figure 6.38. Use three-winding transformer to connect two 100 MW WFs.

   The rated capacity of AC connected compensation reactor is 70 Mvar. In DC and AC connection schemes, the rated capacity of the transformer is 240 MVA, and it is allowed to operate at a power factor less than 1.0.

   The reactive power generated by 150 kV AC system is absorbed by the onshore reactor with a fixed capacity of 52 Mvar, which is the average value of the reactive power generated by the cable at no-load and full load.

3) **500 MW WF**

   For 500 MW case, three transmission schemes are considered: AC connection scheme, two 150 kV lines; AC connection scheme, one 400 kV line; and DC connection scheme, two VSC-HVDC lines of 150 kV.

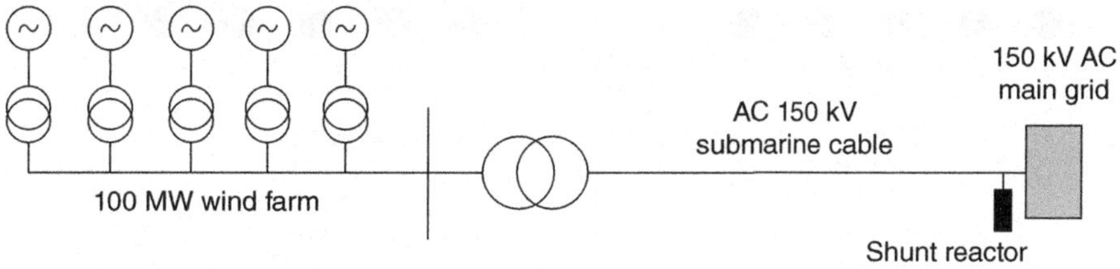

Figure 6.36 100 MW WF AC connection system.

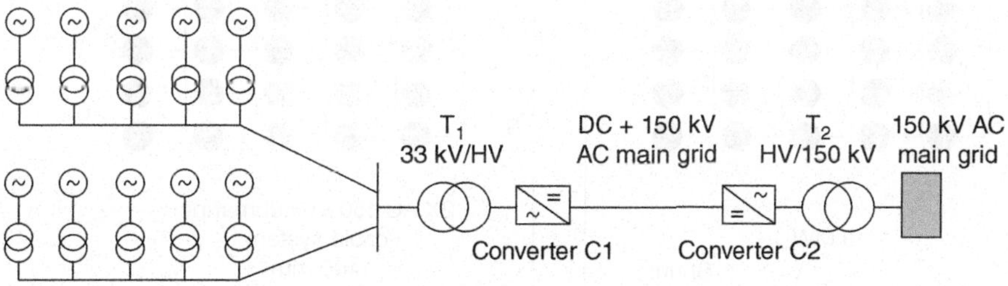

Figure 6.37 100 MW WF DC connection system.

Table 6.10 Performance of DC and AC cables.

Cross-sectional area	Type	Rated voltage	Rated current (65°C)	DC resistance (20°C)	Inductance	Capacitance	Dielectric loss	Charging current
mm²		kV	A	Ω/km	mH/km	μF/km	W/km	A/km
				AC cable				
95		33	250	0.193	0.38	0.17	11.6	0.90
150		33	315	0.124	0.36	0.2	13.7	1.10
240		33	410	0.0754	0.33	0.23	15.7	1.30
300		33	460	0.0601	0.32	0.25	17.1	1.40
400		150	595	0.0470	0.6	0.15	212	4.10
				DC cable				
240	DC	±150		0.0754				

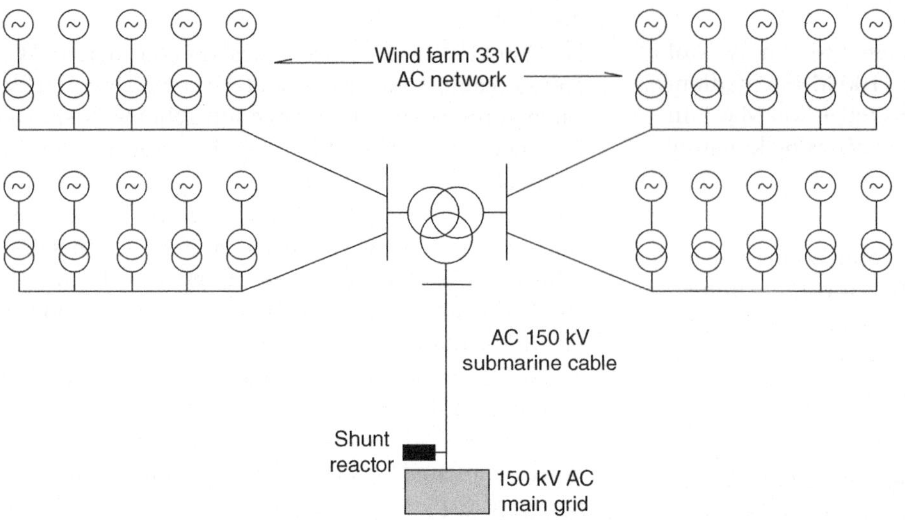

Figure 6.38 200 MW WF AC connection system.

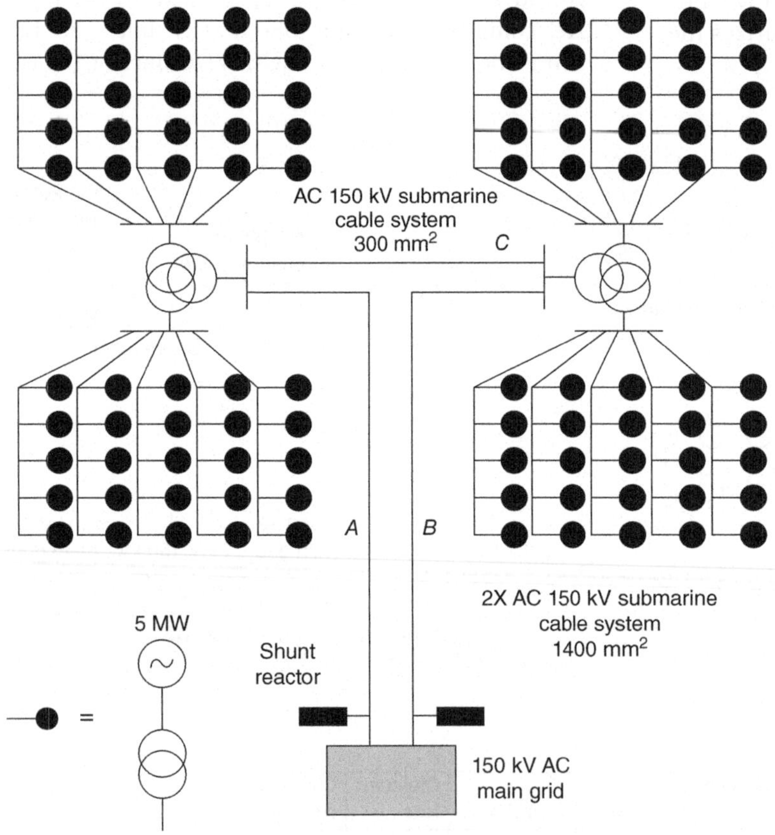

Figure 6.39 Electrical system of a 500 MW WF using two 150 kV submarine cables for AC connection.

In the AC, 150 kV connection scheme, as shown in Figure 6.39, the rated capacity of each line is half of the rated capacity of the WF maximum power (250 MW); cable C between the two systems is used to enhance system reliability. As long as the connection cost is less than the net present value of increasing energy output, the investment in improving reliability is positive and effective. Detailed analysis can be quantitatively explained by calculating the increasing energy output and the unavailability of connection to the shore, and the increasing energy output can be obtained from the WF annual output curve. Figure 6.40 shows the wind power that can be transmitted in single-loop or two-loop under the conditions with or without cable system C. If the out-service time of one of the two cables is greater than 1.13%, the selection of a scheme with cable C is cost effective. Two 150 kV AC lines are connected to onshore with reactors of 82 Mvar.

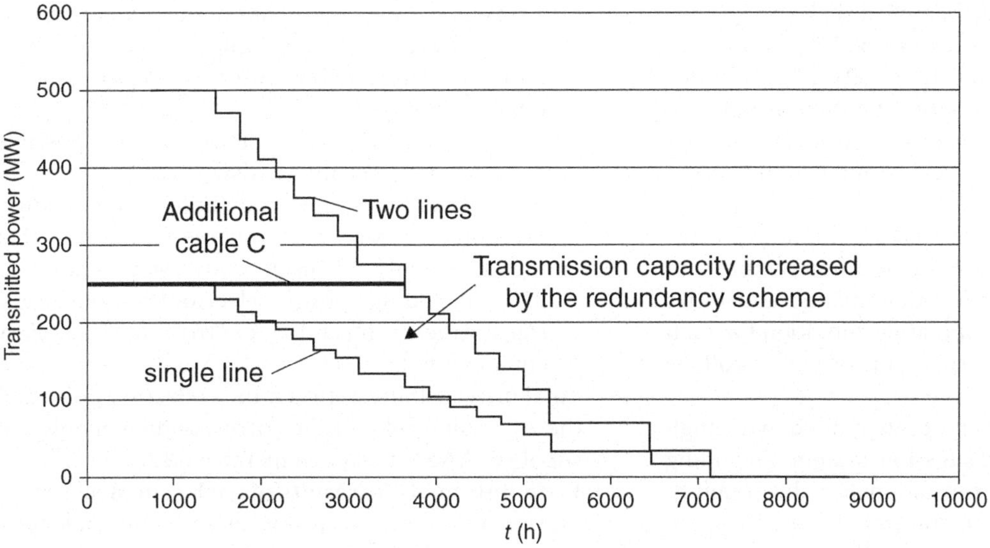

Figure 6.40 Annual power distribution transmitted to the shore and gain of additional cable.

For the AC connection scheme of one AC 400 kV, reactive power compensation is from offshore substation, and the other is from the onshore substation. The rated absorption of offshore reactor is 240 Mvar, equal to half of the total reactive power produced by the cable at no-load. The onshore reactor is rated at 222 Mvar, which is equal to the average value of the maximum and minimum reactive power provided to the substation onshore by the cable when the offshore reactor is in service.

The 150 kV DC connection scheme of a 500 MW WF includes two VSC-HVDC lines, of each the rated capacity is 250 MW, and they are connected to a 150 kV AC grid. In addition, a 150 kV cable is connected to the two halves of WFs. The scheme of using a VSC-HVDC line with the rated capacity of 500 MW is not considered in the study, because such VSC-HVDC technology is not available at present.

6.6.4.2 Comparison of Technical and Economic Analysis of the Connection Schemes

1) **Comparison of annual losses**

   The total annual energy loss of the electrical system between the WTGs and main grid connection point can be evaluated through repeated power flow calculations. The results are shown in Table 6.11 (including absolute value and percentage value). Annual variable energy losses are calculated according to the 18 operating segments on the production duration curve, and for the annual constant energy loss, the annual operating time is assumed to be 8,400 h. Comparing these three cases, the loss of HVDC-VSC is higher than that of AC scheme, and this is mainly because the loss caused by the converter is higher.

2) **Unavailability and outage power loss**

   In order to compare different schemes, the investment costs, the operating and maintenance costs (O&M),

Table 6.11 Comparison of annual energy losses of the three cases being studied.

Case	Transmission type	Annual variable energy losses		Annual invariant energy losses		Annual total energy losses	
		MWh	%	MWh	%	MWh	%
100 MW	AC	8923	2.33	4444	1.16	13367	3.49
	DC	20482	5.36	2993	0.78	23475	6.14
200 MW	AC	14739	1.93	7158	0.94	21897	2.87
	DC	39220	5.13	5482	0.72	44702	5.85
500 MW	AC 150 kV	36142	1.89	17264	0.90	53406	2.79
	AC 400 kV	28962	1.51	30173	1.58	59135	3.09
	DC	98054	5.13	13387	0.70	111441	5.83

the loss and the energy not supplied shall also be considered. Investment costs include 33 kV AC offshore grid cables and their installation costs, HV cables for connecting the WF to the shore and their installation costs (AC or DC), the costs of transformers, converters, AC switch stations, reactors, and support structure of offshore platforms.

Cable installation costs depend on the laying technology, the conditions of existing infrastructure and the machinery protection requirements, and other parameters, so the basic operating condition for each marine cable is about € 50,000/km (related to voltage levels).

The assessments of O&M costs are limited to the high voltage part of the transmission system (including submarine cables and substations). Cable maintenance costs for AC and DC transmission are assumed to be the same as € 200,000/year. The annual substation maintenance costs are estimated to be 0.4% of the total investment of transmission lines.

According to the government arrangements, they commit 10-year fixed subsidies for each kilowatt-hour of wind power production, until the maximum number of 18,000 full load hours is reached. The economic value of the loss and ENS at the beginning 5 years is € 97/MWh, and in the next 5 years is € 40/MWh.

ENS costs are assessed by reliability analysis of the transmission system. The collection system of 33 kV AC system is considered always available (irrelevant in the comparison), therefore, only the transmission system connected to the shore is considered. The equipment reliability data used in the assessment are as follows.

a) Availability of AC and DC high voltage cables are the same. Assume that within the connection lifetime (such as 20 years), each cable system may experience one failure. Assume that the mean time to repair is 30 days.
b) Transformer failure rate has nothing to do with its voltage levels (400 kV or 150 kV) and the power rating, and is 3. 44 × 10^{-2}/year. Mean time to repair is 21 days.
c) For the VSC-HVDC system, it is impossible to determine reliability data based on operational experience, because there is no publicly available recorded data. Therefore, assume that the availability of each conversion station is 99% (including power electronics converters, transformers, reactors, control and auxiliary equipment).

Total unavailability between the collection point and the AC main grid and the corresponding energy not supplied (ENS) are shown in Table 6.12.

3) **Economic analysis results of each schemes**

The economic analysis results are listed in Table 6.13. The assumed financial parameters are: nominal interest rate of 7%, inflation rate of 2%, and economic lifetime of 20 years.

In each scheme, the costs are mainly investment costs. For the AC connection scheme, the biggest investment is high-voltage cables; for the DC connection scheme, the biggest investment is the converter station. The main benefits of the DC scheme are lower cable manufacturing and maintenance costs; however, these savings are insufficient to cover the costs of converter station.

Power loss of VSC-HVDC connection is significantly higher than that of the AC connection, which, particularly affects the economy of 100 MW WFs. ENS fees between the two schemes are very different, because the assumed unavailability for each converter station is 1%. In any case, these expenses have no big impact on the final economic evaluation. In general, the study of each scheme has shown that: the VSC-HVDC scheme is more expensive than the AC scheme.

Table 6.12 Unavailability of transmission connection to the shore and the corresponding annual energy not supplied.

Case	Transmission type	Failure mode	Unavailability of the connection	Energy not supplied (GWh)	
100 MW	AC	Connection	0.0061	2.33	
	DC	Connection	0.0241	9.22	
200 MW	AC	Connection	0.0061	4.66	
	DC	Connection	0.0241	18.43	
500 MW	AC 150 kV	HVAC cable	0.0082	4.95	8.73
		Transformer	0.0040	3.78	
	AC 400 kV	HVAC cable	0.0041	7.85	11.64
		Transformer	0.0040	3.78	
	DC	Converter station/cable	0.0443	26.63	30.42
		Transformer	0.0040	3.78	

Table 6.13 Economic analysis results of all schemes.

Case	Transmission type	Investment costs [M€]	Annual zero loss costs [M€]	Annual zero O&M costs [M€]	Annual zero ENS costs [M€]	Annual zero total [M€]	Annual zero total [M€/MW]
100 MW	AC	58.21	10.03	5.03	1.75	75.02	0.7502
	DC	60.39	17.61	5.14	6.91	90.05	0.9005
200 MW	AC	78.86	16.43	5.43	3.49	104.21	0.5211
	DC	105.27	33.53	6.75	13.83	159.38	0.7969
500 MW	AC 150 kV	181.66	40.06	11.44	6.55	239.71	0.4794
	AC 400 kV	209.36	44.36	10.32	8.73	272.77	0.5455
	DC	260.74	83.59	15.41	22.82	382.56	0.7651

In the 100 MW scheme, the costs of two transmission options are slightly different, and the impact of wind turbine types on economic outcomes shall also be considered. Roughly take the price of the constant speed wind turbine as €889.3/kW, and that of the variable-speed wind turbine as €950/kW. The economic advantages of fixed speed wind turbine are insufficient to balance the higher costs of HVDC system. In fact, the total cost of power generation and transmission of AC transmission scheme is 172.32 million euro, slightly lower than the total costs of VSC-HVDC scheme (181.28 million euro).

4) **Sensitivity analysis**
Economic research results of the 100 MW OWF show that actual costs of HVDC scheme is 20% higher than the costs of AC scheme. In order to assess significance of the impact of related economic parameters on the economic assessment and the economic indicator conditions supporting the selection of DC schemes. Sensitivity analysis on the following parameters is carried out; transmission distance: basically 60 km, varying from 20 to 120 km; costs of cable installation and protection: basically €50,000/km, varying from €50000 to 200,000/km; costs of converter station: basically €110,000/MW varying from € 90,000 to 130,000/MW; loss level of converter station: basically 1.8% at full load of each converter station, varying from 1.2% to 2.41%; availability of converter station: basically each converter station is 0. 99, varying from 0.985 to 0.995.
For the 100 MW WFs, sensitivity analysis results show that when the connection length changes, the cost break-even point of AC and DC schemes is at about 90 km from the OWF to the onshore substation. After the sensitivity analysis carried out on cable laying and protection costs, the converter station costs, the losses in the substation, and substation availability, and so on, the total costs of the DC and AC schemes are converging. Finally, when considering the effect of different types of wind turbines, the comparison results become more favorable to DC schemes. Because HVDC allows the use of constant-speed wind turbines, and power electronic converters are not required, so the costs are less than that of variable-speed wind turbines.

5) **Comparison results**
A detailed technical-economic analysis is carried out on different transmission systems (150 kV AC, 400 kV AC and VSC-HVDC) for connection the three offshore WFs of different sizes (100 MW, 200 MW, and 500 MW) to the shore. The total results are the HVDC scheme is more expensive than the AC scheme, partly because the transmission distance considered in the analysis is 60 km, and the total investment costs of VSC-HVDC are higher. Moreover, VSC-HVDC is also much higher than AC in term of costs associated with the power loss. However, the results of sensitivity analysis on different parameters (transmission distance, cable protection, converter station costs, and reliability) show that, when the transmission distance to the shore is further than 90 km, HVDC seems the cheapest option.

6.6.5 Power Collection System Optimization

So far, most OWFs built in the world are of small scales, are electrically connected using simple connection, and usually follow some of the practices of onshore WFs. However, with further increase in the size of the OWFs, a suitable WF electrical system structure becomes critical, because the efficiency, costs, reliability, and performance of the entire WF depend largely on the structure of the electrical system. Electrical system involves all components between wind turbines and the power grid point of supply (point of common coupling between the WF and the power grid), including generators, switchgears, transformers, transmission cables, power electronic converters, and reactive power compensation equipments, and so on. The whole function of OWF electrical system is to bring together the power of all wind turbines, take it to the shore, and convert it into the appropriate voltage and frequency. The WF collection system completes the task of converging the electricity generated by all wind turbines in accordance with a certain way to the substation bus bars. For a given WF, the structure of the collection system is not unique, you can have different designs, depending on the size of the WF and the reliability requirements on the collection system.

OWFs are much more expensive than onshore WFs in terms of equipment cost and installation and maintenance costs. Section 6.2 lists some of the basic forms of the WF collection system structure, the economy and the steady-state performance of different system schemes are different, which may result in different costs, reliabilities, and efficiencies of system. Therefore, the optimization design of the structures of OWF collection systems is very important. When optimizing the collection system design of a specific WF, consideration is made from the technical and economic aspects, which mainly assesses the power losses of the collection system and changes in the collection system bus voltage under different operating conditions, and so on.

6.6.5.1 Power Loss

Key issues associated with the power loss is the energy loss value and the power factor in operation of the cable. In addition, introduction of more loss to the collection system is related to the wind turbine technology applied.

The MVA loss of each wind turbine in the power collection system series can be calculated in accordance with Formula (6.19). The formula estimates the total rated power of all wind turbines in the same series based on the simulated active and reactive power (P and Q) from the series to the offshore main substation.

$$S_L = \frac{\sqrt{(P_E - P_F)^2 + (Q_E - Q_F)^2}}{N_{WT}} \tag{6.19}$$

Where, S_L is the MVA loss of each wind turbine under the predetermined wind speed and power factor; P_E and P_F respectively represents the total power of all wind turbine in the same series connected to the offshore main substation and the active power obtained through simulation; Q_E and Q_F respectively represents the total reactive power of all wind turbine in the same series connected to the offshore main substation and the reactive power obtained through simulation; and N_{WT} is the number of wind turbines in the same series.

Collection system losses are related to the wind turbine operating power factor. Obviously, in operation under the unit power factor, the losses are minimal, and the operating power factor of uncompensated cage asynchronous generator is low, so the losses are greatest.

Literature [5] carries out simulation calculation for 1 GW WFs, and the results show that, the losses produced by unilateral circuit are minimal, followed by bilateral circuit and star structure, and the losses produced by radial structure are greatest. The study on the collection system of the Shanghai Donghai Bridge 100 MW Wind Farm in the literature [8] has obtained similar results.

6.6.5.2 Steady-State Voltage Level

Under normal operation, voltage levels of the entire collection system must be within the range allowed by the power grid specification. GB12325-1990 "Power Quality - Admissible Deviation of Supply Voltage" provides for the supply voltage of 35 kV or above, the sum of absolute values of positive and negative deviations of the supply voltage shall not exceed 10% of the rated voltage. Voltage regulation device shall adjust the bus voltage within its reactive power range by continuously adjusting the reactive power of the WF, or adjusting through the transformer tap in the collection center. A study on the collection system structure (with single hub connection) shows that voltage levels do not significantly change with the structure types, the voltage difference between different structures is less than 0.3%, and differences of the four schemes are in line with provision [5]. Simulation results of the literature [8] conclude that under normal system operation, the voltage difference of radial structure is 1.03%, the voltage difference of unilateral circular is 0.96%, the voltage difference of bilateral circuit is 1.00%, and the voltage difference of composite circuit is 0.98%. Unilateral circular scheme is optimal, and the radial scheme is worst.

6.6.5.3 The Impact of Redundancy Design

WF collection system redundancy design aims to keep as many wind turbines connected to the system as possible during emergencies. Experience has shown that the repair time of OWFs is much longer than that of onshore installation. For example, submarine cable failure repair time is 720 h in summer and 2,160 h in winter.

Collection system with or without redundancy are not terribly different in normal operation, but will be very different in failures. Redundancy is considered by ring structures. During emergency, such as breaker tripping and main cable disconnection caused by disturbance, power flow will be re-distributed, and the loads of some lines are heavy, resulting in increase of collection system losses and voltage deviations. From the view of power loss, the unilateral circuit is lowest, followed by composite circuit, and bilateral circuit is worst. From the view of bus voltage deviations of the three structures: unilateral circuit, bilateral circuit, and composite circuit all are conform to the prescribed standards.

6.6.5.4 Economy

The costs of typical collection system take up the smaller part in the total investment costs of the WF, particularly small-scale WF using the collection system of radial series. However, large-scale OWFs could significantly change the situation. Presently, in the new design, in order to obtain the maximum wind energy, redundancy requirements have been put forward for the collection system. It is clear that this will increase the costs of submarine cables for the increase of cable length or the use of cables of higher rated capacity, which will increase the investment costs of WFs.

The economic assessment contents involve all devices of an electrical system—wind turbines, 35 kV switchgear (such as isolators for remote operation, generator incomers, hub transformer feeders, hub bus tie lines and various devices), the 35 kV cables (cross-sectional area of 95, 185, 500, 630, or 2,000 mm^2, determined by the design) and buried cables, 35/110 kV transformer, and installation costs of all equipments.

The capital costs of the entire collection system and the WF can be estimated in accordance with Formulas (6.20) to (6.22) [4]:

$$C_{CB} = C_{CA}(\theta)N_{CA}I_{CA} \tag{6.20}$$

$$C_{CS} = C_{SW}N_{SW} + C_{TX}N_{TX} + C_{CB} \tag{6.21}$$

$$C_{WF} = C_{WT}N_{WT} + C_{CS} \tag{6.22}$$

Where, C_{CB}, C_{CS}, and C_{WF} respectively represents the costs of the cable system, the entire collection system and the WF. $C_{CA}(\theta)$ represents the submarine cable costs (i.e., costs of per km cable and its installation costs), which is the function of cross-sectional area θ. N_{CA} and I_{CA} respectively represents the total number and the length of the cable of each type (i.e., a specific cross-sectional area). C_{SW} and C_{TX} respectively represents the costs of switchgears and transformers (including equipment and installation). N_{SW} and N_{TX} respectively represents the total numbers of switchgears and transformers. C_{WT} represents the costs of wind turbine equipment (unit). N_{WT} represents the total number of wind turbines in the WF.

The literature [8] compares the construction cost, and divides them into three parts for calculation (Formula (6.23)):

$$C_{wf} = C_{wt} + C_{cs} + C_{cb} \tag{6.23}$$

Where C_{wf}—the total construction costs of the OWF; C_{wt}—the total costs of wind turbines (including blades, generators, gearboxes, control systems, and other auxiliary devices, towers etc.); C_{cs}—switchgear costs; C_{cb}—submarine cable costs.

The literature [5] compares per kW costs of the four concept collection systems. Results showed that the loop structure and non-loop structure are significantly different in economy. The investment costs of bilateral structure are high, the costs for installation of per kW wind power is £175.3, which is twice as great as that of a typical radial collection system (£83.3/kW). The cost of a unilateral circuit is £131.4/kW, which is 58% higher than that of a typical radial collection system. The cost of star structure is £81.3/kW, which is 3% lower than that of radial structure. For the costs is only determined by submarine cables, the differences of the four collection systems are obvious. Costs of bilateral and unilateral circuits are £120.2/kW and £74.2/kW respectively. Cables costs of radial and star structures are £30.4/kW and £24.8/kW respectively. Thus, the cable investment of bilateral circuit is about four times as great as that of the collection system, and cable investment of unilateral circuit is 2.5 times as great as that of radial system. The results obtained by the literature [8] show that the cost of submarine cables of radial structure is the lowest, and the cost of submarine cables of unilateral structure is 2.3 times as great as that of the radial structure, bilateral structure 1.2 times, and composite structure 1.4 times; from the view of total construction costs of the OWF, chain structure is the lowest, as 6556 yuan/kW, the unilateral circuit is 2.65% times greater than that of chain structure, and the bilateral structure is 0.4% greater, and the composite structure is 1% greater.

It should be noted that the above economic assessment has its limitations. In the WF planning process, not only the construction costs, but also the costs of the entire system in the life cycle shall be considered, and for transmission system, the operation and maintenance costs shall be considered as well. As the analysis above, although loop structure is not dominant in construction costs, its redundant structure can reduce the time to repair, making its operation and maintenance costs lower, and its operational reliability and efficiency higher, therefore, it may become the optimal choice for large-scale WF collection systems.

Optimum design of OWF collection systems mainly aims to select the optimal electrical connection under given wind turbine coordinates, in order to get the optimal economic and technical indicators. The investment costs of the electrical part of OWFs mainly depend on the numbers of wind turbine, location distribution, link topology of low-voltage collection system between wind turbine, voltage classes of high-voltage transmission lines, line length, type of cables, number of substations, and other factors. These factors are discrete and nonlinear, and interrelated and interacting with each other, so solving the investment optimization problems of the electric part of OWFs is complex.

The main structure of OWF electrical wiring optimization platform applied in the literature [19] is shown in Figure 6.41. Improved genetic algorithm is used for optimization. The platform includes several functional modules: cost calculation, reliability assessment, network loss calculation, AC-DC integrated power flow algorithm, and so on. An optimization goal is to consider the average production cost of reliability index (LPC). LPC is the levelized cost of per unit of electricity produced in discounted life cycle. It takes into account

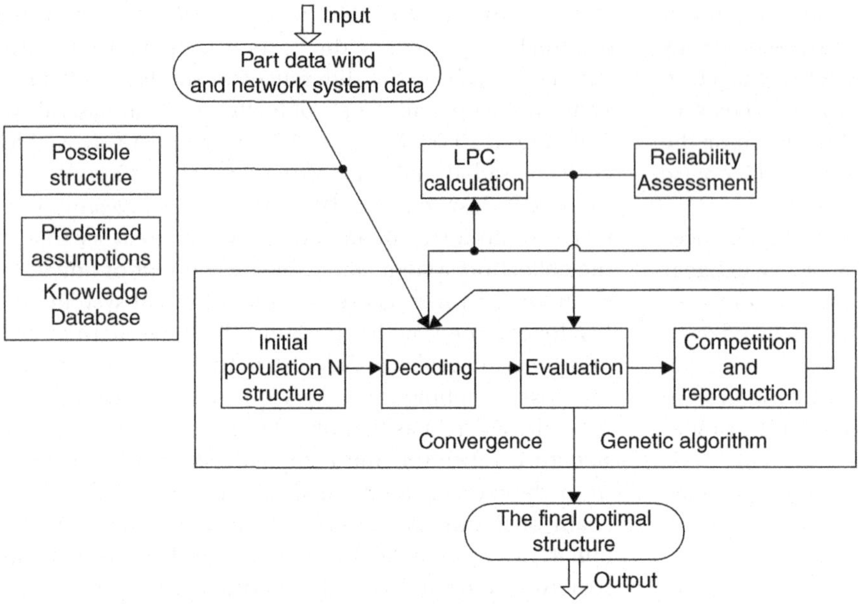

Figure 6.41 Main structures of optimization platform.

the investment costs, power loss, and electricity generated. The new reliability index-loss of generation ratio probability (LOGRP) is put forward when evaluating the electrical system. An AC/DC comprehensive power flow program is applied to variable-speed wind turbines. An AC/DC converter is integrated into basic DC flow algorithm by modifying Jacobian matrix. The two iterative methods have considered separately control strategies and power losses in the power electronic converter.

Conditions considered by OWF collection system optimization assessment is the selection of wind turbine types (3 MW, 3.6 MW, 4.5 MW, and 5 MW, etc.); the possible voltage classes at PCC (132 kV or 400 kV, etc.); cable costs (including installation costs); the methods of constituting a series (e.g., number of wind turbines in one series). Required data tables include wind speed distribution; data sheet of possible wind turbine types, including the power curve, rated voltage, cost, and available parameters; data sheet of possible types of cables used at sea; data sheet of possible cable types use on shore; data sheet of possible transformer types, and so on. These data can be used to form a private data sheet available for the optimization procedure. Taking into account the optimization search space is quite large, the genetic algorithm (GA) can be used for optimization.

6.7 Examples of Typical OWF

6.7.1 Development of OWFs

According to statistics, at the end of 2008, the installed global offshore wind power capacity was 1.473 GW, of which 99% is from Europe, and the offshore wind power installed capacity was 1% higher than the world's total wind power installed capacity (121.2 GW). The offshore wind power installed capacity newly built in 2008 was 350 MW, of which the growth rate was 30%. Britain is a leading country in offshore wind power, with the capacity of seen wind power farms in operation reaching 530 MW, the total capacity of the six projects under construction is 1.2 GW, and another 900 MW was added in 2012. The capacity of Denmark's Horns Rev and Nysted OWFs is 500 MW, and its offshore wind power capacity reached 1 GW as of 2012. Germany only installed three offshore wind farms in 2004 to 2008, including the Ems Emden wind farm installed in 2004 with a capacity of 4.5 MW, and the Hooksiel wind farm installed in 2008 with the capacity of 5 MW. The capacity of the first WF—Alpha Ventus WF—is 60 MW. In 2012, the installed capacity of offshore wind power reached 1.4 GW. Table 6.14 shows some of the offshore wind farms that have been put into operation.

At present, manufacturers that already have commercial production capacity of offshore wind power equipment include Vestas (Denmark), Bonus (Denmark), NEG-Micon (Denmark), GE Wind Energy (USA), Nordex (Germany), Enercon (Germany), REpower (Germany). The rated power of single turbine includes 2 MW, 2.3 MW, 3.6 MW, 4.2 MW, 4.5 MW, and 5 MW. Impeller diameters include 80 m, 85.4 m, 100 m, 110 m, 114 m, 116 m, and 126 m. The trend of using large-scale and mega-scale wind turbine generators has been very obvious. Table 6.15 shows the production status of offshore wind turbines.

6.7.2 Introduction of Several OWFs

A few early OWFs with some characteristics are briefly introduced here [2, 6, 20–22].

Table 6.14 Part of the OWFs have been put into operation in Europe [2–4].

Country	WF name	Number/model of wind turbine (capacity)	Putting into operation date	Offshore distance	Water depth	Wheel hub height	Foundation type
Denmark	Vindeby	11 Bonus (5 MW)	1991	1.5–3 km	2.5–5 m	37.5 m	Concrete Caisson type
Denmark	Tuno Knob	10 Vestas V39 (5 MW)	1995	6 km	3–5 m	40.5 m	Concrete Caisson type
Sweden	Bockstigen	5 Wind World 37 (2.8 MW)	1998	4 km	6 m		Single pile foundation Seat type
Sweden	Utgrundon	7 Enron Wind 70 (10.5 MW)	2000	12 km	7~10 m		Single-pile foundation Seat type
United Kingdom	Blyth	2 Vestas V66 (4 MW)	2000	1 km	6 m	58 m	Single-pile foundation Seat type
Denmark	Middelgrunden	20 Bonus 76 (40 MW)	2001	2–3 km	2~6 m	60 m	Concrete Caisson type
Sweden	Yttre Stengrund	5 NEG-Micon 72 (10 MW)	2001	5 km	8 m	60 m	Single-pile foundation Seat type
Denmark	Horns Rev	80 Vestas V80 (160 MW)	2002	14–20 km	6–14 m	70 m	Single-pile foundation Seat type
Denmark	Samso	10 Bonus 82 (23 MW)	2003	3.5 km	11–1 8 m	61 m	Single-pile foundation
Denmark	Nysted	72 Bonus 82 (165.6 MW)	2003	9 km	6–10 m	70 m	Gravity-based foundation
Ireland	Arklow Bank	7 GE 3.6 (25 MW)	2003	7 12 km	5 m	74 m	Single-pile foundation Seat type
United Kingdom	North Hoyle	30 Vestas V80 (60 MW)	2003	7–8 km	12 m	67 m	Single-pile foundation
United Kingdom	Scroby Sands	30 Vestas V80 (60 MW)	2004	2.3 km			Single-pile foundation
United Kingdom	Kentish Flats	30 Vestas V90 (90 MW)	2005				
United Kingdom	Barrow	30 Vestas V90 (90 MW)	2006				
United Kingdom	Beatrice	2 Repower 5 MW (10 MW)	2006				
the Netherlands	OWEZ	36 Vestas V90 (108 MW)	2006				
Sweden	Lillgrund	48 Siemens 2.3 MW Mk II (110 MW)	2007	7 km	4–8 m		Concrete

Table 6.15 Development status of offshore wind turbines.

Model (Country)	Rated output (kW)	Blade diameter (m)	Key Technologies
A4B182 (Germany)	2000	80	
CD87 = (Denmark)	2300	80, 90	Stall control, double-speed, asynchronous generator
EFG 01HD8 (Denmark)	2750	80, 90	Gear transmission, variable speed, adjustable blade inclination
I4 = JK = (Denmark)	3000	90	Gear transmission, variable speed, adjustable blade inclination, asynchronous generator
GF $ 182 (US)	3600	100	Gear transmission, variable speed, adjustable blade inclination
F846HD8 (Germany)	4500	112	Direct-drive, variable speed, adjustable blade inclination, synchronous generator
ED624L (Germany)	5000	115	Gear transmission, variable speed, adjustable blade inclination, asynchronous generator
MF NDB46 (Germany)	5000	115	Direct-drive, variable-speed, adjustable blade inclination
NO3424646 (Germany)	5000	115	Hybrid transmission, variable speed, adjustable blade inclination

6.7.2.1 Arklow OWF

Arklow is the first OWF invested in and constructed by the GE Company. The WF is located in southeast Ireland waters, and is installed with GE's seven wind turbines of 3.6 MW, with a total installed capacity of 25.2 MW, and a total investment of about 38 million pounds. Wind turbines of the WF are layout of sea bottom sand dunes, about 12 km away from the shore line, the mounting height was about 70 m, and the annual average wind speed of the farm is about 8.5 m/s.

The project construction period was about six months, of which the wind turbine hoisting and cable laying began in July 2003, the construction was completed in December 2003, and the WF started to produce power to the grid from March 2004. After being put into operation, except for the occurrence of one fault, of which the submarine cable was pulled off by ship, and nearly one month was spent for maintenance, the unit is operating normally in the rest of the time. The main problem now with the engineering is the wind turbine foundation has been severely corroded, but the unit operation has not been affected.

6.7.2.2 Horns Rev OWF

Horns Rev was built in Jutland in the North Sea by Elsam (the WF owner in Denmark) and Eltra Power Company since 2001; as the equipment supplier, Vestas had completed the installation of the WF (160 MW) in December 2002; the test items of the final batch of the WF was completed, and put in operation in 2003.

The installed capacity of Horns Rev OWF is 160 MW, the offshore substation includes the following facilities: 36 kV switch stations; 36/150 kV transformers; 150 kV switch stations; control and instrumentation, and communication systems; an emergency diesel generator, including 2 × 50 tons of fuel; employees and service facilities; helicopter platform; crawler cranes; an MOB (man overboard) yacht.

The whole project encountered a lot of problems in the early stage of operating: 16 units were scheduled to be put into use in July 2002, but because the main cable was in fault, a new cable of 150 meters long had to be connected, resulting in the delay of the construction period. In November 2002, two test units were damaged due to security system fault, resulting in the suspension of the development work of the entire WF. It was the blade being damaged. After verified, the cause of fault was human factor: the staff made mistakes during pre-commissioning, and interfered with the control system, causing the blade to operate over-speed, and over-speed could not be controlled due to security system failure, so the standby emergency system took the measure of emergency brake to stop the wind wheel, but the wind wheel could not withstand the high load, which resulted in damage of all the three blades. After the accident, Vestas decided to change unit braking procedures, so, all wind turbine was not put into trial operation until February 2003. And in the next one and a half years, the simultaneous operation time of the 80 units was only half an hour, but the number of total maintenance was 75,000 times, averaging two times for each wind turbine every day, of which the maintenance rate was relatively high.

In addition, offshore transformer was also a frequent fault source. In August 2003, technical faults started to occur in part of the transformers, and by the end of the year, about 20% to 30% of the transformers had encountered problems. There are a variety of factors that caused the faults, but the most common causes were the transformer being impacted by sea waves, insufficient resistance to salt spray corrosion, and seawater often entered into the transformer.

6.7.2.3 Denmark Samsø OWF

Samsø OWF was put into operation since February 2003, and had generated a total electricity of 159,246,260 kWh in two years. The original estimated annual electricity generation of the WF was 77,650,000 kWh, but the actual electricity generation was 79,623,1000 kWh, which is 102.54% of the expected capacity.

In the two years of operation, the problems of Samsø OWF mainly concentrated on the transformers and connection cables: because the bending of the power cable between the transformer and the unit main switch was too small, the cable was in long-term fatigue state, causing damage in the cable end sleeve and the insulation, thus resulting in overheating and smoking accidents. For these reasons, two transformers were respectively damaged in 2004 and 2005. The only solution was to replace the cable and transformers, which cost about 200,000 to 240,000 euros, equivalent to 0.07% of the total construction cost.

6.7.2.4 Denmark Nysted OWF

Nysted OWF was completed in 2003, and put into operation in December 2003, and its installed capacity was 165.6 MW. The cost of the entire project was under budget, and was put into operation a month before schedule.

The operation data of Nysted OWF in the first two years was very satisfactory: the equipment utilization rate was up to 97%, due to good accessibility of the units, the maintenance was convenient, so the equipment fault rate was low, and the WF was operating in good condition.

For the unit, problems mainly occurred in the gear box, if the high-speed bearings and intermediate bearings of the gear box were damaged, replacement was usually carried out during the periods of low wind speed, which would take 48 hours on average. In addition, due to insufficient quenching, the gear was of low intensity, as the operating time increased, cracks between teeth gradually expanded and eventually caused tooth fracture. It was recorded that 13 gear boxes had been damaged, which had seriously affected the generation capacity.

According to this, the measure of installing vibration monitoring system for units have been taken, this system is sensitive enough to detect any damage of the gear box before it is damaged. At present, two gear boxes have been detected by the monitoring system before the damage, so measures can be taken in advance to avoid the electricity generation being affected severely.

6.7.2.5 United Kingdom North Hoyle OWF

The installed capacity of North Hoyle OWF is 60 MW. Starting construction in April 2003, the WF was putting into operation in July 2004, and the equipment supplier Vestas was responsible for the operation and maintenance of the WF within the first five years.

In the construction phase of North Hoyle OWF, at first, the large vessels used for the hoisting of the units had problems, so most of the hoisting was completed by small vessels. Thus, the grid-connection project was delayed, and the first trial operation of the wind turbines was postponed to November 21, when the most severe whether occurred in the waters, and sudden storms could damage the installed machinery, which greatly increases the difficulty of construction. By the end of 2003, 27 of 30 wind turbines were installed, while the remaining three units were not installed until the middle of March 2004

North Hoyle OWF, as the United Kingdom's first major OWF, had carried out a lot of test projects about offshore wind power security and environmental factors, for example, the impacts of OWFs on the sea transportation, radio, mobile phone, radar, and satellite navigation and so on. Extensive research shows that these effects can be neglected.

6.7.2.6 Sweden Lillgrund OWF

Sweden Lillgrund OWF has 48 variable-speed cage asynchronous generators, of which the single-unit rated power is 2.3 MW. The electricity generated the generators is transmitted by voltage-source converter (VSC) to the OWF electrical system. As shown in Figure 6.42, the WF electrical system also includes a transformer

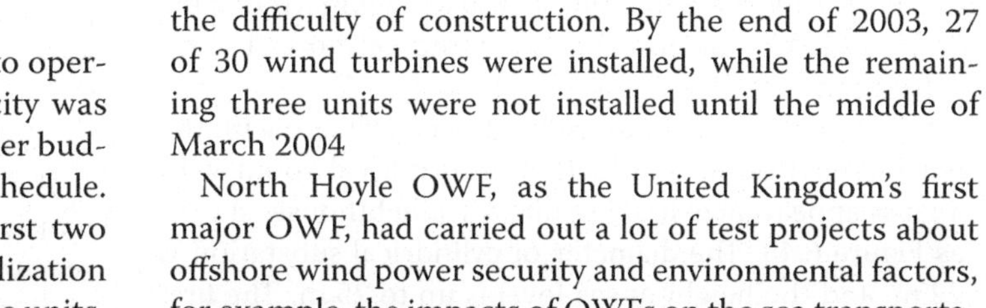

Figure 6.42 Cage asynchronous generators of Lillgrund OWF was grid-connected through VSC.

Figure 6.43 Lillgrund OWF substation.

and a 33 kV switch station. The switch station and the transformer are located in the bottom of the tower.

Lillgrund OWF has five feeders connected to the 33 kV switch stations of the offshore substation, and each feeder is connected to nine or ten turbines. The offshore substation includes control system and auxiliary power supply. The offshore substation is connected through a three-phase 130 kV submarine cable and a 130 kV onshore cable to the onshore 130 kV substation.

Offshore substation is designed to coordinate with the marine environment, which is a cylindrical room with a glass appearance, looking like a sea lighthouse, shown as Figure 6.43. The diameter of cylindrical substation is 22 m, and the height at sea level is up to 25 m. The first layer is an open cable deck. The second layer includes the transformer which is in the middle, the 33 kV switch room, the diesel generator and the transformer room for local power supply. The third layer includes the main transformer, UPS, battery backup system, and the control and monitoring equipment.

The offshore substation electrical equipments are same with mentioned in 6.2.2.

The offshore substation only has a main transformer, so the operational risk is high. Transformer fault may cause outage of several months. To reduce risk, emphasis has been put on extended factory acceptance tests, and besides the standard protection, monitoring, and control equipment, the transformer is also installed with gas sensor in transformer oil, an intelligent fault monitor that can read out the composite values of the gases. The gases generated by the flashovers caused by transformer internal fault and the composition of the gas will indicate the types of faults, so that early warning can be realized, thus informing the operators to developing fault conditions that could lead to transformer failure and unscheduled outages.

The 130 kV overland cable consists of three single-phase cables and one fiber-optic cable, whereas the 130 kV submarine cable consists of three-core copper conductor cables, each conductor is shielded with lead to avoid water penetration, and a fiber-optic cable is integrated in it. The 33 kV submarine cable consists of three-core copper conductor cables, each conductor is also shielded with lead, and the communication optical fiber cable is also integrated in it.

The 33 kV internal network of Lillgrund offshore substation is shown in Figure 6.44, and it is a single radial structure. The cross-sectional area of 33 kV feeder cables in the figure can be selected based on the magnitude and distribution of power flow. As shown in Figure 6.45, different conductor cross-sectional areas are selected for feeders connected to different wind turbines: the cross-sectional area of the feeders connecting the first to sixth WT is 95 mm^2, the cross-sectional area of the feeders connecting WTs 7 to 9 is 185 mm^2, and the cross-sectional area of the feeder between the tenth WT and the offshore substation is 240 mm^2.

The offshore substation is connected through 7 km submarine cable and 2 km overland cable to the onshore 130 kV substation, as shown in Figure 6.46. The submarine cable is directly connected to the offshore platform's main transformer, and there is no switch or circuit breaker at the high voltage side of the main transformer. The main circuit breaker is located in the onshore substation.

The reactive power produced by the 130 kV cable system is about 10 MVar. The system is not installed with reactors to absorb capacitive reactive power, in order to maintain the power factor at one at the point of common coupling, that is, exchange of reactive power is zero, so the reactive power can only be controlled by the VSCs in the wind turbines of the OWF.

When the 130 kV main circuit breaker of Lillgrund switch station switched on, the Bunkeflo substation on shore will go through serious transient oscillations. In case that the switching transient causes oscillation, which is composed of a high-frequency voltage (around 650 Hz) will overlapping the fundamental frequency (50 Hz). This oscillation is caused by the resonant circuit set up by the capacitance of the 130 kV cable and the inductance of main transformer. For special circumstances for the Lillgrund WF 130 kV system (with both capacitive load and inductive load), in order to reduce switching transients, the scheme of inserting resistor in advance is used, as shown in Figure 6.47. As shown in the figure, it has two circuit breakers A and B. When the switch is closing, circuit breaker A will close first, and the transient inrush current is limited by resistor R. After a few cycles, circuit breaker B is closed, then circuit breaker A is opened or the resistor is short connected, and then the closing process is completed.

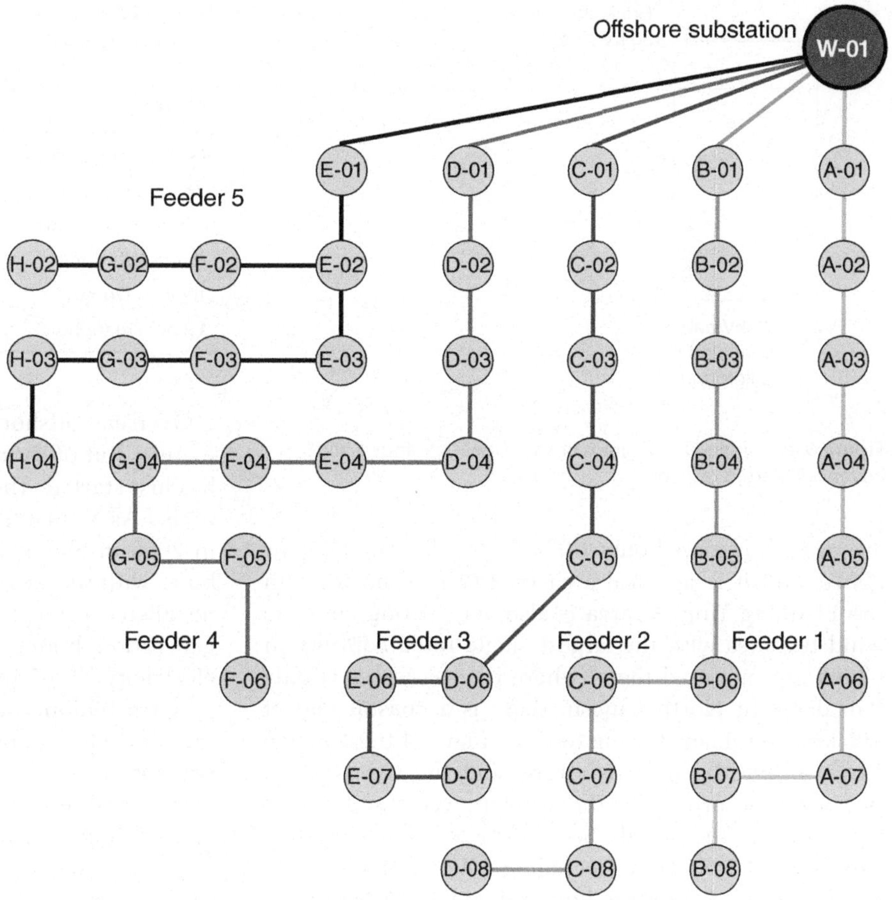

Figure 6.44 33 kV internal network of Lillgrund offshore substation.

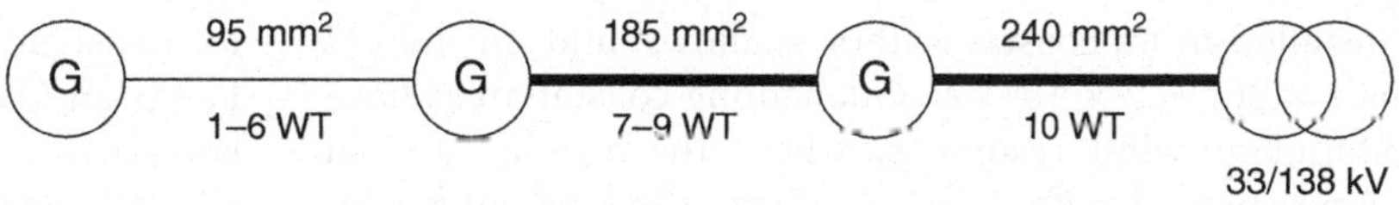

Figure 6.45 Different cross-sectional areas are used to feeder cables in the WF.

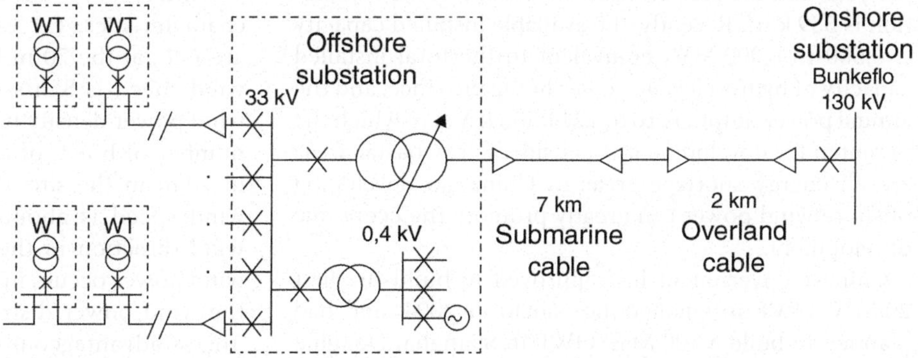

Figure 6.46 Single-line diagram of Lillgrund WF electrical system.

6.7.3 Prospects of China's Development in OWFs

China has a wealth of offshore wind resources. The depth of the vast waters along China's eastern coast is 2 m to 15 m, of which the available wind resource is about three times that of on land, that is, 700 GW, and is close to the power load center.

Along Tianjin and Hebei Huanghua coastal lines, the waters area within 20 km is only 10 m deep, the beaches are very vast, and the wind is very strong there, so the

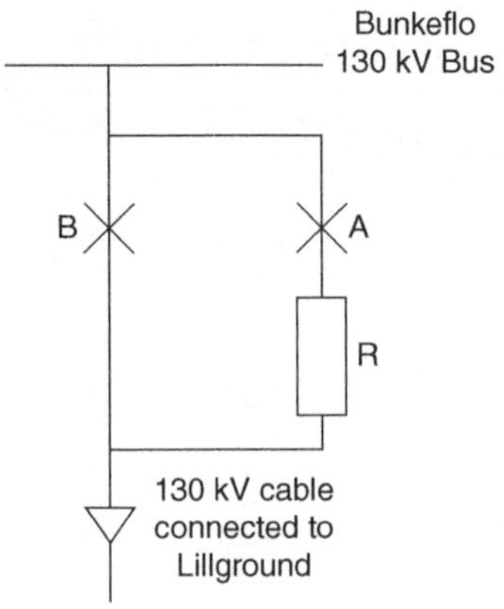

Figure 6.47 Single-line diagram of switchgear equipped with advance inserted resistor.

average utilization hours of WFs can be 3000 h/a or above, and it is an ideal area for OWFs. The wind in the Liaoning Yingkou area is also very strong, and the wind turbines with the largest single rated capacity in China are mounted there, where the sea water is only 7 m deep. In North Jiangsu, there is a coastal line of 700 km from Lianyungang to the estuary of the Yangtze River, where the wind is strong and the sea water is not deep, according to the wind power standards of Germany in the North Sea (6 km × 6 km, installed capacity: 50 × 104 kW, averaging 1.4 × 104 kW/km^2), wind power of nearly a hundred million kW can be installed within the offshore waters of 10 km. Fujian Pingtan is an island, where the wind power can be installed in its coastal waters, shallows, and on the hill is $1 \times 10^6 \sim 2 \times 10^6$ kW. Guangdong coastal areas have abundant wind resources, where the average annual wind speed is 6 to 7 m/s or above, the efficient power generation time is about 7500 h, taking up 85% of time per year, and the area available for wind turbine installation is 539 km^2. Recently, the available installed capacity is 5,500 to 6,000 MW, equivalent to the total installed capacity of hydro-electric power in the province, and the annual power output is 10 to 12 billion kWh, of which the potential for development is considerable. Coastal areas are all energy shortage areas in China, so developing offshore wind power can greatly promote the economic development.

Chinese government had approved to build the first 20 MW OWF in Guangdong Nanao in 2004, and had planned to build a 100 MW OWF in Shanghai Donghai Bridge (on side of Shanghai). In addition, the construction of OWFs in Zhejiang Cixi, Linhai, Daishan, Jiangsu Yancheng, Shandong Qingdao were in planning, and the OWFs under construction in China are shown in Table 6.16.

Table 6.16 Planned OWFs in China.

Location	Installed capacity (MW)
Guangdong Nanao	20
Shanghai	600
Zhejiang Daishan	200
Jiangsu Rudong	200
Jiangsu Dongtai	200
Hebei Huanghua	200

Overseas offshore wind power technology has been mature. But offshore wind energy development in China has just started. The construction of China's first, which is also Asia's first OWF—the Donghai bridge WF began in 2006 in Shanghai. The WF consists of 34 wind turbines with the single rated capacity of 3 MW, so its total installed capacity is 102 MW, the designed annual power generation hours is 2,624 hours, the annual on-grid electricity is 267 million kWh, and its total investment is 2.365 billion yuan. At 21:15 on September 4, 2009, the first three units were grid-connected for power generation.

The wind resources in Shanghai are abundant, and the area of Yangtze River offshore water is about 12,000 km^2, with the water depths of 5 to 15 m, which is available for wind energy development. The offshore wind energy reserved here is initially estimated to be 47 GW. According to the results of the feasibility study, the sites available for construction of large-scale WFs in Shanghai include Fengxian OWF, Nanhui OWF, and Huangsha OWF. By comparison of the construction conditions, Shanghai was intended to construct Yangshan port (Donghai Bride) OWF first.

Wind resources in Donghai Bridge have the following characteristics: (1) wind energy resources in the site are rich, at the 70 m height of the site area the average wind speed is 8 m/s or above, and the annual average wind power density is more than 500 W/m^2. (2) A larger number of hours of effective wind speed: at the height of 70 m in the site area, the annual hours of effective wind speed is about 8,000 h or above. (3) Prevailing wind direction in the site is stable, and the prevailing wind power occurs in south-east-southeast direction, so the wind power distribution is relatively concentrated. This is advantageous to the wind turbine arrangement, which can reduce electricity losses caused by the wake influence between wind turbines. (4) Small intensity of turbulence, the intensity of turbulence at the height of 70 m in the WF is about 0.12, which has little effect on wind turbine performance and service life. (5) Good distribution of daily wind speed and wind power, and the high values of daily wind speed and wind power in the WF occur in 15:00 to 23:00, matching well with the load curve, which indicates that the WF output power is close to the changes of load demand, and is conductive to the scheduling and regulation of power grid. (6) The number of annual equivalent load hours is relatively high, the WF annual hour of equivalent load is about 2,500 h.

According to the analysis of topographic features of the site area, submarine pipe line location and direction, channel location and direction, and other influencing factor analysis, Donghai Bridge OWF is arranged in parallel along two sides of the bridge, and the WF is about 1 km away from Donghai bridge at both sides, and is about 6 to 14 km away from Lingang New Town coastline. The rated capacity of each wind turbine is considered as 2 MW, 50 units are installed, so the total installed capacity is 100 MW. Fifty wind turbines are divided into four sets, and are connected to the self-constructed 110 kV substation of the WF through the four-loop 35 kV submarine cables, with a demonstration project of HVDC transmission system based on the modular multilevel voltage source converters (VSC-HVDC). Finally, the WF is connected through 110 kV overhead line or cable to Lingang New Town 220 kV substation.

References

1 Jie, Ren. 2006. New trend of offshore wind power. *Perspective of Science and Technology Economy*, **10**, 52–54.

2 Bin, Lu and Beiwen, Zhang. 2007. European offshore wind farm construction. *Shanghai Electric Power*, **2**, 129–134.

3 Chu, Song and Han-zhong, Liu. 2006. The development and trend of wind power plant at sea. *Electric Power Survey & Design*, **2**, 55–58.

4 Ackermann, Thomas. 2005. *Wind Power in Power Systems*. New York: John Wiley Sons.

5 Quinonez-Varela, G., Ault, G.W., Anaya-Lara, O., and McDonald, J.R. 2007. Electrical collector system options for large offshore wind farms. *IET Renew. Power Gener.*, **1**(2), 107–114.

6 Li, Zheng, Zhao, Menghua, and Chen, Zhe. 2006. Efficiency evaluation for offshore wind farms. International Conference on Power System Technology, October 22–26, pp. 1–6.

7 Larsson, Ake. 2008. Practical experiences gained at Lillgrund offshore wind farm. Seventh International Workshop on Large Scale Integration of Wind Power and Transmission Networks for Offshore Wind Farms, May 26–27, Madrid, Spain.

8 Jing, Jin, Qian, A.I., Lingling, Xi, and Xin, Zhang. 2007. Internal electrical wiring systems of off-shore windfarms. *East China Electric Power*, **35**(10), 20–23.

9 Ackermann, Thomas. 2002. Transmission systems for offshore wind farms. *IEEE Power Engineering Review*, **35**(10), 20–23.

10 Gengyin, Li, Peng fei, Lu, Guangkai, Li, and Min, Zhou. 2003. Development and prospects for hvdc light. *Automation of Electric Power Systems*, **27**(4), 77–81.

11 Martinez de Alegria, Iñigo, Martin, Jose Luis, Kortabarria, Iñigo, Andreu, Jon, Ibañez Ereño, Pedro. 2009. Transmission alternatives for offshore electrical power. *Renewable and Sustainable Energy Reviews*, **13**, 1027–1038.

12 Lu, Weixing and Ooi, Boon-Teck. 2003. Optimal acquisition and aggregation of offshore wind power by multi-terminal voltage-source HVDC. IEEE Transactions on Power Delivery, **18**(1), 201–206.

13 Jovcic, Dragan. 2008. Offshore wind farm with a series multi-terminal CSI HVDC. *Electric Power Systems Research*, **78**, 747–755.

14 Koutiva, X.I., Vrionis, T.D., Vovos, Nicholas A., and Giannakopoulos, Gabriel B. 2006. Optimal integration of an offshore wind farm to a weak AC grid. *IEEE Transactions on Power Delivery*, **21**(2), 987–994.

15 Prasai, Anish, Yim, Jung-Sik, Divan, Deepak, Bendre, Ashish, and Su, Seung-Ki. 2008. A new architecture for offshore wind farms. IEEE Transactions on Power Electronics, **23**(3), 1198–1204.

16 Hopewell, P.D., Castro-Sayas, F., and Bailey, D.I. 2006. Optimising the design of offshore wind farm collection networks. Universities Power Engineering Conference. Proceedings of the Forty-First International, September 6–8, pp. 84–88.

17 Scutariu, M. 2007. Techno-economical optioneering of offshore wind farms electrical systems. Power Tech, IEEE, Lausanne, July 1–5, pp. 2195–2200.

18 Bresesti, Paola, Kling, Wil L., Hendriks, Ralph L., and Vailati, Riccardo. 2007. HVDC connection of offshore wind farms to the transmission system. IEEE Transactions on Energy Conversion, **22**(1), 37–43.

19 Zhao, M., Chen, Z., and Blaabjerg, F. 2009. Optimization of electrical system for offshore wind farms via genetic algorithm. *IET Renew. Power Gener.*, **3**(2), 205–216.

20 Zhiqiang, Wang. 2006. Development of oversea wind farm and its inspiration. *Energy Technology*, **27**(6), 252–253, 256.

21 Zuoxia, Xing, Lei, Chen, Chao, Wang, Cheng, Cheng. 2006. Operation status and experience of European offshore wind park. *Renewable Energy*, **3**, 98–101.

22 Feng, Deng and Tieqiang, Wang. 2007. Analysis of the current situation of offshore wind farm construction and lessons learned. Proceedings of the Thirteenth China Marine (Offshore) Engineering Symposium, Nanjing, Jiangsu.

7

Analysis of Power Systems Containing Wind Power

7.1 Overview

In order to assess the impact of grid-connected wind farms (WFs) and to ensure that the network conditions can allow the effective operation of wind turbines, analysis of the power system is needed. In the past, we simply used hand calculations. Now we have computer programs for calculations. The pre-existing programs are generally developed on behalf of conventional generators (synchronous generators). For systems that use asynchronous generators, the programs need to be reconsidered. Commonly used power system analysis includes power flows calculation, fault calculation or short-circuit calculation, stability calculation, and electromagnetic transient calculation.

Due to the randomness and variability of the wind power, the calculation of the uncertainty associated with these characters is also needed, such as the probabilistic power flow, and the generating capacity credibility, etc. In addition, the economic assessment of WFs is different from that of conventional power plants, which requires special methods and calculation programs.

International Wind Agency (IEA) organized the comprehensive research on impacts of wind power integrated into power systems in 2006 to 2008 and drew the spatio-temporal map of effects of wind power on the system, as shown in Figure 7.1 [1].

The aim of this study is to analyze and further develop the method for evaluating the effects of wind power on the power system, focusing on power system reliability and efficiency (loss) and the analysis of system-wide effects rather than local effects.

The space-time range of different impact studies on grid-connected wind power is shown in Figure 7.1. Since the power quality mainly studies the voltage quality output by WFs, its space-time range is minimal, which is a local matter. The power system stability studies the possibility of wind power supporting system voltage and frequency when the system is disturbed, of which the time range is from seconds to minutes. The standby/management and dispatch study shows how the uncertainty of wind power injection affects system configuration and management of system backup. Unexpected changes in large WFs shall be combined with any other unexpected changes (e.g., load changes), as the time range considered is from a second to half an hour. The system efficiency and optimal unit commitment studies how wind variations and prediction errors change the operation modes of other power plants and the power flows between regions. Benefits of the WF grid-connection depend on the quantity of the wind, its relative location to the load center, and the system operation practice, of which the studied time range is from hours to days; power system adequacy studies effects of wind power grid-connection on system adequacy; wind power generating capacity credibility represents that, at the same level of reliability, to meet the same load, the ratio of the capacity of conventional unit required to the WF installed capacity. When determining the network adequacy, analysis on static power flow and dynamic system stability shall be carried out, and the studied time range is from hours to years.

Wind power brings more uncertainty to power system operation. The change of wind power covers the entire time span and is predictable, but the accuracy is worse than load forecasting. For large power systems, the cumulative effects of the geographically distributed wind power can be smooth, rapid fluctuations, and natural diversity power changes is a great advantage. However, the impact of wind power on the power system will rise with the increase of the propotion of the wind power. What's more, although it is technically possible to integrate large wind power into the power system, limited to socially and economically acceptable costs, the actual utilization of wind power scale is far below the scale of available wind resources.

In the electric power systems containing WFs, electromagnetic transient calculations shall describe the generators and power electronics converters in detail, using the detailed time domain simulation to simulate the electromagnetic transient process and quasi-steady-state from microseconds to several

Integration of Large Scale Wind Energy with Electrical Power Systems in China, First Edition. Zongxiang Lu and Shuangxi Zhou.

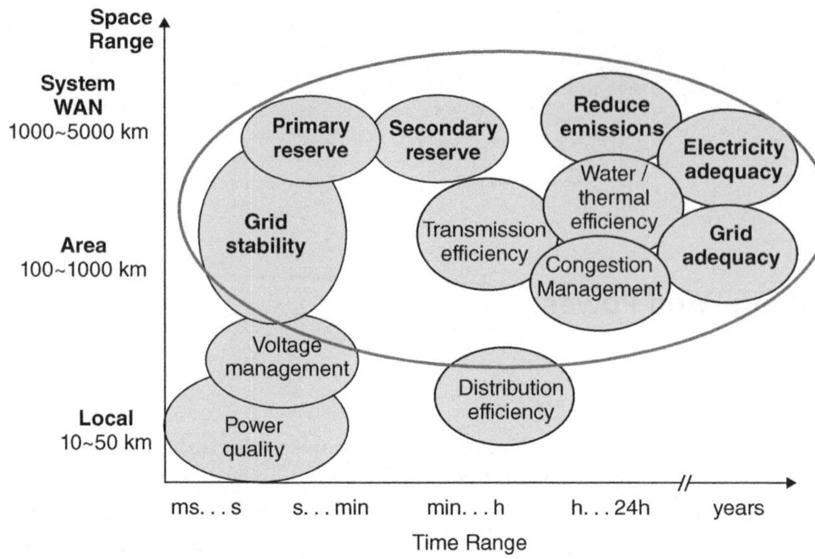

Figure 7.1 Spatio-temporal map of study on grid-connected wind power.

seconds. The electromagnetic transient calculation is not a routine calculation in the design of WFs. It is only used for studies on particular problems, such as characteristics of converters under the control of double-fed generators and direct drive synchronous generators, which is usually analyzed by softwares like EMTP and PSCAD.

This chapter establishes basic mathematical models required for analysis and study of power systems containing the large-scale wind power and make the power flow analysis (probabilistic power flow analysis), the fault analysis, the voltage stability analysis, the large disturbance (transient) stability analysis, the small disturbance (dynamic) stability analysis, and the frequency stability analysis.

7.2 Mathematical Models of Wind Power Generation System

Wind power generation is a method to convert wind energy into electricity, which is often referred to as the wind energy conversion system (WECS). WECS mainly includes absorption and conversion equipments of wind energy—wind turbines, connecting link, the hubs, gearboxes, couplings—as well as the conversion equipment for converting mechanical energy into electrical energy. When studying issues related to wind power connected to the system, in addition to appropriate models established for the above aspects of the single wind turbines, wind speed and wind power distribution models of WFs shall be established according to the need for resolution. In addition, because the capacity of a single wind turbine is low, WFs are usually large scale, and some even have hundreds or thousands of wind turbines. Thus, in order to correctly describe the characteristics of WFs, research on equivalence and simplification on WFs shall be carried out.

Mathematical models of wind turbine generator systems (WTGS) to be established for simulation calculation are related to simulation accuracy requirements and the object of study (steady-state analysis/transient analysis). The main objects of simulation analysis are power quality and stability. Besides the wind power model, the mathematical models of WTGS usually used in the simulation calculation include wind turbines, drives, and generator models.

7.2.1 Wind Speed Models

Wind speed is the driving force of wind turbines, yet its model is relatively independent against that of the WTGS. In the study of the steady state of the power system, in order to accurately describe the wind characteristics of randomness and variability, four-component wind model is usually used at home and abroad [2–4], that is, the wind is decomposed into four components: basic wind v_A, gust wind v_B, ramp wind v_C, and random wind v_D.

The wind components are calculated as follows:

1) Basic wind: basic wind always exists in the normal operation process of wind turbines. It determines the size of the rated power delivered by the wind turbine to system, and basically reflects the changes in average wind speeds of the WF. It can be approximately determined by the Weibull distribution parameters obtained from wind measurement of the WF (Formula (7.1)).

$$\overline{v} = A \cdot \Gamma\left(1 + \frac{1}{K}\right) \tag{7.1}$$

Where $\overline{v}$ is the basic wind speed, m/s; A and K are the Weibull scale parameter and shape parameter; $\Gamma(\cdot)$ is a gamma function.

2) Gust wind: Gust wind describes the characteristic of wind speed's suddenly changing (Formula (7.2)).

$$v_{WG} = \begin{cases} 0 & t \le T_{1G} \\ v_{cos} & T_{1G} \le t < T_{1G} + T_G \\ 0 & t \ge T_{1G} + T_G \end{cases}$$

$$v_{cos} = (v_{WGmax}/2) \times \{1 - \cos 2\pi[(t/T_G) - (T_{1G}/G_G)]\} \tag{7.2}$$

Where v_{WG} is the gust wind speed, m/s; T_G is the cycle; s; T_{1G} is the start-up time, s; and v_{WGmax} is the maximum value of gust, m/s. In the dynamic analysis of wind power systems, it can usually be used to assess dynamic characteristics of the system under the conditions of larger velocity changes.

3) Ramp wind: ramp characteristics of wind speed can be simulated by the ramp wind component (Formula (7.3)).

$$v_{WR} = \begin{cases} 0 & t \le T_{1R} \\ v_{ramp} & T_{1R} \le t < T_{2R} \\ v_{WRmax} & T_{2R} \le t < T_{2R} + T_R \\ 0 & t \ge T_{2R} + T_R \end{cases}$$

$$v_{ramp} = v_{WRmax}\{1 - (t/T_{2R})/(T_{1R} - G_{2R})\} \tag{7.3}$$

Where v_{WR} is the ramp wind speed, m/s; T_R is the holding time, s; T_{1R} is the start time, s; T_{2R} is the end time, s; v_{WRmax} is the maximum value of ramp wind, m/s.

4) Random noise wind: random noise wind speed reflects the random characteristics of wind speed changes. Random noise wind component is represented as (Formula (7.4)):

$$v_{WN} = 2\sum_{i=1}^{N}[S_V(\omega_i)\Delta\omega]^{1/2}\cos(\omega_i + \phi_i) \tag{7.4}$$

$$\omega_i = \left(i - \frac{1}{2}\right)\cdot\Delta\omega$$

$$S_V(\omega_i) = \frac{2K_N F^2|\omega_i|}{\pi^2[1 + (F\omega_i/\mu\pi)^2]^{4/3}}$$

Where ϕ_i is the random variable uniformly distributed between 0 to 2π; K_N is the surface roughness coefficient (generally preferable to 0.004); ω_i is the frequency of each frequency band; F is the disturbance range, m^2; μ is the average wind speed at the relative height, *m/s*; and N is the number of frequency spectrum sampling points.

To sum up, the wind velocity simulation acts actually on the wind turbine is (Formula (7.5))

$$v_W = \bar{v} + v_{WG} + v_{WR} + v_{WN} \tag{7.5}$$

Where v_W is the wind speed acting on the wind turbine, *m/s*.

In the study of transient state, because the power fault time is short, it can be considered that wind speed acting on the wind turbine remains the same in the process of transient state.

Since the wind speed accepted by the wind turbine is mainly the wind speed v_w at the hub height H, the wind speed from wind measuring point H_0 to the hub height H must be modified. The modified formula is (Formula (7.6))

$$v_w = v_{w0}\left[\frac{H}{H_0}\right]^\alpha \tag{7.6}$$

Where is wind speed of wind measuring point, *m/s*; α is the altitude correction factor, ranging from 0.10 to 0.40, and 1/7 is applied for general engineering.

7.2.2 Model of Wind Turbine System

The wind turbine mainly consists of blades, hubs, gearboxes, couplings, and other transmission devices, as shown in Figure 7.2 [5, 6]. Wind turbine model is expressed as a physical process of the blades extracting power from the wind, and the wind power is converted into the mechanical torque acting on the hub, which is transmitted through a gearbox from the low-speed shaft of the wind turbine to the torque of the high-speed shaft of wind generator.

1) **Wind turbine torque model.**
Wind turbine blades extract wind power from the wind, that is, the relationship between the power and wind speed is (Formula (7.7))

$$P_{wt} = \frac{1}{2}\rho A v_w^3 C_p(\lambda, \theta) \tag{7.7}$$

The relationship between the wind speed and wind turbine torque is (Formula (7.8))

$$T_w = \frac{1}{2}\pi\rho C_p \frac{R^3 v_w{}^2}{\lambda}\frac{\Omega_N}{P_N} \times 10^{-3} \tag{7.8}$$

Where P_{wt} is the power extracted from the wind, W; ρ is the air density, kg/m^3; A is the rotor blades swept area, m^2; T_w is the torque generated by wind turbine blades, kg-m; R is the blade radius, m; v_w is the wind speed acting on the blade, m/s; $\lambda = \Omega R/v_w$ is tip speed ratio; Ω is the blade mechanical angular speed, rad/s; Ω_N is the wind turbine rated mechanical angular speed, rad/s; and P_N is the wind turbine rated power (MW).

C_p is wind energy utilization factor (i.e., within a unit time, the ratio of the wind energy absorbed by the entire wind turbine to all the wind power passes through the wind wheel rotating surface). According to the Bates theory, at a high tip speed ratio of $\lambda > 3$, if the blade is airfoil shape optimized, the eddy current losses are small, and the C_p maximum value is 0.593. This means that even if all the energy of the wind is

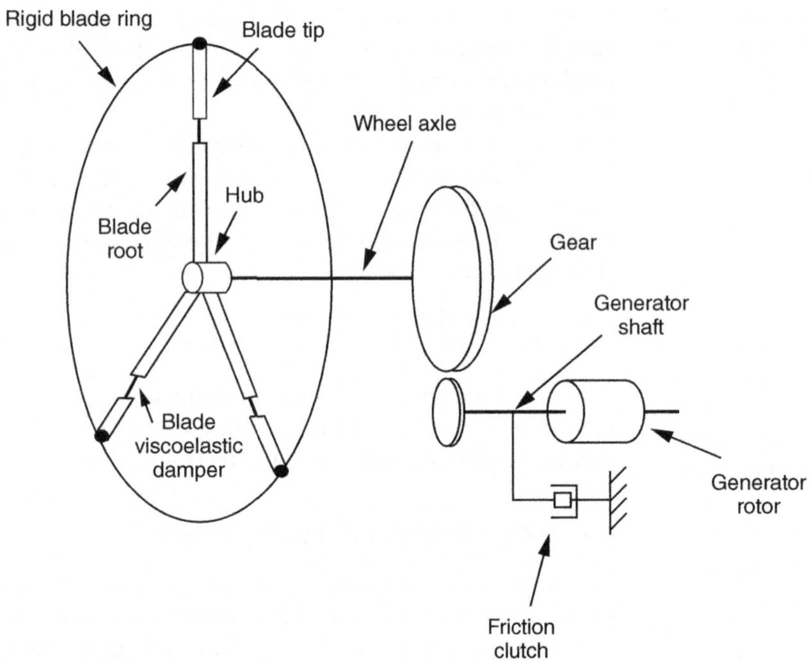

Figure 7.2 Diagram of wind power generation system.

absorbed without loss, only 59% of the energy is used. But Bates theory has not considered the eddy current losses. Wind turbine tip speed ratio λ is the ratio of wind turbine blade tip speed to wind speed, which is an important design parameter of the wind turbine. λ directly affects energy capture of the blade, and affects the wind energy utilization factor. Only when λ is at a certain value C_p will be maximal. In constant-speed wind turbines, due to the constant rotor speed and often changes of wind speed, λ cannot be maintained at an optimum value (even if it is variable pitch blade). C_p value is usually different from its maximum value, making wind turbine operating on an inefficient state. While variable speed wind turbine generators use regulating mechanisms to maintain λ at the optimum value, so as to improve the efficiency of wind turbines. In practice, relationships between the tip speed ratio λ, pitch angle β and wind energy utilization factor C_p can be expressed with Formula (7.9)

$$C_p = c_1(c_2 - c_3\beta - c_4\beta^\alpha - c_5)e^{-c_6} \tag{7.9}$$

Where $c_1 = 0.5$, $c_2 = \lambda$, $c_3 = 0$, $c_4 = 0.22$, $c_5 = 5.6$, $c_6 = 0.17\lambda$, and $\alpha = 2$ are the functions fitted by actual measuring data, which can satisfy the engineering applications.

In actual operations of wind turbines, restricted by wind speed, the torque of the wind turbine can be represented by segmentation according to the wind speed (Formula (7.10)):

$$T_w = \begin{cases} 0 & v_w < v_{in} \\ \frac{1}{2}C_P\rho A{v_w}^3/\Omega_N & v_{in} \le v_w \le v_R \\ P_N/\Omega_N & v_R \le v_w \le v_{out} \\ 0 & v_w \ge v_{out} \end{cases} \tag{7.10}$$

Where $v_{in,}$ v_R, and v_{out} respectively represent the cut-in wind speed, rated wind speed, and the cut-out wind speed.

2) **Transmission system model**
According to different nature and objectives of the research questions, the wind turbine shaft system modeling methods are different. Usually, if the stress distribution analysis and the design of mechanical strength are not required, the concentrated mass of block equivalent method can be used.
   a) If the wind turbine, the gearbox, the drive shaft and the generator rotor are equivalent to a concentrated mass of block, the motion formula of the shafting model will be (Formula (7.11)):

$$2H_M\frac{d\omega_M}{dt} = T_w - T_e - D_M\omega_M \tag{7.11}$$

   Where H_M is the inertia time constant of the equivalent mass of block; D_M is the damping coefficient of the equivalent mass of block; ω_M is the electric angular speed of the equivalent mass of block, that is, the electric angular speed of the wind turbine and the generator; T_w and T_e respectively represents the mechanical torque of the wind turbine and the generator electromagnetic torque.
   b) If the wind turbine and the generator rotor is respectively equivalent to a mass block, a two-mass block wind turbine shaft model can be established. Converting various amount of the wind turbine low-speed shaft to the generator rotor side, the shaft motion formula can be expressed with per-unit value as (Formula (7.12)):

$$2H_w\frac{d\omega_w}{dt} = T_w - K_s\theta_s - D_s(\omega_w - \omega_g) - D_w\omega_w$$

$$2H_g \frac{d\omega_g}{dt} = K_s\theta_s - T_e + D_s(\omega_w - \omega_g) - D_g\omega_g$$
$$\frac{d\theta_s}{dt} = \omega_0(\omega_w - \omega_g) \quad (7.12)$$

Where H_w and H_g respectively represents the inertia time constant of the wind turbine and the generator rotor (including gearbox); ω_w and ω_g respectively represents the electric angular speed of the wind turbine and the generator rotor; θ_s is the wind turbine angular displacement relative to the generator rotor; D_s is the damping coefficient between the wind turbine and the generator; D_w and D_g respectively represents the damping coefficient of the wind turbine and the generator; K_s is the rigidity coefficient of the transmission shafting, of which the value can be estimated based on the shafting torsion frequency.

c) If the flexibility of the blade is considered, and the blade is equivalent to a mass block, then the shaft motion formula will be changed from Formulas (7.12) to (7.13)

$$\begin{cases} 2H_b \frac{d\omega_b}{dt} = T_w - K_{bh}\theta_a - D_{bh}(\omega_b - \omega_h) - D_b\omega_b \\ 2H_h \frac{d\omega_h}{dt} = K_{bh}\theta_a - K_{hg}\theta_b + D_{bh}(\omega_b - \omega_h) \\ \qquad -D_{hg}(\omega_h - \omega_g) - D_h\omega_h \\ 2H_g \frac{d\omega_g}{dt} = K_{hg}\theta_b - T_g + D_{hg}(\omega_h - \omega_g) - D_g\omega_g \\ \frac{d\theta_a}{dt} = \omega_0(\omega_b - \omega_h) \\ \frac{d\theta_b}{dt} = \omega_0(\omega_h - \omega_g) \end{cases} \quad (7.13)$$

Where H_b, H_h, and H_g respectively represents the per-unit inertia constant of the concentrated equivalent mass of blocks the three blades of the wine turbine, the hub and the generator rotor (including the gearbox); ω_b, ω_h, and ω_g, respectively, represents the electric angular speed of the three mass of blocks; $\omega_0 = 2\pi f$ is the base value of system electric angular speed; θ_a and θ_b respectively represents the wind turbine blade angular displacement relative to the hub, and the hub angular displacement relative to the generator rotor; D_{hb} and D_{hg}, respectively, represents the damping coefficient between the wind turbine blade, the hub, and the generator; D_b, D_h, and D_g, respectively, represents the damping coefficient of the equivalent three equivalent mass of blocks; K_{bh} is the equivalent rigidity coefficient between the wind turbine blade and the hub; K_{hg} is the equivalent rigidity coefficient of the low-speed shaft between the hub and the generator, of which the value can be estimated based on the torsion frequency of the blade and the shafting.

In the electromechanical transient simulation of power systems, the establishment of a detailed mathematic model for each link of the transmission mechanism of the mechanical part is not required, and transmission losses can be neglected. The expansion of the wind turbine transmission system is shown in Figure 7.3. Two methods are used here for the simulation of the transmission part. One is the simplified description method that considers the transmission system as a rigid device, and use first-order inertia link to represent the time lag of wind power going through the blades to reach the generator for working. The simplified transmission model is (Formula (7.14)) [7–10]:

$$\frac{dT_m}{dt} = \frac{1}{\tau_h}(T_{ae} - T_m) \quad (7.14)$$

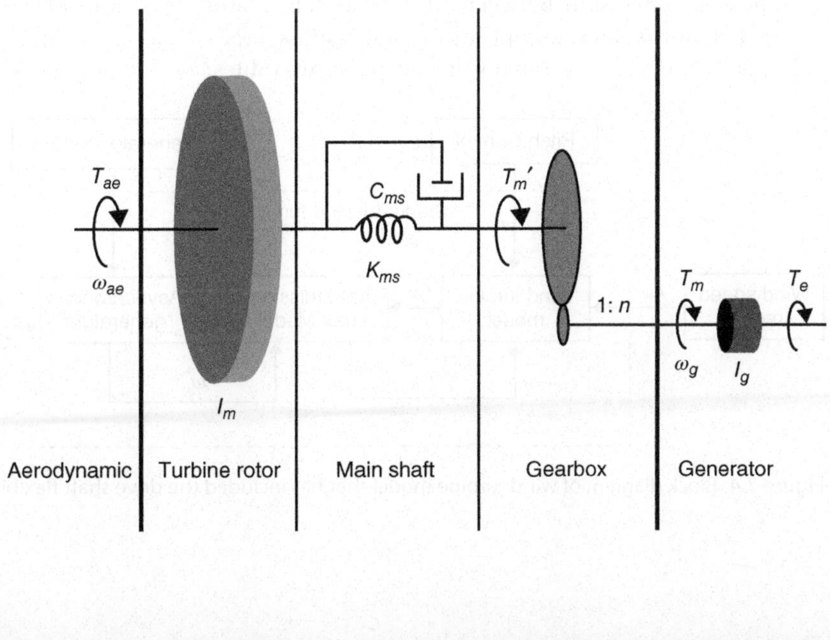

Figure 7.3 Wind turbine transmission system.

Where T_{ae} is the wind turbine output torque; T_m is the mechanical torque input to the generator; τ_h is the inertia time constant of the transmission part.

Another is the description method that considers the drive shaft flexibility. Because the inertia of the wind turbine in the wind power system is large, and the inertia of the generator rotor is small, in order to connect two devices with large inertia difference, the drive shaft must be flexible. Shaft of the conventional synchronous generator is much longer than that of the small asynchronous generator, but its rigidity is about 20 to 80, so it can be considered as a rigid shaft. And the rigidity of the wind turbine transmission shaft is 0.15 to 0.40. Also, the inherent oscillation frequency and torsion frequency of wind turbine systems that have considered the transmission shaft flexibility is about 1 Hz, which is in the same range with the typical characteristic frequency of the power system. Thus, in order to reflect the wind turbine dynamic process after disturbances more accurately, flexibility of the drive shaft shall be considered in the wind turbine model. Block diagram of the wind turbine model which has considered the drive shaft flexibility is shown in Figure 7.4 [8–10].

The drive shaft flexibility is indicated in the low-speed part of the shaft. Its characteristic parameters include the rigidity coefficient k_m and the damping coefficient D_m. Select the torsion angle difference θ_k between both ends of the flexible shaft and the wind turbine mechanical angular speed ω_{ae} as state variables, generator rotor angular speed ω_g and the wind turbine mechanical torque output T_{ae} as inputs, and the mechanical torque input to the wind turbine T_m and the wind turbine rotor angular speed ω_g as outputs. The inertias of the wind turbine rotor and the generator rotor is respectively represented by J_m and J_g, which is respectively converted to $\tau_m(s)$ and $\tau_g(s)$ by using the per-unit value. In order to interface better with the electrical parts and to avoid repeated conversion between the actual values and the per-unit values, except for θ_k (Rad) and k_m (N · m/rad), the above variables are all per-unit values.

The state formula and output formula are listed below (Formulas (7.15) and (7.16)) [8]:

$$\begin{bmatrix} \dot{\theta}_k \\ \dot{\omega}_{ae} \end{bmatrix} = \begin{bmatrix} 0 & \Omega_B \\ -\dfrac{k_m\Omega_B}{\tau_m S_B} & -\dfrac{D_m}{\tau_m} \end{bmatrix} \cdot \begin{bmatrix} \theta_k \\ \omega_{ae} \end{bmatrix} + \begin{bmatrix} 0 & -\Omega_B \\ \dfrac{1}{\tau_m} & \dfrac{D_m}{\tau_m} \end{bmatrix} \cdot \begin{bmatrix} T_{ae} \\ \omega_g \end{bmatrix} \tag{7.15}$$

$$\begin{bmatrix} T_m \\ \omega_{ae} \end{bmatrix} = \begin{bmatrix} \dfrac{k_m\Omega_B}{S_B} & D_m \\ 0 & 1 \end{bmatrix} \cdot \begin{bmatrix} \theta_k \\ \omega_{ae} \end{bmatrix} + \begin{bmatrix} 0 & -D_m \\ 0 & 0 \end{bmatrix} \cdot \begin{bmatrix} T_{ae} \\ \omega_g \end{bmatrix} \tag{7.16}$$

Where S_B is the wind turbine rated power; Ω_B is the rated mechanical angular speed.

3) **Pitch control system**

The blades of variable pitch wind turbine can rotate around the blade central shaft, its pitch angle β changes between 0° to 90°, which can adjust the wind energy captured by the wind turbine in a certain range. When the unit fails, the blades can be feathered to improve the force of the unit, and ensure the safety of the system. Under normal operating conditions, the variable pitch wind turbine also has better aerodynamic characteristics, it maintains the wind turbine to follow the maximum value of C_p under rated wind speed by adjusting the blade pitch angle, so as to improve the efficiency of the wind energy utilization, and maintain a constant power output of the wind turbine above the rated wind speed. Variable pitch wind turbines are suitable for regions with higher frequency of wind speeds higher than the rated wind speed, which can significantly increase the power generation capacity. However, since a variable pitch mechanism is added to the variable pitch wind turbine, the failure rate of the unit is also increased, making the process of bearing design and troubleshooting more complex.

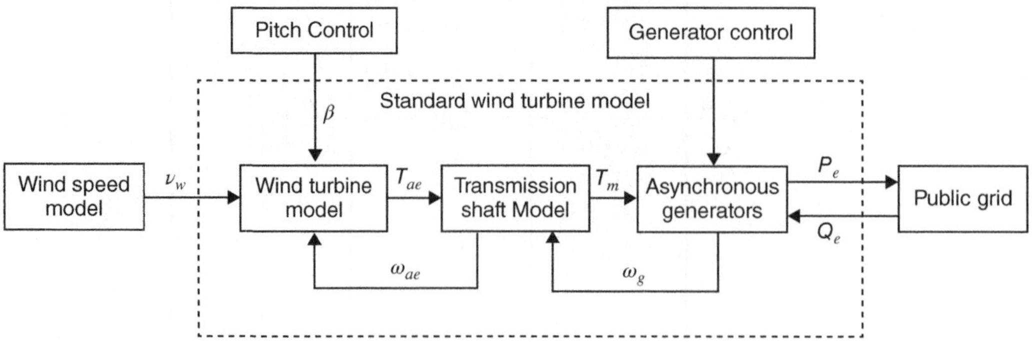

Figure 7.4 Block diagram of wind turbine model that has included the drive shaft flexibility.

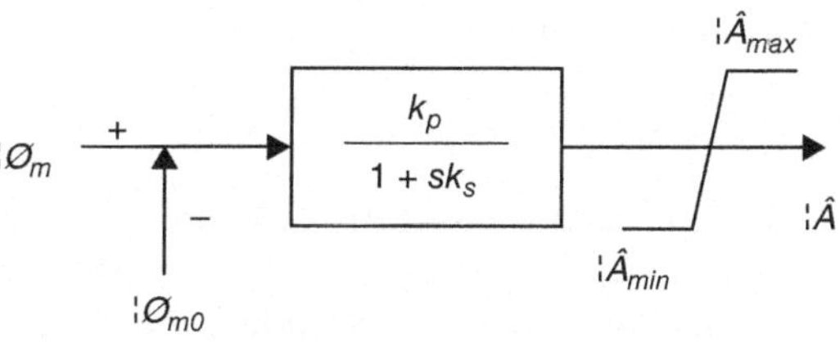

Figure 7.5 Regulating links of variable speed wind turbine pitch angle.

The pitch angle of the variable speed wind turbine can be adjusted by adjusting the speed deviation of the wind turbine rotor, which is called the rotor speed control. It can also be adjusted by controlling the wind turbine output power, which is called the power control. Because the inertia of the wind turbine blade is large, it is quite difficult to make the power regulation speed catch up with the changes in wind speed. In fact, if there are no other measures (such as the control of generator rotor current), the power regulation of the variable pitch wind turbine will be powerless to the high frequency wind speed variation. See Formula (7.17) for the variable pitch wind turbine speed control model (Formula (7.17)) [11, 12]:

$$\frac{d\beta}{dt} = \frac{1}{k_s}(k_p \Delta\omega_m - \beta) \tag{7.17}$$

Where k_s is the inertia time constant of the regulating device (including the delayed effect of measurement links); kp is the gain coefficient; $\Delta\omega = \omega - \omega_{m0}$, is the wind turbine speed deviation. Its control block diagram is shown in Figure 7.5.

The power regulation system of a variable pitch wind turbine is shown in Figure 7.6.

When the wind turbine is operating normally, the power control is mainly used, the control of the pitch angle is determined by Formula (7.18)

$$\begin{cases} \dfrac{d\beta}{dt} = \dfrac{1}{\tau}(\beta_c - \beta) \\ \dfrac{dY}{dt} = k_{pi}(P_e - P_{ref}) \\ \beta_c = k_w v_w + k_{pp}(P_e - P_{ref}) + Y \end{cases} \tag{7.18}$$

Where Y is the intermediate state variable introduced in the integration element; τ is the time constant of the controller servomechanism; k_w, k_{pi}, and k_{pp} are controller parameters; P_e and P_{ref} respectively represents the active power output by the wind turbine and the rated active power.

7.2.3 Cage Asynchronous Generator Model

7.2.3.1 Steady-State Model of Cage Asynchronous Wind Turbines

Grid-connection system of cage asynchronous wind turbines is shown in Figure 7.7 [13]. The cage asynchronous generator (CAG) is directly connected to the power grid, and coupled with the wind turbine rotor through a gearbox. This generator system is constrained by the grid frequency, of which the speed change is small; it can be considered that this wind turbine operates at a fixed speed. At high wind speed, the wind turbine rotor uses the stall action (the wind turbine stall regulation) or the blade pitch angle (the wind turbine pitch regulation) to limit the power extracted from the wind. CIG needs to absorb reactive power to establish a magnetic field, so a parallel capacitor is installed at the generator terminal to provide the asynchronous generator with the excitation current, thus improving the power factor.

The CIG steady state equivalent circuit is shown in Figure 7.8. Wherein, subscript 1 represents the physical quantity on the stator side, subscript 2 represents the physical quantity on the rotor side. Only the quantity on one side is considered and all quantities are converted to the stator side.

Ignore the stator resistance and core power losses, and because $x_m \gg x_1$, asynchronous generator Γ-type equivalent circuit can be obtained through simplifying, which gives the following relationship (Formulas (7.19–7.22)) [14]:

$$P_e = -\frac{U_1^2 R_2/s}{(R_2/s)^2 + X_k^2} \tag{7.19}$$

$$s = -\frac{U_1^2 R_2 - \sqrt{U_1^4 R_2^2 - 4P_e^2 X_k^2 R_2^2}}{2P_e X_k^2} \tag{7.20}$$

$$\varphi = \arctan\left(\frac{R_2^2 + X_k(X_k + X_m)s^2}{R_2 X_m s}\right) \tag{7.21}$$

Figure 7.6 Variable pitch power regulation system.

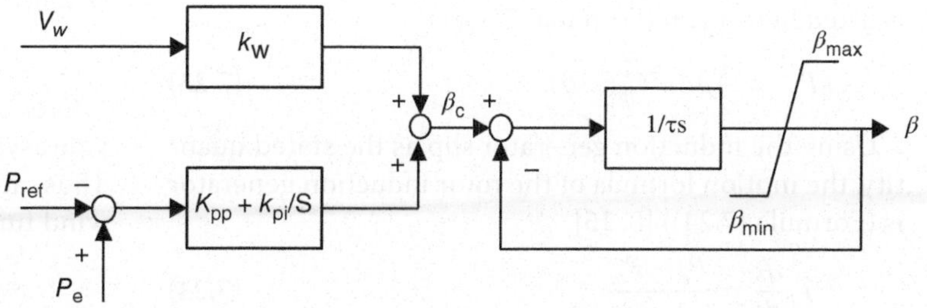

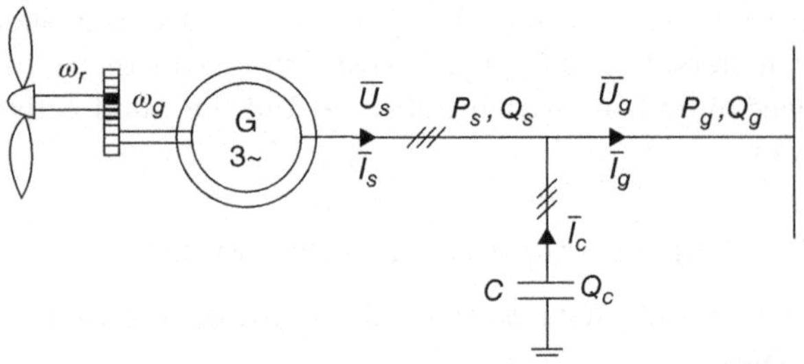

Figure 7.7 CIG wind power generation system architecture.

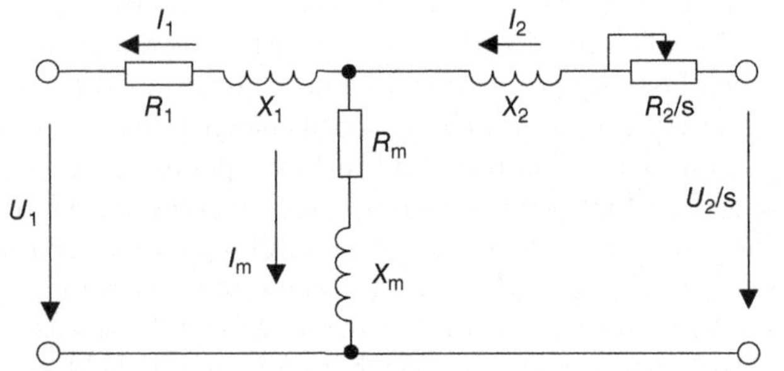

Figure 7.8 Asynchronous generator steady state equivalent circuit.

$$Q_e = -\frac{R_2^{\,2} + X_k(X_k + X_m)s^2}{R_2 X_m s} P_e \tag{7.22}$$

Where $X_k = X_1 + X_2$; s is the slip; φ is the power factor angle; P_e and Q_e respectively represents the active power produced and reactive power absorbed by the generator.

In power flow calculations containing asynchronous wind turbines, taking into account the steady-state model of induction generators, the iterative solution method is used for calculations. First, the active power P_e output by the wind turbines can be obtained based on the wind speed power. Set the initial voltage at wind generator bus as U_1, use P_e and U_1 to calculate the slip s according to Formula (7.20); and use P_e and s to calculate Q_e according to Formula (7.22); consider the wind power bus as PQ node, and use conventional power flow calculations to calculate the power flow of the entire system.

7.2.3.2 Dynamic Model of Cage Asynchronous Generator

The wind power extracted by the wind turbine from the wind can be calculated according to Formula (7.7), which is listed here again (Formula (7.23)):

$$P_{wt} = \frac{1}{2}\rho A v^3 C_p(\lambda, \theta) \tag{7.23}$$

Using the induction generator slip as the stated quantity, the motion formula of the rotor induction generator is (Formula (7.24)) [8, 15]

$$T_J \frac{ds}{dt} = \frac{P_e - P_m}{1 - s} \tag{7.24}$$

That is (Formula (7.25)),

$$\frac{ds}{dt} = \frac{1}{\tau_j}(T_e - T_m) \tag{7.25}$$

Where s is the slip, according to the generator convention, $s = (\omega_0 - \omega_r)/\omega_0$, ω_r, and ω_0 respectively represents the rotor speed and the rated (synchronous) speed; P_m is the input mechanical power, ignoring the wind turbine and gearbox losses, $P_m = P_{wt}$; P_e is the output electromagnetic power; T_m is the input mechanical torque; T_e is the induction generator electromagnetic torque, $T_e = \psi_d i_q - \psi_q i_d = E'_q i_q + E'_d i_d$; τ_j is the inertia time constant of the induction generator.

When the shaft flexibility is considered, using the wind turbine speed and the generator speed (ω_m and ω_g) as state variables, the motion formula of the drive shaft is (Formula (7.26))

$$\begin{cases} T_{wt} - T_m = J_m \dfrac{d\omega_m}{dt} \\ T_m = D_m(\omega_m - \omega_g) + k_m \displaystyle\int (\omega_m - \omega_g)dt \\ T_m - T_e = J_g \dfrac{d\omega_g}{dt} \end{cases} \tag{7.26}$$

Where T_{wt} is the mechanical torque of the wind turbine rotor shaft; $2H_m = J_m$ is the wind turbine rotor inertia; $2H_g = J_g$ is the generator rotor inertia. $T_e =$ is the generator electrical torque; K_m and D_m respectively represents the rigidity and damping of the mechanical coupling system.

On the d-q coordinate axis of synchronous speed ω_s rotation, the CIG electromagnetic transient formula is (Formulas (7.27) and (7.28)):

$$\frac{de'_d}{dt} = -\frac{1}{T'_0}(e'_d - (X_s - X'_s)i_{qs} + s \cdot \omega_s \cdot e'_q \tag{7.27}$$

$$\frac{de'_q}{dt} = -\frac{1}{T'_0}(e'_q + (X_s - X'_s)i_{ds} - s \cdot \omega_s \cdot e'_d \tag{7.28}$$

Where assuming x_1, x_2, x_m respectively, represents the stator leakage reactance, rotor leakage reactance, and excitation reactance of the induction generator in per-unit value, then

$$X_s = x_1 + x_m, X' = x_1 + \frac{x_2 x_m}{x_2 + x_m}, T'_0 = \frac{x_2 + x_m}{r_2}.$$

The electromagnetic power formula is (Formula (7.29)):

$$P_e = T_e = \mathrm{Re}(\dot{E}'\hat{I}_1) \tag{7.29}$$

CIG uses power factor correction capacitors to provide asynchronous generator with the excitation current. Thus, at the generator node, the current injected by the wind turbine is (Formulas (7.30) and (7.31))

$$i_{dg} = i_{ds} + i_{dc} = i_{ds} + \frac{1}{X_c}u_{qg} \tag{7.30}$$

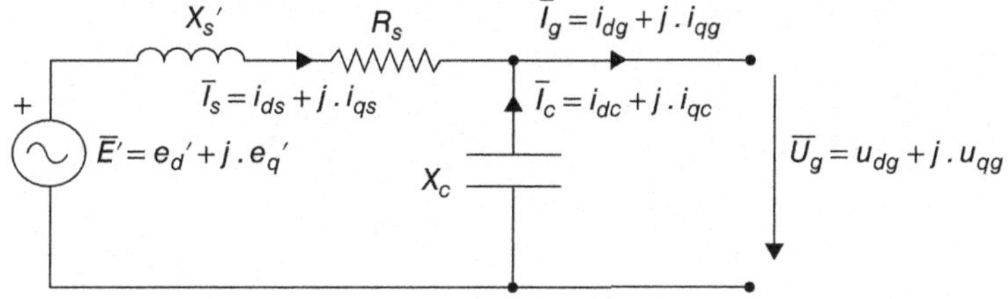

Figure 7.9 CIG wind turbine dynamic equivalent circuit.

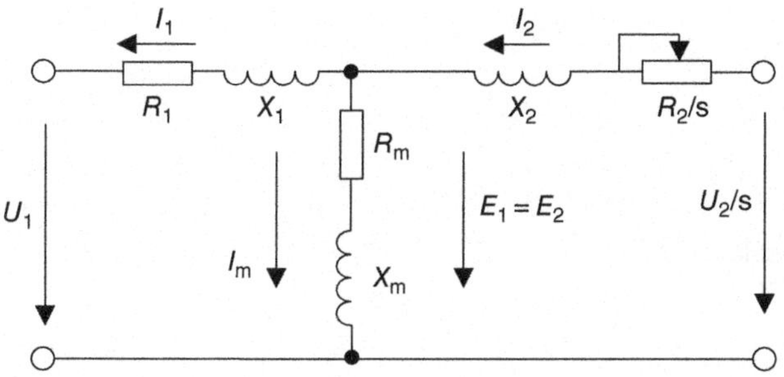

Figure 7.10 Double-fed asynchronous generator steady state equivalent circuit.

$$i_{qg} = i_{qs} + i_{qc} = i_{qs} - \frac{1}{X_c} u_{dg} \tag{7.31}$$

Where X_c is the reactance of the power factor correction capacitors; i_{dc} and i_{qc} are current components of the capacitor; u_{dg} and u_{qg} are voltage components at wind turbine power generation nodes. The model of CIG wind turbine power generation system can be illustrated by the equivalent circuit shown in Figure 7.9 [13].

7.2.4 Double-Fed Asynchronous Generator Model

7.2.4.1 Steady-State Model of DFIG

The steady-state equivalent circuit of the double fed asynchronous generator (DFAG, DFIG) is shown in Figure 7.10. Wherein, the subscript 1 represents the physical quantity on the stator side, the subscript 2 represents the physical quantity on the rotor side. Only the quantity on one side is considered and all quantities are converted to the stator side.

As shown in Figure 7.10 (Formula (7.32)):

$$\begin{cases} \dot{U}_1 = \dot{E}_1 - \dot{I}_1(R_1 + jX_1) \\ \dfrac{\dot{U}_2}{s} = \dot{E}_2 + \dot{I}_2\left(\dfrac{R_2}{s} + jX_2\right) \\ \dot{E}_1 = \dot{E}_2 = (\dot{I}_2 - \dot{I}_1)(R_m + jX_m) \end{cases} \tag{7.32}$$

Total active power injected in the system by the double-fed asynchronous generator consists of two parts: one is the active power P_1 generated by the stator winding, and the other is the active power P_2 produced or absorbed by the rotor winding. Ignore the stator winding resistance, after the calculation, the following results can be obtained at steady state (Formula (7.33)):

$$P_2 = \frac{R_2 X_{11}^2 ({P_1}^2 + {Q_1}^2)}{X_m^2 |\dot{U}_1|^2} + \frac{2R_2 X_{11}}{X_m^2} Q_1 - sP_1 + \frac{R_2 |\dot{U}_1|^2}{X_m^2} \tag{7.33}$$

Where $X_{11} = X_1 + X_m$ (Formula (7.34)).

$$P_e = P_1 + P_2 = \frac{R_2 X_{11}^2 ({P_1}^2 + {Q_1}^2)}{X_m^2 |\dot{U}_1|^2} + \frac{2R_2 X_{11}}{X_m^2} Q_1 + (1 - s)P_1 + \frac{R_2 |\dot{U}_1|^2}{X_m^2} \tag{7.34}$$

Slip s in the formula can be obtained according to the speed characteristic curve of the wind turbine.

The reactive power produced or absorbed by the convertor is very small that can be ignored. Then the VSCF wind turbine reactive power Q_e can be approximated as the stator winding reactive power Q_1. When the generator uses the constant power factor operation mode, if the value of the wind turbine power factor is set by cos φ, then $\tan\varphi = Q_1/P_1$, and the power formula of wind turbines will be (Formula (7.35)):

$$P_e = \frac{R_2 X_{11}^2 {P_1}^2}{X_m^2 |\dot{U}_1|^2}(1 + \tan^2\varphi) + \left(\frac{2R_2 X_{11}}{X_m^2} \tan\varphi + 1 - s\right) P_1 + \frac{R_2 |\dot{U}_1|^2}{X_m^2} \tag{7.35}$$

In the power flow calculation containing double-fed wind turbines, taking into account the steady-state model of the double-fed induction generator, the iterative solution method is used for calculation. First, the active power Pe output by the wind turbines can be obtained based on the wind speed-power; set the initial voltage at the wind generator node as U_1, calculate the slip s according to the law of the speed control of the wind turbines, and calculate P_1 according to Formula (7.35), then $Q_1 = P_1 tan\varphi$ is obtained; consider the wind power node as a PQ node, and use the conventional power flow calculation to calculate the power flow of the entire system. When the generator is using constant voltage operation mode, the WF node can be processed as a PV node.

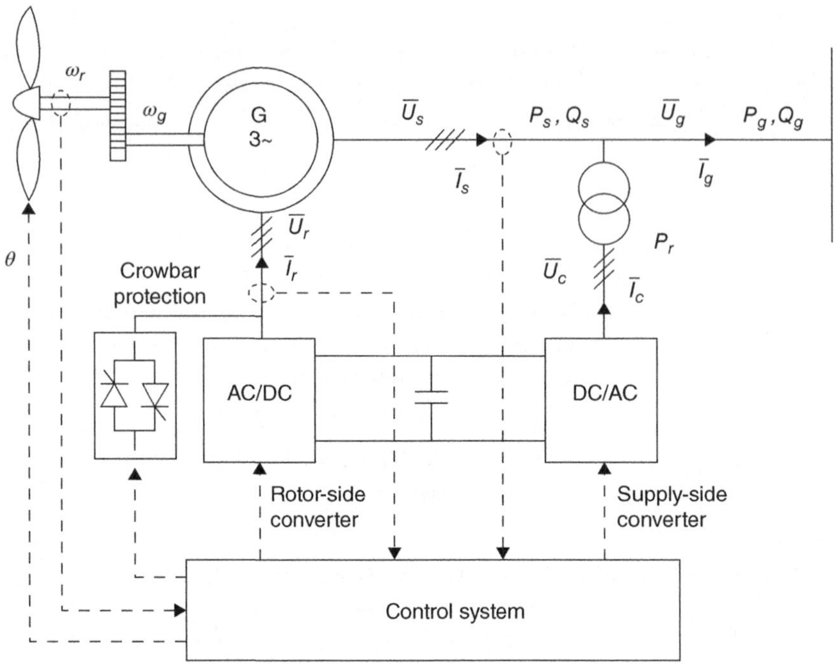

Figure 7.11 DFAG wind power generation system structure.

7.2.4.2 Double-Fed Asynchronous Generator Dynamic Model

The detailed model of wind turbines containing double-fed asynchronous generator includes wind turbine rotor, a power generation system consists of the generator and converter, and the control system, as shown in Figure 7.11 [16, 17].

The DFAG wind turbine rotor model is the same as the CIG wind turbine rotor model. However, in order to limit wind power obtained and the speed at strong wind, the DFAG wind turbine rotor model includes the control of blade pitch angle θ. Therefore, the power extracted by the DFAG wind turbine is expressed by Formula (7.23), and DFAG rotor motion formulas are expressed by Formula (7.24), Formula (7.25), or Formula (7.26).

The wound type rotor of the DFAG wind turbine is expressed with the third-order asynchronous generator model, and represented by the d-q coordinate axis of synchronous speed ω_s rotation. Since d-axis is at the same direction as the stator flux maximum value, such conversion makes the active and reactive power of the DFAG wind turbine able to be controlled independently (Formulas (7.36), (7.37), and (7.38)).

$$-\frac{de'_d}{dt} = -\frac{1}{T'_0}(e'_d - (X_s - X'_s)i_{qs} + s \cdot \omega_s \cdot e'_q - \omega_s \frac{X_m}{X_r} u_{qr} \tag{7.36}$$

$$-\frac{de'_q}{dt} = -\frac{1}{T'_0}(e'_q + (X_s - X'_s)i_{ds} - s \cdot \omega_s \cdot e'_d + \omega_s \frac{X_m}{X_r} u_{dr} \tag{7.37}$$

$$-T_e = \frac{X_m}{\omega_s}(i_{ds} \cdot i_{qr} - i_{qs} \cdot i_{dr}) \tag{7.38}$$

It consists of bidirectional transducers connected by two IGBT bridges through the DC bus. One transducer is connected to the rotor winding (the rotor side or the generator side converter), another transducer is connected to the grid (the supply side or grid side converter). The rotor side converter controls the rotor voltage, to make the wind turbine operate at an optimum efficiency below rated wind speed, or limit the output power of the wind turbine within the rated power when operating above the rated wind speed, and has an expected power factor. The converter on power supply side maintains power exchange from the rotor circuit to the grid and makes the system operate at the unit power factor. Assume that these converters are ideal, the DC voltage between the converters is constant, the converter can achieve decoupling control, and only the fundamental frequency component is considered.

The rotor side converter is simulated as controllable voltage source, use the q-axis rotor voltage u_{qr} to control the rotor speed, and the d-axis rotor voltage u_{dr} to control the reactive power. The converter at the power supply side represents controlled current source, providing the active power exchange from the rotor circuit to the grid with unit power factor. The DFAG generator dynamic equivalent circuit is shown in Figure 7.12.

Figure 7.12 DFAG dynamic equivalent circuit.

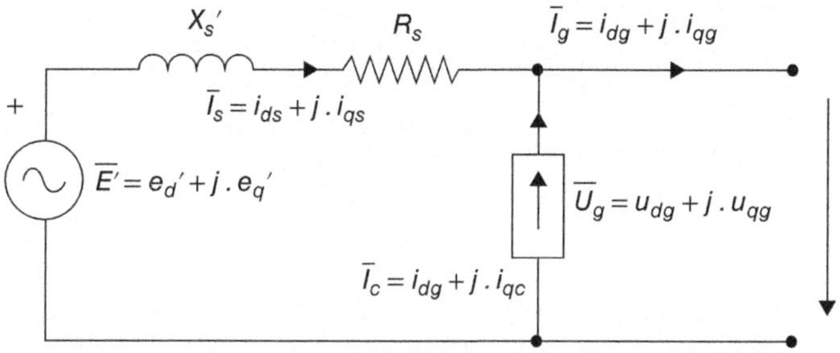

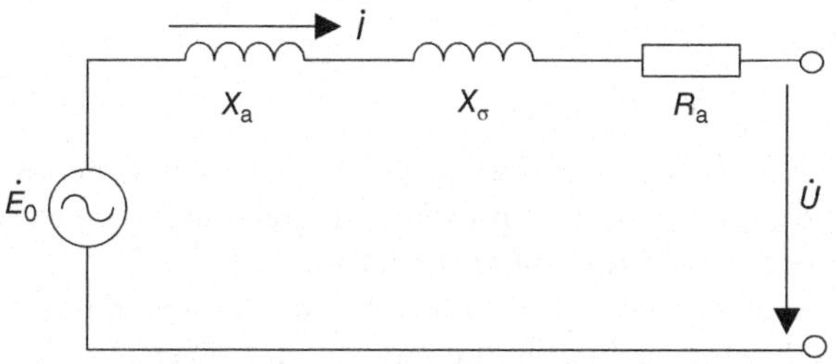

Figure 7.13 Equivalent circuit of PM synchronous generator.

7.2.5 Direct-Drive Permanent-Magnet Synchronous Generator Model

7.2.5.1 Steady-State Model of Permanent-Magnet Synchronous Generator

The equivalent circuit of PM synchronous generator is the same with that of conventional synchronous generators, as shown in Figure 7.13 [18, 19].

The generator terminal voltage formula (Formula (7.39)):

$$\dot{U} = \dot{E}_0 - \dot{I}(R_a + jX_s) \tag{7.39}$$

Where $X_s = X_a + X_\sigma$; $E_0 = \omega_e \Psi_f$; E_0 is the internal EMF of the generator; Ψ_f is the rotor flux, of which the value is constant.

7.2.5.2 Dynamic Model of Direct-Drive Permanent-Magnet Synchronous Wind Turbine

The wind speed model and wind turbine torque model are the same as the above, but because the wind turbine drives the synchronous generator directly, so the model of the transmission part is different from the above, as $\omega_g = \omega_{ae}$, that is, the mechanical speed of the generator rotor ω_g is equal to the wind turbine speed ω_{ae}.

The mathematical model for permanent magnet synchronous generators is (Formula (7.40–7.42)) [20, 21]

$$\begin{cases} u_d = -R_a i_d + p\psi_d - \omega_e \psi_q \\ u_q = -R_a i_q + p\psi_q + \omega_e \psi_d \end{cases} \tag{7.40}$$

$$\begin{cases} \psi_d = -L_d i_d + \psi_f \\ \psi_q = -L_q i_q \end{cases} \tag{7.41}$$

$$\tau_j \frac{d\omega_g}{dt} = T_{ae} - T_e \tag{7.42}$$

Where u_d, u_q, i_d, i_q, ψ_d, and ψ_q respectively, represents the d, q-axis component of the stator voltage, current, and flux linkage; R_a is the stator resistance; ω_e is the rotor electric angular speed, $\omega_e = n_p \omega_g$ and n_p are numbers of pole-pairs of the generator rotor; L_d and L_q, respectively, represents stator coil inductance of d- and q-axises; ψ_f is the flux linkage of the permanent magnet; τ_j is the inertia time constant of the permanent magnet synchronous generator; T_{ae} is the mechanical power output by the wind turbine.

Substitute Formula (7.41) into Formula (7.40), we can obtain (Formula (7.43)):

$$\begin{cases} u_d = -R_a i_d + \omega_e L_q i_q - L_d \dfrac{di_d}{dt} \\ u_q = -R_a i_q - \omega_e L_d i_d - L_q \dfrac{di_q}{dt} + \omega_e \psi_f \end{cases} \tag{7.43}$$

The equivalent circuit under d, q coordinates is shown in Figure 7.14.

Electromagnetic torque formula (Formula (7.44)):

$$T_e = 1.5 n_p (\psi_d i_q - \psi_q i_d) \tag{7.44}$$

Set d-axis at the direction of the rotor voltage, $u_q = 0$. Then, $P_1 = u_d i_d + u_q i_q = u_d i_d$, $Q_1 = u_q i_d - u_d i_q = -u_d i_q$, there is no coupling between active power and reactive power, which is respectively determined by the stator current components on d-axis and q-axis. Set $u_d' = -R_a i_d - L_d \frac{di_d}{dt}$, $u_q' = -R_a i_q - L_d \frac{di_q}{dt}$, $\Delta u_d = \omega_e L_q i_q$, and $\Delta u_q = -\omega_e L_d i_d + \omega_e \psi_f$, Formula (7.43) can be changed to (Formula (7.45)):

$$\begin{cases} u_d = u_d' + \Delta u_d \\ u_q = u_q' + \Delta u_q \end{cases} \tag{7.45}$$

Where u_d' and u_q' are the decoupling terms for achieving decoupling control of the stator voltage and current; Δu_d and Δu_q are the compensation terms for eliminating cross coupling of the rotor voltage and current.

7.2.6 WF Equivalent Model

The equivalent model of the WF containing CAG and DFAG are built on the equivalent wind turbines with remarked power capacity. All wind turbines of the WF are integrated and are equivalent to one (or several) wind

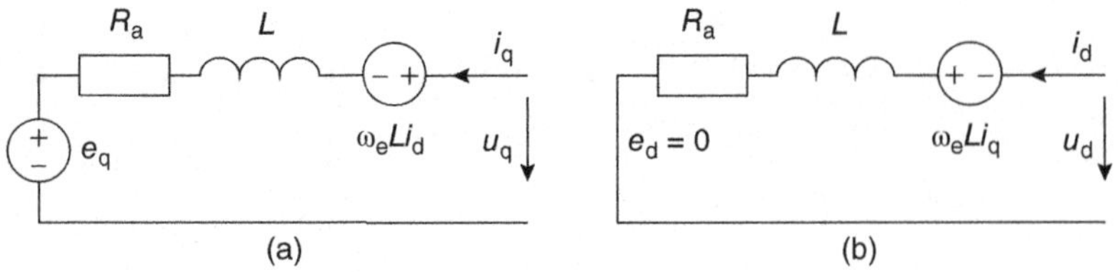

Figure 7.14 PMSG dynamic equivalent circuit under dq coordinates. (a) *q*- axis equivalent circuit; (b) *d*-axis equivalent circuit.

turbine(s), and the wind captured by each wind turbine is considered the same. The equivalent models of each part of the WF is described below [13].

7.2.6.1 Equivalent Wind

Wind distribution of the WF depends on its location. For WFs located on flat land and at sea, the wind turbines are arranged in row. Wind accepted by the wind turbines on the same row is similar, but due to the shielding effect between wind turbines, wind is different between rows. For WFs located in rough land (mountain land) or WFs of which the wind turbines are widely distributed, the winds accepted by the wind turbines are different. Therefore, in some WFs, the group of wind turbines accept similar winds, in other WFs all of the wind turbines receive different winds.

In fact, the most widely used equivalent WF model is based on an equivalent wind turbine that have a set of wind turbines that receive the same wind integrated. The equivalent wind received by the equivalent wind turbine equals the wind received by a group of wind turbines. For a set of wind turbines that accept similar wind (with little different in wind speed), the average value of wind speeds acting on the wind turbine set is used as the equivalent wind speed acting on the equivalent wind turbine, that is (Formula (7.46)),

$$v_{eq} = \frac{1}{n}\sum_{j=1}^{n} v_j \tag{7.46}$$

Where v_{eq} is the wind speed acting on the equivalent wind turbine; v_j is the wind speed acting on a single wind turbine, the subscript *j* represents the corresponding single wind turbine; *eq* represents the equivalent wind turbine; *n* is the number of wind turbines in one set.

The number of equivalent wind turbines shown in a WF equals the number of wind turbine groups accepting similar wind.

Because the generated power of wind turbines accepting different winds is different, the integration of these wind turbines requires an equivalent wind turbine that considers the wind speed acting on each wind turbine. This equivalent model is based on an equivalent wind turbine with remarked power capacity. Equivalent winds of different wind turbines that accept different winds are derived from the wind turbine power curve, for which the method is as follows:

1) The wind turbine power approximation. The power generated by each wind turbine is derived from the wind turbine power curve according to the wind received (Formula (7.47)).

$$P_j^{wt} = PC_{wt}(v_j) \tag{7.47}$$

Where PC_{wt} is the wind turbine power curve function, the superscript *wt* represents the per-unit value with a single wind turbine as the base value.

2) The total power. The sum of the power generated by each wind turbine is the equivalent power P_{eq} (Formula (7.48))

$$P_{eq}^{wt} = \sum_{j=1}^{n} P_j^{wt} \tag{7.48}$$

3) The equivalent wind turbine power curve PC_{EWT} is obtained from the sum of power of the corresponding wind turbine of each wind speed. The power curve generated with the equivalent wind turbine as the base value is the same as the power curve of the single wind turbine, [Formulas (7.49) and (7.50)].

$$P_{eq}^{ewt} = PC_{ewt}(v_{eq}) \tag{7.49}$$

$$PC_{ewt} = PC_{wt} \tag{7.50}$$

Where the superscript ewt represents the per-unit value with the equivalent wind turbine as the base value.

4) The equivalent wind is derived from the inverse function PC_{ewt}^{-1} of the equivalent wind turbine power curve and the power of the equivalent wind turbine obtained according to Formula (7.48), which is represented by the equivalent base value (Formula (7.51)):

$$v_{eq} = PC_{ewt}^{-1}(P_{eq}^{ewt}) \tag{7.51}$$

5) If the wind speeds accepted by all DFAG wind turbines are above the rated wind speed, then the equivalent wind is the average wind. Because the power curve is limited to the rated power, any wind speed above the rated wind speed can be the solution of Formula (7.51).

7.2.6.2 Equivalent Wind Wheel of the CIG Wind Turbine

The equivalent power of the CIG wind turbine set is the sum of rated power of the wind turbine set (Formula (7.52))

$$S_{eq} = \sum_{j=1}^{n} S_j = nS_j \tag{7.52}$$

The equivalent wind turbine accepts the equivalent wind as its wind. Using the single equivalent wind acting on the equivalent wind turbine means that wind turbines of all sets are accepting the equivalent wind as their winds, therefore, they all create the same mechanical torque. Thus, when using per-unit value based on the single wind turbine, the equivalent generator mechanical torque is n times as great as that of the single wind turbine, that is, (Formula (7.53)):

$$T_{mec.eq}^{wt} = \sum_{j=1}^{n} T_{mec.j}^{wt} = nT_{met.j}^{wt} \tag{7.53}$$

The generator mechanical torque of a single wind turbine is calculated according to the same rotor and drive chain model, because any single wind turbine has the same parameters in Formulas (7.23) and (7.26).

When the equivalent generator mechanical torque is represented by the equivalent wind turbine base value, it is equal to the generator mechanical torque of the single wind turbine that is represented by the equivalent wind turbine base value (Formula (7.54)).

$$T_{mec.eq}^{wwt} = T_{mec.j}^{wt} \tag{7.54}$$

CIG uses its equivalent circuit (Figure 7.9) to represent a set of wind turbine power generation system, it means that the models of a single wind turbine and the equivalent asynchronous generator are the same, which have the same mechanical and electrical parameters (per-unit values) when expressed by Formula (7.26) Formula (7.28).

The application of power factor correction capacitor set can obtain an equivalent capacitor with the same reactance (per unit) as that of a single wind turbine. However, the use of the capacitor with constant reactance may cause errors in reactive power approximation of the wind turbine set. Because the reactive power of CIG depends on the active power and generating voltage, these variables are different in each wind turbine when the winds are different. A better approximation of the reactive power can be obtained from the equivalent capacitor with variable reactance. The variable reactance provides reactive power balance between the wind turbine set and the equivalent wind turbine without capacitors. The calculation method of equivalent capacitor is described as follows:

1) Appropriation of reactive power Q_j of each wind turbine is obtained from the simplified asynchronous generator steady-state model (Figure 7.15), the electric power $P_{e,j}$ and the generating voltage $U_{g,j}$ in accordance with Formula (7.55), that is:

$$Q_j = U_{g.j}^2 \cdot \frac{X_c - X_m}{X_c \cdot X_m} + (X_{\sigma s} + X_{\sigma r}) \cdot \frac{U_{g.j}^2 + 2(R_r + R_s)P_{ej}}{2((R_r + R_s)^2 + (X_{\sigma s} + X_{\sigma r})^2)} - (X_{\sigma s} + X_{\sigma r}) \times \frac{\sqrt{(U_{g.j}^2 + 2(R_r + R_s)P_{ej})^2 - 4P_{eg}^2((R_r + R_s)^2 + (X_{\sigma s} + X_{\sigma r})^2)}}{2((R_r + R_s)^2 + (X_{\sigma s} + X_{\sigma r})^2)} \tag{7.55}$$

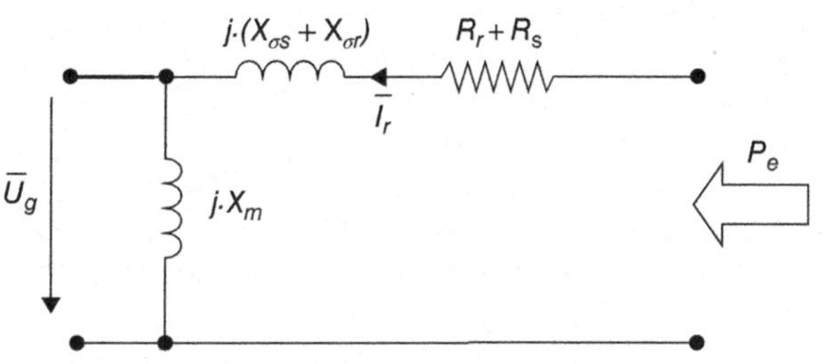

Figure 7.15 Simplified asynchronous generator steady-state model.

Electric power of each wind turbine is derived from the power curve (as shown in Figure 7.16) and the wind. The generating voltage $U_{g,j}$ of each wind turbine is the approximation of the generating voltage $U_{g,eq}$ of the equivalent wind turbine operating in the equivalent WF network.

2) Reactive power Q_{awt} of wind turbine set is equal to the sum of the reactive power of each wind turbine (Formula (7.56))

$$Q_{awt} = \sum_{j=1}^{n} Q_j \tag{7.56}$$

3) Reactive power $Q_{eq,wc}$ of the equivalent wind turbine without power factor correction capacitors is obtained from the steady-state model of the equivalent generator without capacitors (Formula (7.57)).

$$Q_{eq.wt} = \frac{U_{g.eq}^2}{X_{m.eq}} + (X_{\sigma s.eq} + X_{\sigma r.eq})$$

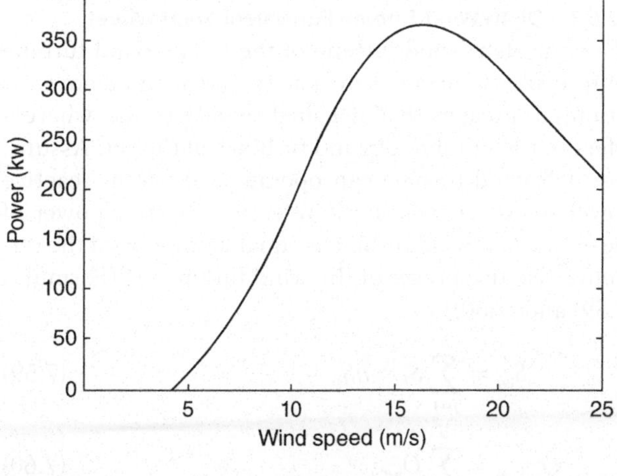

Figure 7.16 CIG wind turbine power curve.

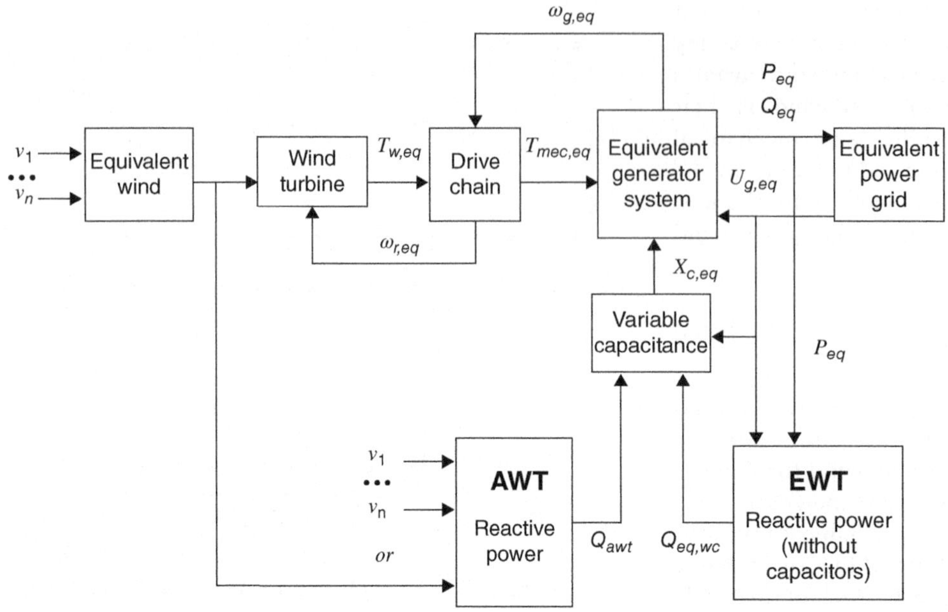

Figure 7.17 Equivalent mode of CIG wind turbines of the WF.

$$\cdot \frac{U_{g.eq}^2 + 2(R_{r.eq} + R_{s.eq})P_{eq}}{2((R_{r.eq} + R_{s.eq})^2 + (X_{\sigma s.eq} + X_{\sigma r.eq})^2)} - (X_{\sigma s.e} + X_{\sigma r.e})$$
$$\times \frac{\sqrt{\begin{array}{c}(U_{g.eq}^2 + 2(R_{r.eq} + R_{s.eq})P_{eq})^2 \\ -4P_{eq}^2((R_{r.eq} + R_{s.eq})^2 \\ +(X_{\sigma s.eq} + X_{\sigma r.eq})^2)\end{array}}}{2((R_{r.eq} + R_{s.eq})^2 + (X_{\sigma s.eq} + X_{\sigma r.eq})^2)} \tag{7.57}$$

4) Variable reactance of the capacitor is calculated as follows (Formula (7.58))

$$X_{c.eq} = \frac{U_{g.eq}^2}{Q_{awt} - Q_{eq.wc}} \tag{7.58}$$

The equivalent model of CIG wind turbines of the WF is shown in Figure 7.17.

7.2.6.3 DFAG Wind Turbine Equivalent Wind Wheel

The equivalent wind turbine of the DFAG wind turbine set re-calibrates its power capacity, and its rated power is n times as great as that of a single wind turbine, where n refers to the number of wind turbines of the set. Assuming each wind turbine can operate at different reactive power reference values; the reference reactive power of the equivalent wind turbine is equal to the sum of the reference reactive power of the wind turbine set (Formulas (7.59) and (7.60)).

$$S_{eq} = \sum_{j=1}^{n} S_j = nS_j \tag{7.59}$$

$$Q_{ref.eq} = \sum_{j=1}^{n} Q_{ref.j} \tag{7.60}$$

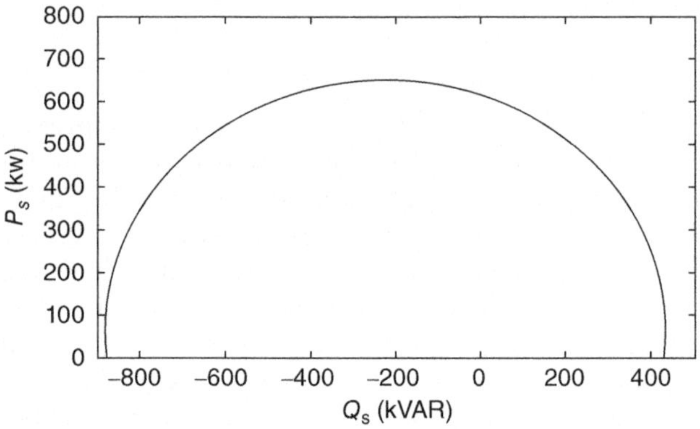

Figure 7.18 DFAG wind turbine generation P_s–Q_s limit curve.

The WF control level provides each wind turbine with the reactive power control command Q_{com}. The reactive power of DFAG wind turbine is limited (Figure 7.18 shows DFAG power P_s–Q_s limit curve), therefore, the reactive power reference value Q_{ref} of each wind turbine can be different from the reactive power Q_{com} required by the WF control level. The maximum reactive power of DFAG wind turbine, that is, the limited stator reactive power $Q_{s.lim}$, depends on the active power P_s, the stator voltage U_s and the maximum rotor current $I_{r.max}$, and is calculated according to Formula (7.61)

$$Q_{s.\lim} = -\frac{U_s^2}{X_s} \pm \sqrt{\left(\frac{X_m}{X_s}U_s I_{r.\max}\right)^2 - P_s^2} \tag{7.61}$$

Since DFAG independently transfers reactive power into the grid through the stator windings, the reactive power maximum reference value of the reactive power controller can be forced to limit the stator reactive power according to formula (7.61).

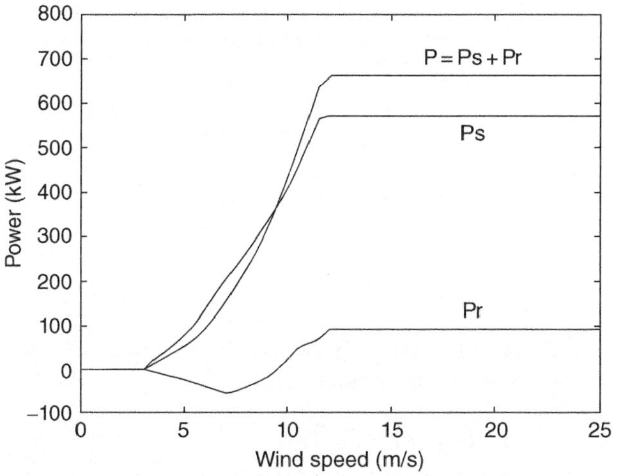

Figure 7.19 DFAG wind turbine power curve.

The active power of the stator is calculated according to the power curve (Figure 7.19 describes the power curve of a kind of DFAG) and the wind. The generating voltage of each wind turbine approximates to the generating voltage of the equivalent wind turbine.

In addition, the equivalent wind turbine receives the equivalent wind, and equivalent wind is used to the single wind turbine rotor and drive chain model, in order to calculate its mechanical torque of the generator. In order to use this method to calculate the equivalent generator mechanical torque, it is assumed that all integrated wind turbines are subjected to the equivalent wind same as the wind, so they all produce the same mechanical torque as the generator. This generator mechanical torque, when represented by the single wind turbine base value, is the same as the generator torque of the equivalent wind turbine represented by the equivalent base value.

Moreover, for the power generation system (asynchronous generators and frequency converters), the control program and the controller (the speed, the reactive power, and the pitch angle controller) and the protection system, models of the equivalent wind turbine are the same as those of the single wind turbine. The model configuration of the equivalent wind turbine of the DFAG wind turbines is shown in Figure 7.20.

7.2.6.4 The Equivalent of the WF Collection System

Equivalent wind turbines operate in the equivalent internal electric network, so the public network integrated with wind turbine shall be replaced with the equivalent impedance. When calculating the equivalent impedance, it is assumed that the short-circuit impedance of the equivalent WF must be equal to the short-circuit impedance of all WFs. There are different methods for the collection system equivalent. Below is the equivalent parameter deduction of the collection system based on circuit analysis [22].

1) **General assumptions**
   The power generated by the wind turbine generator is usually of low voltage level (typically 575 V or 690 V). The capacity of a single wind turbine ranges from hundreds of MW to 5 MW. Usually the transformer matching with the wind turbine will boost the voltage up to medium voltage level, typically 35 kV. As shown in Figure 7.21, the equivalent of aWF collection system includes wind turbine transformers and the equivalent of 35 kV collection networks.
   Feeders can be cables or overhead lines. Feeders are connected to the 35 kV bus of the WF substation transformer, and the transformer boosts the voltage up to the secondary transmission voltage level (e.g., 66 kV or 110 kV). The following assumptions are

Figure 7.20 DFAG wind turbine generator equivalent model.

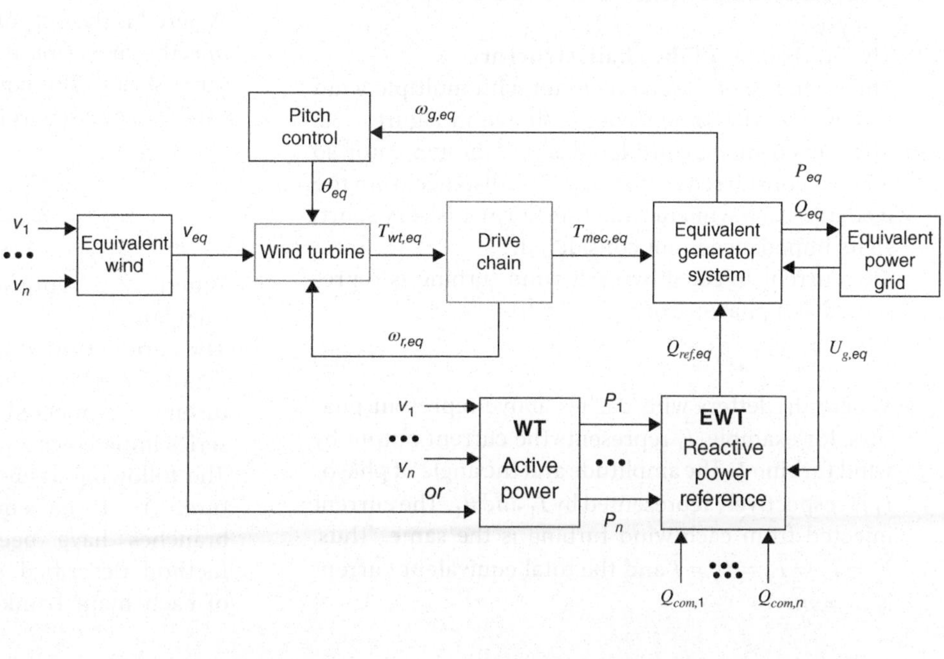

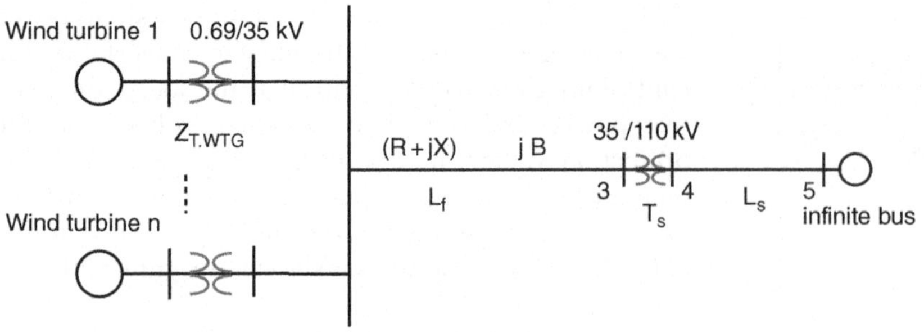

Figure 7.21 A single-line diagram of an equivalent wind turbine connected to the substation.

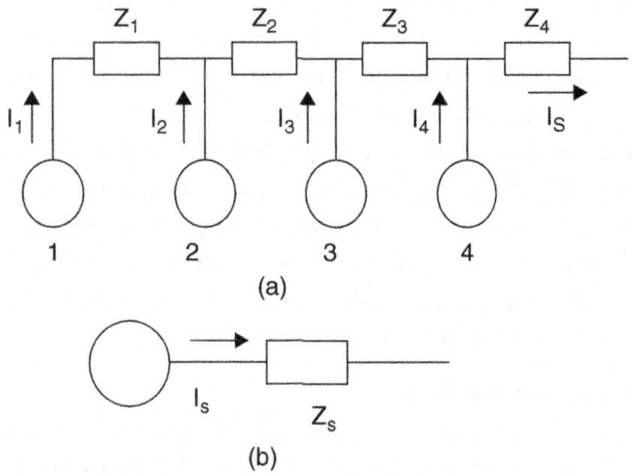

Figure 7.22 Chain structure circuit of multiple wind turbines on a feeder. (a) Single-chain wiring (b) the equivalent circuit of Figure (a).

made for the internal circuits of the WF: ① amplitudes and phase angles of the currents injected by all wind turbine generators are the same; ② the reactive power generated by the line parallel capacitor is based on the bus voltage per-unit value of 1.0.

The following equivalent circuit parameters are based on the apparent power loss. Equivalent parameters of the collection system are obtained from the circuit analysis.

2) **The equivalent of the chain structure**

The structure of the chain circuit with multiple wind turbines on the same feeder is shown in Figure 7.22. The transformer is considered part of the generator, so it is not considered in this step. The distance from the feeder to each wind turbine transformer is very short, so its impedance is not considered.

The current injected by each wind turbine is represented as a phasor (Formula (7.62))

$$\bar{I}_m = I_m \angle \theta_m \tag{7.62}$$

Where the letters with dashes above represent phasors, for example, $\bar{I}_1$ represents the current output by wind turbine 1. The amplitude and the angle of phasor $\bar{I}_1$ is respectively represented by I_1 and θ_1. The current injected from each wind turbine is the same. Thus, $\bar{I}_1 = \bar{I}_2 = \bar{I}_3 = \bar{I}_4 = \bar{I}$ and the total equivalent current is $\bar{I}_S = n\bar{I}$.

The voltage drop between both ends of each impedance is (Formula (7.63))

$$\begin{cases} \Delta U_{Z1} = \bar{I}_1 Z_1 = \bar{I} Z_1 \\ \Delta U_{Z2} = (\bar{I}_1 + \bar{I}_2) Z_2 = 2\bar{I} Z_2 \\ \quad \cdots\cdots \\ \Delta U_{Z4} = (\bar{I}_1 + \bar{I}_2 + \bar{I}_3 + \bar{I}_4) Z_4 = 4\bar{I} Z \end{cases} \tag{7.63}$$

The power loss on each impedance is (Formula (7.64))

$$\begin{cases} S_{Loss,Z1} = \Delta U_{Z1}\, \bar{I}_1^{*} = \bar{I}_1\, \bar{I}_1^{*} Z_1 = \bar{I}\, \bar{I}^{*} Z_1 = \bar{I}^2 Z_1 \\ S_{Loss,Z2} = \Delta U_{Z2}\, (\bar{I}_1 + \bar{I}_2)^{*} = 2\, \bar{I}\, Z_2\, (2\, I)^{*} \\ \qquad = 2^2\, \bar{I}^2\, Z_2 \\ \quad \cdots\cdots \\ S_{Loss,Z4} = \Delta U_{Z4}\, (\bar{I}_1 + \bar{I}_2 + \bar{I}_3 + \bar{I}_4)^{*} = 4^2\, \bar{I}^2\, Z_4 \end{cases} \tag{7.64}$$

Since the total current $I_S = n\,I$, the total power loss is (Formula (7.65))

$$\begin{aligned} S_{Tot,loss} &= \bar{I}^2\,(Z_1 + 2^2 Z_2 + 3^2 Z_3 + 4^2 Z_4) \\ &= \bar{I}^2 \sum_{m=1}^{n} m^2 Z_m \end{aligned} \tag{7.65}$$

Where I is the output current of a single wind turbine; m is the subscript; n is number of wind turbine in the same series. The equivalent of the simplified circuit can be expressed as (Formulas (7.66) and (7.67))

$$S_{Totloss} = \bar{I}_s^2 Z_s \tag{7.66}$$

$$Z_S = \frac{\sum_{m=1}^{n} m^2 Z_m}{n^2} \tag{7.67}$$

Where Z_m represents the series impedance of each line.

3) **The connection at the feeder levels**

Figure 7.23 shows the conditions of three sets of wind turbines connected in parallel through their own series impedance Z_1, Z_2, and Z_3 to the same bus.

The following assumptions are made for this structure: ① Three equivalent circuits of the three branches have been obtained according to the method described above; ② first, the equivalents of each main trunk circuit, respectively, consist of

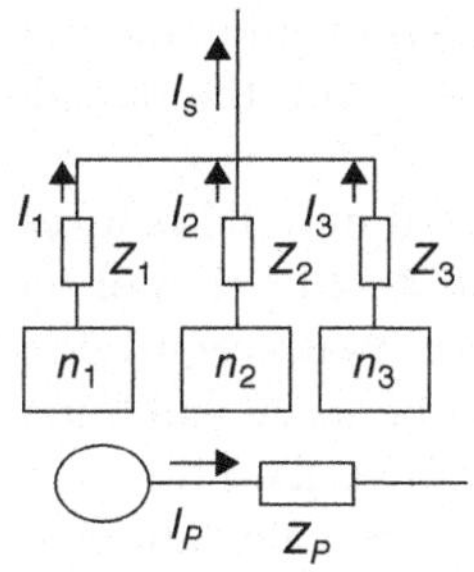

Figure 7.23 Three wind turbine sets connected in parallel through the series impedance Z_1, Z_2 and Z_3.

n_1, n_2, and n_3 generators; ③ outputs generated by all generators are of the same amplitude and phase angle; ④ the impedance of each set connected to the interconnection point is Z_1, Z_2, and Z_3.

The output current of each set is

$$\bar{I}_1 = n_1\bar{I}, \bar{I}_2 = n_2\bar{I}, \bar{I} = n_3\bar{I}$$

The total output of the parallel branch is

$$\begin{aligned}\bar{I}_P &= \bar{I}_1 + \bar{I}_2 + \bar{I}_3 = n_1\bar{I} + n_2\bar{I} + n_3\bar{I}\\ &= (n_1 + n_2 + n_3)\bar{I}\end{aligned}$$

The loss of each branch

$$S_{Z1} = \bar{I}_1^2 Z_1, S_{Z2} = \bar{I}_2^2 Z_2, S_{Z3} = \bar{I}_3^2 Z_3$$

The total loss (Formulas (7.68) and (7.69))

$$\begin{aligned}S_{ZP} &= \bar{I}_1^2 Z_1 + \bar{I}_2^2 Z_2 + \bar{I}_3^2 Z_3\\ &= \bar{I}^2 n_1^2 Z_1 + \bar{I}^2 n_2^2 Z_2 + \bar{I}^2 n_3^2 Z_3 = I_P^2 Z_P\end{aligned} \tag{7.68}$$

$$Z_P = (n_1^2 Z_1 + n_2^2 Z_2 + n_3^2 Z_3)/(n_1 + n_2 + n_3)^2 \tag{7.69}$$

The further simplification is (Formula (7.70))

$$Z_P = \frac{\sum_{m=1}^{n} n_m^2 z_m}{\left[\sum_{m=1}^{n} n_m\right]^2} \tag{7.70}$$

Z_m represents the equivalent impedance of each parallel set.

4) **The connection on the 35 kV main bus**
Figure 7.24 shows the system wiring of several feeders connected to the 35 kV collection line. Following assumptions are made on the circuit: ① the size and phase of the output from all generators are the same ② there is a series impedance Z_{1S}, Z_{2S}, or Z_{3S} between two points of interconnection; ③ each wind turbine set has a series impedance Z_{1p}, Z_{2p}, or Z_{3p}; ④ each branch has its own impedance and is connected to the wind turbine set; ⑤ at first, the equivalent circuit of each main trunk circuit consists of n_1, n_2, or n_3 wind turbines, respectively (Figure 7.24).

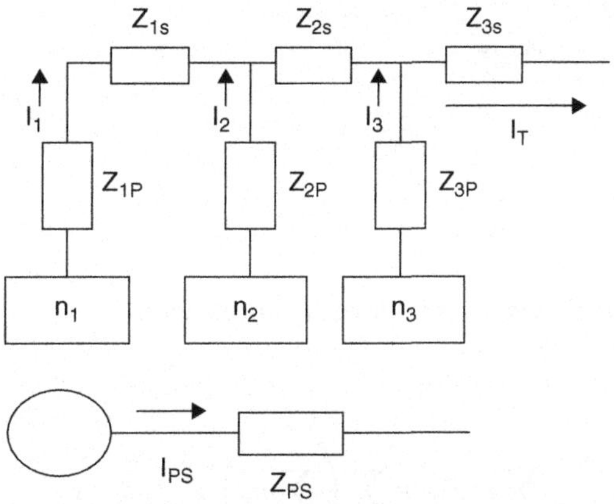

Figure 7.24 Parallel series connection of three wind turbine sets.

The current output from each wind turbine set is

$$\bar{I}_1 = n_1\bar{I}, \bar{I}_2 = n_2\bar{I}, \bar{I} = n_3\bar{I}$$

The total output of the parallel branch is

$$\bar{I}_T = n_1\bar{I} + n_2\bar{I} + n_3\bar{I} = (n_1 + n_2 + n_3)\bar{I}$$

The voltage drop between two ends of the impedance is

$$\begin{aligned}\Delta U_{Z1P} &= \bar{I}_1 Z_{1P} = n_1\bar{I}Z_{1P};\\ \Delta U_{Z1S} &= \bar{I}_1 Z_{1S} = n_1\bar{I}Z_{1S};\\ \Delta U_{Z2P} &= \bar{I}_2 Z_{2P} = n_2\bar{I}Z_{2P};\\ \Delta U_{Z2S} &= (\bar{I}_1 + \bar{I}_2)Z_{2S} = (n_1\bar{I} + n_2\bar{I})Z_{2S};\\ \Delta U_{Z3P} &= \bar{I}_3 Z_{3P} = n_3\bar{I}Z_{3P};\\ \Delta U_{Z3S} &= (\bar{I}_1 + \bar{I}_2 + \bar{I}_3)Z_{3S}\\ &= (n_1\bar{I} + n_2\bar{I} + n_3\bar{I})Z_{3S}\end{aligned}$$

Ibid, according to the loss in each impedance, the total equivalent loss can be obtained from the formula below (Formula (7.71)):

$$S_{ZT,LOSS} = I_T^2 Z_T \tag{7.71}$$

Based on that the total loss is equal to the sum of losses of all impedances, the equivalent impedance of the series-parallel circuit shown in Figure 7.24 can be obtained (Formula (7.72)).

$$\begin{aligned}\bar{I}_T^2 Z_T = \bar{I}^2[&(n_1^2 Z_{1P} + n_2^2 Z_{2P} + n_3^2 Z_{3P})\\ &+ (n_1^2 Z_{1S} + (n_1 + n_2)^2 Z_{2S}\\ &+ (n_1 + n_2 + n_3)^2 z_{3S}]\end{aligned}$$

$$\begin{aligned}Z_T = \frac{1}{(n_1 + n_2 + n_3)^2}[&(n_1^2 Z_{1P} + n_2^2 Z_{2P} + n_3^2 Z_{3P})\\ &+ (n_1^2 Z_{1S} + (n_1 + n_2)^2 Z_{2S}\\ &+ (n_1 + n_2 + n_3)^2 z_{3S}]\end{aligned}$$

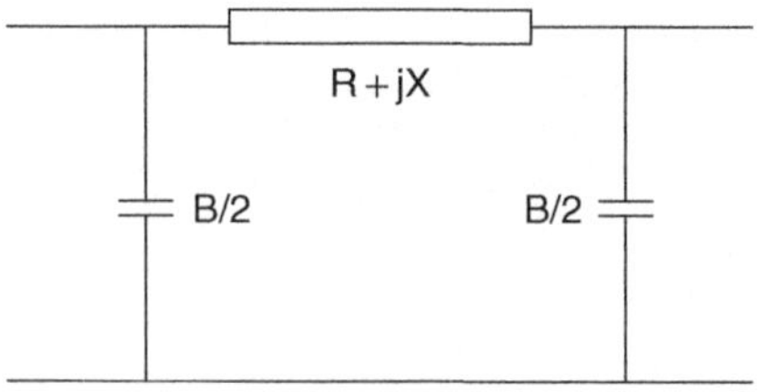

Figure 7.25 Representation of WF circuit susceptance.

that is,

$$Z_T = \frac{\sum_{i=1}^{np} n_i^2 Z_{iP} + \sum_{i=1}^{ns}\left(\sum_{j=1}^{ns} n_j\right)^2 Z_{iS}}{\left(\sum_{i=1}^{nP} n_i\right)^2} \tag{7.72}$$

5) **The presentation of the shunt susceptance**
The equivalent circuit of WF circuit shunt susceptance is shown in Figure 7.25. Since the reactive power generated by the capacitor is proportional to the square of its voltage, considering that bus voltage is close to one under normal conditions, the shunt susceptance can be represented as the sum of all shunt susceptances in the network, that is, the total shunt susceptance of the WF is (n represents number of branches of collection line in the WF) (Formula (7.73)).

$$B_{tot} = \sum_{i=1}^{n} B_i \tag{7.73}$$

6) **The representation of transformer**
For the system shown in Figure 7.21, the transformer impedance (leakage reactance) will produce a voltage drop and apparent power loss, which is equal to the losses of all wind turbines. The equivalent impedance $Z_{\mathrm{T,WF}}$ of wind turbine transformers in the same series is represented as impedance $Z_{\mathrm{T.WTG}}$ of the single wind turbine divided by the number n_{T} of the total wind turbines (Formula (7.74))

$$Z_{\mathrm{T,WF}} = Z_{\mathrm{T,\ WTG}}/n_{\mathrm{T}} \tag{7.74}$$

7.2.7 The Dynamic Equivalent Algorithm for Wind Turbines

In order to save simulation time, an appropriate algorithm is required to simplify the simulation process, which is the necessity of a dynamic equivalent algorithm. The commonly used dynamic equivalent methods at present include the homological equivalent method based on relevant concepts, the mode method based on eigenvalue analysis and the identification method based on online measurement. Compared to the first two methods, the advantage of the identification method is that it does not need detailed data from the outside system,

The identification method described here takes a reference from the concept of the transfer function of the homological equivalent method, which makes equivalent the inertia time constant, and the impedance and initial slip of the wind turbine generator set based on the asynchronous machine model. In equivalent calculations of the rotor winding time constant and the asynchronous generator reactance, the least square fitting is carried out for the frequency response of the overall transfer function of the entire WF, to obtain equivalent parameters of the best-fit point corresponding to the actual response curve.

1) **Wind turbine grouping principle in the equivalent model**
In the equivalent process, if it is required to equivalent to multiple wind turbine generators, we will have the problem of wind turbine grouping. When the problem is relatively simple, we can simply group the wind turbines based on their types and installation locations. When the installation location of wind turbines is irregular, the grouping of asynchronous generators can be based on the characteristic roots of the mechanical transient mathematical model formulas, which determines the dynamic characteristics of the generator in occurrence of small voltage disturbances. Of course, the grouping principle must also consider the actual installation location of each wind turbine. When only the mechanical transient is counted, the wind turbine mathematical model is of first order. See Formulas (7.75) and (7.76) the rotor motion formula and the electromagnetic torque formula [23].

$$T_j\frac{ds}{dt} = M_e - M_m \tag{7.75}$$

$$M_e = \frac{U^2/x_k}{\omega x_k\left(\frac{r_r}{sx_k} + \frac{sx_k}{r_r}\right)} \tag{7.76}$$

Where $x_{\mathrm{k}} = x_{\mathrm{s}} + x'_{\mathrm{r}}$, ignoring the stator resistance r_{r}, and taking the correction factor as $C = 1$, then the critical slip $s_{\mathrm{cr}} = \frac{r'_{\mathrm{r}}}{x_{\mathrm{k}}}$, approximately considered as $\omega = 1$, and the electromagnetic torque is given by (Formula (7.77)):

$$M_e = \frac{U^2/x_k}{\frac{s}{s_{cr}} + \frac{s_{cr}}{s}} \tag{7.77}$$

Linearize Formula (7.77) at rated slip s_{N} point, we can obtain (Formula (7.78)):

$$\Delta M_e = \frac{U_N^2}{x_k} \times \frac{s_{cr} - s_N^2/s_{cr}}{(s_{cr} + s_N^2/s_{cr})^2}\Delta s \tag{7.78}$$

The rated slip s_N is generally small, and can be ignored. Assume that the mechanical torque M_m is constant, i.e., $\Delta M_m = 0$. From Formulas (7.75) and (7.78) we can obtain (Formula (7.79)):

$$\Delta \dot{s} = -\frac{U_N^2}{T_j r_r}\Delta s \tag{7.79}$$

In addition, its characteristic root can be determined as $\lambda = -\frac{U_N^2}{T_j r_r}$. Group asynchronous generators with approximate characteristic roots together as one set, and each set can be equivalent to one generator.

2) **Wind turbine models and physical quantities to be equivalent**
When making equivalent on the asynchronous wind turbine set, we first study the physical quantities needed to be equivalent according to the mathematical model of a single wind turbine. In the WF simulation, the equivalent physical quantities needed for the equivalent wind turbine set include the rotor inertia time constant T_J, the initial slip s, the synchronous reactance x and the transient reactance x'. In addition, the rotor winding time constant T_0' can be obtained through the stator and rotor impedances, but the simulation results show that the error of the equivalent amount obtained from the identification method is smaller, and therefore, the equivalent rotor time constant can be obtained from the identification method.
3) **The principle of the least square method used by parameter identification**
The measurement data (t_i, y_i), $i = 1,\ldots, m$(m is the sampling point) is provided. The fitting curve selected is (Formula (7.80))

$$y(t) = \varphi(t, x) \tag{7.80}$$

Where, the parameter $x = (x_1, \ldots, x_n)^T \in R^n, n < m$

$$\text{If} \begin{cases} \varphi_i(x) = \varphi(t_i, x) - y_i, i = 1, \ldots m \\ F(x) = [\varphi_1(x), \ldots, \varphi_m(x)]^T \end{cases}$$

Then x is the solution of Formula (7.81)

$$\min f(x) = \sum_{i=1}^{m} [\varphi_i(x)]^2 = F(x)^T F(x) \tag{7.81}$$

Generally, the above problem is referred as nonlinear least square problem [24].
4) **The equivalence principle of the rotor time constant, the synchronous reactance and the transient reactance**
The premise of such equivalence is that all asynchronous generators are connected to the same bus through the transformer at the outlet before the equivalence. The capacitor bank for improving power factor is installed at the outlet.

In the electromagnetic loop convergence of the synchronous generator, the fourth-order model can be used to derive its multi-machine equivalent circuit. The electromagnetic loop admittance matrix formula obtained can be described by Formula (7.82) [23]:

$$\begin{bmatrix} i_d \\ i_q \end{bmatrix} = \begin{bmatrix} 0 & Y_{dq} \\ Y_{qd} & 0 \end{bmatrix} \begin{bmatrix} u_d \\ u_q \end{bmatrix} + \begin{bmatrix} Y_{df} \\ 0 \end{bmatrix} E_f \tag{7.82}$$

The definition formula of each element is as follows (Formula (7.83)):

$$\begin{cases} Y_{dq} = -\dfrac{1 + pT'_{d0}}{X_d + pX'_d T'_{d0}} \\ Y_{qd} = -\dfrac{1 + pT'_{q0}}{X_q + pX'_q T'_{q0}} \\ Y_{df} = \dfrac{1}{X_d + pX'_d T'_{d0}} \\ E_f = \dfrac{X_{ad}}{r_f} u_f \end{cases} \tag{7.83}$$

Where T'_{d0} and T'_{q0} respectively represents the transient time constant of d-, q-axis; r_f is the excitation winding resistance; u_f is the excitation winding voltage; E_f is the generator no-load electromotive force. Asynchronous generators can be seen as a special case of synchronous generators, that is, the excitation voltage is identically equals zero, d-, q-axis parameters are equal, and the speed is non-synchronous speed. Therefore, when Formula (7.82) is used to indicate the asynchronous generator electromagnetic circuit, the following formula will be obtained (Formula (7.84)):

$$Y_{\mathrm{dq}} = Y_{\mathrm{qd}} \tag{7.84}$$

There the equivalent d-, q-axis determination problem in synchronous generator equivalent is not exist in the equivalent process of asynchronous generators. Assume that the outlet voltages of the asynchronous generators are equal, the following electromagnetic formula can be obtained (Formula (7.85)):

$$\begin{bmatrix} \sum_j i_{Dj} \\ \sum_j i_{Qj} \end{bmatrix} = \begin{bmatrix} 0 & Y_{DQ} \\ Y_{DQ} & 0 \end{bmatrix} \begin{bmatrix} u_D \\ u_Q \end{bmatrix} + \begin{bmatrix} Y_{DF} \\ Y_{QF} \end{bmatrix} E_{feq} \tag{7.85}$$

Where $\sum i_{Dj}$ and $\sum i_{Qj}$ respectively represents the d-, q-axis current component flowing in the bus; u_D and u_Q respectively represents d-, q-axis component of the bus voltage;
Y_{DQ} and Y_{QD} respectively represents the sum of Y_{dq} and Y_{qd} of each wind turbine; E_{feq} represents the equivalent quantity corresponding to E_f in Formula

(7.82); Y_{DQ} is the generator transfer function, which is used to calculated the rotor time constant, and is defined as follows (Formula (7.86)):

$$Y_{DQ} = -\frac{1 + pT_0'}{X_d + pX_d'T_0'} \tag{7.86}$$

Set $p = j\omega$, then the frequency response of the transfer function can be obtained. The form of the frequency response of the transfer function of the equivalent asynchronous generator can also be obtained by the analogy of Formula (7.86), of which the formula is as follows (Formula (7.87)):

$$Y_{DQ}^* = -\frac{1 + j\omega T_0'^*}{X_d^* + j\omega X_d'^* T_0'^*} \tag{7.87}$$

Where the superscript * represents the physical quantity is equal to the corresponding equivalent quantity of the equivalent asynchronous generator in Formula (7.83).

The solution of the equivalent model refers to Gauss Newton algorithm in the solution of the least square method, fitting the amplitude-frequency response curve of the transfer function. The specific solving process is as follows (Formula (7.88)) [24]:

Set the actual amplitude-frequency response as:

$$G(j\omega) = R_k + jI_k \tag{7.88}$$

In addition, the unknown amplitude-frequency response curve is (Formula (7.89)):

$$G'(j\omega) = \frac{C(jw)}{D(jw)} = \frac{A_0 + j\omega A_1}{1 + j\omega B_1} \tag{7.89}$$

Set $e_k = (G(j\omega_k)D(j\omega_k) - C(j\omega_k)) = e_{Rk} + je_{Ik}$

Where $\begin{cases} e_{Rk} = -A_0 - \omega_k I_k B_1 + R_k \\ e_{Ik} = -\omega_k A_1 + \omega_k R_k B_1 + I_k \end{cases}$

Set

$$E = \sum_{k=1}^{N} (e_{Rk}^2 + e_{Ik}^2)$$

E respectively seeking partial derivative to A_i and B_i, and make all the partial derivatives be zero, the matrix can be obtained as follows (Formula (7.90)):

$$PX = Q \tag{7.90}$$

Where $P = \begin{bmatrix} V_0 & 0 & T_1 \\ 0 & V_2 & -S_2 \\ T_1 & -S_2 & U_2 \end{bmatrix}$, $X = [A_0 \ A_1 \ B_1]^T$, $Q = [S_0 \ T_1 \ 0]^T$

The calculation formula of elements of Matrix$P$: V_0, V_2, T_1, S_2, and U_2 is provided in Formula (7.90). In order to eliminate the impact of ω, the improved algorithm of the frequency domain response identification method presented in the literature [24] is used as follows:

$$e_k' = \left[G(j\omega_k) - \frac{C(j\omega_k)}{D(j\omega_k)}\right]\frac{D(j\omega_k)_l}{D(j\omega_k)_{l-1}}$$

$$E' = \sum_{k=0}^{N} |e_k'|^2$$

where N is the number of sampling points, l is the number of iterations.

Elements in matrix P are improved as (Formula (7.91)):

$$\begin{cases} V_i = \sum_{k=0}^{N} \omega_k^i W_l(\omega_k) \\ S_i = \sum_{k=0}^{N} \omega_k^i R_k W_l(\omega_k) \\ T_i = \sum_{k=0}^{N} \omega_k^i I_k(\omega_k) \\ U_i = \sum_{k=0}^{N} \omega_k^i (R_k^2 + I_k^2) W_l(\omega_k) \end{cases} \tag{7.91}$$

where $W_l(\omega_k) = 1/|D(j\omega_k)_{l-1}|$, subscript $i = 0, 1, 2$.

The iterative calculation is as follows:

1) set the initial value $D(j\omega_k)$;
2) Determine V_i, S_i, T_i, and U_i according to (7.91);
3) Substitute in $PX = Q$ matrix to solve A_0, A_1, and B_1;
4) Use the factor A_0, A_1, and B_1 to calculate $D(j\omega_k)_l$;
5) Repeat the calculations in step (2) to step (4), until $E(j\omega)$ is smaller than the given error indicator.

Substitute A_0, A_1, and B_1 obtained from the above method in the formula of the asynchronous generator transfer function, then we can obtain the reactance X, the transient reactance X' and the rotor winding time constant T_0', see Formula (7.92).

$$\begin{cases} X = 1/A_0 \\ T_0' = A_1/A_0 \\ X' = B_1/A_1 \end{cases} \tag{7.92}$$

5) **The equivalent calculation method of the wind turbines multi-machine** [23, 25]

For wind turbine multi-machine dynamic equivalent, the physical quantities required include the rotor inertia constant, the initial operate slip, the synchronous reactance, the transient reactance, and the rotor time constant.

a) **Equivalent inertia time constant.** Similar to asynchronous generators, the equivalent method of asynchronous generator rotor inertia time constant may be equivalent by means of weighted capacity (Formula (7.93)).

$$T_J = \frac{1}{S_M}\sum_{j=1}^{n} T_j S_j \tag{7.93}$$

Where S_j and T_j respectively represents the capacity and inertia time constant of asynchronous generator No. j, S_M is the sum of the capacity of all asynchronous generators, and T_J is the inertia time constant of the equivalent asynchronous generator.

b) **Equivalent initial operate slip.** First, parallel the Γ-type equivalent circuit of the asynchronous generator, use Thevenin equivalent method to obtain equivalent resistance r_M of the two generators, and use Formula (7.94) to obtain slip s_M of the equivalent generator under initial state (Formula (7.94)):

$$s_M = \frac{R_M\left(\frac{R_1}{s_1}+\frac{R_2}{s_2}\right)^2 + R_M(x_{l1}+x_{l2})^2}{\left(\frac{R_1}{s_1}\right)\left[x_{l2}^2+\left(\frac{R_2}{s_2}\right)^2\right]+\left(\frac{R_2}{s_2}\right)\left[x_{l1}^2+\left(\frac{R_1}{s_1}\right)^2\right]} \tag{7.94}$$

Where R_1 and R_2 respectively represents the sum of stator resistances and the sum of rotor resistances of the two generators; X_{l1} and X_{l2}, respectively, represents the sum of stator leakage reactances and the sum of rotor leakage reactances of the generators; s_1 and s_2, respectively, represents the initial operate slip of the two generators, wherein each physical unit is per-unit value.

The equivalent of the initial operate slip of multiple wind turbines can be completed by continuous use of the method.

c) **Equivalent reactance, equivalent transient reactance and equivalent rotor time constant.** First calculate the Formula (7.95) synchronous reactance X, transient reactance X' and rotor time constant T_0'.

$$\begin{cases} X = X_s + X_m \\ X' = X_s + X_r X_m/(X_r + X_m) \\ T_0' = (X_r + X_m)/(2\pi R_r) \end{cases} \tag{7.95}$$

where the physical meanings of R_s, R_r, X_s, and X_r are explained in Formula (7.82), X_m is the excitation reactance, of which the unit is per-unit value. Then use the method described in Section 7.4 to obtain its equivalent amount.

The step completed identification of the three parameters at the same time. The equivalent time constant of rotor windings can be obtained by the calculation of the equivalent impedance, but the error of the identification value obtained from the least square fitting is smaller, and more suitable as the equivalent parameter in simulation.

7.2.8 Per-Unit Value System

Mathematical models for WTGS can be divided into electrical part and mechanical part. For the model of the electrical part and the network formula interface, in order to facilitate the calculation, the variables shall be converted to the per-unit value under a unified base value of the system. In the model of the mechanical part, if actual values are used for each of the parameters and state variables, its physical meaning will be clearer, but frequent conversion between the actual values and per-unit values is inevitably required in the calculation process. In addition, mechanical characteristics (such as rotor diameter, rated speed of rotor) of wind turbines of various types in the WF are different, the rated speeds of the low-speed shaft and the high-speed shaft on both sides of the gearbox of a wind turbine are also different, so unified per-unit value system is applied for the variables here.

Set the system power base value as S_B; the voltage base value as the average value U_B of all voltage levels; $S_B = \sqrt{3}U_B I_B$, I_B as the base current; and the impedance base value as $Z_B = U_B^2/S_B$. The speed levels on both sides of the gearbox are different, so we can choose the actual mechanical speed Ω_B corresponding to the synchronous speed ω_0 as the speed base value of the high-speed shaft and the low-speed shaft by the analogy with conversion of voltage classes by the transformer; based on $P = T \cdot \Omega$, the torque base value is $T_B = S_B/\Omega_B$. In a per-unit value system like this, the gearbox speed rising ratio γ will no longer apear in the mathematical model.

The per-unit value system is a method used to simplify the calculation, it expresses all values as ratios:

$$\textit{Per unit value} = \frac{\textit{Actual value (any unit)}}{\textit{Base value or reference value (the same unit)}}$$

The advantages of using per-unit values in the power system include: ① reduce the appearance of $\sqrt{3}$ in calculation; ② similar per-unit values can be used in systems of different sizes; ③ by carefully selecting the voltage base value, the process of solving networks containing of transformers becomes easy.

For simple calculations, such as calculations for WF assessment in simple distribution network require the following steps:

1) Assume an arbitrary power base value S_B (e.g., 10 MVA).
2) Select the base value U_b of each voltage level in the network (e.g., 121 kV, 11 kV, and 38.5 kV). These voltage base values shall be related to the rated transformer turns ratio.
3) Calculate the suitable active and reactive power flow per-unit values at the generator terminal. If the power base value (S_b) is selected as 10 MVA, per-unit value of active power of 5 MW is $P = 5/10$ or 0.5 p.u., also, the per-unit value of the reactive power of 1 MVAr is $Q = 1/10$ or 0.1 p.u.

4) If necessary, the ohm impedance of the circuit is also converted into per-unit value, and the impedance base value is $Z_B = U_B^2/S_B$.
5) The calculation formula of the base value current at any specific voltage class is $I_B = S_B/(\sqrt{3}U_B)$.

7.3 Power Flow Analysis of Power System Containing Wind Power

7.3.1 Overview

Operation requirements of the power system under normal balanced three-phase steady-state conditions: ①the generator output is equal to the load demand plus the system losses; ② the bus voltage is maintained close to the rated values; ③ the generators operate within specified active and reactive power limits; ④ transmission lines and transformers are not overloaded. The power flow calculation of the power system is the basic tool to study these requirements. Its basic task is to obtain voltage and angle at each bus of the power system, the active power, reactive power, and equipment losses of all the equipements connected to each bus under balanced three-phase steady-state conditions.

The nodal or loop analysis of conventional circuit theory is not suitable for the power flow studies, because the load data is usually given in the terms of power rather than impedance, and the power plant is used as a power source rather than a voltage or current source. Therefore, in terms of mathematics, the power flow problem is transformed into a set of nonlinear algebraic formulas suitable for computer solving.

The input data of the power flow calculation includes the network topology and wiring, cables, transformers parameters (R, X, G, Band transformer ratio, etc.), the user load and the generator output (PQ or PV); the data to be calculated includes the static characteristics of the power grid, that is, the voltage amplitude and angle, the active power or reactive power (or current) flows through the equipments, as well as various losses in the system.

In the power flow calculation including WFs, the suitable simulation is required for the WF collection system and all kinds of wind turbines. In the power flow calculation, a WF consists of n wind turbines can be modeled as: ① indicate each wind generator in detail; ② equivalent to one wind generator; ③ simulate as several equivalent wind generators, but each one shall be a single model. In conventional power flow calculation, the bus is divided into three categories: PQ node, PV node, and $V\theta$ node. Because of the particularity of the wind power system, the characteristics of the wind turbine itself must be taken into account at the time of calculation. This chapter describes power flow calculations of power systems containing cage induction generators, DFIG, and direct drive permanent magnet synchronous generators.

7.3.2 Power Flow Calculation of Power Systems Containing Cage Induction Generators

The constant speed induction generator has no excitation device, it relies on the reactive power provided by the grid to create a magnetic field, so it is unable of voltage regulation and cannot be regarded as a constant voltage amplitude PV node like a conventional synchronous generator; asynchronous generators will absorb a certain reactive power from the system while outputing active power, the size of its reactive power is closely related to the sizes of the slip s and the node voltage U, and therefore cannot be simply regarded as a negative constant load node, that is, the PQ node. When a WF containing cage induction generators is considered in the power flow analysis, usually the PQ or RX node models are used [26, 27].

It is considered as a PQ node in study of the wind generator steady-state problem, that is, the wind generator active and reactive power are calculated according to the given wind speed and power factor. When the conventional PQ node model is used, the active and reactive power are constant. The improved PQ node model is a correction of these constants, such as P and Q is represented as related to the voltage or frequency, or relevant characteristics of the wind turbine are considered in order to improve P and Q.

When the WF is using PX node model, ignoring the stator leakage reactance and rotor leakage reactance of the generator, the reactive power absorbed by the asynchronous generators depends on the excitation reactance, that is, $Q_g = U^2/X$ wherein, U is the generator terminal voltage amplitude, X is the excitation reactance. This reactance can be expressed as constant excitation reactance at rated voltage, or as a nonlinear function of the terminal voltage in order to consider the saturation characteristics.

RX model is used for asynchronous generator in the following two cases:

a) If the generator parameters and the slip are already known, the generator can be used as an impedance in the steady-state simulation study.
b) If the power factor curve of the wind generator and the wind speed are known, then the slip of the generator can be calculated.

Wind turbine characteristics are assumed already known in both cases. Wind turbine power curve is generally provided by the manufacturer. When the asynchronous generator parameters are unknown, the estimation must try to be made.

1) **The PQ model of the cage asynchronous generator**
   The way of establishing the WF PQ model is to assume that the generator active power and power factor have been given, the generator reactive output can be obtained from the power factor. If we consider

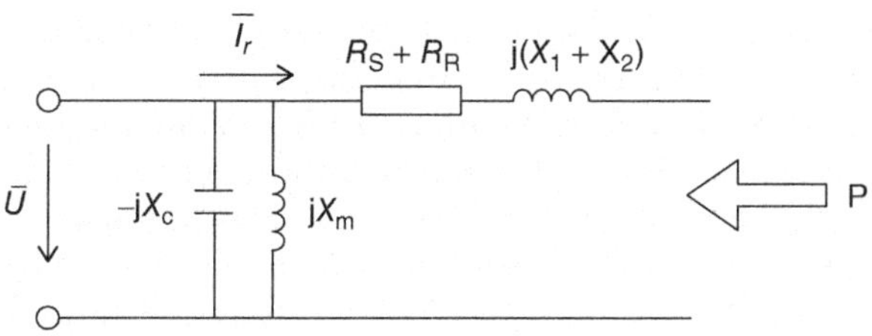

Figure 7.26 The steady-state model of the cage asynchronous generator (with terminal capacitor compensation).

the steady-state model of the cage asynchronous generator, and include the terminal capacitance compensation, the steady-state model shown in Figure 7.26 will be obtained [27].

In Figure 7.26, if it is a PQ node model, then $P = f(U)$; if it is a RX node model, then $P = -I_r^2 R_r \frac{1-s}{s}$.

From Figure 7.26, the reactive power of the asynchronous generator (positive values indicates reactive power absorption) can be calculated according to Formula (7.96) by using complex power (Formula (7.96)).

$$Q = U^2 \frac{X_C - X_m}{X_C X_m} + X \frac{U^2 + 2RP}{2(R^2 + X^2)} - X \frac{\sqrt{(U^2 + 2RP)^2 - 4P^2(R^2 + X^2)}}{2(R^2 + X^2)} \quad (7.96)$$

where U is the voltage; P is the active power (positive value indicates injection into the power grid); X is the sum of stator and rotor leakage reactance; X_m is the excitation reactance; X_c is the capacitor bank reactance; R is the sum of stator resistance and rotor resistance.

The reactive power of asynchronous generatosr can also be calculated as a function of active power using Formula (7.97):

$$Q = -Q_0 - Q_1 P - Q_2 P^2 \quad (7.97)$$

where Q_0, Q_1, and Q_2 are obtained through experiments.

If the wind speed is an input data, then the active power can be obtained by using the wind turbine power curve and Formula (7.7).

2) **The RX model of the asynchronous wind turbine**

When using the RX model, the power flow iterative process can be divided in two steps: the conventional power flow iterative calculation and the slip iterative calculation of the asynchronous wind generator, which are specially carried out in the following steps [27]:

a) According to the power factor curve of the wind turbine, calculate the power extracted by each wind turbine from the wind at the given wind speed and rotor speed.
b) According to the result of the power flow analysis and the rotor speed given in step a, calculate the power generated by each wind turbine.
c) Compare the two kinds of power, calculate the slip value to make the electromagnetic power at this slip be equal to the mechanical power at the given speed.

The RX model is built on the steady-state model of the induction generator, and is represented in the way of impedance: $\overline{Z} = R_s + jX_1 + (jX_m || (R_R/s) + jX_2)$, the susceptance of the shunt capacitance is Bc. The excitation branch is between the stator and the rotor, which is a good approximation of the actual generator.

The wind turbine mechanical power is determined by the wind speed [Formula (7.7)], and if the wind turbine mechanical power is assumed to be constant, the slip can be calculated from Formula (7.98):

$$s = \frac{-U^2 R_R + \sqrt{U^4 R_R^2 - 4PR_R^2(PX^2 + U^2 R_R)}}{2(PX_2 + U^2 R_R)} \quad (7.98)$$

Formula (7.98) is based on the generator steady-state model (see Figure 7.26), wherein the stator resistance R_S is ignored.

The iterative process is as follows: when the process begins, first determine the initial value of slip, and generally the rated slip or the slip obtained according to Formula (7.98) is used. Substitute the value of s in $\overline{Z}$, the power flow analysis initialization can be performed. With the outcome, the generator mechanical power can be calculated by using formula $P_m = -I_R^2 R_R((1-s)/s)$, in which I_R is the effective value of the rotor current, which is calculated from the effective value of the stator current I_s, and the calculation formula is: $\overline{I}_R = (jX_m/(R_R/s) + j(X_2 + X_m))\overline{I}_S$, wherein $\overline{I}_R$ and $\overline{I}_S$ are the phase values of the rotor current and the stator current.

On the other hand, taking into account the wind speed and the generator slip, the value of the power factor can be calculated by using formula $C_p = f(\lambda)$, and the wind power extracted can be calculated by using Formula (7.7). If the two power values (mechanical power and the wind power extracted) are not equal, the convergence process of the both values will be started. In this case, the slip is improved as follows (Formula (7.99)):

$$s_k = s_{k-1} + \Delta s \quad (7.99)$$

where s_{k-i} and s_k respectively represents the present value and the new value of the slip (the slip must be considered in the power flow analysis in the next step), and (Formula (7.100)):

$$\Delta s = J^{-1} \Delta P_m \quad (7.100)$$

where ΔP_m is the difference between the two power values. If the operation is close to the working point, thenconsider as $(1 - s)/s \approx 1/s$, J can be calculated by using the following formula (Formula (7.101)):

$$J = \left(\frac{S}{U}\right)^2 \frac{R_R X_m^2 (R_R^2 - s^2(X_2 + X_m)^2)}{(R_R^2 + s^2(X_2 + X_m)^2)^2} \tag{7.101}$$

where $S = \sqrt{P_g^2 + Q_c^2}$, in which P_g and Q_c respectively represents the active power generated by and the reactive power absorbed by the generator. These power value can be calculated from $P_g = -(U^2/Z^2)\mathrm{Re}\{\overline{Z}\}$ and $Q_c = (U^2/Z^2)\mathrm{Im}\{\overline{Z}\}$ respectively. When the new values of S and J is obtained, carry out the power flow analysis again, when the mechanical power P_{mk} and the extracted wind power P_k are equal, or their difference is within an acceptable range, the calculation process is completed. The calculation error is $\sum_{k=1}^{n} (P_{mk} - P_k)^2$, wherein n is the number of wind turbines in the WF.

In short, the algorithm of WFs simulated as the RX node is carried out according to the following steps:

a) Start calculating with the same slip $s = s_{nom}$ of each wind turbine, substitute the value in the formula, calculate impedance $\overline{Z}$;
b) Use the above values to establish the RX model, in which the admittance matrix of generator and the corresponding item of Jacobian matrix J in the admittance matrix are included.
c) As the result of the first power flow calculation, the bus voltage can be calculated. Then, use the following expression to calculate the mechanical power: $P_m = -I_R^2 R_R[(1 - s)/s]$
d) Use the obtained values of s to calculate λ and C_P, and use Formula (7.7) to calculate the wind power extracted.
e) Compare the both power values, if not equal, use Formula (7.99) to calculate the value of s again, and go back to step b). If both power values are equal, then the calculation process is completed.

3) **Convergence properties of the two methods**

The PQ model has the advantages that, in the first iteration of the power flow analysis, the active power can be calculated as a function of wind speed, and its value remains the same in process later. The reactive power depends on the active and the bus voltage. Since the active power is considered a constant, the only variable is the bus voltage. There are two possible methods for the calculation of reactive power: one is to consider the voltage as a constant. In this case, in the first iterative computation, the active power and reactive power has been determined, and remain the same in the later iterations. Thus, the bus is treated as a traditional PQ bus. This simplification does not bring about a significant margin of error. The second method is that the reactive power is calculated as a function of the voltage. In this case, the reactive power is updated in each iteration. Whatever the method, the convergence characteristic of the PQ bus model is similar to those of the conventional PQ model.

When using RX model, there are two iteration processes: the system power flow analysis and the calculation of the slip of induction generator. From the standpoint of power flow analysis, RX bus is operating as the PQ bus of which the active power and reactive power are zero. Therefore, more power flow iterations are not required. However, to obtain the final slip, usually two to six iteration processes are required.

Therefore, when calculating the power flow of power system containing WFs, the wind turbine mathematical model must be taken into account, and extended to the power formula of the system, for simultaneous solution; or make alternate iteration of the power formula of the system and the formulas describing the wind generator mathematical model.

4) **Calculating process for solving the power flow using alternate iterative method**

The Generator equivalent circuit and power transfer relations are show in Figure 7.27. The asynchronous motor operates as a generator, of which the power transfer relation is in contrast with that of the motor. Natural wind blows the turbine blades, converting wind energy into mechanical energy, and after deducting the mechanical loss, the mechanical power obtained here is the mechanical power P_Ω transmitted to the asynchronous generator rotor, which corresponds to the electric power on the variable resistance $r_2(1-s)/s$, $s < 0$ here. Deduct the rotor copper loss P_{Cu2} and iron loss P_{Fe}, the electromagnetic power P_M input into the stator winding can be obtained, and then deduct the stator copper loss P_{Cu1}, the electric power P_e injected into the power grid is obtained.

In the equivalent circuit shown in Figure 7.27, if we ignore the stator resistance and the core power loss, and due to $X_m \gg X_1$, the magnetizing branch can be moved to the circuit head end, to obtain a simplified Γ-type equivalent circuit of the asynchronous generator, as shown in Figure 7.28. At the positive direction as shown in the diagram, the power injected into the grid P_e is the electromagnetic power P_M, that is, the electrical power on the resistance r_2/s.

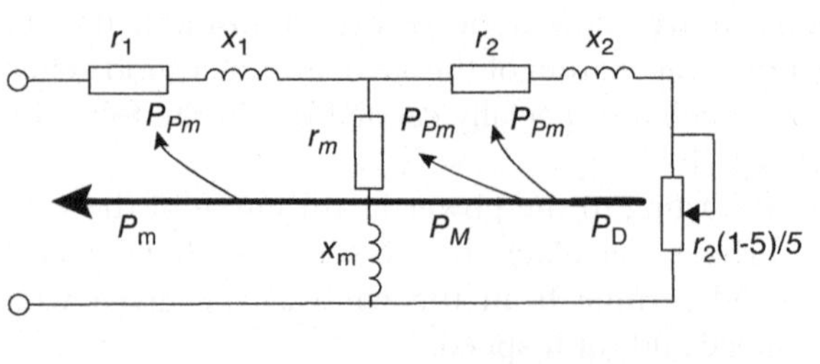

Figure 7.27 Induction generator-power relationship.

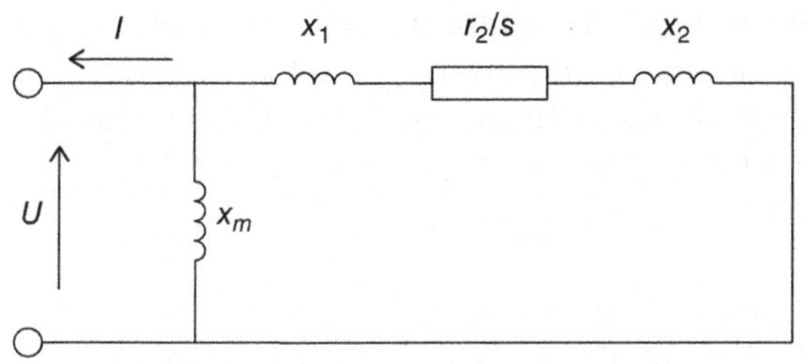

Figure 7.28 Simplified equivalent circuit of the asynchronous generator.

Obtain from the circuit relations (Formula (7.102)):

$$P_e = -\frac{U^2 r_2/s}{(r_2/s)^2 + {x_k}^2} \tag{7.102}$$

where $x_k = x_1 + x_2$. The generator slip s is obtained by calculation (Formula (7.103))

$$s = -\frac{U^2 r_2 - \sqrt{U^4 r_2^2 - 4P_e^2 x_k^2 r_2^2}}{2P_e x_k^2} \tag{7.103}$$

From the equivalent circuit, it can be seen that, the angle of the power factor of the asynchronous generator is related to the value of the slip (Formula (7.104))

$$\varphi = \tan^{-1}\left(\frac{r_2^2 + x_k(x_k + x_m)s^2}{r_2 x_m s}\right) \tag{7.104}$$

The relationship between the reactive power absorbed by the asynchronous generator and the active power (Formula (7.105)):

$$Q_e = -\frac{r_2^2 + x_k(x_k + x_m)s^2}{r_2 x_m s} P_e \tag{7.105}$$

It can be seen from the above formula that, when the active power P_e output by the asynchronous generator is constant, the reactive power Q_e absorbed by it has close relationships with the node voltage U and the size of slip s.

In the power flow calculation containing wind turbines, taking into account the steady-state mathematical model of the asynchronous generator, the iterative calculation method is used, of which the calculation process is as follows [14]:

a) The active power P_e of the WF is given, set the initial voltage of the WF node at U;
b) Use P and U to calculate the wind turbine slip s according to Formula (7.103);
c) Use P and the slip s obtained from calculation to calculate the reactive power Q_e according to Formula (7.105);
d) Consider the WF node as a PQ node to solve the power flow of the entire system, to obtain the new WF node voltage U';
e) If $U' \neq U$, then make $U = 0.5\ (U + U')$, return to step b, continue to perform step b through d, until the difference between the two voltage values obtained is within the predetermined error range.

5) **Calculating process for solving the power flow using joint iterative method** [16, 28]
The use of the iterative method of alternating between the conventional power flow calculation and the wind turbine internal circuit calculation, although increases the accuracy, brings a large computation. The joint iterative calculation model for the WF power flow not only takes into account the terminal voltage of the wind turbine, the active power, the reactive power, and the coupling of the slip, but also shorten the calculation time. The joint iterative calculation model also uses the simplified equivalent circuit model of asynchronous generators shown in Figure 7.28, of which calculations of the electric power P_e, slip s and reactive power Q_i are respectively the same as Formula (7.102), Formula (7.103), and Formula (7.105).

After the (generator) slip is eliminated, the calculation formula of the reactive power of the asynchronous generator is obtained as (Formula (7.106)):

$$Q_e = \frac{2P_e^2 X_k^2 \left(-r_2^2 - \frac{\left(-U^2 r_2 + \sqrt{U^4 r_2^2 - 4P_e^2 r_2^2 X_k^2}\right)^2 (X_k + X_m)}{4P_e^2 X_k^3}\right)}{r_2\left(-U^2 r_2 + \sqrt{U^4 r_2^2 - 4P_e^2 r_2^2 X_k^2}\right) X_m} \tag{7.106}$$

It can be known from Formula (7.106) in the power flow calculation of power system, if the active power of the wind turbines is known, the reactive power can be expressed as a function of voltage and active power. In the Newton-Raphson power flow calculation, the active and reactive power of the WF node can be calculated in accordance with Formulas (7.102) and (7.106), for partial derivative of the reactive increment of the WF node to voltage in power flow Jacobian matrix, the following values shall be added (Formula (7.107)):

$$\begin{aligned}\frac{\partial Q_e}{\partial U} = &-\frac{\left(-2Ur_2 + \frac{2U^3 r_2^2}{\sqrt{U^4 r_2^2 - 4P_e^2 r_2^2 X_k^2}}\right)(X_k + X_m)}{r_2 X_k X_m} \\ &- \frac{2P_e^2 X_k^2 \left(-2Ur_2 + \frac{2U^3 r_2^2}{\sqrt{U^4 r_2^2 - 4P_e^2 r_2^2 X_k^2}}\right)}{r_2\left(-U^2 r_2 + \sqrt{U^4 r_2^2 - 4P_e^2 r_2^2 X_k^2}\right)^2 X_m} \\ &\times \left(-r_2^2 - \frac{\left(-U^2 r_2 + \sqrt{U^4 r_2^2 - 4P_e^2 r_2^2 X_k^2}\right)^2 (X_k + X_m)}{4P_e^2 X_k^3}\right)\end{aligned} \tag{7.107}$$

The driving force behind wind power was uncontrollable, the electricity generation of the wind turbine and the size of its output depend on wind conditions. The output power of the wind turbine can be approximately obtained from its power curve. The formula given by reference [29] is (Formula (7.108)):

$$P_e = P_m = \sum_{i=0}^{8} a_i v^i \tag{7.108}$$

where a_1 to a_8 are parameters of the power curve of a wind turbine, which are slightly different from wind turbines, v is the wind speed.

To sum up, for power systems containing cage asynchronous wind turbines, its calculation models can be divided into the following calculation steps [28]:

a) Given wind speed of the WF;
b) Given initial value of the WF node voltage;
c) Determine the active and reactive power of the WF based on the given wind speed and Formulas (77.108) and (77.106).
d) Call the conventional Jacobian matrix computation program;
e) Modify corresponding elements in the Jacobian matrix of the WF node according to Formula (77.107)
f) Solve the modified formula, and modify the voltage and angle of each node
g) Verify whether it is converged, if yes, the calculation is completed; otherwise, use the modified node voltage as the initial value, go to step c for the next iteration.

7.3.3 Power Flow Calculation of Power System Containing Double-Fed Asynchronous Generators

In the power flow calculation, a WF with n double-fed asynchronous generators (DFAG) can also be modeled as: ① indicate each wind turbine in detail; ② equivalent to one wind turbine; ③ simulate as several equivalent wind turbines, but each one shall be a single-machine model. The equivalent impedance reflects the aggregate impact of the collection system can be included in the impedance of a step-up transformer, as shown in Figure 7.29 [30]. n wind turbines are connected on the same bus, the total input generator power P_{gen}, Q_{max} and Q_{min} are obtained by accumulating according to the actual power of each wind turbine, or approximated as n times of a single generator capacity.

7.3.3.1 Steady-State Equivalent Circuit and Power Calculation of a Double-Fed Asynchronous Generator

The steady state equivalent circuit of double-fed asynchronous generator is shown in Figure 7.30: $R_S + R_R$where r_s and x_s respectively represents the stator winding resistance and impedance, r_r and x_r respectively represents the rotor resistance and impedance, x_m is the excitation reactance (the excitation resistance is ignored), s is the slip.

When the wind speed is known, the total active power P_e injected by the wind turbines into the system at the wind speed can be obtained according to the wind turbine characteristics, P_e consists of two parts, one is the active power P_s generated by the stator winding, and the other is the active power P_r generated or absorbed by the the rotor winding. If the speed is higher than the synchronous speed, the rotor winding will generate active power $P_r > 0$; and if the speed is lower than the synchronous speed, the rotor winding will absorb active power $P_r < 0$.

In case that the stator winding resistance is ignored, the active power generated by the rotor winding can be represented as (Formula (7.109)) [31, 32]:

$$P_r = \frac{r_r x_{SS}^2 (P_s^2 + Q_s^2)}{x_m^2 U_s^2} + \frac{2 r_r x_{SS}}{x_m^2} Q_s - sP_s + \frac{r_r U_s^2}{x_m^2} \tag{7.109}$$

where $x_{SS} = x_s + x_m$, $U_s = |U_s|$.

The active power injected by the wind generators into the system is (Formula (7.110)):

$$\begin{aligned} P_e &= P_s + P_r \\ &= \frac{r_r x_{SS}^2 (P_s^2 + Q_s^2)}{x_m^2 U_s^2} + \frac{2 r_r x_{SS}}{x_m^2} Q_s + (1 - s)P_s + \frac{r_r U_s^2}{x_m^2} \end{aligned} \tag{7.110}$$

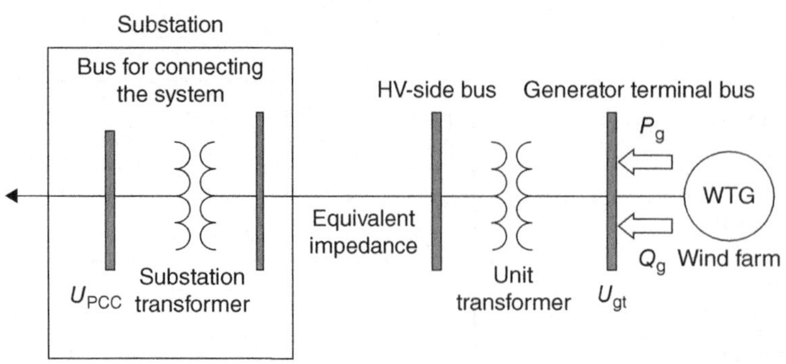

Figure 7.29 DFAG power flow model.

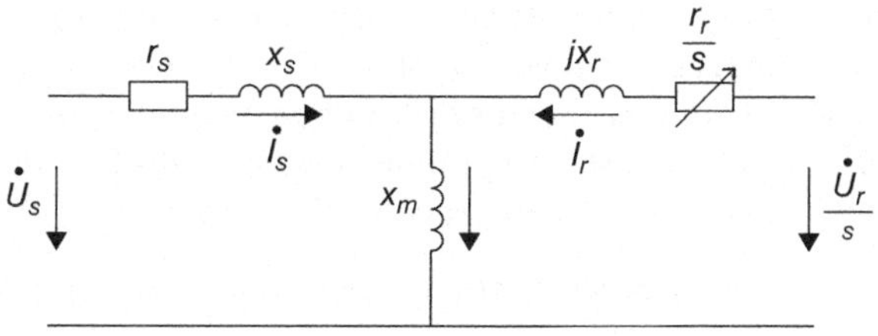

Figure 7.30 Steady state equivalent circuit of double-fed asynchronous generator.

where slip s can be obtained according to the control strategy of the rotor speed of VSCF wind turbines.

7.3.3.2 Control Strategy of Rotor Speed of DFAG

Control strategy of the rotor speed of DFAG refers to the correspondence between the wind turbine speed and the wind turbine mechanical power P_M, usually the control strategy used is as Formula (7.111), and the curve is shown in Figure 7.31 [31, 32].

P_M in Figure 7.31 represents mechanical power of the wind turbines. ω is the speed of the wind turbines, P_{max} is the maximum capacity of the wind turbines, and P_1, P_2, and P_3 are determined by the wind turbine parameters. And ω_{min} is the lower limit of wind turbine speed, ω_{max} is the upper limit of wind turbine speed, ω_r is the wind turbine speed corresponding to the generator synchronous speed. When the power is less than P_1, the wind speed is maintained at the lower limit ω_{min}, as shown by curve segment ab in the figure 7.31; when the power value is between P_1 and P_2, the relationship between the wind turbine speed and the active power approximates to a cubic curve, as shown by curve segment bc in the figure 7.31; when the power value is between P_2 and P_3, the wind turbines operate at synchronous speed, as shown by curve segment cd in the figure; when the power exceeds P_3, the wind turbine operate at a constant torque, then the torque is maximal, and the relationship between speed and power is linear, as shown by curve segment de in the figure 7.31 (Formula (7.111)).

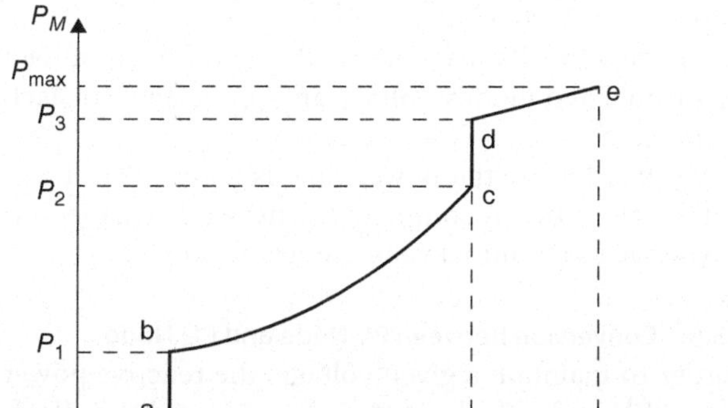

Figure 7.31 Wind turbine speed control law.

$$\begin{aligned}
&\omega = \omega_{min} && 0 < P_M < P_1 \\
&\omega = \sqrt[3]{\frac{P_M}{K_{opt}}} && P < P_M < P_2 \\
&\omega = \omega_r && P_1 < P_M < P_3 \\
&\omega = \omega_r + \frac{\omega_{max} - \omega_r}{P_{max} - P_3}(P_M - P_3) && P_3 < P_M < P_{max}
\end{aligned} \tag{7.111}$$

where k_{opt} is a factor determined by the wind turbine blade parameters [33], which can keep the optimal wind turbine power factor. After the wind turbine speed is determined according to Formula (7.111), the wind turbine slip s can be calculated.

Because double-fed asynchronous generators usually have two operation modes, that is, constant power factor and constant voltage operation modes. Power flow calculation under the two operational modes is respectively discussed below:

7.3.3.3 Constant Power Factor Control Method

If the generator synchronous speed is ω_1, and the rotation speed of the rotor is ω, then $s = (\omega_1 - \omega)/\omega_1$. If the DFAG uses constant power factor control method, namely stator power factor is set as $\cos\varphi$, then $Q_s = P_s \tan\varphi$. Also, because transfer of active power by the converter is relatively small, the reactive power delivered or absorbed by the inverter is very small, so the reactive power Q_e of DFAG can be considered approximately equal to the stator winding reactive power Q_s, when the constant power factor method is used, the following formula can be obtained (Formula (7.112)):

$$Q_e = Q_s = P_s \tan\varphi \tag{7.112}$$

Substitute the above formula in Formula (7.1110), then we can obtain (Formula (7.113)):

$$\begin{aligned}
P_e = P_s + P_r &= \frac{r_r x_{ss}^2 P_s^2}{x_m^2 U_s^2}(1 + \tan^2\varphi) \\
&+ \left(1 + \frac{2r_r x_{ss}\tan\varphi}{x_m^2} - s\right) P_s + \frac{r_r U_s^2}{x_m^2}
\end{aligned} \tag{7.113}$$

The following solution can be obtained from Formulas (7.112) and (7.113) (Formula (7.114)):

$$Q_e = \frac{-bU_s^2 + U_s\sqrt{cU_s^2 + 4aP_e}}{2a}\tan\varphi \tag{7.114}$$

where $a = \frac{r_r x_{ss}^2}{x_m^2}(1 + \tan^2\varphi)$; $b = 1 + \frac{2r_r x_{ss}\tan\varphi}{x_m^2} - s$; $c = (1 - s)^2 + \frac{4r_r x_{ss}\tan\varphi}{x_m^2}(1 - s) - \frac{4r_r^2 x_{ss}^2}{x_m^4}$.

It can be seen from Formula (7.114), when the active power, power factor and slip of DFAG are determined, the reactive power is only a function of the terminal voltage. In general, if wind speed and power factor are given, then the active power of the generator is known, and the slip can be obtained by rules to control the speed.

When considering a WF installed with n wind turbines, if the internal line losses and transformer losses are ignored, assume that all terminal voltages U_s, are the same and equal to the unknown WF bus voltage U_f, then the WF consists of wind turbine generators has the same expression for active power and reactive power (Formula (7.115)):

$$P_f = \sum_{i=1}^{n} P_{ei}(v_i)$$

$$Q_f = \sum_{i=1}^{n} Q_{ei}(P_{ei}, U_f) \quad (7.115)$$

where P_f and Q_f respectively represents the WF active and reactive power, P_{ei} and Q_{ei} respectively represents the active and reactive power injected by wind turbine No. i into the power grid, v_i is the wind speed at the location of wind turbine No. i.

In the constant power factor control mode, the procedure of power flow calculation in power system of a WF consisting of double-fed asynchronous wind turbines is as follows [32]:

1) Wind speed v and initial value of WF node voltage U_f are given;
2) Determine the active power P_{ei} of each wind turbine according to the wind speed-power curve, obtain the slip s_i of each wind turbine according to the speed control law according to Formula (7.111);
3) Calculate the reactive power Q_{ei} of each wind turbine set according to Formula (7.112);
4) Calculate the total active power P_f and reactive power Q_f of the WF according to Formula (7.115);
5) Consider the WF as PQ (of which the value is P_f and Q_f respectively) node, use the amplitude and initial phase angle value of the node voltage to obtain the constant terms of the modified formula and the Jacobian matrix elements;
6) Solve the modified formula, to obtain the modified values of the voltage amplitude and the phase angle, and use them to modify the node voltage.
7) Verify whether the power flow is converged, if yes, the calculation is completed; otherwise, use the modified node voltage as the initial value, go to step (3) for the next iteration.

7.3.3.4 Constant-Voltage Control Mode

Under PV-control mode, DFAG can absorb or emit reactive power in order to maintain a constant terminal voltage. Within the range of wind turbine reactive power regulation, the WF can be considered at PV node. For DFAG, the reactive power control range is limited by the limiting thermal current of stator winding and rotor winding and the maximum current of the inverter, among which the most important limit is the inverter maximum current. The reactive power Q_e of VSCF wind turbine can be approximated as the stator winding reactive power Q_s. When using constant terminal voltage control mode, the power flow calculation process is as follows:

1) Set the WF operating voltage U_s, set the wind speed v;
2) Obtain the active power P_{ei} injected into the system according to the wind speed-pow curve, calculate P_f according to Formula (7.115);
3) Consider the WF as PV node, of which the power is P_f, the voltage is set as $U_f = U_s$, obtain the reactive power Q_f injected by the WF bus by using power flow calculation;
4) Calculate slip s according to the wind turbine speed control law;
5) Use Formula (7.110) to calculate P_s;
6) Substitute P_s, Q_s, and $U_f = U_s$ in Formulas (7.116) and (7.117), to obtain the maximum current limit for the converter and upper and lower limits for the reactive power:

$$P_s^2 + \left(Q_s + \frac{u_s^2}{x_{ss}}\right)^2 \leq \frac{u_s^2 x_m}{x_{ss}^2} I^2{}_{max} \quad (7.116)$$

We can obtain:

$$Q_{s\cdot min} = -\frac{u_s^2}{x_{ss}} - \sqrt{\frac{u_s^2 x_m}{x_{ss}^2} I^2{}_{max} - P_s^2},$$

$$Q_{s\cdot max} = -\frac{u_s^2}{x_{ss}} + \sqrt{\frac{u_s^2 x_m}{x_{ss}^2} I^2{}_{max} - P_s^2} \quad (7.117)$$

where I_{max} is the converter maximum current limit.

7) Verify whether Q_s is beyond limit, if yes, modify the node type. If it is beyond the upper limit, the reactive power Q_s is set as the upper limit value. If it is beyond the lower limit the reactive power Q_s is set as the low limit value. The corresponding WF node is converted to PQ node, go to step 9);
8) If it is not beyond the limits, then calculate the unbalance value in the modified formula, and obtain Jacobian matrix elements;
9) Use Newton-Raphson method to solve the modified formula and modify voltage and phase angle of each node;
10) Verify whether the power flow is converged, if yes, the calculation is completed; otherwise, use the new voltage as the initial value, return to step (2).

7.3.3.5 Conversion Between PV Node and PQ Node

In order to maintain a given voltage, the reactive power of some PV nodes in the systems may exceed the allowable range, especially when the reactive power source of the system is not sufficient, the reactive power of the PV nodes is more likely to overlimit. Thus, in the power flow calculation process, PV node reactive power must be controlled within the allowable range [34].

If the maximum allowable value of the reactive power of PV node is Q_{iM}, its reactive power calculated in an iterative process is Q_i, then the reactive power reserve should be (Formula (7.118)):

$$\Delta Q_i = Q_{iM} - Q_i \tag{7.118}$$

where if ΔQ_i is negative, it means that the reactive power at that node has over-limited, in this case, the iterative process must be interfered, to reduce the reactive power $|\Delta Q_i|$ at this node, and return it to the reactive power upper limit Q_{iM}. Thus, the consequences are bound to change the voltage of each node in the system, but the biggest change is in the voltage of the over-limited point. In fact, at this moment, the node i has converted to PQ from PV.

In the iteration of Newton-Raphson method, if the reactive power of PV node is found overlimited, it shall be converted to PQ node, and $(Q_{iM} - Q_i)$ is considered as the reactive power error of that node. In this case, appropriate changes shall be made to the modified formula. When the polar coordinate expression is used, an formula corresponding to ΔQ_i shall be added in the modified formula, when using a Cartesian coordinate expression, just substitute the modified formula of ΔQ_i for that of ΔU_i^2.

When the voltage of PQ node is lower than the given allowable operating voltage, measures must be taken on the operation mode of the entire system. For this reason, the voltage at this node shall be maintained within the voltage limit, in order to convert it to PV node. When using a polar coordinate system, conversion of node i from PQ node to PV node can be achieved by simply removing the formula of ΔQ_i from the modified formulas, and removing U_i from the unknown quantities. When using Cartesian coordinate method, just change the formula of ΔQ_i into the formula of ΔU_i^2 in the modified formulas.

7.3.4 Power Flow Calculation of Power System Containing Direct Drive Synchronous Generators

Electricity of variable frequency and voltage output by the permanent magnet synchronous wind turbines is converted to the three-phase alternating current with constant frequency and voltage through the full-power converter. It can be known from the wind power characteristics of direct-drive permanent magnet synchronous generators, for a particular wind speed, there is always a constant output power corresponds to it. Therefore, for the power system containing direct-drive permanent magnet synchronous wind turbines, in power flow calculation, the active power and voltage at a particular wind speed can be considered as constant. That is, the synchronous generators operate at constant voltage, and the WF can be considered as a PV node. In this case, we can first set the reactive power initial value Q_e at wind turbine node; obtain the active power P_e output by the wind turbines from the wind speed according to the wind speed-power curve; regard the wind tuebine node as a PV node, solve the power flow of the entire system by using a conventional power flow calculation method. Detailed procedure of the power flow calculation is as follows [35]:

1) Set the wind speed, WF node voltage, and phase angle as well as the initial value of reactive power;
2) Determine the active power P_{ei} of each wind turbine according to the wind speed-power curve;
3) Calculate the total active and reactive power of the WF;
4) Obtain the constant terms of the modified formula and the Jacobian matrix elements based on the voltage amplitude and phase angle of the node;
5) Solve the modified formula to obtain the corrections of the voltage phase angle and the reactive power, and use these corrections to modify the voltage phase angle and reactive power of this node;
6) Verify whether the power flow is converged, if yes, perform step (7); otherwise, return to step (4) for the next iteration.
7) Set the next wind speed value, if the next wind speed reaches the rated wind speed, the calculation is completed; otherwise, substitute the next wind speed value in, set the phase angle of the voltage and the initial value of the reactive power in WF node as those in step (1), and return to step (2) step for calculation.

When the direct-drive synchronous generator is operating with constant power factor, the WF node can be considered as PQ node. First, we can set the initial voltage U of the wind turbine node; and obtain the active power P_e output by the wind turbines based on the wind speed according to the wind speed-power curve, then $Q_e = P_e \tan\varphi$; the WF node is considered as PQ node, use the conventional power flow calculation method to solve the power flow of the entire system.

7.3.5 Probabilistic Power Flow Calculation of Power Systems Containing Wind Power

7.3.5.1 Overview

Probabilistic power flow calculation is a special type of power flow, and a macro statistical method for power system steady-state operation. Conventional power flow is an estimation of voltage and current of a specific location under predetermined power system conditions, of which the results are determined. In fact, conventional power flow calculation conditions are often uncertain, for example, the generator output, node load, system configuration will change over time, random failures and out of service of power generating equipment, transmission equipment, and transformer equipment, and so on. Thus, when the results of power flow study are used for planning and decision making, we often need to know the impact of changes in a known amount of data on system operating conditions, and the corresponding variation range of the power flow results.

In order to ascertain the effect of wind power generation on the system operation, in the study of planning, design, and operation mode, we need to know a number of circumstances (schemes) that may occur as a large number of power flow calculations. Normal power flow calculation is of large calculation amount, and is difficult to fully reflect the overall situation of the system. When using probabilistic power flow calculation method, the input data is the probabilistic statistical characteristics of known random variables (such as the expected value of injected power, variance, and probability distribution functions of a given node, etc.), and the output data is the probabilistic statistical characteristics of the node voltages and branch power flows, including expected values, variances, and probability probability distribution functions, and so on. From these results, we can know the node voltage, branch power, reactive power of the PV node, and average value, value range, and probability of the balancing node power, and so on. Probabilistic power flow can be used for analysis of branch power flow, node voltage probability distribution, expectations, variances, and limit values, in order to give a comprehensive and integrated evaluation on the performance of the entire power grid under various operating conditions, and making a quantitative analysis on the grid weaknesses, giving the line overload probability, probability of voltage overlimit, and probability of the system lost steady stability. Setting the voltage control node in the random power flow calculation can be used to analyze the impact of random fluctuations in node voltage on other node voltages and branch currents of the system. For voltage control node, random distribution of the reactive power injection can also be calculated, in order to determine the reactive power compensation equipment capacity that should be configured on these nodes. These can provide more complete information and scientific basis for the electric power planning and development of operations related decisions by the operation department.

Modeling and solution methods of probabilistic power flow are classified into two categories: analytical and simulation methods. The following introduces the power flow calculation of power system containing wind power based on Monte Carlo simulation.

7.3.5.2 Power Flow Calculation of Power System Containing Wind Power—Monte Carlo Simulation

Since the analytical method needs to conduct more simplifications on linear approximation or quadratic approximation of deterministic flow formulas in solving probabilistic power flow, to accommodate the convolution and other mathematical methods, therefore, in order to simulate in detail the impact of wind power generation on the power flow of the power system, Monte Carlo simulation is often used to calculate the probabilistic power flow of power system containing WFs.

Based on stochastic simulation and statistical experiments, the Monte Carlo method is a means generating a random numerical sequence that matches the probability density characteristics of a certain random variable, and is used as input variables for specific experiments and solutions. In application of the method, the random numerical sequence generated is required consistent with the particular probability distribution of the random variable. And for random numericalsequences of a variety of specific and non-uniform probability distribution generated, the feasible method is to first generate a uniformly distributed random numerical sequence, then try to convert it into the random sequence of specific probability distribution, which can be used as the input variables for digital simulation solution. The basic steps are as follows:

1) Construct probabilistic models; construct a probability model consistent with its characteristics for the problem to be studied.
2) Generate a random number sequence as the sample inputs for a large amount of digital simulation, and then obtain a lot of simulation values.
3) Process the simulation results statistically (such as calculate the frequency, mean value and other characteristics), give solution and the accuracy estimation of the solution to the problem to be solved.

When the probability distribution of each node and the probability distribution of the node generator power are known, the load curve can be converted into a probability model. Probability distribution can be obtained from the frequency histogram of the discrete point load, to produce a set of random numerical sequences in accordance with the distribution. Assume that n samples are produced, use these n sets of data for power flow calculation or use the empirical formulas of unknown quantities and the power injected for calculation, to finally obtain n sets of sample values of the unknown quantities (node voltages, branch power flows, branch network losses), then use probability statistics to obtain the expectations, variances, probability distributions of the unknown quantities.

Generally, the computation of Monto Carlo method is not affected by system size, the frequency in sampling of this method is inversely proportional to the sampling accuracy squared, and in a certain accuracy, reducing variance is effective in reducing the frequency in sampling. Under normal circumstances, the sampling frequency can be 5,000 to 10,000 times, so as to ensure the effectiveness of simulation. The method takes a very long time, and is likely to occur data non-convergence problem of randomly selected nodes. Considering the advantage in accuracy, Monte Carlo simulation method is commonly used as a benchmark to compare.

Conventional probabilistic power flow calculation in power system is based on the probability distribution of each unknown node and the probability distribution of the node generator power. In probabilistic power flow calculation of power system containing wind power, WF node is considered as the power injection node, which must obtain the sampling of WF output power, that is, probability distribution of the output power of the WF node. There are two methods: ① obtained from WF power prediction, ② obtained from wind speed prediction.

The time series of output power of a WF and its probability distribution can be predicted according to the long-term records of WF power output. According to the results of wind speed prediction, Formula (7.119) is used to describe the wind turbine (equivalent WF) power characteristic curve:

$$P_t = \begin{cases} 0 & 0 \le SW_t \le v_{ci} \\ A + B \times SW_t + C \times SW_t^2 & v_{\mathrm{ci}} \le SW_t \le v_r \\ P_r & v_{\mathrm{r}} \le SW_t \le v_{co} \\ 0 & v_{co} \le SW_t \end{cases} \tag{7.119}$$

where A, B, and C are the parameters of wind turbine power characteristic curve, which are different from wind turbines, v_{ci}, v_{r}, v_{co}, and P_r respectively represents wind turbine start-up wind speed, rated wind speed, cut-off wind speed, and the rated power of wind turbine. When the WF wind speed is given, the WF power output and its probability distribution can be obtained according to the wind turbine power characteristic curve.

The use of Monte Carlo method to simulate the probabilistic power flow of power system containing wind power can be divided into the following steps [36, 37]:

1) Simulate the WF wind speed according to a certain wind parameter sampling;
2) Obtain corresponding output power of the WF according to Formula (7.119);
3) Simulate specific load values of different load nodes according to the load parameter sampling;
4) Calculate the power flow (deterministic power flow) of the power system containing wind power according to the method described in Section 7.3, to obtain the voltage amplitude of each node and the active and reactive power of each branch;
5) Repeat the above steps, to obtain multiple data sets of power flow results. Then carry out statistical analysis on the power flow results according to the mathematical statistics method, to obtain the probability distributions of voltage amplitude of each node, the active and reactive power of each branch and other unknown quantities;
6) Assess the impact of wind power generation on power flow of the power system according to certain criteria, such as whether the node voltage or the line power flow is overlimited.

7.3.5.3 Calculation Example

Wiring diagram of power system containing wind power used for analysis of probabilistic power flow calculation is shown in Figure 7.32, it is a typical example of WFs integrated into an 110 kV high-voltage distribution network. WFs are connected to the end of the distribution network, and the high-voltage distribution network is connected to the main power grid through a 220 kV substation. The key observation focus is on the impact of wind power injection on local grid voltage, so external systems that above the voltage class of 220 kV are simplified as equivalent power sources and equivalent impedances. The power voltage, that is, the voltage of Node 1 is denoted as U_{s}, impedance between Node 1 and Node 2 is the equivalent impedance of the external systems, which is denoted as Z_{s}. Node 10 is WF connection point, between Node 2 and Node 10 is the local load of the regions near the WF, the wind turbines selected by the WF are cage asynchronous wind turbines with the rated capacity of 600 kW. The capacity base value used for calculation is 100 MVA. For wind turbine parameters, see reference [14].

In order to maximize the use of wind resources, and get maximum benefit, we need to calculate various impacts of wind power generation on power systems. Here we emphasize on the impacts of wind power on the network power flow, and design the following three calculation schemes: Scheme 1: no wind turbines are connected; Scheme 2: 25 cage asynchronous wind turbines are connected to Node 10; Scheme 3: 50 cage asynchronous wind turbines are connected to Node 10. For results of probabilistic power flow calculation, see Table 7.1 and Table 7.2.

Table 7.1 lists the power flows between Nodes 1 and 2 under the three different schemes. It can be seen that after the wind power connected to the system, energy demand of load areas on conventional power plants is reduced, so the existing electrical equipment can be better used, to reduce electricity costs of users.

Table 7.2 lists the probabilities of voltage over-limit at Node 9 (connection point between the WF and the main grid) under the three different schemes. As can be seen, with the increase in wind power capacity, the probability of voltage overlimit at Node 9 also increases, which may bring some risk to the operation of the system.

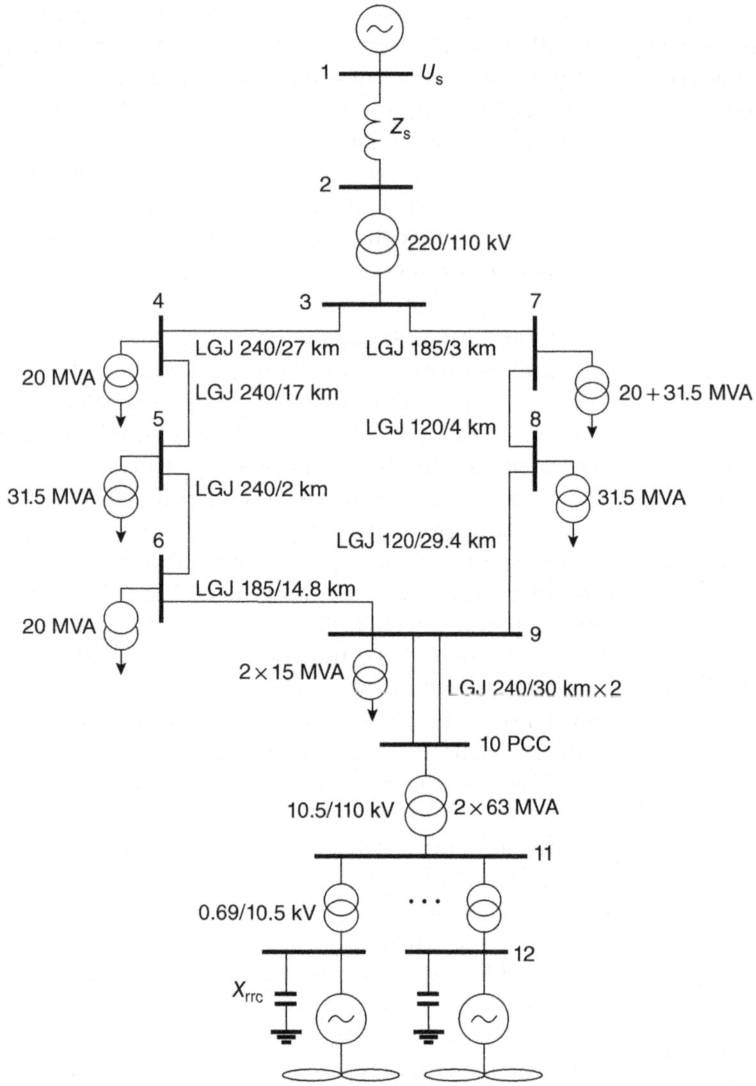

Figure 7.32 System wiring diagram.

Table 7.1 Line power flow between Node 1 and Node 2 under different schemes.

	Active power average value of line between Node 1-2 (p.u.)
Scheme 1	0.558
Scheme 2	0.526
Scheme 2	0.486

Table 7.2 Voltage changes at Node 9 under different schemes.

Voltage of Node 9	Average value (p.u.)	Mean squared error	The total probability of voltage over-limit
Scheme 1	0.99	0.028	0
Scheme 2	0.99	0.031	0.054
Scheme 2	0.99	0.032	0.067

7.4 Short-Circuit Analysis of Power Systems Containing Wind Power

7.4.1 Overview

Power system fault calculation refers to calculation of fault current, voltage, and their distribution in occurrence of power system failures. Short circuit is a type of fault in the power system, which refers to a line or electrical equipment in operating state occuring abnormal connection between phases or phase to ground directly or via an external impedance. The basic forms of short-circuit fault include three-phase short circuit, two-phase short-circuit, two-phase grounding fault, and single-phase grounding short circuit. They can also be divided into symmetrical and nonsymmetrical short circuit.

When a short circuit occurs, the short-circuit current flows through the circuit is much greater than the load

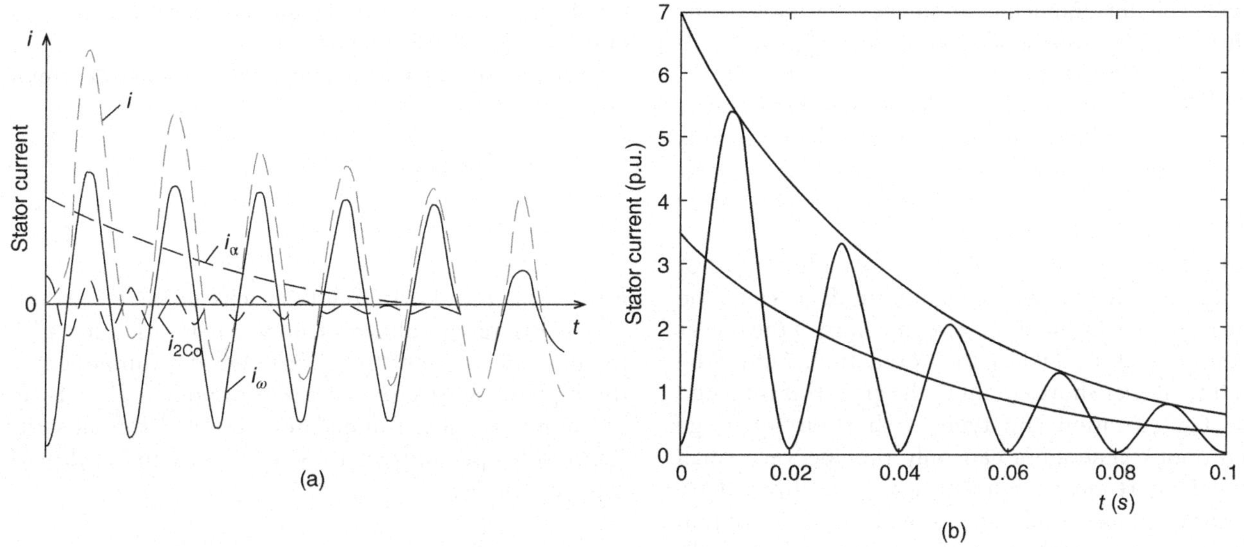

Figure 7.33 Synchronous/asynchronous generator stator short-circuit current curve. (a) Synchronous generator short-circuit current curve (b) Asynchronous generator short-circuit current curve.

current during normal operation, the thermal effect and machinery (force) effect of such a large short-circuit current may cause damage to electrical equipment. Short-circuit fault changes the normal operation condition of the system, may cause cascade accident to the system, resulting in system collapse or breakdown.

The purpose of short-circuit fault calculation is to determine the sizes of the fault current and voltage, their distributions (positive, negative sequence, and zero-sequence current) in the system.

Results of fault current calculation can be used to select electrical equipment (circuit breakers, disconnectors, current carrying conductors, transformers, current limiting reactors, etc.) ratings, set the action value of relay protection, study measures to limit the short-circuit current, and so on.

Short-circuit current usually contains three components: ① Fundamental component, namely frequency periodic component; ② DC component, that is, non-periodic component; ③ Nonperiodic component of other frequencies (see Figure 7.33a). Sizes of the three components and how they change over time are related to the synchronous generator parameters, excitation control system, location, and form of a short-circuit fault, the interaction between the generators and so on.

In occurrence of three-phase short-circuit at generator terminals of the asynchronous generator, short-circuit components in the stator and rotor of the asynchronous generator consist of the following parts [38]:

Steady-state current. The stator current frequency is ω_s (which is a synchronous frequency ω in grid-connection), and the rotor current frequency is $\omega_r = s \cdot \omega_s$.

Stator DC current. When the rotor rotates at $(1-s)\ \omega_s$ (assuming a pair of pole) relative to the stator, the DC current will induce an alternating current with the frequency of $(1-s)\ \omega_s$ in the rotor.

Rotor DC current, this current rotates together with the rotor, and induce an alternating current with the frequency of $(1 - s)\ \omega_s$ in the stator.

The results are shown in Table 7.3., where T_s and T_r respectively represents the decay time constant of stator and rotor DC component currents (also called transient current).

Typically, stator single-phase current waveform is shown in Figure 7.33b. As can be seen from the figure, the periodic component of the short-circuit current and non-periodic component of the generator rotor decay rapidly. If synchronous generator is connected, due to its excitation, the synchronous generator can still provide periodic short-circuit current to the grid during a power failure.

After a half cycle (slightly greater $t = T/2$), the stator current is approximate to the maximum value, so the current at $t = T/2$ can be used as the maximum current approximation.

Various components of the short circuit and their changes can be accurately calculated by using power system fault calculation program. In general, we only

Table 7.3 Current components in the stator and rotor during stator short-circuit.

	Steady state	Transient			
Stator	ω	(dc)	T_s	$(1-s)\ \omega$	T_r
Rotor	$s\omega$	$(1-s)\omega$	T_s	(dc)	T_r

calculate the instantaneous frequency periodic component right after the short circuit occurs ($t = 0^+$) in engineering applications.

Grid short-circuit current is mainly provided by the synchronous generator of the system. When a WF is connected to the power system, calculation of fault current shall take account of asynchronous generators or power electronic inverter model. Since the cage asynchronous generators operating in parallel do not have separate exciters, due to significant decrease of the terminal voltage during short-circuit failure of the power grid, the generators will lose their excitation, and will be difficult to deliver short-circuit to the grid. This is related to the fault location and type of short-circuit. Cage asynchronous generators can only provide very small sustained short-circuit current ($I \approx 0$) in three-phase short-circuit fault, and can provide the maximum short-circuit current in two-phase short-circuit fault. The short-circuit currents provided by the DFAG or direct-drive permanent-magnet synchronous generators are related to the control mode.

Because most of the wind turbine outlet voltages are of low voltage system, such as 0.69 kV, of which the impedance shall be multiplied by $K^2 = (U_{35}/U_{0.69})^2$ when converted to the 35 kV (or higher voltage classes) side, thus, from the point of view of the equivalent circuit on 35 kV side (HV side), the wind turbine and the corresponding low voltage cable are equivalent to a very large current-limiting reactor, which limits the delivery of short-circuit current.

When determining the contribution of generator to the short-circuit current, the most severe condition considered is the occurrence of short circuit at generator terminal, and the short-circuit current obtained at this moment may reach the maximum value. There are many types of modern wind turbines, the following analyzes the failure response characteristics of several mainstream wind turbines and their impacts on power system short-circuit current [39, 40].

7.4.2 Short-Circuit Capacity

An important concept associated with the short-circuit failure analysis is **short-circuit capability.** Short-circuit capacity of a node is equal to the product of three-phase short-circuit current at the node and the node voltage before the occurrence of short-circuit. The size of the short-circuit capacity is an important indicator to measure intensity of the bus, and represents the tightness of the link between that point and the power source. The greater the short-circuit capability at the point of common coupling (PCC) between the WF and the power grid, the closer the WF to the electrical system, the stronger its ability to withstand disturbances, and the greater the accessible wind power capacity. Therefore, the short-circuit capacity is often used as a basic measure of the effects of WFs on power grid.

Short-circuit capacity is calculated as follows (Formula (7.120)):

$$S_{i*} = U_{iN*} I''_{i*} = \frac{1}{Z_{ii(1)*}} \text{ p.u.,}$$

$$S_i = S_{i*} \times S_B = \frac{S_B}{Z_{ii(1)}} \text{ MVA} \tag{7.120}$$

where S_i is the short-circuit capacity of node i (MVA), U_{iN} is the rated voltage of node i before failure; I''_{i*} is the per-unit value of frequency periodic component of three-phase short-circuit current of node i; $Z_{ii(1)*}$ is the per-unit value of self-impedance of node i; the subscript * represents per-unit value; S_B is the system benchmark capacity (MVA).

7.4.3 Fault Response Characteristics of Wind Turbine

7.4.3.1 Fixed-Speed Asynchronous Generator (FSAG) Response to Grid Fault

When a short-circuit fault occurs in the grid, the voltage drop is transmitted to the WF through the grid. The size of the WF voltage drop is related to the location and type of the fault, the number and capacity of synchronous generators close to the fault point. In an area with large wind power and small synchronous generators, a voltage drop can be large; when the voltage drop reaches the wind turbines, the active power produced by them will decline, whereas the mechanical power remains unchanged, so the wind turbines will accelerate. If the wind turbine inertia is relatively small, acceleration of the generator is very large.

When the short circuit is cleared, because of the increased speed, the active power is higher than that before the fault, so there is a larger current, which in turn causes a large voltage drop in the lines and transformers, resulting in a transient period during which the voltage at the asynchronous generator cannot immediately recover to the value before the fault, and the generator is likely to continue to accelerate until the over speed protection trip off. In fact, the closest WF to point of fault will trip off first, other WFs may also trip if the overspeed protection threshold is reached. When some WFs trip off, the voltage tends to recover. Thus, normally cutting off the fault by the transmission system may induce a great number of WFs near the fault to trip. With increase of wind power penetration level, the system had become weak, because when the power transmission line is in fault, a large number of wind turbines will trip due to overspeed protection.

If the WFs have low-voltage protection, at the time of power failure, voltage drop will cause instantaneous trip of all wind turbines with low voltage protection; after the fault is cleared, the voltage can be quickly restored, but the power transmission of the WFs is related to connection of external power grid, if the transmission lines are also tripped due to overload protection, the wind turbines may also be tripped off later due to overspeed protection.

If there is no low-voltage protection for instantaneous trip, at the time of power failure, the wind turbines will have great oscillation, which will continue for seconds, and the long-term disturbance will stimulate tripping off other generators in the region that would have been impossible to trip, which will not be stable until some wind turbines being tripped due to overspeed protection. Based on this, the instantaneous trip is good. But on the other hand, if all wind turbines in the WF have been installed with low-voltage protection for instantaneous trip, in some single-circuit line fault, all wind turbines in the affected area will be tripped off. Then the problem is to trip as quickly as possible those wind turbines that are more affected by the transmission system fault without tripping the rest. This objective can be achieved by using an instantaneous voltage threshold much lower than 85%, or a suitable time delay. Perhaps a scheme combining the voltage threshold and related time delay could be a good solution.

If transmission system fault is cleared quickly, the most affected WFs are tripped successfully in a short time, then the disturbance will probably be harmless from the power system view point. However, this objective is not easily achievable since the compromise between an excess of wind turbine tripping and the lack of it (that would lead to a late overspeed tripping of more WFs) is a difficult task.

The analysis on asynchronous generator respond to its stator terminal symmetric fault can be described by the state space vector. Stator current maximum value approximation can be expressed as (Formula (7.121)) [41]:

$$i_{s\max} = \frac{\sqrt{2}U_s}{X_s'}\left[e^{-T/2T_s'} + (1-\sigma)e^{-T/2T_r'}\right] \quad (7.121)$$

where U_s is the stator voltage (terminal voltage); X_s' is the stator transient reactance; $T_s' = L_s'/R_s$ is the decay time constant of the stator DC component; $T_r' = L_r'/R_r$ is the decay time constant of the rotor DC component; $\sigma = 1 - L_m^2/L_sL_r$ is magnetic (inductive) coefficient; L_m, L_s, and L_r, respectively, represents the excitation inductance, stator leakage inductance, and rotor leakage inductance. $L_s' = L_{s\sigma} + L_{r\sigma}L_m/(L_{r\sigma} + L_m)$ is the stator transient inductance; $L_r' = L_{r\sigma} + L_{s\sigma}L_m/(L_{s\sigma} + L_m)$ is the rotor transient inductance; T is a cycle time, 20 ms.

Reference [42] carries out simulation analysis on cage asynchronous generator transient characteristics under different operating conditions for the testing systems as shown in Figure 7.34 on a real-time simulation platform, including line tripping off island operation, phase a ground fault, phase a, b phase-to-phase fault, and phase a, b two-phase ground fault. The basic conclusions are as follows:

1) Single-phase (phase a) ground fault, the fault phase loses excitation, but the other two phases still have the same level of excitation, in this case, all the three have large currents, of which the sizes are different. The asynchronous generators accelerate during the fault, and then reach a new steady state. During the fault period, since the wind turbine blades get more energy from the wind, the power output will produce abnormal oscillations.
2) Phase to phase fault occurs between phase a and phase b. The asynchronous generator instantaneous phase voltage is $u_a(t) = u_b(t) = -u_c(t)/2$, the three phases all produce large currents. The asynchronous generators also accelerate during the fault, and then reach a new steady state. As can be seen, both phase-to-phase fault and single-phase ground fault can produce oscillations.
3) In contrary, during two-phase ground fault, fault phase a and b completely lose excitation. The loss of excitation of phase a and phase b causes larger current in the non-fault phase, which is larger than that of the fault phase. The asynchronous generators accelerate during the fault, until the fault is cleared, and its output power also produces oscillations.

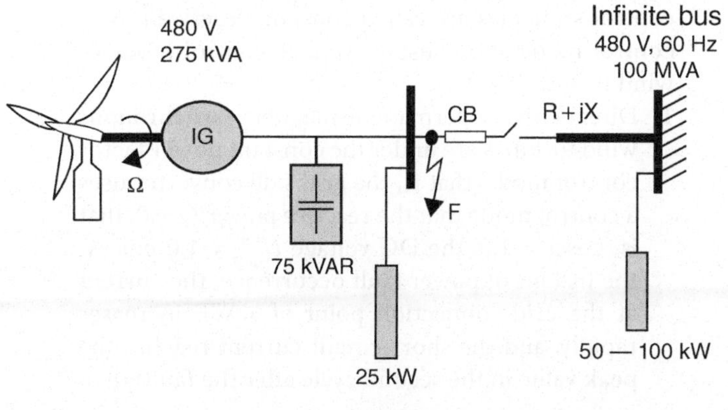

Figure 7.34 The testing systems being studied.

4) When the WF is disconnected from the grid and becomes Island (in Figure 7.34 the system breaker CB is opened), the reactive power required for asynchronous generator excitation is provided by the capacitor bank connected to the generator terminal. In this case, due to the phenomenon of self-excitation, the asynchronous generators may be subject to high overvoltage. The time for the voltage to reach the maximum (theoretical) value depends on the level of reactive power compensation. Voltage with large reactive compensation restores quickly, and the maximum value is high. Asynchronous generators may speed up or slow down, the corresponding stator voltage frequency and output active power may increase or decrease related to the reactive power compensation level. When reactive power compensation is large, the corresponding maximum values of the speed, frequency, and output power of the asynchronous generators are higher.

1) **The response to power fault of direct-drive permanent-magnet synchronous generator (DPMSG)**

In concept, the voltage source converter is used as the direct-drive permanent magnet synchronous generator (DDPMSG) output device, which is similar to that of the synchronous generator, the converter output voltage amplitude and phase control the output of active and reactive power. However, direct-drive permanent-magnet synchronous generator's converter voltage is composed of pulse-width modulation, which has a high level of controllability and rapidity. Unlike the synchronous generator, of which the fault characteristics are dominated by physical processes, with little difference between the different structures. Fault current at full-converter generators is a function of the particular control structure, the fault current characteristics between different control structures differ significantly.

Control strategies of DDPMSG converters on grid side can be divided into two types: DC bus voltage/reactive power, DC bus voltage/converter voltage at AC side, for all the wind turbines, usually these two control strategies are called constant reactive power control mode and constant voltage control mode of wind turbines [21].

1) Direct-drive permanent-magnet synchronous wind turbines are under the constant power factor control mode, that is, the grid side converter uses a control mode that the reactive power $Q = 0$, that is, $\cos\varphi = 1.0$, the DC voltage $U_{dc}^{ref} = 1.0$ p.u. At the instant of power fault occurrence, the current at the grid-connection point of a WF increases rapidly, and the short-circuit current reaches the peak value in the second cycle after the fault; then the short-circuit current at the grid-connection point basically maintains at a constant value, until the fault is cleared. This is because the permanent magnet synchronous generators provide constant excitation current during the fault, and the full power converter plays a certain role of grid fault isolation, the impact of the fault on the generators is relatively small; change in reactive component of converter AC current on grid side during the fault is small, reactive power fluctuations of corresponding wind turbines are also small; and the size of the short-circuit current at grid-connection point during short-circuit fault mainly depends on the active power delivered by the WFs into the grid before the fault and the voltage at the grid connection point; for wind turbines under constant power factor control mode, the short-circuit current at the grid-connection point is mainly active component, the reactive current component is small.
2) Direct-drive permanent magnet synchronous wind turbine under constant voltage control mode, means converters on grid side use the control mode of AC side voltage $U_{ac}^{ref} = 1.0$ p.u., DC voltage $U_{dc}^{ref} = 1.0$ p.u. If the grid fault is not serious, that is, when the ground impedance is large, the fault is not sufficient to cause the wind turbine over-voltage or over-current protection action to lock the converter, the terminal voltage of wind turbines can remain constant with the action of the converter control system, and the short-circuit current at the WF grid-connection point can still stabilize at a constant value after increasing continuously, the short-circuit current is mainly active current component; in case of serious fault in the power grid (the grounding impedance is small), since the reactive power provided by the wind turbines is limited, the terminal voltage of wind turbines is lower than its rating, there are minor fluctuations in short-circuit current at WF grid connection point, it can stabilize at a certain value after several hundred milliseconds.

For the fault at different distances away from wind turbines, or different fault impedances, the short-circuit current provided by the direct-drive wind generator is basically the same.

The IGBTs or IGCTs that constitute the converter are very sensitive to overcurrent. In a typical application, a current larger than 2 or 3 p.u. is easy to destroy IGBTs in a very short time. Thus, a voltage source converter high-speed control must be used to limit the output current, particularly during the fault. Current limit is achieved through high-bandwidth current regulator and "hard-limiting," the IGBT or IGCT device is turned off before the maximum allowable current level is reached [40].

For unbalanced fault, fault current characteristics may be very complex, size of the short circuit is related to the type of fault, converter control strategies protection settings, and other factors, and the objective of the control is to limit the current of power electronic devices.

During grid voltage drop, direct-drive permanent magnet synchronous wind turbines continue to transmit active and reactive power to the grid. When the voltage drop ends, the active power will quickly restore to its original value. When the voltage drop starts, since the output power of the grid side converter is limited by AC voltage drop of the grid, the excess power generated by the permanent magnet synchronous generators will flow to the DC capacitor, resulting in increase of the DC voltage U_{DC}. For this reason, the converter is equipped with unloading resistance protection. When U_{DC} reaches U_{max}, insert the unloading resistance, then U_{DC} will start to drop; when U_{DC} reaches U_{min}, the unloading resistance is disconnected. Figure 7.35 shows the block diagram of DC side unloading control, in which damp load is composed of IGBT devices and unloading resistance R_d. The control formula on DC side is (Formula (7.122)) [43]:

$$\frac{1}{2}C\frac{d(U_{dc}^2)}{dt} = P_s - P_g = \Delta P \tag{7.122}$$

where P_s is the permanent magnet synchronous generator stator output power; P_g is the power delivered by the wind power system into the grid; ΔP is the difference between input and output active power; when the unloading resistance is inserted, the IGBT conduction duty cycle δ_d can be given by (Formula (7.123)):

$$\frac{(\delta_d U_d)^2}{R_d} = \Delta P \tag{7.123}$$

Formula (7.123) demonstrates that when the power on DC side is unbalance, the unloading resistance will absorb the excess power. When the system is in normal operation, U_{dc} and ΔP fluctuate within a limited range, at this moment, $\delta_d = 0$, and the damp load is not involved; when ΔP exceeds the set value, the unloading resistance is immediately inserted, and the conduction duty cycle δ_d is determined by Formula (7.1231). When U_{dc} exceeds the permitted value, makes $\delta_d = 1$, and the IGBT is fully opened, the maximum unloading is reached.

Reference [43] carries out simulation on terminal failure of the PMSG wind power generation system connected back to back to the power grid. When the grid voltage drops, keep constant control of the generator-side converter, active power of the grid side converter is controlled by a d-axis current, and the reactive power is regulated by q-axis current, changing q-axis reference current can easily adjust the output power factor, and the grid-side converter usually operates at unit power factor, the reactive power is set as zero, that is, $i^*_q = 0$. When the grid voltage drops, determine the conduction duty cycle of IGBT by calculating the power needed to be absorbed by the unloading resistance, of which the switching frequency is the same as other converters. Detailed simulation parameters are as follows: the output power is 100 kW, the output line voltage effective value is 600 V, DC side voltage rating is 1200 V, power devices are IGBTs, the switching frequency is 2 kHz, the DC side capacitor is 6800 μF, input inductance is 1 mH, output filtering inductance is 1.2 mH, damp load is $R_d = 5\Omega$, DC bus voltage controller PI parameters are $k_p = 0.002$, $k_i = 0.05$, generator side, and the grid side converters current regulator PI parameters are $k_p = 2.5$, $k_i = 500$.

In the fault simulation, the voltage drop amplitude is 50%, the drop duration is 0.5 s, the drop begins at 1.5 s, and the simulation ends after 5 s. The simulation results show that during the voltage drop, the stator current increases, but subjected to the current order limiter of the grid side converter, the output current is approximately 1.4 p.u. Decoupling control has been achieved for active and reactive power, and the reactive power Q is substantially zero. At the instant of voltage drop, because the current cannot be suddenly changed, the output of active power is quickly dropped to around 60% of

Figure 7.35 DC side damp load control block diagram.

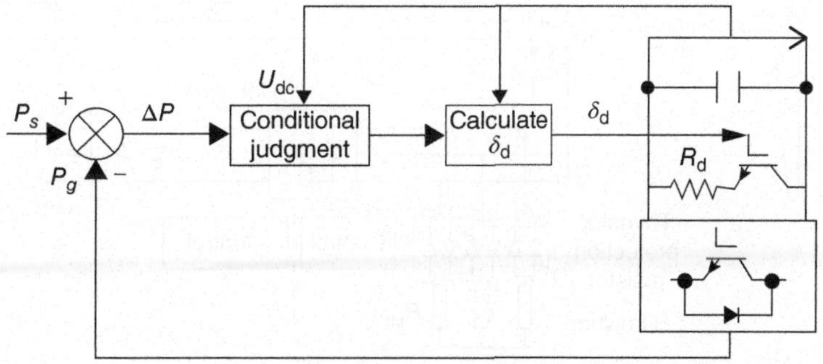

the rating, but with the current increasing, the active power also increases, which is about 0.7 p.u. during the voltage drop. After the voltage is restored, still because the current cannot be suddenly changed, the active power will instantly reach 1.3 p.u., which after a short-term of oscillations, will return to the rated value. The reactive power will have a small fluctuation at the instant of voltage drop and recovery. During the voltage drop, the voltage on DC side has an over-shoot, which after being regulated by the controller, is gradually restored to the rating; and there is a voltage sag when the voltage is restored, which corresponds to the active power waveform. When the grid voltage drops, the power injected into the capacitor instantaneously increases, thus the DC voltage is increased; when the grid voltage restores, the power output by the capacitor instantaneously increases, thus the DC voltage is decreased.

Without the use of damp load, the DC-side voltage will significantly increase, and may exceeds the upper limit of the voltage of the DC-side capacitor. If the current-limiting measures of the grid-side converters are all removed, in case of large voltage drop, in order to maintain the output power, the per-unit value of current becomes very large, can reach 5 p.u. or larger, which will damage the converter power devices.

2) **Response of double-fed asynchronous generators (DFAG) to grid fault**

The double-fed asynchronous generator combines the characteristics of full power converter generator and cage asynchronous generator. DFAG has power electronic converters connected to the rotor circuit, its fault transient is associated with the extreme voltage and current burdens withstood by the converter. Grid voltage drop will cause large (oscillation) current in the DFAG rotor circuit, which may damage the converter. To avoid damage to the converter switch, the rotor crowbar (crowbar) protection is used, as shown in Figure 7.36 [41]. When the rotor current is too large, the thyristor is triggered, a large current will flow through the crowbar resistor instead of the converter.

With rotor crowbar protection, when a bypass resistor (with the resistance of R_{cb}) is connected to protect the converter, the rotor transient time constant will be changed to (Formula (7.124)):

$$T_r' = \frac{L_r'}{R_r + R_{cb}} \tag{7.124}$$

The Formula (7.121) for the calculation of maximum stator current is changed to (Formula (7.125)):

$$i_{smax} = \frac{\sqrt{2}U_s}{\sqrt{X_s'^2 + R_{cb}^2}}[e^{-\Delta t/T_s'} + (1-\sigma)e^{-\Delta t/T_r'}] \tag{7.125}$$

where Δt is the time for the current to reach its first peak value, which depends on the generator slip and the value of the bypass resistor.

Larger bypass resistor will produce smaller T_r'. At the same instant, Δt will decrease.

Therefore, terms in the square bracket of Formula (7.125) are approximately unchanged. As rough approximations, the maximum stator current obtained will be (Formula (7.126)):

$$i_{smax} \approx \frac{1.8U_s}{\sqrt{X_s'^2 + R_{cb}^2}} \tag{7.126}$$

During grid short-circuit faults, DFAG with a crowbar can work in different ways: ① All wind turbines are disconnected from the grid. ② Disconnect the converter from the rotor and wind turbines are not disconnected from the grid. Then the generator is operating as a conventional asynchronous generator with large rotor resistance. ③ Maintain connection of wind turbines with the grid, and the converter is connected to the generator rotor. With this control mode,

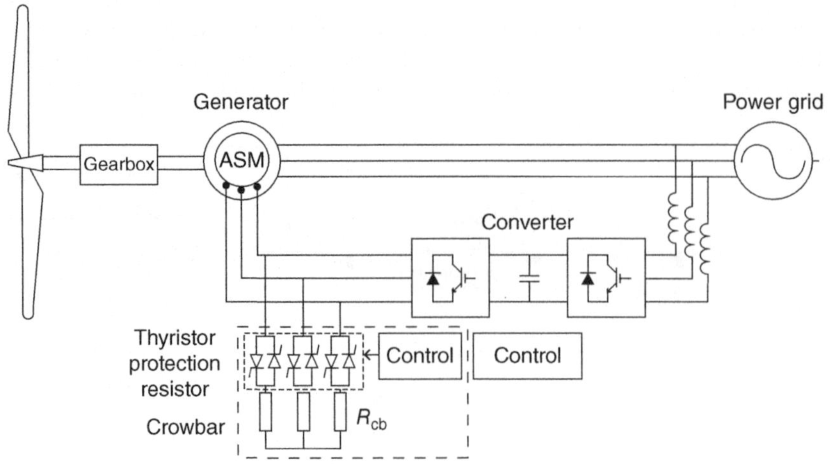

Figure 7.36 Crowbar rotor protection in DFAG rotor circuit.

normal operation may be recovered immediately after the fault is cleared. When voltage drops last for more than hundreds of milliseconds, wind turbines can even support the grid during voltage drop.

Fault response of DFAG is the same as that of cage asynchronous generator to a great extent. But the two assumptions made in analysis of cage asynchronous generator are incorrect for DFAG. One is that, in case of fault, when the bypass resistor is connected to the rotor, the resistance can no longer be ignored. Another is that when the DFAG is used as an asynchronous generator, its slip(s) is not always close to zero.

After the short-circuit fault, the phase difference between the rotor flux and stator flux after a half cycle is 180 degrees, this means that when $t = T/2$, the current reaches its maximum value. This assumption is approximately valid for cage asynchronous generator (where the slip is small, the stator and rotor flux are approximately in same-phase) at the instant of fault. However, double-fed asynchronous generators can operate under big slip, this means at the instant of fault, two flux vectors are not in same phase. When the DFAG operates at super-synchronous speed, the rotor flux is advanced ahead of the stator flux, it will go through less than a half cycle before the phase difference between the two fluxes reaches 180 degrees (when the maximum current is obtained); when DFAG operates below synchronous speed, the rotor flux will lag behind the stator flux, and it will go through more than a half cycle before the phase difference between the two fluxes reaches 180 degrees (when the maximum current is obtained). When the stator terminal voltage drops to zero, the stator, and rotor flux vectors will stop rotating. In fact, they will slowly rotate, depending on the resistances of the rotor and the stator. The greater the resistances, the faster they rotate. For the asynchronous generator, this rotation can be ignored, because the resistance is small. For the DFAG with bypass resistors, this rotation can no longer be ignored. This is why it takes less than a half cycle of time before the current of the double-fed asynchronous generator reaches its first peak.

The short-circuit current generated by the DFAG after grid fault is related to the distance between the generator and the fault location, the short-circuit current may reach five to eight times of the stator rated current. Therefore, the mechanical drive and the axis system will be subjected to a great stress. Moreover, the current flows through the converter may cause overvoltage in the DC connection. Most of the DFAG equipped with CrowBar (CB) protection can protect the converter from these impacts. When CB is triggered, the converter is disconnected from the rotor, and the rotor is short connected through the CB. The measure is enough for the protection of power electronics, but cannot significantly change the generator current and torque characteristics. Because when CB is inserted, as a user of reactive power, the wind turbine will further reduce the voltage level of the power grid. After the fault is cleared, the voltage will jump up to the normal value, which may cause a second trigger of the CB resulting in adverse effect on the voltage distribution.

Fault simulation is carried out for three DFAG with different capacities (3 MW, 2.75 MW, and 660 kW) that have crowbar protection, the crowbar makes the rotor voltage reach its approximately rated voltage, and the phase A stator current generated in the short-circuit fault is shown in Figure 7.37. The main results obtained from the simulation are as follows [41]:

1) In order to describe the generator failure characteristics and the contribution of wind turbines to the short-circuit current in grid fault, the

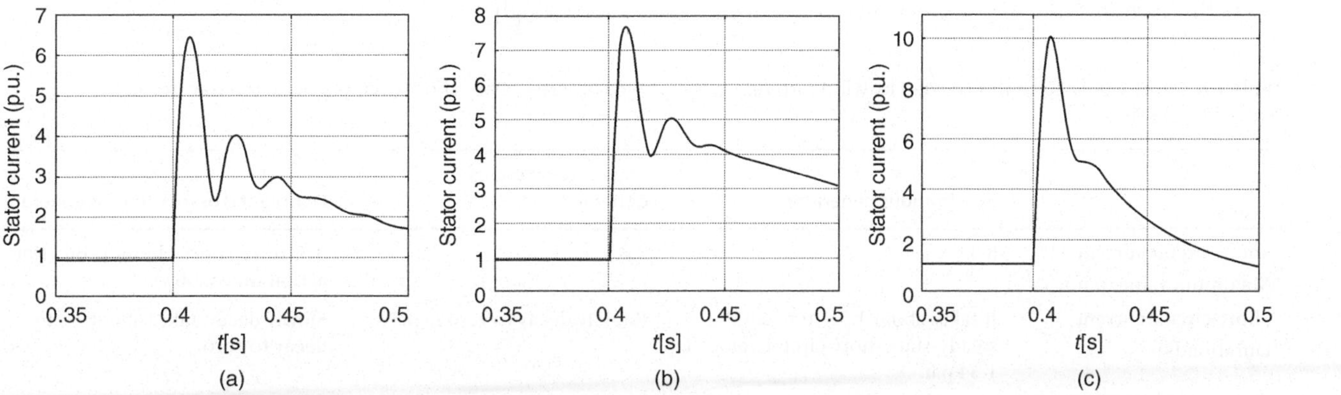

Figure 7.37 When with crowbar protection, DFAG stator phase A current in stator terminal short circuit. (a) 3 MW wind turbine; (b) 2.75 MW wind turbine; (c) 660 k wind turbine.

short-circuit current formulas shall be listed, but usually many simplifications and approximations are made. Compare the results obtained from the approximate formula with those obtained from the time domain simulation, the maximum error rate is about 15%.

2) The worst case considered is that the fault occurs in the wind turbine stator end, but most faults usually occur in distant. When distant short circuit occurs, the impedance between DFAG and the fault point shall be added to the stator impedance, then the original formula can be used. The short-circuit current provided by the WF in distant short-circuit is relatively small.
3) Size of the short-circuit current provided by the DFAG is closely related to whether there is crowbar protection and the protection action time. According to the maximum allowable rotor voltage, the determined crowbar resistance value is relatively small, which means wind turbines provide large short-circuit current to the grid. It is observed that values of the short-circuit current from the wind turbines to power grid is between 6 to 10 p.u. of which the maximum current decreases with the rated power of the wind turbines increasing.

3) **Impact of wind turbines on power grid fault current**

A commercial power system fault calculation program containing wind turbines can be used to calculate each short-circuit current component and its changes in short-circuit fault. Reference [44] carries out comparative study on the short-circuit currents provided by the conventional synchronous generator, the cage asynchronous generator and the double-fed asynchronous generator (with the same capacity of 2 MW) in occurrence of symmetric fault in a power grid, and the results are presented in Table 7.4 and 7.5.

As can be seen from Table 7.4 and Table 7.5: ① the short-circuit current provided by synchronous generators is the largest, followed by the cage asynchronous wind turbines and double-fed asynchronous wind turbines, of which the short-circuit currents are almost of the same size; ② the short-circuit current provided by the cage asynchronous wind turbines decays the fastest, followed by the double-fed asynchronous wind turbines, of which the decay rate is related to the rotor short-circuit protection action time, but the short-circuit currents provided by the cage asynchronous wind turbines and the double-fed asynchronous wind turbines will soon decay to zero. The short-circuit current provided by synchronous generators decays the slowest and only decay to steady state short-circuit current.

If the short-circuit current in the WF power generation unit is unknown, we can estimate its effective value using the following method: for synchronous generators, take eight times of the sum of the generator rated currents; for asynchronous generators, take six times of the sum of the generator rated currents; for generators with power electronic converter, take the sum of the generator rated currents.

Long duration (three-phase) short circuit is significantly different from the short circuit lasts for several cycles, depending on whether it is powered by a synchronous generator or an asynchronous generator. If the excitation provided has nothing to do with the grid, the synchronous generator can continuously provide short-circuit current. However, if the excitation current is extracted from the grid, the short circuit will cause voltage collapse. On the other hand, the asynchronous generator cannot provide short-circuit current for a long time, so in this case sustained short-circuit current is zero. For the short circuit with some distance away from generators, impedances of lines and transformers between the generator terminal and the short-circuit point shall be considered. Depending on the distance and impedance, lower amplification factor is generated accordingly.

Table 7.4 Short-circuit currents provided by wind turbines in occurrence of three-phase short-circuit fault on the transformer high-voltage side.

	Synchronous generator	Cage asynchronous generator	Double-fed asynchronous generator
Short-circuit current Maximum value (p.u.)	8.0	3.8	3.6 (rotor short-circuit protection action time 0.06 s)
Short-circuit current Duration (s)	It takes about 1.5 s to reach the steady-state short-circuit value of 1.34 p.u.	Rapidly decay to zero in 0.5 s	Slowly decay, takes about 1 s to decay to zero
Short-circuit current Effective value (p.u.)	4.6	2.4	2.1

Table 7.5 Comparison of short-circuit currents provided by WFs in occurrence of short-circuit in distant WFs.

No.	Fault point	Wind generator type	i_N/kA	i_F/kA	Increased value %
1	Station No. 1 220 kV bus	Synchronous generator	11.04	13.71	24.2
		Asynchronous generator		13.13	18.9
		Double-fed generator		12.78	16.1
2	Station No. 4 220 kV bus	Synchronous generator	9.49	10.15	7.0
		Asynchronous generator		10.10	6.4
		Double-fed generator		9.99	5.3
3	Power plant No. 4 220 kV bus	Synchronous generator	9.23	9.81	6.3
		Asynchronous generator		9.73	5.4
		Double-fed generator		9.66	4.7
4	Power plant No. 8 220 kV bus	Synchronous generator	23.96	25.00	4.3
		Asynchronous generator		24.67	3.0
		Double-fed generator		24.55	2.5
5	Station No. 2 220 kV bus	Synchronous generator	21.09	21.82	3.5
		Asynchronous generator		21.65	2.7
		Double-fed generator		21.47	1.8

Note: i_N–short-circuit current without wind turbines; i_F–short-circuit current with wind turbines

7.5 Voltage Stability Analysis of Power System Containing Wind Power

7.5.1 Overview

Voltage stability refers to the ability the power system to maintain a steady state voltages of all nodes in the system after suffering disturbances under a given initial operating state. It depends on the ability of the power system to maintain/restore balance between load demand and load supply. Voltage instability refers to occurrence of continues voltage drop or voltage increase at a certain node. The main affecting factor of voltage instability is the power system structure, power generation, and load mode. The possible outcome of voltage instability is loss of local loads, or tripping off transmission lines and other elements by the protection system, resulting in cascading outages. The term "voltage collapse" refers to the process of power system outage or abnormal low voltage accompanying the sequence of events of voltage instability.

Voltage problems often occur in power systems in faults of heavy load and/or in reactive power shortage. Among different factors that affect voltage instability, there is a special association between voltage instability and insufficient reactive power reserve. Voltage collapse is related to unmet load reactive power needs due to availability of reactive power reserve and limitations of reactive power transmission.

There are two types of tools for voltage security assessment: dynamic and static.

Dynamic analysis uses time-domain simulation to solve differential algebraic formulas of nonlinear systems, providing the most accurate reproduction of power system time response. However, the time-domain simulation consume time and is computationally intensive. These restrictions usually make the determination of the inspection and the stability limit of the system in wind range by dynamic analysis unpractical.

Static analysis is based on conventional or improved power flow formula solving. Its objective is to determine how bus voltage in the system changes with the wind power capacity integrated into the system (including different types of wind turbines), or to determine the maximum power that can be transmitted within the allowable voltage limit (the voltage stability limit). Based on power flow calculation, static analysis only includes algebraic formula solving, and therefore is much higher computational efficiency than dynamic analysis. Voltage stability analysis usually requires the testing of system status and many chance events, therefore, methods established on the basis of static analysis is more feasible. Within the voltage stability limit, the maximum allowable system load is usually determined based on the well-known P-V curve or Q-V curve. P-V curve is drawn according to the constant power factor, and Q-V curve is drawn according to the constant power. In order to produce this family of curves, it requires a series of computer simulations. Other methods for solving voltage stability limit include multiple power flow, and Jacobian singularity criteria, and so on [45, 46].

For the voltage stability study in power system, the MW distance for system load gradually and slowly increase (in one direction) to the voltage collapse point is a good measure of the voltage stability limit.

Problems caused by wind power integration into the system are mainly voltage stability problems, of which the impacts on system voltage stability include the following aspects:

1) Reactive compensation method of cage asynchronous generators is capacitor compensation, the compensation amount is proportional to the square of the voltage at the connection point, when the system voltage level decreases significantly, the reactive compensation amount also decreases a lot, whereas demand of the WF for reactive power is increasing, so the voltage level will further deteriorate, and will collapse in severe case, forcing the wind turbines to shut down.
2) Although double-fed asynchronous generators can use power electronics technology to control reactive power, but influenced by the wind turbine rotor speed and rotor-side active power changes and the constraints of the rotor-side power electronic converter, the stator-side reactive power can only be adjusted within a certain range.
3) In the case of fault and the operation of the power grid does not lose stability, some wind turbines are shut down due to their own protection, resulting in corresponding reduction in active power output of the WF.
4) If grid fault is not cleared timely, transient voltage instability will occur.
5) Fluctuations in wind speed and randomness cause the wind turbine output to change over time, which lead to random fluctuations of the system voltage.

For voltage stability analysis of power systems containing wind power, usually static analysis method is used to check how system critical bus voltage changes with the wind power penetration levels (including systems containing WFs connected with different types of wind turbine); dynamic stability analysis is used to observe the fault critical clearing time when different faults are applied to the system and the ability of the system critical bus to restore to the original voltage level after clearing the fault when the WF is connected with different types of wind turbines.

7.5.2 Impacts of Wind Power on the Power System Voltage Stability

The WF consists of different wind energy conversion systems, thus impacts on system voltage stability are also different. Impacts of different types of WFs on the voltage and voltage stability are explained in detail below.

7.5.2.1 WFs with Cage Asynchronous Wind Generators

When the cage asynchronous generator is connected to the grid, it is very different to operate as a generator and a motor. If it operates as a motor, when the external circuit impedance is $R + jX$, the voltage drop (per unit) on the impedance is approximately $\Delta U = PR + QX$ (per-unit value); when it operates as a generator, the voltage drop is approximately $\Delta U = -PR + QX$.

During the normal operation, the cage asynchronous generator terminal voltage drop or voltage rise depends on the active power it delivered. Voltage drop is also a function of the active power. Voltage drop only occurs when little active power is delivered, and the voltage will increase with the active power increases. The relationship between the voltage drop and the active power also associated with the generator parameters and feeder R/X ratio, which is different under different system conditions.

During asynchronous generator-terminal short-circuit fault, due to decrease of the terminal voltage and generator power output, the short-circuit current will rise. Due to this voltage drop, the electric power and torque output decreased significantly, and only a small amount of electric power is fed into the grid. Because the mechanical power remains unchanged, the generator will accelerate.

Electromagnetic torque built in the asynchronous generator at any given wind speed is proportional to the square of its terminal voltage, $T_e = K(s)U^2$ (where, T_e = electromagnetic torque; K = a constant related to the generator parameter; s = generator slip; U = terminal voltage). As can be known, voltage drops by 30% will cause the electromagnetic torque to drop by 50%. Before the fault, the electromagnetic torque and mechanical torque is balanced, after the fault, 50% of the mechanical torque is suddenly used to accelerate the generator [47].

Dynamic characteristics of the generator rotor are controlled by the swing formula $Jd\omega/dt = T_m - T_e$ (where J = rotary inertia; ω = rotor angular speed; T_m = mechanical torque applied to the rotor; T_e = electrical torque).

After the fault is cleared, the generator terminal voltage is restored. Before any energy can be converted between the rotor and the stator (and thus between the generator mechanical and electrical side), an air gap field must be established. Establishment of this field requires a large excitation current (reactive current), which is called "excitation surge." Such current causes the terminal voltage to drop, and thus the voltage and the electrical torque cannot be fully restored. If during the restoration, the electrical torque is lower than the mechanical torque, the generator will keep accelerating, and will lose synchronization. For the power system, the result is a voltage collapse. Unless the protection acts to disconnect the generator at this moment, the rest of the system would gradually be destroyed. Even after the re-establishment of the field, the operating speed of the generator is still much higher than the rated speed, this will require more reactive power, namely, the so-called "mechanical surge," causing voltage continues to decrease, which is, "voltage drop after the fault."

Overall, when short-circuit fault or voltage drop occurs in the power grid voltage, the active power output by the cage asynchronous wind turbine WF will decrease quickly, and the reactive power absorbed will surge. Due to unbalance between the input mechanical torque and output electromagnetic torque, the asynchronous generator is accelerated, and the reactive power absorbed is increased. In addition, reactive power of the compensation capacitor at a low voltage decreases significantly, which impedes the restoration of asynchronous generator speed and voltage, resulting in destruction of the voltage stability in power grid.

From the stability point of view, asymmetric fault is not as serious as symmetry fault. But we cannot ignore the asymmetrical fault analysis in process of design, because asymmetry fault is much more widespread than the symmetrical fault. Moreover, the asymmetric fault clearing time may be longer than that of symmetrical fault.

In the case that wind speed fluctuates, especially when the gust is of large intensity and lasts for long, the WF voltage may continue to fall and eventually collapse. When gusts occur, in the process of the wind speed increases and then decrease in a short time, the pitch angle adjustment mechanism has a mechanical lag time, and has certain limitations in speed adjustment, when the wind speed suddenly increases, it cannot adjust in a timely manner, which causes increase of the generator torque and the speed, thus the asynchronous generator reactive power demand is increased, the voltage continues to drop, and the active power cannot be delivered, resulting in the unbalanced torque on wind turbines increases, the speed continues to increase. When the system is unable to provide adequate reactive power support, the terminal voltage cannot be rebuilt, which ultimately leads to voltage collapse. Of course, if the gust duration is very short, although there is no pitch adjustment, because the change in the kinetic energy absorbed by the wind turbines is relatively small in a short time, the wind turbines can be maintained stable after the gust.

In systems containing synchronous generators near the cage synchronous generator WF, during voltage drops caused by grid faults or disturbances, due to the increase of reactive power absorbed by the wind generators or local equivalent motors (industrial or resident load), the nearby synchronous generator transient overexcitation limiter may be triggered, limiting the reactive power output. This eventually results in excessive acceleration and tripping of the asynchronous generators. This instability is driven by the asynchronous motors, wind generators, or the equivalent motors [48]. Therefore, in this case, the cause of instability is transient excitation limit of synchronous generator, unstable driving force result in asynchronous generators dynamic condition, and eventually results in excessive acceleration and tripping of the asynchronous generators, which, however, has no further impact on the grid.

7.5.2.2 WFs with Double-Fed Asynchronous Generators

Double-fed asynchronous generators can achieve decoupling control of active and reactive power, so the generators can have a certain reactive power regulation capability. WFs based on double-fed wind turbines can achieve zero reactive power exchange with the systems, or even deliver reactive power to the grid, so in normal operation, double-fed asynchronous generators can maintain a proper level of WF voltage, and help to improve the level of static voltage of local power grid.

After a system fault, double-fed asynchronous generators are capable of controlling the reactive power and make the power factors of the wind turbines quickly restore to the 1.0. The wind turbines will not absorb reactive power from the grid that just recovered from the fault, thus the required reactive reserve of the system is reduced, which is conducive to the regional power grid voltage stability; although randomness and volatility of the wind speed have some impact on the system voltage with WFs integrated into, for WFs with double-fed generators, due to the reactive power regulation capability of double-fed wind turbines, in wind speed fluctuations, the wind turbines are able to control reactive power exchange with the grid around zero, and will not absorb reactive power from the grid, thus, although fluctuations in wind speed may have an impact on voltage of the system, but it is small, and will not destroy the voltage stability of the local grid.

For WFs with cage asynchronous generators (CAG) and double-fed asynchronous generators (DFAG), in a dynamic process of grid fault, as the proportion of DFAG to CAG increases, the voltage drop will decrease, and the fault critical clearing time will increase. The ratio will stop increasing after risen to a certain level, the role of DFAG on dynamic voltage stability is limited.

7.5.2.3 WFs with Direct Drive Synchronous Generators

For WFs with direct drive synchronous generators, because the full power converters have a role of isolating grid faults, in occurrence of large-scale disturbances on the grid side, such as symmetrical three-phase short-circuit faults, the generators are less affected by the power faults. Since the direct-drive permanent magnet synchronous generators can achieve decoupling control of active and reactive power through the full power converters, the voltage/reactive characteristics of WFs are depending on the control of direct-drive permanent magnet synchronous wind turbines, which is especially related to the control strategies of the full power converter control systems. In the operation of direct-drive permanent magnet synchronous wind turbines, four control modes are usually used for the control of the converters (P-Q, P-Uac, U_{DC}-Uac, and U_{DC}-Q). Control of grid-side converters of the wind turbines on DC bus voltage U_{DC}/reactive power Q or DC bus voltage U_{DC}/converter AC side voltage U_{DC}, are usually referred to as constant power factor and constant voltage control modes of the wind turbines [43].

1) **No additional control**
When no additional control is added to the wind turbines, during grid short-circuit fault, the terminal voltage of wind turbines will significantly drop. At the instant of fault, the terminal voltage of the wind turbines and the active power output of the WF decrease rapidly, and the reactive power required by the WF rapidly increases, the DC bus voltage is gradually showing a rising trend. Because the mechanical torque is greater than the electromagnetic torque of the generator at the instant of fault, the speed of permanent magnet synchronous generator is therefore showing a rising trend, and the AC side current of grid-side converter is increasing rapidly. In an instant when the fault is cleared, the DC bus voltage reaches the maximum value. In the role of the converter control system, the relevant parameters can quickly decay, and restore to a stable state after the fault is cleared.
2) **Constant power factor control**
WFs consisting of direct-drive permanent magnet synchronous wind turbines can maintain zero reactive power exchange between its outlet and the grid, that is, the whole WF does not deliver nor absorb reactive power, therefore, the reactive load on the lines connecting the WF and the system is the only reactive power of the system, of which comparing to the WFs with cage asynchronous wind turbines, the voltage stability will be significantly improved. However, because the wind turbines cannot provide dynamic reactive power to support the voltage, cutting off a fault line will lead to a weaker grid structure, and lower terminal voltage, active power of wind turbines cannot be completely delivered, and the full power converter DC bus voltage rises to over-voltage protection limit, these may cause protection operation of all wind turbines to be triggered, and all wind turbines trip off the grid, which will affect the operation of the WF and the safety of the grid.
3) **Constant voltage control**
Wind turbines can control the reactive power delivered by the WF to compensate increasing reactive power demand of the lines, so as to further improve the voltage stability of wind power systems. During grid faults, the terminal voltage of wind turbines will significantly decrease. At the instant of fault, the terminal voltage of the wind turbines and the active power of the WF is decreasing rapidly, and the reactive power of the WF is rapidly increasing; because at the instant of fault, the mechanical torque is greater than the electromagnetic torque of the generator, the speed of permanent magnet synchronous generator is therefore showing a rising trend; because the grid-side converter is using a transient voltage control mode during a fault, it can provide reactive power to support the terminal voltage, playing a supporting role on recovery of the grid voltage level; maximum DC voltage value is about 1.18 p.u., of which the generator unit characteristics have been improved in some degree compared with those of the generator units without additional controller; under the role of the control systems, the relevant parameters can quickly decay, and restore to a stable state after the fault is cleared. Meanwhile, grid faults have little effect on the generator-side converter and the permanent magnet synchronous generator.
4) **Additional DC voltage coupling control**
Since the generator-side converter and the grid-side converter use separate control systems, the DC link can play a certain role in grid fault isolation, so grid faults have little effect on the generator-side converter and the permanent magnet synchronous generator. However, when the terminal voltage of wind turbines is reduced, the generator-side active power and the grid-side active power are unbalanced, which will lead to a sharp rise in the DC bus voltage, and may trigger the protection system action, thus affecting the operation of the entire WF and the safety of the grid, and therefore, when using grid-side converter constant-voltage-control mode, additional DC voltage coupling control shall be used. In steady-state operation, the DC bus voltage is at its rated value, the additional DC voltage coupling control does not work, and the generator-side converter control system is used for achieving wind turbine maximum power tracking. When larger DC voltage fluctuations of wind turbines are caused by grid faults, the DC bus voltage measurement is compared with its nominal value to get an error signal, which will pass through the PI controller to generate the t reference value of d-axis component of the generator-side converter stator current, and then achieve restrictions on active power of the generators through inner loop current control system, enabling coordinated control on the DC bus voltage by the generator-side converter and the grid-side converter, until the DC bus voltage is restored to its nominal value, the generator-side

converter control system will restore its maximum power tracking function.

When the same short-circuit fault occurs in the power grid, due to the use of an additional DC voltage controller, the wind turbine control system can better control the changes of the parameters of each unit caused by a short-circuit fault of the power grid, and the entire WF can restore to a steady state after the fault is cleared; at the instant of fault, changes of relevant parameters of the wind turbines and the WF are similar to changes in those when the DC voltage coupling control system is not used. During the fault, the permanent magnet generator speed is showing a tendency to decay, with coordinated control of DC voltage coupling control, the control system can keep the changes of active power in the AC side of the generator side converter basically consistent with the changes of active power in the AC side of the grid-side converter, so as control the DC bus voltage within the reasonable range; because the generator-side converter control system is involved in the DC bus voltage regulation, so the impact of grid faults on the generator-side converter of the permanent magnet generator is greater than that without a coupling control system.

7.5.3 A Method of Assessing Voltage Stability

This section describes a method for assessing voltage stability, which can determine the PV surface of the safe operation of the wind power within the entire range of variation. The surface established is called voltage secure region of operation (VSROp) [49].

7.5.3.1 P-V Analysis

P-V voltage stability analysis is a static tool, which builds the relation curve of a node (or nodes) with the local load. The advantage of this method is that, it provides instructions near voltage collapse throughout the load level range. The input required is the standard power flow input data, and the output is the PV curves for all of the specified nodes. Voltage distribution of the system is shown by the P-V curve, which is drawn by the power flow program. In the curve, the load value changes from the basic operating condition to the point of collapse.

Figure 7.38 shows the P-V curve of the system. The three curves correspond to the basic conditions (BC) and two accident conditions (C1, C2). In accident condition, PV margin is reduced (PV Margin BC > PV margin C1 > PV margin C2), therefore, inclusion of accident conditions in the voltage stability margin estimation is critical.

7.5.3.2 Combination of Wind Power Changes to the Maximum Power Estimation

Traditional PV curve is used to study voltage distribution characteristics on specific load bus at the assumed conventional power generation growth direction. Existing static voltage stability analysis assumes that all generators can be dispatched, but the wind power generation is generally not considered to be dispatched. Therefore, it requires a different approach to understand the impact of wind power variability on voltage stability limit (margin). In systems of high levels of wind power penetration, dispatching of wind power will play a key role.

Wind power is highly variable in several different time ranges (hour, day, and season). In order to maintain grid stability, momentary balance must be kept between power generation and power consumption. Variability of wind power give a challenge to the integration of a large number of wind power into the grid. When a large number of wind power is integrated into existing power grids, conventional PV curve cannot capture the voltage stability margin of systems with high penetration of wind power (~20%).

In order to include wind variability, reference [49] proposes the P-V surface for evaluation of voltage stability secure operation, which is called voltage secure region

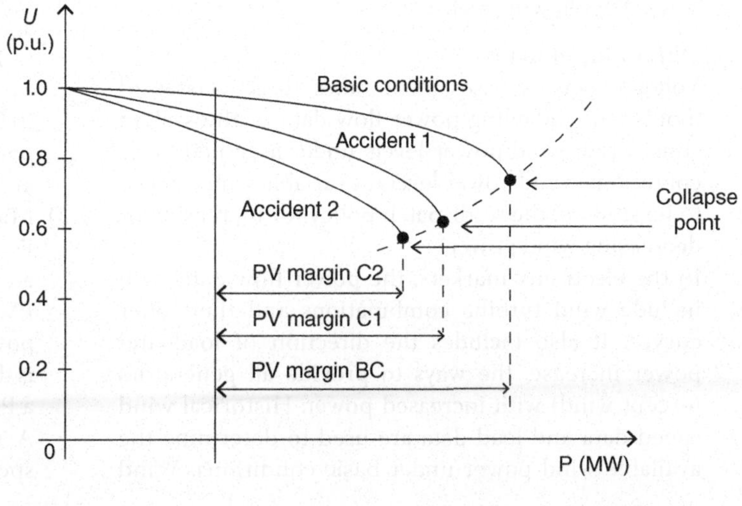

Figure 7.38 P-V curve: Basic working conditions and accident conditions.

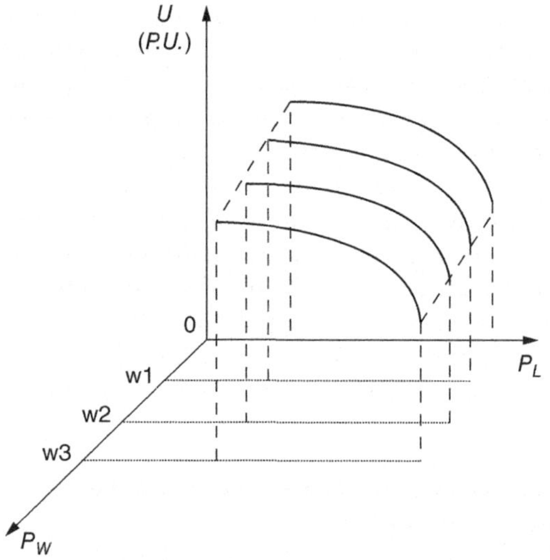

Figure 7.39 Voltage secure region of operation domain (VSROp).

of operation (VSROp). The surface combines the wind power of different levels, and gives three-dimensional voltage secure region of operation domain by drawing different PV curves at different levels of wind power. three-dimensional voltage secure region of operation domain consists of non-dispatched wind power (z-axis), the existing conventional power generation, including system losses (x-axis) and per-unit voltage (y-axis), as shown in Figure 7.39. The four four PV curves in the figure correspond to no wind, and three levels of wind power penetration W1, W2, and W3 (W3 > W2 > W1). For each PV curve, wind power generation remains as a constant, load and power output are increased in accordance with the set load and power generation increasing way, which remains unchanged relative to all the PV curves. Another calculation algorithm input to the PV surface the re-dispatch policy for increasing or decreasing wind power.

Figure 7.40 shows the calculation process of voltage security assessment. The recommended method includes the following steps:

1) **Obtain input data**
   Voltage security assessment takes three classes of input data, including power flow data of the system considered; wind power levels under assumed basic conditions, wind power levels at variable wind speeds to be studied; the re-dispatch policy for increasing or decreasing wind power.
   In the electricity markets, the power flow data shall include wind turbine combinations and their offer curves. It also includes the direction of load and power increase, the ways to provide all generators (except wind) with increased power. Historical wind speed data and load data are used to determine the available wind power under basic conditions. Wind speed maximum change forecast is used to determine at which wind change value the PV curve shall be divided.
   The change rate of wind speed and ramp rate of available generation can be used to establish power generation re-dispatch policy to compensate wind power changes in the system.
2) **Optimal power flow of basic conditions**
   The optimization objective of basic OPF is to minimize system costs, to constrain power flow, power generation, and node voltage limits, the formula used are as follows:
   The objective function (Formula (7.127)):

$$\min f(P_{gk}) = \sum_{k=0}^{n} C(P_{gk}) \tag{7.127}$$

   Constraints (Formulas (7.128–7.133)):

$$\sum_{k=1}^{n}(P_{gk}) - \sum_{i=1}^{m} P_{di} - \sum P_{loss,j} = 0 \tag{7.128}$$

$$\sum_{k=0}^{n}(Q_{gk}) - \sum_{i=1}^{m} Q_{di} - \sum Q_{loss,j} = 0 \tag{7.129}$$

$$P_{gk,\min} \le P_{gk} \le P_{gk,\max} \tag{7.130}$$

$$Q_{gk,\min} \le Q_{gk} \le Q_{gk,\max} \tag{7.131}$$

$$U_{b,\min} \le u_b \le u_{b,\max} \tag{7.132}$$

$$-F_{lin} \le F_j \le F_{\lim} \tag{7.133}$$

   where $b \in (1, 2, \cdots, q)$, q–total number of nodes $j \in (1, 2, \cdots, p)$, p–total number of branches; $k \in (1, 2, \cdots, n)$, k–total number of generators $i \in (1, 2, \cdots, m)$, m–total number of load nodes. u_b–node voltage; F_j–branch power flow. Formula (7.127) represents the total generation costs, where $C(P_{gk})$ is the price of power P_{gk} of generator unit k; Q_{gk} is the reactive power injection of generator unit k; P_{di} and Q_{di} respectively represents the active power and reactive power demands of bus i; $P_{loss,j}$ is the active loss on line j, and $Q_{loss,j}$ is the reactive power absorbed; Formula (7.128) and Formula (7.129) are active and reactive power balance formulas. Formula (7.130) and Formula (7.131) are limits of generator active and reactive power; the node voltage limit is (0.9 to 1.1 p.u.), and the line capacity limit is (<1.0 p.u.), which is respectively given by Formula (7.132) and Formula (7.133).
3) **Margin estimate based on all accidents**
   For a fixed wind dispatch, power flows of all $(n-1)$ accidents are used to draw the PV curves. Record the PV curve corresponding to the accident of minimum power margin and save. A collection of all PV curves is drawn in three-dimensional space in order to obtain a PV surface of voltage security.
   A series of PV curves on different surfaces corresponding to a particular wind power penetration

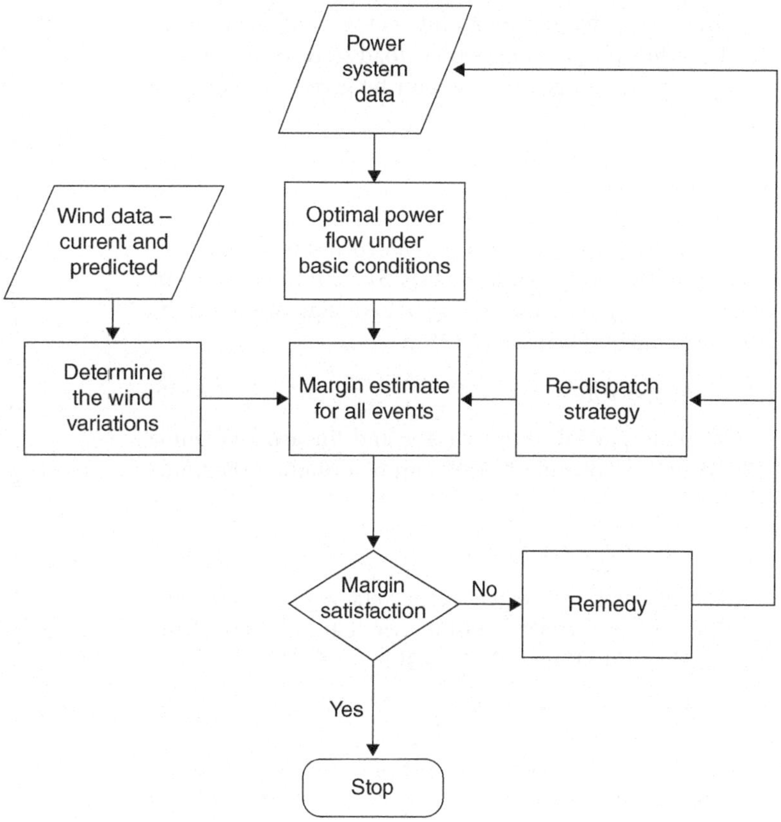

Figure 7.40 Voltage security assessment power flow diagram.

levels forms a multi-dimensional space, which represents the stable voltage operation area. Then the dispatch of basic conditions is used to estimate the smallest available margin of PV surfaces.

4) **Margin check and remedies**
   Margin obtained in Step (3) is used to determine the power margin requirement to be met. If the power margin requirement is not met, remedial measures will be taken to increase the margin: the modified power flow data is fed back to the Step (1), and the entire process is repeated until the desired margin is obtained.

Remedial measures can be taken include, but are not limited to switching capacitor, increasing generator combination or dumping load, and so on.

This multi-dimensional space provides the power system operator with a specific area, which may be too conservative, but is a completely secure region of operation. It also gives the current wind power dispatching and estimation of changes in the next hour, so that operators will be able to determine the available margin quickly.

7.5.4 Probabilistic Voltage Stability Margin of Systems Containing WFs

Aiming at wind power uncertainty, Reference [50] proposes an algorithm to calculate the probability distribution of voltage stability margin of systems containing WFs. This method first calculates the sensitivity matrix between voltage stability margin and wind speed, then calculate central distance of wind speed distribution of each WF, as well as the all-order semi-invariant, and finally use Gram-Charlier series inverse transform to calculate the voltage stability margin probability distribution.

7.5.4.1 Voltage Stability Load Margin Sensitivity Model

Conventional power system power flow formulas (Formula (7.134))

$$\begin{cases} P_{is} - V_i \sum_{j \in i} V_j (G_{ij} \cos\theta_{ij} + B_{ij} \sin\theta_{ij}) = 0 \\ Q_{is} - V_i \sum_{j \in i} V_j (G_{ij} \sin\theta_{ij} - B_{ij} \cos\theta_{ij}) = 0 \end{cases} \tag{7.134}$$

which can be summarized as (Formula (7.135)):

$$f(x, \lambda, p) = 0 \tag{7.135}$$

where x is the system state value, λ is the system load column vector, p is a variable parameter of the system, which is the wind speed of each WF here, so the active power output of wind turbine can be given by the wind speed higher-order functions (Formula (7.136)):

$$P_{wind} = c_1 p + c_2 p^2 + \ldots + c_8 p^8 \tag{7.136}$$

See reference [29] specific system parameters.

If the system load growth unit vector $\hat{k}$ and the current load condition λ_0 are known, the load state vector at critical point of voltage collapse can be given by (Formula (7.137)):

$$\lambda = \lambda_0 + \hat{k}L \tag{7.137}$$

L is the load margin and $L = |\lambda - \lambda_0|_1$.

Due to singularity of Jacobian matrix f_x nearby the critical point of voltage collapse [46, 51, 52], the left eigenvector $\omega(x, \lambda, p)$ corresponding to zero eigenvalue can be found to make (Formula (7.138)):

$$\omega(x, \lambda, p) f_x(x, \lambda, p) = 0 \tag{7.138}$$

Calculate partial derivative for and linearize Formula (7.135), then Formula (7.139) can be obtained (Formula (7.139)):

$$f_x \Delta x + f_\lambda \Delta\lambda + f_p \Delta p = 0 \tag{7.139}$$

Multiply the left eigenvector $\omega(x, \lambda, p)$ to Formula (7.139), at the critical point, we can obtaine from Formula (7.138) (Formula (7.140)):

$$\omega f_\lambda \Delta\lambda + \omega f_p \Delta p = 0 \tag{7.140}$$

Due to $\Delta\lambda = \hat{k}\Delta L$, the relationships between load margin and system parameters can be expressed as (Formula (7.141)):

$$\Delta L = -\frac{\omega f_p}{\omega f_\lambda \hat{k}} \Delta p \tag{7.141}$$

If the the distribution function of wind speed of each WF in the system is known, we can obtain the probability distribution of ΔL through convolution operation based on the probability distribution of random variable Δp.

1) **Calculation method for semi-invariant of random variable convolution operation**
Conventional convolution and calculation of random variables are very complicated, with large calculation amount. Using the Gram-Charlier series based on semi-invariant is a means to approximate continuous random variable distribution by normal distribution, which still has a good result when used to describe multiple discrete random variables [53, 54].
When distribution of the random variable is known, its order moments can be obtained. Set the density function of random variable x as $g(x)$, its v order moments a_v can be given by the following formula (Formula (7.142)) [54]:

$$a_v = \int_{-\infty}^{\infty} x^v g(x) dx \tag{7.142}$$

The order moments of a random variable are its numerical characteristics, which represent the characteristics of the random variable to some extent. Semi-invariant is also a kind of numerical characteristics of the random variable, and it can be obtained from each order moment not higher than the corresponding order. Relationships between the semi-invariants of the first eight orders of the random variable and the order moments can be given by Formula (7.143):

$$\left.\begin{aligned}
K_1 &= a_1 \\
K_2 &= a_2 - a_1^2 \\
K_3 &= a_3 - 3a_2a_1 + 2a_1^3 \\
K_4 &= a_4 - 4a_3a_1 - 3a_2^2 + 12a_2a_1^2 \\
K_5 &= a_5 - 5a_4a_1 - 10a_3a_2 + 20a_3a_1^2 \\
&\quad +30a_2^2a_1 - 60a_2a_1^3 + 24a_1^5 \\
K_6 &= a_6 - 6a_5a_1 - 15a_4a_2 + 30a_4a_1^2 - 10a_3^2 \\
&\quad +120a_3a_2a_1 - 120a_3a_1^3 + 30a_2^2 - 270a_2^2a_1^2 \\
&\quad +360a_2a_1^4 - 120a_1^6 \\
K_7 &= a_7 - 7a_6a_1 - 21a_5a_2 + 42a_5a_1^2 - 35a_4a_3 \\
&\quad +210a_4a_2a_1 - 210a_4a_1^3 + 140a_3^2a_1 \\
&\quad +210a_3a_2^2 - 1260a_3a_1^2 + 840a_3a_1^4 \\
&\quad -630a_2^3a_1 + 2520a_2^2a_1^3 - 2520a_2a_1^5 + 720a_1^7 \\
K_8 &= a_8 - 8a_7a_1 - 28a_6a_2 + 56a_6a_1^2 - 56a_5a_3 \\
&\quad +336a_5a_2a_1 - 336a_5a_1^3 - 35a_4^2 + 560a_4a_2a_1 \\
&\quad +420a_4a_2^2 - 2520a_4a_2a_1^2 + 1680a_4a_1^4 \\
&\quad +560a_3^2a_2 - 1680a_3^2a_1^2 - 5040a_4a_2^2a_1 \\
&\quad +13440a_3a_2a_1^3 - 6720a_3a_1^5 - 630a_2^4 \\
&\quad +10080a_2^3a_1^2 + 25200a_2^2a_1^4 \\
&\quad -20160a_2a_1^6 - 5040a_1^8
\end{aligned}\right\} \tag{7.143}$$

In order to simplify the form of the series, define (Formula (7.144)):

$$g_v = K_v/\sigma^v = K_v/K_2^{\frac{v}{2}} \tag{7.144}$$

where g_v is the normalized semi-invariant of v order, σ is standard deviation.

As an important numerical characteristic of the random variable, semi-variable has the following important properties:

a) All-order semi-invariant of the sum of independent random variables is equal to the sum of all-order semi-invariant of each variable;
b) If random variable y is the linear function of random variable x, and $y = ax + b$, then all-order semi-invariants of random variable x is $g_v^{(x)} (v = 1,2, \ldots, k)$, and the calculation formula of all-order semi-invariant $g_v^{(x)}$ of the random variable y is (Formulas (7.145) and (7.146)):

$$g_1^{(y)} = ag_1^{(x)} + b, \text{ where } v = 1 \tag{7.145}$$

$$g_v^{(y)} = a^v g_v^{(x)}, \text{ where } v > 1 \tag{7.146}$$

c) By using the normalized all-order semi-invariant g_v, the distribution functions of arbitrary random variables can be expressed as a series consisting all-order derivatives of the normal random variables, the Gram-Charlier series expansion formula commonly used in the power system is as follows (Formula (7.147)):

$$\begin{aligned} f(x) = & \int_x^{\infty} N(x)dx + \frac{g_3}{3!}N^{(2)}(x) - \frac{g_4}{4!}N^{(3)}(x) \\ & - \frac{10g_3^2}{6!}N^{(5)}(x) + \frac{g_5}{5!}N^{(4)}(x) \\ & + \frac{35g_3g_4}{7!}N^{(3)}(x) + \frac{280g_3^3}{9!}N^{(8)}(x) \\ & - \frac{g_6}{6!}N^{(5)}(x) - \frac{35g_4^2}{8!}N^{(3)}(x) \\ & - \frac{56g_3g_5}{8!}N^{(7)}(x) - \frac{2100g_3^2g_3^3}{10!}N^{(9)}(x) \\ & - \frac{15400g_3^4}{12!}N^{(11)}(x) + \dots \end{aligned} \tag{7.147}$$

where $f(x)$ is the probability of the random variable value greater or equal to x, $N(x)$ is the standard normal density function, $N^{(r)}(x)$ is the r-order derivative of $N(x)$.

Due to the above characteristics of the semi-invariant, convolution operation of random variables can be simplified as summation operation of the semi-invariant. For voltage stability problems of systems containing wind power, if the wind speed distribution function or statistical distribution of each WF in the system is known, semi-invariant method based on sensitivity matrix can be used to calculate probability distribution of voltage stability margin of the system containing wind power.

2) **Calculation process of voltage probabilistic stability margin of systems containing wind power**

To sum up, the calculation process of voltage probabilistic stability margin of systems containing WFs is as follows:

a) Use the continuation method based on expended power flow to calculate the voltage stability load margin L of the system containing WFs [46];
b) Linearize the critical point power flow formula, to obtain the left eigenvector $\omega(x, \lambda, p)$ corresponding to zero eigenvector;
c) Use Formula (7.134) to Formula (7.141) to calculate the sensitivity matrix between the voltage stability margin of the system containing WFs and the wind speed;
d) Calculate the all-order semi-invariants of the wind speeds of different WFs;
e) Carry out convolution computation according to Formula (7.145) and Formula (7.146) and the sensitivity matrix, to obtain the all-order semi-invariants of the load margin variation ΔL;
f) Carry out Gram-Charlier series inverse transform to obtain the probability distribution function of voltage stability margin.

3) **Calculation Example**

Use the IEEE39 node system as an example. The node system 39 consists of 10 generators, 39 nodes, 12 transformers, and 34 lines. As shown in Figure 7.41, to examine the impact of wind power on the voltage stability of power system, 300 MW conventional generators at Node 30 and Node 35 are respectively replaced by WFs of the same capacity, of which the wind turbine control mode is constant power factor mode. Data of load node is listed in Table 7.6. See references [29, 55] for the detailed wind speed distribution and wind turbine parameters.

For any of a random variable ξ, the probability density function $f(x)$ of its normalized random variable $\overline{\xi}$ is expanded according to Gram-Charlier series, and has the following form (Formula (7.148)):

$$\begin{aligned} f(x) = & \phi(x) + (c_1/1!)\phi'(x) + (c_2/2!)\phi''(x) \\ & + (c_3/3!)\phi'''(x) + \dots + \end{aligned} \tag{7.148}$$

$\phi(x)$ is the probability density function of the standard normal distribution, the coefficient c_k is defined as (Formula (7.149)):

$$c_k = (-1)^k \int_{-\infty}^{\infty} H_k(x)f(x)dx \tag{7.149}$$

$H_k(x)$ is the k order Hermite polynomial [54].

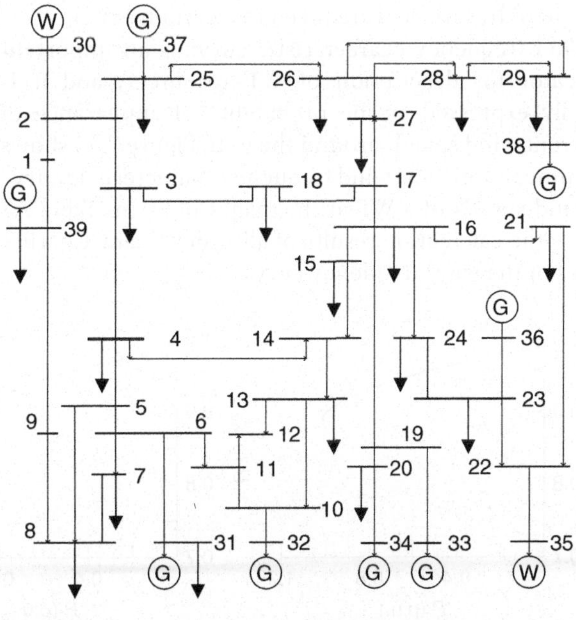

Figure 7.41 Node 39 test system wiring diagram.

Table 7.6 Data of load buses in the simulation system.

Bus	Active power (P_{Li0})		Active growth factor (K_{PLi})	
	Expectation value	Mean square error(%)	Expectation value	Mean square error(%)
03	3.220	0.09	1.0	0.05
04	5.000	0.07	1.0	0.05
07	2.338	0.13	1.0	0.05
08	5.220	0.11	1.0	0.05
12	0.085	0.10	1.0	0.05
15	3.200	0.08	1.0	0.05
16	3.290	0.09	1.0	0.05
18	1.580	0.08	1.0	0.05
20	6.800	0.12	1.0	0.05
21	2.740	0.11	1.0	0.05
23	2.375	0.10	1.0	0.05
24	3.086	0.07	1.0	0.05
25	2.240	0.10	1.0	0.05
26	1.390	0.09	1.0	0.05
27	2.810	0.11	1.0	0.05
28	2.060	0.05	1.0	0.05
29	2.835	0.06	1.0	0.05
31	0.092	0.05	0.0	0.00
39	11.04	0.08	1.0	0.05

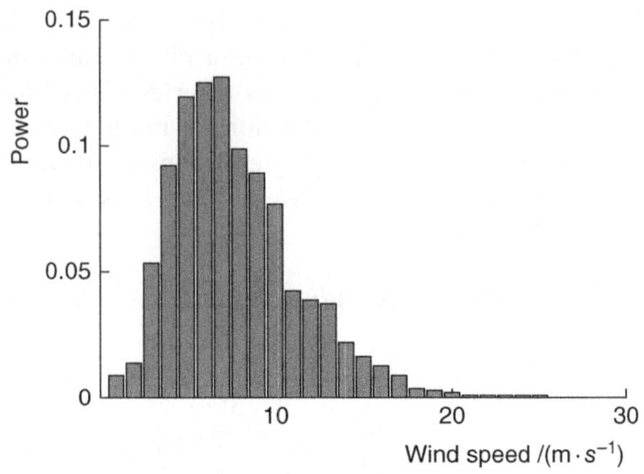

Figure 7.43 Wind speed probability map of WF at the height of 55 m.

Table 7.7 Gram-Charlier series of the WF.

K1	7.7893
K2	13.0984
K3	46.5296
K4	242.9919
K5	1.2678×10^3
K6	$1.5191e \times 10^3$
K7	-1.5650×10^5
K8	-4.3204×10^6

Figure 7.42a and b respectively shows the PV curve before and after the conventional generators in the system are replaced with the wind turbines, as can be seen from the figure, after wind turbines of the same capacity replace the conventional generators, voltage stability margin of the system is reduced to a certain extent.

Wind frequency characteristic curve is an important indicator for description of WF resources, and it is usually expressed as the probability of appearance of different wind speeds around the year. Figure 7.43 shows the annual statistical wind frequency characteristic curve of wind speeds of a WF at the height of 55 m, Table 7.7 shows the calculation results of all-order Gram-Charlier series of the wind frequency curve.

Figure 7.44 shows the probability distribution curve of fluctuations of voltage stability margin of system in the calculation example caused by wind speed fluctuations. As can be seen from the above results, using the above algorithm can quickly calculate the probability distribution of voltage stability margin and determine the probability of the stability margin in a certain range. These results can help operating and planning personnel intuitively detect the potential risks and weaknesses in system operation, and then economically and rationally select the maximum installed capacity of the WF and the grid connection schemes.

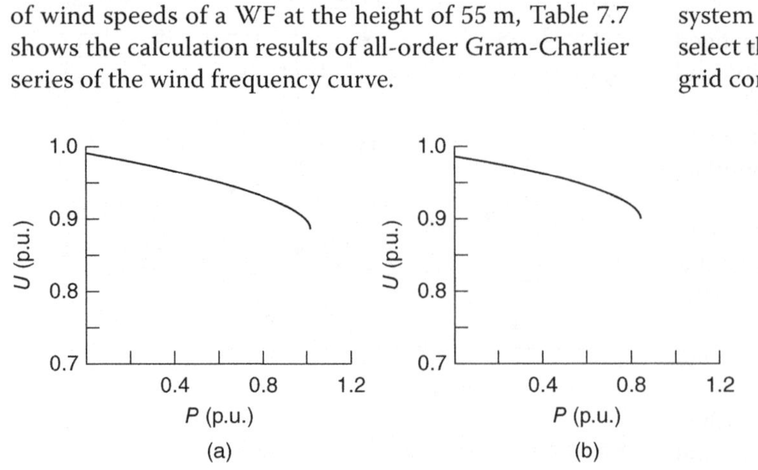

Figure 7.42 IEEE39 node system PV curve. (a) system PV curve of Node 39 without WFs (b) system PV curve of Node 39 with WFs.

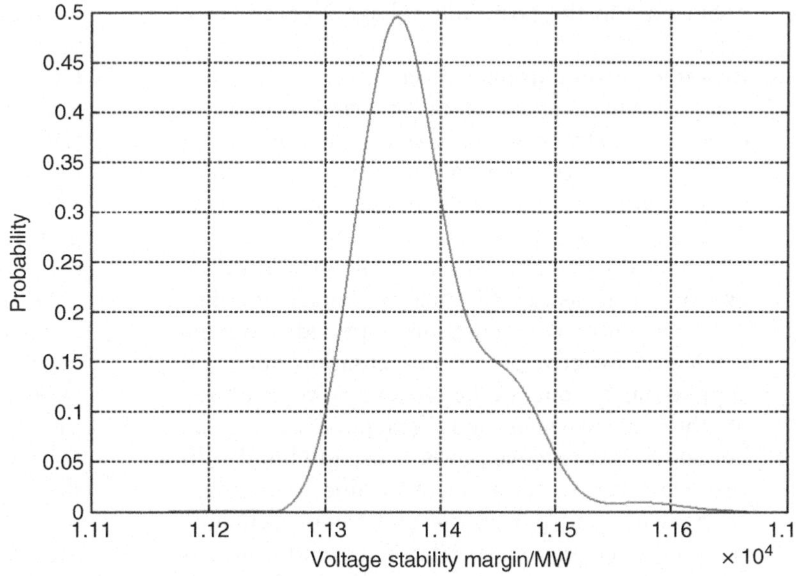

Figure 7.44 Probability distribution of voltage stability margin.

7.5.5 The Support for Voltage Stability of the Power System by Enhancing Wind Power

The objectives of voltage stability analysis are: when the WF is at full power, the existing lines will reach thermal limit; the total production of active power by WFs is as much as possible; the WF voltages are distributed around $\pm 5\% U_N$ deviation of the rated voltage (refer to 35 kV voltage level); no reactive power exchange between the local power grid and the main transmission grid; for all wind power output levels, maintain acceptable power quality of all buses connected to common load; minimize electrical losses of the network. Among which the main objective is to make full use of existing transmission lines, that is, to make the transmission capacity of the lines as close to the thermal limit as possible, while maintaining acceptable power quality of each load points in the system. When the voltage stability may become the constraint on maximum power output of the WF, it is necessary to take measures to enhance the voltage stability, make the voltage stability transmission power limit be equal to or greater than the thermal limit [56].

The size of the wind power unit capacity, fault location and fault clearing time all have significant impacts on the voltage stability of the grid: the greater the capacity of wind turbines close to the fault point, the easier the loss of system stability; the closer the short-circuit point to point of common coupling of the plurality of WFs, the easier the loss of system stability; and the longer the fault duration, the easier the loss of system stability. Therefore, to improve voltage stability of the power grid after large-capacity WFs are integrated into, the following measures can be taken [46, 54, 57, 58]:

1) **Limit the capacity of the wind power unit**
The two specific measures can be taken: limiting the wind power capacity connected to a certain point, so in occurrence of faults, its impact on other WFs can be minimized, that is, use the principle of "decentralized connection" to connect large-capacity WFs to as many points of common coupling as possible; carrying out splitting operation for wind turbines, to reduce the unit capacity.
2) **Increase the triggering speed of protection, and reduce the fault duration**
Shortening fault time can reduce the velocity increment of wind turbines in the process of short circuit, and significantly improve voltage stability.
3) **Selection of wind turbine types**
Efficiency drooping characteristic of the wind turbines after overspeed can prevent acceleration of the wind turbines, which significantly improves the stability of the system.
4) **Selection of wind turbines**
Cage asynchronous generator is most unfavorable to voltage stability. When the voltage U is lower than the threshold voltage U_{cr}, the reactive power absorbed by the asynchronous generators will increase dramatically, thus from the prospective of improving voltage stability, asynchronous generator of lower U_{cr} shall be chose. Double-fed asynchronous generators and direct-drive synchronous generators are capable of voltage/reactive power control, which are generally favorable to the system voltage stability.
5) **Improve wind turbine control**
Improve reactive power control and fault (low voltage) ride-through capability of wind turbines. Pitch angle control of fixed-speed wind turbines can effectively reduce the mechanical power input by wind turbines in grid faults and avoid generator overspeed and voltage instability caused by wind turbine power

imbalances in the post-fault voltage recovery process.

6) **Reactive power compensation**
For variable characteristics of wind power, in long-distance transmission systems with high penetration of wind power, in order to maintain voltage of the entire system within the acceptable range, additional reactive power support is necessary.
Mechanically switched shunt capacitors (MVC) is one of the measures to enhance voltage stability. Reactive power compensation is provided on the 690 V side. Reactive power compensation amount is approximately equal to the reactive power required by the cage asynchronous generator in no-load operation. The reactive power is used to build voltage. However, in terms of WFs, due to frequent changes in wind power, the number of switching required is large; voltage related to wind power is instable, the time range is within a few seconds, the mechanical switching is slow, so its effect is small; the characteristic of shunt capacitor reactive power output proportional to the square of the voltage is unfavorable to the reactive power compensation required by the wind power.
Taking into account the SVC with the response level of 1.5 to 2 cycles, it is able to provide rapid reactive support to help the network voltage recovery. Installation of static Var compensator (SVC) at points of common coupling of WFs, can suppress voltage fluctuation and flicker caused by wind disturbance, and improve the voltage stability of the grid after a short circuit. Smaller SVCs distributed on key buses throughout the system is in favor of the dynamic voltage support close to WF access points. SVCs selected for this purpose are directly connected to 110 kV, or to the third winding of 35 kV autotransformer. SVC can use secondary voltage control strategy, that is, measure line reactive power, control the controller setpoint voltage, and adjust its reactive power output, so that the reactive power exchange between the high-voltage and low-voltage grids is within a set range. SVC is an excellent device for WF reactive power compensation and voltage support. It advantages include continuously adjust the susceptance (reactive power output); rapidly responds; within the range of operating limit, can well maintain the voltage level at its connection point or at the remote node.
When researches show that the WF requires dynamic reactive power compensation, usually combination of fixed switched capacitor bank and SVC dynamic reactive compensation is used, to effectively coordinate for the control of system voltage regulation (such as the process of changing wind, unplanned events, and other processes), support wind power generation and reduce the user's voltage offset.

7) **Operate according to capacity curve of wind arms with double-fed asynchronous generators (DFAG)**
Reference [49] studies and compares DFAG-based WFs operating (dispatching) according to the limited power factor and according to the DFAG capacity curve. From the prospective of voltage stability, operating according to the limited power factor may cause overly conservative estimation of transmission capability of system power; utilization of DFAG capacity curve makes the power transmission capability of the system increase significantly.
With continuous increase of number of DFAG in the WF, correct expression of reactive power of these generators is important. The expression shall include wind power variability and its impact on active and reactive power dispatch of WFs with DFAGs. An accurate estimate of the available reactive power plays an important role in assessment of voltage stability margin of the power system. Capacity curve of the generator rightly expresses its reactive power, thus helping to determine accurately the active power transmission margin.
In operation of stand WFs, an important requirement is to maintain the power factor at grid connection point within the range of ±0.95. This standard requirement is because of comparing with conventional generators (with inherent reactive capacity), the reactive capacity of WFs requires additional costs. Presence of power electronics control in DFAGs, will provide them with a fast-acting reactive power as the synchronous generators directly connected to conventional power plants.
Single DFAG capacity curve is scaled to integrate capacity characteristics of WFs with DFAGs. Figure 7.45 shows that the DFAG is operating within the power factor range of ± 0.95. The curves in the figure correspond to the DFAG capacity curves at variable power output (different wind speeds). The figure shows the capacity curves at the slip rate of 0.25, 0.1, −0.05, −0.15, and −0.25. It covers the entire wind speeds from the cut-in wind speed at slip rate of 0.25 to the cut-out wind speed at the slip rate of −0.25.
WFs rarely operates at 100% output continuously, so important additional reactive power is available in the time of less than 100% output. Achieving operation according to the capacity curve control by WFs with DFAGs, can contribute to lower system losses, and improve recovery of fault voltage after disturbances, and improve system performance, especially at low output power levels of the WF.

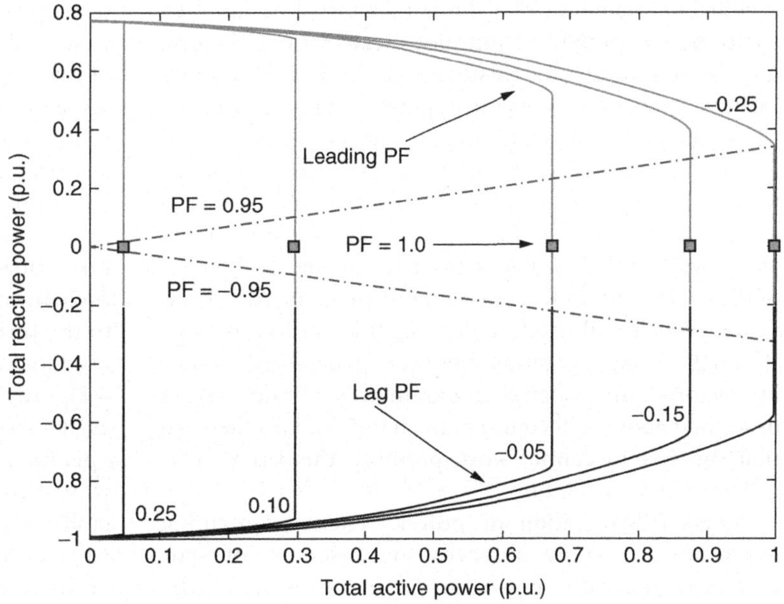

Figure 7.45 The static power capacity curves of WFs with DFAG.

8) **Low-voltage load shedding**
   Installation of low-voltage shedding devices (including removal of wind turbines) in WFs is very effective to prevent vicious voltage drops.
9) **Using SVC-HVDC for grid-connection of WFs**
   Dynamic compensation of cage asynchronous generator (WFs) transient reactive power demand through converter stations, can raise the terminal (WFs PCC) voltage, and enable asynchronous generators to recover the power output capacity rapidly. After the fault is cleared, rapid increase the power output of the asynchronous generators through the active regulation of SVC-HVDC, thus reducing generator acceleration and the increase in reactive power demand caused by unbalanced moment of the generator, is favorable for post-fault voltage recovery of wind turbines and improvement of transient voltage stability of WFs with asynchronous generators.

7.6 Transient Stability Analysis of Power Systems Containing Wind Power

7.6.1 Description of Transient Stability

Transient stability refers to the ability to maintain synchronous operation after large disturbance of the power system in a certain operating state. Large disturbances encountered in power system operation include the occurrence of a variety of short-circuits and disconnection faults, integrating into or removal of large-capacity generators and other important transmission equipment, and integrating into or removal of large capacity users, and so on. In addition, occurrence of some disturbances may be accompanied by a series of operations, such as removal of fault lines by the action of the protection devices and switches. In addition, in order to maintain or improve stability of the power system, the disturbance may be accompanied by removal of generators, load-shedding, putting into forced excitation, fast valving, and other control measures. In the role of large disturbances and the subsequent series of operations, state variables of the power systems vary widely, thus linearizion of state formulas of the system is not allowed, and nonlinearity of the state formulas of the power system must be considered. After large disturbances, frequency of synchronous generator will temporarily deviate from the synchronous frequency (50 Hz/60 Hz), and the generator power angle is changing. The purpose of transient stability study is to determine whether the generators are able to return to synchronous frequency in the new steady-state power angle, as well as to pay close attention to the changes in power flow and bus voltage.

According to the length of time, stability problems of the power system can be divided into short-term stability, mid-term stability, and long-term stability as needed, and the models of system components they used in the analysis are different. Typical time range includes: short-term stability of 0 to 10 seconds; mid-term stability of 10 seconds to a few minutes; and long-term stability of several minutes to tens of minutes [59].

In modern large-scale power systems, for complex networks or networks containing a large number of synchronous generators, transient stability studies are generally achieved by using digital computer programs. For a given disturbance, the program uses algebraic power flow formulas that represent the network and the non-linear difference formulas that represent the synchronous generator for joint and alternating iterative solutions, to achieve continuous calculations before, during, and after the disturbance. Program outputs include changes in generator power angle, frequency, node voltage and line and transformer power flows, and so on, over time. If synchronization between the generators cannot be maintained in the power system, then the electric power delivered by the generators is not a fixed value, and the node voltages, branch currents, and power of the entire system will no longer be maintained at fixed values, which will result in very significant fluctuations. If synchronous operation between generators cannot be restored, the operating parameters of the system described above will continue to change and oscillate, so that the system cannot work properly, this situation is called system instability.

Transient simulation of power systems containing wind power mainly observes the response of synchronous generator power angle. In many cases, the transient stability depends on the magnitude of the generator power angle during the first swing after the disturbance. During the first swing that usually lasts for 1 s, the mechanical power output and the internal EMF of the generators are generally assumed as constants. But taking into account the multi-swing for a few seconds, the models of the generator and the excitation system, as well as more detailed model of the generator shall be considered, in order to obtain accurate results of transient stability for longer period. To simplify the transient stability studies, we usually assume that: ① only the balanced three-phase systems and balanced disturbances are considered. Therefore, only the positive sequence network is used; ② deviation of the generator frequency from the sync frequency is small, the DC offset and harmonics are ignored. Therefore, the network constituted of transmission lines, transformers, and load impedance is substantially in steady state; voltage, current, and power can be calculated by algebraic power flow formulas.

Fault critical clearing time (CCT)is often used to measure transient stability margin of conventional power systems, CCT also applies to transient stability analysis of power systems containing wind power.

Speed (transient) stability of grid-connected asynchronous generators can also be determined by their respective CCT. Determination of CCT of the conventional asynchronous generator is through a lot of trial and error, and repeatedly operating of time domain simulation program. References [60, 61] use the well-known torque-slip and generator speed-time characteristics to analyze transient stability of asynchronous generators and to represent torque-slip characteristics as slip-time characteristics. Stability limit at the three-phase asynchronous generator fault means that asynchronous generator CCT can be obtained according to the torque-time characteristics, which saves the time used for repeated domain simulation.

Figure 7.46 shows the torque-slip characteristics of asynchronous generators. Assume that the steady-state operating point is Point 1, the generator speed is s_f, which also represents the generator speed when the fault starts. In case of asynchronous generator terminal voltage drop, for example, by 50%, due to network fault or disturbances, the corresponding torque-slip characteristics will change from T_{em} to T_{em}' as shown in Figure 7.44. At this moment, because the mechanical torque T_m and electromagnetic torque T_{em} are unbalanced ($T_m > T_{em}$), the generator will accelerate, of which the operating point will move along the torque-slip characteristic curve T_{em}'. If the generator operating point moves beyond point 2, the input mechanical torque T_m will be greater than the electromagnetic torque T_{em}', and the generator will further accelerate. Therefore, point 2 or speed s_{lim} is the stability limit or the asynchronous generator. That is, if clearing time of the fault that causes the voltage drop is relatively short ($s < s_{lim}$), the asynchronous generator is stable; if the fault clearing time is longer ($s > s_{lim}$), the asynchronous generator will go into an unstable mode.

Torque-time characteristics are shown in Figure 7.47. In the figure, by using $s = (w-w_f)/w$ and the speed formula, the torque characteristics of the asynchronous generator is converted from the slip domain to the time domain. In Figure 7.45, t_f and t_{lim} respectively corresponds to s_f and s_{lim} in Figure 7.44. In Figure 7.45, t_{lim} (where $T_{mech} = T_{em}$) is the critical fault clearing time where the asynchronous generator can stably operate after the fault is cleared, that is, CCT. Stability margin of the asynchronous generator in three-phase fault

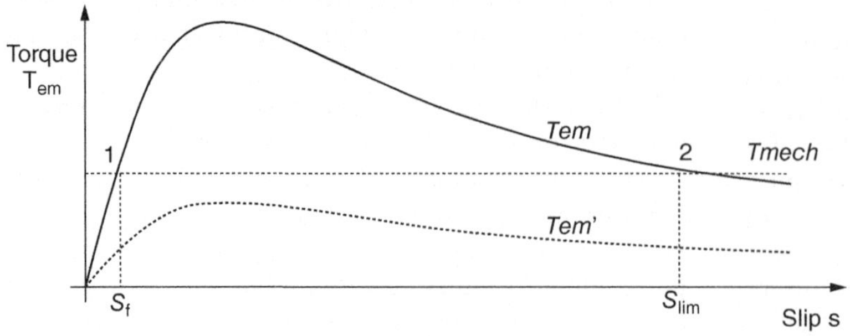

Figure 7.46 Torque-slip characteristics.

Figure 7.47 Torque-time characteristics.

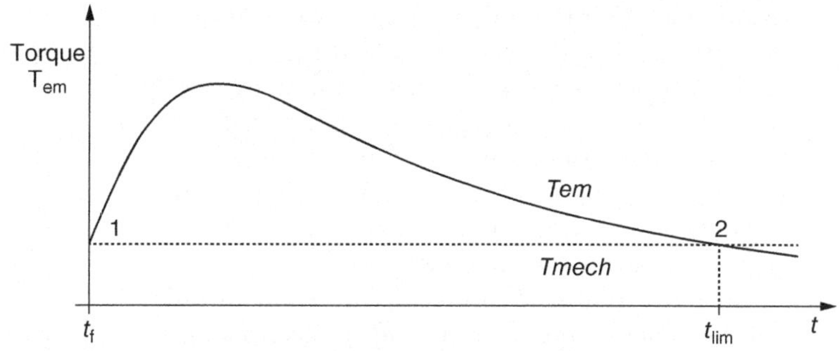

conditions can be determined by using the slip-time characteristic curve shown in Figure 7.45.

CCT is represented as (Formula (7.150)):

$$CCT = t_{cl} - t_f \tag{7.150}$$

where t_f and t_{cl} respectively represents the fault occurrence and clearing time.

Extensive studies have been carried out on transient stability of power systems containing wind power at home and abroad [62–69]. It is generally believed that the impact of wind power on the power system transient stability can be characterized by the fault limit clearing time, and can be tested by the most serious three-phase short-circuit fault; compared with cage asynchronous generators, DFAG equipped with four-quadrant DC/AC converters increase the transient stability limit of the power grid; DFAG equipped with power electronic converters and fault ride-through capability can be interconnected with weak grids without reducing stability; optimization of variable speed wind turbine operation mode can enhance the transient stability nearby conventional generators; various types of wind turbines are not directly involved in the power system oscillations, specifically, the wind power penetration has a damping effect on reduction of synchronous generator capacity (participation in the power system oscillations) due to integration of wind power; simulation of synchronous generator and DFIG with the same capacity in the same position in the power system showed that, through appropriate arrangement of operation and control of the DFIG PWM converter, DFIG can successfully pass through serious grid faults; comparison of effects of WFs consisting of constant speed asynchronous wind turbines, double-fed asynchronous wind turbines and direct-drive permanent magnet synchronous wind turbines on the power grid showed that, constant speed asynchronous wind turbines are of poor stability, double-fed asynchronous wind turbines and direct-drive permanent magnet synchronous wind turbines are capable of raising short-term voltage stability of synchronous generators after a power fault, reducing reactive power reserves required by the system, helping stabilize the grid voltage.

7.6.2 Transient Characteristics of Wind Turbines and Their Impact on Power Grid Transient Stability

1) **Cage asynchronous wind turbines**

Impacts of WFs with cage asynchronous generator (CAG) on transient stability can be analyzed from two aspects: one is, in changes of wind speeds or occurrence of power faults or disturbances, CAG may be subjected to transient instability; the other is, impacts of CAG WFs on transient angle stability of other conventional synchronous generators.

When a fault occurs close to the grid-connected CAG, due to severe voltage drop, the generator can be accelerated to a very high speed, which is far away from the system frequency. This phenomenon is related with the power system stability, which, however, is not included in conventional concepts (such as the rotor angle stability, voltage stability, and frequency stability). Because it is an asynchronous generator, so the stability of CAG cannot be classified as the rotor angle stability phenomenon. After the fault is cleared, the system voltage may be restored to a new allowable value, but the speed of the asynchronous generator may rise to an unacceptable value, so it cannot be classified as voltage stability phenomenon either. After the fault is cleared, the system frequency may be acceptable, so it cannot be classified as a frequency stability phenomenon either. Thus, these stability phenomena are known as the rotor speed stability. Rotor speed stability refers to the ability of asynchronous (induction) machines to maintain connection with the power system and operate at a mechanical speed, which is close to a speed corresponding to actual system frequency after suffering from disturbances [70, 71].

The factors affecting the CAG speed stability include: ① flexibility and inertia of shafting. Because gearbox transmission is added between the wind turbine and the generator, and rotary inertia of the wind turbine is relatively much larger than the generator inertia, shafting between the wind turbine and generator is very flexible. Increase of inertia and rigidity coefficient of the transmission shaft can improve the speed stability of the generator unit. ② mechanical disturbance, that is, random variation of wind speed; ③ electrical disturbances, faults, and their locations suffered by grid-connected wind turbines, especially the power grid voltage drop; ④ the operating point of the generators. Among these factors, the active power output (operating point) can be controlled by adjusting the pitch of wind turbine blades. By reducing the power output of the constant-speed wind turbines, the reactive power absorbed by the wind turbines will be reduced (because the slip is decreased), so the rotor speed stability margin of the wind turbines can be improved. The use of constant speed wind turbines with active stall can also enhance rotor speed stability. ⑤ Equivalent model of asynchronous wind turbines selected in the simulation calculation is different choices. In terms of the generator models, there is a detailed model that has considered the stator electromagnetic transient, a simplified model that has ignored the stator electromagnetic transient, and a steady-state model that has ignored the stator and rotor transient electromagnetic transients at the same time. The use of steady-state model will lead to considerable error in the analysis results, and considering or ignoring the stator electromagnetic transient has little impact on the transient stability analysis results of the generators. In terms of wind turbines, there are different wind turbine shaft equivalent models adopting one or two masses, or detailed models adopting three to five masses, the model considering one mass may get inaccurate transient stability analysis results. Therefore, when analyzing the wind turbine transient stability, the use of wind turbine models combining the generator model ignoring stator electromagnetic transient and the shaft model of two masses is appropriate.

Impact of CAG WTs on the conventional synchronous generator transient stability depends mainly on wind power penetration levels, grid structures, and characteristics of grid faults. If the penetration level of wind power is not high, the dynamic behavior of wind turbines has no significant effect on the transient stability of synchronous generators. With increasing levels of wind power penetration, the power flow will change significantly, and the conventional generator rotor angle will increase. Increase of the first rotor angle swing value can be explained as follows: increase of cage asynchronous wind turbines will require more reactive power. Large reactive power demand will increase the line current and voltage drop, and lead to a greater voltage phase angle. With increasing levels of wind power penetration, conventional generators and infinite bus will provide more reactive power and less active power. Fault clearance will cause increases of reactive power, the cage asynchronous generators will absorb more reactive power from the grid due to their high speed, and the bus voltage drop is so serious it cannot be restored to normal level.

Factors affecting the impacts of CAG WFs on the transient stability include: ① short-circuit power at PCC. The short-circuit power is large, that is, the short-circuit ratio is large, the bus is stronger, and WFs have no significant effect on transient stability; ② reactive power compensation. Sufficient reactive power compensation and high-power factor has some good effect on transient stability; ③ wind power penetration levels. Generally, the penetration level of wind power is high, and has great impact on system transient stability. ④ locations of WFs in the system. Especially when the WFs are located in a particular area which may result in significant changes in power flows (including increase of tie-line power), then the fault critical clearing time may be significantly reduced [69, 71–74].

2) **Double-fed asynchronous generator**

After the large disturbances, the recovery mechanism affecting the synchronous generators returning to synchronization is related to the interaction between the synchronizing force and the inertia of the generators in the system. However, in the case of DFAG, the inertia of the wind turbines is decoupled with the power system. In DFAG, the power electronic converter controls the double-fed machine performance and serves as an interface between the generator and the power grid. With conventional control, rotor current is always controlled to extract the maximum energy from the wind. Thus, with increasing the penetration of WF based on DFAGs, the effective inertia in the system will be reduced, and reliability of the system after large disturbances will be significantly affected.

In DFAG cases, the generator load disturbances also cause changes in the rotor speed and position. However, because it contains asynchronous operations, the position of the rotor flux vector is independent of the physical position of the rotor, there is no synchronous torque-angular characteristics [69].

Because DFAG is an asynchronous generator, its performance is mainly determined by the converter and its control. DFAGs themselves are not involved in the electromechanical oscillation mode of the conventional synchronous generator, their influence on transient angle stability is due to the substitution of some of the synchronous generators, which may increase the electrical distance between the rest of the synchronous generators and change the normal operating point; output increase of WFs with double-fed induction generators changes the power flow distribution of the system, and even make some lines overloaded; conventional synchronous generators are replaced by wind turbines, so the active regulation ability of the system gets worse.

Because AC excitation control is used for wind turbines based on DFAG, they will have the characteristic of variable speed during operation, the power angle is not strictly relevant to the rotor speed, but is determined by its control characteristics, so its coupling to the grid is flexible. Its unique speed-torque characteristics determine that its prime mover mechanical torque will reduce during a grid fault. Compared to the synchronous generator, its acceleration area has decreased, while the deceleration area has increased, so its transient stability characteristics are better than those of the synchronous generator.

From the point of view of energy balance, during a system fault, due to the unbalance between the electromagnetic torque and the mechanical torque, the double-fed wind turbines will accelerate, some of which are temporarily stored in the kinetic energy as wind turbine rotor blades spining up. This part of the staging of energy can reduce the impact of wind turbines in the transient process of the grid. Therefore, transient stability of power systems connected with WFs with DFAG is better than transient stability of power systems with synchronous generators of the same capacity connected to the same point [69, 76].

Adding DFAG to low-voltage systems can considerably improve the voltage and rotor angle stability. The improvement is mainly due to the DFAG absorbs less reactive power than the CAG in transient fault. Although the DFAG is also operating as a conventional asynchronous generator during transient disturbances due to crowbar action, its absorption of reactive power is less than CAG, because crowbar works only when the rotor current exceeds 1.5 p.u., and will be released quickly when reaching normal operating conditions. Therefore, the time of the DFAG operating as asynchronous generator is much less than the total acceleration time of a conventional asynchronous generator. Furthermore, the additional resistance applied by crowbar during short-circuit is connected in series with the rotor resistance, which also reduces the DFAG reactive current.

In addition, since the DFAG grid side converter is used as STATCOM, it can provide voltage support for local WFs during a transient disturbance, which will further improve the voltage and rotor angle stability of the system, especially when DFAG is connected to the low voltage network.

Although the above-mentioned various factors have improved voltage and transient rotor angle stability at low levels of wind power penetration, at high wind penetration (such as greater than 50%), the stability of the system is still significantly decreased, because in transient disturbance period, DFAG wind turbines also absorb a lot of reactive power.

3) **Direct-drive permanent magnet synchronous generators**

   In a WF, due to different wind speeds, the power output by each wind turbine before the grid fault voltage drop is different; during disturbances (voltage drop), because the output current is limited by the transformer, all wind turbines deliver the same electric power, so the total wind power cannot be fully transferred to the grid. To this end, part of the wind power is stored as kinetic energy, so that the wind turbine speed is increased; the other part is stored as electric energy, which makes the DC link capacitor voltage increase. After the end of the voltage drop, the wind turbine speed, capacitor voltage, and the electric power delivered will slowly return to their initial values.

The direct-drive permanent magnet synchronous generator (DDPMSG) is excited by the permanent magnet, there is no reactive power exchange between the generator and the stator-side rectifier, which is similar to the power output of the synchronous generator acting on the grid through the VSC-HVDC, and is able to overcome the negative impact of grid-connected WFs on grid stability and power quality.

During and after a grid fault, the permanent magnet synchronous generator (PMSG) does not absorb reactive current from the grid, and the grid-side converter (inverter) of PMSG can control the active and reactive power exchanged between the inverter and the grid by controlling its d, q-axis current components, and is able to participate in reactive power and voltage regulation of the system within a certain range, so it can maintain the grid's short-term voltage stability, which can help more to improve the system stability compared with CAG and DFAG.

Reference [77] caries out grid fault simulation analysis on the model system shown in Figure 7.48. Rated speed of the permanent magnet synchronous generator (PMSG) is the maximum rotor speed $\omega_{r.max}$. Pitch angle controller will act when the rotor speed exceeds the maximum rotor speed. Using DDPMSG concept and full power converter. Generator-side converter uses vector control, its q-axis stator current can control the active

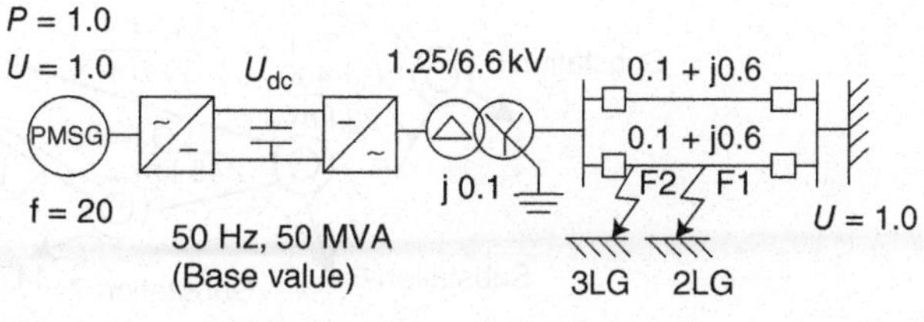

Figure 7.48 Model system.

power, of which the reference value can be determined by the maximum power provided by the wind turbine to the grid. *d* -axis stator current can control reactive power, of which the reference value is set to zero.

In Figure 7.48, symmetrical three-phase ground fault (3LG) and two-phase (phase B, C) to ground fault (2LG) in the grid shown in Figure 7.48 occur in transmission lines of different points of faults. The fault occurs at 0.1 s. The circuit breaker (CB) opens the fault line at 0.2 s, and recloses at 1.0 s. For transient stability analysis, the wind speed is maintained at the rated speed, and then the PMSG reference power is at the rated level. Wind speed is considered constant within a short transient time. The simulation step and simulation time is respectively selected as 0.00001 s and 10 s. The simulation program uses PSCAD/EMTDC, by observing the terminal voltage, PMSG rotor speed, active power of the grid, DC link voltage, wind turbine blade pitch angle, in order to inspect the transient stabilities of the wind power system under three fault conditions:

For 3LG faults, the points of fault is at the middle of line F1 in Figure 7.48. During a network fault, the grid side converter can provide the necessary reactive power, therefore, the terminal voltage can return to the level before the fault. Pitch controller acts when the rotor speed exceeds the rated speed of PMSG. As can be seen from the wind turbine blade pitch angle, the DC link voltage U_{dc}, grid active power and PMSG rotor speed response, and so on, when 3LG fault occurs in distance away from the wind turbines, the transient stability of wind power systems can be enhanced by controlling.

When 3LG fault occurs in the transmission line of the sending end (at F2 in Figure 7.48), that is, when 3LG fault occurs near wind turbines, the transient stability of wind power systems can also be enhanced by controlling, but fluctuations are more serious compared with the fault occurs at the middle of the line.

When 2LG fault occurs at the sending end of a transmission line (F2 in Figure 7.48). As can be clearly seen from the simulation results of the grid side converter active power, reactive power, terminal voltage of the grid, PMSG speed, and the response of U_{dc}, even under the condition of asymmetric fault, the transient stability of wind power systems can still be enhanced by controlling.

In short, for DDPMSG, the use of appropriate controls, can enhance the transient stability of wind power systems, increase DDPMSG's low-voltage ride through (LVRT) capability, and reduce the phenomenon of wind turbine outage. Faults occuring near generators are more serious than faults occur in distance away from the generators.

7.6.3 Impact of WF on Power System Transient Stability

Impacts of WFs integrated into the power system on transient stability of power systems are associated with network topological structure, operation mode of power grid, and location of the WF in the system, wind power technology type, network fault conditions, and many other factors. Integration of wind power into the power system may improve transient stability of the power grid, or reduce the transient stability of the power grid, and it must be determined by simulation calculation of specific cases. Wherein the common practice is to check the synchronous generator speed, rotor angle, voltage, and other state variables through time-domain simulation of dynamic model containing WFs, or determine the impact of WFs on transient stability through solving of critical fault clearing time under the conditions of grid transient fault.

Take the planning grid shown in Figure 7.49 as an example, to examine the different effects of WFs consisting of different wind turbines on the power grid

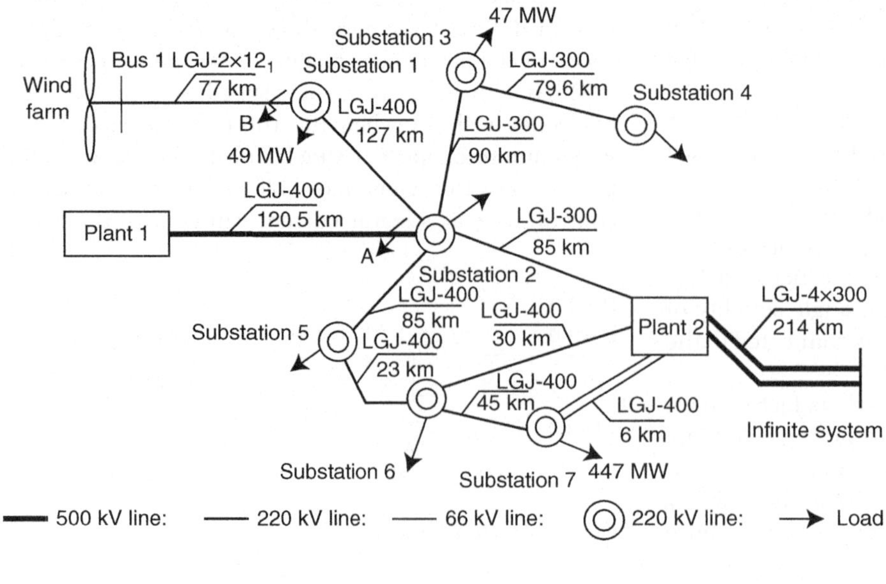

Figure 7.49 Structure of a local planning grid.

transient stability. In the simulation, we assume that bus 1 is connected to a WF with the capacity of 50 MW respectively consisting of cage asynchronous wind turbines, double-fed asynchronous wind turbines, and direct-drive AC synchronous wind turbines. Set two grid faults: ① three-phase short-circuit fault occurs at point A close to the thermal power plant 1; ② three-phase short-circuit fault occurs at point B near the WF. Use the response of rotor angle and speed of conventional synchronous generator, WF outlet voltage, reactive power and speed of the wind turbine, and other state variables to a grid fault to measure the impact of WFs with different wind turbines on the grid transient stability, and we get the following results [69].

1) **Short-circuit fault occurs at point A**
   When three-phase short-circuit fault occurs at point A in the system, if cage asynchronous wind turbines, double-fed asynchronous wind turbines, and direct-drive AC synchronous wind turbines are respectively used on bus 1 of the WF, the critical fault clearing time of the synchronous generator in the system is respectively 250 ms, 285 ms, and 300 ms. If the fault duration of the system is longer than the critical fault clearing time of the system, the synchronous generator will be unstable. The simulation results show that, compared with constant speed cage wind turbines, the use of variable speed wind turbines can increase the stability margin of the system.
2) **Short-circuit fault occurs at point B**
   If three-phase short-circuit fault occurs at point B, and the fault is cleared after 0.15 s, CAG speed will gradually increase, and eventually lose stability; double-fed generators use the converter to control rotor current and maintain balance between mechanical torque and electromagnetic torque of the generator unit through rotor output power, thus to maintain stable operation of the unit; direct-drive wind turbines control the stator current through an converter in order to maintain stable operation of the unit.

When a short-circuit fault occurs in the power grid, since the input mechanical torque of the wind turbine remains relatively stable while the generator output electromagnetic torque is decreasing, this results in acceleration of the generator and makes a lot of kinetic energy accumulated by the unit shaft of low rigidity (the wind turbine is connected to the generator through a gearbox of relatively large speed rising ratio). Because the constant speed cage asynchronous generators are not capable of controlling reactive power and voltage, they will need to absorb large amounts of reactive power from the grid for excitation after clearance of grid faults, which reduces the recovery rate of the generator voltage at the outlet. When wind turbines are connected to weak grids, the energy stored by the newly built magnetic field of the generator is less than that stored in the unit shaft, the generator speed continues to increase, the wind turbine loses stability, then the reactive current absorbed from the grid is also increasing, and the outlet voltage of the unit will continue to decrease; since the unit shaft is relatively "soft," the generator outlet voltage will oscillate, showing transient voltage instability. Double-fed units and direct-drive wind turbine units can utilize converters to involve in the reactive power and voltage of the control system, so that unit outlet voltage can be quickly restored after clearance of the grid fault, and oscillations of the unit outlet voltage can be reduced.

When double-fed wind turbines and direct-drive wind turbines are connected to bus 1, reactive power of the synchronous generator can be restored to the pre-failure value after the fault is cleared. When constant speed cage asynchronous generators are connected by bus 1, to maintain the dynamic stability of the wind turbine outlet voltage, the synchronous generator is required to output a certain reactive power for compensation, thereby increasing the reactive power output of the synchronous generator. After the clearance of the fault, the system has a reactive transient process with a larger magnitude. Therefore, grid-connection constraint of constant speed wind turbines is the reduction of network voltage stability of local power grid in the system.

The results showed that:

1) When the external conditions are the same, in occurrence of short-circuit fault in a certain point in the grid, impacts of WFs of the same capacity consisting of the three types of wind turbines on the stability limit of synchronous generator are different, compared with the constant speed wind turbines, variable speed wind turbines improve the stability of the system.
2) When the grid fault is cleared, cage asynchronous generators need to absorb large amounts of reactive power from the grid for excitation, whereas the dual-fed units and direct-drive wind turbine units can use the restarted converters to control active and reactive power. When WFs are connected to a weak power grid, the speed of the constant speed wind turbine will gradually increase, and eventually lose stability, whereas variable speed wind turbines can maintain stable operation of the unit by controlling the converter.
3) In order to maintain the outlet voltage level of the wind turbine, after the grid fault is cleared, compared with the variable speed wind turbine, constant speed wind turbines require the synchronous generator to output a certain reactive power, and thus the reactive power output of the synchronous generator is

increased. When the grid is weak (transmission reactive power is limited) and its reactive power reserve is small, it may lead to the loss of voltage stability of the local power grid.

Reference [78] studies the impacts of fixed-speed asynchronous generator (FSAG) and double-fed asynchronous generator (DFAG) respectively connected to high-voltage/low-voltage (HV/LV) buses on power grid transient stability based on the IEEE-14 bus system shown in Figure 7.50. WFs are connected to the low-voltage system bus 9, 10, 11, 12, 13, and 14, and high-voltage system bus 4 and 5. Increased load is distributed around the WF to consume the increased power output of the WF.

Figure 7.51 and Figure 7.52 respectively shows how fault critical clearing time (CCT) changes with wind power penetration levels in occurrence of three-phase short-circuit on the high-voltage bus 5 and low-voltage bus 14. For each case, with increasing levels of wind power penetration, voltage, and rotor angle stability of the system are reduced. When the penetration level of wind power is low, the system provides a voltage of higher stability than the rotor angle stability, but when the penetration level of wind power is increasing, the system voltage stability is dramatically decreased. This is because, as more FCAG added into the system, the reactive power demand increases considerably, thus the system becomes increasingly vulnerable until the voltage collapses. Impact of high-voltage system faults on system stability is greater than that of low-voltage faults, that is because, compared to the low-voltage system fault, the WF can maintain a longer period of time at low voltage, therefore, an asynchronous generator in a low-voltage system will accelerate for a longer time and therefore, absorb more reactive power from the system. It can be assumed that, this situation becomes more serious when the penetration level of wind power is high, and

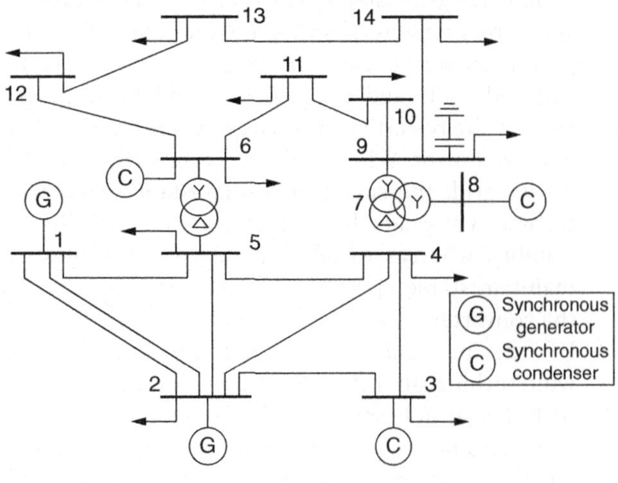

Figure 7.50 IEEE-14 bus system.

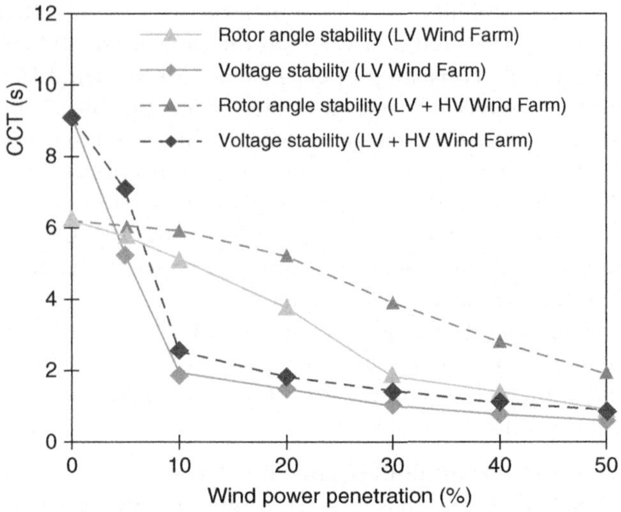

Figure 7.51 Changes in CCT when faults occur on the high-voltage bus 5 (LV/HV network connection).

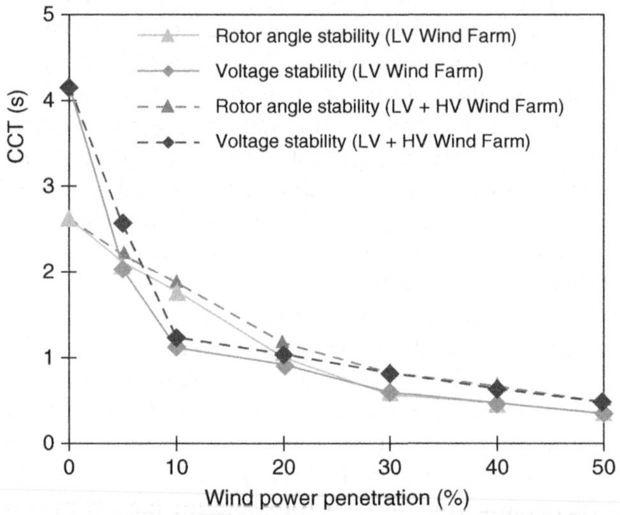

Figure 7.52 Changes in CCT when faults occur on the low-voltage bus 14 (LV/HV network connection).

ultimately lead to voltage collapse. It also makes the synchronous generator tend to be unstable due to loss of a considerable number of electromechanical torque.

When FSAG are connected to the high-voltage system, the adverse effect of absorption of large amounts of reactive power by the asynchronous generator will be reduced. This is because, compared with low-voltage bus, the short-circuit impedance of high-voltage bus is lower, that is, the high-voltage bus is stronger, and is supported by the voltage provided by the synchronous generator.

Because the bus voltage of the high-voltage WF can recover quickly, the acceleration of FCAG is less, and therefore the reactive power absorbed is less. In addition, due to the improvement of local voltage, the capacitor

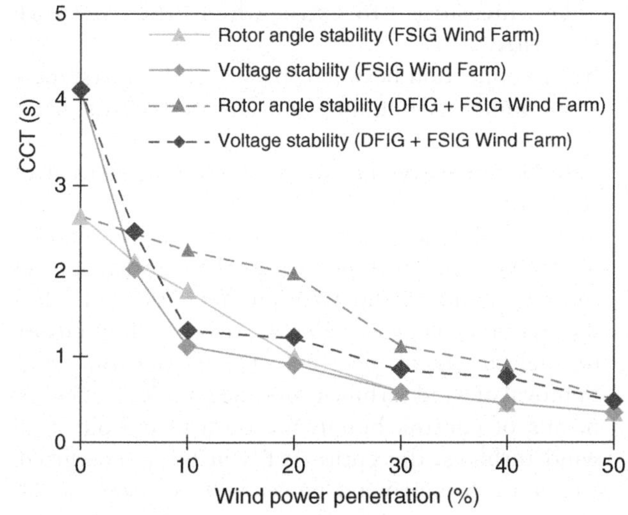

Figure 7.53 Changes in the CCT when faults occur on high-voltage bus 5.

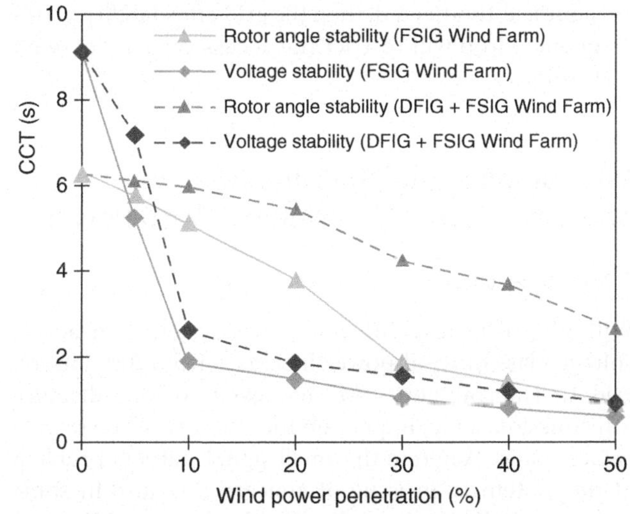

Figure 7.54 Changes in the CCT when faults occur on high-voltage bus 14.

bank can also provide better reactive power support. These factors have an important role to improve voltage stability, and, due to the improvement of electrical and mechanical torques, the stability of the rotor angle is also increased. Although the distribution of FCAG WF improves voltage and rotor angle stability, this improvement has been overpassed by the adverse effects of the wind power penetration level increased from 5% to 50% of FCAG.

FSAG and DFAG WFs of equal capacity are connected to the low-voltage system. The FSAG WFs installed on bus 10, 13, and 14 are replaced by DFAG WFs. Changes in the grid CCT with the penetration levels of wind power in occurrence of faults on bus 5 and 14 are shown in Figures 7.53 and 7.54.

As can be seen from the figures, compared with FSAG WFs, DFAG WFs have considerably improved the voltage and rotor angle stability of the system. The improvement is mainly due to the DFAG absorbing less reactive power than the FSAG during transient faults. Strengthen the system voltage stability helps to quickly restore synchronous generator rotor angle, because the electromechanical torque is improved.

7.6.4 Dynamic Models Required in Transient Stability Analysis

Establishment of reasonable wind turbine models is the basic of wind power system transient stability studies. The models required for transient stability studies mainly include the following main components [79–83]:

1) Wind turbine aerodynamics model;
2) Wind turbine model, there are usually two equivalent modeling methods. One is that the wind turbine and generator units are equivalent to a mass, to establish a rigid model for the unit. The second is that the wind turbines and generator rotors are respectively considered as a concentrated mass, to study transient stability of the wind turbine two-mass shaft model. In addition, considering the impact of blade flexibility on transient stability of the unit, the wind turbine blade is divided into blade tip and blade root along the point of break in the middle of the blade, the blade tip is considered as a mass, and the blade root including the hub, shaft, and gear box is equivalent to another mass, to establish an equivalent two-mass wind turbine model which has considered the blade flexibility, so that the shaft is modeled as a two-mass axis, of which one mass represents the rotor/wind turbine blades, and the other represents the generator.
3) When studying the wind power system transient stability, for modeling of cage asynchronous generator, some literatures apply the fourth-order detailed model, some literatures apply the second-order model that ignores the stator electromagnetic transient, and some literatures apply the transient model that completely ignores stator and rotor electromagnetic transient. Of course, the lower the order, the simpler the model, and the faster the simulation speed, but the accuracy is also lower, which may lead to incorrect transient stability analysis. Therefore, analysis shall be made in order to choose a reasonable model.

Electrical characteristics of double-fed asynchronous generators—because the rotor side converter can control the rotor current quickly, rotor flux dynamics can be ignored, and its characteristics can be modeled as a controllable current source.

The direct-drive permanent magnet synchronous generator does not absorb reactive current from the grid during grid faults and after clearance of faults, and the grid-side converter of the unit participates in system reactive power and voltage regulation, its characteristics can be approximately modeled as a controllable voltage source.

1) Electrical control model—three controllers are used to provide frequency/active power, voltage/reactive power and pitch angle/mechanical power controls;
2) The relay setting values are mainly overspeed protection and low voltage protection.
3) WF models in transient stability analysis. Once the wind turbine and generator models are determined, they can be composed of WF models. Usually, all wind turbines in the WF (or connected to the same bus) are of the same type, so single equivalent model with the same rated capacity of the WF can be used as a model of the entire WF. This method has some limitations, because the input wind speed of each wind turbine in the WF is not the same, these wind turbines are not homologous, and so they cannot be accurately described with one machine. A large WFs can be divided into several equivalent generators (group the wind turbines based on similar wind speed input). However, this method makes no big difference, and only be considered in a detailed study. Wind turbine layout of the WF shall also be considered, especially those with variable speed wind turbines. WF layout changes the regulation capacity of these wind turbines and all of their dynamic characteristics. Therefore, dynamic equivalent of each WF shall be carried out, which makes the power system model is very inconvenient to use. Thus, we should consider a compromise between accuracy and practical application of the model.

7.6.5 Support for System Transient Stability by Enhancing Wind Power

1) **Use dynamic reactive power compensation device**
If dynamic reactive power compensation device SVC is added to the WF substation low-voltage side of a system containing WFs, during the system fault, it can support PCC voltage of the WFs and help to improve the transient stability of the system and the recovery after the fault. But the effect of dynamic reactive power compensation device SVC on improving the transient characteristic of the system is limited, because the reactive power output of the dynamic reactive power compensation device SVC is proportional to the voltage squared, during a fault, if voltage of the SVC at the installation point is relatively low, the reactive power provided by the SVC is limited, and it will lose the role of voltage support.
2) **Take measures to strengthen the network structure**
If the connection between the WF and the system is weak, additional outgoing lines shall be considered for the WF.
3) **Take the measure of removal of wind turbines after the fault**
Removal of wind turbines after the fault can quickly block the wind turbines to absorb reactive power from the grid in the transient process, and allow the system voltage to quickly restore, thus ensure normal supply of electricity to the network load. Removal of wind turbines after the fault is an effective means of control, but in the off-grid period of all wind turbines, the entire WF will be de-energized. Generally, loss of wind power will not cause a significant impact on the frequency of the grid, but if a large number of wind turbines are distributed in the system, and a large number of wind turbines are off-grid at the same time, the grid frequency will be significantly affected, then the grid affordability of frequency drop will restrict the access capacity of wind turbines.

7.7 Small Signal Stability Analysis on Power System Containing Wind Power

7.7.1 Overview

With increasing levels of wind power penetration, access voltage class of wind power increases from distribution grid to transmission grid, and even to long-distance transmission, so its impact on the stability of the system is increasing. Wherein the small signal stability problem of the system is an issue of concern, because in some systems, small signal stability problem may also become a limiting factor in wind power development.

Study on Conventional Power System Small Signal Stability is more concerned about low frequency power oscillation caused by a lack of damping of the system, which is a synchronous generator angle stability problem. The analysis methods applied are mainly linearized system state matrix eigenvalue analysis (frequency domain) and time-domain simulation method. When using eigenvalue analysis, the eigenvalues are calculated according to the system state matrix A, wherein the real eigenvalues σ, is corresponding to the non-oscillation mode, and the negative real eigenvalue indicates decay mode. Complex eigenvalues appear in the form of conjugate character pairs $\sigma \pm j\omega$ and each pair corresponds to an oscillating mode. If the negative σ represents the real eigenvalues of complex pairs, the oscillation mode of the system is damped oscillation. The real part gives the damping, while the imaginary part ω gives the oscillation frequency. Damping ratio is given by the Formula (7.151).

$$\xi = \frac{-\sigma}{\sqrt{\omega^2 + \omega^2}} \tag{7.151}$$

Size of the damping ratio characterizes the speed of oscillation attenuation, when $\xi = 3\%$, after five periodic oscillations, the amplitude attenuates approximately to 39% of its initial value; when $\xi = 5\%\xi = 5\%$, after five oscillation cycles, the amplitude attenuates approximately to 21% of its initial value. Generally, it is required that the damping ratio of each oscillation mode is greater than 3%–5%, otherwise it is a weak damping.

Characteristic analysis method provides system properties of each oscillation mode, and can also be used to adjust the controller parameters. When comparing results of eigenvalue analysis and dynamic time domain simulation, we can find that eigenvalues are not sufficient to determine behaviors of the system after disturbance, and therefore the dynamic time domain simulation is always needed. When using time-domain simulation, the simulation model of the entire power system can be obtained by combining the relevant model described in Section 7.2 and the power system model, then by using a variety of commercial power system analysis software, the curve of system state variables of interest following the change of time can be obtained through calculation. The negative or weak damping mode obtained from eigenvalue analysis can be further tested by time domain simulation. Power oscillation cannot be seen in power systems under normal and stable operation. In order to obtain the required power system oscillations, excitation (interference) is required to be applied to the power system. Three-phase short-circuit fault and line switching are often used for excitation. In the simulation model, the fault time (such as line disconnection time) Tex is determined based on the oscillation frequency to be observed, preferably a half cycle of the oscillation frequency observed half cycle [84], that is (Formula (7.152)),

$$T_{\text{ex}} = \frac{1}{2f} \tag{7.152}$$

where f is the observed oscillation frequency of power system. When the simulation is completed, the representative part of the curve for small signal analysis is obtained through calculation, based on the curve obtained in the simulation, the eigenvalues of the simulation, including the oscillation frequency and damping ratio, can be obtained by using prony analysis.

Asynchronous generators are used for most of the wind power generation, and WFs using synchronous generators is also connected to the grid through power electronic converters, so they are not directly involved in the electromechanical oscillation of the synchronous generator in the system, but after large-scale integration of wind power into the power system, the size and direction of the power flows in the power system are changed, which will have an impact on small signal stability of the system. Because the wind turbines used are of different types, and are connected to different locations in the system, so they have different effects on small signal stability of the system.

When using eigenvalue analysis method to analyze a small signal stability of wind power systems containing wind power, a variety of scenarios shall be set [85], which usually include: (1) keep the WF capacity constant, when changing the WF output, other units within the grid will contribute to keep the system load balancing by reducing output. (2) keep the WF capacity constant, changing the WF output, other units within the grid will contribute to keep the system load balancing by downtime. (3) Change the installed capacity of WFs, the wind turbines are at full-power, other units within the grid will contribute to keep the system load balancing by downtime; (4) at the same power supply node, the conventional synchronous generators are replaced by different types of wind turbines, and so on.

Some studies of the impact of wind power on small signal stability of power systems have been carried out at home and abroad [85–97], and considerable progress has been made regarding modeling of small signal stability analysis for different wind turbine systems and impacts of wind power on the small signal stability of the power systems. This section describes wind turbine (WFs) models used for small signal stability analysis, the impacts of WFs with different wind turbines on small signal stability of the system, and measures for improving small signal stability of power systems through the WF.

7.7.2 Wind Turbine System Models for Small Signal Stability Analysis

The method commonly used for small signal stability analysis are eigenvalue analysis and time domain simulation. When using time-domain simulation, the simulation model of the entire power system can be obtained by combining the relevant model described in Section 7.2 and the power system model, then by using a variety of commercial power system analysis software, the curve of system state variables of interest following the change of time can be obtained through calculation. The following only describes the linearized models used for small signal stability analysis.

7.7.2.1 General Linearized Model

In the Power System Small Signal Stability eigenvalue analysis, by linearizing the dynamic models (differential formulas) describing each dynamic device in the system at the operating point, the linearized model of the system in general formula is given as (Formula (7.153)):

$$\begin{cases} \frac{d}{dt}\Delta x_i = A_i\Delta x_i + B_i\Delta U \\ \Delta i_i = C_i\Delta x_i - Y_i\Delta U \end{cases} \tag{7.153}$$

wherein, x_i is the state variable of a single dynamic device; i_i is the current injected into the network by the device; U is the network node (bus) voltage vector. The state formulas of all dynamic devices in the system can be combined into the following form (Formula (7.154)):

$$\begin{cases} \frac{d}{dt}\Delta x = A_D\Delta x + B_D\Delta U \\ \Delta i = C_D\Delta x - Y_D\Delta U \end{cases} \tag{7.154}$$

Node formula of the interconnected system can be expressed as (Formula (7.155)):

$$\Delta i = Y_N\Delta U \tag{7.155}$$

Substitute Formula (7.155) into Formula (7.154) and eliminate the voltage vectors, the following formula can be obtained (Formula (7.156)):

$$\frac{d}{dt}\Delta x = A_D\Delta x + B_D(Y_N + Y_D)^{-1}C_D\Delta x = A\Delta x \tag{7.156}$$

Formula (7.156) is small signal first-order model, which is a multi-input and multi-output system. Where x is the system-wide state vector; A_D and C_D are the diagonal matrices of matrix A_i and C_i associated with each device. $A = A_D + B_D(Y_N + Y_D)^{-1}C_D$ is augmenter state matrix for the system, including a variety of dynamic components and their network connection relations, deciding the small signal stability characteristics of the system, and Small Signal Stability information of the system can be obtained by solving it.

Calculate eigenvalues of the state matrix, left/right eigenvector and participation matrix (participation factor) based on the system state matrix. The participation factor is considered zero when it is less than 10^{-7}. Contribution of state variable in each oscillation mode of the system can be determined by using the participation factor.

7.7.2.2 Small Signal Analysis Model of Fixed-Speed Asynchronous Generator System

Grid connection of FSAG systems is shown in Figure 7.55, the power system is simply expressed as the equivalent impedance $R_e + jx_e$ and infinite bus. In order to build the model of the single machine infinite bus system (SMIB), which is expressed by synchronous reference axis as shown in Figure 7.55, we assume that: (a) the stator and rotor currents flow at positive direction into the generator; (b) the formula is derived from the synchronous coordinate represented by the direct axis (d) and the horizontal axis (q); (c) all the system parameters and variables are per-unit values and imputed to the stator side of FSAG [93].

FSAG model needs to consider the capture mechanism of wind energy, rotating shaft, asynchronous generator, WF collection system, and transmission system, but does not need to consider the impact of the rotor blade dynamics, tower structure, and yaw movement on entire system dynamics.

1) **Electrical dynamic model**

FSAG electrical formula consists of the stator and rotor voltage formulas, the flux formula and the electromechanical torque formula (Formulas (7.157–7.163)).

$$u_{sd} = R_s i_{sd} - \omega_s \psi_{sq} + \frac{1}{\omega_b}\frac{d\psi_{sd}}{dt} \tag{7.157}$$

$$u_{sq} = R_s i_{sq} + \omega_s \psi_{sd} + \frac{1}{\omega_b}\frac{d\psi_{sq}}{dt} \tag{7.158}$$

$$u_{rd} = R_r i_{rd} - \omega_2 \psi_{rd} + \frac{1}{\omega_b}\frac{d\psi_{rd}}{dt} \tag{7.159}$$

$$u_{rq} = R_r i_{rq} + \omega_2 \psi_{rq} + \frac{1}{\omega_b}\frac{d\psi_{rq}}{dt} \tag{7.160}$$

$$\psi_s = L_s i_s + L_m i_r \tag{7.161}$$

$$\psi_r = L_m i_s + L_r i_r \tag{7.162}$$

$$T_e = \frac{L_m}{L_s}(\psi_{sq} i_{rd} - \psi_{sd} i_{rq}) \tag{7.163}$$

According to Formulas (7.157) to (7.162), as the stator flux and rotor current functions, the stator and rotor state formulas can be written as (Formulas (7.164–7.167)):

$$\frac{1}{\omega_b}\frac{d\psi_{sd}}{dt} = -\frac{R_s}{L_s}\psi_{sd} + \frac{R_s L_m}{L_s} i_{rd} + \omega_s\psi_{sq} + u_{sd} \tag{7.164}$$

$$\frac{1}{\omega_b}\frac{d\psi_{sq}}{dt} = -\frac{R_s}{L_s}\psi_{sq} + \frac{R_s L_m}{L_s} i_{rq} - \omega_s\psi_{sd} + u_{sq} \tag{7.165}$$

$$\frac{L'_r}{\omega_b}\frac{di_{rd}}{dt} = -R'_r i_{rd} + \frac{R_s L_m}{L_s^2}\psi_{sd} - \frac{L_m}{L_s}\omega_r\psi_{sq} + L'_r\omega_2 i_{rq} - \frac{L_m}{L_s}u_{sd} \tag{7.166}$$

$$\frac{L'_r}{\omega_b}\frac{di_{rq}}{dt} = -R'_r i_{rq} + \frac{R_s L_m}{L_s^2}\psi_{sq} + \frac{L_m}{L_s}\omega_r\psi_{sd} - L'_r\omega_2 i_{rd} - \frac{L_m}{L_s}u_{sq} \tag{7.167}$$

where $L'_r = L_r - \frac{L_m^2}{L_s}$; $R'_r = R_r + \left(\frac{L_m}{L_s}\right)^2 R_s$.

The state formulas corresponding to the capacitor voltage (i.e., the generator terminal voltage)

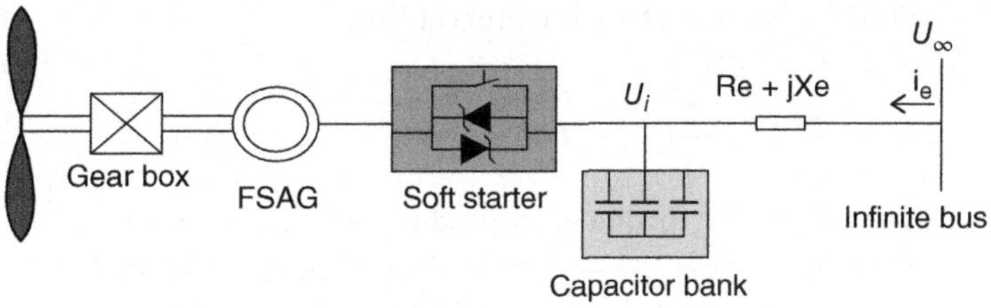

Figure 7.55 FSAG connected to infinite bus.

and network current can be written as (Formulas (7.168–7.171)):

$$\frac{C}{\omega_b}\frac{du_{sd}}{dt} = -\frac{1}{L_s}\psi_{sd} + \frac{L_m}{L_s}i_{rd} + i_{ed} + C\omega_s u_{sq} \tag{7.168}$$

$$\frac{C}{\omega_b}\frac{du_{sq}}{dt} = -\frac{1}{L_s}\psi_{sq} + \frac{L_m}{L_s}i_{rq} + i_{eq} - C\omega_s u_{sd} \tag{7.169}$$

$$\frac{L_e}{\omega_b}\frac{di_{ed}}{dt} = -u_{sd} - R_e i_{ed} + L_e\omega_s i_{eq} + u_{\infty d} \tag{7.170}$$

$$\frac{L_e}{\omega_b}\frac{di_{eq}}{dt} = -u_{sq} - R_e i_{eq} - L_e\omega_s i_{ed} + u_{\infty q} \tag{7.171}$$

where u_s, u_r–stator and rotor voltage amplitudes; L_s and L_r–stator and rotor self-inductances; ψ_s and ψ_r–stator and rotor flux; R_s and R_r–Stator and rotor resistance; u_{sd} and u_{rd}–d-axis components of stator and rotor voltage; L_m–excitation inductance; u_{sq} and u_{rq}–dq-axis components of stator and rotor voltage; H_r and H_t–generators and wind turbine inertia constants; i_{sd} and i_{rd}–d-axis components of stator and rotor current; n_p–number of generator pole pairs; i_{sq} and i_{rq}–q-axis components of stator and rotor current; K_s–shaft coefficient of rigidity; u_∞–infinite bus voltage; ω_b–angular frequency base value (rad/s); $u_{\infty d}$, $u_{\infty q}$–d-axis and q-axis components ofinfinite bus voltage; ω_s and ω_r –synchronous and rotor angular frequency (p.u.); T_m, and T_e–mechanical and electrical torques; s and ω_2–rotor slip and slip frequency; n_{gear}–gearbox transmission ratio; θ–shaft torsional angle; D_{Tm}–mechanical torque fluctuation amplitude; a_{kj}–element of state matrix A at column j, row k; 0–subscript corresponding to the balance amount.

2) **Drive chain process**

Because the inertia of the wind turbines in the wind power system is large, and the generator rotor inertia is small, in order to connect two devices with large inertia difference, the drive shaft must be flexible. Shaft of conventional synchronous generator is much longer than that of the small asynchronous generators, but its rigidity is about 20 to 80, so it can be considered as a rigid shaft. The size of the shaft rigidity of wind turbine is approximately 0.15 to 0.40. Moreover, given that natural oscillation frequency and torsional vibration frequency of the wind turbine system, which has considered the transmission axis flexibility is about 1 Hz, which is within a range with the typical characteristic frequency of the power system, in order to reflect the wind turbine dynamic process after disturbance more accurately, transmission shaft flexibility shall be considered in the wind turbine model. Wind turbine drive chain is usually treated as two masses in dynamic studies. Two-mass model is as follows (Formulas (7.172–7.174)):

$$\frac{d\omega_g}{dt} = \frac{T_e + K_s\theta + (\omega_t - \omega_g)}{2H_g} \tag{7.172}$$

$$\frac{d\omega_t}{dt} = \frac{T_m - K_s\theta - D(\omega_t - \omega_g)}{2H_t} \tag{7.173}$$

$$\frac{d\theta}{dt} = \omega_b(\omega_t - \omega_g) \tag{7.174}$$

where ω_g and ω_t respectively represents the generator and wind turbine speeds, p.u.; ω_b is the synchronous speed, p.u.; θ is the torsion angle between two shaft ends, rad; H_g and H_t, respectively, represents the inertia constant of the wind turbine and the generator; K_s is the shaft coefficient of rigidity, p.u./elec.rad; D is the coefficient of damping, p.u.; T_e and T_m, respectively, represents generator electrical torque and wind turbine mechanical torque, p.u.

The dynamic formula of the single machine infinite bus system (SMIB) described by Formula (7.164) to (7.174) in Figure 7.55 can be written in the form of formula (7.175) as:

$$\begin{cases} \dot{x} = f(x, u) \\ 0 = g(x, u) \end{cases} \tag{7.175}$$

where x and u respectively represents state variable and input variable vector $x = [\psi_{sd}, \psi_{sq}, i_{rd}, i_{rq,}\omega_r, \theta, \omega_t, i_{ed}, i_{eq}, u_{sd}, u_{sq}]^T$ and $u = [T_m, u_{\infty d}, u_{\infty q}]^T$. Linearize and rearrange Formula (7.175), and the system linearized model can be obtained as (Formula (7.176))

$$\Delta\dot{x} = A\Delta x + B\Delta u \tag{7.176}$$

The small signal stability model established in reference [89] does not consider the drive chain flexibility, but takes the wind turbine first-order inertia, pitch angle control, and the asynchronous generators third-order model into consideration (Formula (7.177)).

3) **Wind turbine first-order inertia link**

$$\frac{dT_T}{dt} = \frac{1}{\tau_h}(T_w - T_T) \tag{7.177}$$

where T_w is the torque generated by the wind turbine blade, $T_w = \frac{1}{2}\pi\rho C_p \frac{R^3 v_w^2}{\lambda}\frac{\Omega_N}{P_N} \times 10^{-3}$; ρ is the air density (kg/m3); R is blade radius; v_w is the wind speed acting on the blades (m/s); $\lambda = \frac{\Omega R}{v_w}$ is the blade tip speed ratio; Ω is the blade mechanical angular speed (rad/s); Ω_N and P_N respectively represents wind turbine rated mechanical angular speed and power; C_p is the conversion efficiency factor of the wind turbine, which is a function of the wind turbine blade tip speed ratio λ and the pitch angle β. T_T is the torque on gearbox side; τ_h is the hub inertia time constant.

4) **Gearboxes and couplings**
Gearboxes and couplings are transmissions to transmit torque between the wind turbine and the asynchronous generator, of which the dynamic formula can be expressed as (Formula (7.178)):

$$\frac{d\omega}{dt} = \frac{1}{\tau_\tau}(T_T - T_m) \tag{7.178}$$

where ω is the wind turbine speed; T_m is the torque on gearbox side; τ_τ is the inertia time constant of hub. Due to the wind turbine speed remains unchanged during stable operation, it is considered $d\omega/dt$ is zero. So T_m and T_T can are considered approximately equal.

5) **Pitch control system**
Wind turbine wind energy conversion efficiency factor C_p is a function of tip speed ratio and the pitch angle. When the wind speed changes, the operating point of the wind turbine will change, in order to ensure stable wind energy conversion efficiency and output of the wind turbine, the pitch of the wind turbine shall be adjusted. The pitch control system is typically expressed by Formula (7.179):

$$\frac{d\beta}{dt} = \frac{1}{\tau_\beta}(\beta_0 - \beta) \tag{7.179}$$

Here, other aspects of pitch control are not considered, such as wind speed control link and power output control link. Thus, as a separate control link, the following state formula will not consider the pitch angle state variable.

6) **Asynchronous generator**
Rotor motion formula (Formula (7.180)):

$$\frac{ds}{dt} = \frac{1}{\tau_j}(T_m - T_e) = \frac{1}{\tau_j}(T_T - T_e) \tag{7.180}$$

where $s = \frac{\omega_r - \omega_0}{\omega_0}$, is the asynchronous generator slip; $T_e = \psi_d i_q - \psi_q i_d = E'_q i_q + E'_d i_d$ is the asynchronous generator electromagnetic torque.

The electromagnetic transient formulas are as follows (Formulas (7.181) and (7.182):

$$\frac{dE'_d}{dt} = \frac{1}{T'_{d0}}(-E'_d + (x - x')i_q - sT'_{d0}E'_q) \tag{7.181}$$

$$\frac{dE'_q}{dt} = \frac{1}{T'_{d0}}(-E'_q - (x - x')i_d + sT'_{d0}E'_d) \tag{7.182}$$

where set x_1, x_2 and x_m as the per-unit values of the asynchronous generator stator leakage reactance, rotor leakage reactance, and excitation reactance.
Then

$$x = x_1 + x_m, x' = x_1 + \frac{x_2 x_m}{x_2 + x_m}, T'_{d0} = \frac{x_2 + x_m}{r_2}.$$

From Formula (7.177) to (7.182), the linearization results of the nonlinear formulas are obtained as follows:

$$\begin{bmatrix} \Delta\dot{T}_T \\ \Delta\dot{\beta} \\ \Delta\dot{s} \\ \Delta\dot{E}'_d \\ \Delta\dot{E}'_q \end{bmatrix} = \begin{bmatrix} \frac{-1}{\tau_h} & 0 & 0 & 0 & 0 \\ \frac{-1}{\tau_\beta} & 0 & 0 & 0 & 0 \\ \frac{1}{\tau_j} & 0 & 0 & \frac{E'_{q0} - i_{d0}x'}{\tau_j x'} & \frac{-E'_{d0} - i_{q0}x'}{\tau_j x'} \\ 0 & 0 & -E'_{q0} & \frac{-x}{T'_{d0}x'} & -s_0 \\ 0 & 0 & E'_{d0} & s_0 & \frac{-x}{T'_{d0}x'} \end{bmatrix} \times \begin{bmatrix} \Delta T_T \\ \Delta\delta \\ \Delta s \\ \Delta E'_d \\ \Delta E'_q \end{bmatrix} + \begin{bmatrix} 0 & 0 \\ 0 & 0 \\ \frac{-E'_{q0}\sin\delta + E'_{d0}\cos\delta}{x'} & \frac{E'_{q0}\cos\delta + E'_{d0}\sin\delta}{x'} \\ \frac{(x - x')\sin\delta}{T'_{d0}x'} & \frac{-(x - x')\cos\delta}{T'_{d0}x'} \\ \frac{(x - x')\cos\delta}{T'_{d0}x'} & \frac{(x - x')\sin\delta}{T'_{d0}x'} \end{bmatrix} \times \begin{bmatrix} \Delta u_R \\ \Delta u_I \end{bmatrix} \tag{7.183}$$

The formation of multi-machine system state formulas also needs the terminal current formula (Formula (7.184)):

$$\begin{bmatrix} \Delta i_R \\ \Delta i_I \end{bmatrix} = \begin{bmatrix} 0 & 0 & 0 & \frac{-\sin\delta'\sin(\delta' - \delta)}{x'} & \frac{\sin\delta'\cos(\delta' - \delta)}{x'} \\ 0 & 0 & 0 & \frac{\cos\delta'\sin(\delta' - \delta)}{x'} & \frac{-\cos\delta'\cos(\delta' - \delta)}{x'} \end{bmatrix} \times \begin{bmatrix} \Delta T_T \\ \Delta\delta \\ \Delta s \\ \Delta E'_d \\ \Delta E'_q \end{bmatrix} + \begin{bmatrix} 0 & \frac{-1}{x'} \\ \frac{1}{x'} & 0 \end{bmatrix} \begin{bmatrix} \Delta u_R \\ \Delta u_I \end{bmatrix} \tag{7.184}$$

7.7.2.3 Small Signal Analysis Model of Double-Fed Asynchronous Generator System

1) **DFAG system model**

DFAG are basically asynchronous generators. Similar to the asynchronous generator, their equivalent circuit is shown in Figure 7.56a [94]. Define:

$$E'_d = -(\omega_s L_m/L_{rr})\psi_{qr};\ E'_q = -(\omega_s L_m/L_{rr})\psi_{dr};$$
$$X_s = \omega_s L_{ss}\ X'_s = \omega_s[L_{ss} - (L_m^2/L_{rr})];$$
$$T'_0 = L_{rr}/R_r.$$

DFAG detailed dynamic model can be written as (Formulas (7.185–7.188)):

$$\frac{X'_s}{\omega_s}\frac{di_{ds}}{dt} = u_{ds} - \left[R_s + \frac{1}{\omega_s T'_0}(X_s - X'_s)\right] i_{ds} - (1-s_r)E'_d - \frac{L_m}{L_{rr}}u_{dr} + \frac{1}{\omega_s T'_0}E'_q + X'_s i_{qs} \tag{7.185}$$

$$\frac{X'_s}{\omega_s}\frac{di_{qs}}{dt} = u_{qs} - \left[R_s + \frac{1}{\omega_s T'_0}(X_s - X'_s)\right] i_{qs} - (1-s_r)E'_q - \frac{L_m}{L_{rr}}u_{qr} - \frac{1}{\omega_s T'_0}E'_d - X'_s i_{ds} \tag{7.186}$$

$$\frac{dE'_d}{dt} = s_r\omega_s E'_q - \omega_s\frac{L_m}{L_{rr}}u_{qr} - \frac{1}{T'_0}[E'_d + (X_s - X'_s)i_{qs}] \tag{7.187}$$

$$\frac{dE'_q}{dt} = -s_r\omega_s E'_d + \omega_s\frac{L_m}{L_{rr}}u_{dr} - \frac{1}{T'_0}[E'_q + (X_s - X'_s)i_{ds}] \tag{7.188}$$

where ψ_{dr} and ψ_{qr} respectively represents direct (d) axis and horizontal (q) axis rotor flux; L_{ss} is the stator self-inductance; L_{rr} is the rotor self-inductance; L_m is the mutual inductance; R_r is the rotor resistance; ω_s is the synchronous angular speed; s_r is the rotor slip; X_s is the stator reactance; X'_s is the stator transient reactance; E'_d and E'_q respectively represents d-axis and q-*axis* potential after transient reactance; T'_0 is the rotor circuit time constant; i_{ds} and i_{qs} respectively represents d-axis and q-*axis* stator current; u_{ds} and u_{qs} respectively represents d-axis and q-*axis* terminal voltage; u_{dr} and u_{qr} respectively represents d-axis and q-*axis* rotor voltage.

2) **Drive chain model**

The drive chain consists of wind wheels, shafts, gearboxes, and other mechanical wind turbine components, it is usually represented by two-mass model. The generator rotor shaft is flexibly connected and coupled to the wind wheel through the gearbox, the two-mass model is as follows (Formula (7.189–7.192)):

$$2H_t\frac{d\omega_t}{dt} = T_m - T_{sh} \tag{7.189}$$

$$\frac{d\theta_{tw}}{dt} = \omega_t - \omega_r = \omega_t - (1-s_r)\omega_s \tag{7.190}$$

$$2H_g\frac{ds_r}{dt} = -T_{em} - T_{sh} \tag{7.191}$$

$$T_{sh} = K_{sh}\theta_{tw} + D_{sh}\frac{d\theta_{tw}}{dt} \tag{7.192}$$

where H_t H_g and respectively represents the inertia constant of wind turbine and generator; ω_t is the wind turbine angular speed; ω_r is the generator rotor angular speed; θ_{tw} is the shaft torsion angle; K_{sh} is the shaft coefficient of rigidity; D_{sh} is the coefficient of damping; T_{sh} is the shaft torque; T_m is the wind wheel torque, and the wind turbine power output; T_{em} is the electromagnetic torque. T_m and T_{em} is respectively given by Formula (7.193) and Formula (7.194):

$$T_m = \frac{1}{2}\rho\pi R^2 C_p v_w^3/\omega_t \tag{7.193}$$

$$T_{em} = P_s/\omega_s \tag{7.194}$$

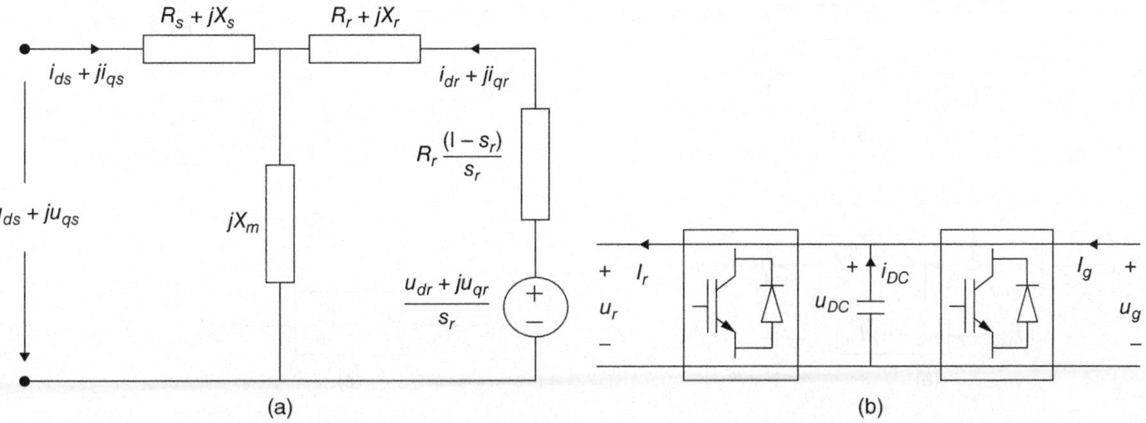

Figure 7.56 DFAG system equivalent circuit. (a) DFAG equivalent circuit; (b) feedback converter equivalent circuit.

where (Formula (7.195)):

$$C_p = \frac{1}{2}\left(\frac{RC_f}{\lambda} - 0.022\beta - 2\right)e^{-0.255(RC_f/\lambda)} \tag{7.195}$$

where: ρ is the air density; R is the blade radius; v_w is the wind speed; β is the blade pitch angle; C_p is the wind turbine rotor power coefficient, and according to Betz theory, the maximum of C_p is 0.593. The largest value of C_p can be obtained by controlling wind speed to follow the maximum wind power. Λ is the wind turbine tip speed ratio; C_f is the blade design constant. P_s is the stator active power.

3) **Converter and DC capacitor model**

The converter structure is shown in Figure 7.56b, the active power passes through the converter is balanced, if taking the current direction in Figure 7.56, the power balance formula will be (Formula (7.196)):

$$P_r = P_G + P_{DC} \tag{7.196}$$

where P_r is the AC terminal active power of rotor side converter; P_G is the AC terminal active power of grid-side converter; P_{DC} is the active power of DC link. They are calculated separately as (Formula (7.197–7.199)):

$$P_r = u_{dr}i_{dr} + u_{qr}i_{qr} \tag{7.197}$$

$$P_G = u_{dg}i_{dg} + u_{qg}i_{qg} \tag{7.198}$$

$$P_{DC} = u_{DC}i_{DC} = -Cu_{DC}\frac{du_{DC}}{dt} \tag{7.199}$$

where i_{dr} and i_{qr} respectively represents d-axis and q-axis rotor current; i_{dg} and i_{qg}, respectively, represents d-axis and q-axis current of grid-side converter; u_{dg} and u_{qg} respectively represents d-axis and q-axis voltage of grid-side converter; u_{DC} is the capacitor DC voltage; i_{DC} is the capacitor current; C is the capacitance of capacitor. When you substitute Formulas (7.197) to (7.199) into (7.196), converters and DC link model can be derived as follows (Formula (7.200)):

$$Cu_{DC}\frac{du_{DC}}{dt} = u_{dg}i_{dg} + u_{qg}i_{qg} - (u_{dr}i_{dr} + u_{qr}i_{qr}) \tag{7.200}$$

4) **Rotor-side converter controller model**

Carry out decoupling control on active and reactive power of the DFAG wind turbine system, take d-axis direction of the d–q coordinate as the stator voltage direction, $u_{qs} = 0$, and u_{ds} is equal to the amplitude of the terminal voltage, so the grid-side active and reactive power can be respectively controlled by i_{dg} and i_{qg}. The stator active power and voltage can also be independently controlled by using decoupling control technology.

The objective of rotor-side converter is to control the active power output of the DFAG to follow the wind turbine torque input, and maintain the generator terminal voltage to the set value. Active power and voltage are independently controlled by u_{qr} and u_{dr}. The control block diagram is shown in Figure 7.57a, and the control formulas are as follows (Formulas (7.201–7.208)):

$$\frac{dx_1}{dt} = P_{ref} + P_s \tag{7.201}$$

$$i_{qr,ref} = K_{p1}(P_{ref} + P_s) + K_{i1}x_1 \tag{7.202}$$

$$\frac{dx_2}{dt} = i_{qr,ref} - i_{qr} = K_{P1}(P_{ref} + P_s) + K_{i1}x_1 - i_{qr} \tag{7.203}$$

$$\frac{dx_3}{dt} = u_{s,ref} - u_s \tag{7.204}$$

$$i_{dr,ref} = K_{p3}(u_{s,ref} - u_s) + K_{i3}x_3 \tag{7.205}$$

$$\frac{dx_4}{dt} = i_{dr,ref} - i_{dr} = K_{P3}(u_{s,ref} - u_s) + K_{i3}x_3 - i_{dr} \tag{7.206}$$

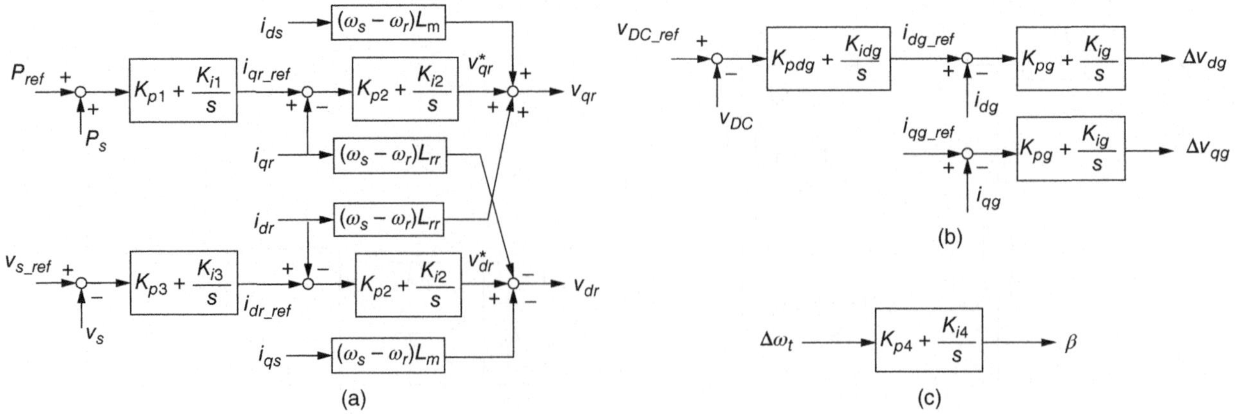

Figure 7.57 DFAG control block diagram. (a) rotor-side converter control block diagram; (b) grid-side converter control block diagram; (c) pitch control block diagram.

$$u_{qr} = K_{p2}(K_{p1}\Delta P + K_{i1}x_1 - i_{qr}) + K_{i2}x_2 + s_r L_m \omega_s i_{ds} + s_r L_{rr}\omega_s i_{qr} \quad (7.207)$$

$$u_{dr} = K_{p2}(K_{p3}\Delta u + K_{i3}x_3 - i_{dr}) + K_{i2}x_4 - s_r L_m \omega_s i_{qs} - s_r L_{rr}\omega_s i_{dr} \quad (7.208)$$

where x_1, x_2, x_3, and x_4 are the introduced intermediate variables; K_{p1} and K_{i1}, respectively, represents the Prop Gain and Integ Gain of the power regulator; K_{p2} and K_{i2}, respectively, represents the Prop Gain and Integ Gain of the rotor-side converter current regulator; K_{p3} and K_{i3}, respectively, represents the Prop Gain and Integ Gain of the grid-side voltage regulator; $i_{\mathrm{dr,ref}}$ and $i_{\mathrm{qr,ref}}$, respectively, represents d-axis and q-axis current control reference value of the generator-side converter; $u_{\mathrm{s,ref}}$ is the predetermined terminal voltage reference value; Pref is active power reference value, of which the value is given by Formula (7.209).

$$P_{ref} = P_B \frac{\omega_t}{\omega_{tB}} \quad (7.209)$$

where ω_{tB} is the wind turbine speed base value; P_{B} is the maximum active power output at ω_{t}.

5) **Grid-side converter controller**
The objective of the grid-side converter controller, as shown in Figure 7.57b is to maintain the DC voltage and the terminal voltage of the wind turbine system. DC voltage is controlled by i_{dg}, and reactive power is controlled by i_{qg}. By introducing intermediate variables x_4, x_5, and x_6, the control formula of the controller shown in Figure 7.57b can be obtained as follows (Formulas (7.210–7.215)):

$$\frac{dx_5}{dt} = u_{DC,ref} - u_{DC} \quad (7.210)$$

$$i_{Dg,ref} = -K_{pdg}\Delta u_{DC} + K_{idg}x_5 \quad (7.211)$$

$$\frac{dx_6}{dt} = i_{dg,ref} - i_{dg} = -K_{Pdg}\Delta u_{DC} + K_{idg}x_5 - i_{dg} \quad (7.212)$$

$$\frac{dx_7}{dt} = i_{qg,ref} - i_{qg} \quad (7.213)$$

$$\Delta u_{dg} = K_{pg}\frac{dx_6}{dt} + K_{ig}x_6 = K_{Pg}(-K_{Pdg}\Delta u_{DC} + K_{idg}x_5 - i_{dg}) + K_{ig}x_6 \quad (7.214)$$

$$\Delta u_{qg} = K_{pg}\frac{dx_7}{dt} + K_{ig}x_7 = K_{Pg}(i_{qg,ref} - i_{qg}) + K_{ig}x_7 \quad (7.215)$$

where K_{pdg} and K_{idg}, respectively, represents the Prop Gain and Integ Gain of the DC bus voltage regulator; K_{pg} and K_{ig}, respectively, represents the Prop Gain and Integ Gain of the grid-side converter current regulator; $u_{\mathrm{DC,ref}}$ is the DC voltage control reference value; $i_{\mathrm{qg,ref}}$ is the control reference value of the grid-side converter current q-axis component.

6) **Pitch controller**
Blade pitch control can maintain optimum speed of the wind turbine, of which the control block diagram is shown in Figure 7.57c, and the control formula is given by Formula (7.216).

$$\frac{d\beta}{dt} = K_{P4}\frac{T_m - T_{sh}}{2H_t} + K_{i4}\Delta\omega_t \quad (7.216)$$

where K_{P4} and K_{i4} respectively represents the Prop Gain and Integ Gain of the wind turbine speed regulator; $\Delta\omega_{\mathrm{t}}$ is the wind turbine speed error.

7) **DFAG wind turbine system dynamic model**
Formula (7.185) to (7.191) and Formula (7.200) represent the dynamic models of wind turbine systems with DFAG, wherein Formula (7.192) and (7.193) describe the stator current dynamic, which can be ignored in electromechanical transient stability analysis, so long as the differential on the left of the formula is made to zero. Formula (7.201), (7.203), (7.204), (7.206), (7.208), (7.210), and (7.212) to (7.216) form the controller model. Dynamic model of wind turbine system with DFAG can be written in a compact form as (Formulas (7.217) and (7.218)):

$$\dot{x} = f(x, z, u) \quad (7.217)$$

$$z = g(x, u) \quad (7.218)$$

where x, z, and u respectively represents the state variable, control output variable, and input variable of the DFAG. They can be expressed as:
$\boldsymbol{x} = [\omega_{\mathrm{t}}, \beta, \theta_{\mathrm{tw}}, \mathrm{s_r}, i_{\mathrm{ds}}, i_{\mathrm{qs}}, E'_{\mathrm{d}}, E'_{\mathrm{q}}, x_1, x_2, x_3, x_4, u_{\mathrm{DC}}, x_5, x_6, x_7]^{\mathrm{T}}$, $\mathbf{z} = [u_{\mathrm{dr}}, u_{\mathrm{qr}}, u_{\mathrm{dg}}, u_{\mathrm{qg}}]^{\mathrm{T}}$, $\boldsymbol{u} = [u_{\mathrm{ds}}, u_{\mathrm{qs}}, i_{\mathrm{dg}}, i_{\mathrm{qg}}]^{\mathrm{T}}$.

8) **Interface of DFAG and the grid**
Take single machine infinite bus system (SMIB) shown in Figure 7.58a, for example, the voltage formula can be obtained as (Formula (7.219)):

$$U_s\angle\varphi - U\angle 0 = -jX_{TL}(I_s + I_g) \quad (7.219)$$

where U is infinite bus voltage; $U_s\angle\varphi$ is the DFAG stator winding terminal voltage; X_{TL} is the transformer and transmission line reactance $X_{\mathrm{TL}} = X_{\mathrm{T}} + X_{\mathrm{L}}$. Network voltage Formula (7.219) and wind turbine model is respectively represented in the x–y and d–q coordinate systems. The relationship between the two

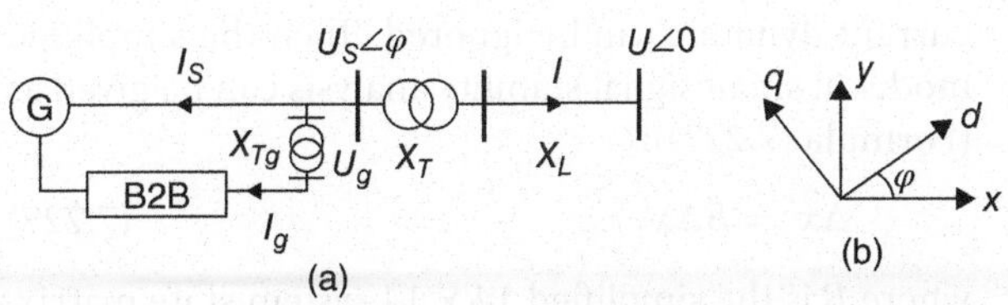

Figure 7.58 DFAG and grid interface. (a) SMIB system; (b) d–q coordinate to x–y coordinate.

coordinate systems is shown in Figure 7.58b, the conversion formula between the two coordinates is (Formula (7.220)):

$$\begin{bmatrix} f_d \\ f_q \end{bmatrix} = \begin{bmatrix} \cos\varphi & \sin\varphi \\ -\sin\varphi & \cos\varphi \end{bmatrix} \begin{bmatrix} f_x \\ f_y \end{bmatrix} \tag{7.220}$$

where f_d and f_q respectively represents d- and q-axis component; f_x and f_y respectively represents x and y-axis component. Used Formula (7.220) in Formula (7.219), the network voltage formula can be written as (Formula (7.221)):

$$\begin{bmatrix} u_{ds} \\ u_{qs} \end{bmatrix} = \begin{bmatrix} \cos\varphi \\ -\sin\varphi \end{bmatrix} U - \begin{bmatrix} 0 & -X_{TL} \\ X_{TL} & 0 \end{bmatrix} \times \left(\begin{bmatrix} \Delta i_{ds} \\ \Delta i_{qs} \end{bmatrix} + \begin{bmatrix} \Delta i_{dg} \\ \Delta i_{qg} \end{bmatrix} \right) \tag{7.221}$$

The voltage formula of grid-side converter is (Formula (7.222)):

$$\begin{bmatrix} u_{ds} \\ u_{qs} \end{bmatrix} - \begin{bmatrix} u_{dg} \\ u_{qg} \end{bmatrix} = \begin{bmatrix} 0 & -X_{Tg} \\ X_{Tg} & 0 \end{bmatrix} \begin{bmatrix} i_{dg} \\ i_{qg} \end{bmatrix} \tag{7.222}$$

X_{Tg} is the grid-side transformer reactance. Linearize Formula (7.221) and (7.222), we can get (Formulas (7.223) and (7.224)):

$$\begin{bmatrix} \Delta u_{ds} \\ \Delta u_{qs} \end{bmatrix} = \begin{bmatrix} 0 & X_{TL} \\ -X_{TL} & 0 \end{bmatrix} \left(\begin{bmatrix} \Delta i_{ds} \\ \Delta i_{qs} \end{bmatrix} + \begin{bmatrix} \Delta i_{dg} \\ \Delta i_{qg} \end{bmatrix} \right) \tag{7.223}$$

$$\begin{bmatrix} \Delta u_{ds} \\ \Delta u_{qs} \end{bmatrix} - \begin{bmatrix} \Delta u_{dg} \\ \Delta u_{qg} \end{bmatrix} = \begin{bmatrix} 0 & -X_{Tg} \\ X_{Tg} & 0 \end{bmatrix} \begin{bmatrix} \Delta i_{qg} \\ \Delta i_{qg} \end{bmatrix} \tag{7.224}$$

Combine Formula (7.223) and Formula (7.224), we can get (Formula (7.225)):

$$\Delta u = K_1 \Delta x + K_2 \Delta z \tag{7.225}$$

9) **DFAG small signal stability analysis model**
Linearize Formula (7.217) and (7.218), substitute Formula (7.218) and (7.225) into Formula (7.217), we can get a small signal stability analysis model with double-fed asynchronous generators (Formula (7.226)):

$$\Delta \dot{x} = A \Delta x \tag{7.226}$$

where A is the 16 × 16 system state matrix. In the small signal stability analysis, by setting the stator current differential in Formula (7.226) to zero, the stator current dynamic can be ignored. Then the simplified model of small signal stability analysis can be given as (Formula (7.227)):

$$\Delta \dot{x}' = B \Delta x' \tag{7.227}$$

where B is the simplified 14 × 14 system state matrix, x′ is the simplified state vector $x' = [\omega_t, \beta, \theta_{tw}, s_r, E'_d, E'_q, x_1, x_2, x_3, x_4, u_{DC}, x_5, x_6, x_7]^T$

7.7.2.4 Small Signal Analysis Model of Direct-Drive Permanent Magnet Synchronous Generator System

In establishing direct drive-permanent magnet **synchronous** generator (DDPMSG) system small signal stability model, in addition to the dynamic formulas of the synchronous generator, the drive chain model of the wind turbine, transformer model and converter controller model shall also be considered.

1) **DDPMSG model**
Basically, DDPMSG is a synchronous generator with constant rotor flux. Its equivalent circuit is shown in Figure 7.59. Its full power to back converter is shown in Figure 7.60.
Select the flux to align with d direction of d−q coordinate, DDPMSG model can be expressed as (Formulas (7.228) and (7.229)):

$$\begin{cases} L_s \dfrac{di_{ds}}{dt} = -u_{ds} - R_s i_{ds} + L_s \omega i_{qs} \\ L_s \dfrac{di_{qs}}{dt} = -u_{qs} - R_s i_{qs} - L_s \omega i_{ds} + \omega \psi \end{cases} \tag{7.228}$$

$$\begin{cases} P_s = u_{ds} i_{ds} + u_{qs} i_{qs} = u_{ds} i_{ds} \\ Q_s = u_{qs} i_{ds} - u_{ds} i_{qs} = -u_{ds} i_{qs} \end{cases} \tag{7.229}$$

where u_{ds}, u_{qs}, i_{ds}, and i_{qs} respectively represents stator voltage, current, and dq-axis component; R_s is the stator resistance; ω is the generator electrical angular speed; L_s is the stator winding self-inductance; ψ is the flux of the permanent magnet; P_s and Q_s, respectively, represents the active and reactive power output by the generator.

2) **DDPMSG drive chain model**
Because in normal state, the wind turbine with DDPMSG decouples with grid through the full power back to back converter, grid transient is invisible on DDPMSG-side and the axis system, so for the wind

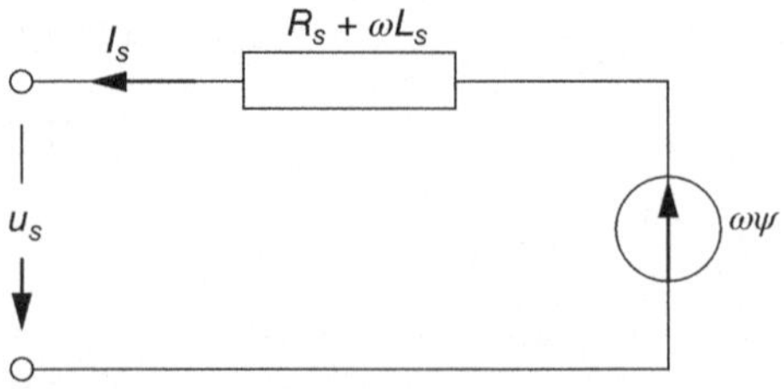

Figure 7.59 DDDPMSG equivalent circuit.

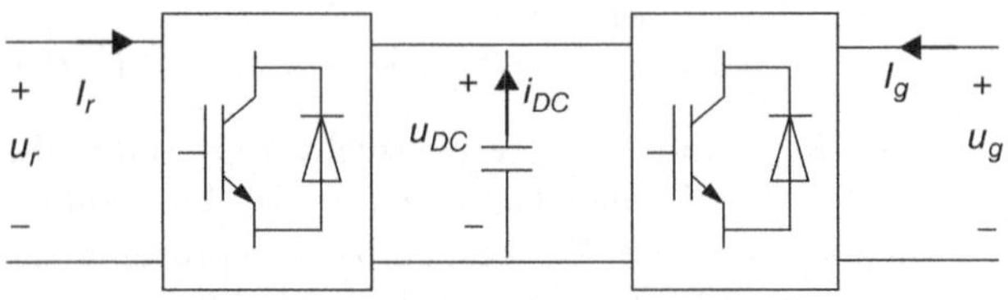

Figure 7.60 Full power back to back converter.

turbine with DDPMSG integrated into the grid, its mechanical system can be simply simulated by using a single-mass model, as (Formula (7.230)):

$$H_{\text{tot}}\frac{d\omega_{\text{t}}}{dt} = T_{\text{m}} - T_{\text{e}} \tag{7.230}$$

where $H_{tot} = H_t + H_g$, H_{tot} are the inertia constants of the entire drive chain; H_t and H_g respectively represents the inertia constant of the wind turbine and the generator; $\omega_t = \omega/p$ and ω_t are wind turbine speeds; p is the number of pole pairs; T_{m} is the wind turbine mechanical torque; T_{e} is the generator electromagnetic torque. T_{m} is the same as Formula (7.186), and T_{e} is given by Formula (7.231):

$$T_e = -p\psi i_{qs} \tag{7.231}$$

3) **Converter model**
DDPMSG wind turbine interfaces with the grid via the full power back to back converter, of which the configuration is shown in Figure 7.64. Because the active power passes through the converter is balanced by a DC link, with the current positive direction predetermined in Figure 7.64, the power balance formula can be expressed as (Formula (7.232)):

$$0 = P_{DC} + P_g + P_s \tag{7.232}$$

where P_{g} is the active power at the grid-side converter AC terminal; P_{DC} is the DC power; they are respectively the same as Formula (7.198) and (7.199) mentioned above.
Because the AC system of the generator and grid is decoupled by the DC link, they can operate at different frequencies. Thus, d–q and D–Q coordinates are respectively used in the generator and grid models. Converter and DC link model can be derived as follows (Formula (7.233)):

$$Cu_{DC}\frac{du_{DC}}{dt} = u_{Dg}i_{Dg} + u_{Qg}i_{Qg} + u_{ds}i_{ds} + u_{qs}i_{qs} \tag{7.233}$$

where i_{Dg} and i_{Qg} respectively represents the grid-side converter D and Q-axis current; u_{Dg} and u_{Qg} respectively represents the D- and Q-axis current of grid-side converter; u_{DC} is the capacitor terminal voltage; $C$is the capacitance of the capacitor.

4) **Generator-side converter control model**
The objective of generator-side converter is to control the generator active power output to follow the wind turbine torque input, and minimize the generator power losses. Active power is controlled by u_{qs}. The generator power losses are minimized by controlling $i_{\text{ds}} = 0$. Control is achieved through u_{ds}. The control block diagram is shown in Figure 7.61, and its control formula is as follows (Formulas (7.234–7.237)):

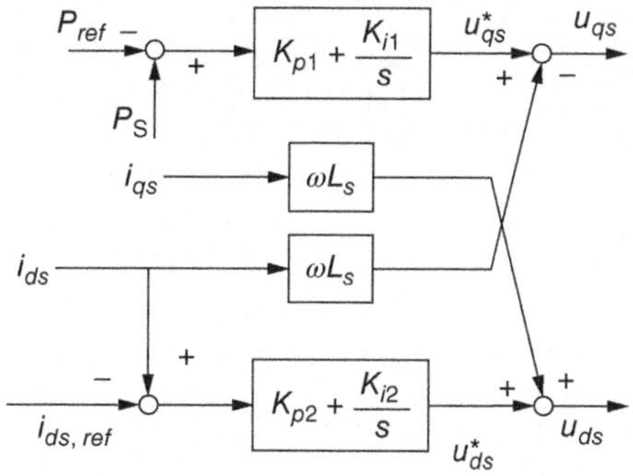

Figure 7.61 Generator-side converter control block diagram.

$$\frac{dx_1}{dt} = P_s - P_{ref} \tag{7.234}$$

$$\frac{dx_2}{dt} = i_{ds} - i_{ds,ref} \tag{7.235}$$

$$u_{qs} = K_{p1}\Delta P + K_{i1}x_1 - L_s\omega i_{ds} + \omega\psi \tag{7.236}$$

$$u_{ds} = K_{p2}\Delta i_{ds} + K_{i2}x_2 + L_s\omega i_{qs} \tag{7.237}$$

where x_1 and x_2 are intermediate variables; K_{p1} and K_{i1}respectively represents the Prop Gain and Integ Gain of the power regulator; K_{p2} and K_{i2}, respectively, represents the Prop Gain and Integ Gain of the d-axis stator current regulator; $i_{ds,ref}$ is the predetermined d-axis stator current reference value; P_{ref} is the active control reference value.

5) **Grid-side converter controller**
The objective of the grid-side converter controller, as shown in Figure 7.62, is to maintain the DC voltage and the terminal voltage of the wind turbine system. Align the direct axis (D) of the D-Q reference axis with the wind turbine system terminal voltage, the DC voltage and the terminal voltage of the wind turbine system can be respectively controlled by u_{Dg} and u_{Qg}. By introducing intermediate variables x_3, x_4, x_5, and x_6, the control formula of the controller shown in Figure 7.62 can be obtained as follows (Formulas (7.238–7.243)):

$$\frac{dx_3}{dt} = u_{DC,ref} - u_{DC} \tag{7.238}$$

$$\frac{dx_4}{dt} = i_{Dg,ref} - i_{Dg} = -K_{p3}\Delta u_{DC} + K_{i3}x_3 - i_{Dg} \tag{7.239}$$

$$\frac{dx_5}{dt} = u_{t,ref} - u_t \tag{7.240}$$

$$\frac{dx_6}{dt} = i_{Qg,ref} - i_{Qg} = -K_{p4}\Delta u_t + K_{i4}x_5 - i_{Qg} \tag{7.241}$$

$$\begin{aligned} u_{Dg} &= K_{p5}\frac{dx_4}{dt} + K_{i5}x_4 \\ &= K_{p5}(-K_{p3}\Delta u_{DC} + K_{i3}x_3 - i_{Dg}) \\ &\quad + K_{i5}x_5 + X_C i_{Qg} \end{aligned} \tag{7.242}$$

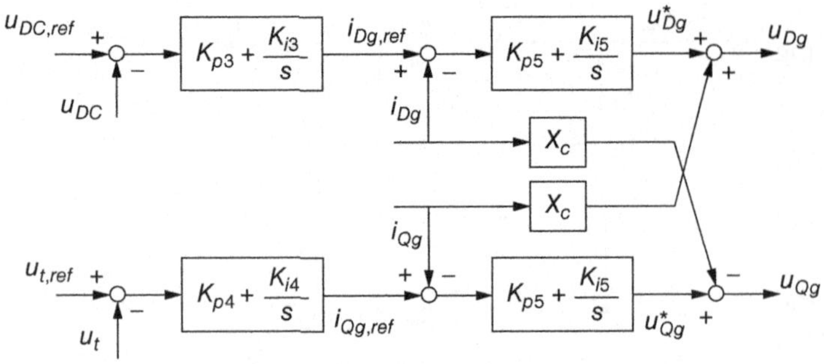

Figure 7.62 Grid-side converter control block diagram.

$$\begin{aligned} u_{Qg} &= K_{p5}\frac{dx_6}{dt} + K_{i5}x_6 \\ &= K_{p5}(-K_{p4}\Delta u_t + K_{i4}x_5 - i_{Qg}) \\ &\quad + K_{i5}x_6 - X_C i_{Dg} \end{aligned} \tag{7.243}$$

where K_{p3} and K_{i3} respectively represents the Prop Gain and Integ Gain of the DC bus voltage regulator; K_{p4} and K_{i4}, respectively, represents the Prop Gain and Integ Gain of the terminal voltage regulator; K_{p5} and K_{i5}, respectively, represents the Prop Gain and Integ Gain of the grid-side converter current regulator; $u_{\mathrm{DC,ref}}$ is the DC voltage control reference value; u_t is the amplitude of the wind turbine system terminal voltage $u_t = \sqrt{u_{Dl}^2 + u_{Ql}^2}$, u_{Dl} and u_{Ql}, respectively, represents D- and Q-axis component; $u_{t,ref}$ is the control reference value of the wind turbine system terminal voltage; X_c is the filter reactance of the grid-side converter.

6) **Dynamic model of DDPMSG wind turbine system**
Formula (7.228), (7.230), and (7.233) represent a dynamic model of the wind turbine system with DDPMSG, wherein Formula (7.228) and (7.233) describe the electrical dynamics, and Formula (7.230) describes the mechanical dynamics. Formulas (7.234) to (7.243) describe the control dynamics. Dynamic model of the wind turbine system with DDPMSG can be written in a compact form as (Formulas (7.244) and (7.245)):

$$\dot{x} = f(x, z, u) \tag{7.244}$$

$$z = g(x, u) \tag{7.245}$$

where x, z, u, respectively, represents the state variable, control output variable, and input variable of DDPMSG. They can be expressed as:
$\boldsymbol{x} = [\omega_t, i_{\mathrm{ds}}, i_{\mathrm{qs}}, u_{\mathrm{DC}}, x_1, x_2, x_3, x_4, x_5, x_6]^{\mathrm{T}}$, $\mathbf{z} = [u_{\mathrm{ds}}, u_{\mathrm{qs}}, u_{\mathrm{Dg}}, u_{\mathrm{Qg}}]^{\mathrm{T}}$, $\boldsymbol{u} = [u_{\mathrm{Dl}}, u_{\mathrm{Ql}}, i_{\mathrm{Dg}}, i_{\mathrm{Qg}}]^{\mathrm{T}}$.

7) **Analysis model for DDPSMG small signal stability analysis**
Interface of the wind turbine with DDPMSG and the grid is shown in Figure 7.63.
Taking the single machine infinite bus system shown in Figure 7.63 as an example, the network voltage formula can be written as (Formula (7.246)):

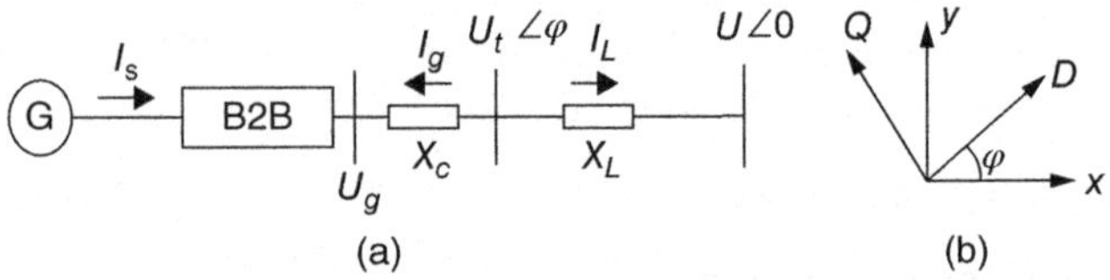

Figure 7.63 DDPMG and grid interface.

$$U_t\angle\varphi - U\angle 0 = jX_L I_l = -jX_L I_g \tag{7.246}$$

where U is the infinite bus voltage; $U_t \angle \varphi$ is the terminal voltage of wind turbine system; X_L is the transmission line reactance. The network voltage Formula (7.246) and the wind turbine system grid-side converter model is respectively represented in the x–y and D–Q coordinate system. The relationship between the two coordinate systems is shown in Figure 7.63b, the conversion formula between the two coordinates is as Formula (7.247).

$$\begin{bmatrix} f_D \\ f_Q \end{bmatrix} = \begin{bmatrix} \cos\varphi & \sin\varphi \\ -\sin\varphi & \cos\varphi \end{bmatrix} \tag{7.247}$$

where f_D and f_Q, respectively, represents D- and Q-axis component; f_x and f_y, respectively, represents the x- and y-axis component. Applying Formula (7.247) to Formula (7.246), the network voltage formula can be written as (Formula (7.248)):

$$\begin{bmatrix} u_{Dl} \\ u_{Ql} \end{bmatrix} = \begin{bmatrix} \cos\varphi \\ -\sin\varphi \end{bmatrix} U - \begin{bmatrix} 0 & -X_L \\ X_L & 0 \end{bmatrix} \begin{bmatrix} \Delta i_{Dg} \\ \Delta i_{Qg} \end{bmatrix} \tag{7.248}$$

The grid-side converter voltage formula is (Formula (7.249)):

$$\begin{bmatrix} u_{Dl} \\ u_{Ql} \end{bmatrix} - \begin{bmatrix} u_{Dg} \\ u_{Qg} \end{bmatrix} = \begin{bmatrix} 0 & -X_C \\ X_C & 0 \end{bmatrix} \begin{bmatrix} \Delta i_{Dg} \\ \Delta i_{Qg} \end{bmatrix} \tag{7.249}$$

Linearize Formula (7.248) and (7.249), and get (Formulas (7.250) and (7.251)):

$$\begin{bmatrix} \Delta u_{Dl} \\ \Delta u_{Ql} \end{bmatrix} = \begin{bmatrix} 0 & X_L \\ -X_L & 0 \end{bmatrix} \begin{bmatrix} \Delta i_{Dg} \\ \Delta i_{Qg} \end{bmatrix} \tag{7.250}$$

$$\begin{bmatrix} \Delta u_{Dl} \\ \Delta u_{Ql} \end{bmatrix} - \begin{bmatrix} \Delta u_{Dg} \\ \Delta u_{Qg} \end{bmatrix} = \begin{bmatrix} 0 & X_C \\ -X_C & 0 \end{bmatrix} \begin{bmatrix} \Delta i_{Dg} \\ \Delta i_{Qg} \end{bmatrix} \tag{7.251}$$

Combine Formula (7.250) and (7.251), and we can get (Formula (7.252)):

$$\Delta u = K_1 \Delta x + K_2 \Delta z \tag{7.252}$$

Linearize Formula (7.244) and (7.245), and we can get (Formulas (7.253) and (7.254)):

$$\Delta \dot{x} = A' \Delta x + B' \Delta z + C' H u \tag{7.253}$$

$$\Delta z = D' \Delta x + E' \Delta u \tag{7.254}$$

Substitute Formula (7.252) and (7.254) into (7.253), we can get a small signal stability analysis model for DDPMG (Formula (7.255)):

$$\Delta \dot{x} = A \Delta x \tag{7.255}$$

where A is 10 × 10 system state matrix $\Delta x = [\Delta\omega_t, \Delta i_{ds}, \Delta i_{qs}, \Delta u_{DC}, \Delta x_1, \Delta x_2, \Delta x_3, \Delta x_4, \Delta x_5, \Delta x_6]^T$.

7.7.3 Impacts of WFs with Different Wind Turbine Units on Power System Small Signal Stability

Power system oscillation is caused by the working principle of synchronous generators, coupling between the synchronous generator rotor speed and electrical torque forms a second-order weak damping system. However, this concept does not apply to the types of generators generally used for wind turbines. Because the wind turbine is an asynchronous machine or is decoupled from the grid via the power electronic converter, and wind turbine itself will not cause new oscillation modes of the system. Cage asynchronous generators use constant speed wind turbines, so in substance, they have better damping. Generators of variable speed wind turbines are decoupled from the grid via power electronic converters, which control the power output, to avoid power system oscillations. In addition, after replacing conventional generators with wind power, the number of the rest of the synchronous generator is reduced, and system impedance is not changed, so decoupling between synchronous generators may become relatively stronger. But when large synchronous generation capacity is replaced by wind power, the oscillation mode of the system (the oscillation type and the oscillation mode) will be changed greatly, and the damping situation will also became complicated.

7.7.3.1 Impact of WFs with Cage Asynchronous Generators on Power System Small Signal Stability

The cage asynchronous generator used by the constant speed wind turbines represents the relationship between the rotor slip (i.e., rotor speed) and the electrical torque, rather than the rotor angle and the electrical torque, so the mechanical part is first-order, and will not render oscillation characteristics, which is the opposite of synchronous generator. While rotor transient is included in the model, an oscillation mode can be found, which is an increase of the order of model. The amplitude of the oscillation mode is small, with good damping. Thus, essentially, cage asynchronous generators have better damping, and will not cause power system oscillations.

Mechanical disturbances, such as changes in wind speed, and other electrical disturbances of grid faults may excite torsional dynamic of constant frequency wind turbine systems, which due to direct grid-connection of cage asynchronous generators, may be transmitted between wind turbines, and is a specific issue of constant frequency wind systems.

References [90, 91] draws the following conclusions through small signal stability characteristic analysis of large capacity and high proportion WFs integrated into actual large grids:

1) When cage asynchronous wind turbines are connected to the grid, oscillation modes in strong correlation with the WF are added in the system. These oscillation modes have better damping characteristics;
2) Due to randomness of changes in WF output, conventional units in the grid maintain the grid power balance by reducing output or shutdown, which would result in increase or decrease of the number of local oscillation modes;
3) Integration of large-capacity and high-proportion WFs into the system does not change oscillation modes between large generator units, but the oscillation characteristics will change. Within a certain range, this impact is limited. Only at high levels of wind power penetration, and when the system power flow distribution is changed due to large WF output, the adverse changes in system oscillation mode damping will appear;
4) With the increase in the proportion of wind power, outgoing power increases, the damping characteristics of the external power grid may make a downward trend;
5) Studies show that the internal power oscillation mode feature is related to the generator unit dispatching mode. If the operation mode that increases the system spinning reserve by reducing output of generator units in the grid is used, its damping characteristics will be increased; and if the operation mode of disabling generators is applied, its damping characteristics will be decreased;
6) The introduction of pitch angle control of the input signal of system frequency deviation, can enhance system damping, reduce low frequency oscillation amplitude, shorten the time of low-frequency oscillation, and improve the system dynamic stability.

Reference [96] introduces the frequency response method to establish small-signal dynamic model for grid-connected WFs consist of cage asynchronous wind turbines. This model represents the capture mechanism of wind energy, the rotating shaft, the asynchronous generator circuit, the collection system of WFs, and the public grid. In the frequency range of a few tenths Hz to 50/60 Hz, according to the model established, we introduce two norm-based performance indexes: PIr and PId, which can be used to test the sensitivity of system performance to parameters and assess the ability of the system to withstand electrical/mechanical disturbances. The model and performance indexes are used to analyze the torsional dynamics of the constant frequency wind turbine system, of which the results are confirmed by time domain simulation of the system to be studied. The results prove that:

1) Electrical disturbances, such as instantaneous three-phase fault, may excite the wind turbine torsional mode, so the WF reduced-order model established based on the wind turbine shaft represented by single mass may cause incorrect results.
2) Low-frequency mechanical disturbances, such as wind gusts, only excite the inertial mode, not the torsional mode. Thus, for low frequency (about 0.1 Hz) disturbances, it is appropriate to represent the wind turbines with a single mass.
3) Electromechanical dynamics of the wind turbine (collection systems and public networks) are coupled by the electrical network, and therefore, any mechanical or electrical disturbances occuring on a wind turbine may excite oscillation model of other wind turbines. For radial grid structure, the degree of coupling by the electrical network is related to the grid rigidity. The greater the rigidity, the less coupling between the wind turbines.

7.7.3.2 Impact of WFs with Double-Fed Asynchronous Generators on Power System Small Signal Stability

Double-fed asynchronous generators (DFAG) are decoupled from the power system via power electronic converters, by using the converter to control rotor speed and electrical power, any rotor speed oscillation that may occurs can be damped. Therefore, DFAG will not has adverse effects on the rotor speed oscillations of the power system that may occur, because these oscillations do not pass through the converter, and the generator cannot feel them. If we assume that the wind power replaces the power generated by synchronous generators, contribution of synchronous generators to all damping provided by the power will be smaller. However, the system topology does not change, so the impedance of synchronous generators against the network is smaller, which improve the damping of any oscillations that may occur between the synchronous generators in most cases.

Reference [97] uses modal analysis and time domain simulation method to calculate and analyze the impacts

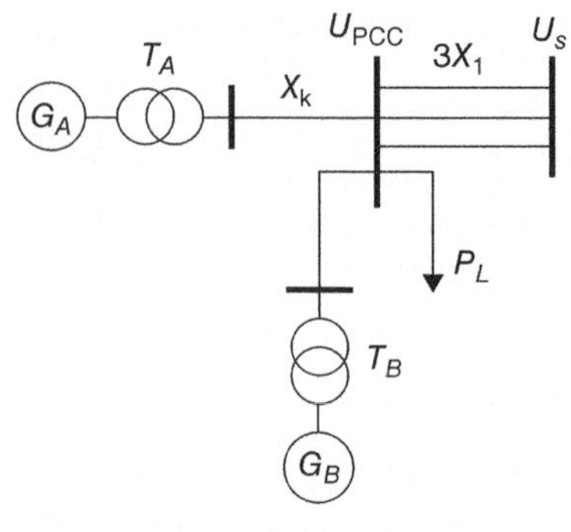

Figure 7.64 The simple system studied.

of DFAG on small and large disturbance stability. The simple system studied is shown in Figure 7.64, generator G_A is a collection of conventional power plants or power plants, and its dynamics are described by the synchronous generator model. Generator G_B is a large WF, its dynamics are described by double-fed asynchronous generator model. Using appropriate models, both of the generators can be represented by the internal EMFX'after the transient reactance E'. Generator G_A is connected to the common bus through short lines (the reactance of each line is X_k), then connected to the larger transmission system via three long lines (the reactance of each line is $3X_l$). The transmission system is represented as an infinite bus, of which the voltage is$_{Us}$. The load is connected to the common bus (PCC). To examine the effect of the presence of WFs on G_A dynamics, it is necessary to know the behaviors of DFAG in both cases with and without G_A. For comparison, a synchronous generator can be used to replace G_B, of which the rated capacity is equal to the rated capacity of the WF, but its dynamics are described with the synchronous generator model.

In small disturbances, the state before disturbance is the same as that after disturbance. Calculating the four operating conditions of different combinations of controllable or uncontrollable G_A and G_B as synchronous generators with DFAG, the system eigenvalues obtained show that when G_B is DFAG instead of synchronous generators, the damping ratio ζ is maximal. However, although the damping ratio increases when containing DFAG, some eigenvalues are closer to the imaginary axis.

In large disturbances, such as a line connected to the infinite bus is cut off, the balance points are different before and after the disturbance, eigenvalue analysis is insufficient to describe the system behavior after disturbance, which needs to be tested by dynamic simulation. If G_A and G_B are the conventional synchronous generators, the system can maintain stability. If G_B is replaced with DFAG, because large disturbances cause DFAG to absorb additional reactive power (system faults, the rotor side converter may lose the ability to control or out of operation, DFAG become cage asynchronous generators, at low voltage, asynchronous generators absorb more reactive from the system), the voltage is instable and cannot be restored. If the sync generator is controllable, the system still able to maintain stability. If the sync generator is uncontrollable, the angle will be unstable.

Participation factor analysis shows that, the state variables δ_B, E'_A and E'_B will affect the high-frequency mode, but these high-frequency eigenvalues have very high damping.

With increasing system load P_L, the track of system eigenvalues moving on the complex plane can be obtained by calculation. When G_B is DFAG, an eigenvalue generated by the system will appear on the real axis, which is very close to the imaginary axis. When PL is large, that is, when the system is overloaded, the eigenvalue will be moved to the right plane. Participation factor analysis shows that, G_B's state variables ω_B has the greatest impact on this eigenvalue, which means that, when the system is overloaded, DFAG adversely affect the small signal stability.

DFAG is very useful for suppressing initial oscillations in G_A caused by system small disturbances, but has less effects on large interferences. That is, when there are WFs near the power system, the system angle stability under small disturbances will be improved, but it may not be conducive to voltage stability in large disturbances. Large disturbances make DFAG absorb more reactive power, and result in voltage unstable and recoverable, leading to instability in the angle.

7.7.3.3 Impacts of WFs with Direct-Drive Permanent Magnet Synchronous Generators on Power System Small Signal Stability

When the WF uses direct drive permanent magnet synchronous generators (DDPSMG), the DDPMSG is connected to the grid via the full-power converter, to achieve a complete decoupling control between the wind turbine and the grid, switching frequency of the converter is much higher than the system oscillation frequency, the direct-drive synchronous generator stator side almost has no response to power oscillations of the power system, and the wind turbine itself, has no damping effect on power oscillations of the system. The absence of any means of control of the system oscillation, may reduce the system damping after DDPSMG WFs are integrated in. In fact, decoupling control can be applied to the permanent magnet direct-drive variable speed wind turbine output power P and Q if DDPMSG generator-side converter uses voltage-oriented vector control, the d-axis of the rotating coordinate system is oriented in the stator voltage d-axis, the electromagnetic power output can be controlled by the d-axis component of stator current. Thus, under normal operation, the control system active power reference value is mainly determined by the wind turbine speed; when power oscillation occurs in the system, the control system active power reference value can be determined by the output of the wind turbine stable control model introduced, which can control the active input reference value to control the output of electromagnetic power output by the wind turbines just by controlling the active power input reference value. If the system frequency signal is used to control the active power reference value, when power oscillation occurs in the system, the system frequency will change with the oscillation and generate active power control reference value of the oscillation in the wind turbine stability control model, thus, to generate the stator current d-axis component that uses the system frequency as its input signal. After stability control model is used by the DDPMSG, no matter how much the WF output is, when power oscillation occurs in the system, the damping power associated with system oscillations will obviously appear in the WF output, and show a trend of mutual offset with the system power oscillations.

Single machine infinite bus system and four-machine system simulations in Reference [95] shows that, wind turbine systems with DDPMSG are stable without a controller, and the dynamic characteristics of the wind turbine system can be improved when using the controller. Dynamic response of the wind turbine system with DDPMSG is simulated based on a single machine infinite bus system and a small system, and is compared with the response of the double-fed asynchronous generator, the simulation results show that, compared with the wind turbine with double-fed asynchronous generators, wind turbines with DDPMSG have better dynamic response, and rapidly decayed oscillations. In disturbances, the DDPMSG wind turbines will generate more reactive power support the terminal voltage. Critical clearing time (CCT) analysis further shows that, both the two wind turbines can improve the power system stability margin. Between these two wind turbines, CCT of the wind turbine with DDPMSG is longer, indicating a better transient stability.

7.8 Frequency Stability Analysis of Power System Containing Wind Turbines

7.8.1 Overview

Frequency stability refers to the ability of the system to maintain stable frequency after severe disturbances in the power system is severely disturbed, and major imbalances between the power generation and the load demand. Frequency stability may be a short-term or long-term phenomenon.

The frequency of the power system is one of the indicators of power quality. In normal operation of the power system, the power supply and the load active power are balanced, and the frequency is within the normal range, the nominal frequency of the grid in China is 50 Hz. Changes in power output and electricity load will cause frequency offset, and therefore, adjustment shall be made at any time according to the frequency deviation.

The amount of unbalance of the power system power is changing, and the change in frequency is a dynamic process. Changes in dynamic response of the system frequency over time at different power shortages can be obtained through simulation. Frequency change not only adversely affects the load operation, but also affects the generators and the power system. When the frequency decreases, mechanical outputs of the forced draft wind machine, induced wind machine, cited wind turbine, feed water pumps, water circulating pumps coal mill, and other equipment for power plant will be reduced due to the decrease of generator speed, resulting in reduction of boiler and turbine output, thus the frequency may continue to decline and create a vicious circle. In particular, when the frequency is reduced to 47 to 48 Hz or less, the above vicious circle will be very intense, which may make the system frequency drops to an unallowable level in a few minutes. This phenomenon is called frequency collapse, of which the consequences will cause large-area blackout, and even the collapse of the entire system. It is the most serious accident in the power system. In addition, the frequency drop will cause increase in the system reactive power, resulting in decrease of system voltage levels.

Frequency deviation caused by power system normal load is generally compensated by the generator power regulating system. In great load changes, the generator speed control system will automatically adjust the active power delivered by the generator, to maintain the system frequency within the range of 50 ± 0.2 Hz, whereas in the case of frequency automatic adjusting device is used, frequency error shall not be more than 0.05 to 0.15 Hz. When the system is in serious power shortages, power growth of the generator speed governing system cannot immediately compensate for these power shortages, so in order to prevent frequency drop and frequency collapse, automatic low-frequency load shedding devices (referred to as UFLS) must be equipped in the system. When the frequency reduces to a certain extent, part of the load is automatically removed by grading of the frequency levels, to make the system frequency restore to 49.5 Hz or above as soon as possible.

Active output of WFs varies with wind speeds, and has randomness and limited controllability. As an unstable power supply of the system, grid-connection and off-grid of wind power are difficult to predict, considering from this point, wind power is actually a source of interference of the system. When wind power capacity shares a larger proportion in the system, the impact of random fluctuation of its output power on grid frequency is significant, which affects the grid power quality and the normal operation of some frequency sensitive loads. This requires that other conventional units in the grid have higher frequency response capabilities, be capable of follow-up control, and suppress the frequency fluctuations. Taking into account the fluctuations in wind power, when the wind power loses its output due to off-blast or wind stall, the grid frequency will be reduced, particularly when proportion of the wind power is large, the system frequency stability will be affected, so to ensure that wind power frequency safety and frequency stability of the power system in continuous operation after wind power is grid connected is also an important issue for wind power research. Only when the power systems are large and closely connected, the frequency issue is not significant.

Steady-state method and time-domain simulation method can be used for frequency stability analysis. Steady-state frequency analysis has nothing to do with the grid structure, and is only related to the load level of the grid, the nature and composition of the power, and it is used for the determination of whether the system frequency deviation under different conditions is within an allowable range. Time-domain simulation can simulate changes in frequency over time caused by a variety of large disturbances in the system. There are a variety of tools available for research on the system frequency stability. PSS/E. Dynamics (PTI), Eurostag (RTE), PSLF (GE), and TSAT (Powertech) are some tools available on many industrial software platforms.

7.8.2 Power System Frequency Characteristic

When the power system is in steady-state operation, the characteristics of system active power varies with frequency is called active-frequency static characteristic, which indicates the frequency variation characteristic in power system power imbalance. The frequency characteristics of the entire power system is the result of integration of power frequency, load frequency characteristics, and voltage influence. When the system frequency changes, the generator output power will change, which is the frequency characteristic of power generation systems or so-called power frequency regulation effect. When the system frequency changes, the power consumption of the load will change, this is the frequency characteristic of the system load or is called load frequency adjustment effect; changes in the system frequency also cause changes in grid voltage, and voltage change in turn causes changes in the generated power and load power [98].

7.8.2.1 Active Power-Frequency Static Characteristics of Load

Among total active loads of the power system, some loads have nothing to do with the frequency change, such as lighting, rectifier equipment, and so on; some loads are proportional to the first power of the frequency, such as the a ball mill, reciprocating pump, and so on; and some load are proportional to the second power of the frequency, such as eddy current loss of the transformer; and some load are proportional to the third power of the frequency, such as the ventilation wind machine, circulating pump, and so on; and some loads are proportional to higher math power of the frequency, such as the water feed pump, and so on.

Relationship between the active load and the frequency of the entire system can be expressed as (Formula (7.256))

$$P_{\mathrm{LD}} = a_0 P_{\mathrm{LDN}} + a_1 P_{\mathrm{LDN}}\left(\frac{f}{f_{\mathrm{N}}}\right) + a_2 P_{\mathrm{LDN}}\left(\frac{f}{f_{\mathrm{N}}}\right)^2 + a_3\left(\frac{f}{f_{\mathrm{N}}}\right)^3 + \cdots \tag{7.256}$$

where P_{LD} is the system active power when the frequency is equal to f; P_{LDN} is the system active load when the frequency is equal to the rated frequency f_N; coefficient a_i ($i = 0,1,2,\ldots$) is the share of the load in P_{LDN}, which is proportional to i power of the frequency, and $a_0 + a_1 + a_2 + \cdots = 1$. Divide both side of the formula by P_{LDN}, the load power-frequency static characteristics represented per-unit value will be obtained (Formula (7.257)):

$$P_{\mathrm{LD}*} = a_0 + a_1 f_* + a_2 f_*^2 + a_3 f_*^3 + \cdots \tag{7.257}$$

When the frequency deviation Δf is not large, the load power-frequency static characteristics can be commonly approximated by a straight line, that is (Formula (7.258))

$$K_{\mathrm{LD}} = \frac{\Delta P_{\mathrm{LD}}}{\Delta f} \tag{7.258}$$

where K_{LD} is called frequency regulation effect coefficient of the load, which is also called unit power regulation of the load, in MW/Hz. Its per-unit value is (Formula (7.259)):

$$K_{\mathrm{LD}*} = \frac{\Delta P_{\mathrm{LD}}/P_{\mathrm{LDN}}}{\Delta f/f_{\mathrm{N}}} = K_{\mathrm{LD}}\frac{f_{\mathrm{N}}}{P_{\mathrm{LDN}}} \tag{7.259}$$

Frequency regulation effect coefficient of the load characterizes the load frequency regulation characteristics, that is, the increased or decreased amount of the power consumption of the load with the increase or decrease of frequency. The size of $K_{\mathrm{LD}*}$ depends on the share of each load in the entire system, $K_{\mathrm{LD}*}$ of different systems at different times may be different. In real systems, $K_{\mathrm{LD}*}$ is 1 to 3. Load frequency regulation effect coefficient is an important data reflecting performance of primary frequency regulation of the system and is a calculation basis for the dispatching department to determine the load-shedding scheme based on the frequency or to determine load shedding in low-frequency accidents to recover the frequency.

7.8.2.2 Active Power-Frequency Static Characteristics of the Generator

When the power system load changes break the balance between the input mechanical power and output electromagnetic power of the generator, the generator speed will change, the governor in the generator speed governing system will be triggered, changing the quantity of air (water) intake of the prime motor, making the input/output power of the generator reach a new balance, to achieve the adjustment of the speed or frequency. Therefore, according to the regulation principle of the generator speed governing system, that is, when the load increases, the generator output will increase, and the frequency will decrease; when the load decreases, the generator output will decrease, and the frequency will increase. The generator active power-frequency static characteristics can be approximated as a broken line, as shown in Figure 7.65.

Line segment 2 to 3 shows in the figure indicates that the generator output has reached its maximum value, and the governor will no longer work. The oblique line in the figure indicates the regulating process of generator frequency, of which the slope is (Formula (7.260)):

$$K_G = -\frac{\Delta P_G}{\Delta f} \tag{7.260}$$

where K_G is called the generator unit power regulation, in MW/Hz. Its per-unit value is (Formula (7.261)):

$$K_{G*} = -\frac{\Delta P_G/P_{GN}}{\Delta f/f_N} = K_G\frac{f_N}{P_{GN}} \tag{7.261}$$

The unit power regulation of generator characterizes the frequency regulation characteristics, that is, the increased or decreased amount of the generator active output with the increase or decrease of frequency.

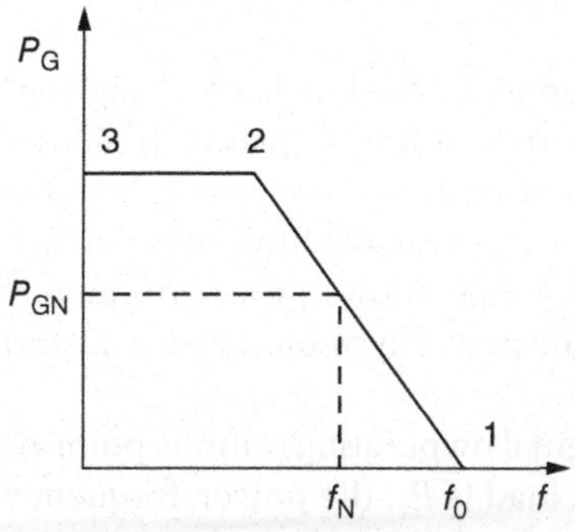

Figure 7.65 The active power-frequency static characteristics of generator.

In addition to the unit power regulation, parameters characterizing the frequency regulation characteristics of system power supply also include the static difference coefficient, which is defined as (Formula (7.262)):

$$\delta_* = \frac{f_0 - f_N}{f_N} \tag{7.262}$$

where f_0 in the formula represents the frequency of the generator in no-load operation. Regulation coefficient indicates the amount of frequency deviation between no-load operation and rated operation, and its relationship with the unit power regulation of generator is as follows:

$$\delta_* = \frac{f_0 - f_N}{f_N} = \frac{f_0 - f_N}{P_{GN} - 0}\frac{P_{GN}}{f_N} = \frac{1}{K_G}\frac{P_{GN}}{f_N} = \frac{1}{K_{G*}}$$

That is (Formula (7.263))

$$K_G = \frac{1}{\delta_*}\frac{P_{GN}}{f_N} \quad \text{and} \quad K_{G*} = \frac{1}{\delta_*} \tag{7.263}$$

In the load and generator unit power regulation derivation formula, power increases and frequency increases are positive, so minus sign in Formula (7.260) indicates the generator output decreasing (increasing) with the increase (decrease) of the frequency, that is, the generator unit power regulation is positive. Similar with K_{LD}, K_G is also an important data reflecting performance of primary frequency regulation of the system, and an important basis of load distribution of parallel units. Different from the frequency regulation effect coefficient of load, it can be set. However, limited by the operation stability of the generator governor mechanism, its setting range is limited. Within the setting range, the greater K_G, the smaller δ_*, the smaller the frequency offset of the adjustment result. Generally, value range of δ_* is: steam turbine generation set = 0.04 to 0.06, water-turbine generator set = 0.02 to 0.04. When the generator set is at full load, limited by the speed-regulating mechanism, the generator set no longer has the ability of frequency regulation, that is, K_G is zero, and δ_* infinite now.

7.8.2.3 Active Power-Frequency Static Characteristic of Power System

In case of frequency changes caused by load changes of the power system, regulation of the frequency is shared by the regulation effects of both the load and generator set, and its adjustment process is called the primary frequency regulation. Figure 7.66 shows power-frequency static characteristics of the power system (take a load and a generator as example).

In Figure 7.66, the original owperating point is point A, that is, when the system load is P_1, the power-frequency static characteristic of the load is $P_{LD}(f)$, and meets with the power-frequency static characteristic of generator $P_G(f)$ in point A, then the system frequency is f_1. When

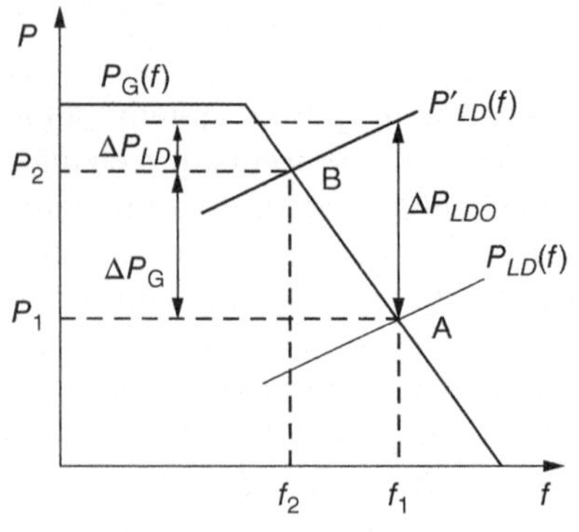

Figure 7.66 Active power-frequency static characteristics of the power system.

the system increases by ΔP_{LD0}, the load power-frequency static characteristic will jump from $P_{LD}(f)$ to $P_{LD}'(f)$. In this case, on the one hand, the input and output power balance of the generator set is broken, the speed changes will cause frequency decreases, trigger the governor, and increase the generator output ΔP_G; on the other hand, due to frequency caused by speed changes, the load demand is decreased according to its own regulating effect ΔP_{LD}. $P_{LD}'(f)$ and $P_G(f)$ meet at point B again, that is, under the combined action of the generator and load regulating effects, the operating point will move from point A to point B, and the operating frequency will fall from f_1 to f_2, then we have (Formulas (7.264) and (7.265)):

$$\Delta P_{LD0} = \Delta P_G - \Delta P_{LD}$$

Also

$$\Delta P_{LD0} = -(K_G + K_{LD})\Delta f \tag{7.264}$$

or

$$\Delta P_{LD0} = -K\Delta f \tag{7.265}$$

where $K = (K_G + K_{LD}) = -\frac{\Delta P_{LD0}}{\Delta f}$ is called unit power regulation of the power system, and its per-unit value is (Formulas (7.266) and (7.2.67)):

$$K_* = -\frac{\Delta P_{LD0}/P_{LDN}}{\Delta f/f_N} = \frac{P_{GN}}{P_{LDN}}K_{G*} + K_{LD*} \tag{7.266}$$

or

$$K_* = K_r K_{G*} + K_{LD*} \tag{7.267}$$

where K_r is called reserve factor, which represents the ratio of generator rated capacity to the total active load of the system at rated frequency. Under normal circumstances, the system always has some spare capacity, so $K_r > 1$.

Unit power regulation of the power system characterizes the frequency regulating characteristic, that is, when the amount of the system frequency increases or decreases, the increase or decrease of system load under combined action of the generator and load regulation effects.

Adjustment of the unit power regulation of system K can only be achieved by regulating the generator unit power regulation K_G. The greater K_G, the larger K, the smaller the frequency change caused by increase or decrease of the load, and the more stable of the system frequency. When the generator is at full power (i.e., $K_G = 0$), $K_* = K_{LD*}$, that is, in increase of the load, its increment can only be borne by its own regulating effect, and the value of K_{LD*} is very small, so the frequency drop caused by load increase is very serious. This requires the generator output should not only meet active power demand of the system at nominal frequency, but also be provided with a certain degree of spare capacity.

7.8.2.4 Frequency Regulation of the Power System

The frequency deviation caused by the first load change will be regulated by the governor of the generator speed governing system, which is called primary frequency regulation; frequency deviation caused by the second load change is regulated by the frequency regulator of generator speed governing system, which is called the secondary frequency regulation; for the frequency deviation caused by the third load, based on active power balance, each generator is instructed to allocate the active power according to the optimal (economic) allocation principle, which is called the tertiary frequency regulation, and is carried out by the energy management system (EMS) of local control center. The following describes the primary and secondary frequency regulations related to the frequency control of wind power.

1) **Primary frequency regulation**

Primary frequency regulation is as the regulating process reflected by the active power-frequency static characteristic of the power system shown in Formula 7.66, that is, the load increment ΔP_{LD0} is balanced by two aspects: the power output increase of the generator regulated by the speed governor and auto reduction of the load power with the frequency decrease. Regulation characteristic formula of the primary regulation is expressed as Formula (7.264) or Formula (7.265).

When there are n units equipped with governors operating in parallel, the unit power regulation of the equivalent unit is (Formula (7.268)):

$$K_{G\Sigma} = \sum_{i=1}^{n} K_{Gi} \tag{7.268}$$

The regulation characteristic formula of primary regulation is changed to (Formula (7.269)):

$$\Delta P_{LD0} = -(K_{G\Sigma} + K_{LD})\Delta f \tag{7.269}$$

or (Formula (7.270))

$$\Delta P_{LD0} = -K\Delta f \tag{7.270}$$

It should be noted that, the n generator units in Formula (7.268) are not at full load, once the units are operating at full load, $K_{Gi} = 0$ shall be taken. Therefore, the more the underloaded units participate in paralleling, the greater the power unit regulation K of the system.

The regulation characteristic formula shows that, for a certain amount of load change ΔP_{LD0}, the greater the system unit power regulation K, the smaller the frequency change Δf. However, affected by the stability of the speed-regulating mechanism, the unit power regulation of the generator cannot be set too large (or the regulation coefficient is set too small). Suppose the regulation coefficient is set to zero, then, although the load change will not lead to changes in the frequency, which seems that a constant frequency is maintained, this will make allocation of load variations cannot be fixed among various units, thus lead to the problem that the speed regulating system of each generating set cannot work stably.

Loads among generator sets with primary regulation are naturally allocated according to the unit power regulation or the regulation coefficient. That is, the increments of power of each unit is (Formula (7.271))

$$\Delta P_{Gi} = -K_{Gi}\Delta f = -\frac{1}{\delta_*} \times \frac{P_{GN}}{f_N}\Delta f \tag{7.271}$$

That is, the smaller the regulation coefficient, the larger the active power change assumed by the generators (with respect to their rated capacity), the increase amount is related to the rated power of the generator.

It can be known from the above analysis that, the primary frequency regulation is droop control, that is, the frequency after regulation is impossible to return to the original value. Therefore, primary frequency regulation can only accommodate to irregular changes with small amplitude and short period. The more the generators participate in the primary regulation, the greater the adjustable capacity of the generator, and the better the system frequency regulation.

2) **Secondary frequency regulation**

For circumstances with large load fluctuations and longer periods, primary regulation alone does not necessarily guarantee the frequency deviation within the allowable range. At this point, the frequency regulator of the generator speed regulating system shall act in order to carry out secondary regulation.

The secondary frequency regulation means, in case of frequency changes caused by load changes, the frequency regulator act to make the generator power-frequency static characteristic move in parallel, thus changing the active output of the generator, to keep the system frequency constant or in the allowable range. For example,

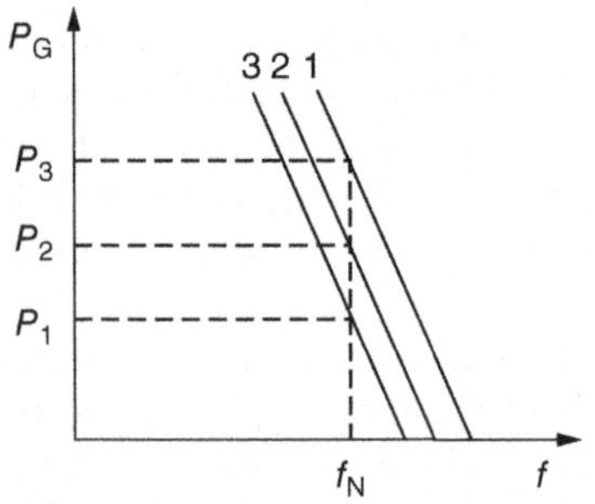

Figure 7.67 Parallel move of active power-frequency static characteristic.

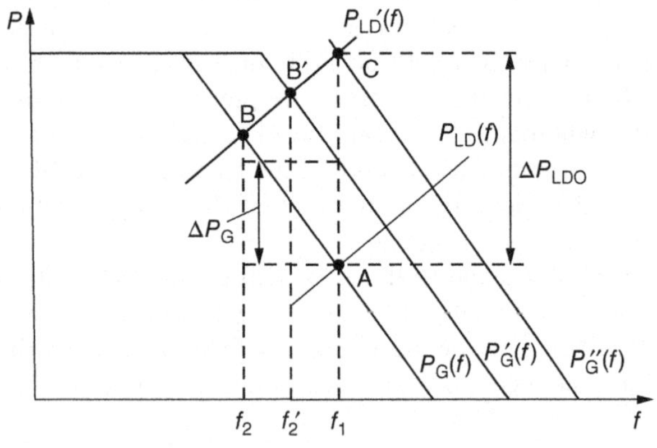

Figure 7.68 Secondary frequency regulation.

in Figure 7.67, the load increases reduce the frequency, the frequency regulator acts to increase the quantity of air (water) intake of the prime motor, making the active output of the generator and the frequency increase, and the power-frequency static characteristic moves in parallel to the right from 2 to 1; conversely, if the load decreases increase the frequency, the frequency regulator will act to reduce the quantity of air (water) intake, to make the frequency reduce, and the power-frequency static characteristic moves in parallel to the left from 2 to 3.

The secondary frequency regulation process is shown in Figure 7.68. The original operating point of the system is point A, and the system frequency is f_1. When the system load is increased by ΔP_{LD0}, if there is only primary regulation, the operating point will move to point B, and the system frequency is f_2. Under the action of the frequency regulator of the generator speed regulating system, the power-frequency static characteristic of the generator will move in parallel to $P_G'(f)$, the operating point is moved to point B', and the system frequency will be f_2'. The frequency regulation characteristic formula is (Formula (7.272)):

$$\Delta P_{LD0} - \Delta P_G = -(K_G + K_{LD})\Delta f \tag{7.272}$$

As can be seen from Formula (7.272), load increment ΔP_{LD0} is shared by three parts, one is the additional power ΔP_G generated by the generator in the secondary regulation, the second part is the additional power $K_G\Delta f$ generated by the generator in the primary regulation, and the other part is the load power $K_{LD}\Delta f$ reduced by the load regulating effect. In addition, because the generator produces additional power in the second regulation, in the same load change, the system frequency deviation is smaller. If the load increment [as $P_G''(f)$ shown in the figure] is all borne by the additional power produced by the generator in the secondary regulation, that is, $\Delta P_{LD0} = \Delta P_G$, then $\Delta f = 0$, and zero droop control of the system frequency is achieved.

Thus, secondary regulation can achieve droop and zero droop controls of the frequency. But secondary regulation cannot change the value of the unit power regulation of the system.

In systems with multiple generators equipped with speed regulators operating in parallel, when the load changes, as long as the generators are underloaded, they all participate in primary frequency regulation; and the secondary frequency regulation is generally undertaken by one or a few generators (power plants).

7.8.3 Frequency Characteristics of Wind Power Systems

The frequency characteristics of wind power systems are for study of how frequency variations in power system cause changes in the wind turbine rotary speed and the kinetic energy, namely how wind turbines respond to changes in grid frequency. Because the generator works in accordance with the principle of electromagnetic torque in the opposite direction of mechanical torque, changes in the rotary speed is possibly only the result of one or two changes among these variables. If during the frequency event, the mechanical torque provided by the prime motor is considered constant, then the electromagnetic torque is only affected by changes in the system frequency, which can affect the generator's inertia response. If the electromagnetic torque is affected by changes in system frequency, then the inertia response can be observed. In conventional synchronous generators, electromagnetic torque is sensitive to changes in system frequency, so the inertia response is easy to be observed. In determining the impact of growth of wind power on the power system, wind turbine inertia response is a key factor. In high wind speed or low speed with a lot of wind power capacity, the wind turbine will replace part of the conventional generators in the system. If the wind turbine cannot demonstrate its inertia effect, the effective inertia mass of the whole system will be reduced. When increasing the proportion of wind power, its adverse effects on the system inertia will also increase. For power systems with higher proportion of wind power under low load conditions and low system capacity, reduction of system inertia is very unfavorable to the grid frequency stability.

7.8.3.1 Inertia Time Constant of Wind Turbines

Inertia time constant is an important characteristic parameter of the wind turbines, it reflects the energy stored in the wind turbine rotating mass. Energy storage of any rotating mass (the kinetic energy) is generally expressed as (Formula (7.273)) [99, 100]:

$$E = \frac{1}{2}J\omega_m^2 \tag{7.273}$$

where J is the mechanical inertia, ω_m is the rotary speed. Divide Formula (7.265) by the generator rated apparent power S_n, the inertia time constant is obtained as (Formula (7.274)):

$$H = \frac{J\omega_m^2}{2S_n} \tag{7.274}$$

The inertia of a given object is (Formula (7.275)):

$$J = \sum m_i r_i^2 \tag{7.275}$$

where m_i is the mass of object i, r_i is the radial distance of the inertia axis. In the wind turbine, the center of mass of the blade is typically one-third of the wind turbine radius r, the mass of each blade is m_b. Then the expression of the wind turbine inertia is obtained as (Formula (7.276)):

$$J = 3m_b\left(\frac{r}{3}\right)^2 = \frac{1}{9}m_r r^2 \tag{7.276}$$

where m_r represents the mass of the blade and the rotor structure, which is equivalent to $3m_b$. Asynchronous generator rotor inertia is usually represented by the inertia time constant H, of which the value range is 4 to 8 s. Inertia time constant of generators of large conventional power plant is within the range of 2 to 9 s [100].

7.8.3.2 Wind Turbine Inertia Response

In occurrence of significant frequency reduction in the grid, the system inertia plays a decisive role in reducing of frequency. The lower the inertia, the faster the system frequency decreases. In the power system, for serious frequency accidents, any reduction in inertia response is dangerous. For generators with different wind turbine technology, the frequency response characteristics are not the same.

Synchronous generators of conventional power plants are connected directly to the power grid, which means the power system frequency and generator electromagnetic torque (and the electrical power output generated) are coupled. Because of this coupling, in case of sudden drop in the power system frequency, active power will suddenly (transient) increase. This is called "natural" inertia response of the generator, which is advantageous to the frequency stability.

Each generator in the power system is modeled as a droop controller, which is used to help stabilize the power system frequency. The typical droop rate of reheat thermal power plants is set to about 4% [99, 101]. And it is assumed to be ideal droop, namely the value of the droop rate unchanged. Thus, conventional synchronous generators are ideal conventional generators (ICSG).

For fixed-speed generator (FSG) based on cage asynchronous machine, due to the coupling of the rotor speed and the system frequency is strong, when the frequency of the power system is reduced, the constant-speed wind turbine speed will also reduce to release the corresponding part of rotational kinetic energy, in order to provide inertia response, of which the response rate (rpm drop speed) and amplitude (minimum speed) depend on the rotational kinetic energy stored in the wind turbine blades, wind turbine rotor, and the generator rotor, as well as the rate of change of the grid frequency. Determination of inertia time constants of FSG and any other asynchronous generators is based on the wind turbine blades and the combination characteristics of generators. Typically, the inertia time constant of FSG is small, so it can respond to rapid change in frequency.

The disadvantage of FSG is that it will cause instability when operating at changing wind speed. Because the time constant of variation of blade pitch angle is very long, and the blades are directly coupled with the rotor, the response of FSG to a sudden change in wind speed is particularly slow. Therefore, power impact changes can often be seen in large gusts. This characteristic requires FSG to be connected to strong power systems. Here, strong refers to large power systems that can effectively deal with the impact of changes, and appropriately distribute this impact to the entire network. Therefore, a large number of FSG integrated in a small isolated system is very difficult.

Because the structure of double-fed asynchronous generator (DFAG) requires the rotor be decoupled from three-phase power system, so when DFAG wind turbine are connected to the system, the mechanical inertia of the blade cannot be used. Inertia response of DFAG must be simulated by changes in the control scheme. DFAG wind turbines are capable of decoupling control between its active and reactive power, but due to its complete decoupling control between the variable speed wind turbine speed and grid frequency, it is impossible for it to provide a frequency response on the grid when the grid frequency is changed. Therefore, in grid frequency changes, the inherent inertia of variable speed wind turbines based on the double-fed generator is manifested as a "hidden inertia" on the grid, and is unable to help reduce the frequency rate of change. The additional converter control loop compensates for low inertia rendered by DFAG wind turbines. By controlling with the use of wind turbines "hidden inertia" to support the primary frequency control of the system for some time, so that the inertia of DFAG can also be comparable to the inertia of FSG.

7.8.4 Support of Wind Turbines on System Frequency Control

The loss of large generators by the system will cause power system frequency drops. Frequency drop causes the generator speed regulator to act, which, however, is limited by the mechanical time constants of the steam or gas turbines associated with event response. These time constants are typically in the range of 1 to 2 s, during which the rate of decline in the frequency will determine the lowest frequency of the system before the action of the speed regulator is triggered. Frequency rate of decline is mainly determined by the size of the rotary inertia available in the system. When conventional power generation is replaced by wind power, the frequency rate of change of the system will increase. Maximum frequency rate of change after loss of power is related to the wind turbine technologies. When fixed-speed wind turbines replace conventional synchronous generators, there is no significant changes in the minimum frequency reached after loss of power. This is due to the inertia response provided by the fixed-speed wind turbines. When DFAG wind turbines replace conventional synchronous generators, the lowest frequency after the loss of power will be reduced this is because the inertia response provided by the DFAG wind turbines is negligible [102]. In systems with higher levels of wind power penetration, the inertia levels may be reduced, depending on whether or not to remove the combinations or structures of wind turbines. The following analyzes the inertia response capabilities of a fixed-speed asynchronous generator (FSAG), a double-fed asynchronous generator (DFAG), and a full converter variable speed wind turbine generator (FCWTG).

7.8.4.1 Frequency Response of Fixed-Speed Wind Turbines

Conventional synchronous or asynchronous generators are directly connected to the grid, and will automatically release the energy rotating masses for grid frequency drops, that is, respond to grid frequency drops. When the gird frequency changes from f_0 to f_1, and it is assumed that the rotor speed is changing from ω_0 to ω_1 in direct ratio, then the kinetic energy released by the fixed-speed wind turbine is ΔE_{k} (Formulas (7.277) and (7.278))

$$E_{k0} = \frac{1}{2}J\omega_0^2 \tag{7.277}$$

$$E_{k0} - \Delta E_k = \frac{1}{2}J(\omega_0 - \Delta\omega)^2 \tag{7.278}$$

where $\Delta\omega = \omega_0 - \omega_i$. Substitute $\Delta\omega$ in Formula (7.269), we can obtain (Formula (7.279)):

$$\Delta E_k = E_{k0}\left(1 - \frac{\omega_1^2}{\omega_0^2}\right) = E_{k0}\left(1 - \frac{f_1^2}{f_0^2}\right) \tag{7.279}$$

Figure 7.69 shows the response of 2 MW cage asynchronous wind turbines connected to an infinite bus to system frequency changes. The infinite bus frequency drops from about 50 Hz to 49.75 Hz. For generators of 2 MW and abrupt changes in frequency from 50 Hz to 49.75 Hz, the rotating mass expected by Formula (7.278) will release the kinetic energy of 40 kW, which can also be determined by the area of the output power curve shown in Figure 7.69.

Rotary speed of the cage asynchronous generator decreases with the frequency. The reducing of rotary speed and the inertia response generated are due to the increase of deceleration electromagnetic torque, whereas the deceleration electromagnetic torque is caused by the decrease in system frequency.

Impact of reducing of system frequency on the electromagnetic torque can be seen from Figure 7.69. In can be seen from the figure the increase of deceleration torque in

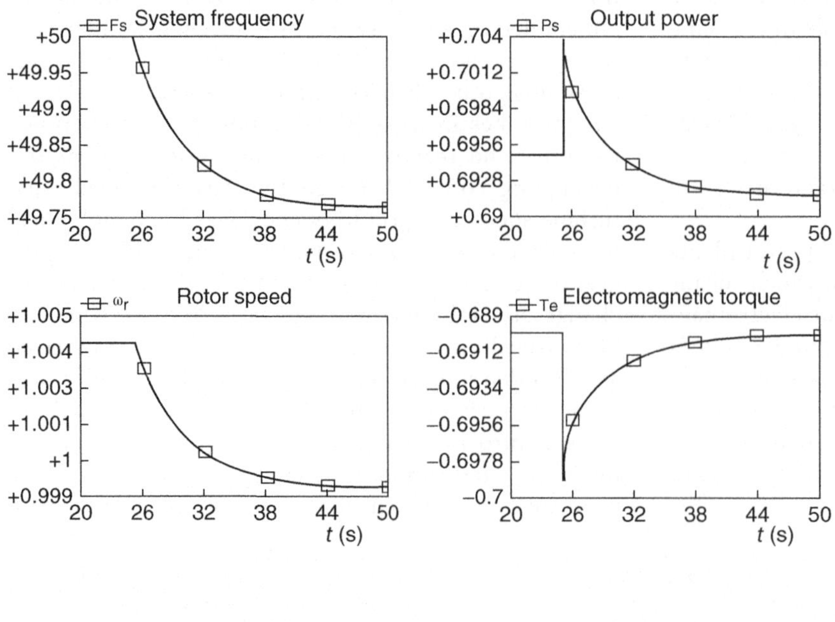

Figure 7.69 Response of cage asynchronous generators to system frequency drop (p.u.).

response to the decrease of system frequency. Reducing the rotary speed generates inertia response and kinetic energy release, and causes an immediate increase in electrical power. If electromagnetic torque is insensitive to changes in system frequency, then the inertia response is unobservable. Torque transients at the beginning and the end of frequency disturbances are associated with the shaft rigidity and damping.

7.8.4.2 Speed and Inertia Control of DFAG

There are many ways for the control of DFAG speed. Basic speed control is achieved by two control loops; this is because the mechanical and electrical systems are operating on very different periods. Electrical system dynamic is much faster than that of the mechanical system. Primary control loop achieves the generator and converter control, which is the control of the active and reactive power generated by the wind turbine. Secondary control loop is used in mechanical systems, to control the pitch angle of the blades and the rotary speed.

Application of primary loop control to the converter is common in the entire power system. For mechanical control, especially for speed control, DFAG is unlike FSAG. By speed control, wind turbines can capture maximum power at different wind speeds.

The main task of the speed control is to maintain the tip speed ratio of the wind turbine at an optimal level. Realization of speed control is shown in Figure 7.70 [98, 99, 103]. The use of steady-state generator speed as the reference value (reference value is determined from the manufacturer's power curve) can establish the speed control loop. At low wind speeds, the generator is kept at a low fixed speed; at high wind speeds, that is at wind speeds higher than the rated wind speed of the wind turbine, the blade pitch is gradually increased to reach the optimal level for maintaining power generation.

In Figure 7.70, all of the components in DFAG that play an important role are marked: ① generator reference speed (ω_{ref}) is determined by the predetermined wind turbine P-ω characteristic curve; ② calculate the error between the speed measurement (ω) and the reference speed ($\Delta\omega$); ③ the error is fed to the PI controller, generating a torque speed reference value($T_{\omega,ref}$); this value, which is caused by the imbalance between the wind turbine torque and the generator torque, will result in an acceleration or deceleration torque, until the desired speed is reached; ④ in the last part of the control, the final speed control torque (T_{ref}) is the combination of speed controller torque ($T_{\omega,ref}$) and inertia response torque ($T_{inertia}$), the additional inertia response torque($T_{inertia}$) is obtained through the secondary control loop.

As can be known from Figure 7.70, speed control relies on the inertia response torque term ($T_{inertia}$), which is usually realized by using in one of the following two controls [98]. The first method is to try to follow the inherent characteristics of the synchronous generator, obtaining power captured by the wind turbine by the available kinetic energy differential at any speed ω_m (Formula (7.280)):

$$P = \frac{dE_k}{dt} = J\omega_m \frac{d\omega_m}{dt} \tag{7.280}$$

Next, replace inertia J with the combined inertia time constant H. This value is derived from Formula (7.273) to (7.276). Then the following formula is obtained (Formula (7.281)):

$$\frac{P}{S} = 2H\frac{\omega_m}{\omega_s}\frac{d(\omega_m/\omega_s)}{dt} \tag{7.281}$$

where ω_s is the grid frequency. By using the per-unit value, Formula (7.281) can be expressed as (Formula (7.282)):

$$\overline{P} = 2H\overline{\omega}\frac{d\overline{\omega}}{dt} \tag{7.282}$$

The per-unit torque obtained from Formula (7.282) is (Formula (7.283)):

$$\overline{T} = 2H\frac{d\overline{\omega}}{dt} \tag{7.283}$$

The control loop in Figure 7.71 is obtained [98, 99]. In order to minimize the impact on the mechanical drive chain loads, a first-order filter (low pass filter) improves the power injection rate, which produces two effects: reducing the rate of electromagnetic torque and reducing peak torque. This is represented by $T_{interia}$ in Figure 7.70, which combines with the speed control reference torque.

The second method is to obtain the inertia response torque term by using the proportional control. The torque term is calculated according to the absolute deviation of the nominal frequency of system, that is

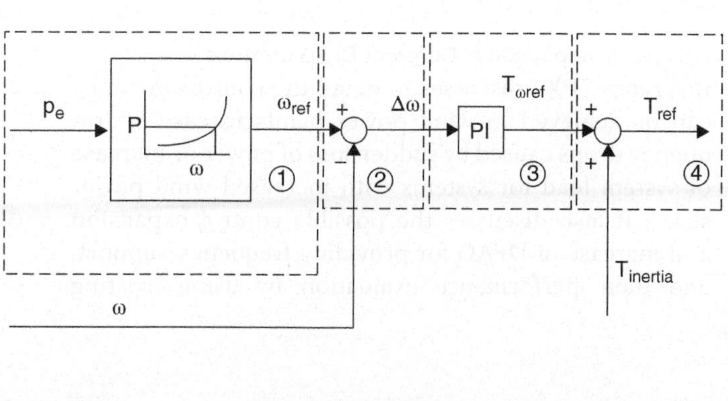

Figure 7.70 DFAG wind turbine speed control block diagram.

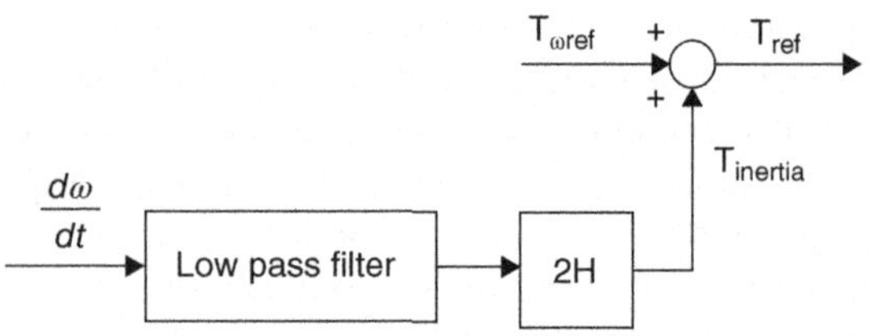

Figure 7.71 Realization of inertia response.

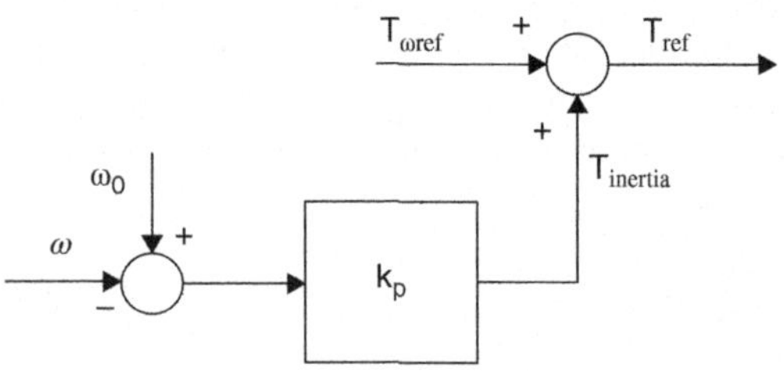

Figure 7.72 DFAG wind turbine droop characteristic control block.

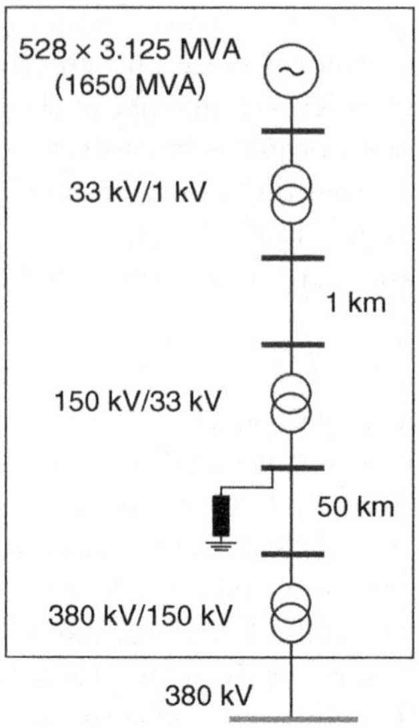

Figure 7.73 WF structure diagram.

(Formula (7.284)):

$$T_{inertia} = k_P(\omega_0 - \omega_{ms}) \tag{7.284}$$

where ω_0 is the rated speed; k_p is the proportionality constant. Such control is called droop control (droop control), as shown in Figure 7.72 [100, 101]. This is equivalent to the primary frequency control applied to a conventional synchronous generator. Droop control generates inertia response torque term, this term is added to the torque terms of speed control.

Due to the results of the two inertia response controls, DFAG can make use of the mechanical inertia stored in wind turbine blades. This allows DFAG be capable of functioning in a manner similar to FSG, having the ability to deal with changes in system frequency, and this is also the capacity, which enables DFAG to maintain the operation at variable wind speed, and generate more power. However, realization of this additional control is slightly delayed for response to system frequency changes. Moreover, the inertia response that may be achieved by DFAG is limited by the operating constraints, such as current limit. In fact, realization of additional control loop may be conflict with maintaining operating constraints within the limits, which requires some design modifications, such as the increase of rated capacity of DFAG converter [102].

7.8.4.3 Comparison of Different DFAG Controls

Reference [100] discusses in detail the control methods can be achieved in wind power plants in case of frequency drops caused by sudden loss of power or increase of system load for systems with increased wind power share. It also discusses the possible control expansion and increase of DFAG for providing frequency support, and their performance evaluation by using existing control systems of wind turbine as the baseline. The WF structure used is shown in Figure 7.73.

The most suitable control structures for wind turbines to provide frequency support during system faults are: ① constant P and Q control; ② constant power and constant voltage control; ③ constant power P and voltage U control, with additional elements controlled by the frequency; ④ constant power P and voltage U control, with additional elements controlled by static and dynamic frequencies; ⑤ frequency-dependent power P and constant Q control. The first two controls are already commonly used by DFAG.

In all cases, assuming the loss of 2.5% of the total generating capacity, the system without wind power is compared with the system with wind power accounted for 51.9% of the total generating capacity of the system. The frequency deviation characteristics obtained are summarized in Figure 7.74. The main results are as follows:

1) Expansion of the existing control structures by wind turbines are possible. Compared with conventional power plants, WFs consisting of wind turbines that are controlled by the power (P) and frequency-dependent voltage (U) may significantly improve the frequency stability after large power imbalances. The downside of this method is that, it contains acceptable short-term network voltage drops, especially voltages of buses nearby the WF.
2) Increasing levels of wind power penetration results in a higher frequency gradient, with the increase of wind power, the system initial frequency drops faster after disturbances;
3) When wind power is not connected to the system, the steady-state frequency drop after losing 2.5% of total generating capacity is 1,000 mHz (curve 0). It has been

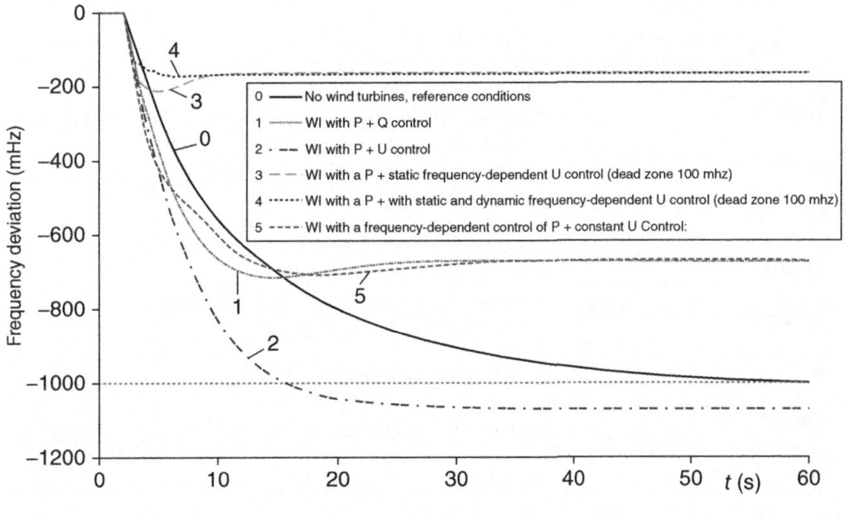

Figure 7.74 Comparison of frequency deviation after loss of 2.5 % power by the system.

compared with all results obtained from using different controls. Results obtained from the use of active power and static and dynamic frequency-dependent voltage control are the best (curve 4). Similar results can also be obtained from the use of active power and static frequency-dependent voltage control (curve 3). If 100 mHz is a dead zone, the steady state frequency deviation for the selected trials network is 164 mHz. When implementing P and Q control (this is a control strategy used by DFAG under any circumstances at present), the steady-state frequency deviation obtained is 669 mHz (curve 1). Similar results can be obtained from frequency-dependent P control, but the starting frequency gradient tends be steep (curve 5). It should be emphasized that the final choice requires good coordination between power and frequency controllers. If this is not the case, the system characteristics may be adversely affected. On frequency stability, the smallest benefit seems to be the choice of active and constant voltage control. In this case, the final frequency deviation is 1070 mHz, lower than the frequency relay start setpoint (curve 2). This is because the voltage controller used here is very strong for wind turbines, the voltage is always quickly stabilized at the reference value. Thus, the voltage controller is designed not only to maintain a constant voltage, but also to consider the impact of softer-voltage characteristics on frequency stability. On the other hand, the voltage control of wind power plants in the future is necessary because voltages of all the conventional wind power replaced by wind power are controllable. In the case of wind turbines under constant voltage control, all the entire reactive power supply (with or without wind power plants) is almost constant. However, for constant Q control, all the reactive power output is slightly declined, which is due to reduction of network voltage (especially buses near the WF near) and decrease of reactive power absorbed by the load as a result. Speaking on the reactive power, decrease of maximum output of wind power plants occurs in the case of frequency-dependent voltage control.

4) Having a frequency-dependent voltage control can not only maintain the grid voltage, but also have considerable contribution to the frequency stability by using the load voltage characteristic.

During disturbances, the network voltage distribution under each control is different. Generally, the use of voltage and reactive power control can cause smaller decrease in voltage. The voltage drop caused by the frequency-dependent U controller depends on the level of frequency drop. The bar graph shown in Figure 7.75 summarizes the changes in the system voltage under different controls. Figure 7.75 shows that the effective frequency support of wind turbines includes balance between voltage and frequency, especially in bus near the WF. In other words, we need to reduce the voltage deliberately to reduce the power absorbed by the load, thereby reducing the pressure on the frequency. However, in all cases, the voltage drop does not exceed 2%; this voltage drop can still provide favorable support level for a given frequency. Further improvement requires not only the application of frequency-dependent voltage control by all wind turbines, but also the application by conventional generators, in order to reduce the voltage drop.

Compared with FSG wind turbines, DFIG wind turbines have larger frequency drops. In systems, which small disturbances may have significant impacts on system frequency, installation of FSG wind turbines may have more benefits. However, it must operate at fixed speeds, so the wind turbines must be installed in suitable locations where average wind speed will produce the most stable power generation. In addition, FSG must be installed in systems without voltage recovery problems, because the voltage stability problems presented by

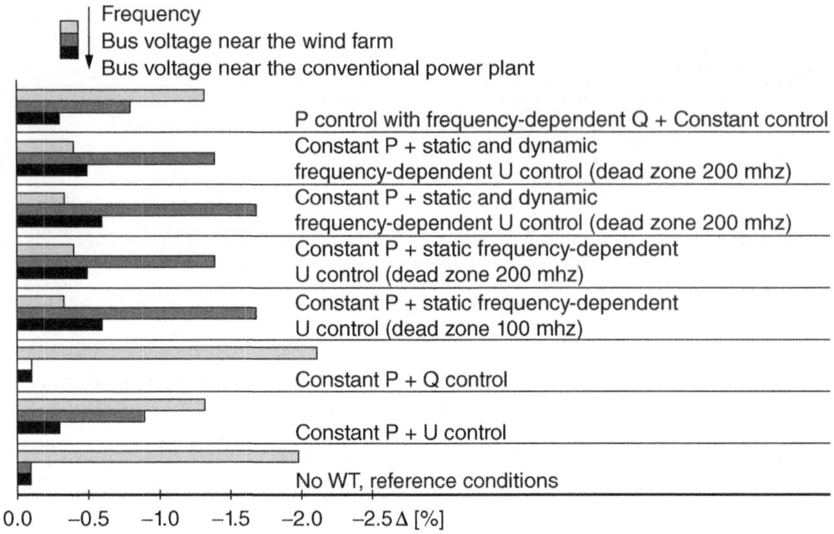

Figure 7.75 Comparison of control schemes.

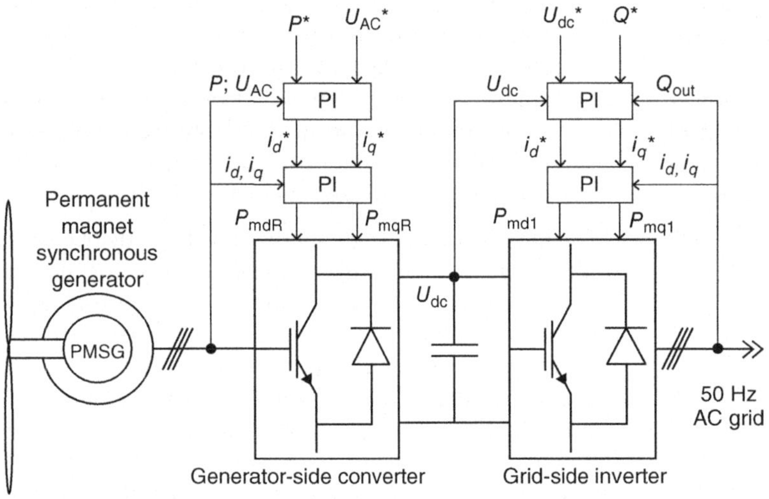

Figure 7.76 Full converter variable speed wind turbine and converter control system.

FSG are more prominent than the frequency stability problems.

7.8.4.4 Frequency Control Strategy of FCWTG

Here are varieties of choices for realizing full converter variable speed wind turbine generators (FCWTG), that is, direct-drive permanent magnet synchronous wind turbines. Figure 7.76 shows the block diagram of control of FCWTG and its converter. The generator-side rectifier controls the active power P and the AC voltage U_{AC} of the generator, it uses two external PI control loops to determine reference value i_d^*,i_q^* of the two inner current loop controls, which control the d, q -axis current components. Inner-loop control determines the modulation index P_{mdR} and P_{mqR}. d and q axises are oriented by the generator AC voltage. Grid-side converter (inverter) is used to control the DC voltage U_{dc} and reactive power output Q_{OUT}, and it also uses two inner current loop controls. The reference axis is oriented by the AC voltage of the converter.

Figure 7.77 shows a control system model of the blade pitch angle. In normal operation, the signal S is zero. Reference pitch angle β^* is determined by the PI controller. When the speed of the permanent magnet synchronous generator (PMSG) ω_G is lower than the maximum speed ω^*_G, output β_N of the PI controller is zero; when ω_G exceeds ω^*_G due to the excess wind power, the PI controller will increase its output β_N to decelerate the PMSG to return to ω^*_G. The actuator simulates the physical process of the blade pitch angle changing from β to β^*, which is limited by the pitch angle, and is limited by the pitch angle β and the rate of change .β of the pitch angle.

The active power output of FCWTG is controlled by adjusting the modulation index of the converter. In case of frequency events, the frequency change will lead to rapid adjustment of the modulation index in order to maintain the active power output within the value provided by the controller. The use of frequency control system to adjust the active reference value of the

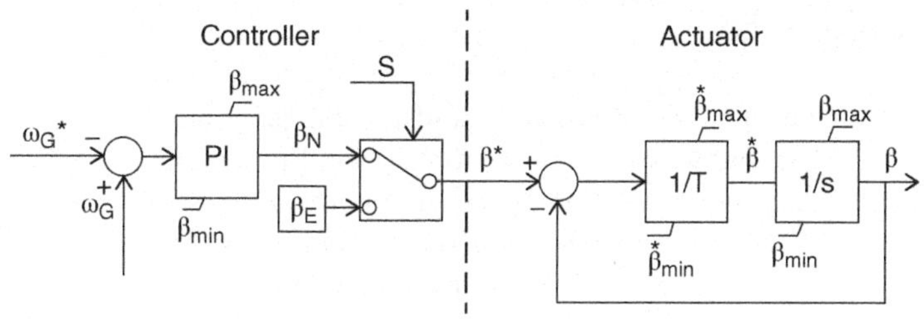

Figure 7.77 Pitch angle control system.

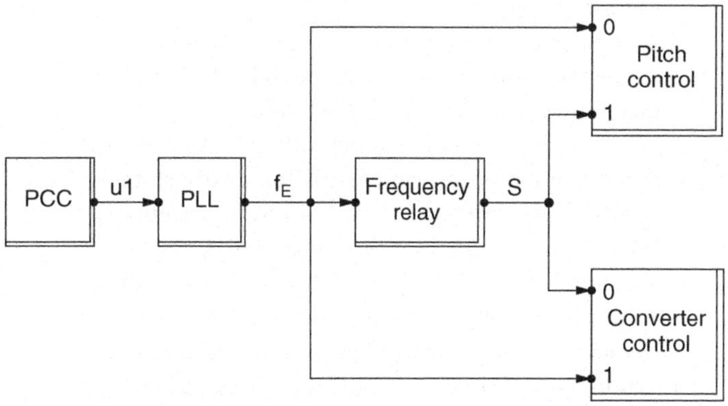

Figure 7.78 A complete frequency control system.

converter controller in response to changes in the power system frequency, emulates the frequency response of ideal synchronous generator.

Figure 7.78 shows a complete block diagram of the improved frequency control system for FCWTG. Measure the positive sequence voltage U1 at WF point of common coupling (PCC), and measure the power system frequency via phase-locked loop (PLL). Frequency relay is used to detect abnormal power system frequency. Signal S of abnormal marking frequency is sent to the pitch controller and the converter controller.

When the signal S changes from 0 to 1 (abnormal marking frequency), the pitch controller PI control loop (Figure 7.77) will fail, the pitch control reference value is directly determined as β_E. If the wind speed of FCWTG is greater than the rated value, the pitch controller can reduce the pitch angle to increase the active power output. To do this, in FCWTG, the active reference value of the converter also must be increased effectively. Figure 7.79 shows how active power reference value P^* in the expanded FCWTG control system is generated.

In Figure 7.79, the maximum power point tracking (MPT) produces the active power reference value P_M^* (MPT optimizes the power of the function of PMSG speed ω_G). Frequency support component of the active power is expressed as P_F^*, which is zero when S is zero. When the frequency relay detects abnormal conditions, set S to 1, P_F^* becomes nonzero, thus providing frequency response. Thus, the active reference value P_U^* is determined by the sum of P_M^* and P_F. In addition to the initial frequency control, signal S limits the reference power P_U^* between 0 and P_{MAX}^*. Where low frequency trip is set at 49.8 Hz, which continues for 0.1 s.

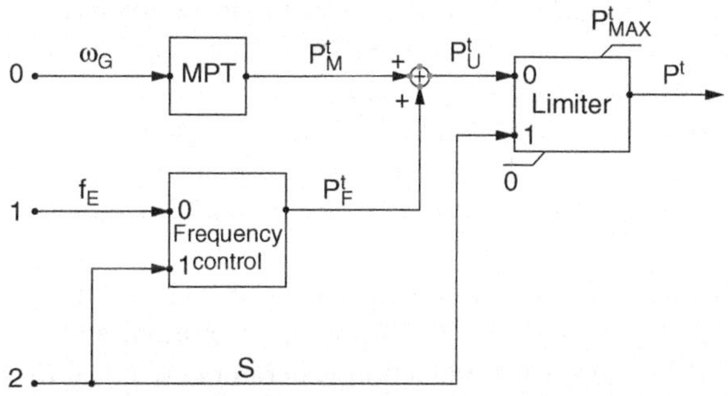

Figure 7.79 General FCWTG droop control structure.

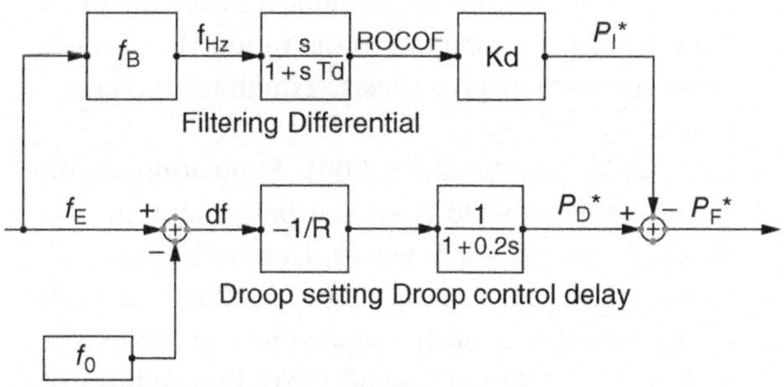

Figure 7.80 Frequency support control system.

Figure 7.80 shows all strategies that produce the signal P_F^* in Figure 7.78. Signal P_F^* has two components, one is P_I^*, which is proportional to the frequency change after filtered (ROCOF), simulating the "natural" inertia response of ideal synchronous generator (ICSG). A low-pass filter must be connected to the differentiator so that the frequency measurement noise does not cause a problem. Another component P_D^* is the droop component, similar to the ICSG droop response.

References

1 Holttinen, H., Meibom, P., and Ensslin, C. et al. 2008. Impacts of large amount of wind power on design and operation of power systems, results of IEA collaboration. Seventh International Workshop on Large Scale Integration of Wind Power and Transmission Network Offshore WFs, May 26–27, Madrid, Spain.

2 Wasynezuk O. and Krause P.C. 1985. Theory and comparison of reduced order models of induction machines[J]. *IEEE Trans,* **3**, 598–606.

3 Feljoo, A.E., Cidras J., Domelas, J.L.G. 1999. Wind speed simulation in wind farms for steady-state security assessment of electrical power systems[J]. *IEEE Trans on Energy Conversion,* **14**(4), 1582–1588.

4 Yunping, Chen, and Wenxia, Pan. 2001. A study on wind power system voltage behaviour. *Journal of Hehai University (Natural Sciences),* **29**(1), 88–92.

5 Trudnowski, Daniel, J., Gentile, Andrew, Khan, Jawad M., and Petritz, Eric M. 2004. Fixed-speed wind-generator and wind-park modeling for transient stability studies. *IEEE Transactions on Power Systems,* **19**(4), 1911–1917.

6 Bin, Zhao. 2008. Studies of transient stability for grid-connected wind turbine with induction generator [D]. Unpublished master's thesis, Chongqing University.

7 Xueguang, Wu, Xuecheng, Zhang, Yonghua, Yin, and Huizhu, Dai. 1998. Application of models of the wind turbine induction generators (WTIGs) to wind power system dynamic stability analysis. *Power System Technology,* **22**(6), 68–72.

8 Junling, Wu. 2004. Some technical problems of large scale wind farm integration operation [D]. Unpublished master's degree thesis, Tsinghua University, Bejing.

9 Risø National Laboratory. 2001. Simulation of interaction between wind farm and power system. Pitney Bowes Management Services, Denmark A/S.

10 Salman, S.K., Teo, A.L.J., Rida, I.M. 2000. The effect of shaft modeling on the assessment of fault CCT and power quality of a wind farm. Proceedings of Ninth International Conference on Harmonics and Quality of Power, Oct 1–4, pp. 994–998.

11 Yuan, Zhang and Chengxu, Wang. 2003. *Wind Power* [M]. Beijing: China Electric Power Press.

12 Xueguang, Wu, Weisheng, Wang, and Huizhu, Dai. 1998. Research on voltage fluctuation characteristics of wind power system. *Wind Power Generation,* **4**, 29–35.

13 Fernández, L.M., García, C.A., Saenz, J.R., and Jurado F. 2009. Equivalent models of wind farms by using aggregated wind turbines and equivalent winds. *Energy Conversion and Management,* **50**, 691–704.

14 Junling, Wu, Shuangxi, Zhou, Jianfeng, Sun, et al. 2004. Analysis on maximum power injection of wind farm connected to power system. *Power System Technology,* **28**(20), 28–32.

15 Fernandez, Luis M. and Ramon Saenz, Jose. 2006. Francisco jurado. Dynamic models of wind farms with fixed speed wind turbines. *Renewable Energy,* **31**, 1203–1230.

16 Yichun, Wu, Ming, Ding, and Lijun, Zhang. 2005. Power flow analysis in electrical power networks including wind farms. *Proceedings of the CSEE,* **25**(4), 36–39.

17 Fernandez, Luis M., Jurado, Francisco, and Ramon Saenz, Jose. 2008. Aggregated dynamic model for wind farms with doubly fed induction generator wind turbines. *Renewable Energy,* **33**, 129–140.

18 Zhefeng, Feng, Enxing, Yang, and Guozhu, Chen. 2007. Unit power factor control of direct-drive permanent magnet wind generator. *Mechanical & Electrical Engineering Magazine,* **24**(9), 62-–65.

19 Wei, Wang. 2008. The steady state model of synchronous wind generator and the research of the grid connected [D]. Unpublished master's degree thesis, Beijing Jiaotong University.

20 Ming, Yin, Gengyin, Li, and Jiancheng, Zhang, et al. 2007. Modeling and control strategies of directly driven wind turbine with permanent magnet synchronous generator. *Power System Technology,* **31**(15), 61–65.

21 Mei, Zhang. 2008. Modeling and control system simulation of permanent magnet synchronous wind turbine generator [D]. 2008. Unpublished master's thesis, Xi'an University of Technology.

22 Muljadi, E., Butterfield, C.P., Ellis, A., Hochheimer, J., Young, R., Miller, N., Delmerico, R., Zavadil, R., and Smith, J.C. 2006. Equivalencing the collector system of a large wind power plant. *Power Engineering Society General Meeting,* **2009**, 9.

23 Sun Jianfeng, Sun, Jiao Lianwei, Jiao, Wu Junling, Wu, Shuangxi, Zhou, and Shousun, Chen. 2004. Research on multi-machine dynamic aggregation in wind farm. *Power System Technology,* **28**(7), 58–61.

24 Shande, Shen. 1993. *Identification in Power System.* Beijing: Tsinghua University Press.

25 Dawei, Ma. 2003. Electromechanical transient process of electric power system [M]. Beijing: Water Conservancy and Electric Power Press, China.

26 Pecas Lopes, J.A., Maciel Barbosa, F.P., and Cidras Pidre, J. 1991. Simulation of MV distribution networks with asynchronous local generation

sources. Proceedings of the Electrotechnical Conference, May 22–24, pp. 1453–1456.

27 Feijóo, Andrés E. and Cidrás, José. 2000. Modeling of wind farms in the load flow nalysis. *IEEE Transaction on Power System,* **15**(1), 110–115.

28 Shuangxi, Zhou, Haichao, Wang, Zongxiang, Lu, and Junling, Wu. 2005. A joint iteration method for power flow calculation of power system containing unified wind farm and its application. *Power System Technology,* **29**(18), 59–62.

29 Feijoo, A. and Cidras, J. 2000. Modeling of wind farms in the load flow analysis[J].*IEEE Trans. on Power Systems,* **15**(1), 110–115.

30 Miller, Nicholas W., Price, William W., Sanchez-Gasca, Juan J. 2004. Modeling of GE wind turbine-generators for grid studies. GE WTG modeling version 3.2.

31 Hong, Shen. 2003. Integrated VSCF wind turbine [D]. Unpublished doctoral thesis, China Electric Power Research Institute.

32 Weihong, Hu, Wei, Wang, Yinglin, Wang, and Haibo, Xiao. 2006. Power flow analysis in electrical power system including wind farms. *North China Electric Power,* **10**, 12–15.

33 Pena, R., Clare, J.C., and Asher, G.M.1996. A doubly fed induction generator using back-to-back PWM converters supplying an isolated load from a variable speed wind turbine[J]. *IEE Proc-Electr Power Appl,* **143**(5), 380–387.

34 Hong, Shen, Weisheng, Wang, and Huizhu, Dai. 2003. Reactive power limit of variable–speed constant–frequency wind turbine [J]. *Power System Technology,* **27**(11), 60–63.

35 Bai Wang, Jun, Wang, Wei, and Chen, Wei. 2009. Researches on power flow algorithm of electric power systems including wind farm of wind turbine with synchronous generators. *Microcomputer Information,* **25**(31), 96–97.

36 Ming, Ding, Shenghu, Li, and Kai, Huang. 2001. Probabilistic power flow analysis based on monte-carlo simulation. *Power System Technology,* **25**(11), 10–14, 22.

37 Haichao, Wang. 2006. Research on grid connected wind power generation. Unpublished doctoral dissertation, Tsinghua University.

38 Vicatos, M. S. and Tegopoulos, J. A. 1991. Transient state analysis of a doubly-fed induction generator under three phase short circuit. *IEEE Transactions on Energy Conversion,* **6**(1), 62–75.

39 Usaola, J., Ledesma, P., Rodriguez, J.M., Fernandez, J.L., Beato, D., Iturbe, R., and Wilhelmi, J.R. 2003. Transient stability studies in grids with great wind power penetration. Modelling issues and operation requirements. *Power Engineering Society General Meeting, IEEE,* **3**, 13–17.

40 Walling, R.A. and Reichard, M.L. 2009. Short circuit behavior of wind turbine generators. 62nd Annual Conference for Protective Relay Engineers, March 30–April 2, pp. 492–502.

41 Morren, Johan and de Haan, Sjoerd W. H. 2007. Short-circuit current of wind turbines with doubly fed induction generator. *Ieee Transactions on Energy Conversion,* **22**(1), 174–180.

42 Ouhrouche, Mohand. 2009. Transient analysis of a grid connected wind driven induction generator using a real-time simulation platform. *Renewable Energy,* **34**, 801–806.

43 Shuju, Hu, Jianlin, Li, and Honghua, Xu. 2008. Modeling on converters of direct-driven wind power system and its performance during voltage sags. *High Voltage Engineering,* **34**(5), 949–954.

44 Hongliang, Guan, et al. 2008. Symmetrical short circuit analysis of wind turbine generator. *Electric Power Automation Equipment,* **28**(1), 61–64.

45 Tung Linh, Nguyen. 2009. Voltage stability analysis of grids connected wind generators. 4th IEEE Conference on Industrial Electronics and Applications, May 25–27, pp. 2657–2660.

46 Shuangxi, Zhou, Lingzhi, Zhu, et al. 2004. *Voltage Stability of Power System and its Control.* Beijing: China Electric Power Press.

47 Martins, Marcia, Sun, Yu, and Bollen, H.J. 2004. Voltage stability of wind parks and simularities with large industrial systems. Nordic Wind Power Conference March 1–2, Chalmers University of Technology, Göteborg, Sweden.

48 Potamianakis, Emmanuel G., and Vournas, Costas D. 2006. Short-term voltage instability: Effects on synchronous and induction machines. *IEEE Transactions on Power Systems,* **21**(2), 791.

49 Vijayan, P., Sarkar, S., and Ajjarapu, V. 2009. A novel voltage stability assessment tool to incorporate wind variability. Power & Energy Society General Meeting, July 26–30, pp. 1–8.

50 Haichao, Wang, Jianfeng, Dai, Shuangxi, Zhou, et al. 2006. Voltage stability margin model for power systems containing wind energy resources. *Journal of Tsinghua University (Science and Technology),* **46**(7), 1185–1188.

51 Dobson, Ian, and Lu, L.1992. Computing an optimum direction in control space to avoid saddle node bifurcation and voltage collapse in electric power systems[J]. *IEEE Transactions on Automatic Control,* **37**(10), 1616–1620.

52 Greece, Scott, Dobson, Ian, and Fernando, L. 1997. Sensitivity of the loading margin to voltage collapse with respect to arbitrary parameters [J]. *IEEE Transactions on Power System,* **12**(1), 262–270.

53 Chengshan, Wang and Xuyang, Yu. 2004. A method for computing the probability distribution of fault

critical clearing time[J]. *Proceedings of the CSEE,* **24**(11), 6–10.

54 Xifan, Wang and Xiuli, Wang. 1988. Probabilistic power flow analysis in power systems[J]. *Journal of Xi'an Jiaotong University,* **22**(2), 87–97.

55 Feijoo, A., Cidras, J., and Dornelas G.1999. Wind speed simulation in wind farms for steady-state security assessment of electrical system[J]. *IEEE Transactions on Energy Conversion,* **14**(4), 1582–1558.

56 Pálsson, Magni þ., Toftevaag, Trond, Uhlen, Kjetil, and Giæver Tande, John Olav. 2002. Large-scale wind power integration and voltage stability limits in regional networks. *Power Engineering Society Summer Meeting,* **2002**(2), 762–769.

57 Li, Lin, Caixin, Sun, Yongping, Wang, Haitao, Wang, and Dezhou, Yang. 2008. Calculation analysis and control strategy for voltage stability of power grid with large capacity wind farm interconnected. *Power System Technology,* **32**(3), 41–46.

58 Xiaoguang, Wei, Yongning, Chi, Xiaoyun, Wei, and Guang, Tang. 2008. Transient voltage stability enhancement of induction generator-based wind farm by VSC-HVDC. *Transactions of China Electrotechnical Society,* **23**(5), 100–107, 147.

59 Kundur, P., Paserba, J., Ajjarapu, V., Andersson, G., Bose, A., Canizares, C., Hatziargyriou, N., Hill, D., Stankovic, A., Taylor, C., Custem, T. V., and Vittal, V. 2004. Definition and classification of power system stability. *IEEE Trans. Power Syst.,* **19**(2), 1387–1401.

60 Senjyu, T., Sueyoshi, N., Uezato, K., Fujita, H., and Funabashi, T. 2003. Transient stability analysis of induction generator using torque-time characteristic. The Fifth International Conference on Power Electronics and Drive Systems, November 17–20, pp. 760–765.

61 Hui, Li, Bin, Zhao, and Li, Han. 2008. Assessment of transient models and stability analysis methods for wind power generation units. *Journal of Chongqing University,* **31**(5), 481–485, 494.

62 Nunes, M. V. A., Lopes, J. A. P., Züurn, H. H., Bezerra, U. H., and Almeida, R. G. 2004. Influence of the variable-speed wind generators in transient stability margin of the conventional generators integrated in electrical grids. *IEEE Trans. Energy Convers.,* **19**(4), 692–701.

63 Muljadi, E., Butterfield, C. P., Parsons, B., and Ellis, A. 2007. Effect of variable speed wind turbine generator on stability of a weak grid. *IEEE Trans. Energy Convers.,* **22**(1), 29–35.

64 Ullah, N. R., and Thiringer, T. 2007. Effect of operational modes of a wind farm on the transient stability of nearby generators and on power oscillations: A Nordic grid study. *Wind Energy,* **11**, 63–73.

65 Sanchez Gasca, J. J., Miller, N. W., and Price, W. W. 2004. A modal analysis of a two-area system with significant wind power penetration. *Proc. 2004 IEEE PES Power Systems Conf. Expo,* **2**, 1148–1152.

66 Slootweg, J. G., and Kling, W. L. 2003. The impact of large scale wind power generation on power system oscillations. *Elect. Power Syst. Res.,* **67**, 9–20.

67 Qiao, Wei and Harley, Ronald G. 2008. Effect of grid-connected DFIG wind turbines on power system transient stability. Power and Energy Society General Meeting - Conversion and Delivery of Electrical Energy in the 21st Century, July 20–24, pp. 1–7.

68 Yongning, Chi, Weisheng, Wang, Yanhua, Liu, and Huizhu, Dai. 2006. Impact of large scale wind farm integration on power system transient stability. *Automation of Electric Power Systems,* **30**(15), 10–14.

69 Na, Cao, Yanchun, Li, Haixiang, Zhao, and Huizhu, Dai. 2007. Comparison of effect of different wind turbines on power grid transient stability. *Power System Technology,* **31**(9), 53–57.

70 Samuelsson, O. and Lindahl, S. 2005. On speed stability. *IEEE Transactions on Power Systems,* **20**(2), 1179–1180.

71 Kanabar, M.G., Dobariya, C.V., and Khaparde, S.A. 2006. rotor speed stability analysis of constant speed wind turbine generators. International Conference on Power Electronics, Drives and Energy Systems, 12-15 December 12–15, pp. 1–5.

72 Tamura, J., Yamazaki, T., Ueno, M., Matsumura, Y., and Kimoto, S.I. 2001. Transient stability simulation of power system including wind generator by PSCAD/EMTDC. Power Tech Proceedings, IEEE Porto, September 10–13, pp. 5.

73 Folly, K.A. and Sheetekela, S. 2009. Impact of fixed and variable speed wind generators on the transient stability of a power system network. Power Systems Conference and Exposition. 15-18 March 15–18, pp. 1–7.

74 Usaola, J., Ledesma, P., Rodriguez, J.M., Fernandez, J.L., Beato, D., Iturbe, R., and Wilhelmi, J.R. 2003. Transient stability studies in grids with great wind power penetration. Modelling issues and operation requirements. Power Engineering Society General Meeting, IEEE, July 13–17, pp. 1534–1531.

75 Gautam, Durga, Vittal, Vijay, and Harbour, Terry. Impact of increased penetration of dfig-based wind turbine generators on transient and small signal stability of power systems. *IEEE Trans. Power Syst,* **24**(3), 1426–1434.

76 Hughes, F. M., Lara, O. A., Jenkins, N., and Strbac, G. 2005. Control of DFIG based wind generation

for power network support. *IEEE Trans. Power Syst.*, **20**(4), 1958–1966.

77 Muyeen, S. M., Takahashi, R., Murata, T., Tamura, J., and Ali, M. H. 2007. Transient stability analysis of permanent magnet variable speed synchronous wind generator. Proceeding of International Conference on Electrical Machines and Systems, October 8–11, Seoul, Korea.

78 Meegahapola, L., Flynn, D., and Littler, T. 2008. Transient stability analysis of a power system with high wind penetration. 43rd International Universities Power Engineering Conference.

79 Gautam, Durga, Vittal, Vijay, and Harbour, Terry. Impact of increased penetration of dfig-based wind turbine generators on transient and small signal stability of power systems. *IEEE Trans. Power Syst.*, **24**(3), 1426–1434.

80 Sánchez, J.A., Veganzones, C., Martínez, S., Blázquez, F., Herrero, N., and Wilhelmi, J.R. 2008. Dynamic model of wind energy conversion systems with variable speed synchronous generator and full-size power converter for large-scale power system stability studies. *Renewable Energy,* **33**, 1186–1198.

81 Wu, Feng, Zhang, Xiao-Ping, and Ju, Ping. 2008. Modeling and control of thewind turbine with the direct drive permanent magnet generator integrated to power grid. DRPT2008, April 6–9 Nanjing, China.

82 Kima, Hong-Woo, Kimb, Sung-Soo, and Koa, Hee-Sang. 2010. Modeling and control of PMSG-based variable-speed wind turbine. *Electric Power Systems Research,* **80**, 46–52.

83 Kazachkov, Y. A., Feltes, J. W., and Zavadil, R. 2003. Modeling wind farms for power system stability studies. *Power Engineering Society General Meeting,* **3**, 1526–1533.

84 Hagstrøm, Espen, Norheim, Ian, and Uhlen, Kjetil. 2005. Large scale wind power integration in norway and impact on damping in the nordic grid. *Wind Energ.,* **8**, 375–384.

85 Mendonca, Â. and Lopes, J. A. P. 2005. Impact of large scale wind power integration on small signal stability. Proceedings of the 2005 International Conference on Future Power Systems, Amsterdam, The Netherlands, November 16–18, 2005.

86 Tabesh, A. and Iravani, R. Small-signal dynamic model and analysis of a fixed-speed wind farm - a frequency response approach. *IEEE Trans. Power Delivery,* **21**, 778–787.

87 Rouco, L. and Zamora, J. L. Dynamic patterns and model order reduction in small-signal models of doubly fed induction generators for wind power applications. IEEE Power Engineering Society General Meeting, June 18–22, 2006.

88 Wu, Feng, Zhang, Xiao-Ping, Godfrey, K., and Ju, Ping. 2006. Modeling and control of wind turbine with doubly fed induction generator. *IEEE Power Systems Conference and Exposition,* **1**(5), 751–760.

89 Hong, Tang, Junling, Wu, and Shuangxi, Zhou. 2004. Modeling and simulation for small signal stability analysis of power system containing wind farm. *Power System Technology,* **28**(1), 38–41.

90 Hongguang, Zhang, Lizi, Zhang, Shuyong, Chen, and Ning, An. 2007. Influence of large scale wind farms on power system small signal stability and damping characteristic. *Power System Technology,* **31**(13), 75–80.

91 Hongliang, Guan, Yongning, Chi, Huizhu, Dai, and Yihan, Yang. 2008. Small signal stability and control of wind turbine with asynchronous generator integration into power system. Automation of Electric Power Systems, Feb.25, 2008, **32**(4)

92 Hagstrøm, Espen, Norheim, Ian, and Uhlen, Kjetil. 2005. Large scale wind power integration in norway and impact on damping in the nordic grid. *Wind Energ.,* **8**, 375–384.

93 Rahimi, Mohsen and Parniani Mostafa. 2009. Dynamic behavior and transient stability analysis of fixed speed wind turbines. *Renewable Energy,* **34**, 2613–2624.

94 Wu, F., Zhang, X.-P., Godfrey, K., and Ju, P. 2007. Small signal stability analysis and optimal control of a wind turbine with doubly fed induction generator. *IET Gener. Transm. Distrib.,* **1**(5), 751–760.

95 Wu, F., Zhang, X.-P., and Ju, P. 2009. Small signal stability analysis and control of the wind turbine with the direct-drive permanent magnet generator integrated to the grid. *Electric Power Systems Research,* **79**, 1661–1667.

96 Tabesh, Ahmadreza and Iravani, Reza. 2006. Small-signal dynamic model and analysis of a fixed-speed wind farm-a frequency response approach. *IEEE Trans. on Power Delivery,* **21**(2),778–787.

97 Elkington, K., Knazkins, V., and Ghandhari, M. 2007. Modal analysis of power systems with doubly fed induction generators. 2007 iREP Symposium - Bulk Power System Dynamics and Control - VII, Revitalizing Operational Reliability August 19–24, Charleston, South Carolina,.

98 Xiaolin, Su and Xiaoxia, Yan. 2007. *Power System Analysis.* Beijing: China Electric Power Press.

99 Vittal, Eknath, McCalley, J.D., Ajjarapu, V., and Harbour, Terry. Wind penetration limited by thermal constraints and frequency stability. 39th North American Power Symposium, September 30–October 2, pp. 353–359.

100 Vittal, Eknath, McCalley, J.D., Ajjarapu, V., and Harbour, Terry. 2007. Wind penetration limited by

thermal constraints and frequency stability., 2007. NAPS '07. 39th North American Power Symposium September 30–October 2, pp. 353–359.

101 Morren, J., Pierik, J., and de Haan, S.W.H. 2006. Inerial response of variable speed wind turbines. *Electric Power System Research,* **76**, 980–987.

102 Conroy, James F. and Rick Watson, Rick. 2008. Frequency response capability of full converter wind turbine generators in comparison to conventional generation. *IEEE Transactions on Power Systems,* **23**(2), 649.

103 Lalor, G., Mullane, A., and O'Malley, M. 2005. Frequency control and wind turbine technologies. *IEEE Transactions on Power Systems,* **20**(4), pp. 1905–1913.

104 Morren, Johan, de Haan, Sjoerd, W. H., Kling, Wil L., and Ferreira, J. A. 2006. Wind turbines emulating inertia and supporting primary frequency control. *IEEE Transactions on Power Systems,* **1**(1), 433–434.

8

WF Power Quality and Its Improvements

8.1 Overview

As we know, wind turbines have an impact on the power grid, and are also under the impact of the power grid when connected to the public grid. Interactions between wind turbines and the grid include the change of short-circuit power, frequency and voltage fluctuations, as well as the flicker effect caused by fluctuations of output power, voltage asymmetry, harmonics, and interharmonics. The cage-type asynchronous generators in the wind farm (WF) with capacitance compensation may result in self-excitation and so on.

The public grid has to avoid devastating effects caused by wind power. For this reason, it is equipped with over-voltage, short circuit, and generator protection. When the voltage and frequency devite from normal operating condition, a quick disconnection from the grid of wind turbines must be ensured. When the wind turbine is operating as a motor, only a very short time is allowed. The power factor of the wind turbine generator must be kept within the range predetermined by the power grid.

Power quality problems of WFs (including single wind turbines) connected to the distribution network have special significance. The capacity of large wind turbines has reached a level of several megawatts (MW), which feeds to the distribution network with a high source impedance, and directly affects the power supply quality of end users. For variable speed wind turbines using power electronic converters, switching the converter will produce harmonics. For constant speed wind generators, regardless of whether the generator is connected to a stationary wind turbine or a wind turbine running at a certain speed, the starting current should be as small as possible. In the former case, the impact of a large impact load of starting current on each component of the wind energy converter shall be avoided; in the latter case, the emission interference (flicker) and grid voltage drop shall be prevented.

Wind turbines under normal operation will produce a continuously varying power output and this power variation is caused by disturbances, wind shear, tower shielding, and operations of the control system. These impacts result in periodic power fluctuations, and the fluctuation frequency is the frequency of the blades sweeping over the tower (for large wind turbines, the frequency is typically about 1 Hz), which is superimposed on a slow power variation caused by the wind speed. Power fluctuations of higher frequency (a few Hz) may also be caused by wind turbine dynamics. The use of fixed-speed cage-type asynchronous generators can cause periodic changes in the output power and voltage. Because of the smoothing effect of rotor flywheel, many fast power variations generated by the variable speed wind turbine are not transmitted to the network.

In terms of voltage quality, the WF and the grid interact with each other. To identify the origin of some power quality in the grid is easy, but the other is difficult. Voltage change of each wind turbine is independent, its wind speed, line impedance, and compensation are different from the others. The most important is the voltage quality at the grid connection point of the WF or the point of common coupling (PCC).

When a large WF is studied, it cannot be expressed as a wind turbine. A large WF should be divided into several groups according to its characteristics (geographic location, wind turbine type, control settings, and line impedance, etc.).

Many countries have paid great attention to power quality problems caused by WFs, and thus have developed a number of criteria in this regard. The following data listed in IEC61400-21 (2001) is related to wind turbine generators power quality [1]:

8.1.1 Rated Power (P_n, Q_n, S_n), Rated Voltage (U_n), and Rated current (I_n)

Rated power $\boldsymbol{P}_n$ is designed achievable maximum continuous electrical power output of the wind turbine under normal operating conditions; rated reactive power $\boldsymbol{Q}_n$ is the reactive power of the wind turbine when generating rated active power at normal voltage and frequency;

rated apparent power S_n is the apparent power of the wind turbine when generating rated active power at normal voltage and frequency; rated current I_n is the current of the wind turbine when generating rated active power at normal voltage and frequency.

8.1.2 The Maximum Allowable Power Output P_{mc}

It is the possible average power output of the wind turbine within 10 minutes, depending on the design of wind turbine, and its value can be over-rated. The maximum allowable power output P_{mc} can be used to provide a desired maximum average power output of the wind turbine within 10 minutes.

Wind turbines with the control of active power output (by a pitch control and/or speed control) can typically provide $P_{mc} = Pn$. Wind turbines with passive control of output power (such as constant speed or stall-controlled wind turbine) can be set P_{mc} higher than P_n by about 20%.

8.1.3 The Maximum Measured Power P_{60} and $P_{0.2}$

The maximum measured power P_{60} is measured average value within 60 s, the maximum measured power $P_{0.2}$ is the measured average value within 0.2 s. They are used for two purposes, the first is for considering the coordination of relay protection settings; the second is that they have a special relationship with the wind turbine during the islanding operation.

Variable speed wind turbine can typically provide $P_{60} = P_{0.2} = P_n$. For stalled or pitch controlled constant speed wind turbine, $P_{0.2}$ is usually greater than P_n.

8.1.4 The Reactive Power Q as a Function of the Active Power (10 Minute Average Value)

The reactive power of the wind turbine is defined as the average value within 10 minutes, which is a function of the average output power of 0.1%, 90%, and 100% of the rated power within 10 minutes. It also provides the reactive power respectively corresponds to P_{mc}, P_{60}, and $P_{0.2}$.

The reactive power of the cage-type wind turbine directly connected to the grid is a function of the active power output. Typically, the capacitor compensation of grouping switching is used. Wind turbines using modern frequency converters can usually control the reactive power to zero, or can deliver or absorb reactive power as needed, which of course, is limited by the converter rated capacity.

8.1.5 The Flicker Coefficient for Continuous Operation $C_e(\psi_k, v_a)$ Is a Function of the Network Impedance Angle ψ_k and Annual Average Wind Speed v_a

Flicker is defined as a feeling of visual instability stimulated by light, of which the brightness or spectral distribution fluctuates over time. Voltage fluctuations may cause flicker. With flicker tester, the time series value of the voltage fluctuation measured or simulated within 10 minutes can be converted to the short-term flicker value Pst.

Fluctuations in the power output of wind turbines may cause voltage changes. Therefore, the wind turbine is a source of flicker. Generally speaking, flicker may be caused by continuous operation and switching operation of a wind turbine, such as cutting in a wind turbine.

Flicker emission of a wind turbine during continuous operation can be characterized by a flicker coefficient. Flicker coefficient is a normalized measure of the flicker emission of a wind turbine. Therefore, the relationship of the flicker value Pst measured at the terminal of the wind generator operating in the grid, in the absence of any other source of disturbance, and the flicker coefficient $C_e(\Psi_k, v_a)$ is shown in Formula (8.1).

$$C_e(\Psi_k, v_a) = P_{st}\frac{S_{kt}}{S_n} \tag{8.1}$$

Where Ψ_k is the grid impedance angle; v_a is the annual average wind speed at hub height of the wind turbines; S_{kt} is the short circuit apparent power at the wind turbines terminal; S_n is the wind turbine rated apparent power.

Since the flicker coefficient C_e of the wind turbine is a function of the grid impedance angle and the annual average wind speed, as shown in Formula (8.1), C_e must be determined by using a series of Ψ_k and v_a. The flicker coefficient of variable speed wind turbine is relatively lower, and the flicker coefficient of the constant speed wind turbine ranges from average (stall control) to a higher value (pitch control).

Flicker emission caused by switching operations can be characterized in a similar manner, but some additional parameters shall be introduced

8.1.6 The Maximum Number of Times of Switching Operations N_{10} and N_{120} Within 10 Minutes and 120 Minutes

There are several switching conditions: the wind turbine starts at cut-in speed; the wind turbine starts at rated wind speed; and in the most severe cases, to switch between the generators (only used for more than one generator or a generator with multiple windings). Because these are likely to cause significant voltage fluctuations.

The acceptability of switching operation depends not only on its effect on the voltage, but also on how frequently they occur. Therefore, the maximum number of times of switching operations within 10 minutes N_{10} and 120 minutes N_{120} is set via the control system of the modern wind turbine.

8.1.7 The Flicker Step Factor When Wind Turbine is Switching $k_f(\psi_k)$

The flicker step factor $k_f(\psi_k)$ and the voltage change factor are both functions of the network impedance angle ψ_k. The flicker step factor is the flicker normalized measure caused by a single switch operation of the wind turbine. It is calculated according to Formula (8.2):

$$k_f(\psi_k) = \frac{1}{130}\frac{S_k}{S_n}P_{st}T_p^{0.31} \tag{8.2}$$

where T_p is the duration of voltage fluctuation caused by switching operations; P_{st} is the flicker caused by the wind turbine; S_n is the rated apparent power of the wind turbine; S_k is the short circuit apparent power of the network.

Flicker step factor must be given for the specified network impedance angle (30°, 50°, 70°, 85°) and the specified types of switching operations (see section 8.1.5).

The variable speed wind turbine usually gets fairly low flicker step factor, whereas the flicker step factor of the constant speed wind turbine ranges from average value (pitch control) to a high value (stall control).

8.1.8 The Voltage Change Factor $k_u(\psi_k)$

Voltage change factor is the normalized measure of voltage changes caused by a single switch operation of the wind turbine. It is calculated according to Formula (8.3):

$$k_u(\psi_k) = \sqrt{3}\frac{U_{\max}-U_{\min}}{U_n}\frac{S_k}{S_n} \tag{8.3}$$

where $U_{\max}$ and $U_{\min}$, respectively, represents the maximum and minimum voltage (effective value RMS); U_n is the normal line voltage; S_n is the wind turbine rated apparent power; S_k is the network short-circuit apparent power.

Voltage change factor must be given for the specified network impedance angle value (30°, 50°, 70°, 85°) and the specified types of switching operations (see section 8.1.6).

Voltage change factor k_u is similar to the impact current factor k_i. k_i is the ratio of the maximum impact current as well as the rated current. Although k_u is a function of the network impedance angle ψ_k, the value of the maximal k_u will be close to k_i.

The variable speed wind turbine usually expects relatively low voltage change factor, and the constant speed wind turbine voltage change factor ranges from average value (pitch control) to a high value (stall control).

8.1.9 Maximum Harmonic Current I_h, Average of Each Harmonic Current During the Period of Continuous Operation for 10 Minutes, Until the 50th Harmonic

The injection of AC current containing high-order harmonics and interharmonics into the grid may interfere with the voltage waveform. Therefore, the high-order harmonic and the inter-harmonic current shall be limited, so that the voltage waveform distortion will not exceed the prescribed limits.

A wind turbine directly connected to the grid without any power electronic converter, under normal circumstances, continuous operation does not cause significant harmonic or inter-harmonic distortion. When starting, due to the effect of the thyristor used for soft start, there may be a sudden distorted pulse for a short time. As long as this pulse does not occur frequently, and its duration time is less than tens of seconds, this pulse is generally not a problem.

Wind turbines with power electronic converters in variable speed operation will produce a limited amount of harmonic and inter-harmonic distortions. Therefore, each harmonic current must be specified, until it reaches the allowable value of fiftieth harmonic current distortion. When the maximum value of each harmonic current is given by the output power, the harmonic current should be represented by the average data of the harmonic within 10 minutes.

Basically, the integration of wind power into the grid does not seriously reduce the distribution network power quality. The general experience is that when a WF with a large number of generators (more than 10) is connected to the grid, the main power quality problem is the increase of static voltage; for hundreds of wind turbines in the weak grid, the limiting factor is often the transient voltage change. In the design and development of new large-scale wind turbines, the power quality is an important issue to be considered, and a factor to improve variable-speed operation and "active stall" control requirements, because the variable speed operation and "active stall" can both reduce the transient change in the output power.

This chapter describes the power quality problems, especially power quality problems related to wind power. In this chapter, we describe the characteristics of power quality and its relevant standards; frequency quality and frequency regulation; voltage quality and voltage/reactive power adjustment; three-phase voltage and current unbalance; power quality associated with WFs; measurement and assessment of power quality of WFs connected to the grid; and measures to improve power quality of WFs.

8.2 Power Quality and Its Characteristics

The term "power quality" is used to describe how close the power supplied to the user to the appropriate quality standards, so that the terminal device can operate properly. Although power quality is affected by the transmission and distribution network operations, but it is essentially a measure of users centered. Continuous power supply complete interrupt is considered an electric network reliability problem rather than a power quality problem.

The three basic properties of electricity are voltage amplitude, voltage waveform, and frequency. An ideal three-phase AC system must continuously supply power to the load at a constant frequency (50/60 Hz) according to a predetermined voltage class with a whole sinusoidal voltage waveform. In the three-phase AC power system, the phase voltages and phase currents should be equal in amplitude, and the phase difference shall be 120 degrees. Therefore, power quality includes the three aspects of voltage quality, current quality, and frequency quality: ① voltage quality is also known as voltage amplitude quality, which is generally believed to be influenced by the supply side, and using generalized deviations between actual and ideal voltage to reflect the level of power supply; ② current quality is mainly influenced by the user, and the current quality issue generally refers to harmonics; ③ frequency quality generally refers to the specification that the synchronous frequency of system power supply is not within the range of the system rated deviation. In regions of weaker power supply, along with the rapid changes in large-capacity active load, periodic or non-periodic deviations will appear in the system frequency. Current frequency regulation control technology and power generation management have been able to better control the frequency change.

Compared with general product quality, power quality has different characteristics. Power quality does not completely depend on the electricity production department. The electrical energy is made a whole from production to consumption, power generation, supply, and use of electricity in the power system are always in dynamic balance, in which any link will have an impact on power quality. And some quality indicators (e.g., harmonic, voltage fluctuation, flicker, and three-phase voltage imbalance) often depend on the load disturbance of the user. Comprehensive safeguards of power quality are both the responsibility of the power companies, and the obligations of users (disturbing loads). Power companies should take the responsibility, ensure the power quality by management (including oversight of the user) and take technical measures. Article 28 of China's "Electricity Law" stipulates: "Power supply enterprises shall guarantee that the quality of the electricity they supply to the users is in conformity to the standards of the State, and shall solve the power quality problems caused by utility power supply facilities without delay." Article 32 stipulates: "Electricity utilization by users shall not endanger the safety of power supply and utilization, nor disturb the order of power supply and utilization."

Parameters that mark the power quality are different for different power supply (or utilization) locations at different power supply (or utilization) moment. That is, the power quality is in dynamic change in space and time; power quality is proposed to be measured by the results of probability statistics, of which the monitoring site shall be specified. Ninety-five percent of the probability value is usually taken as the measure basis in China. This feature also puts forward corresponding requirements on the performance of electrical equipment, that is, the electrical equipment should not only be able to operate normally within the specified standard value, but also have the ability to withstand over-limited operations for short time. Power quality standards are specific to the general and majority of users and special requirements on power quality of special users can be specified in the power supply agreement. Power quality content and standards will vary with the levels of economic development and techniques of various countries. Power quality control requires considerable investment (including improvements on grid structure, balance between active and reactive power, the use of a variety of frequency regulation, voltage regulation, filtering and reactive power compensation devices, and dispatching and operational technology management, etc.), and the control technology is evolving and improving continuously. In fact, deviations between power quality indicators and the ratings are inevitable. The objective of formulation power quality standards is to determine the appropriate allowable values of power quality indicator deviations. It needs to consider the safety and economy of the power system, and whether users (electrical equipment) can accept these deviations and their consequences, as well as the possibility provided by current technology and the price to be paid in order to meet the standards.

Power quality standards are a reflection of modern technology and economic level of a country. From the early 1980s, China's State Bureau of Technical Supervision has listed the development of national power quality series standards as a key project. It has so far issued five national standards, namely: GB12325-90 "Admissible Deviation of Supply Voltage"; GB12326-90 "Admissible Voltage Fluctuation and Flicker"; GB/T 14549-93 "Harmonics in public supply network"; GB / T15543- 95 "Admissible three-phase voltage unbalance factor"; and T15945-95 "Permissible Frequency Deviation for Power System." In these standards, the definitions of terms and symbols are given, and measuring instruments, methods, and calculation methods of the allowable values are explicitly stipulated. These standards are the basis of current measures of the pros and cons of power quality in China.

The reason for power quality problems being brought up is the increasing use of sensitive load equipment, including the use of various computer-based controllers and power electronic converters, as well as the users' awareness of the adverse consequences of the power system disturbances on equipment operation. With the development of economy, science, technology, and social life, the demands of application of new technologies for power quality is increasingly high. From the available statistics and research results, the problem that will bring more serious damage is the voltage amplitude quality problem, which is also the domestic and international research emphasis in recent years, so the main power quality in the narrow sense refers to the voltage quality. The part most heavily affected by the power quality disturbance is the industrial load in the distribution network, so the distribution network power quality problems is in the most important position in the overall power quality study.

There are four factors affecting the power quality: ① natural phenomena and disasters, such as lightning; ② failure of electrical equipment and device or protection mal-operation; ③ the extensive use of nonlinear, and impact pollution loads by end users; ④ man-made events. Ensuring of power quality levels needs the combination work of many aspects, in order to achieve a win-win or multi-win results.

8.3 Power Quality Problems of WFs Associated with the Grid

Wind power generation affects the power quality, and grid disturbances also affect equipment and operations of wind power systems, so in terms of power quality, WFs, and the grid interact with each other. Figure 8.1 shows the origin of the power quality when a WF is connected to the public grid.

Figure 8.1a shows the various effects on the WF that may be caused by the transmission and distribution network, for example, voltage sags at the grid-connection point of the WF (i.e., the voltage drops to 0.1 to 0.9 pu, and lasts for 1 minute) is usually caused by transmission/distribution network faults, which should be paid special attention, because in case of the low voltage and generator load reduction, the wind turbine will be over speeded; and when at low voltage, the wind turbine reactive power demand is increasing, resulting in further reduction of the network voltage. This low voltage may cause the contactor and the voltage-sensitive control circuits to trip off. Voltage increase (over-voltage up to 1.1 pu and above) also occurs in WFs, so it should be paid attention.

Harmonic voltage distortion will result in increase in losses of wind generators, and may interfere with the current characteristic of the control system and power electronic converters. Typically, asynchronous wind turbines (fixed speed) are equipped with power factor correction capacitors, which present a low impedance to the harmonic current, and may cause resonance oscillations with other inductive reactance in the network.

Momentary interruption and automatic reclosing of the grid may also bring great harm, because automatic reclosing may be out-of-phase reclosing at asynchronous generator with voltage. Usually, a main protection quick trip is a measure to deal with this situation. In case of power grid interruption, the wind turbines will run off-grid.

Grid voltage imbalance will affect the rotation of asynchronous generators, and causes increased losses in the generator and torque ripple. Voltage imbalance can also make a power converter to inject unwanted harmonic current into the grid, unless consideration of the situation of imbalance in the power supply has be considered in its design. During normal operation of the asynchronous generator (i.e., slip s is close to zero), its effective resistance relative to the rotor with negative sequence (unbalanced) voltage is only $R_r/2$, so any small imbalance in the voltage applied will result in a large current imbalance, which is a particular problem of the wind turbine connected to the 10 kV distribution network. The wind turbine connected to 10 kV three-phase circuit will produce a large imbalance current (up to 15% to 20%).

Figure 8.1b shows how disturbances are introduced into the distribution network by the wind turbine, thus

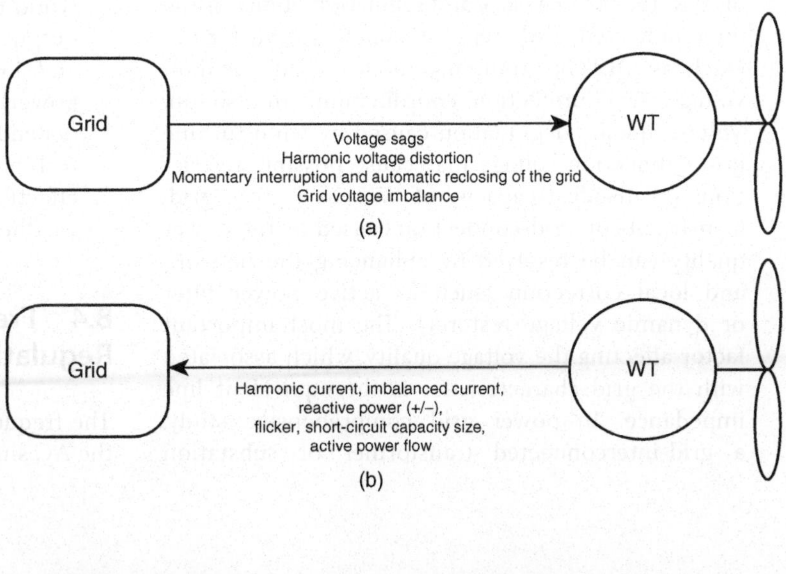

Figure 8.1 Origin of power quality problems of the WF. (a) Impact of the power grid on wind turbines; (b) Impact of wind turbines on the grid.

causing reduction in power quality. For example, constant speed wind turbines could produce or absorb reactive power and output active power, depending on the network, load and power generation levels, which may lead to undesirable steady state voltage deviation; when using a soft starter or interconnected through a power electronic conversion device with the system, the wind turbine will cause harmonics, and may inject harmonic current into the distribution network; wind turbine unbalanced operation will result in a negative sequence current injected into the network, which in turn causes network voltage imbalance; continuous operation of the wind turbine and switching operations may cause voltage flicker; impact of the wind turbine on the grid depends largely on the short-circuit capacity size of grid-connection point of the WF (wind turbines), the greater the short-circuit capacity, the smaller the impact.

Power quality problems have particular significance for the WF. Factors affecting the power quality of WFs can be divided into the following aspects [2]:

1) **Wind turbine technology**
   Includes the different types of generators; with or without gearbox transmission; whether the wind turbine is directly/controllably connected to the power grid. A typical characteristic of the WF is built on the basis of an asynchronous generator (currently mainly double-fed asynchronous generator), which conveys variable power to the grid, and its power flow (and reactive power flow together) may impact the flicker emission and voltage distribution. Although this can be offset by reactive power compensation device, such as double-fed asynchronous generator or the generator with full rated power converter typically provides less active output fluctuations and reactive power compensation, the drawback of using power electronic converters is the possible production of larger harmonics.
2) **PCC grid conditions**
   Includes the PCC short circuit capacity and the ratio of X/R; the voltage class and regulation ability of the interconnected grid; type of coupling transformer (such as OLTC); grounding mode of the medium voltage grid; protection coordination, and so on. Voltage quality degradation caused by wind turbine grid-connection voltage may cause possible restrictions on installed capacity of WFs in a given grid. Constraints on grid-connection caused by the power quality can be resolved by enhancing the network and local correction (such as active power filter or dynamic voltage restorer). The most important factor affecting the voltage quality, which associated with the grid characteristics is the equivalent line impedance. In power grid comprehensive study, a grid-interconnected transformer or substation short-circuit power and phase angle (S_k and ψ_k) are usually introduced. Network enhancement can improve the short-circuit power of PCC, in order to allow integration of large WFs. However, due to economic reasons, this is not always feasible.
3) **WF structure and control**
   Includes the number and rated power of the wind turbines; characteristics of the WF collection system (X/R); impact of the available capacity of the WF internal cable system; power/voltage control and regulation capabilities. WF topology is mainly constrained by the micro address of the wind turbine. For micro-siting of the wind turbine, it shall not be installed in the wake flow of other wind turbines. It is usually believed that the power fluctuations (the International Electrotechnical Commission [IEC] power quality standard is equivalent as flicker) caused by several wind turbines can be offset according to $1/\sqrt{N}$, where N is the number of wind turbines, electric series or regions in the WF. Therefore, in the initial stages of WF construction or electrical design, for the same wind power installed capacity, optimization shall be made among different WF structures or electrical structures, which have different effects on the power quality, this is very important in a weak isolated system (such as islands). Although the WF topology is not stated in any of the existing standard, but its impacts on the dynamic characteristics of the WF, that is, the impacts on the power quality of the WF cannot be ignored.
4) **The local characteristics of wind flow**
   Includes turbulence intensity; the situation of wind turbines operating in wake flows; the wind speed and direction; the spatial variability of the wind. The most typical characteristic of primary energy sources that drive the wind turbine, namely, the wind flow is characterized by its intense non-fixed and non-random characteristics, which is generally commonly called "turbulent flow" or "turbulent current." Atmospheric conditions or physical characteristics of the wind (wind turbulence, etc.) cause fluctuations of the WF output power and fluctuations of WF bus voltage and PCC voltage, thus affecting the voltage (and power) quality of WFs. Of course, this impact also depends on wind turbine technology. Modern wind turbines with reactive power control capabilities can effectively control the voltage quality, even in severe weather conditions.

8.4 Frequency Quality and Frequency Regulation

The frequency of the power system is the frequency of the AC sinusoidal fundamental voltage generated by the synchronous generator in the power system. All generators are in synchronous operation under steady-state conditions, and the frequency of the entire system is equal, which is a consistent system-wide parameter. The rated frequency of the power system is 50 Hz or 60 Hz, and 50 Hz is generally applied in China and Europe, whereas 60 Hz is applied in most areas of Americas. Power generation and electrical equipment in the power system is designed and manufactured in accordance with the rated frequency, and can only function best when operating around the rated frequency. When the system frequency variation is too large, operations of the users and power plants will be adversely affected.

Quality of frequency is an important indicator for measuring power quality. The Chinese national standard "Permissible Deviation of Frequency for Power System" issued in 1995 stipulates: "the normal power system frequency deviation allowable value is ± 0.2 Hz. When the system capacity is small, the deviation allowable value can be "±0.5 Hz," "system frequency variations caused by the user impact load generally shall not exceed ± 0.5 Hz." Frequency deviation allowable values of some provisions of the industrial developed countries are not greater than ± 0.1 Hz.

Frequency of the power system can only remain unchanged when the total active output of all generators is equal to the total active load (including all the grid losses), and when the total active output and total load is imbalanced, the speed of each generator set and its corresponding frequency would change. The load of the power system is constantly changing, changes in any load would cause power imbalance of the entire system, leading to changes in frequency. In power system operations, the output of each generator shall be timely adjusted (by adjusting the prime mover power elements—inputs of vapor or water), in order to keep the frequency deviations within the allowable range. Frequency regulation is mainly related to active power regulation and control. Therefore, in practice, frequency regulation should be combined with power plant active power control, economic distribution of the system active power and tie-line power control.

In order to maintain qualified frequency quality of the power system at any time, dispatch operation management of the system shall also must be prepared: prepare accurate daily load plan; develop and release generating task to each power plant, and ensure the main frequency regulation plant can adjust the system frequency according to the provisions, under the conditions that the economic operations of turbines and boilers of the power plant, system security, and reliability are ensured, and sufficient frequency regulation capacity is left for the main frequency regulation plant.

Frequency regulation of interconnected power systems is more complex, so different frequency regulation modes can be applied: the regulation mode based on the frequency, the regulation mode based on the tie-line exchange power and the regulation mode based on deviations of the frequency and the exchange power. Frequency regulation work using complex frequency regulation modes shall be done through the dispatching automatic system.

In case of large active power shortage occurring in the power system or in local system after splitting, significant frequency drop will cause the output of the turbo generator unit to drop or the turbo generator unit to trip off, and cause a vicious circle of further decline in the frequency of the system and further reduction of the active power, thus results in blackouts of the power system or the local system. This phenomenon is called frequency collapse, and is the most severe fault in the power system, because it makes the whole system or local system after splitting collapse and lose the entire load.

Frequency deviation caused by power system normal load fluctuation is generally compensated by the generator power regulating system. In case of severe power shortage, the power growth of the generator speed regulating system cannot immediately compensate for this power shortage. The main measure to prevent frequency drop and frequency collapse in case of serious power shortage of the system is load-shedding, using chain load shedding for faults that will cause serious power shortage in a certain area, and implementing low-frequency load shedding according to a certain logic based on the degree of frequency drop.

Systems containing a large number of wind power on the one hand can affect the system frequency changes: integrating into or removal of massive wind power may cause system frequency increases or decreases, which may even beyond the frequency limits of power grid operation. WFs are not like conventional power plants, which can increase the output when needed by the system to promote the increase of grid frequency, unless the WF has implemented partial load operation.

Output of wind turbines is closely related to the wind speed and direction. Therefore, changes in wind speed bring great impact on the wind turbine output. The output of the wind turbine may change from zero to full output. Within the corresponding period (10 minutes to 30 minutes) of power system load-frequency control (LFC), wind power is likely to reach full output. Therefore, fluctuation characteristics of wind power must be considered for the study of problems of grid LFC, which contains a large amount of wind power. In the study of static frequency control, wind turbine output power can be seen as relevant only to wind changes, regardless of the system frequency.

Wind power frequency control requirements include frequency response capabilities, active change rate (ramp rate), and active power limitations. The frequency response is the ability to change active output in response to system frequency changes. Frequency response ability of wind power is poor, in the case of high wind power penetration, especially in the "low demand, high wind power," wind turbines usually only rely on conventional generators to provide frequency response or purchase its frequency control obligations from other generators.

Limiting the active power change rate is to limit the system frequency fluctuations. Because extreme change of wind speed and start-up and stopping of the WF will cause large frequency fluctuations, therefore, the active output change rate of the WF shall be limited in positive or even negative, for example, Germany seeks a positive WF active power change rate of 10% of the rated power per minute. Scottish provides that, WFs shall have a negative active power change rate of 3.3% of the rated power output per minute.

The normal frequencies in a power system under synchronous operation are the same, and are generally maintained within a narrow bandwidth. If the frequency deviation exceeds the normal range and endangers the power system reliability, it will endanger the electrical equipment, including power electronic devices. In the past, the grid operator requires wind turbines to disconnect from the grid in case of frequency over-limits. However, if the wind turbine capacity trips from the grid at low frequencies, it may exacerbate the system frequency drops, and affect the restorability of the system. For this reason, some countries and regions have developed new specifications, requiring wind turbines remain connected to the system over a wide frequency band range, see Figure 8.2. The specifications stipulate the outputs of wind turbines connected to the power grid, and the allowed operation time limits, to help restore the grid frequency and make the power system operate stably.

8.5 Voltage Quality and Voltage Regulation

Voltage quality is the specifications of the power system operation voltage and supply voltage. It consists of three elements: deviation of the voltage value, fluctuations and flicker of voltage value, voltage dips or sags, voltage swells or cracks, as well as AC voltage waveforms (harmonics). Voltage flicker and waveforms will be described in another chapter.

8.5.1 Voltage Value Deviation

Deviation of the voltage value refers to the offset of the operating voltage from the rated voltage, which is commonly expressed as a percentage (Formula (8.4)):

$$\text{Voltage deviation } \Delta u\% = \frac{\text{Node actually measured voltage value-rated value}}{\text{rated value}} \times 100\% \tag{8.4}$$

Voltage deviation is an important technical indicator of power quality. The China national standard "Admissible Deviation of Supply Voltage" issued in 1990 provides for supply voltage of 35 kV or above, sum of absolute values of positive and negative deviations of the supply

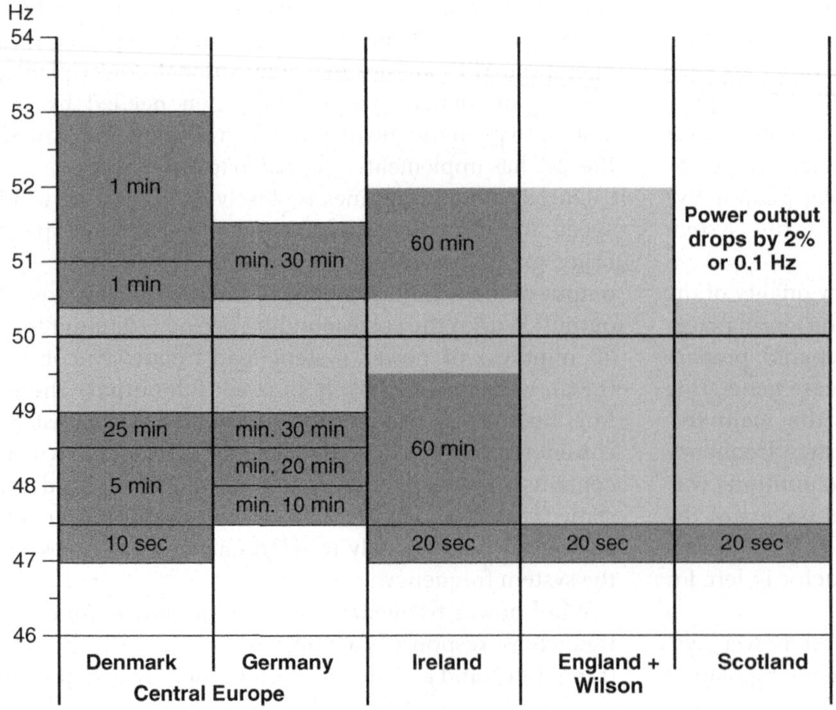

Figure 8.2 Requirements for WF operation frequency range of some countries.

voltage shall not exceeds 10% of the rated voltage. If the upper and lower supply voltage deviations are of the same sign (both positive and negative), the larger absolute value of the deviations shall be used as a measure basis. Three-phase supply voltage admissible deviation of 10 kV and below is ±7% of the rated voltage; single-phase supply voltage admissible deviation of 220 V is +7% and −10% of the rated voltage.

Different from the power system frequency, the voltage level of the power system and the voltage distribution of each bus depend on voltage of each power supply point, load size and nature of each bus, reactive power distribution and operation in the power system, especially the reactive power flow distribution. Load of the power system is changed at any time, and operation and power flow distribution of the power system are also changed at any time, so the voltage level of the power system and the voltage distribution of each bus are functions of time and space, that is, voltages of the same bus are time functions, whereas voltages of different buses at the same instant are different. To make the voltage of each bus in the power system at all times comply with the requirements of power quality, the voltage must be adjusted.

Wind power grid-connection changes the power flow distribution and the bus voltage level of the system. Especially when integrated into a distribution network with local load, due to significant changes in the power flow, or even due to bidirectional flow, the voltage deviation may be very large. When WFs are connected to a weak grid, as the wind power output increases, the voltage deviation may exceed the prescribed limits. For a WF, voltage problems are often the most important factors restricting the integration capacity of wind power. Thus, the WF and its voltage regulation are the basic problems of wind power systems.

In high-voltage grid, the voltage regulation is essentially achieved by adjusting the reactive power distribution of the power system. So first of all, in the planning, design, construction, renovation phases of the power system, the rational allocation of reactive power compensation, and reactive power shall be reasonably resolved based on the hierarchical (voltage) partitioning (power) principles, to create conditions for the voltage regulation measures taken in operation means. The total capacity of reactive power source of the power system shall be able to meet the load (including reactive demands of the power grid) demands and have some spare capacity. Reactive capacity shortage will cause the system voltage levels to drop, so that other voltage regulation measures cannot play their role.

In addition to synchronous generators, reactive power source of the power system also includes shunt capacitors, synchronous condensers, and static compensators and other reactive power compensation equipment, as well as the charge power of high voltage power lines, and so on.

Under the condition of local balance of reactive power hierarchical partitioning, the methods to achieve voltage regulation are adjusting the excitation of the generator and synchronous condenser; adjusting the regulating transformer (including load regulating transformer) tap; switching shunt capacitors and reactors; using static var compensator for automatic voltage regulation, and so on. In voltage regulation of WFs, in addition to full use of the reactive power and voltage regulation abilities of double-fed asynchronous generators and direct-drive permanent-magnet synchronous generators, we shall also draw help from the regulation of the transformer and the control of reactive power compensation equipment.

Because achieving voltage regulation has changed the reactive power flow of the power system, and reactive power flow is closely related to the network power losses and the economy of the power system operation, while adjusting the voltage, economic operation shall also be taken into account. Reactive power flow optimization is to seek for minimum network losses under the premise of meeting voltage requirements.

With the development of power system automation and higher demands for power quality, implementation of centralized voltage control for the power system based on decentralized voltage regulation is a new trend. That is, the system control center (dispatch station) monitors changes in voltage and reactive power of each pilot node in the system at any time, releases voltage regulation instructions to control the equipment of various power plants and substations through online calculation of the computer, in order to minimize the voltage deviation of each pilot bus and the network losses of the entire system. Considering that centralized automatic voltage control (AVC) of reactive power regulation of the WF is a challenging task, the randomness of wind power will make centralized automatic voltage control more difficult and frequent.

8.5.2 Voltage Fluctuations of WFs

Voltage fluctuation is another important technical indicator of power quality. Voltage deviations are slow voltage changes in a long term, whereas voltage fluctuations are short-term fast voltage changes.

Voltage fluctuations are caused by a series of voltages (root-mean-square value) changes, as shown in Figure 8.3. Voltage variation (or commonly known as voltage fluctuation) value is expressed as a percentage of the difference ΔU between two extreme values U_{max} and U_{min} of the root-mean-square values of voltages on the fluctuation characteristic curve to the rated voltage U_N (Formula (8.5)):

$$d(\%) = \frac{U_{max} - U_{min}}{U_N} \times 100\% \tag{8.5}$$

Usually, d value is used as a measure of voltage fluctuations. The root-mean-square values of fluctuation voltage

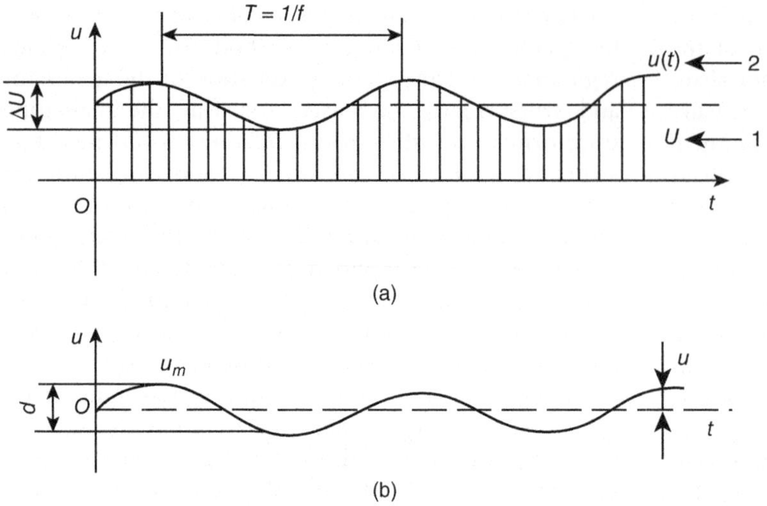

Figure 8.3 Modulation of fluctuation voltage *u* on root-mean-square voltage *U*. (a) Power grid rms voltage *U*(t); (b) Amplitude-modulated volage *u(t)*. 1—Root-mean-square voltage; 2—sine-modulated wave..

u(t) (waveform of each mean square value is represented by a vertical bar) shown in Figure 8.3a can be considered as a modulation result of amplitude-modulated wave voltage u(t) on the normal voltage [U(t) is a constant]. Figure 8.3b shows the sine-modulated wave, the horizontal coordinate is the mean value of U(t), d is the difference of u(t) between peak and valley, which is represented as a percentage of U_N. The number of the voltage changes per unit time is called frequency r, in unit of s^{-1} or min^{-1}.

When calculating the frequency, calculate in ascending order and descending order of the voltage respectively for once, the relationship between the shown frequency r and frequency F is (Formula (8.6)):

$$f(Hz) = \frac{r(s^{-1})}{2} = \frac{r(\min^{-1})}{120} \tag{8.6}$$

The voltage fluctuation limits stipulated in the Chinese national standard "Power Quality - Admissible Voltage Fluctuation and Flicker" (GB12326-90) are shown in Table 8.1.

Note: ① r in the table represents the voltage fluctuation frequency. For low fluctuation frequency r (less than once a day), the range of the value of voltage fluctuation d can also be widened; ② for random irregular voltage fluctuations, measured according to the probability of 95% of the value, values with "*" in the table are its limits.

Table 8.1 Voltage fluctuation limit at all levels.

$r(h^{-1})$	d(%) LV, MV	HV
$r \leq 1$	4	3
$1 < r \leq 10$	3	2.5
$10 < r \leq 100$	2*	1.5*
$100 < r \leq 1000$	1.25	1

If wind-powered loads in the power grid account for a large proportion, then in normal operation power output variations caused by changes in wind speed may result in voltage fluctuations. This is mainly decided by the path that providing the apparent power, and is also related to the short circuit power and the corresponding phase angle of the power grid. Voltage variations can be caused by load changes, putting into or removal of loads or generators, switching between generators, wind speed fluctuations and tower shelter.

When fixed-speed generators, especially large-capacity cage-type asynchronous generators are connected to power grids, because of low slip, power output variations caused by wind speed fluctuations are transferred directly to the grid.

When fixed-speed generators are connected to the power grid, considerable smoothing of power output can be achieved by the short-term energy storage in the rotating mass of the drive chain. With the increase of dimensions and rated power output, the wind turbine will show a high rotor speed raising time constant. Therefore, compared with small wind turbines, large variable speed (especially in a short time scale) wind turbines can obtain smoother power output. The range of wind turbine speed regulation also contributes to the smooth level of power output, wind turbines with large variations in wind speed can better restrain output power fluctuations. For wind turbines coupled via frequency converters, the starting current has little effect, because its short circuit current contribution is less than 1% of the rated current. In this kind of wind turbines, therefore, short-time voltage variation is less important.

Small- and medium-size wind turbines usually have obvious effects on the grid only when they are connected in a large group, but as more wind turbines with large output power fluctuations are connected together, the power output will be more stable. For small wind turbines, since the rotor time constant is small, smooth effects comparable with those of large wind turbines must be obtained by increasing the speed adjustment range. Large wind turbines with high output volatility operating alone can lead to weak power grid voltage changes, so power output fluctuation of large wind turbines should be as low as possible.

In determining voltage changes caused by wind turbines connected to the power grid, the so-called "voltage drop" direction related to the transformer and the line must be taken into consideration. Figure 8.4 shows the simplified equivalent circuit and phasor diagram of asynchronous machine in parallel operation. Figure 8.4a and b respectively shows the equivalent circuit of generator and motor modes, c and d are the corresponding phasor diagram. As can be seen from the figure, under the motor mode, the motor voltage U_{Mot} is usually expressed as the difference between the grid voltage U_{grid} and the voltage drop $dU = (R_{L+T} + jX_{L+T})I_{1Mot}$ (Figure 8.4d). The phasor diagram of the generator mode shows that, the grid voltage U_{grid} and the generator voltage U_{Gen} roughly have the same value (Figure 8.4c). However, the voltage drop between the two voltages have obvious phase displacement, which is known as the line impedance phase angle ψ. Obviously, the generator voltage can be either leading or lagging the grid voltage, depending on location of the generator current and the ratio of the resistance and reactance in the lines (including overhead lines and cables) and the transformer.

Figure 8.5 indicates the typical structure of wind turbines connected to the grid in a simplified form. Assume that the load at the PCC is small relative to the wind turbine power. According to the voltage u_1 at higher grid point (GP) or the rated voltage u_n, voltage drops at the PCC with and without wind turbine power can be calculated, thus the corresponding voltage value U_{PCC} at the PCC can be obtained. ΔUr in the figure characterizes the voltage rise with wind turbines at the PCC, and Δu_{rpern} characterizes its allowed deviation value (such as 2%, in medium voltage networks).

Figure 8.6 shows the relative changes in PCC voltage U_{PCC} with the impedance phrase angle (Figure 8.6a) and the capacity of short circuit (Figure 8.6b).

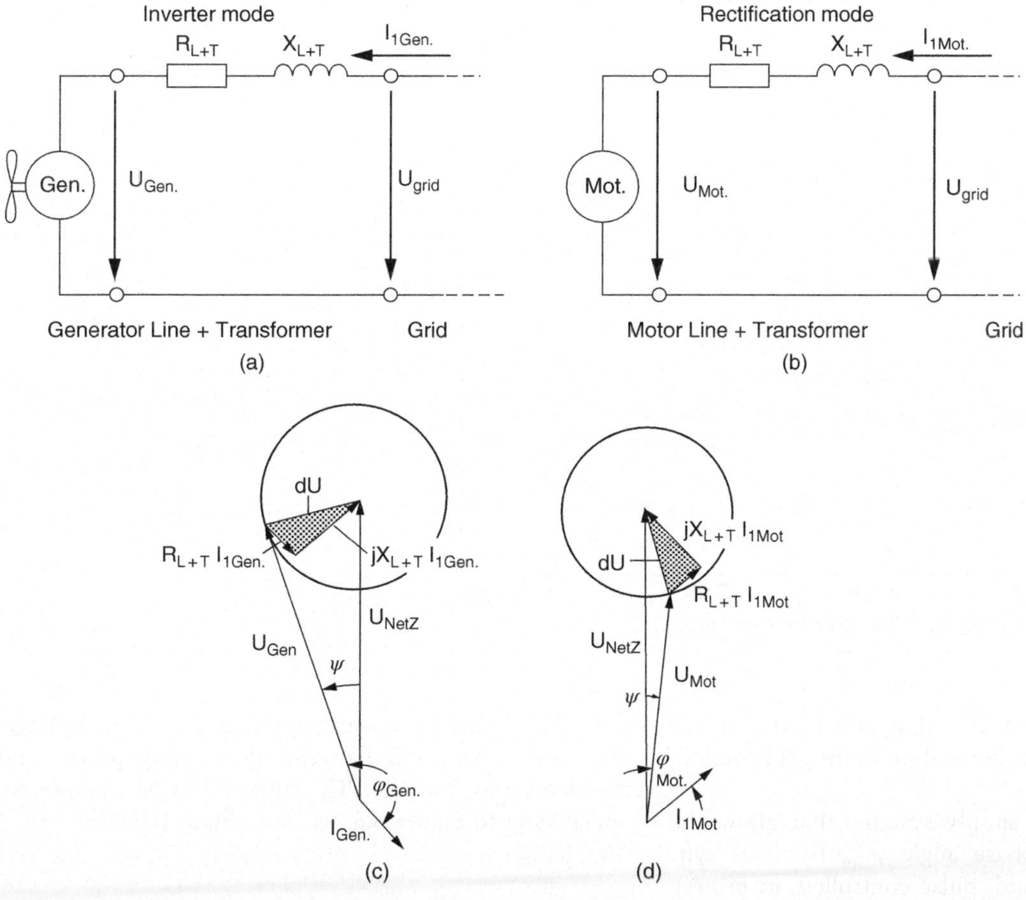

Figure 8.4 Generator/motor equivalent circuits and phasor diagrams at the grid. (a) Generator equivalent circuit, (b) Motor equivalent circuit; (c) Generator phasor diagram; (d) Motor phase diagram.

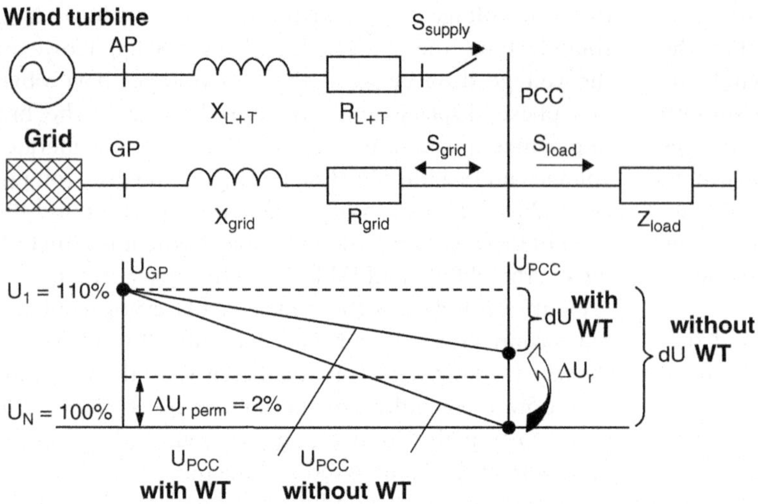

Figure 8.5 Equivalent circuit and voltage variations with and without wind turbines at the PCC.

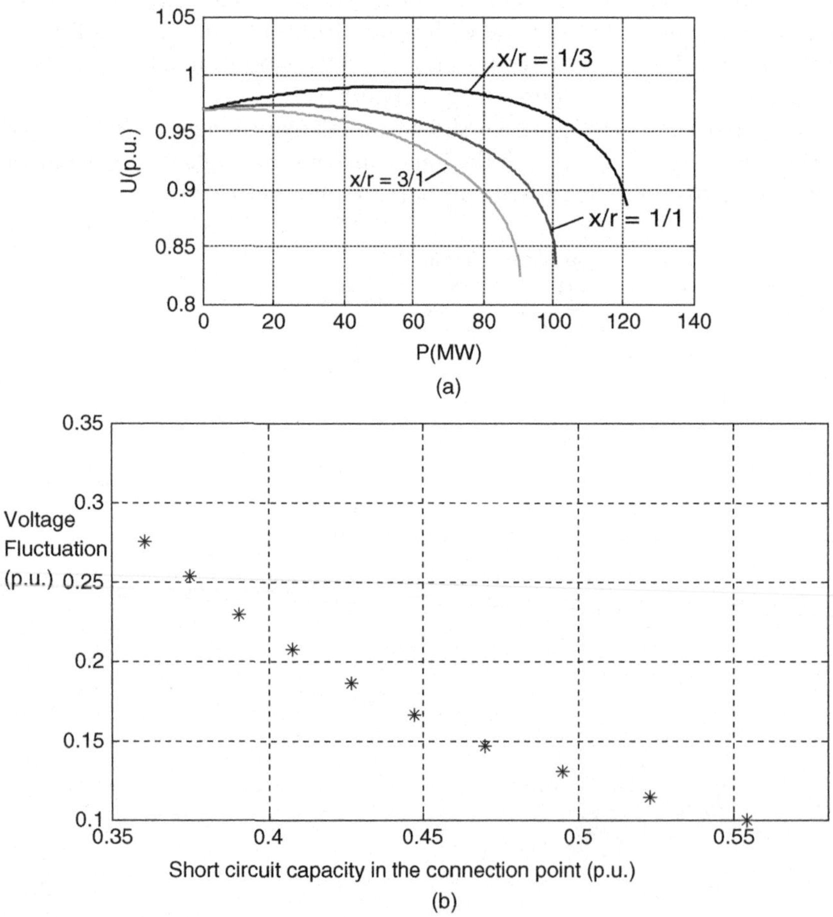

Figure 8.6 The PCC voltage changes with the impedance phrase angle (a) and the capacity of short circuit (b).

Figure 8.6b shows the PCC voltage fluctuation in the case of 20% active power fluctuation in the grid connection point.

In studies of power supply systems that allow the free readjustment of phase angle φ_{Gen} (such as synchronous generators and pulse-controlled inverters), the phase angle of power supply may affect the voltage of grid connection point. As can be known from Figure 8.6, inductive power supply decreases the voltage, whereas capacitive power supply increases the voltage. Moreover, the ratio between the supply power and the load power at the PCC should also be considered. According to Figure 8.6, we can obtain (Formulas (8.7) and (8.8)):

$$U_{PCC} = U_1 - \frac{(P_{load} - P_{supply})R_{grid} + (Q_{load} - Q_{supply})X_{grid}}{U_1} \tag{8.7}$$

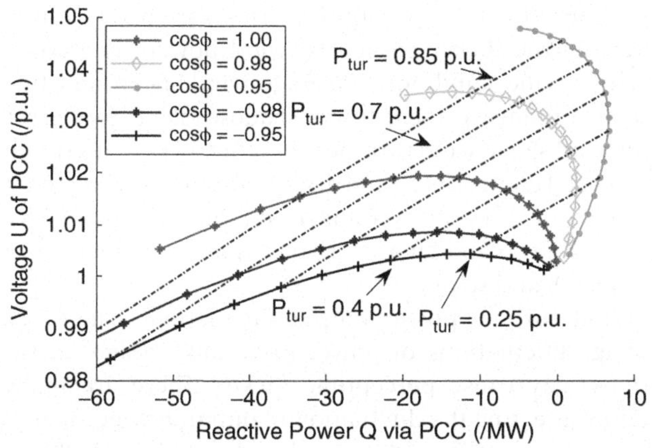

Figure 8.7 The U-Q state curves of PCC in different active and reactive power injection.

$$S_{grid} = S_{load} - S_{supply} = (P_{load} - P_{supply}) + j(Q_{load} - Q_{supply}) \quad (8.8)$$

In the same way, PCC voltage u_{PCC} considering line resistance and reactance can also be obtained according to Figure 8.5.

Figure 8.7 shows the changes of the voltage and reactive power of PCC are simulated by injecting different *active* and *reactive* power. The U-Q curves exhibiting eyebrow-like shape indicate that with the increasing of the *active* power of wind turbines, the PCC states change from right to left. Whereas those nearly parallel lines indicate the PCC states change from left to right with the increasing of *reactive* power at Bus L. (Here, negative power factor denotes inductive injection.) Obviously, the voltage and reactive power of PCC increase linearly with the growth of reactive power injection, and the gradients nearly keep constant within a large range of the operation point.

8.6 WF Voltage Flicker

8.6.1 Overview

Flicker is the intuitive visual experience of human eyes on the lighting anomaly caused by voltage fluctuations of a certain frequency. Flicker is closely related to voltage fluctuations, but is a concept different from voltage fluctuations. Voltage fluctuations are a series of voltage variations or continuous deviations. The IEC has developed strict criteria and indicators on calculation and measurement of flicker, the statistical formula for the flicker perception rate F (%) is (Formula (8.9)):

$$F = \frac{C+D}{A+B+C+D} \times 100\% \quad (8.9)$$

where A is the number of people being unaware; B is the number of people being slightly aware; C is the number of people being obviously aware; D is the number of people feeling unbearable.

Instant sensitivity factor to flicker S is the human visual response to the luminance fluctuations caused by voltage fluctuations, usually 50% of the flicker perception rate F is taken as a unit of measurement of the instant sensitivity factor to flicker, that is the perception unit $S=1$. Usually taken as the voltage fluctuation corresponding to the perception unit $S=1$ as a basis for the study of flicker, and the flicker waveform factor is defined as (Formula (8.10)):

$$R(f) = \frac{\text{Sinusoidal voltage fluctuation at the sensitivity factor of } S=1 \text{ perception unit}}{\text{Rectangular voltage fluctuation at the sensitivity factor of } S=1 \text{ perception unit}} \quad (8.10)$$

The flicker sensitivity factor coefficient recommended by IEC is (Formula (8.11)):

$$K(f) = \frac{\text{8.8 Hz sinusoidal voltage fluctuation at the sensitivity factor of } S=1 \text{ perception unit}}{f \text{ Hz sinusoidal voltage fluctuation at the sensitivity factor of } S=1 \text{ perception unit}} \quad (8.11)$$

$K(f)$ essentially describes the frequency characteristics of the lamp-eye-brain link, the maximum detectable frequency range of human to the luminance fluctuations is 0.05 to 35 Hz, and among this range, human is most sensitive to the luminance fluctuations around 8.8 Hz.

Short-term flicker value P_{st} and long-term flicker value P_{lt} are the statistical characteristics reflecting voltage fluctuations, and their calculation formula are (Formulas (8.12) and (8.13)):

$$P_{st} = \sqrt{0.0314P_{0.1} + 0.0525P_1 + 0.0657P_3 + 0.28P_{10} + 0.08P_{50}} \quad (8.12)$$

$$P_{lt} = \sqrt[3]{\frac{1}{N}\sum_{k=1}^{N} P_{stk}^3} \quad (8.13)$$

where $P_{0.1}$, P_1, P_3, P_{10}, and P_{50} respectively represents the perception value of the flicker sensitivity factor S(t) for over 0.1%, 1%, 3%, and 50% of the time within 10 min.

IEC recommends $P_{ST} = 1$ as a flicker limit of low voltage power supply, known as the unit flicker. P_{St} and P_{lt} are two indicators for determination of the flicker. For P_{St}, the statistical time shall be at least a few minutes, of which the typical value is 10 minutes; the statistical time of P_{lt} is over time 1 hour, and 2 hours is generally taken.

For the total flicker consisting of multiple flicker emission sources, equivalent severity index A_{St} (short time)

Table 8.2 IEC flicker compatible value and flicker planning value.

Parameters	Flicker compatible value LV, MV	Flicker planning value	
		MV	HV
P_{st}	1.0	0.9	0.8
P_{lt}	0.8	0.7	0.6
A_{st}	1.0	0.75	0.5
A_{lt}	0.5	0.3	0.2

Table 8.3 Flicker limits at all voltage levels.

System voltage	LV	MV	HV
P_{st}	1.0	0.9(1.0)	0.8
P_{lt}	0.8	0.7(0.8)	0.6

and A_{lt} (long time) are defined (Formulas (8.14) and (8.15)):

$$A_{st} = \sum_i A_{sti} = \sum_i P_{sti}^3 \tag{8.14}$$

$$A_{lt} = \frac{1}{N}\sum_{k=1}^{N} A_{stk} \tag{8.15}$$

IEC has developed the flicker compatible value and flicker planning value for networks of at all voltage levels (LV: low-voltage network, $U_n \leq 1$ kV; MV: medium-voltage networks, 1 kV $< U_n \leq 35$ kV; HV: high-voltage network, 35 kV $< U_n \leq 230$ kV), as shown in Table 8.2.

The allowable voltage flicker limits stipulated in Chinese national standard "Power Quality - Admissible Voltage Fluctuation and Flicker" (GB12326-90) are listed in Table 8.3.

Note: ① In GB 12326-2000, the measurement cycles of P_{st} and P_{lt} is 10 minutes and 2 h, respectively. ② MV value in brackets is only available for the case of all users are connected to the same voltage level at the PCC. ③ GB 12326-2000 voltage rating: LV ($U_n \leqq 1$ kV), MV (1 kV $< U_n \leqq 35$ kV), HV (35 kV $< U_n \leqq 220$ kV).

8.6.2 Origins of Flicker

Voltage flicker describes the dynamic changes of the network voltage, which is caused by the wind turbines or the changes of load. Flicker standard is usually used to represent the instantaneous voltage changes, and is important to wind power generation, because: ① WFs often use a large single power generating unit; ② may frequently start up and stop; ③ the power input is the wind in random; ④ power fluctuation of each wind turbine in the WF is usually not synchronized. There are many factors affecting the flicker caused by wind power generation, such as wind conditions (average wind speed and turbulence intensity, etc.), tower shadow effect, yaw error and wind shear, wind turbine type, control systems (pitch and speed control, etc.) and grid conditions (short-circuit capacity of WTG grid-connection point or PCC, power line X/R ratio and characteristics of loads connected to the PCC), and so on.

Wind turbine power output fluctuations can cause voltage fluctuations of power grid, and voltage fluctuations may cause perceptible flicker. There are many factors affecting the fluctuation of output power, such as the air density ρ, impeller speed ω, the pitch angle β, and the wind speed v, and so on. Generally, rapid changes in wind speed is at about the magnitude of 0.1 Hz, whereas the perceivable flicker frequency range by human eye is 0.1 Hz to 35 Hz, so the possibility of perceptible flicker caused by rapid changes in wind speed is little. Changes in impeller speed ω and pitch angle β are associated with the wind turbine type and the control system, advanced control systems can reduce the fluctuation of wind turbine power output.

Wind conditions, in particular the average wind speed and turbulence intensity, have a direct impact on the voltage fluctuation and flicker caused by the wind turbine. As the wind speed increases, voltage fluctuations and the flicker of wind turbines increase. When the wind speed reaches the rated wind speed and continuously increases, flicker of constant wind turbines continues to increase, but flicker of variable-speed wind turbines has begun to decrease due to the smoothing effects of the wind turbine on fluctuations in power output. The impact of turbulence intensity on voltage fluctuation and flicker is also significant, and is almost proportional to the growth of the voltage fluctuation and flicker.

Tower shadow effect is the obstruction of wind turbine tower to the air flow. When the blade passes the tower, the torque it generates is reduced. The wind speed far away from the tower is constant, as it approaching to the tower, the wind speed begins to increase, and when it gets closer to the tower, the wind speed begins to decline. Wind turbines down the wind are the most severely affected by tower shadow effect. Tower shadow effect can be represented by the Fourier series of 3p times (p is the blade rotational frequency) of the frequency. Due to the existence of the vertical wind speed gradient distribution in the blade swept area, wind shear can also cause torque fluctuations. Wind shear can be represented by the binomial series under the polar coordinate system using the wind turbine hub as the pole. Fluctuation frequencies of common torque and output power are the same as the frequency of the blade passing the tower. For three-blade wind turbines, the main frequency fluctuation is 3p or integer multiple of

3p, flicker level caused by periodic voltage fluctuations at such frequency, sometimes will reach 90% of the total wind turbine flicker value. Under normal circumstances, the 3p frequency range is 1 to 3 Hz, and the maximum torque fluctuation range is approximately 20% of the mean torque.

In addition to fluctuations of output power caused by the tower shadow effect and wind shear, fluctuation components with the frequency of p can also be detected in the output power of the wind turbine, of which the main reason for its emergence may be the incomplete symmetry of the blade structure or gravity. In addition, fluctuation component with the tower resonant frequency as its frequency is also obvious, which may be due to the lateral oscillation of the hub. Furthermore, the flicker problems caused by different types of turbines are also different, the flicker problems caused by constant speed wind turbines are more serious. Variable speed wind turbines can smooth the power output. In general, the intensity of flicker caused is only about a quarter of that of fixed-speed wind turbines.

Grid-connected wind turbines generate voltage fluctuations and flicker not only in the course of continuous operation, but also in the course of switching operation. Typical switching operations include startup and stopping of wind turbines, and switching between generators, wherein switching between generators only applies to wind turbines with multiple generators or with multiple generator windings. These switching operations can cause power fluctuations, and further cause voltage fluctuations and flicker at wind turbine terminals and other corresponding nodes.

Apart from the form of wind itself and the wind turbine characteristics, the network structure of the system, which the wind turbines are connected to also has a greater impact on the voltage fluctuations and flicker caused by it. WF PCC short-circuit capacity (short-circuit ratio), the impedance size, and the ratio of resistance and reactance (R/X) of the grid lines reflect the strength of the network, and are important factors affecting voltage fluctuations and flicker caused by wind turbines. The greater the PCC short-circuit capacity (short ratio), the smaller the voltage fluctuations and flicker caused by wind turbines. Suitable R/X value allows the voltage fluctuations caused by the active power to be compensated off by the voltage fluctuations caused by the reactive power, so that the overall average flicker value will be decreased a little. Voltage fluctuations and flicker caused by grid-connected wind turbines has non-linear relationship with line X/R ratio, when the corresponding line impedance phase angle is between 60° to 70°, the voltage fluctuation and flicker are minimal.

For wind turbines operating in a WF, as there are differences in wind speed and wind changes in time among different wind turbines, and the different rotation characteristics of wind turbine blades, the output power of each wind turbine and the voltage of each network bus have different volatility patterns, therefore, the total output power fluctuation and total voltage fluctuation of the WF can be not simply obtained from summing up the output power of all wind turbines or the voltage fluctuations caused by them. In a WF, many wind turbines operating at the same time can produce a smooth effect on fluctuations in the power output of the WF. When n wind turbines are operating at the same time, the amplitude of power fluctuation will be reduced to $1/\sqrt{n}$ of the sum of power fluctuations of all wind turbines, that is, if the number of wind turbines is subject to the non-associated changes in the torque, its power output and its impact on flicker will be reduced to (Formula (8.16)) [3]:

$$\frac{\Delta P}{P} = \frac{1}{\sqrt{n}} \frac{\Delta p}{p} \tag{8.16}$$

where n is the number of generators; P and p respectively represents the power of the WF and the wind turbine; ΔP and Δp are the corresponding amplitudes of their power fluctuations.

Of course, when multiple units are operating simultaneously, the power injected into the power grid is increased, that is, the total amount of power fluctuation is increased, so the amounts of voltage fluctuation and flicker caused thereby are also increased.

8.6.3 Calculation and Assessment of WFs Flicker

China has not developed relevant standards for calculation and assessment of WF flicker, so national standard GB/Z 19963-2005 has proposed to temporarily perform in accordance with IEC 61400-21. In December 2001, the IEC enacted the IEC 61400-21 standard "Measurement and Assessment of Power Quality Characteristics of Grid Connected Wind Turbines," which mainly proposed the principles and methods for measurement and assessment of power quality of wind turbines and WFs. At that time, most of the WFs were connected to the distribution network. The following describes some methods for flicker calculation and assessment.

8.6.3.1 Flicker Calculation and Assessment in the IEC 61400-21 Standard

1) **Assessment of flicker in the process of WF continuous operation**

   Flicker coefficient of wind turbines in continuous operation can be divided into long-term flicker coefficient and short-term flicker coefficient. According to the IE C61000-3-7, long-term flicker value can be calculated from the cubic root of the average cubic value of 12 consecutive short-term flicker values.

It is generally believed that flicker of the wind turbine is a function of wind speed, and generally the wind condition is likely to last for a long time, so the 12 consecutive short-term flicker values may be equal. Therefore, for the wind turbine, the long-term flicker coefficient is equal to the short-term flicker coefficient.

IEC 61400-21 firstly gave an assessment formula for single wind turbines. The 99th quantile point-flicker value caused by the grid-connected single wind turbine during continuous operating period can be estimated according to Formula (8.17) below:

$$P_{st} = P_{lt} = c_e(\psi_k, v_a)\frac{S_n}{S_k} \quad (8.17)$$

where $c_e(\psi_k, v_a)$ is the wind turbine flicker coefficient, which is a function of the annual average wind speed v_a at the given PCC grid impedance angle ψ_k and the given wind turbine hub height, S_n is the wind turbine rated apparent power; S_k is the PCC short-circuit capacity; P_{lt} is the long-term flicker value.

The flicker coefficient of wind turbine corresponding to the actual ψ_k and v on-site, can be obtained by using linear interpolation based on the data sheet obtained from the measurement results. Table 8.4 shows $C_e(\psi_k, v_a)$ of a single wind turbine,

Multiple wind turbines are connected to the PCC, thus, the flicker value is estimated according to Formula (8.18).

$$P_{st\Sigma} = P_{lt\Sigma} = \frac{1}{S_k}\sqrt{\sum_{i=1}^{N_{wt}} (c_i(\psi_k, v_a)S_{n,i})^2} \quad (8.18)$$

where c_i (ψ_k,v) is the flicker coefficient of a single wind turbine; $s_{n,i}$ is the rated apparent power of a single wind turbine; n_{WT} is the number of wind turbines connected to the PCC.

2) **Assessment of flicker during switching operation**
Switching operations of the wind turbine can be divided into starting up at cut-in wind speed, starting up at rated wind speed and starting up under the worst conditions. During a switching operation given in the IEC 61400-21, flicker caused by a single wind turbine can be separately estimated by using Formulas (8.19) and (8.20):

$$P_{st} = 18N_{10}^{0.31}k_f(\psi_k)\frac{S_n}{S_k} \quad (8.19)$$

$$P_{lt} = 8N_{120}^{0.31}k_f(\psi_k)\frac{S_n}{S_k} \quad (8.20)$$

where k_f (ψ_k) is the flicker step factor of ψ_k given at the PCC. The wind turbine flicker step factor corresponding to the actual ψ_k on-site can be obtained by using linear interpolation based on the data sheet obtained from the measurement results.

Table 8.4 Flicker coefficient of a wind turbine (the value in the table is only for example).

The grid impedance phase angle ψ_k (degrees)	30	50	70	85
Annual average wind speed v(m/s)	Flicker coefficient $C_e(\psi_k, v_a)$			
6.0	7.5	3.2	0.8	3.8
7.5	7.7	3.3	1.2	4.0
8.5	7.9	3.4	1.8	4.5
10.0	8.0	3.5	2.0	5.2

If more wind turbines are connected to the PCC, the total flicker can be estimated by using Formulas (8.21) and (8.22):

$$P_{st\Sigma} = \frac{18}{S_k}\left[\sum_{i=1}^{N_{wt}} N_{10,i}(k_{f,i}(\psi_k)S_{n,i})^{3.2}\right]^{0.31} \quad (8.21)$$

$$P_{lt\Sigma} = \frac{8}{S_k}\left[\sum_{i=1}^{N_{wt}} N_{120,i}(k_{f,i}(\psi_k)S_{n,i})^{3.2}\right]^{0.31} \quad (8.22)$$

where $n_{10,i}$ and $n_{120,i}$ respectively represents the number of switching operations of i th wind turbine within 10 minutes and 120 minutes; $k_{f,i}(\psi_k)$ is the flicker step factor of the i th wind turbine; $S_{n,i}$ is the rated power of i th wind turbine.

The formula for estimating voltage variation $d(\%)$ caused by switching operations of single wind turbines given in the IEC61400-21 is as follows (Formula (8.23)):

$$d(\%) = 100 \times k_u(\psi_k)\frac{S_n}{S_k} \quad (8.23)$$

where K_u (ψ_k) is the voltage variation coefficient of wind turbine corresponding to ψ_k given at the PCC.

8.6.3.2 Flicker Transmission Calculation

The installed capacity of WFs in China is growing, the WF locations are more remote, so the electric power is usually transmitted to the transmission grid via long transmission lines. In order to establish consistent flicker limits for different voltage levels in the power system, it is important to know how the voltage fluctuation spreads through the network. Flicker disturbances from various sources spread to the network according to the network impedance.

In terms of flicker transmission analysis, knowledge about the system structure is important. Flicker transmission coefficient ($k_{f,ij}$) represents the relationship of the two flicker values at the two measurement points (i and j)

Voltage fluctuations and flicker produced by fluctuating loads can be measured at the grid PCC, the flicker p_{St} produced at the same network node by the flicker sources in different locations is the superposition of the flicker $p_{St,i}$ transmitted to that node by such flicker sources (Formula (8.24)) [4, 5]:

$$P_{st} = \sqrt[m]{\sum_i P_{st,i}^m} \tag{8.24}$$

where m is the composite coefficient. m=1 is the linear superposition of voltage fluctuations of the same waveform; m=2 is the frequency-dependent non-independent random voltage fluctuations superposition; m=3 is the frequency-independent independent random voltage fluctuation superposition.

Flicker changes in the transmission process in the power grid, which is defined as (Formula (8.25)):

$$k_{v,ji} = u_i / u_j \tag{8.25}$$

is the voltage fluctuations transmission coefficient for voltage fluctuation at node j transmitting to node$i$. Definition (Formula (8.26))

$$k_{f,ji} = P_{st,i} / P_{st,j} \tag{8.26}$$

is the flicker transmission coefficient for flicker at node j transmitting to node i. If $k_{v,ji}$ is frequency-independent, then (Formula (8.27))

$$k_{f,ji} \approx \sqrt{k_{v,ji}} \tag{8.27}$$

Calculation of radial grid transmission coefficient is relatively simple. As shown in Figure 8.8, voltage fluctuations are regarded as a disturbance source, similar to the calculation method of short circuit faults, the following is obtained by passing to the direction of the power supply (Formula (8.28)):

$$k_{v,21} = \frac{Z_{scc}}{Z_{scc} + Z_{12}} \tag{8.28}$$

Because the system short circuit parameters and line parameters are mainly reactances, Formula (8.28) can be approximated to the arithmetic operation of the impedance (without considering the phase angle). Similarly, although the change of fluctuation voltage frequency will cause impedance amplitude changes, due to the characteristic of reactance dominant within the power frequency range, the numerator and denominator in Formula (8.28) are approximately proportional to the changes in frequency, which has little impact on $k_{v,21}$, thus $k_{v,21}$ is approximately treated as a

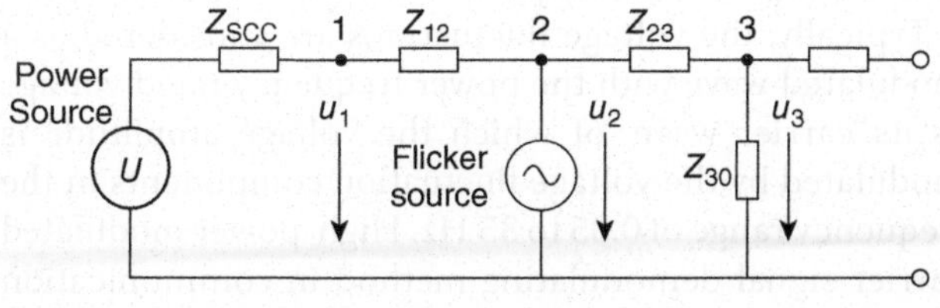

Figure 8.8 Flicker passes along the transmission line.

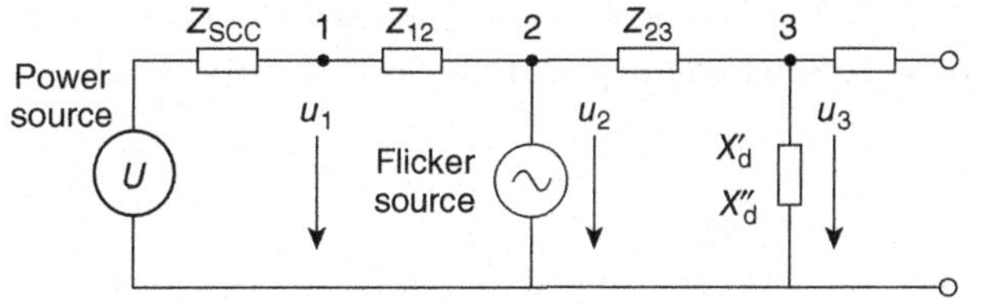

Figure 8.9 Transmission of flicker along the transmission line with distributed power generation units.

frequency-independent quantity. The above impedance parameter can use the calculation values at the power frequency. As shown in Figure 8.9, when voltage fluctuations are passed to the power supply side, the voltage fluctuation transmission coefficient is (Formula (8.29))

$$k_{v,23} = \frac{Z'_{30}}{Z'_{30} + Z_{23}} \tag{8.29}$$

where Z'_{30} is the equivalent impedance of the power supply side at node 3. $|Z'_{30}| \gg |Z_{23}|$, and impedance calculation in Formula (8.29) can still be replaced by arithmetic operation, and $k_{v,23}$ is frequency-independent. If there are distributed power generation units on power supply side at node 3, the value of Z'_{30} shall be adjusted, the literature [4] proposes that, when the flicker frequency is lower than 2 Hz, replace Z'_{30} with transient resistance x'_d; when the flicker frequency is higher than 2 Hz, replace Z'_{30} with subtransient reactance x''_d.

For mesh network, grid admittance matrix, and the principle of superposition can be used. This method is similar to the power grid short-circuit current calculation, the flicker source is replaced by u_a, and flicker current i_a is gnerated at corresponding node. Based on the fluctuation frequency, the synchronous machine is replaced by transient reactance or subtransient reactance, and other components are also replaced by impedance or reactance. Except for the flicker source, all other voltage sources short circuit to ground, and power sources are open circuit to ground. Thus, the relationship between the flicker source u_a and the voltage fluctuation at each node in the grid can be described by the node admittance matrix (Formula (8.30)).

$$\begin{bmatrix} 0 \\ \vdots \\ i_a \\ \vdots \\ 0 \end{bmatrix} = \begin{bmatrix} Y_{11} & \cdots & Y_{1a} & \cdots & Y_{1n} \\ \vdots & & \vdots & & \vdots \\ Y_{a1} & \cdots & Y_{aa} & \cdots & Y_{an} \\ \vdots & & \vdots & & \vdots \\ Y_{n1} & \cdots & Y_{na} & \cdots & Y_{nn} \end{bmatrix} \begin{bmatrix} u_1 \\ \vdots \\ u_a \\ \vdots \\ u_n \end{bmatrix} \tag{8.30}$$

Convert Formula (8.30) into impedance matrix (Formula (8.31)):

$$\begin{bmatrix} u_1 \\ \vdots \\ u_a \\ \vdots \\ u_n \end{bmatrix} = \begin{bmatrix} Z_{11} & \cdots & Z_{1a} & \cdots & Z_{1n} \\ \vdots & & \vdots & & \vdots \\ Z_{a1} & \cdots & Z_{aa} & \cdots & Z_{an} \\ \vdots & & \vdots & & \vdots \\ Z_{n1} & \cdots & Z_{na} & \cdots & Z_{nn} \end{bmatrix} \begin{bmatrix} 0 \\ \vdots \\ i_a \\ \vdots \\ 0 \end{bmatrix} \tag{8.31}$$

Voltage fluctuation transmission coefficient from node a flicker source to the grid node i is (Formula (8.32)):

$$k_{v,ai} = \frac{Z_{ai} i_a}{Z_{aa} i_a} = \frac{Z_{ai}}{Z_{aa}} \tag{8.32}$$

where Z_{aa} is the self-impedance at flicker source node a; Z_{ai} is the mutual impedance from flicker source node a to node i. Within the power frequency range, HV grid Z_{aa} and Z_{ai} dominate by reactance, so $k_{v,ai}$ is frequency-independent. The above impedance matrix can still use the values at power frequency. In medium-voltage power grids, with the increase of resistance value of the grid, the error may be larger.

There are two simplified methods to estimate the flicker transmission effect:

a) **Short-circuit capacity method**
For flicker transmission between neighboring nodes, short-circuit capacity method can be used. It is also the method recommended by GB12326-2000. Set S_{sc1} as the short-circuit capability of node 1, S_{sc12} is the short-circuit capacity at node 2 in the direction from node 1, S_{sc21} is the short-circuit capacity at node 1 in the direction from node 2 (see Figure 8.10). Voltage fluctuation transmission coefficient from node 2 to node 1 (Formula (8.33)):

$$k_{v,21} = \frac{S_{sc12}}{S_{sc1} - S_{sc21}} \tag{8.33}$$

Using the same weighting parameter of the flicker caused by voltage fluctuation k, the value of P_{st} at node i is (Formula (8.34)):

$$P_{sti} = k\frac{\Delta U_i}{U_i} \tag{8.34}$$

Then the flicker transmission coefficient from node 2 to node 1 is (Formula (8.35)):

$$k_{f,2\to1} = \frac{\Delta U_1}{\Delta U_2} \approx \frac{S_{sc2}}{S_{sc1}} \le 1 \tag{8.35}$$

where ΔU_1 and ΔU_2 are the node voltage deviations. S_{sc1} and S_{sc2} is the short-circuit capability at node 1 and node 2, respectively. Previous studies have shown that, if there is no large asynchronous motor or generator capable of providing a relatively large amount of short-circuit power, then $k_{f,2\to1}$ is in the range of 0.95 to 1.0.

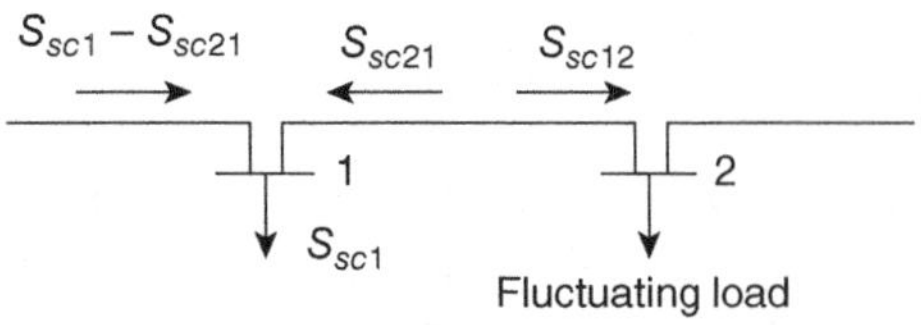

Figure 8.10 Flicker transmission estimation based on short-circuit capability.

b) **Power flow calculation method**
In essence, voltage variation is caused by the fluctuating load connected to the grid, or is equivalent to load periodically switching on and off at the PCC. Ignore the changes in the switching process, the valley and peak of the voltage variation will correspond to value at steady state after inputting or removing the load. Thus, voltage variation at each node in the grid caused by fluctuating load can be estimated by using the changes in power flow distribution under these two states (Formula (8.36)):

$$u = \frac{\Delta U}{U} = \frac{U_0 - U_1}{U_N} \tag{8.36}$$

where U_0 is the voltage value at any one node in the power grid when the fluctuating load is disconnected; U_1 is the voltage value of the node when the fluctuating load is disconnected; U_N is the rated voltage. Since the voltage phase changes little in the two way, the phase angle can be ignored and Formula (8.36) can be directly used for arithmetic operation. After the voltage variation of each node is calculated, the correlation flicker transfer coefficient can be calculated according to the formula (8.35).

8.6.4 Measurement of WF Flicker

8.6.4.1 Flicker Detector

"Flicker" is the subjective visual sense of human to illuminance fluctuations. It is not only related to the size of the voltage fluctuation, but also related to the frequency (i.e., the amplitude-modulated frequency to power frequency voltage) of the fluctuation, lighting performance of the lamp, and human visual sense. The IEC has gave the voltage flicker meter design specifications in accordance with the recommendation of the International Electric Association (UIE) in 1982, of which the block diagram is shown in Figure 8.11.

Block 1 drops the supply voltages of different levels to the voltage value adaptable to the internal circuit of the instrument, and produces standard modulated wave used for its self-test.

Block 2, 3, and 4 comprehensively simulates the response of the lamp-eye-brain link to voltage fluctuations. Block 2 reflects the relationship between light intensity and voltage, demodulates the voltage fluctuation components, and gives a voltage that has linear relationship with the amplitude of the modulated wave.

Typically, the voltage fluctuations are considered as a modulated wave with the power frequency rated voltage as its carrier wave, of which the voltage amplitude is modulated by the voltage fluctuation components in the frequency range of 0.05 to 35 Hz. High-power modulated carrier signal demodulating method in communication theory can be used as the detection method of voltage fluctuation components: multiply the modulated signal by the periodic signal with the same frequency and phase of the carrier signal, separate the voltage fluctuation

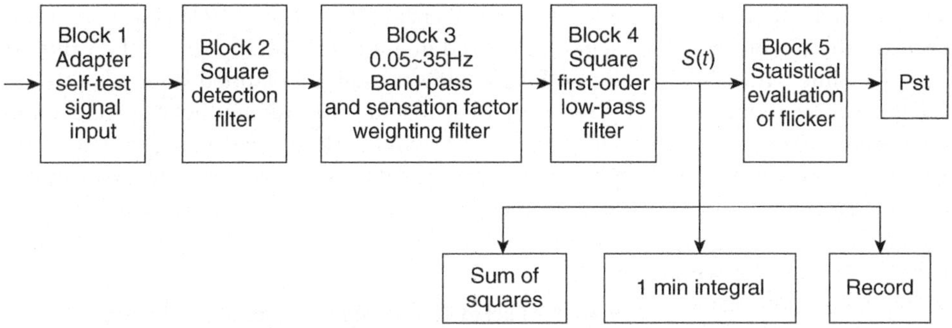

Figure 8.11 Block diagram of Voltage flicker meter design.

component from the power-frequency carrier voltage, and obtain the fluctuation components through the band-pass filter. Power-frequency voltage u(t) can be expressed as: u (t) = A (1+ $m$cosωt)cos ω_0t, where A is the amplitude of the power-frequency carrier voltage; ω_0 is the angular frequency of the power- frequency carrier voltage; m is the amplitude of the modulated wave voltage; $m$cosωt is the fluctuation voltage. There are three commonly used methods for detection of voltage fluctuation: square demodulation detection method, full-wave rectification detection method, and half-wave RMS detection method.

The band-pass and sensation level weighting filter in Block 3 reflects the sensitivity of the human eye to voltage fluctuations at different frequencies, and the passband is 0.05 to 35 Hz;

Block 4 contains a squarer and a first-order low-pass filter (the time constant is 300 ms), which are used to simulate the transient non-linear response and memory effects of the human brain to illumination change, and its output S(t) reflects the instantaneous flicker sensation level of human vision for voltage fluctuations. Different treatments of S (t) may reflect the grid voltage flicker situation.

Block 5 shows the statistical analysis of flicker, that is, carrying out online statistical analysis according to S (t) or carrying out offline statistical analysis (Cumulative probability function-CPF method) for its filtering outputs to obtain the output short-term flicker severity P_{st}, then use the short-term flicker severity value to extrapolate the long-term flicker value P_{lt}.

Based on the CPF curve, short-term flicker value P_{st} is obtained according to the Formula (8.11). Long-term flicker value P_{lt} is calculated from the short-term flicker values included in the measured period according to Formula (8.12). According to this principle and block diagram, a simulation flicker detector and a digital flicker detector can be designed. In the simulation, the filter circuit is realized by using the chip, which has the characteristics of fast processing speed and so on. The filtering operation of the digital detector is realized by software, and the amount of calculation is large, but the structure is simple and it is more flexible.

8.6.4.2 Huitengxile WF Flicker Test

The literature [6] reports the flicker test results of the huitengxile WF. A huitengxile WF is installed with five types of wind turbines currently, which is integrated into the system through two sets of 10 kV/110 kV step-up transformers, the 110 kV bus and 10 kV bus are both divided into two segments and operating in parallel, segment I and segment II both have a six-circuit 10 kV outgoing line respectively, and segment I is connected to 28 wind turbines, of which the total capacity is 16.3 MW; segment II is connected to 35 wind turbines, of which the total capacity is 21 MW; segment I and segment II of 10 kV is respectively connected to one group of 750 kVA compensation capacitors, among which only one group is operating when testing; compensation capacitors are connected to the wind turbine terminals, in actual operation, the control system will control the switching of the capacitors in accordance with the generator power factor. Main wiring diagram of the WF and the Caoduoshan substation is shown in Figure 8.12.

The testing point selected was the 110 kV bus of Caoduoshan substation, the first and second main transformer sides. Continuous testing time was from December 6, 2000 at 17:10 to December 7, 2000 at 13:30. Wherein, during December 6, 17:10 to December 7,

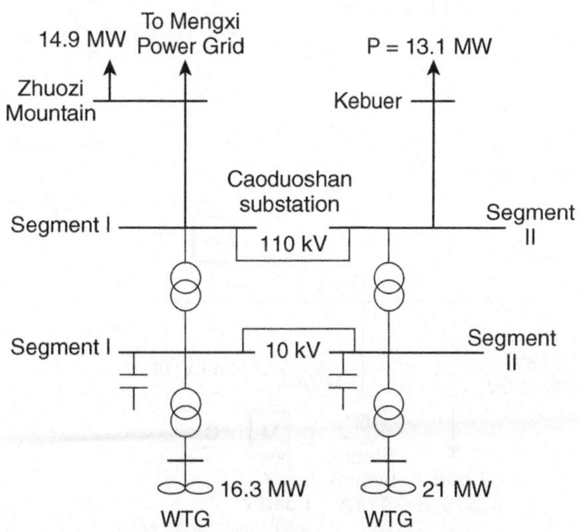

Figure 8.12 Huitengxile WF system wiring diagram.

9:30, the test was carried out on the first main transformer side, and test is carried out on the second main transformer side for the rest of the time. The ABC three-phase voltage of the 110 kV bus was tested. In the test, for short-term flicker p_{St} prediction, the statistical time selected was 10 minutes, and for long-term flicker P_{lt} prediction, the statistical time selected was 1 hour.

Wind speed of the WF experienced the change process from 4 m/s to 13 m/s during the test, some wind turbines started up and stopped, and the terminal capacitors also has automatic switching operations. The test includes different operating conditions of the WF under normal operation, which can basically reflect the flicker generated under the integrated action of different types of wind turbines and multiple flicker sources in the WF.

The test results show that the values of ABC three-phase voltage P_{st}, P_{lt}, A_{st}, and A_{lt} meet the requirements, and the values of P_{st} and A_{st} are slightly higher at some moment, wherein, at 17:50 on December 6, the values of P_{st} and A_{st} had reached the IEC planed value.

Table 8.5 110 kV bus voltage flicker test results (P_{st}, and A_{st}).

	Phase A		Phase B		Phase C	
Time	P_{st}	A_{st}	P_{st}	A_{st}	P_{st}	A_{st}
17:50	0.239	0.014	0.798	0.508	0.346	0.041
18:20	0.293	0.025	0.226	0.012	0.505	0.129
06:20	0.386	0.057	0.492	0.229	0.638	0.260
10:20	0.519	0.140	0.372	0.052	0.465	0.101
10:50	0.452	0.092	0.559	0.174	0.519	0.140

Several larger flicker values of the 110 kV bus voltage are separately listed in Table 8.5, among which the values of P_{St} and A_{st} are lower than the planned value.

From the test results, the voltage flicker produced by Huitengxile WF wind turbines was compliant with the planning requirements. As the flicker value is a result of unified measurement calculation, occasional overlimits can be ignored.

8.6.4.3 Flicker Test of WFs with Double-Fed Wind Turbines

The literature [7] tests the flicker of double-fed wind turbines connected to the medium-voltage distribution system, which is shown in Figure 8.13. In the figure, double-fed asynchronous wind turbines are connected to the WF 11.4 kV medium-voltage network, the flicker measuring points are at Point A, B, and C, and Point A (11.4 kV) is connected to the public grid; Point B (11.4 kV) is the medium-voltage collection system of the WF, in which there is a large motor load and a 9 MW turbo generator. Point C (440 V) is on the load side.

The ADX-3000 and RPM flicker meters are used. The flicker data is sampled once every 10 minutes. The data has been sampled for three months. The values of P_{st} recorded at Point B are between 0.139 to 0.676, and the values of P_{lt} are between 0.209 to 0.475. As can be seen, the wind turbine seems to provide additional support to the network voltage, because the voltage fluctuation is decreased during high power output period of the WF.

The values of P_{st} and P_{lt} recorded during the period of wind power average output power (about 70% of the double-fed asynchronous generator rated power) and the

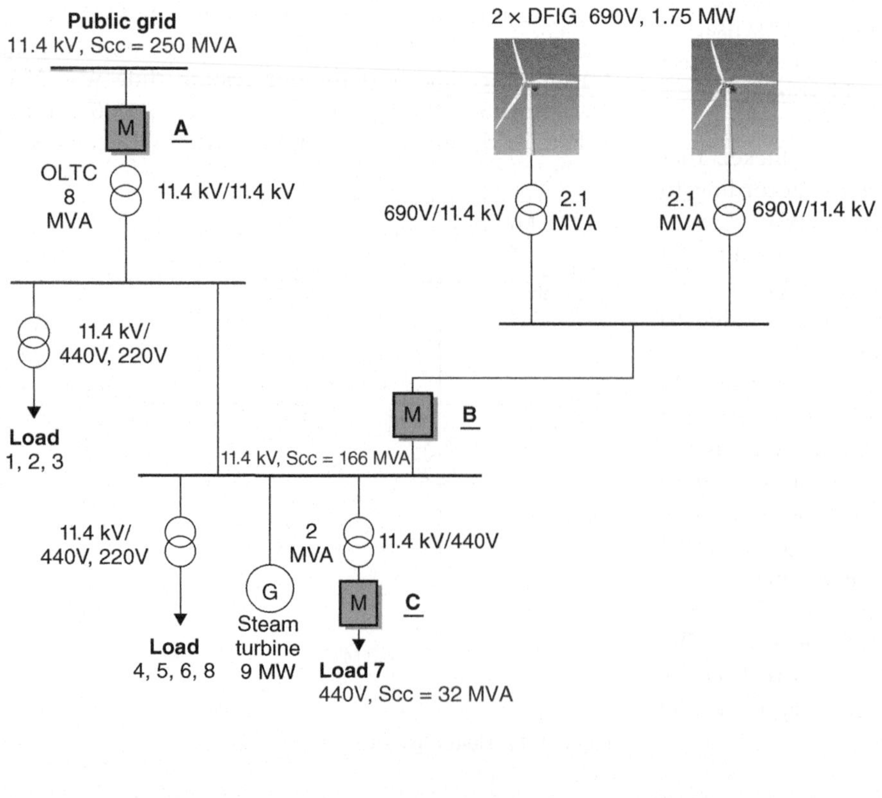

Figure 8.13 Test system single-line diagram.

Table 8.6 Flicker test results during the period of 70% wind turbine power output.

	Average value				95% of the cumulative probability value				99% of the cumulative probability value			
Measuring point	Phase R/P_{st}	Phase T/P_{lt}	Phase R/P_{st}	Phase T/P_{lt}	Phase R/P_{st}	Phase T/P_{lt}	Phase R/P_{st}	Phase T/P_{lt}	Phase R/P_{st}	Phase T/P_{lt}	Phase R/P_{st}	Phase T/P_{lt}
A	0.262	0.262	0.274	0.272	0.379	0.372	0.321	0.317	0.437	0.427	0.343	0.339
B	0.309	0.308	0.331	0.331	0.478	0.484	0.378	0.381	0.563	0.572	0.402	0.407
C	0.384	0.387	0.405	0.409	0.574	0.584	0.451	0.459	0.669	0.682	0.472	0.484

Table 8.7 Flicker test results during the period of no wind turbine power output.

	Average value				95% cumulative probability value				99% cumulative probability value			
Measuring point	Phase R /P_{st}	Phase T/P_{lt}	Phase R /P_{st}	Phase T/P_{lt}	Phase R /P_{st}	Phase T/P_{lt}	Phase R /P_{st}	Phase T/P_{lt}	Phase R /P_{st}	Phase T/P_{lt}	Phase R /P_{st}	Phase T/P_{lt}
A	0.262	0.264	0.275	0.275	0.407	0.400	0.366	0.363	0.479	0.468	0.412	0.407
B	0.301	0.305	0.324	0.329	0.506	0.515	0.441	0.447	0.609	0.619	0.499	0.506
C	0.386	0.389	0.408	0.412	0.589	0.596	0.488	0.492	0.691	0.701	0.528	0.532

period of no wind power output are different. The statistical analysis of the measurement data is listed in Table 8.6 and Table 8.7. In the table, average values of P_{st} and P_{lt}, and 95% and 99% of the cumulative probability values are also listed. The comparison of the statistical data in Table 8.6 and Table 8.7 shows again that, wind turbine output does not deteriorate the original flicker level of the grid, but relieves some voltage fluctuation problems of it.

By simultaneously measuring flicker of different locations, if it can be ensured that other time-variable loads do not obviously action or cause interactions, transmission of the flicker via the network can be determined. According to measurement results, Table 8.8 shows the flicker transmission coefficient of the upstream and downstream directions at about 70% of the double-fed asynchronous generator rated output. The last line in the table is calculated according to the short-circuit capacity method, it can be seen that the average transmission coefficient k_f is less than 1.0. As can be seen, the maximum flicker appears at Point C, this may be caused by many sources of higher power fluctuation, including a lot of rotating machinery and weaker systems at the the measuring location.

Table 8.8 The flicker transmission coefficient obtained at 70% of the rated power.

DFIG bus	Y	B			
Bus	X	A		C	
Phase	-	R	T	R	T
Measured mean value	$TF_{Y\text{-}X}$	0.91	0.92	1.34	1.37
Measured 95% of CPF	$TF_{Y\text{-}X}$	1.54	1.56	2.36	2.44
Measured 99% of CPF	$TF_{Y\text{-}X}$	1.85	1.88	2.87	2.98
Short-circuit-based calculation	$TF_{Y\text{-}X}$	0.66	0.66	5.19	5.19

8.7 Harmonics and Interharmonics of the WF

8.7.1 Harmonics

Ideal waveforms of voltage and current in the AC power system are sinusoidal. Due to the presence of non-linear electrical equipment in the power system, especially due to extensive use of power electronic technology in the power system, a lot of fast time-variable loads and electrical equipment with non-linear characteristics have appeared in the power system, such as HVDC converter devices, electrified locomotives and high-power rectifier, inverter, frequency conversion equipment, voltage regulating equipment, electric equipment with arc non-linear characteristics, as well as a variety of household controllable rectifiers or power-switching devices. These loads will produce a variety of non-sinusoidal currents in the role of sinusoidal voltage. These currents flowing in the network will cause the voltage waveform distortion, which, in addition to the normal fundamental frequency, may also contain a variety of higher harmonics. Power loads of these high order harmonic currents produced are called harmonic sources.

The extent of distorted waveforms deviating from the sinusoidal wave, that is, harmonic content is often expressed as sinusoidal waveform distortion rate. It is a measure of energy quality. Total harmonic distortion (THD) is defined as a percentage of the harmonic ($h \geq 2$) effective value to the fundamental component ($h = 1$)

effective value. If the DC component in the voltage and the current are not considered, then the voltage distortion rate (D_U) and the current distortion rate (D_I) are defined as follows (Formula (8.37)):

$$D_U = 100 \times \frac{\sqrt{\sum_{n=2}^{\infty} U_n^2}}{U_1}(\%)$$

$$D_I = 100 \times \frac{\sqrt{\sum_{n=2}^{\infty} I_n^2}}{I_1}(\%) \quad (8.37)$$

In engineering, a certain harmonic content of voltage or current distorted waveforms is often required to be given in order to monitor or take preventive measures. Defined the nth harmonic content (D_{Un} or D_{In}) as the percentage of the effective value U_n of nth harmonic voltage or the effective value I_n of nth harmonic current to the effective value U_1 of its fundamental voltage or the effective value (I_1) of its fundamental current (Formulas (8.38) and (8.39)):

$$D_{Un} = 100 \times \frac{U_n}{U_1}(\%) \quad (8.38)$$

$$D_{In} = 100 \times \frac{I_n}{I_1}(\%) \quad (8.39)$$

The presence of harmonics not only causes pollution to the power system, but also causes serious harm to electrical equipment and its operation. In order to ensure the normal operation of various electrical equipment, the sinusoidal waveform distortion rates of the current and voltage must be limited in the allowed range. The limits of voltage sinusoidal waveform distortion rate for the power system is provided in the Chinese national standard "Public Supply Network Harmonic," which was issued in 1993 (see Table 8.9).

In order to control the power system harmonic voltage, the harmonic current injected into the power system by each harmonic source must be limited. Therefore, allowable value of the harmonic current injected by nonlinear electrical equipment of users and its calculation methods are also provided in the "Public Supply Network Harmonic."

Table 8.9 Limits of voltage sinusoidal waveform distortion rate.

Power grid nominal voltage (KV)	Voltage total harmonic distortion rate (%)	Containing rate of harmonic voltage of each order (%)	
		Odd order	Even order
0.38	5.0	4.0	3.0
6	4.0	3.2	1.6
10			
35	3.0	3.4	1.2
66			
110–220	3.0	1.6	0.8

Harmonic pollution is one basic problem of the power quality. In WFs, only constant-speed wind turbines inject important harmonic currents into the power grid. Constant-speed wind turbines, especially those with power factor correction capacitors, change the harmonic impedance of the distribution network, which in certain conditions will form a resonant circuit. If constant-speed and variable-speed wind turbines are installed in the same WF, this may be an important issue.

The thyristor soft starter (Figure 8.14) is typically used for grid connection of asynchronous generators driven by constant speed wind turbines. When the wind turbine is connected to the grid, the machine shaft will accelerate, and when the asynchronous generator speed is close to the synchronous speed (about 92% to 99% of synchronous speed), the grid-connected contactor will act. The generator is connected to the grid via a group of bi-directional thyristors, control the thyristor triggering unit, to make the conduction angle of the bi-directional thyristor gradually increase from 0 ° to 180 °, and adjust the thyristor conduction angle open rate to ensure impact current in the grid-connection process is not greater than the predetermined value of the technical conditions. When the transient process ends, the bypass switch is closed, so the thyristor is short connected. In the process of grid connection, the grid voltage is gradually added to the generator, so the current extracted by the generator is controlled, thus the grid voltage fluctuations are lowered. But at the same time, as the trigger angle of the thyristor changes, different harmonics will be produced. Soft starter only works for a few seconds when the wind turbine is connected to the grid, so effects of such short-term harmonics are harmless and can be ignored. If the soft starting unit does not quit (is not bypassed) during normal operation, then the harmonic currents must be assessed.

Modern variable speed wind turbine uses grid side voltage source converter (Figure 8.15). It uses insulated-gate bipolar transistor (IGBT), and the switching frequency is several kHz, so the low-order (e.g., <nineteenth order)

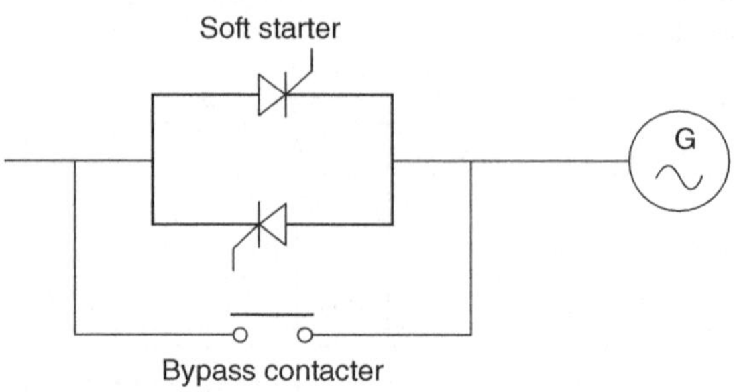

Figure 8.14 Soft starter device (had drawn only single phase).

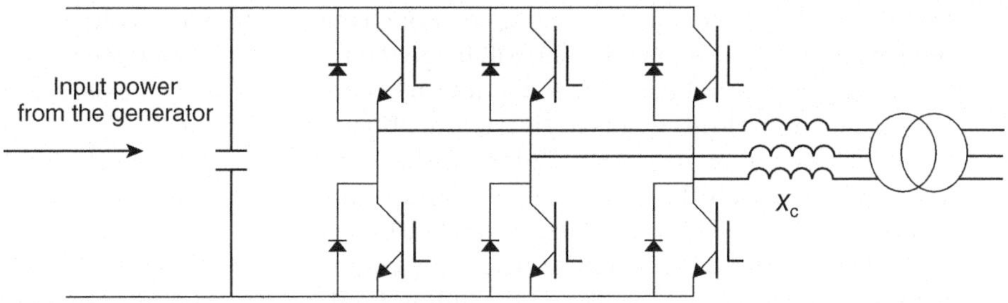

Figure 8.15 Six pulsed two-level IGBT voltage source converter.

harmonics are eliminated from the sinusoidal waveforms produced, but the harmonics of the output current in the vicinity of the switching frequency of device (i.e., the range of 2 to 6 kHz) are obvious. Although this high frequency currents are relatively easy to be filtered, they may affect the selection of coupling reactor X_c, which can be used as a part of the filter to suppress harmonic currents in the vicinity of the switching frequency.

8.7.2 Interharmonics

8.7.2.1 The Concept of Interharmonic

Harmonic usually refers to the waveform component of integral multiple power frequency (fundamental frequency), and waveform component of non-integer multiple fundamental frequency is called interharmonics. Interharmonics can be characterized by the non-integral factor μ (Formula (8.40)):

$$\mu = \frac{f_\mu}{f_1} \tag{8.40}$$

where f_μ is the interharmonic frequency, f_1 is the fundamental frequency (e.g., 50 Hz). Interharmonic frequency mainly appears as the below or above sideband of harmonic frequency, which can be lower than the fundamental frequency, that is, $\mu < 1$.

Studies on interharmonic are deepening. The existing national standard, "Power quality— Public Supply Network Harmonic" (G/B 14549-93) only provides limits and measurement methods for harmonics. Existing research shows that interharmonic exists widely in power systems, and that its amplitude is small and does not cause direct damage to the system or device, but sometimes it will trigger unpredictable resonance between the transformer and the capacitor bank. Interharmonics can also cause flicker. And sub-harmonic currents at frequencies lower than the fundamental frequency can easily cause saturation of power transformer or damage to the synchronous generator.

8.7.2.2 Interharmonic Sources [8–10]

Interharmonics widely present in the power system, various nonlinear fluctuating loads (electric arc furnaces, welding machines, thyristor powered mills, etc.), all kinds of frequency converters, low synchronous cascade control, asynchronous motors and other equipment with core and ferromagnetic oscillations in the distribution network, and so on, may be the sources of interharmonics, modern wind power plants may also be an important source of interharmonics.

1) **Frequency converter**
   AC frequency converter is divided into two major categories, namely, AC-DC-AC converter and AC-AC converter. These two frequency converters both have harmonics in their supply currents in use, and the harmonic frequencies produced have the following expression (Formulas (8.41) and (8.42)):

$$f_{h1} = (p_1 m \pm 1)f_1 \pm p_2 n f_0 \tag{8.41}$$

$$f_{h2} = p_2 n f_0 \pm f_1 \tag{8.42}$$

   where m=1, 2, 3 …; n=0, 1, 2…; f_{h1}- frequency that is dependent of the number of trigger pulses; f_{h2}– frequency that is independent of the number of trigger pulses; p_1- the number of trigger pulses; f_1- fundamental frequency of power input; p_2- factor related to the phase number of a frequency converter load, p_2=6 is the 3-phase loads, p_2=2 is single-phase load; f_0-converter output frequency.
   As can be seen from Formulas (8.41) and (8.42), the harmonic frequency of the supply current is not all the multiples of the fundamental frequency f_1 and the output frequency f_0, that is, it contains interharmonic components.
2) **Fluctuating load**
   Industrial electric arc furnaces, thyristor rectifier powered mills are fast-moving impact loads, of which the electric quantity (voltage or current) changes can be observed within a few ms or more than 10 ms, so interharmonics of continuous and discrete components are produced. Set the system voltage u(t) = sinωt, the load characteristic is r(t) = 1 − r sinω_mt, r < 1, ω_m is load changing frequency, then the load current is (Formula (8.43)):

$$\begin{aligned} i(t) = \frac{u(t)}{R(t)} &= \frac{\sin \omega t}{1 - r\sin \omega_m t} \\ &= \sin \omega t(1 + r\sin \omega_m t + r^2 \sin^2 \omega_m t \\ &\quad + r^3 \sin^3 \omega_m t + \cdots) \end{aligned} \tag{8.43}$$

   As can be found from Formula (8.43), the load current i(t) contains $\omega \pm \omega_m$, $\omega \pm 2\omega_m$, $\omega \pm 3\omega_m$ and other frequency components, so if ω and ω_m are not change in sync, an interharmonic current will be generated. For example, if ω_m=8 Hz, r =0.5, the interharmonic component in the load current $i(t)$ waveform obtained by simulation are 36 Hz, 44 Hz, 52 Hz, 68 Hz, 76 Hz, and 84 Hz, which are the same with the interharmonic values (60 ± 24, 60±16, 60±8Hz) estimated by Formula (8.43). It can be seen as side frequency components are added to each harmonic (including fundamental harmonic) after the input fundamental voltage is modulated by the modulating wave of the load. These side frequency components are interharmonics, which appear in pairs.

3) **Ferroresonance**
Simplified schematic diagram of ferroresonance is shown in Figure 8.16. In the figure, L is the nonlinear inductance with iron core, when voltages at both ends of L increase or have inrush currents, the core may saturate, and the inductive reactance is reduced, and when it is reduced to ω_l=1/ω_c, the series resonance condition is met, the ferroresonance will occur, which is the main source of interharmonics in the power grid.
4) **Synchronous cascade control device**
Low-synchronous cascade control is mainly used for wound rotor asynchronous motor. It adds a rectifier in the rotor circuit instead of the traditional series resistance, to convert the slip power into DC power, which is fed back into the grid through the inverter. Speed control is achieved by changing the slip power. The speed control method is of high efficiency, low loss, wide control range, and good performance, but there will be harmonic currents generated in the inverter and the stator circuit, including interharmonics.
5) **Asynchronous motor**
Asynchronous motor stator and rotor slots produce irregular magnetizing currents due to core saturation, resulting in interharmonics generated in the low voltage power grid. In normal speed, the motor interference frequency is in the range of 500 to 2,000 Hz, which is wider when starting up. This motor will interfere with the grid when connected to longer (>1 km) low voltage overhead line ends. The interharmonic voltage can reach 1% of the rated voltage.

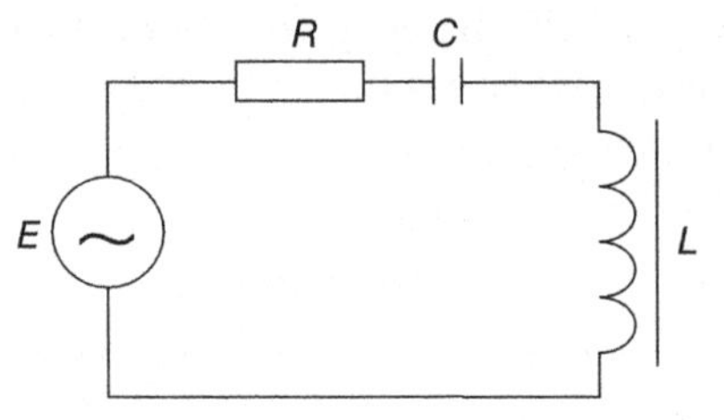

Figure 8.16 Series ferroresonance resonance circuit.

6) **High voltage direct current transmission**
HVDC power transmission is similar to the inverter, which also use converter transformers and inverters for AC-DC-AC conversion. When the two AC systems work on different frequencies, interharmonics will be generated. Modulation theory can also be used for its analytical method, applying switching function method to calculate the interharmonics in the two AC system respectively.
7) **WF**
Interharmonics of the wind power generation system are mainly caused by frequency converters, and may also be caused by the rotary motor.

8.7.2.3 The Harm of Interharmonics

Interharmonic voltages will cause the lights to flicker, and will cause the interference with pulse-controlled audio receivers, TV, radio, or other audio devices, and will cause noise and vibration of asynchronous motor, abnormal operation of the low frequency relay and overcurrent tripping of passive power filter.

1) **Wave distortion**
Interharmonics will cause zero-crossing deviations of the waveforms, and changes in the positive and negative half wave amplitude.
2) **Flicker**
Interharmonics at frequencies below the fundamental frequency can cause voltage flicker. As long as the interharmonics frequency is close to the harmonic frequency or the fundamental frequency, flicker will occur. Harmonics are integral multiples of the power frequency, so harmonics always change synchronously with the fundamental frequency, but interharmonics and the fundamental frequency do not change synchronously, if a signal contains a interharmonic frequency of f_1, that the envelope of the signal waveform will fluctuate in the frequency of f_m, in which (Formula (8.44)):

$$f_m = |f_1 - f_k| \tag{8.44}$$

where f_k is a harmonic or fundamental frequency close to the interharmonic f_1 frequency. Because the human eye is sensitive to flashing lights at the frequency of around 8.8 Hz, if the difference between interharmonics and harmonic or fundamental frequency is around 8.8 Hz, human vision would be severely disrupted, resulting in visual clutter. Interharmonics with frequencies below the fundamental frequency can also cause abnormal operation of the low-frequency relay, as well as the failure, or even the damage of existing harmonic compensation equipment.
3) **Influence on the measuring instrument**
Non-synchronous changes in interharmonics and the fundamental frequency will lead to changes and

zero-crossing deviations of positive and negative half wave amplitude of the waveform, and errors in data sampling or zero-crossing action of the digital relay, or even cause accidents due to misoperation, and will also affect the results and accuracy of traditional harmonics measurement, as well as cause measurement error in the measuring instrument. Interharmonic voltage will also cause TV image scrolling and the noise of radio or other audio equipment.

4) **Influence on the motor**
Influence of interharmonics on the noise and vibration of the motor is very large. The force generated by interactions between the air-gap magnetic flux and the rotor current caused by interhamonics can be decomposed into multi-frequency components. If the component frequency of the force is close to the natural frequency of the motor rotor and is deeply coupled on the "circular vibration mode" order, it will generate strong noises and vibrations. In addition, interharmonics can provoke unpredictable resonance between the capacitor bank and the transformer reactance. Harmonics with frequencies below the fundamental frequency will cause torque torsional vibration of turbo-generator.
5) **Influence on the power factor**
Due to the the waveform distortion caused by interharmonics, the load power factor is reduced, and energy losses are increased, resulting in reduction of the insulation level.

8.7.2.4 Interharmonic Standards

Research shows that interharmonic voltage must be limited to a level low enough [8, 11]:

1) Harmonics below 25 Hz should be limited to less than 0.2% so as not to cause the lights to flash (flicker);
2) For audio ripple control receiver, interharmonic voltage should be limited below 0.3%, otherwise it will be disturbed.
3) Interharmonic voltage below 2.5 Hz should not exceed 0.5%, or it will interfere with the TV, and cause noise and vibration of the asynchronous motor, as well as abnormal operation of the low-frequency relay;
4) If the interharmonics between 2.5 to 5 kHz exceed 0.3%, it will cause radio noise or noises of other audio devices;
5) When there are non-linear loads, interharmonics at the frequency of f_h will produce side frequency components (n=1, 2, …) at the frequency ($f_h \pm 2nf_1$); amplitudes of these side frequencies may be very close to those of the interharmonics, of which in terms of amplitude of the flicker frequency band, have been extended to four times of that of the fundamental frequency, and have also been extended to the same times of amplitude of the audio control frequency. Thus, the influence of interharmonics will be greatly expanded.

In view of the above findings, it is recommended in IEC61000-3-6 that the interharmonic voltage level shall be limited below 0.2%. Currently passive filters consist of capacitance, inductance and resistance are widely used industrially to absorb harmonics and improve the power factor, but these filters will only have obvious effect on major harmonics of several orders (usually harmonics of integral orders, such as 5,7,11), and will probably have amplification effect on interharmonics. In serious cases, the filter will not be able to be put into operation due to harmonic overload, and even be damaged very quickly. To suppress the resonant amplification caused by the filter system to the interharmonics, typically a damping resistor is added to the single harmonic filter.

8.7.3 Harmonic Test of WFs

8.7.3.1 Nanao WF Harmonic Test [12]

When testing, the Nanao WF was installed with 130 units with the capacity 53.8 MW. In the WF, there are wind turbines from the United States, Denmark, and Spain with the capacity of 150 to 600 kW, and China made wind turbines with the capacity of 200 to 300 kW. The WF capacity accounts for 7% of the total grid capacity in Shantou. Terminal voltages of the WF step-up to 10 kV by the box-type transformer, and then converged to Dalankou substation and Niutouling substation. The Zhuzaiao substation and Niutouling substation is respectively connected to Shantou power grid through 35 kV submarine cables and 110 kV submarine cables, and the grid connection point is at Maipu station (see Figure 8.17). WF turbines are asynchronous generators, and are connected to the grid through the soft parallel devices; the WFs are located in the foothills of the island, the wind speed varies randomly, so wind speed of each WF is different; starting time and stopping time of the wind turbines are different; and the power electronic control system (rectifier) will also generate harmonics.

In order to test the effects of harmonics produced by the WF on the power grid, at the three voltage levels of 10 kV, 35 kV, and 110 kV, the 10 kV bus of Niutouling substation, 110 kV bus of Niutouling substation, 10 kV bus of Dalankou substation, 35 kV bus of Zhuzaiao substation, 110 kV Jinling Line, and 35 kV Zhulan Line were selected as the test points for testing. When testing, the wind scale was at about 6~7 levels, and the wind turbine power was about 70% to 80% of the maximum power output. System status: the 2# main transformer at Niutouling had quit operation, and the 35 kV Nanao line at Zhuzaiao substation was in the state of power-off maintenance. The testing equipment used was DZ-4B power quality analyzer; the test was implemented in accordance with the GB T14549-1993.

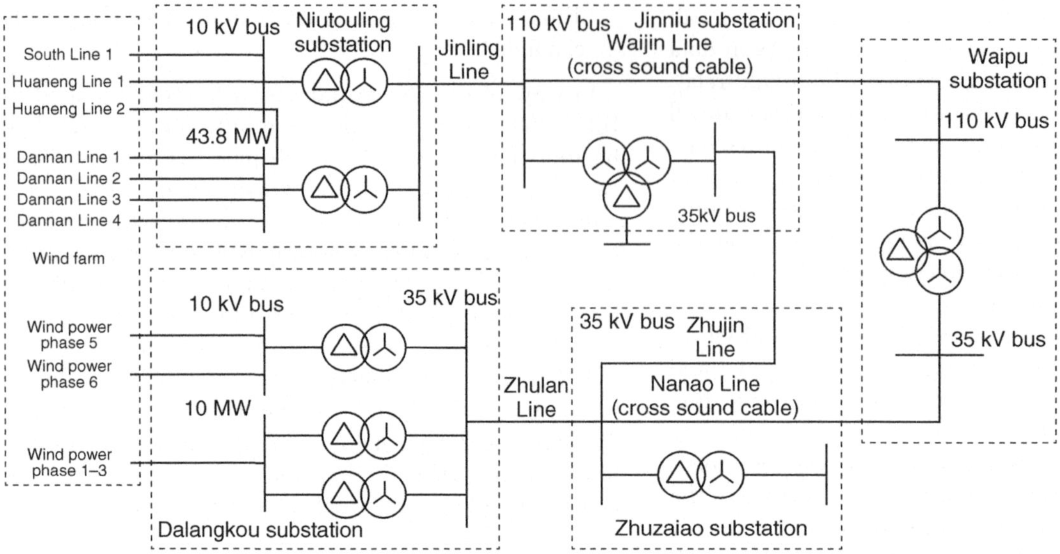

Figure 8.17 Nanao WF grid-connection wiring diagram.

The test results and analysis are as follows:

1) Among the three voltage levels of 10 kV, 35 kV, and 110 kV on Nanao Island, the maximum value of the voltage total harmonic distortion rate was 1% (phase C voltage of 10 kV bus of Dalankou substation), and the minimum value was 0.5% (phase A and phase C voltages of 10 kV bus of Niutouling substation), they were both within the allowable range, and the harmonic component was small.
2) The harmonic currents injected by the 110 kV Jinling Line into Shantou system were all within the allowable range: the fundamental current was 54.4 A, and the maximum (fifth order) harmonic current was 0.5 A, which was far smaller than the allowable value (5.4 A). The content rate of the harmonic distortion current injected into the system by the 35 kV Zhulan Line was relatively high: when the fundamental current was 26.7 A, the fifth order harmonic current was 1.2 A, which was smaller than the allowable value (6.7 A). By comparison of the harmonics of the 10 kV buses of Niutouling and Dalankou substations, it can be seen that the capacity of units connected to Niutouling substation of 110 kV was larger than the capacity of the units connected to Dalankou substation of 35 kV, that is, the harmonic current value injected into the system by Niutouling substation was far larger than that of Dalankou substation, but because the capacity of the 10 kV bus of Niutouling substation was larger, the voltage distortion rate of the 10 kV bus was smaller than that of Dalankou substation. Total harmonic distortion rate of 10 kV bus voltage of Niutouling substation was 0.5%, whereas the total harmonic distortion rate of 10 kV bus voltage of Dalankou substation was 0.6%.
3) Nanao WF indeed had some harmonic components, and the maximum harmonic currents was fifth and seventh orders. The capacity of Nanao WF had reached 7% of that of Shantou grid, but the harmonic current injected into Shantou grid was far smaller than the allowable value of national standard. The total harmonic distortion rate and content rate of each harmonic of the bus voltage of each voltage level connected to the WF were within the allowable range stipulated by GB T14549-1993, so the Nanao WF did not have a bad influence on the power quality of Shantou power system.

8.7.3.2 Harmonic Tests of Different Wind Generator Systems

The literature [13] reports the measurement results of two different systems: the inverter coupled to synchronous generator (SG) and double-fed asynchronous generator (DFAG), as shown in Figure 8.18 and Figure 8.19. Technical parameters of six types of wind

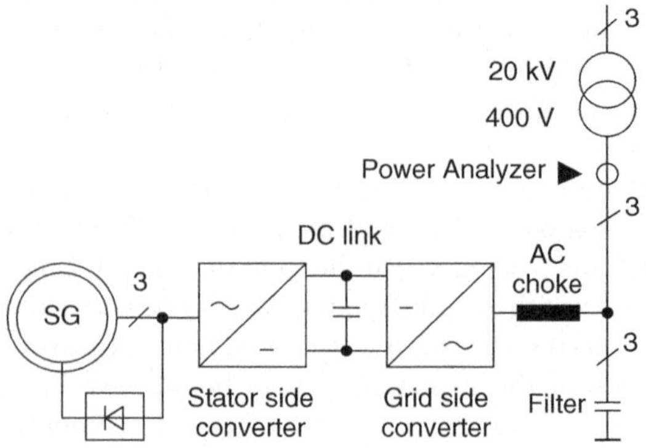

Figure 8.18 Inverter coupled to the synchronous generator (SG).

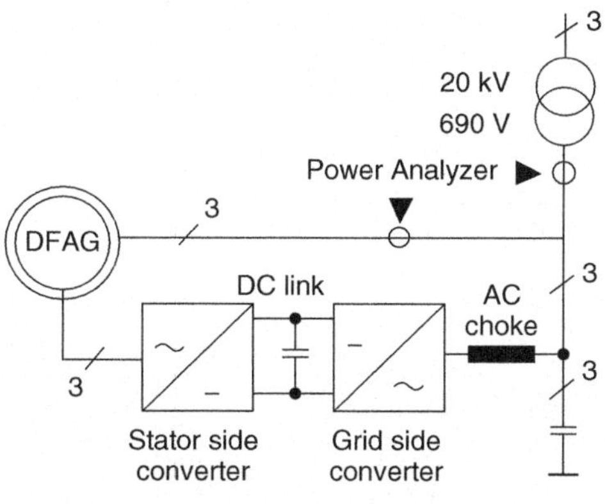

Figure 8.19 Double-fed asynchronous generator (DFAG).

generators are listed in Table 8.10. Current and voltage harmonic parameter on LV side and HV side of the wind turbine transformer and the DFAG stator connection point are measured.

The measurement results are summarized in Tables 8.11. It can be clearly seen from Table 8.11 that the deviations of various wind turbines from the power quality. The differences depend not only on the types of converters, but also on the connection conditions decided by different network impedance and filter structures, of which, the impedance is formed by the transmission lines and transformers in the WF. Although the obtained harmonic parameter range of the current is always different, the obtained curve shape trends to be similar. Table 8.11 shows that the stator current distortion level measured is always higher than the total distortion rate on the grid side, stator distortion of DFAG system is generated by transmission of rotor-side harmonics to the stator. Rotor-side harmonics are generated by the rotor-side converter. Harmonic parameter reaching a very high value (800%), shows a very high current distortion. On the contrary, very low voltage fluctuation, of which the value is between 1% and 4%, shows that it has no direct relationship with the relative power. Large current distortion always happens at low-power range of low rated power. Sampling rate for assessment of the harmonic parameters are 12.8 kHz and 6.4 kHz, and the mean value of 1 minute is obtained.

Distortions of the same shape with low current harmonic parameters can be observed on the 20 kV medium voltage side. Table 8.12 shows the comparison of measurement results on the medium voltage side and the low voltage side. As can be seen, total harmonic distortion of the WF is not the sum of the fluctuations of single wind turbines. The harmonic parameter is within the range of 3% to 60%. This is a result of harmonic damping effects between the different voltage levels. Damping elements are primarily transformers and transmission lines. Compared with a single wind turbine, distortion level on medium voltage side does not change, this is due to the changes of the current that flows through the network impedance between the two voltage levels caused by power fluctuations.

8.7.3.3 Measurement of Interharmonics

The amplitude and (or) frequency of harmonics usually vary, but there are also relatively stable (quasi-steady-state) interharmonics. For interharmonic of quasi-steady state, frequency domain analysis instrument can be used. This instrument uses analog filtering technology to complete signal analysis, and can be tuned to the appropriate frequency to measure the amplitude and frequency of the corresponding signal. Since the measuring instrument is difficult to rapidly change for simultaneous measurement, the frequency domain instrument cannot be used to measure fluctuating or rapidly changing interharmonics, so in this case, special time domain analyzer is generally used to measure interharmonics, and suitable sampling time window width shall be selected

Table 8.10 Technical parameters of six types of wind turbines.

Parameter	Wind turbine type					
Wind turbine type	1	2	3	4	5	6
Generation type	DFIG*	DFIG	DFIG	SG**	SG**	DFIG
Rated power [MW]	1.5	1.5	1.5	1.5	1.8	1.5
Voltage [V]	690	690	690	400	400	690
Number of pole-pairs	2	2	2	36	36	2
Gear ratio	1:90	1:90	1:90	–	–	1:90
Converter type	A	B	A	C	C	A
Pulse frequency[kHz]	3	2.25	3	5–12	5–12	3
Transformer[V/kV]	690/20	690/20	690/20	400/20	400/20	690/20
Rotation speed range[rpm]	11–22	11–22	11–22	8–22	8–22	11–22
Tower height[m]	100	100	100	98	98	100

Table 8.11 Low-voltage current and harmonics of six types of wind turbines.

Parameter/WT type	Measured value of each type of WT [%]					
	1	2	3	4	5	6
THD(I)	8–450	4–90	5–400	5–130	2–30	3–100**
$THD_z(I)$	8–1500	10–60	10–250	10–200	2–35	3–100
*THD(I)_stat**	20–600	10–150	6–800**	-+	-+	3–100
THD_z *(I)_stat*	50–600	10–120**	10–600**	-+	-+	3–40
PWHD(I)	1–50	1–50	5–400	15–200	3–40	5–100**
PWHD(I)_stat	75–600	10–150**	40–800**	-+	-+	3–50
THD(U)	1.5–2.5	1.2–2	1.5–3.5	1–3	1–2.5	1.5–3.5
$THD_z(U)$	2–5	1.25–2.5	2–5	1–3	1–2.5	1–3.5
THD(U)_stat	2.5–5	1.5–2.5	2.5–5	-+	-+	1.5–4
THD_z*(U)_stat*	2–4.5	1.5–2	1–1.5	-+	-+	2–4

* 1, 2, 3, and 6 are DGAG, 4 and 5 is SG; *THD*: total harmonic distortion rate; *THDz*: total harmonic distortion rate containing interharmonics; *P WHD (I)*: weighted harmonic distortion rate of the part that larger than or equal to 14th order; *I*: current; *U*: voltage; stat * value is measured at the DFAG stator connection point;
** are estimated values, because only limited power measurement is available;
+ SG stator measurement alone is not necessary.

Table 8.12 Current and voltage harmonic parameters on low voltage and medium voltage side.

Measuring location	THDz(*I*) (%)	THDz(*U*) (%)
SG 5 (400 V)	2–35	1–2.5
DFAG (690 V)	3–100	1–3.5
WF (20 kV)	3–60	1–3.5

according to the frequency of the interharmonic to be measured [8, 14].

The commonly used time-domain harmonic analyzer generally carries out discrete sampling for the value of a cycle (20 ms) of the power frequency, carries out digital processing for the sampling data, and uses discrete Fourier transform (DFT) or fast Fourier transform (FFT) to determine the size of each harmonic, so the result can only be the harmonics of integral multiples of the power frequency. To analyze the interharmonics, the sampling period must be lengthened (i.e., increase the time window width); for example, take two cycles (40 ms) for sampling and analysis, then the power frequency is twice the harmonics, and the fundamental frequency is one-half times the interharmonic. The corresponding $m/2$ times (m=1,2, …) harmonics can be obtained.

In general, setting the sampling time window width as T_w, all f_m=mT_w can be analyzed. Of course, the longer the sampling period, the more interharmonics obtained from analysis, and the greater the amount of calculation and the longer the corresponding computation time. The appropriate sampling period shall be selected in order to adapt to the frequency of the interharmonic to be analyzed. For example, band width of audio pulse controlled receivers is 5 Hz, in order to analyze the interharmonic of the band width of 5 Hz, window width T_w can be taken as 0.16 s (i.e., 160 ms or eight cycles), the all interharmonic components at the frequency of f_m=6.25 Hz can be measured.

8.8 Measures to Improve Power Quality of WFs

Power quality problems are linked to WF voltage deviations, voltage fluctuations and flicker, harmonics, and so on. Currently, most compensation devices used to improve and enhance the power quality have the function of suppressing voltage fluctuations and flicker, such as static var compensator (SVC), active power filter (APF), dynamic voltage restorer (DVR), and distribution system power quality unified controller (DS-UNICON), and so on.

8.8.1 Improvement of WF Voltage Deviation

When making reactive power compensation for relevant voltage deviations, the following work is required: ① without the installation of any compensation device, use power flow calculation to analyze the reactive power exchanged between the WF and the power grid under different WF output states and the voltage conditions of major nodes, and the main concern is the WFs PCC voltage and power factor levels, whether the specification requirements of the power grid can be met under various WF output states. ② According to the results of power flow analysis, determine the reactive power compensation scheme, including the type and size. Then carry out power flow analysis for WFs with reactive power compensation. ③ Since the voltage and reactive power characteristics of the WF are closely related to the wind turbine types, the actual wind turbine type constitution of the WF shall be analyzed.

Capacitors configured at cage-type asynchronous generator terminals and the charging power of the output cable lines can basically meet the reactive power demand of the wind turbines, so the reactive power absorbed from the grid is not much. But the WF collection system and transformer require a certain reactive power, so when during outage and low output of the wind turbine, the voltage at the PCC may be relatively high. When WF output power is large, the voltage at the PCC is low, reactive power compensation measures needed to be taken for the WF.

Variable speed wind turbines have the capacity of voltage and reactive power control. The role of the parallel capacitor compensation of the WF is to reduce the burden of reactive power control on the wind turbine power electronic converter, and reduce the amount of reactive power exchanged between the WF and the power grid.

For reactive power and voltage regulation of voltage deviations, grouping switching shunt capacitor bank (reactors), transformer tap regulator, WF transformer taps and WF reactive power and voltage automatic control devices with the coordinated regulation of shunt capacitors, and SVC, and so on, can be used. The shunt capacitor bank (reactor) is most commonly used because it is cheap and reliable.

The SVC device automatically adjusts its reactive power output according to the control target based on different active power output of the WF, and always maintains the WF bus voltage at a given value. In the process of zero output to rated output of the WF, SVC device can control the voltage of the controlled bus and the reactive power output of its own, reducing the reactive power exchange between the WF and the power grid, and the regulation is very smooth. Configuration of SVC in large-scale WFs with rapid and frequent voltage changes is very appropriate. When implementing, the fixed reactive compensation device and the dynamic reactive power compensation device are combined to minimize the dynamic reactive power compensation capacity in order to save investment.

8.8.2 WF Flicker Suppression

A variety of measures can be used for the suppression of flicker caused by WFs, such as SVC, active power filter (APF), dynamic voltage restorer (DVR), unified power quality controller (DS-UNICON), and other compensation devices, as well as the control technologies of variable speed wind turbines and the energy storage system of the wind farm can also be used.

1) **SVC**

   Voltage flicker is a special expression form of voltage fluctuations, and the flicker severity is related to the voltage fluctuations caused by load changes. In HV or MV distribution network, voltage fluctuations are mainly related to the change amount of the reactive load and the power grid short-circuit capacity. Under the circumstance that the power grid short-circuit capacity is determined, voltage flicker is mainly caused by dramatic changes in the reactive load. Therefore, the most commonly used method for the suppression of voltage flicker is to install static var compensation device, and now this technology is quite mature. However, because some types of static var compensation devices also produce low-order harmonic currents, they shall be used in parallel with the passive filter. In actual operations, some harmonics may be seriously amplified due to system harmonic resonance. Therefore, when making compensation, a compensators having a short response time and directly compensation for reactive impact current and harmonic current of loads are required to be used.

2) **APF**

   To suppress voltage flicker, in case of sharp fluctuations of load current, real-time reactive current compensation shall be made following the load changes. In recent years, the use of APF consisting of gate turn-off thyristor (GTO, IGBT, etc.) and pulse width modulation (PWM) technology can be used to make real-time compensation for load current. APF works in a completely different way with conventional passive power filters, it uses gate turn-off power electronic devices to carry out compensation control by applying the instantaneous reactive theory based on the principle of coordinate transformation, of which the action principle is the use of power electronic controllers instead of system power supply to provide the required distortion current to the load, so as to ensure that the system only need to provide sinusoidal fundamental current to the load.

   Compared with passive power filter, APF has the following advantages: fast response time, high compensation rate for voltage fluctuations and flicker, the compensation capacity can be reduced; no harmonic amplification and resonance problems, stable operation; strong control capacity, has a role in control of voltage fluctuations, flicker and voltage stability, can effectively filter high-order harmonics, and compensate for the power factor.

3) **DVR**
   In low and medium voltage distribution networks, rapid fluctuations in the active power will also lead to voltage flicker, which requires the compensation equipment to compensate for the reactive power while suppressing voltage fluctuation and flicker, so as to reduce the reactive power fluctuation of the power supply line and provide instantaneous active power compensation. Thus, the traditional reactive power compensation methods cannot effectively improve the quality of this type of energy, and only compensation devices with the energy storage units can meet the requirements.
   DVR is a three-phase converter (inverter) consisting of three single-phase voltage source converters that are connected in series between the grid and the load needed to be compensated. The three single-phase structures used for the inverter here is for the purpose of more flexible three-phase voltage and current control, and providing compensation for system voltage asymmetry. The core part of the device is the synchronous voltage source inverter. When the line-side voltage suddenly changes, the inverter on the DC side of the power source will generate an AC voltage, and then connect in series to the original grid voltage through a transformer to compensate for system voltage drop or offset system voltage surges. Due to its own energy storage unit, it can inject the voltage difference between the normal and fault voltages in order to overcome the influence of system voltage fluctuations on the users. Therefore, it is an effective tool for resolving voltage fluctuations, asymmetry, harmonics, and other dynamic voltage quality problems, and is a compensation device for the load, which plays the role of isolation between the system and the load. The device is only for compensation of specific load, so its capacity depends only on the compensation capacity of the load and the required compensation range.
   Currently, most DVR use capacitors on the DC side to provide DC voltage, which can only provide limited energy. If long-term voltage compensation is required, the voltage output must be perpendicular to the current, so the device does not provide active power and only exchanges reactive power, thus the long-term compensation requirement can be met.
4) **DS-UNICON and other compensation devices**
   Unified power quality controller is an integrated compensation device that combines the characteristics of series and parallel compensation device, and has the function of unified compensation for voltage and current quality problems. User power integrated compensation device containing the combination of series and parallel energy storage units, can be applied for harmonic compensation of the distribution system, as well as solving the problems of momentary power interruptions and voltage fluctuations and other dynamic voltage quality problems, and improve the reliability of power supply.
5) **Control technology for variable speed wind turbines**
   Since variable speed wind turbines are connected to the grid via power electronic converters, on the one hand, the converter will produce harmonic pollution on the grid in operation, on the other hand, the converter is easy to be controlled, by proper control, the fast dynamic characteristic of the converter can ensure the provision of high-quality power to the grid. Specific requirements for converters used for wind turbines: the harmonics generated shall be as few as possible; the output power factor is 1.0 or controllable; controllable output voltage, adapt to changes in grid voltage; stable output power; controllable generator electromagnetic torque; control system should be as simple as possible.
   Based on the generator side converter control, a constant DC bus voltage and torque limited by the strength of the mechanical components are maintained, and the maximum current and power limited by ratings of the generator and the converter are maintained. SPWM rectifier keeps returning power to the grid as a constant power supply. The power reference value is determined by the wind speed and the wind turbine speed, in order to provide more stable power to the grid.
   Power converters can provide unity power factor to the grid. Through control of the network side bridge, the output reactive power (inductive or capacitive) can be adjusted.
6) **Utilization of WF energy storage systems**
   There are many types of WF energy storage systems interfacing with the grid through power electronic converters. At this time, the proper control of the energy storage system and the converter connected to the system, can also suppress power and voltage fluctuations, and improve power quality (discussed in detail in section 10.7).

8.8.3 WF Harmonic Suppression

In order to improve the operating conditions of the power system and to improve the waveform quality, in the availability of technical and economic conditions, measures must be taken to suppress harmonics in the system. For example, increasing the short circuit capacity of the harmonic source bus, and reducing harmonic impedance in order to reduce harmonic voltage of the bus; increasing the number of pulses of the converter unit to reduce the harmonic currents generated; using AC filters to absorb harmonic currents, and so on.

Power system harmonics is caused due to the presence of non-linear devices, including ① ferromagnetic saturated type; ② arc type; ③ electronic switch type.

For wind turbines, harmonics generated by the generator itself is negligible, and major source of harmonic currents are the power electronic components used in the wind turbine. For fixed-speed wind turbines, power electronic devices do not participate in continuous operation, so basically no harmonics are generated; when the wind turbine is connected to the grid, the soft grid-connected device is in working condition and will generate harmonic current, but because its input process is very short, the harmonic injection can be ignored. Variable-speed wind turbines use power electronic components with large-capacity, AC-DC-AC converters of direct-drive permanent magnet synchronous wind turbines use controlled PWM rectifier or non-controlled rectifier and then connected to DC-AC converters, and the grid side uses the PWM inverters to output three-phase AC with constant frequency and voltage; double-fed asynchronous wind turbine stator windings are directly connected to the AC power grid, the rotor winding wiring terminals are led by three slip rings to the bi-directional power converter, and PWM inverter is also used on the grid side. Ports of the stator winding will always deliver power after connected to the grid, and power flow at ports of the rotor winding depends on the slip. Regardless of the type of the variable speed wind turbine, the converter will always be in a working state after connected to the grid. Therefore, the harmonic injection problems of variable-speed wind turbines shall be considered.

Harmonics generated by wind turbines can be determined by actual measurement of the harmonic meter. Measurement of harmonics are general measurement of the output harmonic currents of the wind turbines or the WF. Content of each harmonic can be obtained through a FFT conversion. In WF harmonic measurement, the measuring points shall include the current of the PCC. If the measured harmonic content is overlimited, filtering measures need to be taken.

To analyze transmission of the harmonics generated by the WF in the power grid and the flow direction of the harmonic power, harmonic power flow calculation is required. Basic steps for harmonic flow calculation are as follows: ① obtain the fundamental power flow according to the usual power flow calculation method based on the grid structure and the given operating conditions; ② determine or calculate the injection current of harmonic source of each harmonic network according to the operating conditions of the harmonic source devices; ③ calculate the harmonic parameters of each component, form the equation for each harmonic network; ④ use iterative method to solve the equation of the harmonic network, determine the harmonic voltage of each node in the network, and calculate the harmonic complex power of each branch.

Harmonic suppression of WFs can be started with two aspects: reduce harmonic emission of harmonic sources, and filter out the main harmonics in the network.

1) Use advanced converter topology and control strategy to reduce the generation of harmonics.

   For converter control in wind power generation systems, harmonic elimination (HE), current hysteresis (CH), space vector (SV), and other PWM control strategies can be used. For HE control, a minor phase error or amplitude error in the output voltage of the converter can cause a major error in the output current, so the security of HE control is the lowest for active loads like a power grid. Due to the introduction of current error feedback signals by the CH strategy, the output current is close to a sinusoidal wave. CH also provides power factor correction; SV method can obtain minimal current changes through the selection of IGBT switching states, so as to reduce the harmonic content of the output current, and improve the power quality.

   Using a double-fed asynchronous generator with active filtering function, its straight axis is used to control the reactive power, its horizontal axis is used to control the active power, and the increased signal components are used to control harmonic compensation of the network, in order to achieve decoupling control of active power, reactive power, and harmonic current compensation. The harmonic distortion rate can be reduced from 30% to 4%, whereas the DC bus voltage is maintained at a constant value.

   A high pulse converter is an effective way to reduce harmonics. In terms of harmonic reduction, it can reduce the capacity of the compensator, but has no effect on reactive power compensation.

2) When installing filters at wind turbine terminals or PCC buses, the filter parameters and capacity shall be determined by analysis of the actual measurement results. Passive filters, active filters, and a hybrid filter combining them both, as well as a high pulse converter and other schemes can be used.

   Passive filters can effectively reduce harmonic distortion and provide a certain amount of reactive power compensation, but when the angle of ignition is changing in a wide range, meeting the changing demands of reactive power compensation is difficult. But passive compensation with fixed parameters can effectively reduce the need for active power filters, as shown in the hybrid filtering scheme.

   APF can provide flexible reactive power compensation and harmonic compensation, and can effectively control changes of the reactive power of converter system. In order to achieve reactive control, role of APF to harmonic compensation shall be decreased. APF apply the principle of compensation to supply necessary higher harmonic current to the higher order harmonic load, so the supply current can be of sinusoidal waveform. This device has the features of correct compensation, small space occupancy, and convenience of extending capacity, so it is increasingly applied.

Hybrid compensation system can effectively control reactive power and harmonic compensation and significantly reduce harmonic distortion, because fast AC current changes during phase conversion of the thyristor can be partly absorbed by passive filters, the flexibility of active filters in the hybrid power filter system can be fully played. Also in hybrid power filter system, the required active filter capacity can be significantly reduced, because a large amount of reactive power can be provided by passive filters.

References

1 Ackerman, Thomas. 2005. Wind Power in Power Systems. New York: John Wiley & Sons.

2 Estanqueiro, A. I., Tande, J. O., and Peças Lopes, J. A. 2007. Assessment of power quality characteristics of wind farms. Power Engineering Society General Meeting, IEEE, pp. 1–4.

3 Burton, Tony, Sharpe, David, Jenkins, Nick, and Bossanyi, Ervin. 2001. *Wind Energy Handbook*. New York: John Wiley & Sons.

4 Renner, H. and Sakulin. M. 2000. Flicker propagation in meshed high voltage networks. *Proceedings of Ninth International Conference on Harmonics and Quality of Power*, **3**, 1023–1028.

5 Honggeng, Yang, Xianyong, Xiao, and Junyong, Liu. 2004. Issues and technology assessment on power quality, Part 4: Measurement and analysis of voltage fluctuation and flicker. *Electric Power Automation Equipment*, **24**(1), 1–4.

6 Yongxiang, Bai, Zhiling, Xue, and Wenzhong, Zhang. 2001. The voltage flash testing in huitengxile wind power field. *Inner Mongolia Electric Power*, **19**(3), 10,11,18.

7 Hsu, Y. J. and Lu, C. N. 2006. Flicker measurements at an industrial power network with wind turbines. IEEE Power Engineering Society General Meeting.

8 Haixue, Lin. 2001. Interharmonics in electrical power system. *Distribution & Utilization*, **18**(3), 6–9.

9 Jiangtao, Hao, Nian, Liu, Yu, Xing Jin, Liya, Bo, and Zhuo, Chen. 2004. Study on inter harmonic of power system. *Electric Power Automation Equipment*, **24**(12), 36–39.

10 Zhiqun, Wang, Shouzhen, Zhu, and Shuangxi, Zhou, *et al.* 2004. Basic concepts, indicators and standards for power quality. The second International Symposium on power quality. Shenzhen, pp. 101–111.

11 IEC 61000-3-6. 1996. Assessment of emission limits for distorting loads in MV and HV power systems. Basic EMC Publication.

12 Huang Zhuangsheng, Huang. 2001. Measurement and analysis for harmonics generated by nanao wind power plants. *Power System Technology*, **25**(11), 80–82.

13 Schulz, D., and Hanitsch, R. E. 2003. Investigation of the current harmonic parameters of wind energy converter. IEEE Bologna Power Tech Conference, June 23–26, Bologna, Italy.

14 IEC 1000-4-7. 1991. General guide on harmonics and interharmonics measurements and instrumentations, for power supply systems and equipment connected thereto.

9

Wind Velocity and Generated Power Forecasts in WF

9.1 Overview

Forecast means that people extrapolate and judge future or unknown status of the object with mastered knowledge and means in advance [1]. The forecast is to conjecture the future on the basis of the past, estimate tomorrow in accordance with yesterday, and anticipate the unknown with known information. Whether a forecast result is consistent with the future development or not is determined by two aspects: one is the development of the object itself and the working state of various factors that affect the development process; the other is people's ability to understanding objective things and voluntarily control their trends.

Wind power forecasts differ from those of other power generation modes and loads. A conventional power plant is mainly affected by physical faults. Compared to the conventional power plant, as a natural phenomenon, wind submits to reliable statistical processing and physical forecasts better. Wide experience of forecasting load demand is available, and the consumption of electric energy is easier than wind power in terms of forecast, and its forecast can reach the reasonable accuracy.

Within the time related to power generation plans and transmission capacity, reliable and accurate forecasts of wind power output facilitates system operation and helps dispatchers respond to the variability of the wind power in order to better manage the wind power in the system. The accuracy of the wind power forecasts has important influence on determining the quantity of reserve demand and the economical operation of the electric power system. If relatively accurate forecasts of wind velocities and wind generator's power can be carried out, the dispatch department in the power system is facilitated to adjust dispatch plans in time, as well as to optimize system reserve capacity and make a reasonable control strategy of wind farms (WFs). As a result, the wind power becomes a dispatchable power supply; adverse effects of the variability of the wind power on an electric network, as well as the spinning reserve and operation cost of the electric network can also be reduced, and the penetration power limit of wind power is improved; a foundation of the WF to join the generation bidding can be laid, and the measurement of the capacity credibility of the WF is facilitated so as to make correct electric energy trade plans under the open electricity power market environment. Forecasts of wind velocities and wind generator's power not only provide a reference for power dispatch, but also supply an effective reference for the operation and maintenance of the WF. For example, if timely and accurate forecasts are performed in windy days when damages of wind turbines might occur, protective measures of the wind turbines can be completed in advance so that the loss is lowered. The shutdown of the wind turbines for maintenance can be performed when the wind velocity is low. In this way, the wind power can be fully utilized. Accurate forecasts of wind power have important significance in preventing wind abandonment or reducing its period, saving the total cost of power system.

The forecasts can be divided into qualitative and quantitative forecasts. The qualitative forecast is the qualitative description of the trend of the development and change of the object based on human knowledge and experience. The quantitative forecast is to make quantitative and specific description of the trend of future development of the object based on the mathematical model according to data obtained through investigation. The wind power forecast is required to be quantitative.

According to time span (range) of forecasts, they are generally classified into short-term forecast, medium-term forecast, and long-term forecast. The short-term (including super short-term) wind power forecast is a subset of wind power time forecast (differing from wind power spatial forecast). The time range of the short-term wind power forecast that concerns us is several days, and the time-stepping is from several minutes to several hours. Wind power forecasts within different time spans can be provided according to users' different requirements. Several-hour short time wind power forecasts facilitate the dispatch department in electric networks to adjust the dispatch plans and realize economical

Integration of Large Scale Wind Energy with Electrical Power Systems in China, First Edition. Zongxiang Lu and Shuangxi Zhou.

operation. Moreover, this kind of forecast facilitates the optimization of the system reserve capacity and the transmission system, because the requirement of reserve power is determined by the maximum forecast error; after obtaining hourly accurate wind power forecast for the next day, operators can optimize the dispatch and power balance of the power plant as well as achieve the optimal unit commitment and a lower unit cost; forecast within 12 to 36 hours is in favor of effective electric power market exchange; several-day forecast helps other power plants, such as a large combined heat and power (CHP) plant, operate optimally. Accurate medium-long term wind velocity forecast benefits the plan and design of WFs such as determining the capacity, place, and so on, of a newly installed wind power set; according to the available wind energy resource, the reasonable proportion between the installed capacity of the WF and complementary power generation equipment is determined; a foundation of research on the dynamic interaction of a large-scale WF and the power system is laid, and this kind of forecast improves the design of the wind turbines as well as the connection between WFs and the electricity network; additionally, the mid-term forecast minimizes the adverse effects of the WF on the power network [2–4].

The spatial ranges of forecasts are not uniform. In terms of the spatial forecasts of wind power, the generated power of one set of wind turbines as well as wind power generation in one WF, in a plurality of WFs of one region, and even in one balanced area can be also forecasted, and the forecasts are determined by the operation requirements of the electric network and the WF. Obviously, in the dispatch center of a power grid, all kinds of forecasts of total wind power in its affiliated balanced (controlled) power system are required.

The first report on the wind power forecast, carried out by one research group at the Pacific Northwest Laboratory explaining the importance and benefits of short-term forecasts of the wind power for electric power companies, was issued in 1978. The result obtained by the research group shows that fully reliable forecast of the wind power have the following applications: weekly day-by-day forecast of the wind power output can be applied to planning maintenance; hourly forecast of the wind power in every day can be included in the strategy concerning load plans; hourly expected forecast of the wind power can be used for dispatch [5].

Through plenty of researches and experiments of more than 30 years, especially since 2000, the wind power forecast has reached a certain high level. Multiple models and methods based on physics and statistics have been explored. The wind power forecast has been applied in many WFs and electric power networks in the United States and Europe.

The power generated in the WF is determined by the wind velocity. The operation cost may be increased due to unexpected changes of the power output in the WF (because the electric power system needs to add primary standby) and the reliability of power supply is possibly endangered by this unexpectation. In order to plan heat standby capacity and manage the operation of electric power network, the operators should know the changes of the wind power output in advance. Therefore, tools of wind power forecast are very important. The tools are required by various transmission system operators (TSOs) as well as terminal user-energy exchangers and energy servers, as well as independent power producers (IPPs). The forecast tools supply input information to economical dispatch, electric energy exchange, safety assessment, and so on. As for the electric power system, wind power forecast is one of the most direct and valuable methods for lowering the uncertainty of wind power production plans.

In this chapter, the variability of the wind power and basic methods for forecasting wind velocity and wind power (concentrated in short-term wind power forecast) and the application theory shall be mainly explained.

9.2 Variability of Wind Power

9.2.1 Variability of Electric Power System and Wind Power

The production and consumption of the electric energy or the supply and demand of the electric energy are naturally variable, which are influenced by a great deal of planned and unplanned factors [6]. The power consumption of industry, agriculture, commerce, and citizens is changed at all times, and the great influence of weather changes, large-scale of public activities and holidays and vacations on the consumption of the electric energy is well-known. Therefore, the load demand of one day (week, month, and year) is inconstant, but likes a fluctuant curve. According to the requirement for balancing the operating power in the electric system, the output of power generation system must be adjusted along with changes of the load, that is to say the production of electric energy is changed in pace with the changeable consumption.

The shutdown of power plants, transmission lines, and electric equipment is also variable, follows the change of the loads under control, and besides it has many irregular changes, which are affected by fuel or special weather. All electric power network internal accidents, sudden changes, or interruptions of power transmission with an external electric network, the change of electricity price in a physical market, and the like can immediately affect the decisions on the starting and shutdown of power plants.

The whole electric power system is variable, and both the supply side and the demand side are variable. One electric power system is essentially like a large pool, there are hundreds of water pipes for input (power plants) and millions of taps (consumptions), and both the water pipes and the taps are possibly disconnected and connected at any time, so that the water level is always variable, however, as long as the water level is kept within a set range, the ability to "keep" the safety of the system is provided. In order to keep the electric power system safe, system operators are required to continuously balance the planned and unplanned changes of the power supply and demand. It is realized through the standby. The variability of the power supply is an inherent characteristic of the electric power system, just like that of the demand.

Wind power generation belongs to a power generation technology with variable output. Generally, the wind is changeable, so that the wind energy is naturally unreliable. In fact, it is incorrect sometimes. The wind power is described as "intermittency or intermittent" energy sometimes, and the words "intermittency" or "intermittent" is easy to be misunderstood. In terms of the whole power system, wind power is irregularly turned on/off. Wind power never disappears like power generated by the conventional power plant. It takes several hours to shut down most wind turbines in the system, even during extreme events such as a rain storm. In addition, the time interval of zero wind power can be forecasted. However, zero power output is gradually realized, and the wider the considered region is, the more gradual the conversion process of zero power output is. It is worth considering that, compared to other power generation technologies, the mentioned wind power technology is at a very high availability level (98%). Therefore, as for the wind power in the system, the terminology "intermittency or intermittent" is inappropriate, and the qualifier "variable-output" should be used instead [6].

The variability of wind energy resources is required to be tested under a wide environment of power system, not a single WF or a single set of wind turbine. In a specific place, the wind does not blow continuously. However, as for a large system, even if the wind stops in a certain place, it always blows in other places. Therefore, at the system level, wind can provide relatively stationary output, that is to say the wind can be utilized in a specific place at all time. Generally, the stopping times of different single wind turbines or different single WFs are unrated to a great extent. However, the effect of the wind power on the variability of the system will be non-ignorable when the wind power is changed like many "taps" on a "bathtub."

Because of the variability of the wind resources, the argument about the natural unreliability of the wind energy is carried out sometimes. But it should be pointed out that there is no fully reliable power plant or power generation technology. In fact, because of accidents, natural disasters, planned outage, and so on, instantaneous tripping of a large-scale power plant will occur. Then power will be lost and reserve capacity is required to be applied immediately. However, at the system level, the wind energy is not suddenly tripping. Since hundreds, thousands of wind turbines are distributed in a relatively wide region, wind energy is relatively stationary not as violent as the power failure of several large-scale power plants. Moreover, as for operators, when the changes of power supply present in the whole system, the output power of large-scale power plants are easier to be forecasted and to be managed. Generally, the outage of a 2 MW wind turbine cannot affect the system, but the system must be respond immediately to outage of a 500 MW coal-fired power plant or a 1,000 MW nuclear power plant.

As for a system with the low wind power penetration level, operators will not notice the influence significantly. Under the condition that the wind power penetration level reaches approximate 20% of the power supply, the method established for handling variable demand and supply and the available reserve are better suitable for dealing with an added variable power supply such as wind power, which is related to the characteristics of the specific system. Some changes may be necessary for an electric power system with high wind power penetration level and its operation mode should adapt to connecting wind power. In Denmark, which is the country with the highest wind power penetration level, 21% of electric power demand was satisfied by wind power in 2004 and approximate 25% of the demand for electric power its western transmission system (not connected with the eastern system) is satisfied by wind power in the normal year. In some seasons, the wind power can completely meet instantaneous demand. Experience of Spain, Denmark, and German who have a large quantity of wind power shows that the problem of the wind power capacity contained in the conventional electric network is possibly an economic problem but not a technical problem, even if amount of wind power is connected to national electric network.

The most serious mistake taken in the discussion about the grid connection of the variable wind power is to handle problems in isolation. The wind power has its own characteristics and it is also similar to some other kinds of power supplies, for example, nuclear power and some gas-fired and coal-fired power plants, which are not flexible and should be operated at a constant generation level. The gird connected wind power is very week in flexibility, because its generated power output is determined by natural wind. In isolation, the characteristics of these power plants are unqualified, because the demand for electric power is remarkably and continuously changed in one day. But there are other coal-fired and gas-fired power plants in the power system that have higher flexibility and their output can be quickly changed. Therefore, the wind power should be integrated into a large system with various power plants.

9.2.2 Variability Characteristics of Wind Power Output

The variability of the wind power output is mainly related to the time series of wind velocity. In fact, there are various time series in the world. Some time series are changed slowly and relatively stably along with the seasons. Monthly power demand possibly presents this kind of time series. Other time series are possibly unordered, so that it is difficult to be forecasted. Shown in Figure 9.1, the time series of wind velocity has this kind of characteristic. It is a random process high in nonlinearity and its mean value and standard deviation are changed at any time. In Figure 9.1, the difference between the maximum and minimum wind velocity reaches 15.9 m/s. Therefore, load forecast is more accurate than wind forecast. Generally, the accuracy of 24-hour power demand forecast can reach approximately 98.5%. However, as for wind forecast within 24 hours, the average error of a report is at a 10% to 20% grade [3].

Both the power loads and the wind velocity present variability. For example, very remarkable seasonal characteristics (every day, every week and every season) exist in a load curve. But the randomness of the wind velocity curve is greater than that of the load curve. By spectroscopic analysis of the time series of both the wind velocity and power load, it is clear that the frequency domain of the load is under the control of several specific frequencies, which is the result of the daily and weekly load changes. However, the frequency spectrum of the wind velocity cannot provide us with any important frequency value.

The variability of the wind velocity is affected by many geographic factors. The roughness of the sea level is very low, and the stability of the thermosphere and wind greatly differs from that observed onshore. In addition, the low roughness enhances the effect on the wind velocity distribution, as a result, compared to the wind velocity onshore, the wind velocity offshore is more stationary and durable. Therefore, offshore wind power will be greatly developed in the future and it is necessary to detect offshore wind for a long term and deeply understand the characteristics of the off shore wind.

The wind power is a kind of power supply of variable output and undulates at any time mainly under the influence of meteorological conditions. The wind changes within the full-time scale range, including second, minute, hour, day, month, season and year.

9.2.2.1 Time Characteristic of Variability of Wind Power

The wind power output changes over time. In order to analyze and respond the changing wind power output, the variability of the wind power is classified into short-term variability and long-term variability in accordance to different time ranges. These specific time ranges are differentiated because this kind of information type is corresponding to different types of power balance.

1) **Short-term variability**
   The short-term variability of the wind power means the changes of output power within time ranges, including second, minute, hour, day, and month. The short-term variability of the wind power (several minutes to several hours) is important, because of

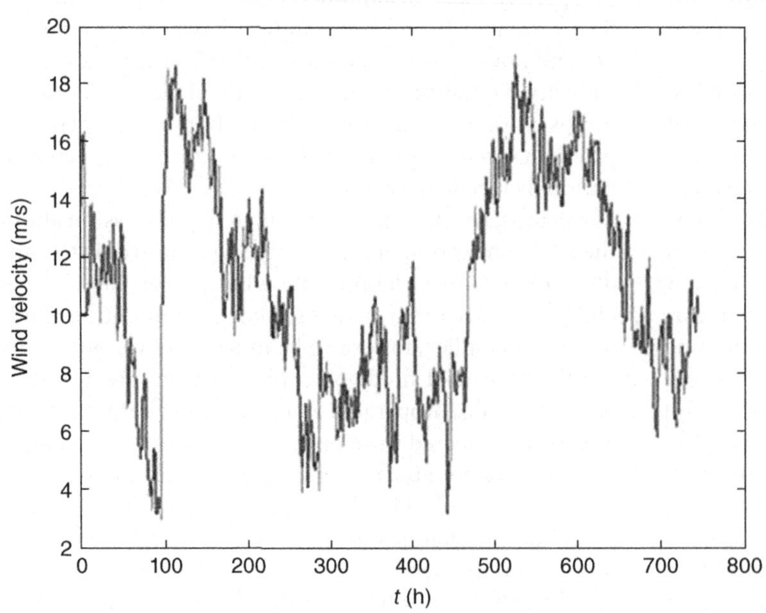

Figure 9.1 A typical time series of wind velocity.

the effect on generating plan, power balance, and determining demand for reserve. The wind power short-term variability bore by the power system is determined by the short-term wind changes (weather mode) and the geographical distribution of WF. The overall variability bore by the electric power system is determined by the simultaneous changes of the loads, all WFs, and other generators. The influence of the short-term variability on power system is determined by the wind power capacity, as well as the combination of different power generation modes and long-distance transmission capacity [6].

On the basis of analyzing the existing data of the operating WF and the meteorological measurement of the position of the typical WF, we can quantify the change of net active power output within a given period, namely a minute, an hour, or several hours. According to the data analysis of the operating wind farm, the below statues can be observed: (1) For large single wind turbine, due to the mean effect of the rotor through which the wind flows and the filter action of the wind turbine inertia, the change of second-grade power output is very little; (2) To a WF, since the output of each set of wind turbine in the WF is equalized, the total power output change within tens of seconds is very little; (3) To multiple WFs distributed in a large region, within a range of several minutes (or less) to tens of minutes, the total power output change of all WFs is little, owing to the "geography diversity."

The short-term wind power variability is as follows: (1) The change within 1 minute is caused by the disturbance or transient events, at the moment, the quick change of grid-connected WF power output (several seconds, minutes…) is quietly little. The reason is that the result of wind power synchronization collection is difficult to be sensed by the system. (2) The change (such as 10 to 30 minutes) within 1 hour has important significance to the system. However, it is considered that these are always related to demand change. This kind of change becomes flat to a great extent due to the geographic variability, and the change of the power output of the geographically decentralized WFs is generally kept within a range of ±5% of the installed wind power capacity. The most remarkable power output change caused by the change of wind velocity is within a range of 25% to 75% of the rated output. The great change occurs before the passing of storm rain, when the wind turbine reaches its storm limit; the output of wind turbine decreased from full capacity to zero within several minutes. Generally, it is important in a relatively small geographical region. As for a large region, it takes several hours for most wind power capacity to disappear during stormy weather. When the wind power penetration reaches a certain level, and the change of the wind power output can compete with the demand (when 5% to 10% of annual power demand is satisfied by the wind power), the power system standby capacity required by system power balance will be a question. (3) Hourly variability means the change from server hours to one day. Several hours ahead change affects power system dispatch, and it can be forecasted by the wind power forecast tools.

2) **Long-term variability**

Slow or long-term change related to the wind power integration to power system includes seasonal change and annual change, caused by climatic influence. The long-term variability of the wind power is not important for daily operation and management of the power network, but plays an important role in system strategic plan.

Monthly and seasonal changes have important significance for electric power traders and the electric system plan, because the electric power traders must deal with transmission contracts, but the wind power capacity has certain influence on electricity price. It is necessary to know the contribution of the wind power to the safety of the electric power system when panning the electric power system, and it is evaluated through the credibility of the wind power capacity. As for a system with low wind power penetration level, the capacity factor is a measurement of capacity credibility. The improvement on technique and percentage increment of the installed offshore wind power can enlarge the capacity factor.

At least 30 years of data analysis of the whole Europe shows the trends of the variability of long-term average wind velocities in the whole Europe are similar, and can be reasonably characterized by normal distribution with standard deviation of 6%. Therefore, small change caused from one place to another place can be speculated. Similar trends can be also observed in other regions of the world, such as Australia, Japan, and Britain. The wind zone in Mongolia, China is in the same status. The result has a great effect on determining the expected change of the long-term wind characteristics and the corresponding wind power project annual energy output. Generally, the annual change of the wind power resources is less than that of water flowing into a hydroelectric station. In terms of the electric power system, the annual change of the wind power is affected by the development of the wind power market and planned land/offshore wind power proportion.

9.2.2.2 Spatial Characteristic of Wind Power Variability

The wind always blows in some certain places and never simultaneously blows strongly everywhere, besides, wind turbines are dispersedly mounted in a wide region, thus the short-term and local wind fluctuations are not related with each other. Therefore, in terms of the system, the wind power output fluctuation is mutually smoothed to a great extend, as a result, the fluctuation amplitude of the power system caused by wind power is lowered. Records of the operating WF have proved the smoothness effect. For a Germany example, the maximum hourly power vibration of a single WF reaches 60% of its installed capacity, whereas the hourly maximum change of a 350 MW Germany synchronized WF will not exceed 20% of its installed capacity. As for a larger region, such as four North Europe countries, the maximum hourly change of the wind power is less than 10% of the installed capacity [6]. In the electric network of Xinjiang, China, only in the WF of Debauching, the effect of the wind power fluctuation on the dispatch can be remarkably felt, and after the system has multiple WFs, less wind power output fluctuation is felt by dispatch.

Geographical spread of wind power causes the effect of "geographic diversity." Besides, it reduces the power fluctuation and increases the total stationary wind power capacity of the system. When the wind power output is aggregated in a large region, the continuous curve of the total power output of the WF becomes flat, and the effective hours of obtained wind power output are increased. At the same time, the maximum power value is reduced. The larger the aggregated WF is, the less the effect of the system operation variability is.

When the WFs mounted in geographic positions have the different weather modes, the synchronization benefit of the WF output is better. The larger the electric power network is, the more remarkable the effect of stabilizing the wind power variability is. Therefore, when building a large-scale wind power plant, it must be integrated into a large-scale power system.

9.2.2.3 Amplitude Characteristic of Wind Power Variability

Generally, the sudden switch in or out of the wind turbines in the WF will cause large impact to the electric power system, which is possible. But in terms of the power system, the capacity of one set of large-scale conventional generator reaches 1,000 MW, the forced outage rate of fossil fuel power plants in China is approximate 1%. Generally, this kind of forced outage cannot be forecasted, and it takes a very short time to realize conversion from full power to zero power. The greatest change of the wind power output occurs before the coming of rainstorm, when the wind turbines reach their rainstorm limit, the output of wind turbines decreased from full power to zero power within several minutes. It is generally important just in a relatively small region. As for a large region, most wind power capacity disappears within several hours during rainstorm. For example, the storm occurred in Denmark on January 8, 2005 is the strongest storm in Denmark in the last 10 years, and the output of a 2,400 MW wind power generator in western Denmark (2,000 km^2) took 6 hours to drop from 2,000 MW to 200 MW. If centralized large-scale offshore WFs are lost, the full output will reduce to zero output within approximately 1 hour.

The maximum change of the wind power output is related to consider the size of the region and different time periods. Wide researches have been carried out in Germany, Denmark, and Britain. Generally, the conclusion that the hourly change degree of the wind power is shown as Table 9.1, where the wind power change is shown as the percentage of installed wind power capacity. Surely, the characteristics of wind in different regions and countries are different, and the hourly change degrees of the wind power are quite different.

9.2.2.4 Wind Power Output Component

The wind power has very strong randomness, and is a non-linear and non-stationary time series signals. During

Table 9.1 Greatest change of wind power output [6].

Region Dimension (km × km)	Greatest Change % (Up/Down)	Example
	Hourly Changes	
100 × 100	50	Britain
200 × 200	30	Denmark
400 × 400	20	Germany, Denmark, Finland
Group countries	10	
	Changes Within 4–12 Hours	
One country	(40–60)/80	Denmark/Germany
Larger region	35	Northern Europe
400 × 400	4h, 80%, 6h, 80%, 12h, 90%	Britain

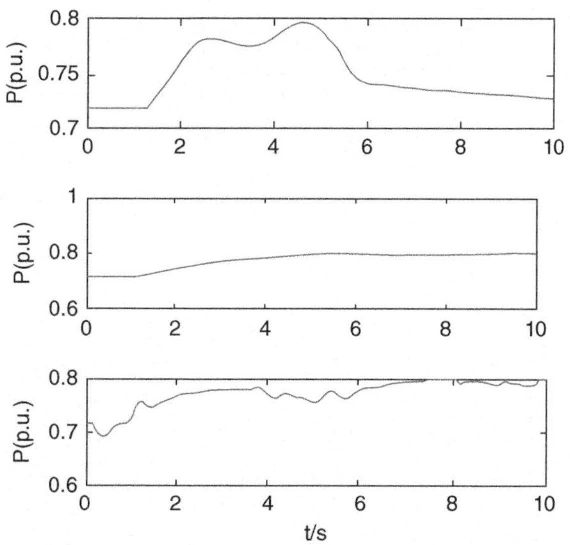

Figure 9.2 Decomposition of WF power output. (a) Power component $P_a(t)$; (b) Power component $P_t(t)$; (c) Power component $P_r(t)$.

the wind power forecast, the nonstationary signals can be disintegrated into several components according to the different scales of fluctuation of different dimensions or trends, such as high-frequency component, low-frequency component and average component.

The power output curve $P_W(t)$ and its components $P_a(t)$, $P_t(t)$ and $P_r(t)$ are shown in Figure 9.2, and they satisfy (Formula (9.1)):

$$P_W(t) = P_a(t) + P_t(t) + P_r(t) \tag{9.1}$$

The power component Pa(t) is the greatest influence on the operation and planning of the power system, shown at Figure 9.2a. It is the mean value of undulating wind power $P_W(t)$, and is calculated on 1-minute time window; the power component $P_t(t)$ is shown in Figure 9.2b. It is a component with high frequency and mean value of zero, and mainly originates the turbulence of the wind velocity; the power component Pr(t) is described at the Figure 9.2c, it is the slope component of $P_W(t)$, and shows the despondence of the WF to the "intermittence" of the wind.

9.3 Wind Power Forecast Model and Method

The change of the wind is random, but the future change trend and the condition of wind are knowable. The development of the weather prediction technique, in particular to the application of numerical weather prediction (NWP), continuously improves the forecast accuracy of the important parts of the weather prediction, including wind velocity and wind direction. Wind power generation depending on the natural wind is also predicable, and future wind power level can be forecasted in accordance with NWP and on-site measurement records, WF power generation models, statistics methods and the like. For decades, multiple wind power forecast tools have been developed in the world, and the precision is continuously improved. The wind power forecast model and method will be described in the section.

9.3.1 Wind Power Forecast Models

Wind power forecast models can be classified according to required input data, shown in Figure 9.3. Two main kinds of model are time series model and NWP-based model. Generally, the developed short-term wind power forecast models can be classified into three types, including time series model, physical model, and statistics model. The future WF power output forecast technology will combine the above three types of models as required in most cases, aiming at improving the forecast precision. The aim of wind power forecasts is to contact the wind forecast of NWP model with wind turbine power output [3, 6–8].

The time series model only utilizes online measured data of wind velocity or wind power, and forecasts wind power generating capacity several hours ago. NWP-based model forecast performance is better than that of the time series model, the forecast time is longer than 4 to 6 hours, and the property can be physical or statistical. The statistics model tries to directly (omitting physical factor) utilize the NWP model output and past NWP output as well as measured data of wind power to calculate wind power output. A statistics/NWP-based model is generally better than the physical model, but under the condition of longer forecast, the difference becomes irrelevant, but the statistics model requires more input data. A physical/NWP-based model only utilizes the output of the NWP model to estimate the local wind velocity of the WF, and then converts the wind velocity into local wind power output. However, in order to calculate model output statistics (MOS) parameter to lower system errors, the past

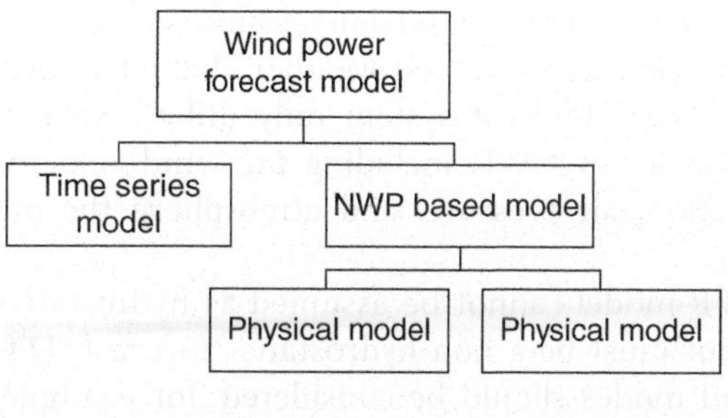

Figure 9.3 Types of models for short-term wind power forecast.

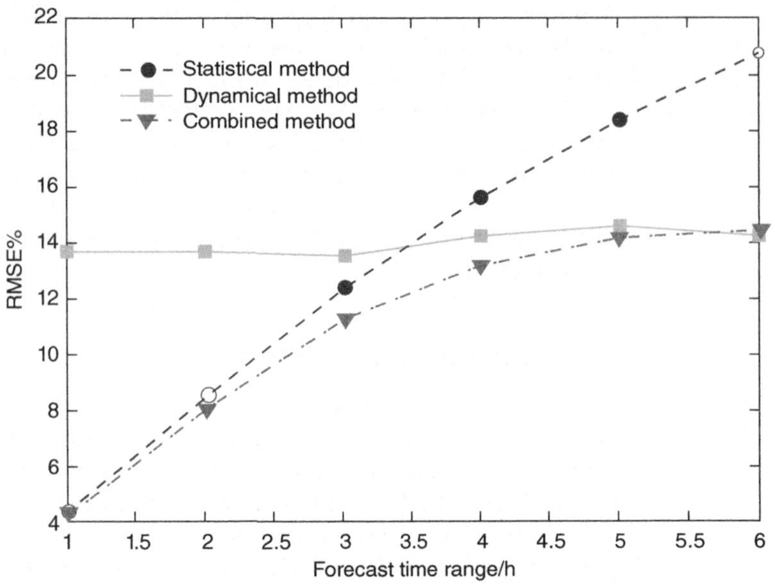

Figure 9.4 Accuracy comparison of time series model with NWP-based physical model of wind power forecast (I) NWP model.

measurement of NWP output and wind power output are required.

As an example, the root mean square error (RMSE) and mean absolute error of the predicted wind power output are shown in Figure 9.4. The forecast result is established by respectively adopting the simplest time series model (continuous method, dashed line) and physical model (predictor, full line) purely based on NWP. The more advanced time series model and a model containing statistics usually can provide better results. The errors shown in Figure 9.4 reflect the current technical level of the wind power forecast system.

The NWP model is used for the computer simulation of terrestrial atmosphere predicted by weather forecast. The most advanced NWP model is operated on the special super computer in the national meteorological center. They predict future atmosphere condition by utilizing as much as possible information on the atmosphere condition at given time points and physical laws. With better understanding of the atmosphere characteristics and improvement of computer speed, the NWP has been improved gradually. The main limit for NWP to accurately forecast atmosphere condition within a very short time range (shorter than 4 hours) is the difficulty in integrating full measurement into the model, whereas the main limit to forecast within longer time (longer than several days) is the disordered weather characteristics. The wind power forecast system only utilizes several output variables of NWP, including the wind velocity, wind direction, air pressure, and atmosphere thermo stability.

The NWP model cannot be assumed as hydrostatics forecast, but must be a non-hydrostatics forecast. Different wind modes should be considered, for example, the breeze in a small region (land/offshore) and the Venturi effect (wind on mountain) require precise digital elevation models (DEMs), to show the topographical conditions of the wind flowing route, therefore, the NWP model is very complex. Furthermore, the forecast of NWP model has two other limit factors, namely data variability and calculation time. The easily changed input data makes data output uncertain, as a result, the forecast accuracy will be reduced, which is difficult to avoid. As a complex mathematical model, the NWP model is usually operated on the super computer, even through, it takes a long time to get the result.

The short-term NWP model mathematically solves the numerical solution of conservation equation (mass, momentum, heat, water, etc.) at a location of given four-dimensional grid (longitude, latitude, height, and time). A coarse grid only provides rough forecast. Generally, a mesoscale weather forecast model provides weather data evaluation in a form of "grid mean value." The terminology "mesoscale" means that the grid is the grid size from $0.5 \times 0.5\ km^2$ to $15 \times 15\ km^2$. Each grid provides the predicted values of wind velocity, wind direction, air pressure and air temperature at a height 50 m above the ground. Because the improvement on the forecast accuracy requires smaller spatial resolution, the mean value of these grids shall be further processed, such as considering characteristic change of a land form in a sub-grid range (for example the hills and mountain). However, meteorology data spatial extrapolation is a challenging task, especially in heterogeneous landscape.

The NWP model has been developed for more than 10 years, and is mainly used for weather forecast. Generally, the NWP model is continuously operated on the super computer, and is in the charge of national meteorological mechanism. However, international private companies can also provide NWP-based weather forecast due to the development of parallel processing.

9.3.1.1 Time Series Model

The wind can be shown as a time series, which can be defined as the observed value of a group of parameters or a group of parameters, and takes the values at many discrete time interval points. These intervals generally take the time length of a certain rule (although not always), and most time series based forecast systems are based on even time interval data, as a result, if time steps between data points are not consistent or the data is default, problems maybe caused. The time series is generally shown as (Formula (9.2)) [8]

$$X = \{\ldots,\ x(t-3),\ x(t-2),\ x(t-1),\ x(t)\} \tag{9.2}$$

As the simplest one, the time series forecast model is the cheapest wind power forecast model. The time series model only requires the latest (several hours) measured data of wind velocity or wind power of the WF, or the measured data of wind velocities or wind power of several typical WFs from a region where the wind power will be forecasted. The simplest time series model is the continuous model, and all of other wind power forecast models must be improved on the basis of the continuous model. The continuous model is established on the basis of the current wind numerical $x(t)$ and the wind numerical $x(t+1)$ of the next forecast time step, having high relevance, namely the forecast value x_{pre} of the next time step (Formula (9.3)).

$$x_{pre} = x(t) = x(t+1) \tag{9.3}$$

Weather scholars as a comparison and supplement of the NWP model established this method. More advanced time series models require more input data (past wind velocities or wind power measurement) and calculation amount. As for short-term forecast, the RMSE of continuous model forecast to about 10% to 30%. This kind of model utilizes time series analysis technologies such as a regression least square algorithm, auto regression model, or artificial neural network model to acquire trends of the measurement of the wind velocity and then extrapolate these trends (future several hours). Many types of time series models have been researched, there is no the simple and the best one, because different models cause different error degrees of different occasions due to different weather conditions and land forms. The forecast before 4 to 6 hours is most accurate, and it is enough for some applications. As for an electric power system having many thermal power plants, the forecast time span is the most critical, because the output of the fossil fuel power plant cannot be quickly changed.

The continuous method is the simplest wind velocity forecast method. For the wind velocity forecast within a very short time range (several minutes to several hours), the simple method is more effective than the NWP model. However, with the increase of forecast ahead of time, the accuracy of the continuous model quickly decreases (see Figure 9.4).

9.3.1.2 Physical Model

The physical method aims to describe the physical process of converting wind into electric power. The physical method comprises a series of models containing different physical processes: the effect of WF shield, the curve of wind turbine power, the model output statistics, and so on.

NWP-based physical model for creating wind power forecast has the following steps shown in Figure 9.5, wherein:

Step 1, obtaining the output result of the NWP model. The NWP model utilizes information about the current atmosphere condition to forecast future atmosphere condition according to known physical laws. These physical laws describe how to change the values with condition variables, such as temperature, air pressure, humidity, and wind velocity at present. The relationship of the variables are described through a set of non-linear partial differential equations not solved in an analytic method. The equation system is numerically solved in each three-dimensional space (longitude, latitude, and height) by the NWP model, and a three-dimensional grid covers a part or all parts of the earth atmospheric layer. The accuracy of the NWP model is limited by incomplete measurement of initial state variable and disordered meteorological characteristics.

Step 2, local refinement. Wind velocity and wind direction output obtained by the NWP model are converted

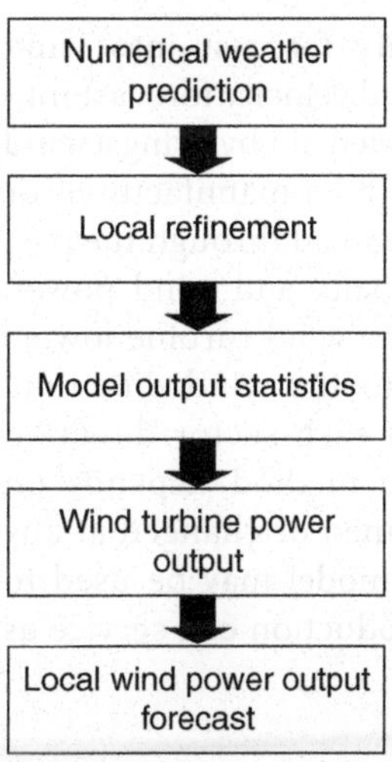

Figure 9.5 Steps of establishing wind power forecast by using NWP-based physical model.

into wind velocity and wind direction at the height of the hub of each wind turbine in the WF requiring wind power output forecast. The simplest method is to utilize the wind velocity and wind direction at the vertical height of the NWP model grid point, which is most close to the WF, in this way, the possibly generated forecast error is smallest. The wind velocity and wind direction can be also obtained through the interpolation of four NWP model grid points around the WF. In order to lower system errors, the influence of landform, ground surface roughness, and existing barriers (trees, building, and the like) should be considered. Those can be simplified into an atmosphere flowing model improved through the ground surface roughness or complex mesoscale numerical atmosphere models, such as an MM5 model or a computational fluid dynamics (CFD) model.

Step 3, MOS. The MOS is used for lowering the system errors as far as possible. The system errors, caused by incorrect roughness parameter, atmosphere thermal stability assumption, neglect of the influence of the complex landform, or excessively low resolution of NWP model and the like, are possibly significant and unavoidable. Fortunately, when the MOS is in use, the reason for the system errors is unnecessary to know and is generally unclear. A mathematical form causing the best error improvement is changed with the time, which is unknown. But a simple linear model can realize the greatest possibility of improvement. The parameters may change over time, and most remarkably change with the seasonal change, aging of wind turbine, and NWP model change. Therefore, the most precise forecast system utilizes the online measurement and a forgetting algorithm to repeatedly calculate the parameters. The MOS can be used before, after, or during converting the wind velocity into the wind power.

Step 4, converting wind velocity forecast into wind power forecast. Converting wind velocity forecast into wind power forecast can be carried out by using a wind turbine power curve provided by a manufacturer, or practical wind power curve obtained through the past measurement of the wind velocity and wind power. Because the wake stream of the wind turbine lowers the total WF power related to the wind direction, the power curve measured in each sector direction should be adopted. Since the method depends on measurement impossibly obtained or changed at any time possibly, a wake stream model may be used to estimate the drop in power production can service as a substitution method.

9.3.1.3 Statistics Model

A statistics model for wind power forecast aims to describe the direct relationship between forecasted wind velocity and electric power output based on statistical analysis of time series of past data.

Statistical models use measured wind power output of a wind farm or region and the corresponding NWP to find statistical parameters that describe the relationship between them. The parameters are possibly changed along with changeable weather conditions, wind turbine aging, and nearby vegetation change, and as a result, the parameters are updated by using weighting of data of the forgetting algorithm. Compared to the physical model, the remarkable advantage of the statistics model includes consideration of all physical aspects. It means that all of complex influence factors including the landform, roughness, wake stream of the wind turbines, and the like are automatically considered. The remarkable disadvantage of the statistic model is that the accuracy of forecast within a longer time span decreases. The past wind power measured value and the result of the NWP model are required to solve the statistics parameters, and in order to make the forecast precise, the parameters are required to be continuously updated.

In wind power forecast, the statistics-based model and another artificial neural network-based method combine to generate a short-term forecast. Most statistics methods adopt an auto regression algorithm, and model parameters are adjusted by using the difference between the forecast and the practical wind velocity at the past close moment. The method based on the neural network adopts considerable quantities of time series data sets to learn the relationship between input data and output wind velocity. The accuracy of these methods will quickly decrease along with the increment of forecast ahead of time.

9.3.1.4 Hybrid Models

Many hybrid models have been used for the wind power output forecast, and the combined types can be the combination of a physical method and a statistics method, the combination of a short-term model and a long-term model, and the combination of a selectable statistics models.

The hybrid models aim to integrate the advantages of each model, thereby obtaining the optimum comprehensive forecast performance. The fact that the adoption of different standards has important influence on model performance evaluation should be stressed, especially the adoption of RMSE and mean absolute error (MAE) standard for the forecast.

9.3.2 Wind Power Forecast Method

Common methods for wind power forecast include the continuous forecast method, the Kaman filtering method, the random time series method, the artificial neural network method, the fuzzy logic method, the spatial relevance method, learning method, the synthesis method, and the like [4, 5, 9].

1) **Continuous forecast method.**
   The continuous forecast method is the simplest method for wind power forecast, which has low requirement on data, and considers that the wind velocity forecasted value is equal to the moving average value of several latest wind velocity values. Generally, the latest wind velocity observed value is simply taken as the wind velocity forecasted value of the next point. The forecast error of the model is greater, and the forecast result is nonstationary but the forecast within a short-term range (several minutes to several hours) is very effective.
2) **Kaman filtering method.**
   The wind velocity services as a status variable to establish a status spatial model, and the wind velocity forecast is realized by using the Kaman filtering algorithm. The Kaman filtering algorithm is obtained under the condition of hypothetically knowing noise statistics characteristics of noise, in fact, estimating the noise statistics characteristic is the difficulty in applying the method. The algorithm is suitable for online wind velocity forecast.
3) **Random time series method.**
   The random time series method adopts a large amount of history data to establish a model, which determines a mathematic model capable to describe the researched time series through model identification, parameter estimation, and model test, further induces a forecast model and finally realizes the purpose of forecast. As long as getting the single wind velocity or power time series, the method can establish a forecast model and obtain a better forecast effect. An accumulated auto regression-moving averaging model is most frequently used in the method.
4) **Fuzzy logic method.**
   Apply fuzzy logic and forecasters' expertise to make data and languages to form a fuzzy rule bank, and then select a linear model to approach the wind velocity with non-linear dynamic change. However, the effect of a simple fuzzy method on the wind velocity forecast is usually poor, the main reason is that the fuzzy forecast learning ability is weaker, the identification of a fuzzy system does not form a complete theory, and selecting the structure of the fuzzy system in the forecast system shall be further researched. Generally, the fuzzy forecast method is required together with other method for use, for example, a fuzzy algorithm and a genetic algorithm are combined to perform short-term wind velocity forecast.
5) **Spatial relevance method.**
   Multiple groups of wind velocity data of the WF and several places close to the WF is required to be considered in the spatial relevance method, and the wind velocity forecast is performed by applying the spatial relevance of wind velocities of several places. In order to obtain required several groups of wind velocity data, several remote wind velocity measurement stations are required to be arranged around the WF. Real time wind velocity data measured in the local WF and a central computer processes each remote wind velocity measurement station, and the wind velocity is forecasted by using the spatial relevance of the wind velocities of the WF and each wind velocity measurement station. In fact, if multiple groups of wind velocity data of the WF and the several places close to the WF can be collected, the method can be used for forecasting the wind velocity. A large amount of original data is collected by the method, but factors considered in the forecast process are increased, so that the forecast effect is better.
6) **Learning method.**
   An artificial intelligence (AI) method is applied to recognize forecasted wind velocity and wind power output from past time series. An artificial neural network (ANN), aiming at simulating the structure and functions of human brain, is formed by connecting large quantities of simple processing elements in a certain type of large-scale topology structure. The ANN has the characteristics of parallel processing, distributed storage, fault tolerance, and the like. Besides, it has the feature of self-learn, self-organization, and self-adaptation ability, and can achieve functions of associative memory, non-linear mapping, sorting and identification, calculation optimization, knowledge processing, and so on, which is very effective in solving complex problem. It can be used for wind velocity forecast, but has problems including slow training speed and the like. When an intelligent forecast system is established, the fuzzy logic system and the neural network are naturally mutual-complementary tools. Although the neural network is a low-grade calculation structure, but can well process unprocessed (raw) data. Fuzzy logic can perform high-grade ratiocination, but it is lack of learning ability, and cannot be adjusted. The forecast result of an individual fuzzy logic method is always bad. Therefore, combining the fuzzy system and the neural network to a comprehensive system to provide a method for establishing very short-term wind power forecast is a viable method. In fact, the neural-fuzzy system is equivalent to the neural network with a fuzzy ratiocination model in function.
7) **Synthesis method.**
   In practical application, the above methods are combined or sometimes hybrid. The physical method nearly always utilizes past data as the input of the

model, or uses model output statistics (MOSs) to correct the result. Statistics method or the AI method usually utilizes knowledge of physical process in the design of the model, such as the power curve shape. The statistics method analyzes the relationship between the weather predicted through the past time series and the generated wind power, as a result, the relationship can be described in other manners in the future. Similar to the statistics method, the AI method is also used to describe the relationship between input data (forecasted by the NWP model) and output data (WF power output), but not explicit statistics analysis, and utilizes an algorithm to implicatively describe the non-linear and high complex relationship between the data. To the statistics method and the AI method, past long-term weather prediction with high quality and the time series of the wind power output are basic and important.

9.3.3 Wind Power Forecast System

Early developed and operated short-term wind power forecast systems have been commercialized, shown as Table 9.2. These forecast tools have been improving [6, 10].

Wind power forecast has been researched in foreign countries early. In early 1990, Landberg developed a set of forecast system, whose main principle is to convert wind velocity and wind direction provided by the NWP to the wind velocity and wind direction at the height of the hubs of the wind turbines in a certain method. Then, the output of the WF is obtained according to a wind turbine power curve and corrected according to the efficiency of the WF. The system adopted a high-resolution limited area model (HIRLAM) of Denmark Academy of Meteorology as the input of the NWP. The WAsP model in Denmark Risø National Laboratory converted the wind velocity and wind direction into the wind velocity and wind direction at the height of the hubs. And the PARK model of Risø considered the weak effect of WF.

The wind power prediction tool (WPPT) was developed by Technical University of Denmark. Since 1994, the WPPT had worked in the electric power system of western Denmark, and begun to work in the electric power system of eastern Denmark from 1999. At first, the system combined the adaptive regression minimum root-squaring method and the exponential forgetting algorithm, and given a forecast result of 0.5 to 36 hours.

Table 9.2 Introduction to early operated short-term wind power forecasts.

Forecast Model	Developer	Method	Operation Condition, Region	Operating Time
Predictor	Risø National Labrotary (Denmark)	Physics	Denmark, Ireland, Germany (America)	1994
WPPT	IMM, Denmark University of Science & Technology	Statistics	≈2.5 GW, Denmark (Eastern and Western)	1994
Previento	Carl von Ossietzky Universität Oldenburg, Energy and Weather System (Germany)	Physics	≈12 GW, Germany	2002
AWPPS (More-Care)	Armines/Ecole des Mines de Paris (France)	Statistics	Crete Island, Madeira Islands	1998
RAL (More-Care	RAL (Britain)	Statistics	Ireland	–
Sipreólico	Universidad Carlos III de Madrid (Spain)	Statistics	≈4 GW, Spain	2002
LocalPred-RegioPred	CENER (Spain)	Physics	Spain	2001
Casandra	Gamesa (Spain)	Physcis	Spain, Portugue and America	2003
GH Forecaster	Garrad Hassan (Britain)	Physics Statistics	Spain, Ireland, Britain, America, Australia	2004
eWind	TrueWind (America)	Physics Statistics	Spain and America	1998
AWPT	ISET (Germany)	ANN Statistics	≈15 GW, Germany	2001
AleaWind	Aleasoft (Spain)	Statistics	Spain	2004
Scirocco	Aeolis (Holland)	Physics	Holland, Spain	2004
Metrological	MBB	Physics	Spain	2004
Meteotemp	Unamed model	Physics	Spain	2004
Zephyr	Risø Technical University of Denmark	Physics Statistics	Spain	

Prediktor is a WF power forecast system developed in Risø National Laboratory, and uses the physical model as more as possible. Large-range air flowing data is provided by the high-resolution limited area model (HIRLAM) of the NWP system. According to geocenter autogration towing law and the logarithm distribution plan of the wind velocity, the wind velocity of upper air is converted into the wind velocity of the ground. As for a specific place, more detailed data is required, so that WAsP program can be used to analyze, and the influence of barriers and roughness, the change of the roughness, the acceleration at mountain, the deceleration in valley, and the like can be considered by the WAsP. The influence of the wake stream of the WF can be considered in a PARK model. At last, two statistics models are used for showing other factors unable to be shown in the physical model.

Risø and Informatics and Mathematical Modeling (IMM) of Technical University of Denmark jointly developed Zephry, a new generation of short-term wind power forecast program. The Zephry integrates the functions of the forecast program Prediktor and the WPPT, and can perform short-term forecast (0 to 9 hours) and day-ahead forecast (36 to 48 hours). The IMM adopts online practically measured data and an advanced statistics method, which can give a very good short-term forecast result, and besides uses weather models such as the HIRLAM so as to greatly improve the precision of the day-ahead forecast.

Previento is a forecast system developed by OldenBurg of Germany, can give a power forecast result within 48 hours in a larger region, and is similar to the Prediktor in forecast method.

AWPT is a part of wind power management system (WPMS) developed by Germany Solar Energy Institute. The WPMS comprises an online monitoring system, a short-term forecast system (1 to 8 hours) and a day-ahead forecast system. The model has the characteristics as follows: forecast is performed according to precise NWP provided by Germany Weather Bureau; the power output of the WF is calculated by using the ANN; the total wind power output injected into the electric power network is calculated in an online extrapolation model.

eWind is a wind power forecast system developed by AWS Truewind of the United States. The eWind mainly is composed of a group of high-precise three-dimensional atmosphere physical and mathematical models—an adaptive statistical model, a wind power output model, and a forecast distribution system.

Sipreolico is a wind power forecast system developed by Universidad Carlos III de Madrid, Spain.

Other forecast systems include More-Care system working in Madeira Island and Crete Island, and Honeymoon system developed by Ireland.

In addition, some companies have begun to handle the short-term forecast model as daily commodities and integrate it to the energy management system, the wind power control system, the SCADA system, and the operation of the WF. Information and communication technologies in the power market framework are expected to play an important role in comprehensive wind power forecast tools.

9.4 Wind Velocity Forecast

Affected by multiple factors such as temperature and air pressure, the wind velocity has very high randomness, so it is difficult to obtain a relatively precise forecast result. The wind velocity forecast is a basis of the wind power output forecast. General models and methods for the wind power forecast are summarized in the previous section. This section describe detailly several wind velocity forecast methods [4, 5, 11–16].

9.4.1 Forecasting Wind Velocity in Time Series Method

Time series analysis (TSA) model proposed by American scholar George Box and Britain statistician Gwilym Jenkins in 1968 is called the Box-Jenkins forecast method, which is widely applied to various forecasts [12, 16], and is one of the main methods for forecasting the wind power. Box-Jenkins forecast method takes the time series i as a random process to research and describe. A large amount of historical data is used to establish a model; a mathematical model capable of describing the researched time series is determined through model identification, parameter estimation and model inspection, and then the model infers a forecast model.

According to the Box-Jenkins method, the random time series models can be assorted into the autoregression model (AR), the moving average model (MA), the autoregression moving average model (ARMA), and the autogression integrated moving average model (ARIMA). As for the AR model, the current observed value is shown by several historical observed values and a current random interference; as for the MA model, the current observed value is shown by the linear combination of white noise series called random interference; the AR model is combined with the MA model to obtain the ARMA model. The time series described by the models AR, MA, and ARMA are called stationary time series.

The wind velocity is a random variable whose time series is a nonstationary series. When forecasted by the time series method, the wind velocity shall be converted into a stationary time series at first, in other words, the wind velocity time series data is subjected to preprocess. At the aspect of mathematics, by unfolding signals in the complete set of a function, in this way, the signals are shown in different methods under any condition. The following part will list several methods of converting nonstationary time series to the stationary time series.

9.4.1.1 Difference Transformation Method

1) **First difference transformation.**
   Ordered difference operator $\nabla = 1 - B$ is introduced, and B is a delay operator. The first difference transformation of the original nonstationary time series $\{y_t\}$ is performed (Formula (9.4))

$$\nabla y_t = y_t - y_{t-1} = (1 - B)y_t \tag{9.4}$$

   ∇y_t is the first difference of y_t. The original time series is subjected to the d ordered difference so as to obtain (Formula (9.5))

$$\nabla^d y_t = \nabla^{d-1} y_t - \nabla^{d-1} y_{t-1} = (1 - B)^d y_t \tag{9.5}$$

   The stationary series obtained after the difference can be described by models AR, MA, and ARMA, and the original time series can be shown as (Formula (9.6))

$$\phi(B)\nabla^d y_t = \theta(B)\alpha_t \tag{9.6}$$

   In this way, an auto regression integrated moving average model (ARIMA) (p, d, q) is obtained.
2) **Seasonal difference transformation.**
   If the researched time series has a seasonal change trend, the seasonal difference transformation of the time series can be carried out, and introduce a seasonal difference operator $\nabla_s = 1 - B^s$, and $\nabla_s^D = (1 - B^s)^D$, wherein S means a period, and the seasonal ARIMA model is (Formula (9.7))

$$\phi(B^s)\nabla_s^D y_t = \Theta(B^s)\alpha_t \tag{9.7}$$

3) **Synthesis difference transformation.**
   By synthesizing the transformation of the above two differences, the obtained model is $(p, d, q) \times (P.D, Q)_s ARIMA$, namely (Formula (9.8))

$$\varphi(B)\phi(B^s)\nabla^d \nabla_s^D y_t = \theta(B)\Theta(B^s)\alpha_t \tag{9.8}$$

In order to determine the sequence of the model, y_t, ∇y_t, ∇_s y_t, or the autocovariance and autocorrelation function of variables after a higher difference are investigated to determine d and D, and then simplify the model into a corresponding AR, MA, or ARMA model, and further p, q, P, Q are determined. After the sequence of the model is confirmed, other parameters of the models can be calculated in moment estimate algorithm or least squares estimation and so son. At last, whether residual error series $\hat{\alpha}_t = y_t - \hat{y}_t$ ($\hat{y}_t$ is a predicated value) is one series sample of the white noise or not is checked, if yes, the established model is suitable, otherwise the established model is not suitable with that the model is required to be modified or identified once again until qualification.

Reference [13] conducts the second ordered difference transformation to the original wind velocity series is performed in, and a $\{\nabla^2 y_t\}$ series and 20 autocorrelation function values thereof are obtained, at the moment, the mean value of $\{\nabla^2 y_t\}$ approximates to zero, which shows that after second difference transformation, the autocorrelation function quickly degenerates to zero, and the series is enough stationary. Therefore, $\{\nabla^2 y_t\}$ can be identified as an ARMA $_{(p, q)}$ model. In view of multiple choices for model identification, in order to obtain the most reasonable model, the parameter estimation and model test of multiple sets of order including p = 0, 1, 2,..., 8; q = 0, 1, 2, 3 is performed, it should be noticed that the model should satisfy stationary condition and reversible condition, finally, the model is determined as ARMA (6, 2), and the original wind velocity series model is ARIMA (6, 2, 2). By using the model to perform wind velocity forecast, the absolute mean error of the obtained wind velocity forecast result is about 26%.

9.4.1.2 Empirical Mode Decomposition Method

In the end of 1990s, America-born Chinese Norden E. Huang and his team proposed the conception of intrinsic mode function (IMF) and gave a method for decomposing any signal into an intrinsic mode function, namely empirical mode decomposition (EMD). The EMD is a new method for processing non-linear and nonstationary signals, integrates the multiresolution advantage of wavelet transformation, besides overcomes the difficult in selecting a wavelet basis in the wavelet transform, and is a self-adaptation wavelet decomposition method. The EMD method decomposes the signals based on the characteristics of the signals, decomposes nonstationary signals into a plurality of intrinsic mode functions or intrinsic mode components in accordance with fluctuation or trends in different sizes, and performs smooth signal processing.

When combining the EMD with the time series method to perform short-term wind velocity forecast, firstly, apply the EMD adaptively to decompose the original series into a series of intrinsic mode function IMFs in different sizes in a adaptive manner, and the decomposed components outstand the partial characteristics of the original series. Secondly, according to the change rule of each IMF, a corresponding time series model is established, and each intrinsic mode component is forecasted. Lastly, the final forecasted value is obtained though adding the forecast results of all components [14].

The intrinsic mode functions (IMFs) obtained by the EMD method for decompose satisfy two conditions: (1) in the whole number set, an extreme point and zero-crossing numbers are equal or the difference between the two is one at most; (2) in any point, the mean value of an envelope function formed by partial maximum and another envelope function line formed by partial minimum is zero.

The steps of performing experience model decompose of time series $X(t)$ are as follows:

1) All partial extreme points of the signals are determined, then all partial maximal points and partial minimum points are connected by a cubic spline to form an upper enveloping line and a lower enveloping line, the upper and lower enveloping lines should envelope all data points, and are respectively marked as $u(t)$, $v(t)$.
2) The mean value of the upper and lower enveloping lines are marked as $m(t)$: $m(t) = (u(t) + v(t))/2$. $X(t) - m(t) = h(t)$ is solved, whether $h(t)$ satisfies two conditions of the intrinsic mode function is judged, if not, $h(t)$ is taken as new $X(t)$, and the above operation is repeated below (Formula (9.9)):

$$\begin{aligned} m_1(t) &= [u_1(t) + v_1(t)]/2 \\ h_2(t) &= h_1(t) - m_1(t) \\ &\cdots \\ m_{k-1}(t) &= [u_{k-1}(t) + v_{k-1}(t)]/2 \\ h_k(t) &= h_{k-1}(t) - m_{k-1}(t) \end{aligned} \tag{9.9}$$

Until $h_k(t)$ satisfies the two conditions of the intrinsic mode function. In this way, the first intrinsic mode function $C_1(t)$ and the rest of the signals are obtained (Formula (9.10)).

$$r_1(t) = X(t) - C_1(t) \tag{9.10}$$

3) EMD decomposition of $r_1(t)$ is continuously carried out according to the method in the step (2) until the obtained rest satisfies the given ending condition, and a decomposition process is finished. Finally, several IMFs components and one margin are obtained (Formula (9.11)):

$$\begin{aligned} r_2(t) &= r_1(t) - C_2(t) \\ &\cdots \\ r_n(t) &= r_{n-1}(t) - C_n(t) \end{aligned} \tag{9.11}$$

The original series $X(t)$ are shown as the sum of all IMFs and the margin (Formula (9.12)):

$$X(t) = \sum_{i=1}^{n} C_i(t) + r(t) \tag{9.12}$$

$r(t)$ is a residual component in the mathematical expression, and the mean trend of the signals is reflected. Each IMF component $C_i(t)$ is a stationary time series.

9.4.1.3 Local Wave Decomposition Method

The basic steps of forecasting the wind velocity based on local wave decomposition (LWD) are as follows: firstly, a local wave method is used to solve the mean value of the signals, and then abstract local wave components in sequence according to the mean value, and decompose nonstationary wind velocity signals very high in randomness into limited relatively stationary signals. In addition, based on decomposition, each local wave component is modeled by using the time series method, short-term change of the wind velocity is forecasted, and finally the short-term changes are overlapped to form a wind velocity forecast curve [15].

The local wave decomposition method is a signal analysis method based on unfolded signals, and is established based on instantaneous frequency concept, whereas the instantaneous frequency is very important physical quantity for researching the transient and nonstationary phenomena.

The process of decomposing the complex nonstationary signals into limited local wave components is called local wave decomposition, also known as a screening process. A corresponding decomposition algorithm is called a local wave decomposition algorithm, comprising two parts, that is, a mean value-solving process and a component extraction process.

1) **Solving mean value**
The first step of the local wave decomposition process is to find the mean value of the original signals. According to different ways to obtain the mean value, the step can be divided into enveloping mean value method, continuous mean value method, adaptive time-variable filtering decomposition method, integral mean value theorem of adaptive filtering, and so on.

Enveloping mean value method. Firstly, find all local extremisms in the signals, then connect all local maximums in the signals by a cubic spline to form the upper envelope. In a similar way, the local minimums generate the lower envelope, and all data shall be contained between the upper envelope and the lower envelope. The mean value of the upper envelope and the lower envelope is the required mean value. The method has a good effect on most occasions, but the problems caused by the cubic spline are overshoot or undershoot, although these problems can be solved in a more complex spline method, the calculation time consumed by the spline interpolation of a higher order is more, and the improvement is not remarkable. In addition, a spline fitting process consumes much time also.

Continuous mean value method. The continuous mean value method means that the mean value of signal continuous extreme points is used, not the envelope mean value. Therefore, the method only requires spine fitting once, not twice. It may be easy to achieve, but the defect is that when the close extreme range is larger, a more serious mean amplitude effect will be caused. This kind of continuous mean value method will enable basic mode components to become signals with constant amplitude, and the characteristics of the signals related to the amplitude are damaged.

Adaptive time varying filtering decomposition algorithm based on signal local characteristics. The steps of the adaptive time varying filtering decomposition algorithm based on signal local characteristics are as follows:

a. All local extreme points of the signals are found and then form a time series $e(t_i)$, wherein t_i represents the time position of the local extreme point i.

b. As for any three continuous local extreme points, $e(t_i)$, $e\ (t_{i+1})$, $e(t_{i+2})$, the following time varying filter is used to calculate the local mean value $m(t_{i+1})$ at the moment of t_{i+1} (Formula (9.13))

$$m(t_{i+1}) = h(t_i) \times e(t_i) + h(t_{i+1}) \times e(t_{i+1}) + h(t_{i+2})) \times e(t_{i+2}) \tag{9.13}$$

in the expression

$$h(t_i) = \frac{1}{2} \times \frac{t_{i+2} - t_{i+1}}{t_{i+2} - t_i}$$

$$h(t_{i+1}) = 0.5$$

$$h(t_{i+2}) = \frac{1}{2} \times \frac{t_{i+1} - t_i}{t_{i+2} - t_i}$$

In order to improve the boundary effect of the decomposition based on an experience mode, the solving of mean values at the two ends of the signals are subjected to special processing, and the following algorithms are adopted (Formulas (9.14) and (9.15)):

$$m(0) = \frac{1}{2}s(0) + \frac{1}{2} \times \frac{t_1}{2t_1 - t_2} \times [e(t_1) - e(t_2) + s(0)] + \frac{1}{2} \times \frac{t_2 - t_1}{2t_1 - t_2} \times e(t_1) \tag{9.14}$$

$$m(T) = \frac{1}{2}s(T) + \frac{1}{2} \times \frac{t_n - t_{n-1}}{T - t_{n-1}} \times e(t_n) + \frac{1}{2} \times \frac{T_2 - t_n}{T - t_{n-1}} \times [e(t_n) - e(t_{n-1}) + s(T)) \tag{9.15}$$

In the expression, $s(0)$ is the start value of the original signals, s(T) is the end value of the originals, and $m(0)$, m(T) are the local mean values of the starting point and the final point of the organ signals.

c. After obtaining the local mean values $m(0)$, $m(t_1)$, …, $m(t_n)$ of the local extreme points from the original signal, the local mean value at other time point is subjected to interpolation with cubic spline, as a result, a local mean value $m1$ is obtained.

Only one time of cubic spline interpolation is required by each decomposition process in the method, thus accelerating the decomposition process. In addition, the selection of the local mean value is not based on the envelop mean value of the local extreme value, but obtained through the time varying filtering algorithm based on the local extreme value, as a result, higher decomposition accuracy is realized, accordingly, the repeat times of each time of decomposition are reduced, and besides higher decomposition precision is obtained.

2) **Extraction of local wave component**

If the solved local mean value of the original signal $S(t)$ is $m_1(t)$, the first component $h_1(t)$ can be obtained by the following expression under the ideal condition (Formula (9.16)):

$$h_1(t) = S(t) - m_1(t) \tag{9.16}$$

Since the envelope mean value is so approximate, some asymmetric waves still possibly exist in h1(t), which need to be filtered continuously, and the process can be repeated many times. In second filtering, h1(t) is required data to be processed, so (Formula (9.17)):

$$h_{11}(t) = h_1(t) - m_{11}(t) \tag{9.17}$$

After the k times of filtering, the following is obtained (Formula (9.18)):

$$h_{1(k-1)}(\mathrm{t}) - m_{1k}(\mathrm{t}) = h_{1k}(t) = C_1(t) \tag{9.18}$$

In the expression, C1(t) is the first local wave component obtained from the original signals. During the processing, it should be extremely careful, because excessive repeats of the processing will cause that the local wave components become pure frequency modulation signals, but their amplitudes are constant. In order to avoid this, a criterion of stopping filtering processing must be determined. The criterion can be achieved through limiting the size of the standard deviation, the standard deviation S_d is obtained through calculating two continuous processing results. The common value of S_d is 0.2 to 0.3.

In one word, the shortest period component of the original signals should be contained in $C_1(t)$, namely the part with the minimum signal size. Signal $R_1(t)$ separated from the original signal $S(t)$ is taken as new data to be processed.

Wherein $R_1(t) = S(t) - C_1(\mathrm{t})$, obviously, R1(t) contains the larger scale part of the signals.

The screening is carried out in sequence (Formula (9.19)):

$$R_2(t) = R_1(t) - C_2(t)$$
$$R_3(t) = R_2(t) - C_3(t)$$
$$\cdots\cdots$$
$$R_n(t) = R_{n-1}(t) - C_n(t) \tag{9.19}$$

So (Formula (9.20)):

$$S(t) = \sum_{i=1}^{n} C_i(t) + R_n(t) \tag{9.20}$$

In other words, the original signals are decomposed into n local wave components ($C_1(t)$~$C_n(t)$) and a trend item $R_n(t)$. Because the scale becomes larger, $R_n(t)$ is relatively slowly changed a message, and is possible a monotonic function or a constant. But mode components $C_1(t)$~$C_n(t)$ are separated messages with different scale.

The processing can stop after the preset condition criterion is satisfied, and the stopping conditions herein can be classified as follows: (1) when the component Cn(t) or residual component Rn(t) becomes less than the predicted value, stopping is performed; (2) when the residual component Rn(t) becomes the monotonic function, the stopping is performed until one local wave component cannot be selected any more. Even if the original signal data has a global mean value of zero, the final residual amount still may not be zero.

Seen from the process, each component can be modulated by amplitude or frequency. Through decomposition, modulations of different amplitudes and frequencies are clearly separated, and can obtain a signal description method of variable amplitude and frequency. The decomposition can effectively separate messages with different time scales,and process nonstationary, non-linear messages. The time and frequency can be simultaneously localized, and the decomposition has very high self-adaptability.

By applying the above local wave algorithm, the local wave decomposition of the wind velocity is performed, and the local wave is decomposed into different basic mode components (relatively stationary random signals), as a result, forecast errors can be reduced to a great extent, and the absolute errors of the forecasts will be lowered to a certain extent because of the reduction of the amplitude value of decomposition amount.

9.4.2 Forecasting Wind Velocity in Time Series Neural Network Method

9.4.2.1 Fundamental Principle of Neural Network Method

An artificial neural network aims to simulate the structure and functions of the human brain, formed by connecting large quantities of simple processing components in a certain large-scale topology structure, and is relatively effective in solving complex problems. Currently, the most widely applied one is a multi-layer feedforward neural network, namely a BP neural network, comprising an input layer, one or more hidden layers and an output layer, inter-laminar neurons are unidirectional connected, whereas intra-laminar neurons are independent of each other. The hidden layer neuron mapping function commonly adopts a hyperbolic tangent S-shaped (Sigmoid) transfer function (tansig function), the output layer adopts a Purelin function, and the output of the whole network can take any value. The learning and training process of the network is just a weight adjustment process, and is realized by a signal forward propagation process and an error backpropagation process. The training is performed to round and round until the errors of the network output are reduced to an acceptable degree, or are performed until the preset training times are up. A grid underwent good training can give proper output for input not interactively trained also, and has generalization ability, which makes the forecast possible [13].

Since the traditional BP algorithm has some inherent defects, such as easiness of falling into local minimum, low learning convergence speed, and the like, so that Levenberg-Marquardt (LM) algorithm is adopted, the weight modified expression of the traditional BP algorithm (Formula (9.21)) is replaced by Formula (9.22).

$$\Delta W = -\eta \frac{\partial E}{\partial W} \tag{9.21}$$

$$\Delta W = (J^T J + \mu I)^{-1} J^T E \tag{9.22}$$

Where W is the weight in the expression; η is learning efficiency; ΔW is weight modification value; E is the error; J is Jacobian matrix of error to weight differential; μ is scalar quantity, which determines that the learning is finished in Newtonian method or gradient method, and μ supports adaptive adjustment. Practice shows that the L-M algorithm can effectively overcome the defects of the traditional BP algorithm, and shortens learning time.

9.4.2.2 Selecting Input Variable

In the three-layered feedforward neural network, as long as the hidden layer adopts a non-linear progressive increment mapping function, and the input layer and the output layer adopt a linear mapping function, any continuous function can be approached by the three-layer network. In this way, the wind velocity forecast network model can be designed to a three-layer feedforward network, wherein the node number of the input layer is equal to the input variable number, the node number of the hidden layer needs to be determined through multi-aspect test, the node number of the output layer takes one, the hidden layer adopts a

Tansig mapping function, and the output layer adopts a Pureline mapping function.

There is no a very good selection standard for the determination of the input variable of the neural network input layer. When using the time series model to select the input variable, the input variable of the network must select those variables that greatly affect output and are capable of being detected or extracted. The research result of the time series method shows that the second-order difference sequence $\{z_t\}$ can be identified as an ARMA (6, 2) model, and forecasted value $\hat{z}_t$ (1) is determined by six historical values closest to $\{z_t\}$ and a linear combination of two most close residual deviation values. Therefore, the input variable of the neural network can be determined as the six historical values closest to the second-order difference sequence $\{z_t\}$ and two closest residual deviation values, the value within the next hour of $\{z_t\}$ can be taken as the target of the network to form a sample pair, and multiple groups of sample pairs are selected to form a training sample set to training the network. $\hat{y}_t(1)$ can be deduced after the forecasted value $\hat{z}_t(1)$ is given by the network (Formula (9.23)). The following can be deduced by $Z_t = \nabla^2 y_t = y_t - 2y_{t-1} + y_{t-2}$

$$\hat{y}_t(1) = \hat{z}_t(1) + 2y_t - y_{t-1} \tag{9.23}$$

9.4.2.3 Adaptive Weight Adjustment

Generally, the weight of the BP network is no longer changed or easily changed after training and determination. Because the wind velocity shows very high randomness, when using the constant weight to forecast, the forecast effect is better when the training sample set is relatively close, and the weight of the network shows inaptitude when training sample set deviates, so that adjustment is required. As a result, the weight requires adaptive adjustment. The specific steps are as follows:

1) Assume the data used for the first group of training sample set is $z_{t+1} \dots z_{t+N}, a_{t+1} \dots a_{1+N}$, after using the group of data to training the network well, $\hat{z}_{t+N+1}$ can be forecasted, and further $\hat{y}_{t+N+1}$ is solved.
2) As time goes on, after obtaining z_{t+N+1}, the data used for training the sample set becomes $z_{t+2} \dots z_{t+N+1}$, $a_{t+2} \dots a_{1+N+1}$. The network is trained once again based on the previous weight, $\hat{z}_{t+N+2}$ can be forecasted after training the network well, and accordingly $\hat{y}_{t+N+2}$ is solved. In a similar way, the previous network weight is inherited, that is the training is identified, the weight can be continuously updated to enable the network to be suitable for the newest change of wind velocity.

We can see that the time series method and the neural network method are the same in input variable, but the neural network method can learn the non-linear law of the changes of the wind velocity, while the time series method cannot do it, which is the reason why the forecast effect of the neural network method is better than that of the time series method.

9.4.2.4 Forecast Correction

The forecasted value of the neural network model can be further corrected in accordance with the distribution statistics characteristics of the wind velocity. We assume that the wind velocity satisfies Weibull distribution, the probability density function of the wind velocity is below (Formula (9.24)):

$$p(v) = \frac{k}{c}\left(\frac{v}{c}\right)^{k-1} \exp\left[-\left(\frac{v}{c}\right)^{k}\right] \tag{9.24}$$

in the expression, k is called shape parameter, and c is called size parameter. According to the practical wind velocity, $k = 1.96$, $c = 7.88$ are determined. Knowing from practical wind velocity data, the wind velocity of the WF is mainly at 2 to 20 m/s, and the probability that the wind velocity is less than 1.5 m/s and greater than 20 m/s is very low, accordingly, the following correction can be performed:

when $\hat{y}_t > 20$, make $\hat{y}_t = 20$;
when $\hat{y}_t < 1.5$, make $\hat{y}_t = 1.5$.

In addition, to draw on the experience of the principle of the continuous forecast method, the relevance between y_t and y_{t-1} is the largest, and it meets the practical experience, so that the correction $\hat{y}_t = (\hat{y}_t + \hat{y}_{y-1})/2$ is performed once again. In this way, the forecast errors can be lowered by 2%. The simulation test in reference [13] shows the final error caused by using the above time series neural network method to forecast the wind velocity, which is about 20%.

9.4.3 Forecasting Wind Velocity in Time Series-Generalized Neural Network Method

Higher forecast precision of the wind velocity series can be obtained by using the time series–neural network method to perform short-term wind velocity forecast, using the time series model to select the input variable of the neural network and selecting and using multi-layer back propagation (BP) neural network and generalized regression neural network (GRNN).

9.4.3.1 Principle of Generalized Regression Neural Network

Radial basis function (RBF) neural network is a three-layered feedforward network with a single hidden layer. An input layer node formed by a radiative function only transfers input signals to the hidden layer, whereas an output node is a simple linear function. The generalized regression neural network (GRNN) can approach any continuous function with precision, which is a local

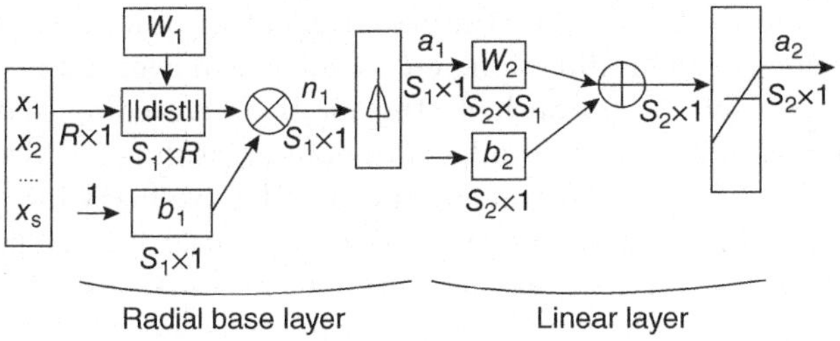

Figure 9.6 GRNN network assumption diagram.

approximation network, namely for a certain local region of the input space, there are only a few neurons decide network output.

The GRNN is a branch of the RBF neural network, of which the method for obtaining the relationship between data differs from interpolation and fitting, and can correct the network with data obtained by sample or calculation in the same structure, and does not require repeated parameter calculation. The inherent advantage of the GRNN is that it has great potential at non-linear fitting. Differing from the typical BP network, the GRNN network is good in simulation effect, quick in calculation, and stationary in result.

GRNN is a feedforward neural network model based on a non-linear regression theory. Differing from the typical BP network, the GRNN approaches to the function through activating neurons, that is to say a function value of the input vector is approached from a function value mapping corresponding to the neuron vector in an adjacent region, whose network structure is shown as Figure 9.6. The GRNN network is formed by an input layer, a hidden layer, and an output layer. Gaussian transformation function is adopted for controlling the output in the hidden layer, thereby restraining the activation of an output unit. In input space, the Gaussian function is symmetrical to a receipt domain. The effect of inputting neuron on the network output shows exponential decay along with the input vector. In the GRNN network, each training vector has a corresponding radial base neuron in the hidden layer, and the neuron stores each training vector. When a new vector is input into the network, the distance between the new vector and each vector in the hidden layer can be calculated by the following expression (Formula (9.25)):

$$dist = |X_j - W_R^j|,\quad j = 1,2,\cdots,S_1 \tag{9.25}$$

In the expression (9.25), S_1 is the hidden layer unit number, X is the input vector, W_1 is a hidden layer unit weight vector, and *dist* is the distance between the input vector and the weight vector. The calculated distance can be adjusted by the following expression (Formula (9.26)):

$$b = 0.8326/s,\quad n_1 = dist \times b_1 \tag{9.26}$$

The output of the Gaussian function is in accordance with the following expression (Formula (9.27)):

$$a_1 = e^{-(n_1)^2} \tag{9.27}$$

in the expression, n_1 is the adjusted distance, and s is the width of the window. If $dist = s$, the adjusted distance $n_1 = 0.8326$, and Gaussian function output value a_1 is 0.5, which is equivalent that the related coefficient is 0.5; if *dist* is far greater than *s*, the Gaussian function output approaches to 0. With the increase of n_1, the output of the hidden layer is gradually decreased. The variable *s* plays a role of the window, that is to say the value *s* plays a role of activating output layer neurons. *s* becomes greater, b is less, and the distance between the hidden layer neurons and the input vector is shortened, so that the number of the neurons activated in the window is increased, otherwise, *s* becomes less while b becomes greater, the distance between the hidden layer neuron and the input vector is prolonged, the hidden layer output is lowered, and the number of neurons activated in the window is decreased.

The output layer of the GRNN network is linear, the output is as follows (Formula (9.28)):

$$a_2 = n_2 = W_2 a_1 \tag{9.28}$$

in the expression, W_2 is the second layer of weight.

Knowing from the above, the characteristics of the GRNN network are that the manually adjusted parameter is less, and only one threshold value exists. The learning of the network completely depends on a data sample. The characteristics determine that the GRNN network avoids the effect of the manmade subjective assumption to the greatest extend.

Two latest historical values and one closest residual deviation value of the original wind velocity series ARIMA (1, 1, 1) are determined as the input variables of the GRNN neural network, and the value of the next 10 min of the original wind velocity is taken as a target output sample pair. The distribution density of the radial base function has important effect on the performance of the GRNN. Theoretically, the distribution density is less, accordingly, the approach of the function is more precise, but the process of the approach is not smooth, the greater the distribution density is, the smoother the process of the approach is, and the greater the error of the approach is. Therefore, in the design process of the network, the value of the distribution density is required to be adjusted until a relatively ideal precision is realized.

9.4.3.2 Wind Velocity Forecast Result

In reference [16], the method of the time series respectively combined with the BP and GRNN neural networks is used to forecast wind velocity 10 minutes in advance. The comparison of forecast effect is shown in Figure 9.7. It is clear that because the lead time of forecast is shorter, the forecast effect of time series-BP and time series-GRNN is not remarkably better than that of a pure

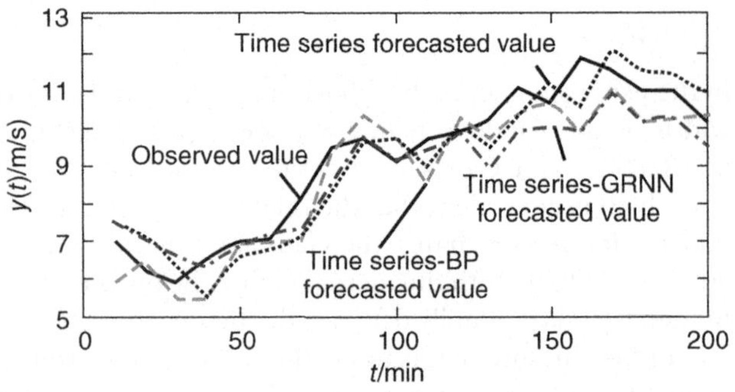

Figure 9.7 Result of forecasting wind velocity in time series combined with neural network method 10 minutes in advance.

time series method. The absolute average error is about 6.7%, if the time series-GRNN neural network is applied to forecast of short-term wind velocity 10 minutes early.

The same method is applied to wind velocity forecast of which the lead time is respectively 20 minutes and 30 minutes. The obtained results are respectively shown in Figure 9.8. Since the lead time of forecast is prolonged, the forecast effects of the time series-BP and the time series-GRNN are remarkably better than that of the pure time series method. As for the forecast result of the time series method 20 minutes early, the absolute average error is about 12.7%. The absolute average error of the time series-BP method to forecast is about 11.8%, and the absolute average error of the time series-GRNN method to forecast is about 11.3%. As for the forecast result with the time series method 30 minutes early, the absolute average error is about 15%; the absolute average error of the time series-BP method is about 13.9%; the absolute average error with the time series-GRNN neural network method to forecast is about 13.5%.

In reference [10], the RBF neural network is adopted to forecast the wind power with the forecast lead time of more than 40 hours. Compared to the continuous method for forecast, both the absolute average error and the root mean square error of the RBF neural network is remarkably lowered.

9.5 Generated Power Forecast in WF

9.5.1 Overview

A wind-generated power forecast is to provide information of the amount of wind power in a time point within future several days. The wind-generated power forecast can be classified into two types, namely, a WF power forecast based on historical data and WF power forecast based on the NWP model.

The power forecast of WF based on the historical data is a method for forecasting WF power according to the historical data, that is to say, a mapping relationship between several historical data (including power, wind velocity, wind direction, etc.) and the power output of the WF is established. The methods are composed Kalman filtering method, continuous algorithm, ARMA algorithm, linear regression model, adaptive fuzzy logic algorithm, and so on. In addition, artificial intelligent methods such as artificial neural network are also adopted. Limited by the accuracy of prediction, the lead time of forecast method is generally within 6 hours or slightly longer.

WF power forecast based on the NWP model is to contact the wind forecast of the NWP model with the wind turbine power output, that is to say, the NWP model provided by the meteorological department is utilized to forecast the weather conditions (mainly including wind velocity, wind direction, temperature, air pressure, and the like) of the WF or a certain point close to the WF; the forecasted value of the NWP model is converted into the power output of the WF through establishing a forecast model and combining other inputs. The time of wind power output forecast based on the NWP can be longer, which is specifically determined by the forecast time length of the NWP, and can be up to 24 hours, 48 hours, even 72 hours or longer. The NWP is provided by the meteorological department. Generally, resolution is scores of square kilometers (such as 60 km^2, 30 km^2, or 20 km^2), and the requirement of direct WT power calculation cannot be satisfied. Therefore, as for the short-term WF power forecast based on the NWP, the WF power is calculated through taking the NWP model as the input of the short-term WF power forecast model. The structure of the WF power forecast system is shown as Figure 9.9.

Like wind velocity forecast previously mentioned, the WF power short-term forecast model can be classified into two types, that is, statistic model and physical model [6, 10, 17].

Statistic model. The statistic model method is to analyze the relationship between the weather prediction from past time series and the generated wind power, that is, a mapping relationship is established between the input (NWP model, measured data of WF, etc.) of a system and the power of the WF. The relationship can be used to forecast further wind velocity and power output. The process comprises the linear method and non-linear method, such as regression least square method and artificial neural network. Knowledge about the physical process, such as the shape of the power curve, is generally utilized in the adoption of the statistic method in the model design.

The advantage of the statistic method is the forecast can voluntarily adapt the position of WF, so that the system errors are automatically reduced. The disadvantage is the method needs long-term data

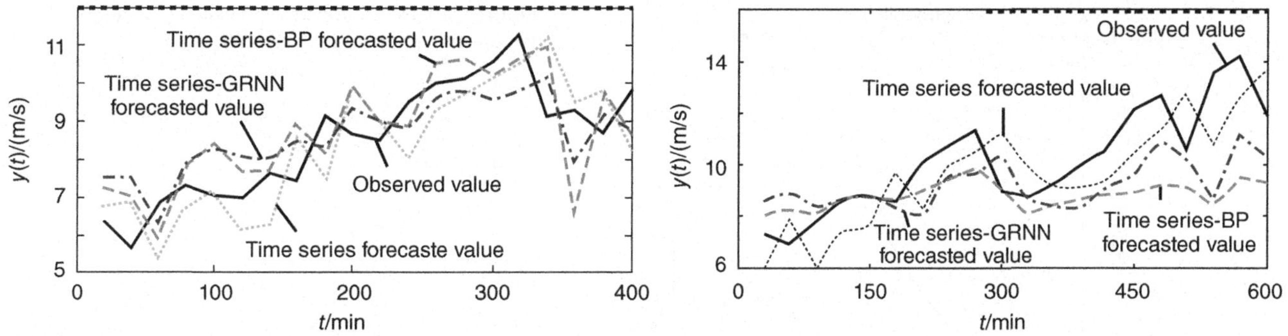

Figure 9.8 Results of wind velocity forecasts with combination method of time series and neural network in different lead time. (a) 20 minutes early; (b) 30 minutes early.

Figure 9.9 Basic structure of WF power forecast system.

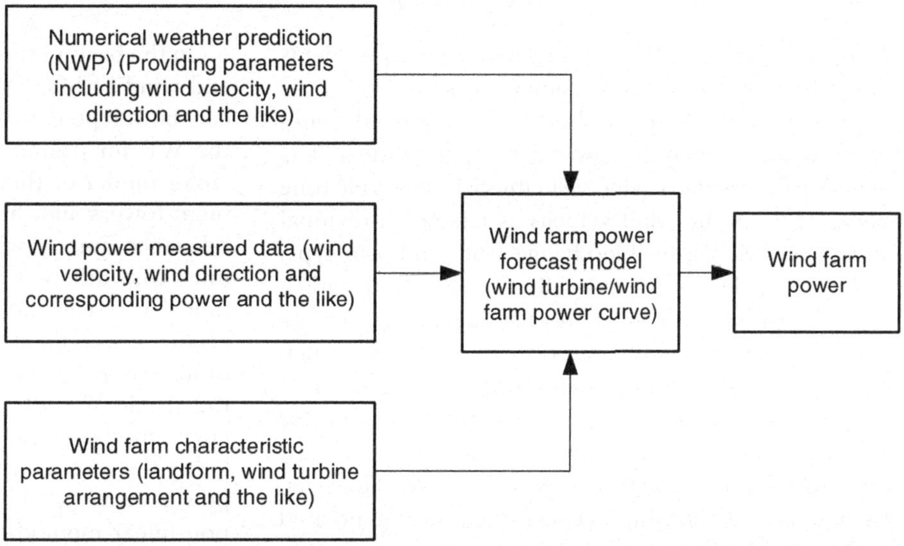

measurement and extra training, in addition, the rare weather conditions almost never occurred at the stage of training, which makes forecast difficult by the system, the corrected forecast of the rare weather conditions is very important, otherwise, very great forecast errors will happen.

Physical model. The physical model describes the physical process of converting wind into electric power. The main model comprises a series of models with different physical processes: condition of wind at the height of the hub in the WT position, influence of WF shading, WT power curve, model output statics, and so on. The method utilizes parameters of detailed physical description of lower atmosphere to solve the basic problem that wind velocity on the rough network from the NWP is converted into wind velocity in the wind turbine position. It is composed of two important steps: the horizontal wind velocity from the network points to the wind turbine coordinates and the wind velocity at the height of atmosphere, provided by the NWP, are converted into the wind velocity at the height of the wind turbine hub, shown as Figure 9.10. Physical factors utilized by the physical method include information on meteorology (NWP) and topology (landform, roughness, and barrier), as well as the technical characteristics of the wind turbine (hub height, wind turbine power curve, and thrust coefficient). The key is to improve the resolution of the NWP model to precisely forecast the weather (wind velocity, wind direction, etc.) of a certain point (e.g., the position of each wind turbine). An NWP model in the local WF is established. The precision of the NWP model is generally from scores of km^2 to 1 km^2 or 2 km^2. The physical model method tries to utilize a mesoscale or micro scale model to establish a relationship between the NWP model and the local wind. After obtaining the best estimation of the local wind velocity, model output statistics (MOS) is used to lower residual errors.

The physical model has the advantage of no requirement for a large amount of data measured for a long term, and it is more suitable for complex landforms. The disadvantage of the physical model is requirement for abundant meteorological knowledge and understanding

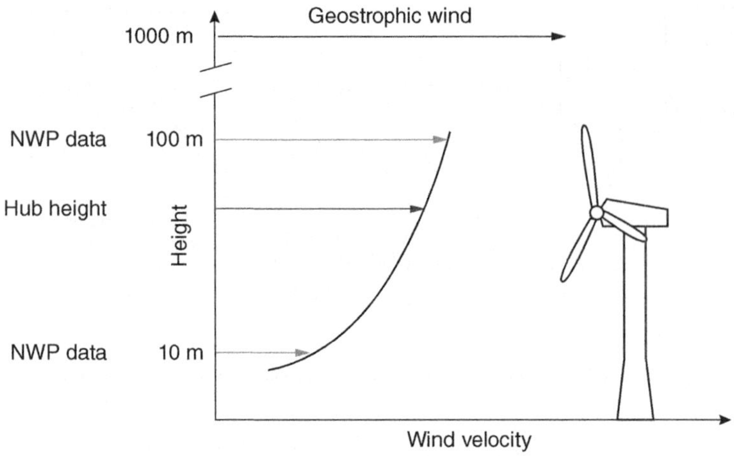

Figure 9.10 Deduction of wind velocity at height of wind turbine hub with given NWP.

of physical characteristics. If a relatively rough model is established, the forecast precision is worse.

The wind power forecast differs from the wind velocity forecast. The electric power generated by the WT is related to difference of wind velocities within a wide time range, whereas the wind velocity is related to regional climate model, seasonable change, and landform type. So far as we know, theoretically, for the wind with a speed of v (m/s), the wind energy passing through a cross section of A(m^2) in unit time is $p = 0.5\ \rho A v^3$, in which ρ is air density (kg/m^3), determined by atmosphere temperature and air pressure. The simplest method for converting the wind velocity into wind power forecast is to utilize a manufacturer's power curves. However, the practical relationship between the power generated by the whole WF and the wind velocity is more complex than a theoretical expression, because of various wind turbines and control modes thereof. Furthermore, power curve can be in the range of a whole WF, a group of wind turbines or a single set of wind turbine. The relationship between the wind velocity and the output power should be treated as non-linear, and is a random time-varied function of the wind velocity. It cannot be described by a certain power curve of the wind turbines. In a WF, multiple wind directions and velocities are utilized to convert the wind velocity into power and obtain a WF wind velocity matrix. Compared to only using power curve from the manufacturer, the power curve generated by measuring the wind velocity is less in forecast errors. Moreover, due to the nonlinearity of the power curves, the wind velocity forecast error is enlarged in the high-slope region between wind turbine cut-in wind velocity and the rated wind velocity state (the error attenuates at the rated wind velocity), in other words, the nonlinearity of the wind turbine power curve further enlarges the error, and small deviation of the wind velocity may possibly cause great power deviation.

On the basis of the measurement of past wind velocity and wind turbine/WF output power, different methods can be adopted to establish a power curve model of the WF, for example: (1) directly establishing a power curve model of the wind velocity of a wind measurement towers and the WF power; (2) on the basis of the relevance among wind turbines, dividing wind power units into different groups, establishing a power curve model between the wind measurement tower and the group power for each group, then adding the results of all groups and obtaining the power curve model of the whole WF; (3) establishing a wind velocity-power curve model of each wind power unit, then accumulating the forecast power of all units, and obtaining the power curve model of the whole WF; (4) adopting a non-linear method to establish a relationship model of the wind velocity of wind measurement tower and the WF power, for example, adopting the neural network method to establish a curve of the relationship between input parameters (wind velocity and wind direction) and output parameter (power).

9.5.2 Wind Power Forecast Based on Physical Model of NWP

The wind power forecast based on the physical model of the NWP comprises the following main steps: NWP, wind power output forecast, and proportional regional increase [4].

Step 1, the weather prediction comprises forecasts of wind velocities and wind directions, some other weather parameters such as air pressure may be required by the wind power forecast, provided by the NWP model.

Step 2, the WF power output forecast requires NWP data as input, meanwhile, the influences of local weather on the wind velocity, wind direction, wind turbine power inversion characteristics, and WF shading, as well as other various factors possibly affecting the total power output, shall be considered. Different methods and combined methods have been developed and used. As for the short-term forecast, online measured wind velocity and/or WF power output are/is taken as the additional input of the forecast. As described in the previous section, different WF power curve models can be established.

Step 3, the forecast is just to summarize the forecast of a representative WF, if a larger region having many WFs or wind turbines is required to be forecasted, the forecast result of the representative WF can be enlarged in proportion and added to regional forecast. In this way, work related to forecast is minimized and the quantity of data required by input from the NWP model is decreased, whereas the forecast accuracy will not be reduced much, because the adjacent WFs present the similar characteristics.

Some problems related to the wind power output forecast based on the physical model of NWP are described below [4].

9.5.2.1 NWP

The data from the NWP model is the most basic input required by all wind power forecasts almost. Generally, cumulative resolution and model chains with different NWP models and different grades are adopted. On the basis of global weather observation and measurement, the model chains are accomplished by meteorologists, meteorological stations, satellites, and the like. All obtained data is taken as the input for calculating the global NWP model (simulating planets and atmosphere). The NWP model calculates the future status of the atmosphere according to physical laws of controlling meteorological phenomena. Because the calculation takes much time, the resolution of the global model must be rough. The global model is only operated in the weather service centers of few countries. In order to provide more precise weather forecast, local area models (LAMs), covering a small part of the earth, are also adopted, but have very high resolution.

These models are used as the input of the forecast of the global model, and calculate weather prediction considering local landform characteristics. One example of the LAM is the LME model in German Weather Service (GWD), it covers the middle Europe, has 105,625 grids (325 × 325), and obtains ground level resolution (size of grid) of 7 × 7 km. An operation model forecast range is 48 hours, and the time resolution is 1 hour. The model is started and operated twice one day, respectively, at AM 0:00 and AM 12:00.

9.5.2.2 Forecast Time Range

The forecast range is a time period between the obtained forecast time and the last forecast point. Different forecasts have different purposes, and the forecast time range is determined by users' demands and forecast feasibility.

As for the current wind power forecasts, deterministic forecast is used within a forecast range of 3 to 5 days. Two main forecast time periods must be differentiated. The day-ahead forecast is mainly used for the day-ahead power trade in the electric power market, so that the forecast time range is determined by a trade organization, for example, trade closing time and trading date. One example of closing time at tomorrow 12:00 is shown as Figure 9.11.

The midnight NWP model is started and operated with day-ahead observed result used. If the calculation is finished at about AM 7:00, the result information is transferred to the wind power forecast system. Generally, the result can be obtained after several minutes of calculation. The result will be analyzed and applied at the AM 12:00 in the next day before the ending of electric power trade. It means that before 48 hours of the forecast, the calculation starts from the NWP model. If there is no trade in weekend and public holidays, the lead time of the calculated "day-ahead" trade is 96 hours or longer in fact.

Very-short-term wind power forecast is mainly used for the intraday trade and the start up of the reserve capacity. Its main characteristic is utilization of the

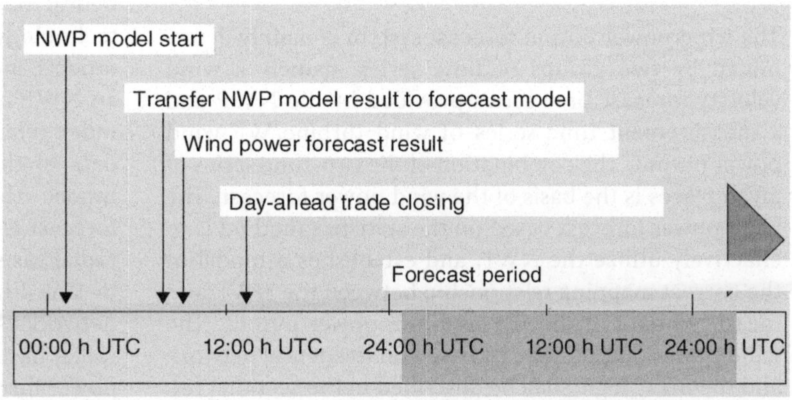

Figure 9.11 Typical wind power forecast schedule for day-ahead trade.

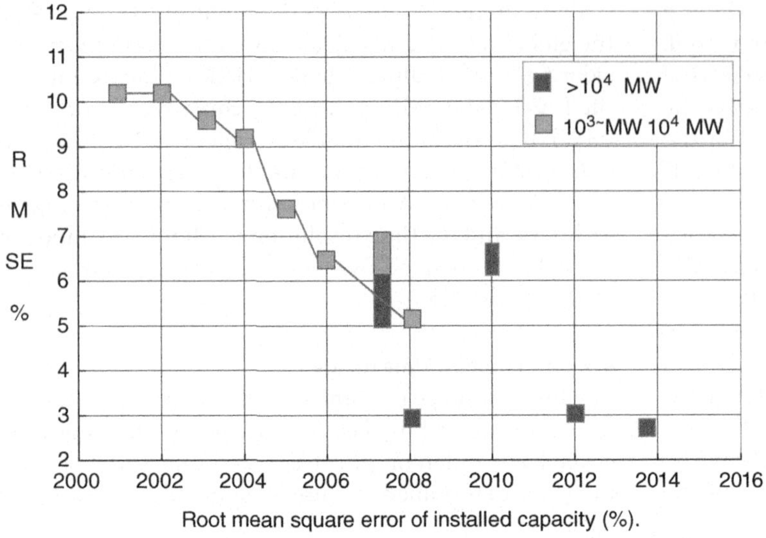

Figure 9.12 Progress of forecast errors of day-ahead wind power forecast operated in one controlled region.

online data of actual power output and/or wind velocity. As for a very short-term forecast time range, the forecast accuracy will be greatly improved. Generally, in the short-term forecast, NWP model data is combined with online measured data. In a longer forecast time range, the NWP data has large weight. And large weight is given to the online measured data in the short-term forecast.

9.5.2.3 Forecast Accuracy

Utilizing wind power management system (WPMS) in the wind power forecast was realized in 2001, then the WPMS has been continuously improving, and accordingly, forecast result errors have been continuously reducing, as shown in Figure 9.12. Wind power forecast errors in one controlled region are shown by comparison between RMSE of the forecasted time series and online monitored time series, which are gradually improved from about 10% realized at the first time to about 6.5% realized in 2005 (reduced to wind power rated installed capacity).

9.5.3 Statistics Method–Based Wind Power Output Forecast

The wind power output forecast system is mainly determined by two groups of time series, namely a wind velocity forecast time series provided by the NWP and a measurement time series of wind turbine/WF wind power output. The combination of the two-time series of given places is the basis of the wind power forecast. The wind power forecast based on the statistics method is to effectively utilize the NWP, and establishes a model of the correct mapping relationship between the NWP and measured data of the WF and WF power output. The wind power output forecast based on the neural network and the fuzzy logic shall be described in the section [18].

9.5.3.1 General Steps of Wind Power Output Forecast Based on Statistics Method

The statistics forecast method herein is established on the basis of the neural network, and utilizes the time series of the wind power output and the NWP as input to estimate future wind power capacity. The two-time series exist in different periods (frequencies), that is to say the value of wind power output per minute is provided by SCADA system, which is online in fact; however, meteorological service delivers several times of NWP every day, covering a range of 48 hours or 72 hours. Generally, the forecast of the NWP model is modified once in 6 hours, whereas the output of the wind power forecast model generally requires hourly modification. Because the weather forecast is related to the space and time resolution of the NWP model, the weather forecast forms the main uncertain source. Inaccurate NWP, especially in a short time range, makes wind power forecast extremely difficult. Surely, it is no doubt that as for short-term and long-term ranges of forecasts, the NWP is necessary, and its accuracy is a key to the accuracy of the wind power output forecast. Therefore, in order to better use the NWP, the artificial neural network (ANN) and fuzzy logic combination technology is adopted. The system comprises three models, including a primary wind power forecast model, an NWP quality evaluation model that provides a fuzzy index related to the reliability of the NWP (main input data of the system), and a final wind power forecast model. The first model providing the primary power forecast is formed by one self-organized map and three radial basis neural networks; the second model is formed by one fuzzy logic model and two radial basis neural networks; the final forecast model is formed by three radial basis neural networks. The whole model of wind power forecast is shown as Figure 9.13.

Figure 9.13 Structure of wind power forecast model.

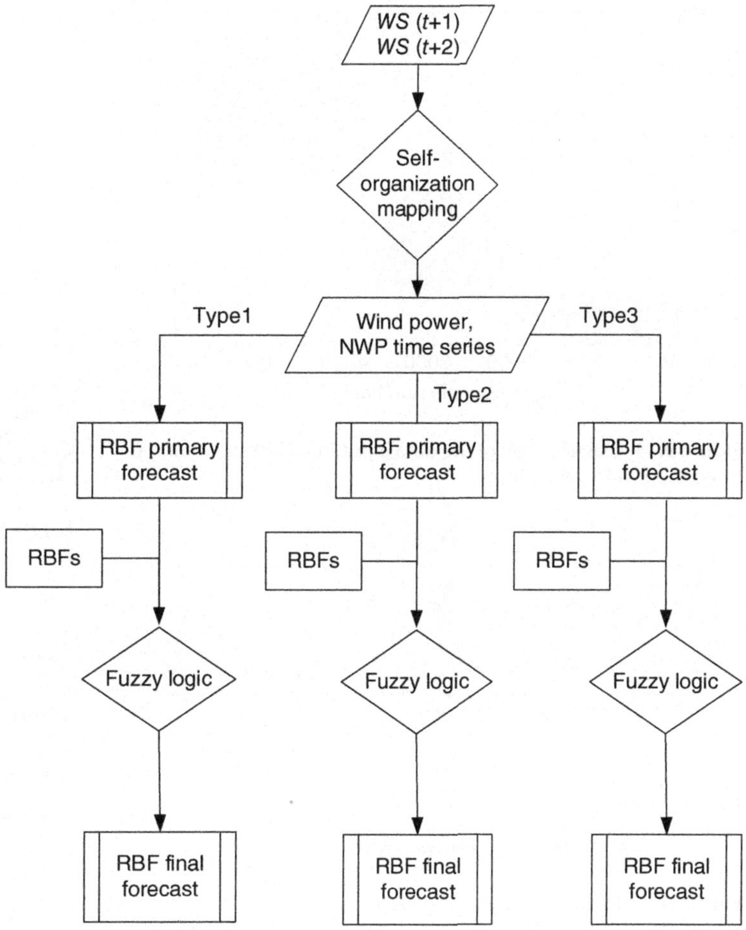

9.5.3.2 Primary Wind Power Forecast

Two forecasted wind velocities WS(t + 1) and WS(t + 2) obtained by the NWP are required by the system as input, and are corresponding to the wind power forecast in the current hour and the wind power forecast in the next hour. The second value of the wind velocity corresponding to the next hour is just used by a forecaster to recognize the increase or deduction trend of the wind velocity. In addition, in order to make an applied radial base function (RBF) have more effective performance, the wind velocity range used is simplified into a cut-in wind velocity threshold value and wind velocity corresponding to the rated power of the wind turbines. The generated power is set to zero if the wind velocity is lower than the cut-in wind velocity threshold value; when the wind velocity is above the rated wind velocity, the generated wind power is set to be the maximum (rated) value. In this way, only wind velocity enough for power generation and the wind velocity status lower than the corresponding constant WF output value are trained by the neural network. Therefore, when the subsets of the wind velocity within a limit range are used, the neural network shows better performance. In the similar way, in order to better utilize the NWP, the time series of the forecasted wind velocity is classified into three sets by the primary forecast model according to the sizes of the wind velocities (small, medium, and large), and the sets are applied to self-organization mapping. Each layer of the Kohonen neural network selected is formed by three neurons, and the wind velocity value can be classified into three types. The classified forecasted values for the wind velocity forecast before 1 hour is shown as Figure 9.14.

Wind velocities based on this classification are reserved as data for wind power forecast, that is to say, past wind power value, wind direction, and one-day forecast time (hour) congregate into three sets according to the types of the wind velocities. Each set trains one individual RBF network. Samples are trained by the radial base function network, because both the non-linear function approach ability of the network and the network convergence speed are better than those of the BP network. The RBF neural network is a multi-input single-output feedforward neural network. As shown in Figure 9.15, the RBF neural network is a three-layer forward neural network formed by an input

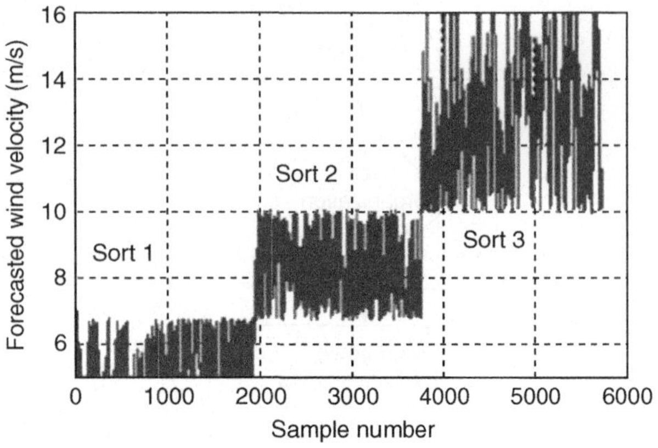

Figure 9.14 Classification of the data input into NWP according to forecasted wind velocities.

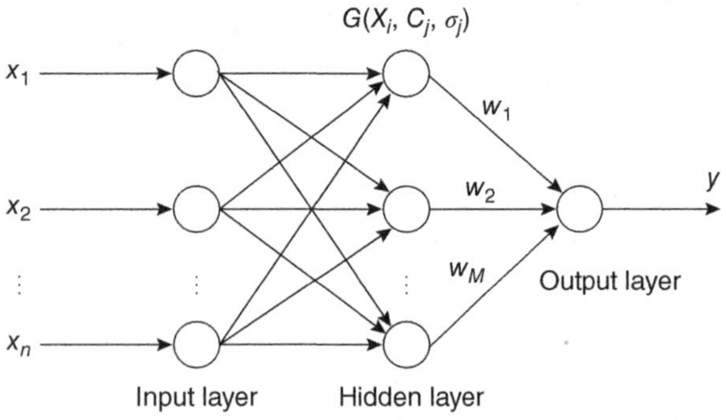

Figure 9.15 Radial base neural network.

layer, a medium layer (hidden layer), and an output layer. The input layer is formed by signal source points that connect the network with the environment; the number of the neurons (units) of the hidden layer is determined by the described problem, its function is to perform the non-linear conversion from input space to hidden layer space, and the input vector is mapped into high-dimensional characteristic space. The hidden layer adopts a radial base function as an excitation function, and locally responds to input signals. The output layer achieves the linear combination of output weights in new space, and adjustable parameters are just the weight of the linear combination and the parameters to control the shape of basis function. The network has been proved in theory, the feedforward network containing one hidden layer can approach any one continuous function. The RBF network is particularly suitable for forecasts of the non-linear time series. The wind velocity and WF power are typical non-linear time series, as a result, selecting the RBF network as one of the forecast methods is suitable.

The structure of the primary wind power forecast model is shown in Figure 9.16. In the figure, the input forms of each RBF network are as follows (Formula (9.29)):

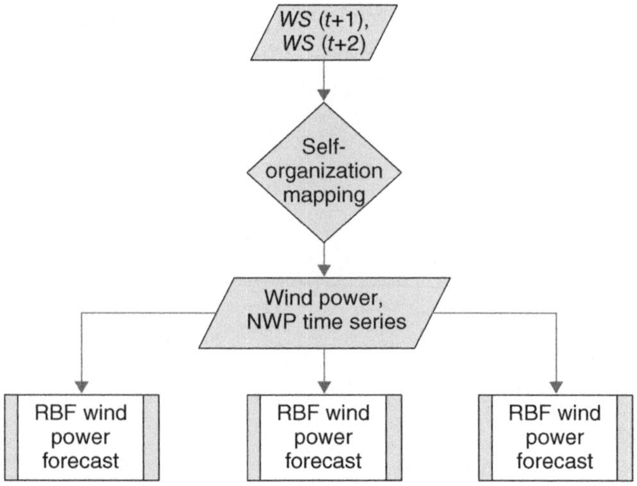

Figure 9.16 Structure of primary wind power forecast model.

$$P(t+1) = f(P(t), WS(t+1),\ WS(t+2),\ WD(t+1),\ H(t+1)) \tag{9.29}$$

where P(t + 1) is the output of the model and is wind power output; *WS* is the wind velocity provided by NWPs; *WD* is the wind direction provided by the NWPs, shown in "arc"; *H* is time (hour) taken by forecast.

The function relationship between the output and input of the radial base function neural network (RBFNN) can be shown as (Formula (9.30)) [19]

$$y_i(X) = \sum_{j=1}^{M} w_{ij}\phi_j(X) + b_j \quad i = 1,2,\ldots,n \tag{9.30}$$

where, $X = (x_1,x_2,\ldots,x_n)^T \in R^n$ is an input vector; y_i is the output value of the unit *i*, and $_M$ is the number of centers; w_{ij} are weights from the neuron *j* to the output unit *i*; b_j is a bias value; $\phi_j(\cdot)$ is the non-linear transfer function in RBF layer. The usual transfer function in RBF layer used, namely the non-linear base function, which is a Gaussian function shown below (Formula (9.31)):

$$\varphi_j = \exp\left(-\frac{\|X - c_j\|^2}{2\sigma_j^2}\right) \tag{9.31}$$

where, $c_j \in R^n (1 \le j \le M)$ is the center of the RBF; σ is the width parameter (threshold value) of the RBF, which can regulate the sensibility of the RBF neurons. Meanwhile, the output function is as follows (Formula (9.32)):

$$y_i(X) = \sum w_{ij} \exp\left(-\frac{\|X - c_j\|^2}{2\sigma_j^2} + b_j\right), \quad i = 1,2,\cdots,n \tag{9.32}$$

According to the output function of the network, the parameter learning of the RBFNN comprises two parts, one is the determination of hidden layer neuron central vector c and normalized parameter vector (width) σ, and the other is the determination of output layer weight w matrix. The effect of the form of the non-linear function used by the RBFNN on the network performance is not very important, the key factor is the selection of a base function center. If the center is not properly selected, for example, some centers are too close, the approximate linear correlation will be generated. As a result, pathological conditions of numerical values are brought. The RBFNN parameters are determined in two methods, wherein one is completed by two steps, firstly, the center and width of the RBF function are determined; secondly, the connection weight of the hidden layer and the output layer is determined. The RBFNN base function center, width and weight adjustment algorithm are determined in many methods. For example, the k-means clustering algorithm is selected for the RBFNN center; the center and width of the RBF function and the connection weight between the hidden layer and the output layer are simultaneously trained in a supervised learning method; the RBFNN center is selected through the orthogonal least square (OLS) algorithm [20].

In the RBF neural network, a key problem is how to determinate the number of the hidden layer neurons. The traditional method is to make the number of the hidden layer neurons equal to that of input vector elements. However, when many vectors are input, the excessive number of hidden layer units is difficult to be accepted by people. The improvement method is to begin the training from zero neuron. Neurons are automatically added to the network through examining output errors. Reuse is carried out, input weight corresponding to the maximum error generated by the network at each time is taken as a weight vector. A new hidden layer neuron is generated. Then the errors of a new network are examined, and the process is repeated until the errors satisfy the requirement or reach the maximum number of the hidden layer neurons.

The structure of the RBF network is the same with that of the generalized regression neural network shown in Figure 9.6. The RBF network can process non-linear problems such as wind power forecast. The effectiveness of the RBF network can be enhanced through allowing some input variables of the RBF network to more greatly influence the output thereof and selecting proper standardized factors.

The information provided by the NWP is the most important, especially within a forecast range of more than 6 hours. Therefore, the input variables of the wind velocity and wind direction are in higher order of magnitude than that used by other variables (past value of wind power and forecast hour). Especially the wind velocity value corresponding to a forecast hour, it is set to be in a higher order of magnitude than other wind velocity variables and wind directions. The fact that the previous wind power in the 6 hour-short-term-prediction is more important than the wind power forecasted within a long-term range should be noticed, so it is expressed by MW. That is the reason why the model operates repeatedly and the previous power forecast is taken as the past wind power. When the forecast time range is enlarged, the power forecast errors increase. Therefore, as for the long-term forecast, the past wind power is expressed by GW. In this way, the same structure of the RBF can be used for short-term and long-term forecasts. Typical normalized input of the RBF has the form below (Formula (9.33)):

$$P(t+1) = f(P(t), WS(t+1)*20, WS(t+2)*10, WD(t+1), H(t+1)/480) \quad (9.33)$$

Hourly errors of forecasts caused by the primary wind power forecast model, which is 1 hour early, are normalized by the installed capacity of the WF, most are within ±20%, and the maximum error is within ±(40%–50%) [4].

9.5.3.3 Quality Assessment of NWP

The accuracy of the wind power forecast is closely related to the quality of the NWP. The status the weather predication departures from the actual value of the wind velocity has great negative effect on forecast quality. When a wind power value is simply calculated out according to wind velocity forecasted by NWP and a wind turbine power curve (wind velocity and power relationship curve), we call the wind power forecast "theoretical value." The maximum error of the simple forecast, which completely established on the basis of hourly wind velocity forecast of the NWP and the power curve, exists in the maximum error in the "theoretical value" of the wind power of the corresponding wind velocity sampling, also exists in the corresponding wind velocity sampling point with the maximum error.

When the "theoretical" wind power is calculated, the effect of the wind direction is not considered, and it is one of the main reasons for errors in this kind of calculation. In order to considering the effect of the wind direction in the wind velocity forecast quality provided by the NWP, two RBF networks are respectively used for each type of wind velocity. One RBF network receives two wind velocities provided by the NWP as the input, and two wind velocities are corresponding to the wind velocity in the forecasted hour and the next hour. The input of other RBF network is the wind direction to be forecasted and forecast hour. Network training is performed so as to obtain the "theoretical value" of the power obtained from the power curve calculation. The output of these RBF networks and the primary wind power forecast model are compared to identify "poor" NWP, in other words, the quality of the NWP is identified by the difference between the primary wind power output and the training output of the first RBF with forecasted wind velocity value, and

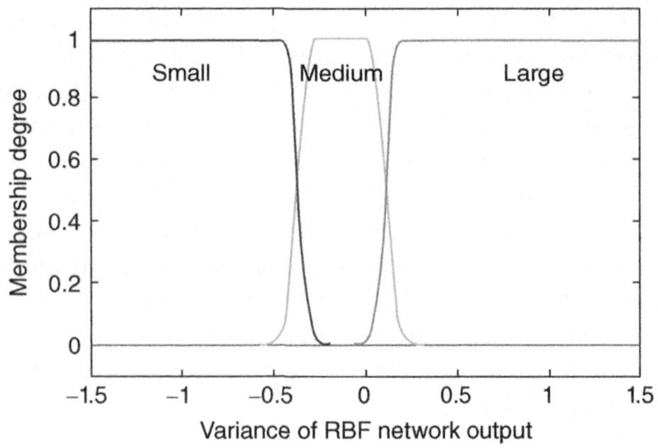

Figure 9.17 Fuzzy set of difference between primary wind power forecast model and output trained by RBF network with forecasted wind velocity value.

the difference between the primary wind power forecast and the second RBF training output with the forecasted wind direction and forecast hour.

In order to use the two differences as parameters to identify inaccurate wind velocity forecast, a fuzzy logic model is introduced. The fuzzy logic model receives the difference between the output of the RBF networks and the primary forecast as the input. The quality of the forecast provided by the NWP is evaluated by using fuzzy rules. When the difference is in a high value and the wind direction is beyond the preset limit value, the quality of the NWP will be estimated by the model. The input (semantic variable) of each fuzzy model is set to be the values of three fuzzy sets (small, medium and large). The participation of the difference between the primary wind power forecast model and the output trained by the first RBF with the forecasted wind velocity value in the three fuzzy sets is shown in Figure 9.17.

The advised fuzzy set model can be explained by 27 rules (Formula (9.34)):

$$\text{If } X_1 \text{ is } A_1, \text{ and } X_2 \text{ is } A_2, \text{ then y is } B_i \tag{9.34}$$

where $X_i (i = 1, 2, 3)$ is an input variable, A_i, B_i are fuzzy sets, and y is the output of model, value within [0, 1]. The fuzzy sets are simulated by the Gaussian function (Formula (9.35)):

$$\mu A_i(x_i) = \exp\left(-\left(\frac{x_i - a_i}{b_i}\right)^2\right) \tag{9.35}$$

Output y of the fuzzy model is between 0 and 1, and provides a quality index of NWP.

9.5.3.4 Final Wind Power Forecast Result

At last, the RBF network is utilized to provide the final forecast. As for each type of wind velocity, the RBF networks are trained by data, same with the RBF network of the primary wind power forecast model, and the output of the fuzzy model. These networks have the structures same with that the RBF networks for providing primary wind power forecast, the training is processed by the same training set, and the quality index of the forecasted wind velocity is added to the training set. As for short-term and long-term forecast, the normalization of the input is realized in the same method.

In reference [19], one wind power system is tested in the above method, the time range is from March 2001 to April 2003, and the weather prediction covers a range of 48 hours and is updated four times every day. The model utilizes wind power output forecast provided by the NWP 2 hours early before each hour, in other words, the NWP provides update before the hour, and NWP can only obtain the forecast before 42 hours. Therefore, hourly wind power forecast covers a range of 41 hours. The distribution of normalization errors before four different observation time (1, 6, 18, and 36 hours) is shown in Figure 9.18. Within a short-term observation time range, greater percentage errors are within ±10%; in a range of longer term, the errors are distributed

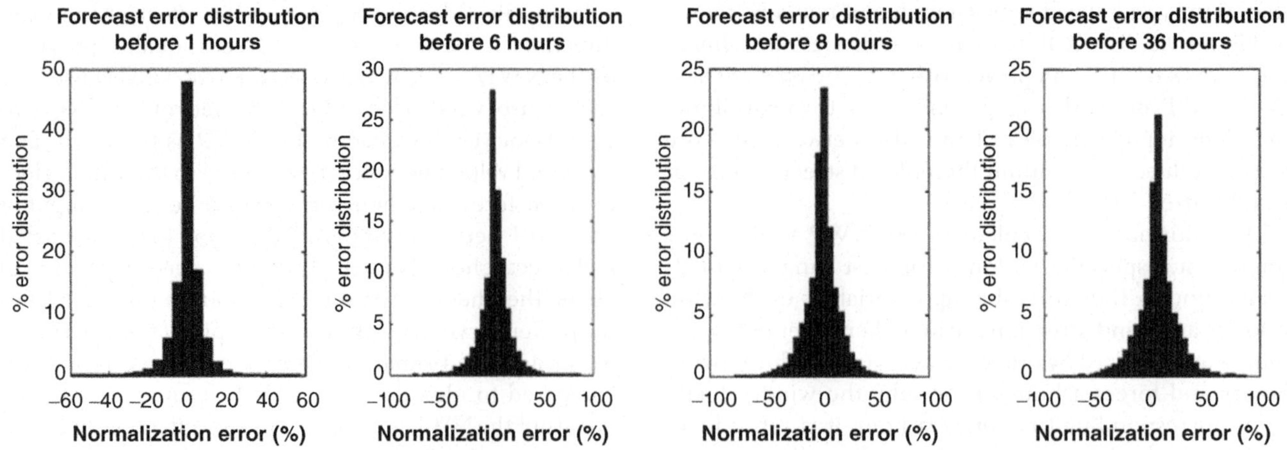

Figure 9.18 Distribution of normalization errors of wind power forecasts before 1, 6, 18, and 36 hours.

within 20% to 25%, where zero error is obtained, and the distribution of the maximum error is less than 1%.

Compared to the continuous forecast method, as for forecasts before four different observation time intervals, the ranges of the normalization mean absolute error (NMAE) of the above statics forecast methods are 5% and 14%, whereas the NMAE of the continuous method reaches 24%. The normalized root mean square error (NRMSE) is always less than 20%, but the NRMSE of the continuous method reaches 34%.

Compared to the result of the primary forecast model, the RBF fuzzy model shows the performance of the primary forecast model can be improved in all observation advanced periods of time through estimating the quality of NWPs, and the improvement of the NMAE reaches 4%.

9.6 Accuracy of Wind Power Forecast

9.6.1 Requirement of System Operation on Accuracy of Wind Power Forecast

An electric power system operation strategy is established on the basis of the future system state forecasted (no matter before season, day, or several minutes). Even if there is no any wind power in the electric power system, much uncertainty is contained in the establishment of operation strategy. As for the wind power generation, in the time frame of load tracking, in other words, as for the load change for further several hours, the variability of the wind power and accuracy of forecast will affect the dispatch and operation of power generation resources capable of being controlled in a balance system, whereas the change cost of the operation plan will affect the cost of load power supply. Uncertainty of the wind power forecast also affects a power generation contract and plan for future one day or several days, as well as electric power market trade [21].

At the aspects of power flow managed by electric network operator and possible electric network congestion, in order to perform real-time processing at the acceptable cost, small forecast errors are required as far as possible. As for the operators, the most urgent situation is to forecast the large variation of wind power output, especially the change is still uncertain 1 to 2 hours before occurrence (shown in Figure 9.19). There is a forecast of high wind power output, however, it does not occur at the predicted moment, and there are "phase errors." Recently, the fact shows the phase errors become less because of the improvement of forecast tools. Extreme storm is an uncommon exception, forecast of occurrence or no occurrence of the wind power output drop is determined by the evolvement of wind velocity. The proper method for handling the challenge is to combine the forecast and control of the WF.

Under the electric power market environment, the forecast accuracy is greatly related to "gate-closure time." It is worth noting that the quality of the short-term wind power forecast should be considered in relation with the electric power market gate-closure time. A common very long closing time is shortened for the forecast accuracy, so that cost of balancing the system is dramatically affected. The total forecast absolute error (based on the wind power output forecast technical level in 1997) is shown in Figure 9.20. It is shown in the percentage of the total actual wind power capacity. According to the Figure 9.20, different absolute errors exist within different absolute time ranges, wherein forecast errors

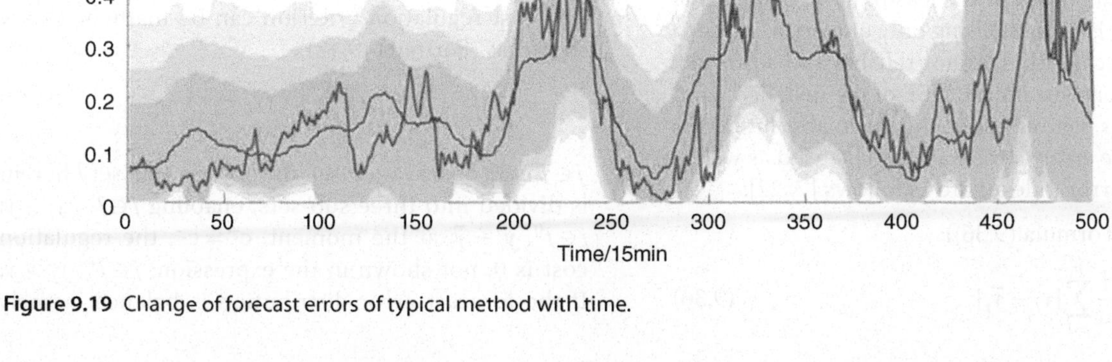

Figure 9.19 Change of forecast errors of typical method with time.

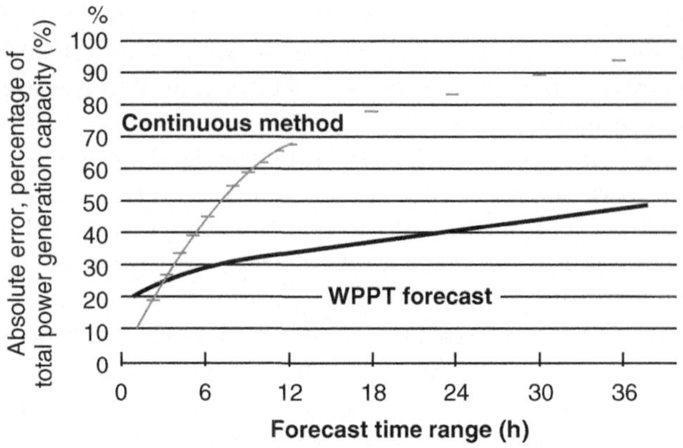

Figure 9.20 Total forecast absolute error within 1 year and forecast time range.

(produced energy) before 36 hours are two times of those before 3 hours.

9.6.2 Evaluation Index of Wind Power Forecast Quality

The wind velocity forecast error of the WF is mainly related to the forecast method, forecast period and the characteristics of the wind velocity of a forecasted place, generally, the shorter the forecast period is, the more moderate of the wind velocity change of the forecasted place is, the less the prediction error is, otherwise, the forecast error is greater.

The forecast is the basis of a decision, and the selection of a wind power forecast evaluation criterion is fundamentally determined by the type of the decision. An electric power company responsible for the power balance of the electric power system is interested in the determination of the most possible wind power generation capacity, while the forecast interest of the wind power company is to obtain the greatest benefit from the wind power generation. The requirements of the two kinds of forecasts are not consistent.

The evaluation of the accuracy and reliability of the forecast result is an important part of forecast analysis. Forecast errors for measuring the accuracy of the wind power forecast can be quantified by the different relation functions. Multiple error indexes are usually used to evaluate the forecast result, in this way, the quality of various forecast methods or the feasibility of one kind of forecast method is judged. Using only one error index to evaluate the forecast result is not reliable and scientific. Generally, the normalization RMSE of the installed wind power capacity is common, and the mean absolute error or mean absolute error percentage can be used as well.

Usually used error indexes are as follows [7, 22]:

1) Mean error (Formula (9.36)):

$$AE = \frac{1}{n}\sum_{i=1}^{n}[y_t - \bar{y}_t] \tag{9.36}$$

2) Mean percentage error (Formula (9.37)):

$$APE = \frac{1}{n}\sum_{i=1}^{n}\frac{y_t - \bar{y}_t}{y_t} \tag{9.37}$$

3) Mean absolute error (Formula (9.38)):

$$MAE = \frac{1}{n}\sum_{t=1}^{n}|y_t - \bar{y}_t| \tag{9.38}$$

4) Mean absolute percentage error (Formula (9.39)):

$$MAPE = \frac{1}{n}\sum_{t=1}^{n}\left|\frac{y_t - \bar{y}_t}{y_t}\right| \tag{9.39}$$

5) Root mean square error (Formula (9.40)):

$$RMSE = \sqrt{\frac{1}{n}\sum_{t=1}^{n}(y_t - \bar{y}_t)^2} \tag{9.40}$$

6) Root mean square percentage error (relative root mean square error) (Formula (9.41)):

$$MSPE = \frac{1}{n}\sqrt{\sum_{t=1}^{n}\left(\frac{y_t - \bar{y}_t}{y_t}\right)^2} \tag{9.41}$$

In the above error indexes, y_t $(t=1, 2,\ldots,n)$ is the observed n values of a certain time series, and $\bar{y}_t(t = 1,2,\ldots,n)$ is the fitting value or forecasted value of a certain model for a time series. The mean error and mean percentage error indexes can be used to measure whether the forecast is unbiased. The mean absolute error and the mean absolute percentage error are generally used to measure the degree of the error deviation. The RMSE and the root mean square percentage error are generally used to measure the distribution status of the errors. These are general methods for defining errors, and are not normalized by installed wind power capacity.

7) Cost regulation criterion [22]. The main issue of the criterion is that the costs of wind power forecast errors are not uniform, that is to say when actual value of the wind power output is above the forecast value, the cost is c^+. And the cost is c^- when the practical value is lower than the forecasted value. Herein, $c^+ \neq c^-$. The costs of upward regulation and downward regulation are respectively c^- and c^+. The cost regulation criterion can be mathematically shown as (Formula (9.42)):

$$\sum_{j\in J^-} C^-(y_t - \bar{y}_t) + \sum_{j\in J^+} C^+(y_t - \bar{y}_t) \tag{9.42}$$

$j \in J$ is an observation set, the observation set J herein is divided into three sub-sets, enabling $j \in J^-$, $y_t - \bar{y}$; $j \in J^0$, $y_t - \bar{y}$, at the moment, $c^+ = c^-$, the regulation cost is 0, not shown in the expression; $j \in J^+$, $y_t > \bar{y}$. If the forecast error distribution is balance, namely

$c^+ = c^-$, so, there is no difference between the forecast error and the mean absolute error. When the cost regulation criterion is applied, the best target is lowest cost of power adjustment to compensate forecast error.

8) Same with the load forecast, the evaluation criterion of the wind power forecast can be defined as following:
   a) Daily wind power forecast accuracy (Formula (9.43))

$$A_i = 1 - \sqrt{\frac{1}{n}\sum_{n=1}^{n} d_{it}^2} \tag{9.43}$$

   Where n is the number of Forecast total points; d_{it}^2 = (forecasted value-practical value)/practical value.
   b) Monthly mean wind power forecast accuracy (Formula (9.44))

$$A = \frac{1}{N}\sum_{i=1}^{N} A_i \tag{9.44}$$

   N is the number of total days in one month.
   c) The highest wind power forecast accuracy and the lowest wind power forecast accuracy of the day i.
   Highest: B_{ih} = 1−(day-ahead actual highest wind power minus the day-ahead highest forecasted wind power)/the day-ahead actual highest wind power;
   Lowest: Bil = 1−(day-ahead actual lowest wind power minus day-ahead lowest forecasted wind power)/the day-ahead actual lowest wind power.
   d) Accuracy of the mean highest and lowest wind power forecasts in the current month (Formula (9.45))

$$B = \frac{1}{2N}\sum_{i=1}^{N}(B_{ih} + B_{il}) \tag{9.45}$$

Considering the characteristics of the wind power, the wind power forecast error is usually deducted to the calculation of the rated power of the WF. For example, the mean percentage error is shown as $APE = \frac{1}{n}\sum_{i=1}^{n}\frac{y_t - \overline{y}_t}{P}$; the root mean square percentage error is shown as $MSPE = \frac{1}{n}\sqrt{\sum_{t=1}^{n}\left(\frac{y_t - \overline{y}_t}{P}\right)^2}$, wherein P is the rated capacity of the installed wind turbines in the WF.

In the above indexes, some are valuable for the improvement of algorithm precision, such as the mean absolute error and the mean absolute percentage error, whereas some are valuable for the dispatch of the electric network, such as the root mean square percentage error (relative to the root mean square error).

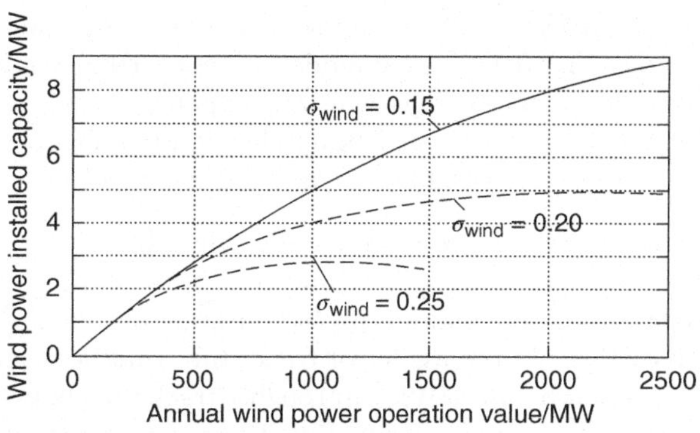

Figure 9.21 Relationship between annual wind power operation value and wind power forecast variance.

The improvement on the forecast precision is very important. Because of the random volatility of the wind velocity, the forecast of the wind energy will not be very precise. Under the condition of the electric power market, the relationship between the precision of the wind energy forecast with the economic benefit of the operation of the WF is very remarkable. When the forecast variance of the wind power generation capacity is respectively 0.15, 0.2, and 0.25, the relationship between the wind power installed capacity and the annual operation value is shown in Figure 9.21. It is shown that the improvement on the precision of the wind power forecast can greatly improve the economy of the system.

Discovered from the Figure 9.21, the installed capacity of the WF is increased from 0 to 2,500 MW when the forecast variance is 0.15, and the annual operation economic value of the installed capacity is almost linearly increased. When the forecast variance is increased to 0.20 to 0.25, the operation value difference of the WF with the capacity of 500 MW is not great. But a very remarkable difference occurs when the installed capacity is 1,000 MW, and the difference of the annual operation values of different forecast variances becomes greater along with the increase of the installed capacity. But when the forecast variance is 0.25 and the installed capacity is over 1,000 MW, the annual operation economic value is reduced. It shows that the increase of the installed capacity of the WF will be seriously restricted if the provided forecast is inaccurate. Therefore, the more the installed capacity of the WF is, the more meaningful the forecast is [11, 12].

9.6.3 Influence Factors of Wind Power Forecast Accuracy

As for the electric power system, the wind power forecast is one of the most direct and valuable solutions to reduction of the uncertainty of the wind energy production plan. The main factors affecting the accuracy of the wind power forecast include accuracy of weather prediction, WF power model, namely model for converting wind energy into electric energy, wind power penetration level of system, electric power market closing time, forecast time period, wind condition of WF, and so on [4, 7].

Weather prediction accuracy. The wind velocity forecast depends on the NWP, and some WFs cannot get data of the NWP temporarily but the common weather prediction. At the aspect of trend, the wind velocity forecasted by the NWP can roughly track the change of the actual wind velocity. But at a specific moment, the deviation fluctuation between the forecasted value and the actual value is still very large, especially when fluctuation which is in large amplitude, of the actual wind velocity occurs in a short time period. The wind velocity forecast is difficult to be tracked. As for the number of the wind turbines in a specific WF, the resolution of the NWP is limited, and approximate wind velocity forecast value of each set of wind turbine is deduced according to the NWP. The power is forecasted by using weather prediction data, and the forecast error is generally greater than that of the weather prediction. The wind energy is proportional to third times of the wind velocity. At a unsaturated section, the wind turbine power is relevant to the wind velocity in a certain proportion. Therefore, the wind velocity error will be enlarged by the power error.

WF power model. The methods for forecasts of the conversion from the wind energy to the electric energy include physical method, statistic method, and learning method. The accuracy of different power forecast methods is different. When a wind velocity-power curve physical model for the wind turbine is adopted, the forecasted power output of WF should be scaled in proper proportion, and the scaled model of different WFs are different.

Wind power penetration level. Generally, in the system, the more the proportion of the wind power is, the more the influence on the electric power system is. But the accuracy of the wind power forecast will be higher.

Forecast time. The wind power forecast is not only related to the forecast method, but also the forecast time (period). Generally, the shorter the forecast period is, that is to say the shorter the forecast lead time is, the more accurate the forecast is. Otherwise, the forecast error is greater.

Wind velocity characteristics. The wind power forecast is closely related to the wind velocity characteristics. Generally, the milder the wind velocity change in the forecast place is, the less the forecast error is, otherwise, the forecast error is greater.

WF spatial distribution. If many WFs are forecasted together, the forecast error will be reduced. The integration of several large regions with several GW wind power farms installed capacity will prompt the reduction of relative forecast error, because the forecast errors of different regions are practically neutralized with each other.

Electric power market closing time. Viewed from the accuracy of wind power forecast, reducing closing time may be favorable, because the short-term forecast is more accurate than the long-term forecast. The electric power market trade rules should be accordingly regulated in consideration of the wind power.

9.6.4 Accuracy Level of Current Wind Power Forecast System

Forecast accuracy is an important issue, because the forecast errors are related to the application, as well as the extra cost, of the system balance power. In Europe, the cost generated by 1 GW forecast errors within several hours is at an order of magnitude about 1 million Euros [23].

No matter what forecast method is adopted, with present tools, within a range of 36 hours, the normalization RMSE of the forecast for one single WF is from 10% to 20% of the wind power installed capacity. After the scale is increased to the grid-tied wind power of the whole region, the errors will decrease below 10% because of the smoothing effect of multiple WFs/wind turbines. The larger the range is, the less the forecast errors are. As for a longer forecast period, the forecast accuracy will be lowered [6].

As for a large region, the forecast mean error is required at an order of magnitude which is 1.5% to 3% of peak load, equal to about 3% to 5% of the total energy of the day-ahead forecast. The accuracy of the day-ahead forecast cannot be obtained by the current available technology. For operators, the comprehensive errors of the forecasts of the wind power and the loads are decisive. When the extra system balance demand of the wind power is determined, only the extra derivation in comparison with the system without wind power is considered, not the total wind power forecast error. Therefore, at the low wind power penetration level, the wind power forecast error only has small influence on the forecast error of the system [6].

Additionally, the accuracy of the wind power forecast is also related to the spatial scale as it shows in Figure 9.22.

The integration of different relative forecasts in the electric power system is described at Figure 9.23. The total forecast error generated by the integration of the different relative forecasts in the electric power system is less than the sum of all forecast errors, because the load is not related to the wind forecast. According to the system power balance, the accuracy related to a balance system (control region) is not the very accuracy

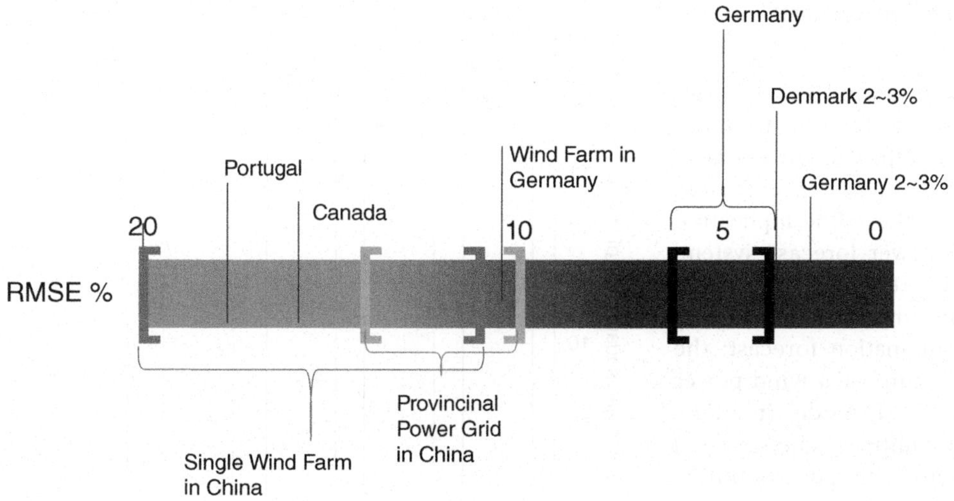

Figure 9.22 Forecast error (MW) in different regional range.

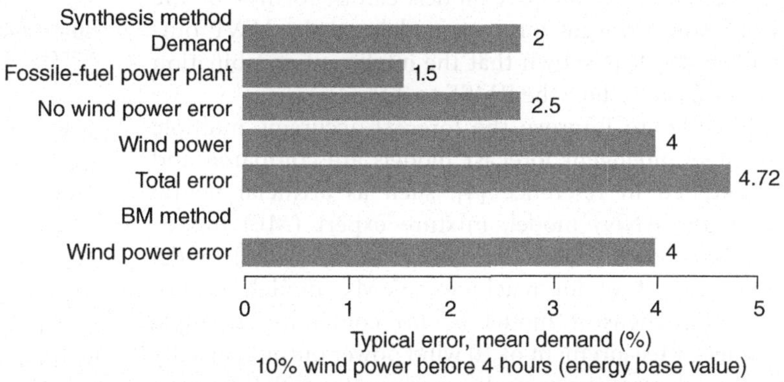

Figure 9.23 Total error of relative planned reserve and error of demand (load) forecast, power plant operation, and wind forecast.

of the wind power forecast, and this is the sum of all demand and supply forecast errors related to the system operation. A balance market (BM) method only observes whether wind power errors are excessive punishment of the wind power forecast errors, because in the system level, the most suitable attached reserve should be established on the basis of combined forecast error. In a real example, the combined forecast error is 2.2%.

In Germany, the accuracy of the wind power forecast is usually assessed by an independent system operator (TSO). The most popular error measurement of forecast quality evaluation is the RMSE. It is established on the basis of the normalization of the deviation between the forecast and measured value for the installed wind power. The RMSE reflects a function of corresponding cost, because it stresses that great forecast errors are disproportionally expensive.

The evaluation of the wind power forecast in Germany and the trade of comprehensive wind power capacity forecast in the whole Germany shows that the forecast accuracy was remarkably improved in past few years. Forecasts from November 2004 to July 2006 are evaluated by a Previento forecast system, including forecast results of storm periods in several winters and springs. The accuracy is shown by the RMSE in the normalization of the installed wind capacity. The RMSE obtained through statistics is 3.1% intraday (0 to 23 hours), 4.4% day ahead (24 to 47 hours), and 5.8% two days ahead (48 to 71 hours) [23].

Typical forecast errors of the Previento wind power forecast system in Germany are higher in winter, because of higher wind velocity in winter and higher uncertainty of storm weather predication, especially in a low air pressure system and a frontal surface region therein. In summer, the forecast errors are mainly controlled by the effect of a temperature layer, especially under the high air pressure condition. The wind velocity at the height of a large hub is rapidly increased after sunset, and it is very difficult to forecast.

Less forecast errors exist in the wind power forecasts of Germany, because of the advantages of good spatial smoothing effects (WFs are relatively scattered), that is to say an error of one place is partly neutralized by errors of remote WFs.

9.6.5 Ways to Improve Wind Power Forecast Accuracy

Along with the continuous increase of wind power capacity, the accuracy of the wind power forecast becomes more important. Since carrying out the wind power forecast, the accuracy has been continuously improved, and the progress will last. According to the online application feedback of the existing wind power forecast system, it is possible to further improve the forecast system technique. Main ways to further improve the accuracy of the forecast include: (1) combination forecast: the combination of different models can use a wind power output forecast model and an NWP model (multiple models and approximating of multiple schemes); (2) proper forecast time range; (3) utilization of smoothing effect of spatial distribution, and so on.

9.6.5.1 Combination of Different Forecast Models

Combination of different models can be formed by the wind power output forecast model and the NWP output model. It is shown that the intelligent combination method can reduce the RMSE to 20% [4].

In order to improve the forecast accuracy, multiple artificial intelligent forecast models are compared and researched in reference [4], such as artificial neural network (ANN) model, mixture expert (ME) model, combination of nearest-neighbor search (NNS) model and support vector machines (SVMs) model, as well as comprehensive model (Com) containing multiple models. Measurement of 10 wind power output capacity in E.on control region and corresponding NWP forecast data of these points from the German Weather Prediction Center are applied in different forecast methods. The comparison between the wind power forecast RMSE of 10 WFs obtained through four different artificial intelligent methods and the forecast errors obtained by combining all methods is shown in Figure 9.24. It is clear that the best result is obtained by SVM under the condition. Compared to each single artificial intelligent forecast model, the synthesis model method can improve the forecast accuracy, even if it is simple synthesis.

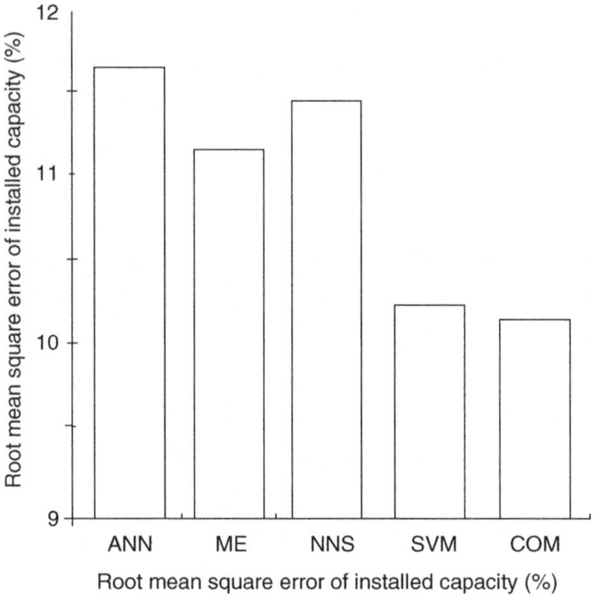

Figure 9.24 Comparison among forecast errors obtained by different artificial intelligent methods.

9.6.5.2 Multi-Model Method in NWP Model

The weather prediction model field is full of the most potential improvement, which is on the accuracy of the wind power forecast. The following aspects can be further improved, including improvement on the spatial resolution of the weather prediction model, improvement on the frequency of each NWP, coupling of the NWP, and the short-term wind power forecast model, as well as acquiring data from multiple NWPs to carry out combination forecast.

Up to now, the common method for calculating the wind power is to only utilize a single NWP model, in this way, high-risks and expensive forecast errors happen, and especially in the case of a special weather event, a single model may be abnormal. It is considered that deviations of forecasts in all NWP forecast models will not simultaneously occur, and each NWP model has its own advantages and disadvantages depending on the weather conditions, so that the forecast accuracy can be improved by the multi-model method in the NWP model.

Germany investigated and researched the effects of three different NWP models used for day-ahead wind power forecast on the forecast accuracy. All the three models are used for the input of wind power management system (WPMS) based on the ANN method. Compared to a single model, even if the forecasts of the three models are averaged, the forecast accuracy is remarkably improved. The RMSE result of the combined model is 4.7%, whereas the forecasted value of each model is respectively within a range from 5.8% to 6.1%. If the optimum combination of a weather model related to different weather conditions is found, and a special program automatically classifying the weather conditions is established, with advantages of different weather forecast models used, as a result, the forecast accuracy can be further improved [4].

In order to find the optimum combination of special weather conditions, one kind of classification scheme is established on the basis of meteorology experience and automatic meteorology classification. By adopting the advanced combination tool, the optimum weight of each weather model can be determined in accordance with weather conditions in the ascendant. As for special condition, selecting the weather model combination with the optimum performance will remarkably improve the forecast of the wind power. Some research results show that the combined wind power forecast is obviously better than the best forecast result obtained through a single weather model and simple fixed combination. Especially under the dynamic weather conditions, the forecast of combination of different models lowers the great and expensive forecast errors, in other words, the combination method has the greatest benefit under extreme conditions, such as storm.

9.6.5.3 Adoption of Proper Forecast Time Range

A shorter forecast range leads to lower forecast errors. However, the forecast time range should be combined with an electric power market mechanism and matched with the dispatch plan of conventional generators.

9.6.5.4 Utilization of Smoothing Effect of Spatial Distribution

The forecast error is determined by the numbers and geographic distribution of the wind turbines and WFs. Integration of WF forecasts in a large region will lower relative forecast errors, because one part of the forecast errors of different regions is mutually neutralized. In Germany, the typical forecast errors (RMSE) of representative WFs are 10% to 15% of the installed capacity, and the typical errors of the controlled region obtained from these representative WFs are 6% to 7%, and there is only 5% to 6% in the whole Germany. If possible, wind power should be integrated in the large area, because the integration promotes remarkable reduction of the forecast error and short-term power fluctuation.

9.7 Application and Challenge of Wind Power Forecast

9.7.1 Expansion of Application Field of Wind Power Forecast

The wind power forecast has different purposes, adapting to the forecast time range. Annual forecasts can be used to make an annual maintenance plan; weekly forecasts can be used for the optimization of power generation charge (cost); hourly or other shorter forecasts are used for optimization dispatch of generators. The wind power forecast provides economic dispatch, electric energy trade, safety evaluation, and the like with basic data input [3, 4, 24, 25]. As for the power system, the wind power forecast is one of the most direct and valuable methods for lowering the uncertainty of a wind energy production plan.

The first and most important application of the wind power forecast is to lower demand for balancing electric energy and backup power. It requires that the wind power is integrated into a power supply system to balance power supply and demand, that is to say, real-time wind power forecast is integrated into DTS/EMS to realize the dispatch optimization of the power plants.

It is considered that the wind power generation dispatch optimization-economic dispatch (including unit optimization combination) of the wind power can lead to lower charge of wind power integration and lower emission of balancing power plant, and finally generates higher value of wind power. In order to optimize power plant dispatch and power balance, the accurate forecast of the wind power generation in the whole region is required. The corresponding time range is determined by technology and a regulatory formation manner, for example, the type of the conventional power plant in the system and the closing time of the electric power market trade.

The effect of the wind power on system backup power is an important problem concerned by people. In a system with a high wind power penetration level, the shutdown of a conventional generator and unexpected load increase or unexpected wind power reduction may cause inadequate power generation of the system. The increase of the wind power possibly requires the increase of the total backup of the system. But the actual changes of the load and the wind power do not affect the backup capacity level of the system, the accuracy of the load and the wind power forecast are related to the system backup level, because they together bring more uncertainty. The greatest forecast error determines the demand of the backup power, in order to minimize the backup power, the improvement on the wind power forecast accuracy is required.

The second application is to provide wind power feed through forecast for electric network operation and electric network security evaluation. Because the WF is always connected to a remote region of the electric transmission network, in order to forecast large congestion and consumption caused by a large power flow, the electric network operators need to know the future and current wind power feedback of each electric network PCC point. High local wind power forecast can forecast possible power transmission bottleneck of a local electric network. In the EMS system of an electric network control center, forecast and uncertainty of the forecast can be used as the input of trend calculation. Detailed view of the expected electric network loads can be obtained. Indication of electric network operation and congestion management decision is provided, and electric network congestion disaster is lowered or avoided. In this kind of application, the current and forecasted wind power generation of the PCC point of each electric network region or WF is required, and forecast of a small region even a single WF is also required.

The third application is energy storage management. Energy storage is very useful for on compensating the fluctuation of wind energy. Different application techniques of restraint for wind energy fluctuation have been raised into the agenda, such as water pumping energy storage, and air compression or hydrogen storage. The wind power forecast can provide the expected energy storage operation with strategy, for example, the state that stored energy is completely emitted before the strong wind is forecasted.

The fourth application is wind energy trade in wholesale market. At present, in some European countries, the wind power is exchanged in a wholesale market. In order to make the wind power to be directly exchanged in the energy market, as long as reliable forecast is provided, the capacity of provided wind power is known. The key point is to utilize the uncertainty of the wind power forecast to estimate various risks related to the wind energy. The wind power forecast is integrated into an electric power market simulator. Training, system test, and analysis tools can be provided to market management and operation personnel.

With the development of the wind power forecast, the application of the wind power forecast will be developed in the field of electric power system planning, operation, maintenance management, market operation, and the like in depth and width.

9.7.2 Challenge to Wind Power Forecast

With the rapid increase of wind power capacity, the accuracy of the wind power forecast becomes more important. The forecast accuracy is the future greatest challenge to the wind power forecast system. In addition, the overall wind power forecast affects many aspects of electric power system planning and operation. How to use the wind power forecast is a challenge to electric power enterprises and wind power generation enterprises.

9.7.2.1 Improvement of Wind Power Forecast Accuracy

The wind power forecast accuracy is improved continuously, and the continuous improvement can be expected. The main ways to lower the wind power forecast error include combination of different forecast models, multiple model method in NWP models, adoption of proper forecast time range, utilization of smoothing effect of spatial distribution, and the like. The specific improvement methods are as follows:

a) The utilization of a combined operation model system, of which the data generated by several NWP models is input, remarkably improves the forecast accuracy. In addition, a method for improving model combination should be developed; the spatial and time resolution of weather service is improved, and local weather data is provided for the WF;
b) The progress of the NWP and more frequent update of the weather prediction will improve the input data of the wind power forecast;
c) The further improvement on the forecast method and the improvement method of the combination of different forecast methods can be expected to further reduce forecast errors;
d) As for short-term wind power forecast, increasing the online measured data of the wind still has potential possibility of the improvement on the forecast;
e) A special offshore wind power forecast model is established. The wind power forecast under the offshore environment may be more reliable than the wind power forecast on the land; as for the weather conditions of the place close to the coast, the atmosphere boundary layer differs from that on the land. The atmosphere stability and the distance to the coast have extremely important effects on the offshore wind power forecast;
f) The forecast of the probability distribution of the forecast errors and the reduction of a large error event provide a chance for reserve capacity required by the reduction and balance of the wind power forecast errors;
g) In order to process congestion management, in an electric network with a high wind power penetration rate, as for each node of a high-pressure electric network, the forecast of high spatial resolution is required;
h) A forecast method more accurate in proportion increase/reduction is established;
i) A new method for the simulation of forecast on a complex landform;
j) Research on the estimation of adaptive parameters, making the model automatically adapt changes of the WFs and the environment;
k) Establishment of a probability forecast model.

9.7.2.2 Expansion of Application of Wind Power Forecast

The wind power forecast is used for application. The effect and value of the wind power forecast are examined by the practical application. Same with the short-term load forecast, the short-term power output forecast of the WF is mainly used for electric network operation, power plant production plan and trade. Important time ranges related to wind power forecast applications include trade time (0 to 23 hours), day-ahead (24 to 47 hours) operation plan, weekend (from Friday to Monday, namely 72 to 96 hours) operation plan, and the wind power plant control time (determined by control requirement, but most are short-term forecast time ranges).

1) **Wind power forecast is integrated into an power grid control center.**
When an electric power company integrates a large amount of wind energy, the wind power forecast is a key factor. If the wind power forecast is not integrated into electric power system operation and electric power market, the forecast is wasted and does not enable the information to be usable and to have an effect. If the operators cannot make sure of the forecast or the forecast information is not presented to the operators in a usable way, the operators maybe operate the system in a more conservative manner, accordingly, the operation cost is increased.
Wind power forecast information is directly integrated into an energy management system (EMS) in the power grid control center. It can realize: (1) reducing risks when the results are not fully known by the electric network control center; (2) improving system operators and dispatchers' confidence in managing the variability of the wind power; (3) as for a whole power system, lowering the cost through improved operation plan and the utilization of power generation and power transmission resource; (4) increasing wind power forecast gain benefit as well as obtaining the complete system value.
The wind power forecast information is integrated into a detailed control room simulation system. The simulation can be used for a dispatcher training simulator (DTS), and photographically simulates the EMS and electric system operation. The DTS provides real simulation of a real-time electric power system and functions of the EMS. As for the electric power system operation and operator training, it is a viable field.
2) **Synthesis of EMS and wind power forecast.**
Same with the load forecast, as for integrated wind power forecast, the first logical step is to directly integrate the real-time wind power forecast into the dispatcher training simulator (DTS) and finally integrated into the energy management system (EMS). The DTS should accept wind power forecast obtained through the calculation of the electric power company with a utility-wide wind power forecast system. The DTS utilizes WF power generation output (measured or forecasted) and drives the corresponding original motivation to approach a forecasted target. Meanwhile, in order to respond the behavior of the WF, all elements of the DTS will simulate the dynamic behaviors of the electric power system and the EMS. During simulation, the key data, such as operation cost deviation, system security, and reliability indices, are all collected and taken as one part of the wind power generation influence analysis.
3) **Synthesis of electric power system and electric power market simulator.**
Under the electric power market environment, the reconstruction of electric power industry has developed new trade entities (independent system operator, power transmission company, and power generation company) utilizing a highly combined operation system with continuously increased complexity. The process means that a high professional level is required, and the quality control is continuously stressed. In order to realize these purposes, simulated environment becomes the key factor of training of management and operation staff, system test, and analysis. As a result, a testing and training simulation environment (TTSE) is needed, comprising the DTS and copies (e.g., a market management system, load forecast, and wind power forecast) of all key factors of a real-time system.
4) **Real-time electric power system forecast with loook-ahead simulation.**
Real-time electric power system forecast is required by the operators. Advanced and real-time operation system forecast can provide conditions for the operators and enable the operators to know requirements of the optimum real-time operation. It may be related to the system with a high wind power penetration level.
The above simulation environment can be operated in the operation environment in which automatic initialization can be performed according to real-time conditions, then the ultrafast simulation of predicting future can be realized, and leveraging wind forecast information is supplemented. The final goal is to provide display and enable the operators to clearly know things will happen for 10 to 40 minutes. It will involve the diversity of indexes and decision-making support tools.
By using this kind of frame work of a real-time operation environment, the operators can complete their duties, and they not only know the wind power forecast in future several hours but also the influence of forecasted wind power on the operation of the electric power system. On the basis of the visualization, the operators can make the most proper decision to supplement leveraging and control wind power generation benefit.
A important issue of future research is analytical method in the simulation environment of study the interactions between statistics output of a wind farm and the probability load of power transmission line and other equipment that connect the wind farm and the load and the probability characteristic of real-time electricity prices. Before the operators can believe a power system forecast containing a wind power forecast, much work is needed for improving the

accuracy of the look-ahead simulation technology. On this point, one important issue is to forecast and provide warning of gale events possibly causing more extreme wind power output fluctuation. This kind of early-warning system provides the operators with continuous state recognition about system health. By utilizing ultra-high-speed simulation, the operators and dispatchers can reduce risks of system faults. They can take prevention solutions according to forecasted system conditions (providing recognition of state improvement and real-time optimum decision-making).

5) **Application of operation and control.** The wind power forecast can be used for determination of the backup capacity required by system balance, arrangement of optimum unit combination (UC), as well as automatic generation control (AGC) and automatic voltage/reactive power control (AV/QC). It requires much work through combining an actual system at these aspects.

References

1 Zheng, Lin. 2007. The Future of Scientific Prediction Studies. Bejing: FangZhi Press.
2 Ahlstrom, Mark L., and Zavadil, Robert M. 2005. The role of wind forecasting in grid operations & reliability. IEEE/PES Transmission and Distribution Conference & Exhibition: Asia and Pacific Dalian, China.
3 Lerner, Jefff, Grundmeyer, Michael, and Garvert, Matt. 2009. The importance of wind forecasting. *Renewable Energy Focus*, **10**(2), 64–66.
4 Ernst, Bernhard, Oakleaf, Brett, Ahlstrom, Mark L., Lange, Matthias, Moehrlen, Corinna, Lange, Bernhard, Focken, Ulrich, and Rohrig, Kurt. 2007. Predicting the wind. *IEEE power & Energy Magazine*, **5**(6), 78–89.
5 Costa, Alexandre, Crespo, Antonio, Navarro, Jorge, Lizcano, Gil, Madsen, Henrik, and Feitosa, Everaldo. 2007. Review on the young history of the wind power short-term prediction. *Renewable and Sustainable Energy Reviews*, **12**(6), 1725–1744.
6 European Wind Energy Association (EWEA). 2005. Large scale integration of wind energy in the European power supply: Analysis, issues, and recommendations. http://www.ewea.org/fileadmin/ewea_documents/documents/publications/grid/051215_Grid_report.pdf
7 Boone, Andrew. 2005. Simulation of short-term wind speed forecast errors using a multi-variate ARMA (1,1) time-series model. [D]Dept of Electrical Engineering Electric Power Systems, Stockholm, Sweden, Royal Institute of Technology.
8 Negnevitsky, Michael and Potter, Cameron W. 2006. Innovative short-term wind generation prediction techniques. Proceedings of IEEE PES Power Systems Conference & Exposition, October 29–November 1 pp. 60–65.
9 Guoyang, Wu, Yang, Xiao, and Sasa, Weng. 2005. Discussion about short-term forecast of wind speed on wind farm. *Jilin Electric Power*, **6**, 21–24.
10 Sideratos, G. and Hatziargyriou, N. 2007. Using radial basis neural networks to estimate wind power production. Power Engineering Society General Meeting, IEEE, pp. 1–7.
11 Shuangxi, Zhou, Haichao, Wang, and Shousun, Chen. 2006. Analysis on operation value of wind power resources[J]. *Power System Technology*, **14**(5), 98–102.
12 Chongqing, Kang, Qing, Xia, and Mei, Liu. 2007. Power system load forecasting. Bejing: China Electric Power Press.
13 Yang Xiuyuan, Yang, Xiao Yang, Xiao, and Chen Shuyong, Chen. 2005. Wind Speed and Generated Power Forecasting in Wind Farm. *Proceedings of the CSEE*, **25**(11), 1–5.
14 Ran, Li, Yue, Wang, and jinyeong, Xiao. 2009. Short-term wind speed forecasting for wind farms based on empirical mode decomposition mode. *Electric Power*, **42**(9), 77–81.
15 Shengli, Guan. 2009. Wind speed forcasting in wind farm based on local wave decomposition and time series. *North China Electric Power*, **1**, 10–13.
16 Kai, Cai, Lunnong, Tan, Chunlin, Li, and Xuefeng, Tao. 2008. Short-term wind speed forecasting combing time series and neural network method. *Power System Technology*, **32**(8), 82–85, 90.
17 Xingkai, Gu, Gaofeng, Fan, Xiaorong, Wang, Haixiang, Zhao, and Huizhu, Dai. 2007. Summarization of wind power prediction. *Technolog. Power System Technology*, **31**, 335–338.
18 Sideratos, George and Hatziargyriou, Nikos D. 2007. An advanced statistical method for wind power forecasting. *IEEE Transactions on Power Systems*, **22**(1), 258–265.
19 Jun, Kang. 2009. Application research Based on radial basis function neural network. Unpublished master's thesis, Hunan University.
20 Liangjun, Zhu and Yanping, Zhang. 2007. Comparison of several radial basis function (RBF) neural networks. *Journal of Fujian Radio & TV University*, **4**, 68–70.

21 Ahlstrom, Mark, Jones, Lawrence, Zavadil, Robert, and Grant, William. 2005. The future of wind forecasting and utility operations. IEEE Power & Energy Magazine, **3**(6), 57–64.

22 Ravn, Hans F. 2006. Short term wind power prognosis with different success criteria. 9th International Conference on Probabilistic Methods Applied to Power Systems KTH, Stockholm, Sweden, June 11–15, 2006.

23 Lange, Matthias and Focken, Ulrich. 2008. New developments in wind energy forecasting. Power and Energy Society General Meeting - Conversion and Delivery of Electrical Energy in the 21st Century, IEEE, July 20–24, pp. 1–8.

24 Watson, S. J., Lanberg, L., and Halliday, J. A. 1994. Application of wind speed forecasting to the integration of wind energy into large scale power system. *Proc. Inst. Elect. Eng., Gen., Transm., Distrib.*, **141**(4), 357–362.

25 Milligan, Michael, Porter, Kevin, and DeMeo, Edgar, *et al.* 2009. Wind power myths debunked. *IEEE Power & Energy Magazine*, **7**(6), 89–99.

10

WF Control and Protection Technologies

10.1 Overview

Modern power plants are highly automatic. Power generator systems can generate active power and reactive power according to the requirements of power systems. Power plants can realize not only coordinated control for individual generator system but also generator system combinations, achieving optimal operation. Hydropower plants in a watershed can also carry out inter-plant coordination and control according to reservoir and hydrological conditions.

The output power of wind turbine generator systems (WTGSs) with uncontrollable motive power (wind) are limitedly controllable, which is different from conventional power plants. The coordinated control of output power of the wind farms (WFs) made up of decentralized WTGSs is more difficult than that of conventional power plants. With the increase of single WTGS capacity and WF capacity, power systems have the following higher requirements for grid-connected operation of wind power: WFs should operate like conventional power plants; WFs should not only operate safely, reliably, and efficiently but also be capable of supporting power systems when necessary. The control and protection of a power system are closely related, and the control system and protection system of a wind turbine generator (WTG) can be integrated into an overall digital control system with technological development.

In the researches about WF control, the objects, places, and methods of WF control should be figured out. For WFs, the first kind of control is realized by WTGs, such as the decoupled active and reactive power control of double-fed asynchronous generators (DFAG); the second kind of control is realized through coordination with other flexible power generation systems, such as the automatic generation control (AGC) of WFs, hydropower plants (pumped storage power stations), and conventional thermal power plants; the third kind of control is realized by means of adding controllable equipment for WFs, such as static or dynamic reactive power compensation equipments and energy storage equipments. This chapter mainly introduces the control and protection technologies of WTGSs and WFs related to power systems.

10.1.1 Control of WTGs and WFs

10.1.1.1 Control Objectives and Functions of WTGSs

Wind power systems are complicated multi-variable non-linear systems characterized by uncertainty, multi-interference, and so on. The basic objects of the control system of a WTGS include ensuring reliable operation, obtaining maximum energy, providing electrical energy with good quality, extending life of the WTGS, and supporting power systems. The control system has the following functions [1, 2]:

1) Realizing stable operation in the operating wind velocity range;
2) Tracking the optimal tip-speed ratio in the low wind velocity mode to capture the maximum wind energy;
3) Limiting the wind energy capture in the high wind velocity mode to keep rated output power of the WTGS;
4) Reducing the changes of torque peak caused by gusts, mechanical stress of the transmission chain, and output power fluctuation;
5) Reducing transient response of the power transmission chain;
6) Realizing a small control cost and limiting the input signal amplitudes, the regulation range, and the rate;
7) Avoiding possible mechanical resonance and self-exciting resonance;
8) Regulating active power and reactive power of the WTGS and supporting and controlling the stability of grid frequency and voltage.

The control system is expected to realize normal operation, regulation, and protection for the whole WTGS. In details:

1) Start control: If the average wind velocity measured in a period by the wind velocity detection system has reached the cut-in wind velocity, and no fault was

Integration of Large Scale Wind Energy with Electrical Power Systems in China, First Edition. Zongxiang Lu and Shuangxi Zhou.

found in the self test, the control system will give a command for brake release and the WTGS will switch from wind waiting state to start with a low wind velocity.

2) Grid connection and disconnection control: When the WTG speed reaches synchronous speed, grid connection will be operated. For reducing impacts on the power grid, a cage asynchronous WTG is usually connected with a thyristor soft starter. During soft start, limit the grid-connected current of the generator and monitor the balance of the three-phase current. If the unbalance is beyond the limit, stop the generator. Additionally, the soft starter can make the WTG start with a low wind velocity. When the wind velocity is lower than the cut-in wind velocity (such as 3 m/s), disconnect the grid-connected generator from the power grid and brake. The grid connection of a double-fed generator or a direct-drive WTG is controlled by a grid side converter and the grid-connected current is small.
3) Yaw and cable untwisting: Yam control refers to automatic wind tracking according to the wind direction. Continuous wind direction tracking may result in cable twisting, so the control system has a cable untwisting function.
4) Speed limitation and braking: If the speed is faster than its upper limit, there will be runaway. The generator will disconnect from the power grid automatically, the blades will get opened for soft braking, and the hydraulic braking system will work to stop the blades. The steering system will make the nacelle deflect to cross the wind by 90° to protect the whole tower.
5) Speed and power control: There will be automatic speed and power control in line with the wind velocity and the power; the corresponding compensation capacitor will be cut in or out automatically according to the power factor; the power grid operation, the WTGS operation, and the wind condition will be detected and recorded automatically.
6) Voltage and reactive power control: The control system will determine control mode of the WTGS and cut the corresponding compensation capacitor in or out automatically according to the terminal voltage and power factor of the generator.
7) Protection function: The control system can judge abnormalities such as grid faults and take the corresponding protective actions automatically. It can also generate charts based on the data recorded to show the performance indicators of the WTGS. If the WTGS is operating in a WF, it should be provided with a remote communication function.
8) The control system can automatically detect and record the power grid operation, the WTGS operation and the wind condition, etc.

10.1.1.2 Control System Structure

At present, the control systems of most of WTGSs are Distributed Control System (DCS) basing on industrial personal computers. Each DCS has many special modules with different functions to choose for convenient local control. Many modules can be directly arranged at the operating points of control objects for local signal acquisition and processing. Meanwhile DCS has great on-site adaptability for convenient on-site debugging of control programs and modifying control parameters of the corresponding WTGS at any time (see Figure 10.1 for structure of the DCS of a WTGS) [3].

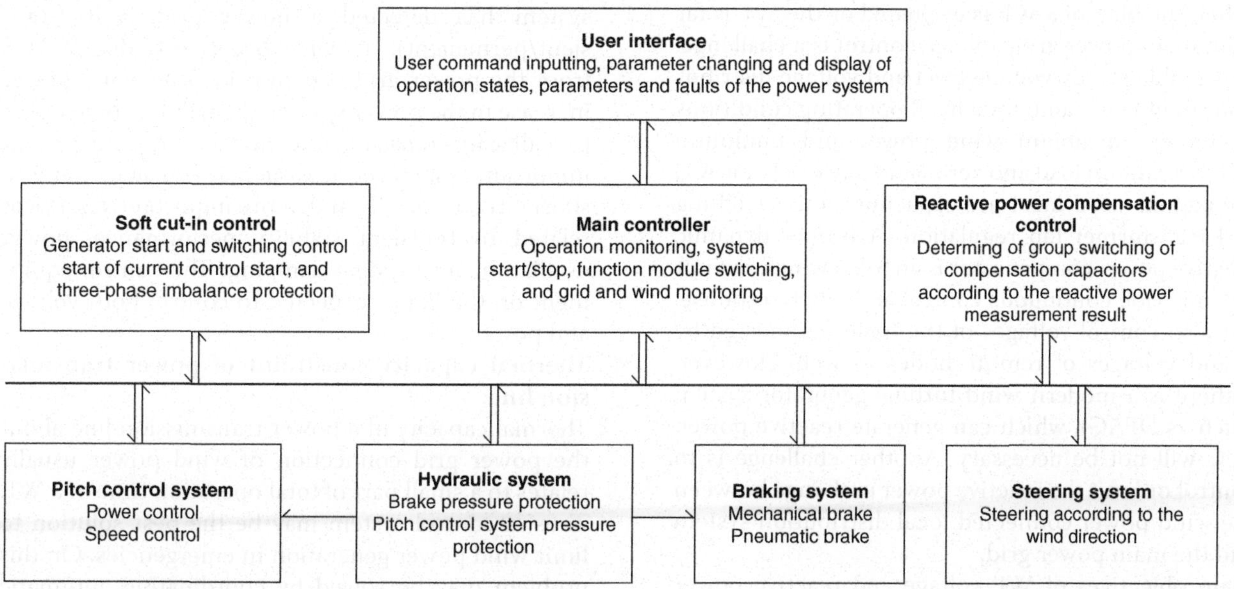

Figure 10.1 Structure of DCS.

10.1.1.3 Main Parameters Monitored By the Control System

During operation, the control system needs to monitor the following main parameters:

1) Electrical parameters—frequency and three-phase voltage of the power grid, three-phase output current and power factor of the generator;
2) Meteorological parameters—wind velocity, wind direction, air temperature and air pressure;
3) WTGS state parameters—speed (of generator and wind turbine), temperature (of generator controller, bearing and speeder oil), cable twisting, mechanical brake, nacelle vibration, and oil level (lubricating oil level and hydraulic system oil level);
4) Feedback signal—inter-blade spoiler recovery, mechanical brake release, yaw brake release, speed signals for the generator before and after grid disconnection, respectively, and commands given by the control system of the WF under intergrated monitoring.

10.1.1.4 Tasks and Objectives of WF Control

From the point view of the operation of a power system, reliable and effective control of the corresponding WFs is necessary for meeting power grid specification and increasing the wind power penetration level. The WF control related to power system operation can be divided into three kinds of control fields as below: ① active power control and grid frequency regulation; ② reactive power control and voltage regulation; ③ Fault Ride Through (FRT) capability and support for grids. The general objective is to maximize the wind energy fed into the power grid while safety operation of the power grid is guaranteed. To realize this objective, the WF may take the following control actions [4–6].

1) **Voltage control and reactive power compensation:** When the plan of a WF is evaluated or the WF is far from main power grid, voltage control is a challenge. Acceptable steady voltage level and voltage distribution should be maintained in all operating conditions (including maximum wind power and minimum load; maximum load and zero wind power). For a WF, the actions used most are capacitor bank switching and transformer tap regulation. A proper dynamic reactive power compensator is an SVC, which can not only provide continuous control and quick response, but also control voltages of the nodes connected by it and voltages of remote nodes as well. However, if there is a modern wind turbine generator system (such as DFAG), which can generate reactive power, SVC will not be necessary. Another challenge is to control or limit the reactive power exchange between the wind power connected local distribution system and the main power grid.
   Main objectives of WF voltage and reactive power control: ① To make the existing circuit reach the thermal limit current I_{th} in all wind power generation conditions; ② To maximize the active wind power fed into the power grid; ③ To limit voltage deviation of the WF within ±5% of the rated voltage (for a medium-voltage power grid); ④ To avoid or minimize net exchange of reactive power between the regional network and the main power grid; ⑤ To keep acceptable electric energy quality for each bus (connected with an ordinary load) in any wind power generation level; ⑥ To minimize electrical loss of the power collection system of the WF.
2) **Voltage stability improvement:**
   Voltage stability is possibly the main constraint for maximum rated capacity of a WF its connecting/disconnecting state. The typical reason of voltage instability or collapse in a power system failing to meet the reactive power requirement is that the circuit is heavily loaded and/or outage. Voltage stability is sometimes called load stability. However, it seems that wind power–related voltage stability is not related to the loads in power systems but to wind turbine characteristics. Wind power–related voltage stability may be regarded as voltage instability in short period because the output power of a wind turbine usually changes greatly in a couple of seconds, which shows the variation of the wind velocity at the wind turbine. Usually, voltage instability may happen when the active power fed into the power grid by the WF is much lower than thermal limit of the circuit. For lower voltage stability constraint, it is necessary to realize sufficient and quick reactive power compensation through the initiative reactive power control of the WTG (such as DFAG) or external reactive power compensation (such as SVC).
3) **Transient stability improvement:**
   Conventionally, after the wind turbine protection system has detected a power grid fault (transient/permanent), it will disconnect the WTGS from the power grid and stop it. With wind power increase in the power system, the wind turbine needs to realize for transient faults ride through and provide important system services such as transient standby, short circuit capacity and, more importantly, services related to transient stability performance, power oscillation, and system damping. The control equipment on the WF is expected to control both voltage and power.
4) **Thermal capacity constraint of power transmission line:**
   Thermal capacity of a power transmission line about the power grid connection of wind power usually relates to a small part of total operation time of a WF. Thus, a control system may be the best solution to limit wind power generation in emergencies. Or, this problem may be solved by coordinating automatic generation control (AGC) on the precondition that

there is a flexibly controllable power plant in the blocked area. For a power system containing both wind power and hydroelectric power, water storage capacity may be used for increasing control flexibility. In this case, AGC equipment may be used to control the power flow to avoid grid congestion (thermal capacity constraint).

5) **Power fluctuation—frequency control:**
Wind energy is essentially a fluctuating power source. For power systems where wind power forms the majority of the electricity generation, their operation problems such as frequency regulation and congestion management are challenging ones because of normal fluctuations of the wind power. To solve these problems, flexible and improved secondary power generation (frequency regulation) control or energy storage system control is necessary.
6) **Adverse effects of the interaction of power electronic converters:**
Modern wind turbines improve their operation flexibility and controllability through power electronic converters. However, adding power electronic applications may result in adverse and uncertain interaction for WFs. Thus, wind turbine converter regulation and filter design are of great importance and must be analyzed as a part of the design and installation of WFs.

10.1.2 WTGS Protection

There are different types of WTGs and their protection configurations and protection schemes are not exactly the same. This section introduces the protection of cage asynchronous WTGSs and double-fed asynchronous power generator systems [7–9].

10.1.2.1 Protection of Cage Asynchronous WTGs

As a typical modern WTG, a cage asynchronous generator can be connected with a power grid synchronously. It is directly connected with a public AC grid with help of a step-up transformer. The terminal voltage of the typical generator is 575 to 690 V (60/50 Hz; three-phase). Typically, the generator terminal voltage may rise to the distribution grid voltage, that is, 35 kV. The generator step-up transformer is a three-phase and oil-cooled (or dry type) one installed at the bottom of a wind turbine tower through gaskets; its primary-side voltage is 35 kV (star-earthed or delta connection); its secondary-side voltage is 575 V or 690 V (the winding is star-earthed). Impedance of the generator step-up transformer must be designed carefully to avoid possible resonance by the harmonic current of asynchronous generator. The typical impedance of the generator step-up transformer is 5% (based on a 750 kVA wind turbine) or 6% (based on 1,500 kVA or another capacity). The asynchronous generator has a variable slip ratio between 1% and 10%.

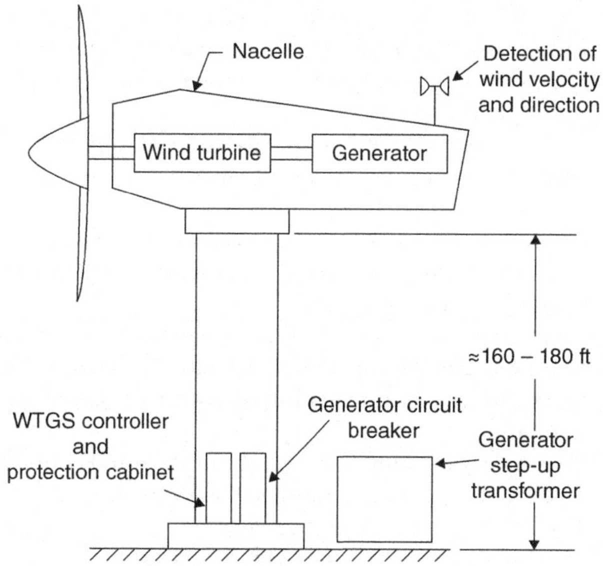

Figure 10.2 Typical wind power system.

A generator controller is a multifunctional digital controller (in the WTG controller and protection cabinet as in Figure 10.2). It has the following control and protection functions.

- Synchronizing the speed accorind to the power grid frequency and limiting the impulse current (cut-in current) in grid connection;
- Limiting the surge current in grid connection through a thyristor circuit;
- Making the cut-in current smaller than the rated load current;
- Switching the capacitor bank and calibrating the power factor according to output power of the WTGS;
- Tripping to protect the wind turbine in a power grid fault or disturbance;
- Providing the wind turbine/generator with relays for undervoltage/overvoltage protection;
- Providing the wind turbine/generator with relays for underfrequency/overfrequency protection;
- Integrating other wind turbine control relays and generator protection relays into the wind turbine/generator control system

The parameters monitored by a wind turbine and a generator include electric quantity, yawer, hydraulic pressure, environment, power transmission line, generator, phase compensation, thyristor, pitch, emergencies, energy input and output, rotor current control, communication, self test, and all kinds of event records. The following parameters are some typical ones:

- Electric quantity: Extremely high/low voltage and time out limits, extremely high/low current and time out limits, high/low frequency and time out limits, reverse (negative) power limit; tripping current of main circuit breaker and time out limit, lagging limit of high/low voltage and current, average voltage and current, leakage current (ground fault); asymmetry between grid voltage and grid current;

- Yawer: maximum/minimum yaw time, automatic yaw limit the generator overspeed when wind velocity is between 0 and 25 m/s;
- Environmental parameters: nacell control temperature, wind velocity attenuation, gearbox temperature, oil temperature, and so on;
- Generator: three-phase sequence, generator temperature (resistance temperature detector (RTDs)), rotor speed (RPMs), and matching of gearbox transformation ratio;
- Phase compensation: capacitor group size and control, power factor compensation feedback circuit;
- Thyristor: maximum temperature;
- Pitch: high/low limit setup, and power error

These relay protection and control functions can protect a wind turbine (or generator) system.

A generator step-up transformer, no matter it is delta-star earthed or star-earthed and-star-earthed, will be provided with a primary-side fuse for short circuit protection. The primary-side fuse is designed based on rated rated load current and surge current of thetransformer and provides fault protection for the secondary side. If the generator step-up transformer is in a nacelle and has a circuit breaker, the circuit breaker must be designed with the possible maximum short-circuit current (including the currents provided by the generators and regional grids connected with neighboring feeders). A single-phase grounding time overcurrent relay is provided to protect the generator step-up transformer. The maximum fault current contributed to the three-phase faults on the primary side of a generator step-up transformer by a typical generator is approximately four to five times of rated load current of the generator step-up transformer (generator).

Because of its characteristics, an asynchronous generator has voltage and frequency limits different from those of a synchronous generator. From the point view of voltage control, the limit for a WTG is ±10% of the generator's terminal voltage (690 V or 575 V).

The undervoltage/overvoltage protection function provided by a WTG system is main protection for the related WTGS and backup interconnection protection for the regional grid WTGS. According to the designs of different manufacturers, generators have different set values of operating voltages range and trip. Possible set values for the undervoltage/overvoltage protection of a WTG are as below:

- $\geq 20\% U_n$: trip after a delay of 0.1 s
- $\geq 10\% U_n$: trip after a delay of 6 s
- $\geq 11\% U_n$: trip after a delay of 0.08 s (internal power factor capacitor)
- $\leq 11\% U_n$: trip after a delay of 1 s
- $\leq 15\% U_n$: trip after a delay of 0.6 s
- $\leq 25\% U_n$: trip after a delay of 0.01 s

Figure 10.3 shows the voltage trip time. Note that system voltage envelope is different from set values of generator voltage protection. Terminal voltage of a WTG is not in the same position as the voltage of a public connection point or a regional grid's interconnection point. Terminal voltage of a WTG at the end of the long collecting feeder is larger than both the voltage of the WTGS at a substation and the voltage close to connection point of a regional system. Thus, some WTGs on a WF may trip upon an undervoltage or overvoltage, but the other ones may operate continually.

The underfrequency/overfrequency protection function of a WTG control system is the main protection for the related WTGS and the backup protection for interconnection between the WTGS and a regional grid as well. Generator frequency protection may comply with the (low-frequency) load-shedding frequency scheme of

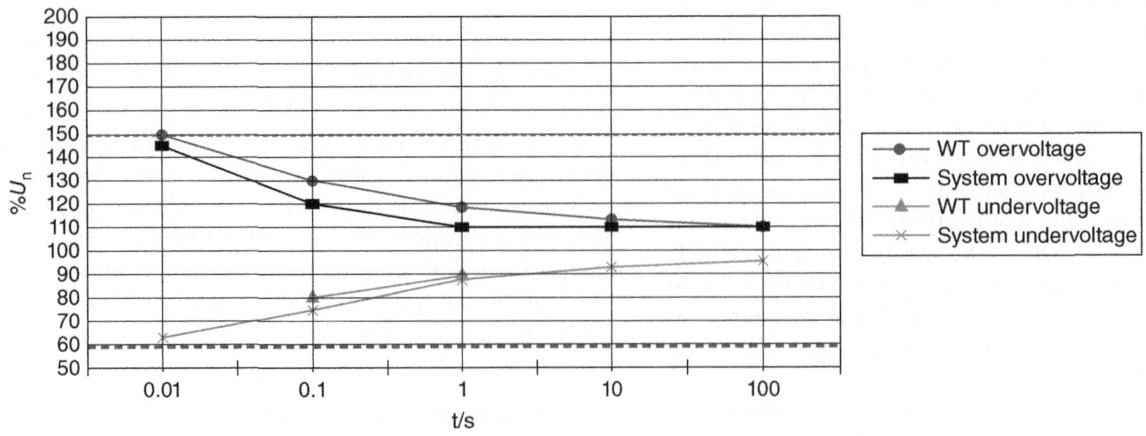

Figure 10.3 Operating voltage of a wind turbine and of a power system.

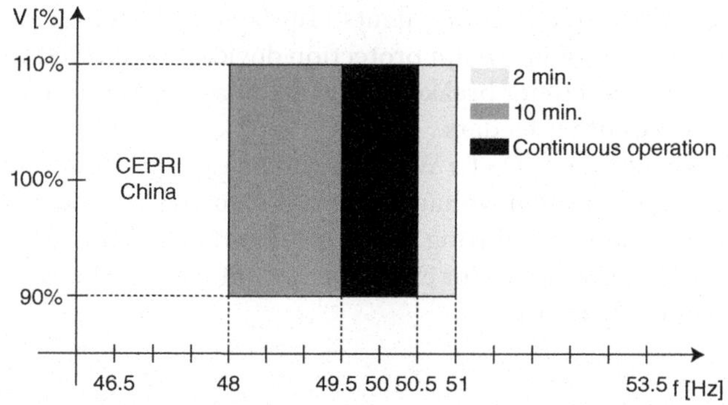

Figure 10.4 Setup of generator frequency protection and requirements of load-shedding frequency (50 Hz system).

a regional grid or not, depending on requirements of the local system (ISO). For a generator control system with a rated frequency of 50 Hz, the typical set values for WTG frequency protection are as below:

- ≥61 Hz: trip with a delay of 0.2 s and disconnect the power factor capacitor within 0.1 s;
- ≤48 Hz: trip with a delay of 0.2 s

These set values above are different from those of a typical synchronous power generator system. Figure 10.4 shows set values of the WTG for trips upon an underfrequency or overfrequency and the set frequency for possible grid load shedding.

WTG frequency protection may be compliant (coordinated) with the load-shedding frequency of a regional power transmission system. Frequency protection differs in the wind turbines designed by different manufacturers.

In interconnected systems, there is undervoltage and overvoltage protection, underfrequency and overfrequency protection (for eliminating power transmission line faults and substation faults, and isolation (islanding)) and other necessary protection for large system integrity.

In fact, the characteristic of WTG protection is backup protection for the equipment interconnected to systems. Thus, the protective devices of interconnected systems should be set more flexibly than WTG protection.

10.1.2.2 Double-Fed Asynchronous Wind Turbine Generator Protection

Figure 10.5 shows the single-line diagram of a double-fed asynchronous wind turbine generator (DFAG), including the configuration of intelligent electronic devices (IED).

A wind turbine with a variable pitch is connected with a wing asynchronous generator through a gearbox, and back-to-back voltage source converters are connected with the rotor circuit of a generator to provide variable speed control. A step-up transformer has three windings, including a high-voltage winding in delta wiring and two low-voltage windings in star wiring and grounding. The high-voltage winding is connected with a power grid through a circuit breaker while the low-voltage ones are connected with a stator winding circuit and a rotor circuit, respectively.

Due to the Delta/Star wound of the wind turbine transformer, ground faults of the stator on the low-voltage side of the wind turbine transformer cannot be detected by upstream protection devices. An IED provides protection for this kind of fault with help of a transient overcurrent component, which can respond to zero-sequence currents, residual (leakage) currents or transformer neutral currents. This component is not expected to cooperate with other protective components and it is allowed to operate with the minimum delay. If it successfully detects a zero-sequence current by the relation between the phase current and the residual current, possible saturation of the current transformer should be considered when the startup set value of an external fault is determined.

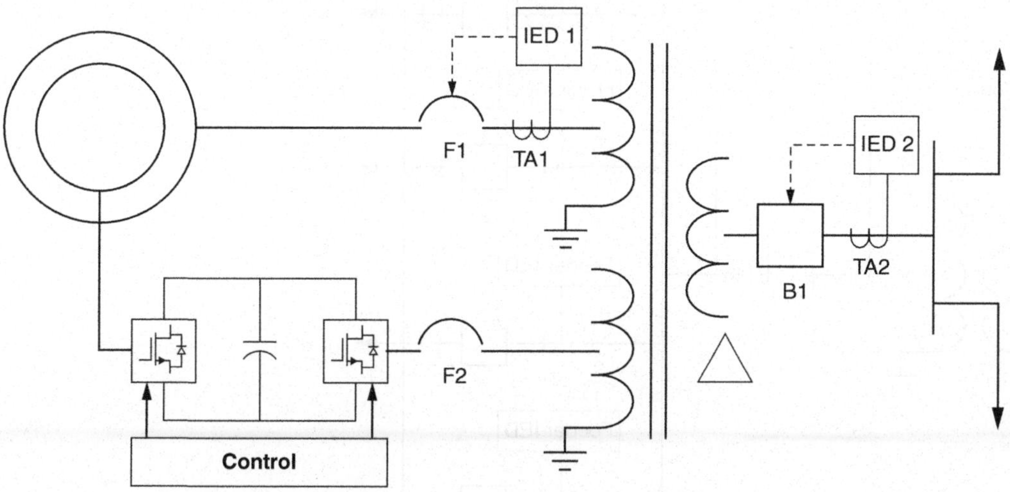

Figure 10.5 Single-line diagram of a DFAG.

IEDs provide low-voltage phase faults protection (interphase fault protection and phase-to-ground fault protection) too. A transient component cuts off a serious fault through a minimum delay. Note that a DFAG provides a short-circuit current for external faults. Current of the transient component should be set as a value lower than the expected minimum phase fault current of the generator terminal and higher than the expected maximum fault current provided by the generator in a power grid fault. A fixed-time overcurrent component detects internal phase faults of the generator, and the upstream fixed-time overcurrent protection devices should cooperate with it.

IEDs with similar protective components can be used in converter circuits too. They can not detect the faults in rotor windings but the faults at converter terminals.

A DFAG need some additional protection functions embedded in WTG controllers or realized in IEDs. These additional protection functions include:

voltage unbalance;
overheat (resistance temperature detector);
inverse phase shift;
out-of-step;
voltage/frequency out-of-limit

A WTG must isolate itself from feeder faults with a minimum delay as much as possible. External fault protection should never be enabled against the faults of neighboring feeders or WTGs. Particularly, it is impossible to realize this performance level through independent measurement of the local voltage and current. Usually, grid fault detection depends on low-voltage components and overvoltage components. These components need to delay until upstream protection devices are allowed to use feeder circuit breakers, in order to avoid trips due to faults of other feeders.

Finally, the IED of a WTG should be capable of capturing the data about voltage waveforms, current waveforms and time series during faults or disturbance. This fault analysis is valuable for ensuring the performance for protection systems.

10.1.3 WF Electrical System Protection

The electrical system of a WF consists essentially of step-up substation, switchgear, and multiple feeders, which is similar to the step-up substation of a conventional power plant. However, its WF wind turbines are more decentralized; its feeders are longer; it may be provided with multiple decentralized medium-voltage collecting buses. The electrical system of a big WF is a medium-voltage (10 to 35 kV) collecting grid. This section mainly introduces the protection of collecting feeders, collecting buses, interconnected transformers, and connection system circuits [7–10].

10.1.3.1 Protection of WF Substation

Figure 10.6 is the single-line diagram of a typical WF. Several feeders are connected with a collecting bus. The power transformer steps up the voltage to the transmission voltage. The WF is connected with a main grid through single-circuit high-voltage power transmission line.

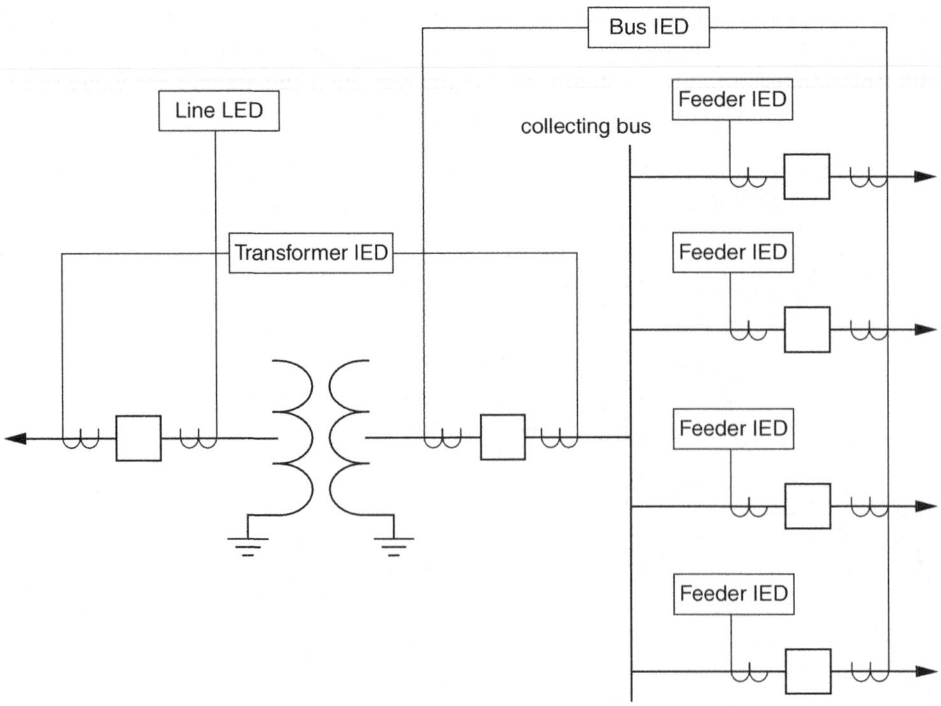

Figure 10.6 Single-line diagram of a typical WF.

The collecting buses need to be protected. Bus faults are eliminated with the minimum time by a high or low impedance differential component. If a low impedance bus differential solution is adopted, the TAs of feeders should not be in parallel connection, otherwise the fault current of the WTG will possibly operate wrongly because of TA saturation in an external fault.

A blocking scheme may be a substitute for the differential bus. Upon a downstream fault, the overcurrent component IED of each feeder will transmit a block signal to the overcurrent component in the transformer breaker IED. Upon a bus fault, no block signal will be transmitted. A substation LAN can provide a method for block information exchange easily.

The power transformer needs to be protected too. A 100% differential component form based on surge constraint is adopted. If there are a few feeders, the bus area and the transformer area may be protected by a multi-constraint transformer differential component combination to avoid using transformer circuit breaker and TAs.

The WF may be connected to a power grid through two output lines or tapped with a multi-terminal circuit too. In any cases, the power transmission line is typically protected by line differential form or a distance component, and the WF needs to be connected with a special communication channel of a remote electrical equipment terminal to obtain the best protection. Through the communication channel, a signal may be transmitted to the electrical equipment terminal to inform it that the WF has been disconnected and reclosing can be done. Out-phase reclosing on the WF must be avoided because it will result in a serious wind turbine torque transient state. Based on single-phase trip, ground faults can be reclosed. In this scheme, the WF is synchronous with the power grid with help of a non-defective phase, which increases the availability of the WF. However, a circuit breaker which can realize IED protection and a single-phase operation is needed.

10.1.3.2 Collecting Feeder Protection

The usual types of electric wiring topology of a WF are as below: radial type, branching radial type, feeder-subfeeder type, ring type, and so on (see Figure 10.7). Each type of the electric wiring topology above has its apparent advantages and disadvantages. When selecting the type of the electric wiring topology, factors such as wind distribution, layout of the available wind turbine towers, and cost must be considered.

Figure 10.7a shows a radial feeder topology. Single-loop feeders from a collecting bus are connected to the wind turbine towers successively. This is the simplest feeder structure. Cable faults or WTG faults inside the wind turbine towers may be isolated to continue the operation. However, this is the most insecure topology because all the generated power on the feeders will lose upon a fault of the circuit breaker B1 of the substation or a cable fault between the substation and the first wind turbine tower.

Figure 10.7b shows a branching radial feeder topology. This topology, similar to the radial feeder wiring topology, has two collecting feeders switched on and off through a collecting bus circuit breaker. It has the lowest installation cost for each feeder; however, it is also the most insecure because the power generated on the two feeders will lose upon a circuit breaker fault or cable fault between the substation and the first wind turbine power.

Figure 10.7c shows a feeder–subfeeder topology. It applies in a typical circumstance that wind turbines are distributed in a large area. It is usually made up of single-loop cable feeders feeding a remote multi-subfeeder switching station.

Figure 10.7d shows a loop feeder topology. Compared with other topologies, it provides high availability. It can operate continually with a single failed component. The faults in or between WTG towers can be isolated and the other WTGs can operate continually.

All types of collecting feeders are not connected with other loads but specially used for WTG connection. Furthermore, the interconnected area usually contributes more to the fault currents on the collecting feeders than the WTGs connected to these collecting feeders. Thus, collecting feeders could be treated as radial feeders. For collecting feeder protection, collecting feeders may be simplified as radial distribution feeders and the typical overcurrent protection may be adopted. However, in setting collecting feeder protection, the fault currents fed by the feeders and the neighboring feeders should be considered.

The typical WF topology above has an inherent limitation. The WF feeders are connected to a collecting bus and their voltages are increased to the transmission voltage through a transformer. The WF feeders are grounded to the reference ground of a medium-voltage collecting system with help of the neutral point of a substation transformer. High-voltage side of the WTG transformer uses delta wiring, so the WTG cannot provide a reference ground for the feeders. If the feeder circuit breaker is switched on during operation, the feeders and the operating WTG will be isolated into an isolated ungrounded system. In the case of a phase-to-ground fault on any of the feeders at this time, the voltage of non-fault phase will increase to the line voltage. For an onshore WF with overhead feeders, ground faults of feeder are the most common faults that may result in a serious transient overvoltage or even insulation failure and equipment damage.

To avoid the limitation above, an independent grounding transformer may be installed on each feeder to control the transient overvoltage condition. The grounding transformer may have a zigzag shape or

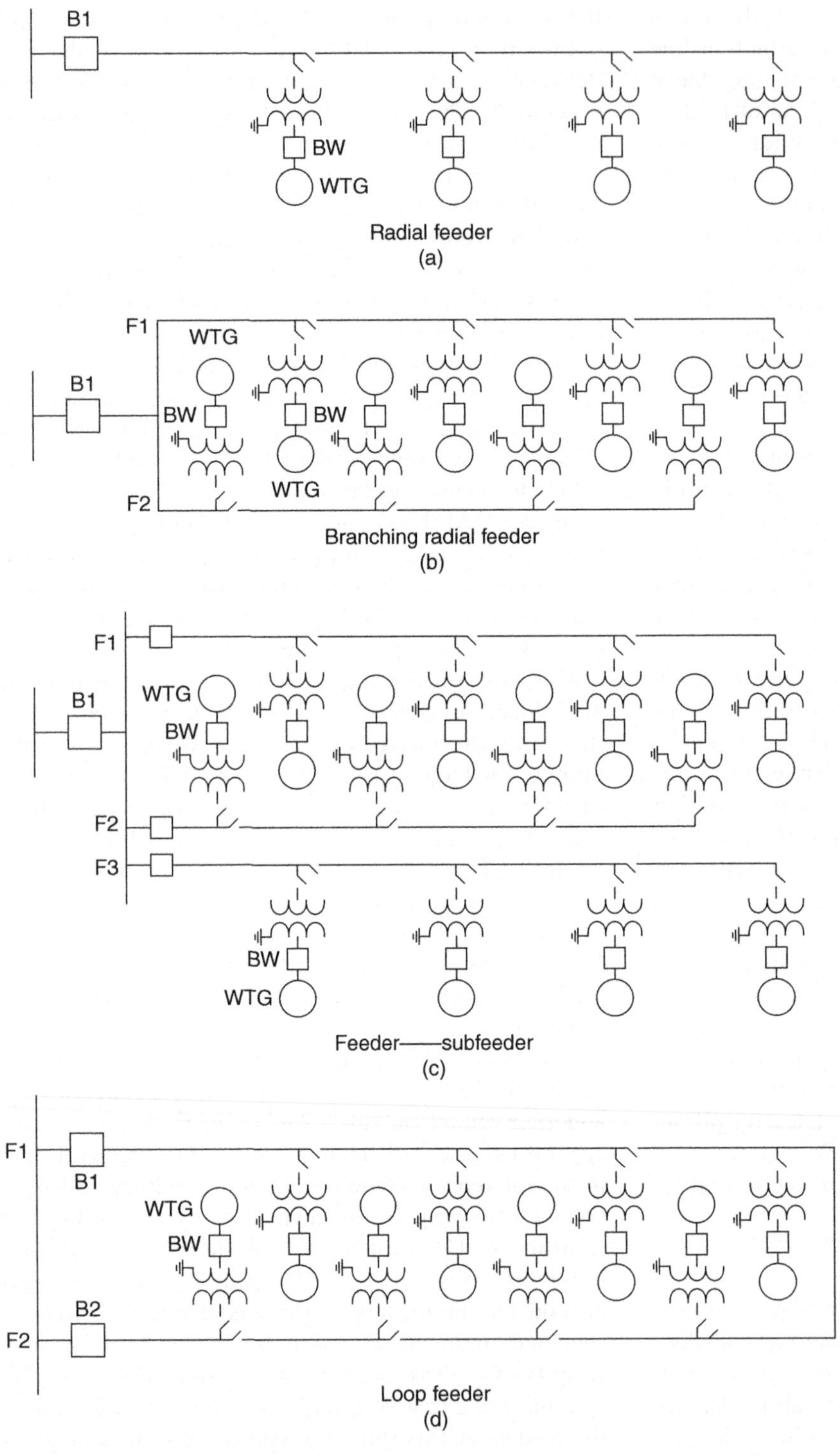

Figure 10.7 Common wiring topologies of feeders.

star-earthed-delta shape. Its ground resistance may be used for limiting the ground currents in feeder faults. However, this will result in higher equipment cost and engineering cost.

A substitute protective method is to make the WTGs disconnect from the feeders before trip of the feeder circuit breaker. However, the IED for feeder protection in the substation is the sole IED for detecting feeder faults selectively. In this case, this IED will transmit a transfer trip signal to all the WTGs on the feeders. After all the WTGs are disconnected, the feeder circuit breaker switched on will not generate a transient overvoltage. For ensuring a coordinated trip, the feeder circuit breaker will have a minimum delay when switched on. In article [7], this action is called transfer trip. Table 10.1 introduces time series of the feeder faults based on

Table 10.1 Time series of transfer trip.

Event No.	*Description*	*Time (ms)*
1	Feeder ground fault	0
2	Fault detection and transfer trip signal transmission by the Feeder IED	32
3a	IED of WTG receiving transfer trip signal and operating	8
4a	Trip of the WTG circuit breaker	60
	Duration of WTG disconnection	**100**
3b	Delay of the feeder IED	30
4b	Trip of the feeder circuit breaker	60
	Duration of feeder disconnection	**122**

transfer trip. In the table, the trip duration in the table is assumed as 60 ms; the duration of information processing and transmission between two ends of the power grid is 8 ms; the delay for trip of the IED from the feeder circuit breaker is about 30 ms to enable all the WTGs to disconnect before fault elimination. Delay of switching of the ethernet is not necessary in Table 10.1 but negligible.

10.1.3.3 Protection of Interconnection Step-up Transformer

The step-up main transformer for interconnection between a 35 kV collecting system and a main grid has a three-winding structure including a low-voltage winding and a high-voltage winding both with star wiring and a third winding with delta wiring. The third winding is used for stabilizing the neutral point of the step-up main transformer and providing a zero-sequence current for the ground faults on two sides of the step-up main transformer. To enable the relay on the power transmission line and the relay on the collecting feeder to operate, the transformer impedor (including its third winding) must be properly designed to provide a sufficient ground fault current. The MVA-rated value of the third winding is usually 35% of capacity of the step-up main transformer.

If the main transformer has delta wiring and star wiring, the grounding transformer may be installed on the delta wiring side of the main transformer to make primary neutral point of the main transformer stable. In another scheme, a zero-sequence voltage relay circuit is installed to detect displacement of the neutral point in a ground fault. However, in this scheme, the system will operate continually following ground protection of the power transmission line, and coordination with backup ground protection and coincidence should be considered. Usually, a main transformer with delta wiring and star wiring and an ungrounded neutral point is installed for each of the WFs of the same size.

10.1.3.4 Protection of Interconnection Power Transmission Line

If a WF is interconnected with a regional grid, its generators should be treated like synchronous generators because synchronous generators can provide a large fault current in several cycles. Protection of interconnection power transmission linemay include the following schemes partially or completely [8].

- Direct transfer trip (transfer trip because of circuit breaker failure);
- Interconnection line differential protection or guide line (differential) protection;
- Segmented distance protection and backup overcurrent delay protection;
- Undervoltage/overvoltage protection;
- Underfrequency/overfrequency protection;
- Circuit breaker failure protection and synchronization

These protection schemes are a part of the compulsory technology requiements for a regional grid. They may differ for different WFs, depending on the related regional grids.

10.2 Active Power and Frequency Control

An important problem for WFs is active power and frequency control. In the active power and frequency control of a conventional power system, output active power of the generators is controlled to track load changes of the power system to maintain the rated system frequency and realize the plan to exchange power with other interconnected power systems. Active power and frequency control is also called AGC (see section 10.5) and load frequency control. In a power system with wind power, particularly a power system with massive wind power, its active power and frequency control is more difficult because of the randomness of wind power. To solve this problem, it is necessary to strengthen the active power control capability of WTGs and realize coordinated active power and frequency control for WFs as well.

10.2.1 Frequency Control

In a power system, its frequency indicates the balance or unbalance between power generation and consumption. For a normally operating power system, its frequency should be close to its normal value. In a power system with a power frequency of 50 Hz, the normal frequency should be usually 50 Hz (deviation: ±0.1 Hz) (the requirement for frequency quality in the related national standard: 50±0.2 Hz) or 49 Hz to 50.3 Hz at most.

If power generation is not balanced with power consumption, first and secondary frequency control should be done. For example, if power consumption is larger than power generation, the rotational energy stored in the large synchronous generator will be used for keeping balance between power generation and consumption and, consequently, the generator speed and the system frequency will both decrease. In a power system, some WTGSs have frequency-sensitive devices. These WTGSs are called first frequency control WTGSs. They are used to increase their power generation until a balance between power generation and power consumption and a stable frequency. The first frequency control lasts for about 1 s to 30 s. The secondary frequency control is for recoverying the frequency to the normal value and release the first backup used, lasting from 10 to 15 minutes. The power generation increase or decrease by the secondary frequency control is slow. In some countries, AGC is adopted; in some other countries, the secondary frequency control is realized manually upon request of a dispatcher.

In normal operation, the output of a WF will change 15% of the installed capacity within 15 minutes, resulting in additional unbalance between power generation and consumption of the power system. In or after an extreme wind condition, the wind power generation possibly changes greatly.

With power generation technologies different from those adopted by conventional power plants, wind turbines have limited capabilities in the first frequency control compared with conventional power plants. However, some electric power company possibly needs a WF to have 3% to 5% capability of primary frequency control (as a thermal power plant does) and another electric power company possibly needs a WF to participate the secondary frequency control. Upon overfrequency, this may be realized by stopping some WTGSs on a WF or controlling the pitch. Wind is uncontrollable, so a WF needs to make its WTG power lower than the possible generation power intentionally under normal frequency. In other words, it needs to take partially loaded operation, in a bid to provide secondary control under low frequency.

10.2.2 Active Power/Frequency Control System of a WF

The active power control of a WF can be realized by several sub-control equipments all characterized by an impact on the set points of WF active power. According to the set point of WF, active power controller of the WF acts on set points of each wind turbine to enable the wind turbines to operate in the active power control range. An active power regulator for WFs developed by Gamesa is introduced below [4]. The active power regulator (APR) has several advanced devices supporting grid stability in aspects such as increasing and decreasing sequcence control that upgrades the climbing capability to two times, and an optimal feedforward regulation for WF power control. As in Figure 10.8, the task of SCADA part of APR is to detect the regutation that is going to be obtained and make sure that this regulation will not pass the limits, before wind turbine releases setting point on a WF; the wind turbines have a task too: to detect and make sure that the set points required could be obtained within the wind turbine limits. The APR is embedded with a wind turbine self-protection mechanism called PQTV limitation (P: active power; Q: reactive power; T: temperature; V: voltage). The measured values in point of common coupling (PCC) include active power (P), reactive power (Q), apparent power (S), and frequency (f).

Active power of WF connected with a large grid or operating independently through regulation can influence the PCC frequency. The active power regulator developed by Gamesa is a special type focusing on frequency control (see Figure 10.9).

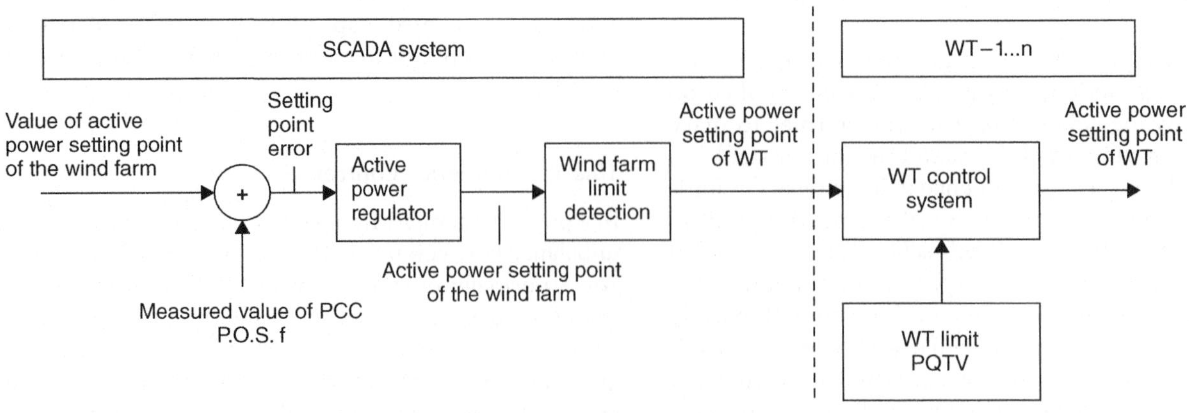

Figure 10.8 Active power regulator and its equipment configuration.

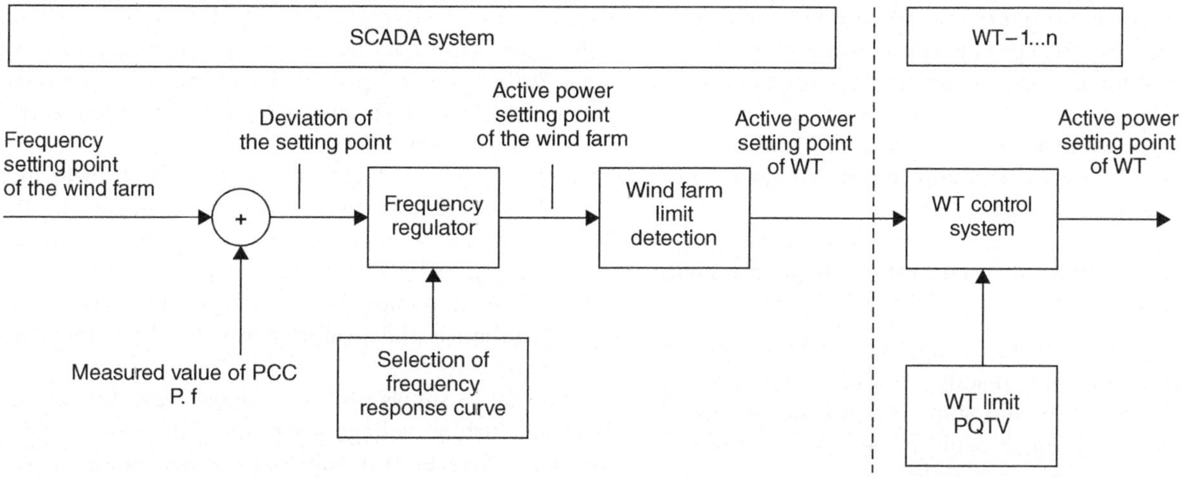

Figure 10.9 Frequency regulator and its equipment configuration.

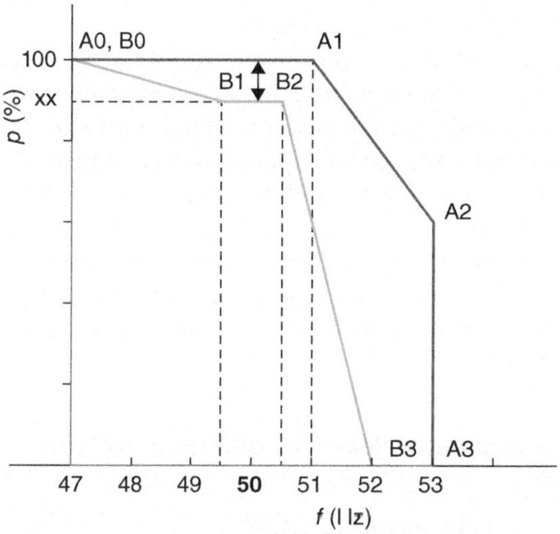

Figure 10.10 Frequency regulator, two examples with frequency response curves.

The frequency regulator needs an input that can respond to the frequency deviation of PCC. Figure 10.10 introduces two examples of frequency response curves. There may be five different frequency response curves at most.

In addition to the necessary frequency and active power control, the active power regulator can also balance control, including control over spinning reserve capacity and control over available active power and gradient (control over active power increase/decrease rate; see Figure 10.11).

The limiting point in different regulation modes is the response time of updating setting points by the WF regulator and the SCADA system. It relates to loads and layout of the communication facilities. In a wind turbine, the frequency operation limits are given by the electrical component used.

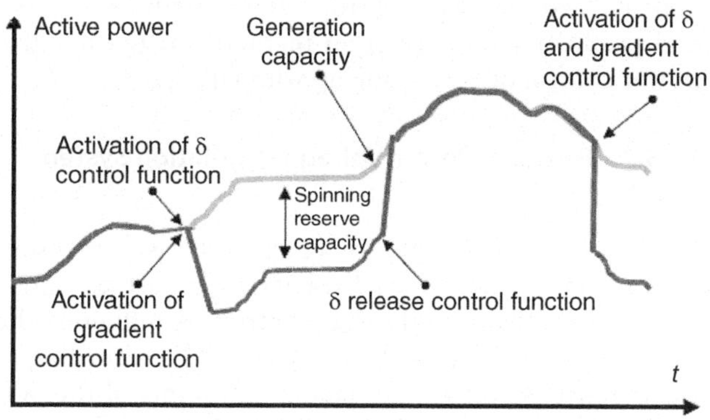

Figure 10.11 Active power regulator, WF balance control.

10.3 Reactive Power and Voltage Regulation

10.3.1 Overview

Another important issue of a WF is cut-in or cut-out reactive power control and voltage regulation. Section 3.4 in Chapter 3 analyzed the voltage control capabilities of three main kinds of WTGs and reaches a conclusion that it may be reasonably believed that the wind turbines of a WF based on DFAG can participate in electrical variable control, particularly reactive power control. For the reactive power output and system voltage control of WFs, different countries and electric power companies have proposed power grid specifications that are not exactly the same. In the power grid specification compiled by ESCOSA (Essential Services Commission of South Australia), the following six requirements are proposed for a WF.

① In the case of full power, high-voltage side of the WF transformer should keep the power factor of ±93%.
② Fifty percent of the power factor correction capacity should be dynamic (with quick response).

③ Reactive power output of the WF should be in direct proportion to active power output of the WF.
④ Voltage of the power transmission system should be adjustable.
⑤ The WF should be kept from trips (LVRT and HVRT) upon faults (voltage drops) and high voltages of neighboring transmission grids.
⑥ Voltage of the power transmission system should be recovered to at least 90% of the rated voltage after fault elimination.

The requirements ① to ④ may be met by means of determination of total reactive power compensation capacity and dynamic and static reactive power compensation of a WF which could be seen in Section 5.5 in Chapter 5. This section mainly introduces the methods to meet the requirements ③ and ④, the ordinary method of coordinated reactive power control for WFs and the two algorithms (proportional-integral method and proportional distribution method) for the reactive power control of WFs made up with DFAGs.

10.3.2 Reactive Power/Voltage Regulation System of a WF

Reactive power control is realized by several sub-controllers of the same characteristics, just as active power control does. The sub-controllers influence the reactive power input and output of a WF. According to the reactive power (power factor) set points of the WF, the reactive power regulator (RPR) distributes reactive power set points to wind turbines within the control range of reactive power controller (see Figure 10.12).

The WF reactive power regulator of Gamesa supports grid stability by means of voltage control and apparent power control. In addition to external reactive power compensation equipment, the reactive power compensators of wind turbines can be used for quick and accurate reactive power control. Economically and technically, the basic reactive power compensation of a big WF should be provided by external equipment such as STATCOM, shunt reactor (SR) or static reactive power compensator (SVC).

Upon a large voltage deviation, the transformer tap will be operated firstly. Voltage regulator of the WF can minimize operation times of the transformer tap and keep the PCC voltage in the specified range strictly, which avoids large voltage fluctuation and grid disturbance and prolongs lives of the connection equipment.

Figure 10.13 shows that the voltage regulator acts on the wind turbine voltage set points. The wind turbine controller detects and controls the set values within the operation ranges of the wind turbines. The voltage regulator is embedded with a self-protection mechanism called PQTV limitation (P: active power; Q: reactive power; T: temperature; V: voltage).

The active and reactive power regulator of a WF has a special operation mode supporting PCC apparent power control or system apparent power tracking control for the WF. Figure 10.14 introduces apparent power control. The control strategy of a WF regulator containing apparent power control has the optimal active power in the reactive input or output range in Figure 10.14. For example, if the wind velocity increases, operating point trace of the WF will move from operating point 1 to operating point 2 along the dot dash.

10.3.3 Coordinated Reactive Voltage Control of a DFAG WF

10.3.3.1 Control Structure of a DFAG

What is important for DFAG modeling based on different wind velocities is to note that a wound-rotor asynchronous generator has a feedback on both its rotor

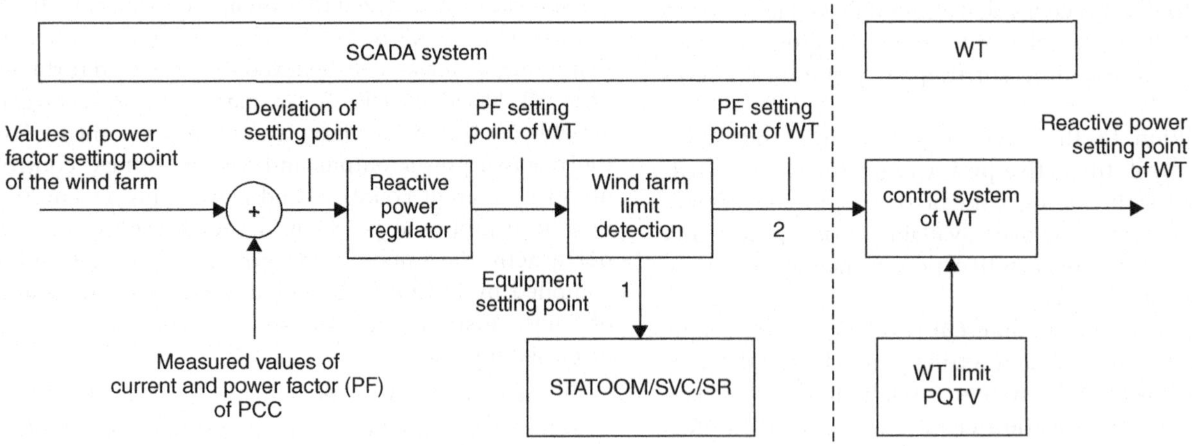

Figure 10.12 Reactive power regulator and its equipment configuration.

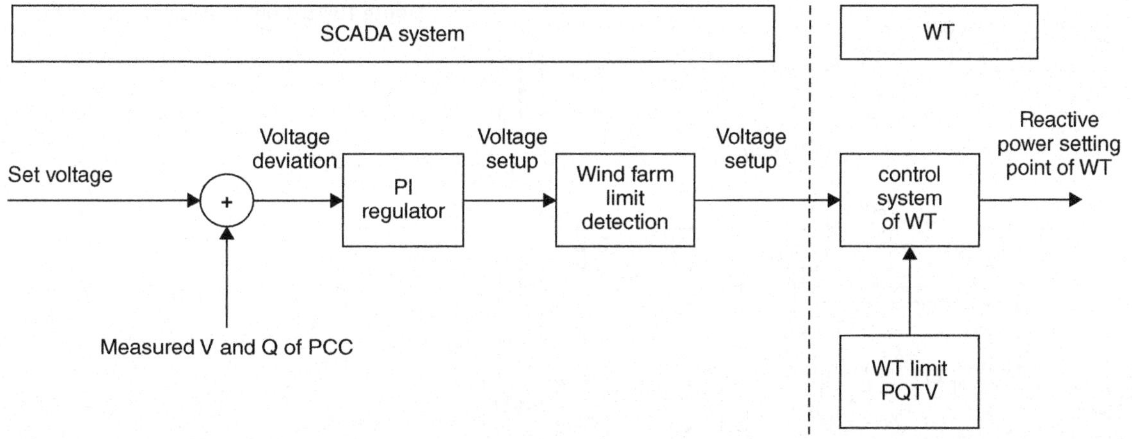

Figure 10.13 Voltage regulator and its equipment configuration.

Figure 10.14 Apparent power control mode of a WF regulator.

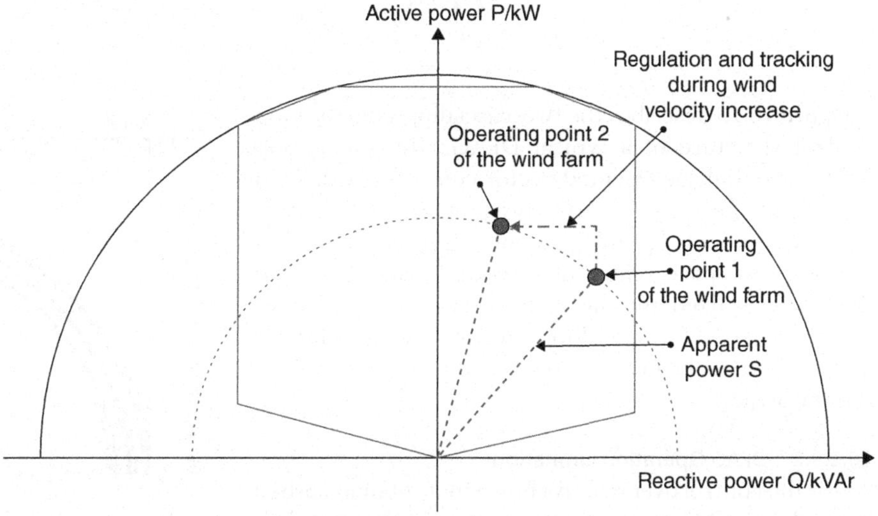

side and stator side. Generally speaking, the stator of a DFAG is connected with the power grid directly and its rotor is connected to grid through variable-frequency power electronic converters. The back-to-back PWM converters can process the power flows in two directions to enable the DFAG to operate between the subsynchronous speed and the super-synchronous speed. We can analyze operating principle of a DFAG through the classical revolving-field theory, the well-known dq model and "3→2" and "2→3" coordinate exchanges.

In order to process dynamic characteristics of the DFAG with the most practical method, both voltage components and current components on the stator side should refer to a fixed coordinate system and those on the rotor side should refer to a coordinate system rotating at the rotor speed during modeling. After that, the three-phase current and the voltage should be transformed into those in a dq form with help of the Clarke transformation method. Thus, starting from the common dynamic three-phase model for generators, the DFAG model can be described with the following matrix formula (Formula (10.1)).

$$\begin{bmatrix} u_{sD} \\ u_{sQ} \\ u_{r\alpha} \\ u_{r\beta} \end{bmatrix} = \begin{bmatrix} R_s + pL_s & 0 & pL_m \cos\theta_r & -pL_m \sin\theta_r \\ 0 & R_s + pL_s & pL_m \sin\theta_r & pL_m \cos\theta_r \\ pL_m \cos\theta_r & pL_m \sin\theta_r & R_r + pL_r & 0 \\ -pL_m \sin\theta_r & pL_m \cos\theta_r & 0 & R_r + pL_r \end{bmatrix} = \begin{bmatrix} i_{sD} \\ i_{sQ} \\ i_{r\alpha} \\ i_{r\beta} \end{bmatrix} \tag{10.1}$$

In the formula: L_s and L_r are self-induction coefficients of the stator side and the rotor side; L_m is the mutual-induction coefficient; R_r and R_s are resistances of the stator side and the rotor side; θ_r is the rotor angle (variable); $p = d/dt$ is a differential operator. In Formula (10.1), voltage component and current component should operate alternately to process the voltage into an independent variable, that is, a system input variable, and the current into a dependent variable, that is, a system output variable, to correspond to a voltage type DFAG.

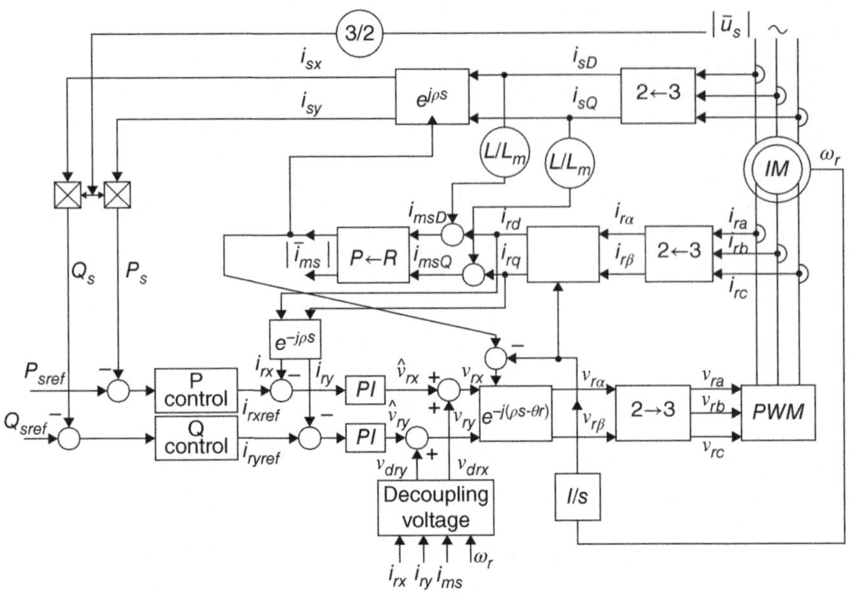

Figure 10.15 Control structure of a DFAG.

Figure 10.15 describes the two-cascading-control-loop control structure of a typical DFAG. The inner loop (stator flux linkage oriented vector control) is expected to control the rotor current direction i_{rx} and orthogonal component i_{ry} independently (represented with reference to space vector of the flux linkage of which the shaft is fixed on the stator). Two same PI controllers are used for inner loop control. The outer loop is expected to control active power and reactive power of the stator side.

10.3.3.2 DFAG Operation Limitation

As for the control over reactive power output or absorbed through a DFAG, there is a problem directly related to the temperature (heat) on the rotor side. In order to obtain the permissible operation limits, there should therefore be a perfect analysis about heat and electromagnetism of the DFAG. Figure 10.16 describes a typical multi-limitation PQ curve of a DFAG.

According to Figure 10.16, the calculation of safe operation limits may be realized through an eccentric semicircular formula. For a consideration of limitation of active power generation capacity and reactive power generation capacity, expressions about the stator side's active power and reactive power directly related to the rotor current component must be established (Formula (10.2)).

$$
\begin{aligned}
P_s &= \frac{3}{2}|u_s|i_{sy} = -\frac{3}{2}|u_s|\frac{L_m}{L_s}i_{ry} \\
Q_s &= \frac{3}{2}|u_s|i_{sx} = \frac{3}{2}|u_s|\frac{L_m}{L_s}(i_{ms} - i_{rx}) \\
&= \frac{3}{2}|u_s|\frac{L_m}{L_s}\left(\frac{|u_s|}{\omega_1 L_m} - i_{rx}\right) \\
Q_s - \frac{3}{2}|u_s|\frac{L_m}{L_s}\frac{|u_s|}{\omega_1 L_m} &= -\frac{3}{2}|u_s|\frac{L_m}{L_s}i_{rx}
\end{aligned}
\tag{10.2}
$$

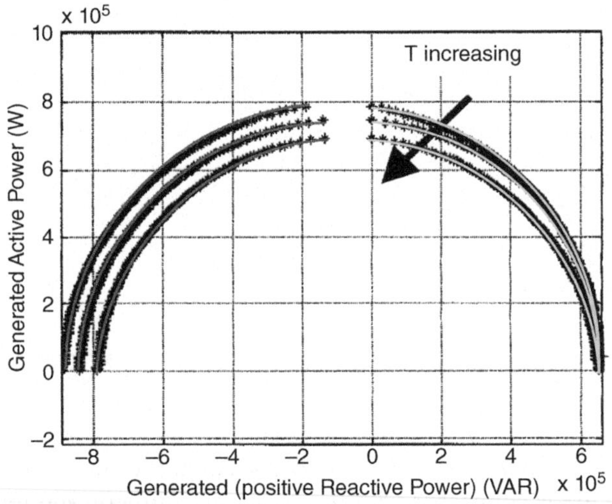

Figure 10.16 Typical active/reactive power limitation.

Formula (10.3) about safe operation limit curve of the generator can be obtained on the basis of Formula (10.2) and eccentricity.

$$
P_s^2 + \left(Q_s - \frac{3}{2}|u_s|\frac{L_m}{L_s}\frac{|u_s|}{\omega_1 L_m}\right)^2 = \left(\frac{3}{2}|u_s|\frac{L_m}{L_s}\right)^2 (i_{r\max})^2 \tag{10.3}
$$

According to Formula (10.3), the operation limits relate to the current on the rotor side directly. According to the safety level needed, the semicircle radius will decrease because the maximum current $i_{r\max}$ on the rotor side has decreased.

Figure 10.17 shows the structure of the WF model with several WTGs (quantity: n) connected with power grid [11].

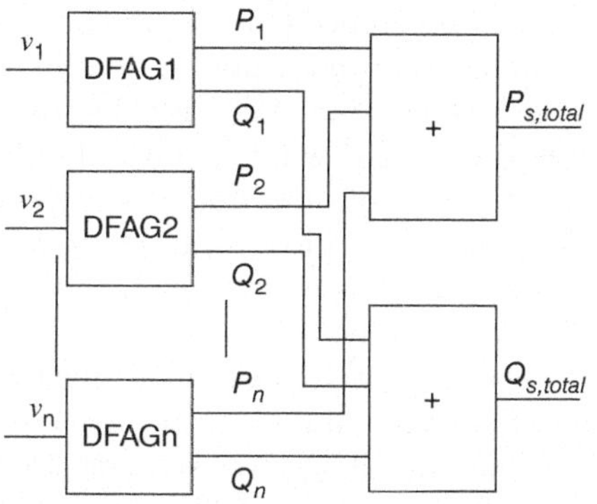

Figure 10.17 WF model.

Each generator has quite different wind velocities (some generators possibly disconnect from the power grid while some other possibly operate with the maximum outputs), total active power and reactive power of the WF is still sum of the active power and reactive power of all wind turbines on the WF (Formula (10.4)).

$$P_{s,total} = \sum_{i=1}^{n} p_i$$
$$Q_{s,total} = \sum_{i=1}^{n} q_i \tag{10.4}$$

In the formula: $P_{s,\ total}$ is sum of the active power transmitted from the WF; $Q_{s,\ total}$ is sum of output/absorbed reactive power of the WF; p_i is active power of the generator i; Q_i is the output/absorbed reactive power of the generator i.

10.3.3.3 Proportional Distribution Algorithm of Reactive Power

In the first algorithm, the reactive power necessary for the WF is, in consideration of the PQ operation limit of each generator (such as that defined in Formula (10.3)), distributed proportionally to all the operating generators. This means that each of the generators operates safely with a specific power factor. Additionally, in this distribution algorithm, the operating point of each active power transmitted by almost each generator has a big gap with the reactive power generation limit. Thus, from the point view of maintenance, the DFAG will be well protected.

In the proportional distribution algorithm, it is necessary to measure the operating state of each generator (the active power transmitted and the P/Q limit) and then the reactive power needed will be obtained. The algorithm is based on the control law in Formula (10.5) [12].

$$q_i = \frac{Q_{SREF}}{\sum_{i=1}^{n} q_{i\max}} \times q_{i\max} \tag{10.5}$$

In the formula q_i is the reference reactive power of each generator calculated; $q_{i\max}$ is the maximum reactive power transmittable by each generator; Q_{SREF} is reference reactive power of the WF.

Steps of the proportional distribution algorithm:

1) Measure the active power transmitted by each generator and consider its P/Q limit, obtain $q_{i\max}$ and read reference reactive power Q_{SREF} of the WF.
2) Calculate the reference reactive power q_{iref} of each generator with the proportional distribution algorithm (Formula (10.6)).

$$q_{iref} = \frac{Q_{SREF}}{\sum_{i=1}^{n} q_{i\max}} \times q_{i\max} \tag{10.6}$$

3) Transmit the reference reactive power q_{iref} to each generator at the same time.
4) Compare the WF's actual reactive power and reference reactive power Q_{SFARM} and then return to step 1).

10.3.3.4 PI Control Algorithm

In the proportional distribution algorithm, the power factors necessary for almost all generators of a WF are within their power generation limits. However, if they are applied in a practical WF, the calculation will be massive. Furthermore, if a wind velocity changes suddenly, the proportional distribution algorithm will be related to the system performance sensitivity.

According to a simulation test based on the actual data of a WF with DFAGs, the WF made a first-order dynamic response upon a power factor step change. Thus, the proportional-integral algorithm is selected in consideration of first-order dynamic response of the WF and improvement of robustness of the control algorithm upon a wind velocity change. For the safe operation of generators, this algorithm actually considerates resisting the impact of wind velocity increase. Steps of the proportional-integral algorithm are as below:

1) The power factor necessary for the WF is calculated as below based on parameters of the power distribution grid and the set active power and reactive power determined by the grid company (it is believed that P_s must be an optimal value and only the reactive power changes with the set value of power factor) (Formula (10.7)).

$$\cos\varphi_{REF} = \frac{P_s}{\sqrt{Q_{SREF}^2 + P_s^2}} \Rightarrow \varphi_{REF} = \arctan\frac{Q_{SREF}}{P_s} \tag{10.7}$$

2) Measure the actual output active power P_s and reactive power Q_s of the WF and calculate actual power factor φ_{FARM} of the WF (Formula (10.8)).

$$\cos\varphi_{FARM} = \frac{P_s}{\sqrt{Q_s^2 + P_s^2}} \Rightarrow \varphi_{FARM} = \arctan\frac{Q_s}{P_s} \tag{10.8}$$

3) Define the power angle deviation (Formula (10.9)).

$$e_{\varphi} = \varphi_{REF} - \varphi_{FARM} \tag{10.9}$$

4) Use the Tustin discrete method (resistation of wind velocity increase is considered) on the PI controller (Formula (10.10)).

$$\varphi_{REF}(k) = \varphi_{REF}(k-1) + k_p \left(1 + \frac{T_s}{2T_i}\right) e_{\varphi}(k) + k_p \left(\frac{T_s}{2T_i} - 1\right) e_{\varphi}(k-1) \tag{10.10}$$

In the formula, T_s is the sampling period; k_p is proportional constant of the PI controller; T_i is integration constant of the PI controller.

5) Make sure the φ_{REF} obtained through calculation is within the PQ limit (Formula (10.11)):

$$\begin{aligned} &if \quad \varphi_{REF}(k) < \varphi_{REF,min} \quad \varphi_{REF}(k) = \varphi_{REF,min} \\ &if \quad \varphi_{REF}(k) > \varphi_{REF,max} \quad \varphi_{REF}(k) = \varphi_{REF,max} \\ &else \quad \varphi_{REF}(k) = \varphi_{REF}(k) \end{aligned} \tag{10.11}$$

6) Transmit $\varphi_{REF}(k)$ to each generator of the WF as its set value.
7) Return to step 5).

According to document [12], a simulation test based on the two control algorithms (proportional distribution algorithm and PI control algorithm) was done on a WF in northern Spain to specially detect robustness of the PI controller upon sudden wind velocity changes. It is believed that each of the algorithms has its advantages and disadvantages. In the proportional distribution algorithm, all the generators operate in similar conditions and are kept from saturation as well as the reactive power limit, but they do not show great robustness upon sudden wind velocity changes. In the control algorithm based on a PI controller, there is great robustness upon sudden wind velocity changes, but some WTGs operate with their limits (saturation) sometimes because the reactive power needed is not distributed proportionally in this algorithm. According to result of the simulation test, the PI control algorithm has better effect and robustness for the WF analyzed.

10.4 Fault Ride-Through (FRT) or Low/High-Voltage Ride-Through (L/HVRT) of WTGS

10.4.1 Overview

According to the power grid specification, a WF should have a function of fault ride through (FRT) or low/high voltage ride through (H/LVRT) and it should be able to respond to a system fault within only several milliseconds. To make FRT function close to power generator system control, FRT can be embedded into the whole generator control system or a special power electronic converter. In addition, well coordination with electrical and mechanical limitations of the wind turbines should be done in a very short period of time. The power electronic converter system has a complete control function for meeting detailed requirements of the existing power grid specification for the FRT function.

Technically, there are several methods for realizing the FRT function. These methods and their respective cost performance vary in different wind power systems. Two common solutions for the FRT function are introduced below. One of the solutions is based on active Crowbar, made up of a power electronic device in the rotor circuit and expected to connect wind turbines with the power grid during low-voltage faults. The other solution is made up of a super big rotor converter and expected to make the rotor converter operate continually during low-voltage faults. The converter system can operate like a synchronous generator to control the excitation and power of the DFAG and its activation or release result in no interruption [4].

The FRT–time curve limits are related to the wind turbines installed, the electric power infrastructure layout, and the electrical equipment connected. The limits must be determined by calculation of the wind turbine control system in different conditions. The voltage to be detected is usually voltage U_{PCC} of the PCC between the WTGS and the power grid.

The FRT–time curve should be between limits of the WF and wind turbines and set values of the circuit breaker and the relay protection. The power grid operator should suggest on the location of the FRT–time curve of each WF, including the minimum voltage and its duration, the voltage recovery speed (creep speed), etc.

In FRT, LVRT is more common and important than HVRT because there are much more voltage drops than voltage rises during grid disturbances or faults.

10.4.2 LVRT of WTGS

LVRT is a course in which a wind turbine can still be connected with the power grid or even provide the power grid with reactive power to support its recovery to ride through the low voltage duration (area) when voltage of the power grid connection point of a wind turbine is dropping. In other words, LVRT characteristic and control objective of the wind turbine during the power grid voltage drop should meet the following requirements: the wind turbine can still be connected with the power grid, provide the power grid with reactive power and support voltage recovery during the fault and restore normal operation soon after fault elimination [13, 14].

A power grid voltage drop will result in some transient state processes for generator operation, such as overvoltage, overcurrent and speed rise, endangering safe operation of wind turbines, and their control system. Generally speaking, if the power grid has a fault, the related wind turbine will disconnect from the power grid immediately to protect itself passively without considering duration and severity of the fault. It is acceptable when the wind power penetration level (proportion of the wind power generation in total power generation of the power grid) is low. However, if the wind turbine still disconnects from the power grid as above in a voltage drop when the proportion of wind power in the power grid is large, the fault will get worse because of combination of the power grid disturbance and the wind power disturbance and the whole system will be more difficult to recover or even other wind turbines of the system will all disconnect from the power grid. Thus, an effective LVRT action must be taken to keep grid stability.

The LVRT capacity is very important for voltage stability keeping, particularly when there is partial, large, and centralized wind power. Premature trips of many WTGs because of partial grid disturbance will possibly increase the risk of system instability and make grid disturbance more influential.

LVRT characteristic can be realized by making the trip conditions variable by means of control modification, so as to make low voltages allowed during grid faults. Besides, wind turbines must be capable of operating for time long enough with a lower supply voltage.

The essential requirement for LVRT is that wind power equipment should keep operating with most of faults which can be eliminated normally. This is particularly important when a WF is connected with a weak grid because wind turbines of the WF need to provide voltage support and power on loads of the line terminals in this case. However, this LVRT requirement will increase the costs of generator systems because the realization of LVRT is usually by adding equipment or modifying control strategy.

At present, there are three types of wind turbines on the market, including fixed speed induction generators (FSIG) with direct grid connection, direct-drive permanent magnet synchronous generators (DDPMSG) and DFAGs. This section introduces the LVRT technology and its requirements, analyzes the transient responses of wind turbines of the three main types above to grid voltage drop faults, summarizes some main LVRT realization schemes and mainly discusses the LVRT schemes of DFAGs.

10.4.2.1 Introduction to LVRT

LVRT is a specific operation function requirement for grid-connected wind turbines to keep grid connection in grid voltage drops. Figure 10.18 shows the typical LVRT rule for a wind turbine. The LVRT characteristic is defined through the minimum voltage in the

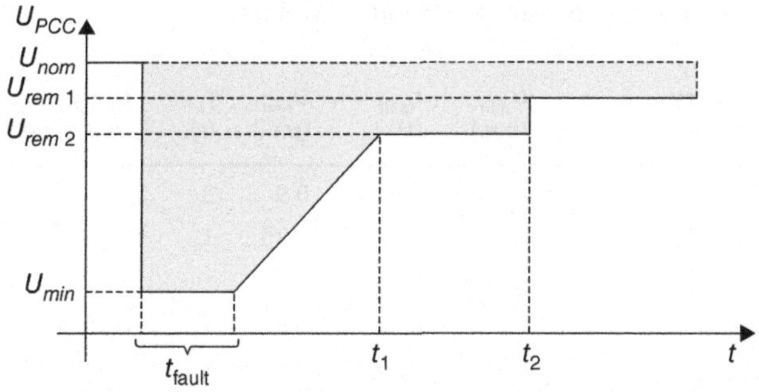

Figure 10.18 Typical LVRT rule of a wind turbine.

whole fault duration and voltage recovery to the normal level after fault elimination. A wind turbine must keep grid-connected operation in the grey area, but it may disconnect from the power grid upon any voltage below the gray area. After fault elimination, the voltage should rise and recover to a voltage close to the voltage before the power grid disturbance. Figure 10.18 specifies: ① the minimum voltage U_{min} and duration t_{fault} that the WF or wind turbine must withstand: value and duration of the minimum voltage depend on the relay protection technology used and both the location and type of fault. Duration of the minimum voltage is the duration when the power grid fault is normally eliminated (10 to 20 cycles). In some power grid specification, the minimum voltage should be zero, in other words, a wind turbine should have the function of zero voltage ride through (ZVRT). ② The speed of voltage recovery after fault elimination: it should be quicker between completion time of fault elimination and t_1 while it should be slower between t_1 and t_2. The voltage recovery speed depends on strength of the power grid and support capability of the reactive power: The stronger the power grid, the quicker the voltage recovery and the lower the ride through requirement for the generators will be. ③ The final voltage after recovery: that is, smaller than the rated voltage by less than 10%. ④ That a WF should provide the PCC with an active power in direct proportion to the voltage and offer the maximum reactive power to avoid overload of the power electronic converters of the wind turbines. Note that the specifications in the figure are suitable for the grid connection points of a WF and that voltage drops are less at the terminals of wind turbines in the WF because of the expected voltage support by them. Thus, the set value for low voltage relay protection of a wind turbine may be $0.3U_n$ (or usually $0.8U_n$ when there is no LVRT).

In the eyes of a power grid operator, the most important parameters for the LVRT rule are the minimum voltage in the grid-connected operation that a wind turbine must keep in a power grid fault and its duration. The specifications about connection of wind power into grids in different countries (regions) have different LVRT requirements (see Table 10.2).

Table 10.2 Comparison of different LVRT rules.

Country (region)	U_{min} (p.u.)	t_{fault} (s)	U_{rem1} (p.u.)	t_1 (s)	U_{rem2} (p.u.)	t_2 (s)
Canada AESO	0.15	0.625	0.9	3	—	—
Canada	0	0.15	0.75	1	0.85	2
(Hydro Québec)	0.25[a)]	0.75[a)]				
Spain	0.2	0.5	0.8	1	0.95	15
Ireland	0.15	0.625	0.9	3	—	—
UK	0.15	0.14	0.8	1.2	0.85	2.5
Germany[b)]	0.0	0.15	0.7	0.15	0.9	1.5
Germany	0.25	0.1	0.75	0.75	1	10
Brazil	0.2	0.5	0.85	1	0.9	5

a) Distant fault
b) WTGS with big symmetric short-circuit current components

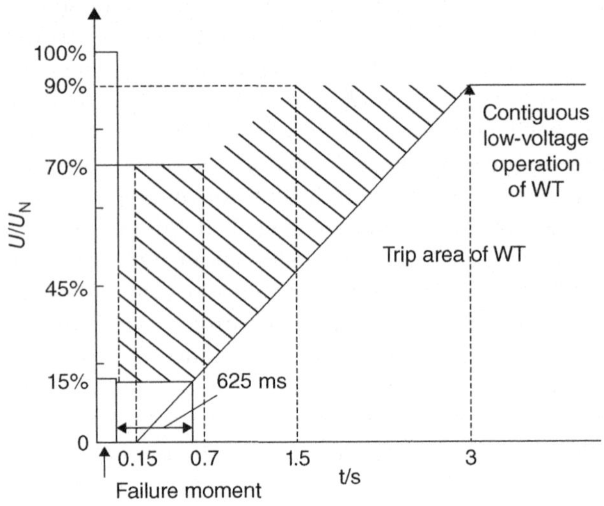

Figure 10.19 LVRT requirements of E.ON Netz.

In northern Germany, the wind turbines are dense. Figure 10.19 shows the LVRT requirements of the power grids here for WFs and WTGSs. According to the figure, when the voltage has dropped by 15% to 45%, a wind turbine should keep providing reactive power support and keep grid-connected operation for at least 625 ms; when the voltage has dropped by more than 90%, a wind turbine should keep grid-connected operation all the time. The speed of reactive power increase is 5% or 20% of the rated power per second, depending on the instant voltage. The speed of voltage rise is approximately 0.227 p.u./s.

10.4.2.2 Response of WTGs to Grid Voltage Drops

The precondition of LVRT studies is analysis of some transient state processes of wind turbines in grid voltage drops and recovery. The wind turbines of different types (such as FSIGs, DDPMSGs, and DFAGs) have different transient states and impacts [13, 14].

FSIGs and DFAGs are all connected with grids directly on stator side. Due to this direct coupling, grid voltage drops are reflected directly in voltages of the stator terminals of generators, resulting in direct current components in stator flux linkage and negative-sequence components in asymmetric faults, which have large slips relative to the rotating rotor (the slip frequencies are close to ω_s and $2\omega_s$, respectively, and ω_s is the synchronous angular frequency), thus resulting in a large rotor electromotive force and a large rotor current and significant increase of voltage and current of the rotor circuit as well.

For FSIGs, their capabilities of providing electricity for grids are limited by grid voltage drops, thus resulting in large currents. Usually, they have great robustness and overload capacity and they can avoid being burned upon short-time large currents. However, the unbalance between their mechanical power and electrical power results in their acceleration. After fault elimination, they possibly cannot return to their speeds before faults because of deceleration failure and then trip because of overspeed protection, depending on their inertias, rated capacities, and fault duration.

For DFAGs, their rotor sides are connected with AC/DC/AC converters and their power electronic devices have limited overvoltage and overcurrent protection capabilities. If no control actions are taken in grid voltage drops to limit the fault currents, the larger transient rotor currents will endanger the weak power electronic devices; however, rotor current control will result in higher direct voltages of the converters and then damage of the converters; additionally mismatching between input power and output power of the converters will possibly make the direct voltages increase or decrease (it is due to the the fact whether DFAGs operate at a super-synchronous speed or a subsynchronous speed during faults). In the transient processes of grid faults (particularly asymmetric faults), the electromagnetic torques of DFAGs change a lot, generating impacts for transmission parts such as wind turbine gearboxes and influencing the operation and lifetime of wind turbines. During stator voltage drops, DFAGs have lower output powers. If the captured power by wind turbine is not restrained, there will be mismatching between the electrical power of grids and the mechanical power input and surely speed rise of DFAGs. In the case of large wind velocities, that is, large driving torques of wind turbines, the electromagnetic torques of DFAGs will increase even after fault elimination and limiting further speed rise of DFAGs as well as increase of absorbed reactive power could be difficult, resulting in decrease of stator terminal voltages, more serious obstruction for rid voltage recovery or even system crash because of grid voltage recovery failure. This relates to inertia, rated values, and fault durations of DFAGs. Thus, the LVRT function of DFAGs, which are restricted by rotor-side converters, are more difficult to be realized than that of other generators in direct grid connection or other direct-drive power generation technologies.

For DDPMSGs, there is no direct coupling between generators and grids and stators are connected with grids with help of AC/DC/AC converters. The transient drops of grid voltage results in decreasing of output power but the output power of DDPMSGs keep constant simultaneously. In this case, the power mismatching will increase the direct voltages of converters and endanger the safety of power electronic devices. The allowed value of direct voltage increase depends on the rated values of converters and the capacitors of direct buses. If control actions are taken to stabilize direct voltages, the currents output to grids will be larger, endangering the safety of converters too. If the direct voltages of converters change in a range, the converters on the generator side can usually be controllable, so converters can realize control over generators during grid voltage drops. DDPMSGs still operate in normal operation mode and have no overspeed because grid faults are almost invisible for them. Therefore, the LVRT functions of direct-drive synchronous generator systems are easier to be realized than DFAGs.

10.4.2.3 Methods of LVRT Realization

1) **LVRT realization for FSIGs**

   During voltage drops, the main problem of FSIGs is quick speed rise as a result of electromagnetic torque attenuation. The simplest method is, after the fault is reliably confirmed, to increase pitch angles quickly by means of quick pitch varying during and after a fault, so as to reduce the input mechanical torques and limit the speed rise. However, wind turbine blades respond slowly because of large inertia, so the effects of pitch control are restricted by the time constants of pitch control related to fault durations.

   Another disadvantage of variable-pitch control is the failure to provide reactive power to support grid recovery because operating FSIGs need to absorb reactive power from grids. The usual method to reduce reactive power absorption is to install capacitor banks in line with maximum output power. However, for the occasions with large output power fluctuations in wind power generation, this method will result in system voltage fluctuations and overvoltage of neighboring buses upon faults. Therefore, a better method is to use a static reactive power compensator (SVC) to realize dynamic reactive power compensation to reduce voltage fluctuations in stable operation and support grid voltages and strengthen FRT during grid voltage drops. Compared with SVCs, static synchronous compensators (STATCOM) have the characteristics of synchronous compensators and their compensation currents are independent from voltages of corresponding connection points and do not decrease during voltage drops when used for reactive power and voltage regulation, so they perform well in support for grid voltages.

2) **LVRT realization of PMSGs**

   During voltage drops, the main problem of permanent magnet synchronous generators (PMSG) is the increase of direct voltages as a result of energy mismatching. These problems can be solved by means of storage or consumption of residual energy through proper actions. The costs, the power grid specifications and the fault depths and durations should be considered for desciding the effect of this kind of design.

   One action is to increase the rated values of converters, in other words, select power electronic devices with higher rated voltages and currents and increase the rated voltages of DC capacitors in converter design. In this way, direct voltage limits can be made higher during voltage drops to store residual energy and currents of the inverters on grid sides can be increased to output more energy. However, there is limited potential for rated value increase of devices because of a cost consideration. Additionally, long-time faults and serious faults may result in serious power mismatching and exceeding of capacities of devices.

   Another action is to decrease the generation power of PMSGs to realize power balance. For example, if grid voltage drop is detected, that is, the set electromagnetic torque of a synchronous machine is reduced, the PMSG will have speed rise so that partial input energy of the wind turbines can be stored by means of temporary speed rise and output power of the PMSG will be lower. Variable-pitch control is not necessary for minor faults but when limits of speed rise or energy storag are reached or speed rise of PMSGs is too large, variable-pitch control can be used directly. Variable-pitch control can reduce the input power of wind turbines fundamentally and benefit the power balance during voltage drops. If used with the method of device capacity increase, this strategy can further improve the margin of FRT.

   For deep faults with longer durations, residual energy can be stored or consumed through additional circuit units. Figure 10.20 introduces two methods to realize LVRT through an external circuit unit. In Figure 10.20a, an energy storage system (ESS) is connected to the DC bus of a converter. In this method, if an excessively high direct voltage is detected, IGBT of the ESS will be triggered to transfer the residual DC energy and the stored energy will be fed into a power grid after fault elimination. In Figure 10.20b, a

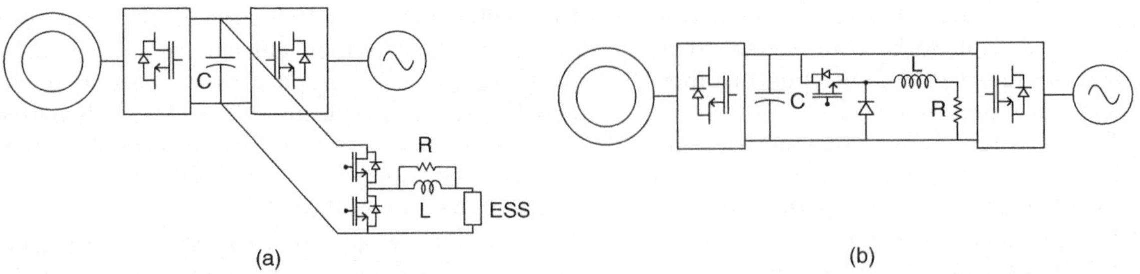

Figure 10.20 Two LVRT realization methods for PMSGs. (a) Connection with an energy storage system (ESS); (b) Connection with a resistor (R) for consumption of residual energy.

Buck converter is used and the residual energy stored in the DC capacitor is consumed by a resistor (R) directly.

3) **LVRT realization of DFAGs**
Compared with FSIGs and PMSGs, DFAGs are endangered most during voltage drops because the transient overcurrents and overvoltages of rotors will damage power electronic devices and the attenuation of electromagnetic torques will result in speed rise.
In consideration of existence of undervoltages and overvoltages on DC buses, ESSs may be connected to DC buses like PMSGs do, so as to keep voltage stable for DC buses. This control scheme based on energy management mainly focuses on keeping direct bus voltages of AC/DC/AC converters and has no direct relation with the LVRT characteristics of DFAGs.
Figure 10.21 introduces two main ride-through technologies for DFAGs, including one based on a short circuit protection circuit (Crowbar) (a) and the other based on a converter with an uninterruptable power supply (UPS) and a lower rated capacity (b). The power flows between DC buses and UPS should be a two-way one. UPSs serve as the "sources" or "pools" of active power, depending on the fact that DFAGs operate at a subsynchronous speed or a super-synchronous speed. This means more complicated structures but more developable control for converters.
If the currents during faults are larger than rated values of the rotor side converters, the Crowbar circuits will provide bypasses. For ride through, Crowbar circuits should be based on forced reversing switches or thyristor switches with active shutdown circuit. If DFAGs operate in a subsynchronous state, it will be more complicated because Crowbar will short rotor windings again and then DFAGs will need rotor side power. In this case, DFAGs will operate in a subsynchronous state like cage asynchronous generators do, in a bid to be accelerated to a super-synchronous speed. Even if there are active Crowbar circuits, DC buses probably still need UPS or DC bus capacity increase to keep DC bus voltages and provide quick recovery after fault elimination.

Several typical LVRT realization schemes for DFAGs are introduced below [14–17]:

1) **Control strategy based on dynamic compensation for the stator voltages of DFAGs**
LVRT characteristic can be realized by making the trip conditions variable by means of control modification. During a power grid voltage drop, the direct current component and the negative-sequence component both in the stator flux linkage will result in large electromotive forces in the rotor circuit and the frequencies will be ω_r and the sum of ω_s and ω_r, respectively (ω_s is the synchronous angular velocity, whereas ω_r is the rotor angular velocity). The rotor circuit has small leakage inductance and small resistance, so a large electromotive force will surely generate a large current in the rotor circuit.
To reduce the impacts of changes of the stator flux linkage on the rotor circuit, dynamic compensation control may be done for the stator flux linkage or, in other words, leakage flux linkage in the DFAG may be controlled to balance out the impacts of the transient current and the negative-sequence component both in the stator flux linkage on the rotor side [7, 18, 19]. The flux linkage relation between the stator and the rotor is as below (Formula (10.12)):

$$\psi_r = \frac{L_m}{L_s}\psi_s + \frac{L_s L_r - L_m^2}{L_s} I_r \approx \psi_s + (L_{s1} + L_{r1}) I_r \tag{10.12}$$

In the formula, ψ_r is the rotor flux linkage; ψ_s is the stator flux linkage; L_m is the excitation inductance; L_s is the stator inductance; L_{s1} is the stator leakage inductance; L_r is the rotor inductance; L_{r1} is the rotor leakage inductance; I_r is the rotor current. Thus, in a fault, the rotor current I_r may be controlled to make the direction of the rotor current be opposite to the direct current component and the negative-sequence component both in the stator flux linkage to reduce to some contents or even avoid impacts of the stator flux linkage on the rotor flux linkage.

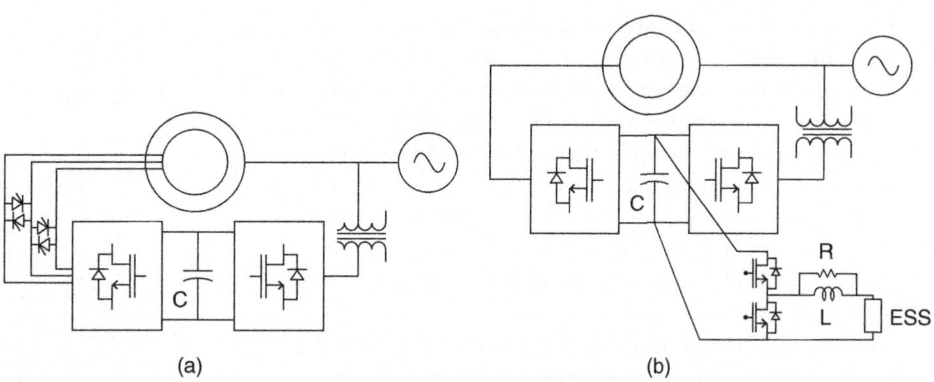

Figure 10.21 Possible LVRT realization of DFAG. (a) Short circuit protection (Crowbar); (b) Connected with an energy storage system (UPS).

Compared with the induced electromotive force caused by a voltage drop, degree of the voltage support provided by the converter is very crucial. In the same conditions, when a voltage drop occurs, a generator at a super-synchronous speed needs a larger rotor terminal voltage than at a subsynchronous speed; the larger the stator voltage drop is, the larger the rotor terminal voltage necessary for rotor current control will be and the more difficult rotor current control will be; in a fault, a heavy load is more difficult to be controlled than a light load, whereas for reactive power control, the control voltage decreases with increase of reactive current. The increased control bandwidth of the current control loop can effectively restrain the peak of rotor current during dynamic transition.

According to Formula (10.12), a converter's compensation capacity for rotor flux linkage changes with help of a rotor current is affected by the stator and rotor leakage inductance: the larger the leakage inductance is, the larger the compensation capacity (the LVRT capacity) will be, and vice versa.

2) **LVRT control strategy based on rotor short circuit protection (Crowbar)**

The rotor short circuit protection (Crowbar) technology began to be early used for the rotor converter protection of DFAGs. It is divided into passive and active Crowbar. For passive Crowbar, which is the so-called "thyristor (SCR)" Crowbar, the rotor Crowbar is triggered upon a specified grid fault limit, thus protecting related wind turbines. It is completely a self-protection Crowbar, providing no support for failed grid voltages or for immediately recoverying power supply to the power grid after grid fault elimination. In active Crowbar, a turn-off device such as GTO or IGBT can be used as a substitute for the SCR in the thyristor Crowbar circuit; when protection is needed, the two-way switch will be triggered to access the bypass resistor to the rotor circuit of the DFAG. Figure 10.22 shows the structure of passive Crowbar or, if the thyristor SCR is replaced by an IGBT, the structure of active Crowbar.

It is important to select a proper resistance value for the bypass protective resistor. If the resistance is too small, the bus voltage will not exceed its limit, but there will be a large current and a large peak of transit torque so that the rotor current will not be limited; if the resistance is too large, there will be a smaller transient current and a smaller torque peak, but the rotor converter will have an overvoltage on its rotor side and not be protected. Furthermore, steady-state operation characteristics of the DFAG after Crowbar are influenced by both the Crowbar resistance accessed to the rotor circuit and degree of the power grid voltage drop.

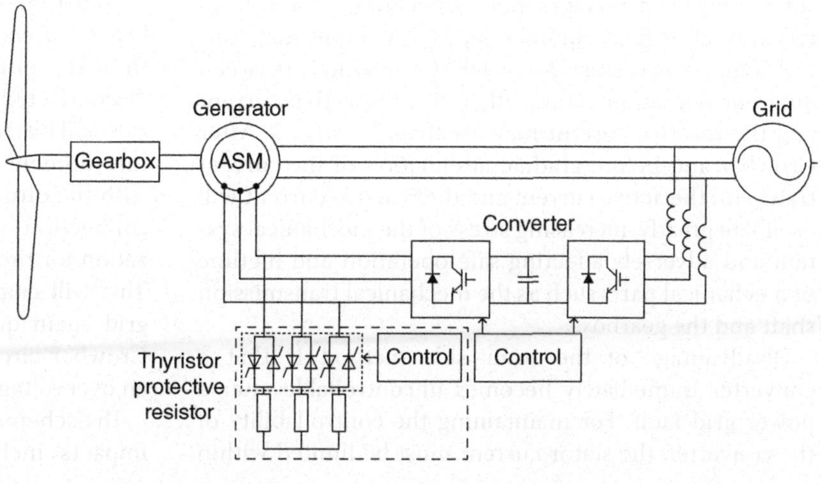

Figure 10.22 Crowbar protection.

A voltage drop will result in rapid speed rise. In this case, an emergency variable-pitch control scheme can be used to increase pitch angle of the wind turbine quickly at a speed of 5° per second, so as to decrease output mechanical torque of the wind turbine quickly, slowing down speed rise of the DFAG, reducing the inductive reactive power absorbed by the DFAG from the power grid during grid voltage recovery after fault elimination and lowering the obstruction for grid voltage recovery.
After Crowbar, operation state of the DFAG is similar to that of a cage asynchronous generator (CAG). When rotor current and rotor converter direct voltage of the DFAG are lower than the set protective values, the control over the DFAG should be recovered to switch from the fixed speed operation state of an CAG to the variable-speed operation state of a DFAG. However, there will be serious current oscillation during Crowbar disconnection and thisswitching needs a logic sequence and a corresponding control strategy, otherwise the Crowbar protection circuit of the DFAG will operate again. Therefore, for smooth switching between the two operation states, a strict logic plan is necessary, including not only control over the active power andthe WTG speed but also the DFAG reactive power for supporting grid stability.

3) **LVRT control strategy for STI**
There are two strategies for LVRT equipment locations, that is, rotor-side method and stator-side method. The rotor-side method is mainly for converter protection. In this system, combination of the Crowbar or Crowbar structure on a rotor side is properly designed to transfer the successive energy on a converter direct bus to adapt to FRT [20].

Rotor Crowbar control can avoid grid disconnection during grid faults. However, this strategy has three main advantages as below: ① After Crowbar operation and before system controllability recovery, a generator with a rotor circuit shorted by rotor Crowbar becomes a load absorbing inductive reactive power as a CAG does, not supporting grid voltages but obstructing grid voltage recovery after fault elimination; ② A complicated control logic is necessary for a DFAG to switch between different operation states, otherwise the active current and the reactive current may oscilliate greatly; ③ After Crowbar and before gradual attenuation of the current transient, the active current and the reactive current will oscilliate greatly, increasing stress of the mechanical system and adversely affecting safe operation and lifetime of mechanical parts such as the mechanical transmission shaft and the gearbox.

Disadvantage of the rotor side method is that a converter immediately becomes uncontrollable after a power grid fault. For maintaining the controllability of the converter, the stator current must be limited within a proper range to ensure the safter of the converter. A method to realize this is to interrupt the stator current completely and immediately after a power grid fault to make the generator effectively disconnect from the power grid for some time. This function is realized through the power electronic switch in the stator circuit. The "short-term interruption (STI)" strategy and scheme is a part of it. As in Figure 10.23, the STI divides LVRT into four areas [17].

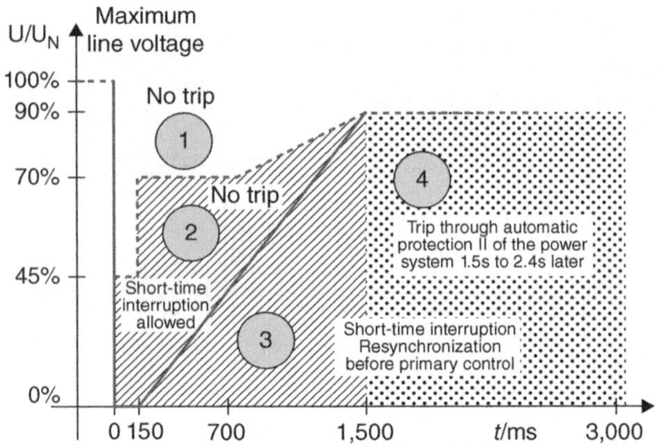

Figure 10.23 LVRT requirements of STI.

A WTG must be in a grid-connected state in areas ① and ②, but it may get disconnected from the power grid for STI control after the power grid voltage has dropped to area ② and a wind turbine faces overload, poor stability or another technical problems; if the power grid voltage has dropped to area ③ because of a power grid fault, then a wind turbine will be influenced significantly and the wind turbine may get disconnect from the power grid temporarily for STI control; if the low voltage duration is longer than 1.5 s, in other words, the power grid voltage is in area ④, the wind turbines are allowed to trip under protection of the system.

Figure 10.24 shows the topology of a DFAG based on STI control scheme. In the figure, a commutation unit (CU) and a stator side switch (SSS) are added. Upon an overcurrent during grid voltage recovery, the SSS will be locked to enable the stator of the DFAG to disconnect from the power grid within 10 s. After the stator has disconnected from the power grid, the CU on the stator side will be triggered and the rotor converter will stop. At this moment, the rotor circuit of the DFAG will connect with the direct bus through the diode in inverse parallel connection with IGBT, inorder to force the demagnetization for the DFAG together with the CU on the stator. This will enable the DFAG to connect with the power grid again quickly after grid disconnection. The rotor Crowbar circuit is used for backup protection to avoid an overvoltage for the direct bus during transition.

The scheme based on STI can avoid a DFAG's adverse impacts including oscillation and absorption of the

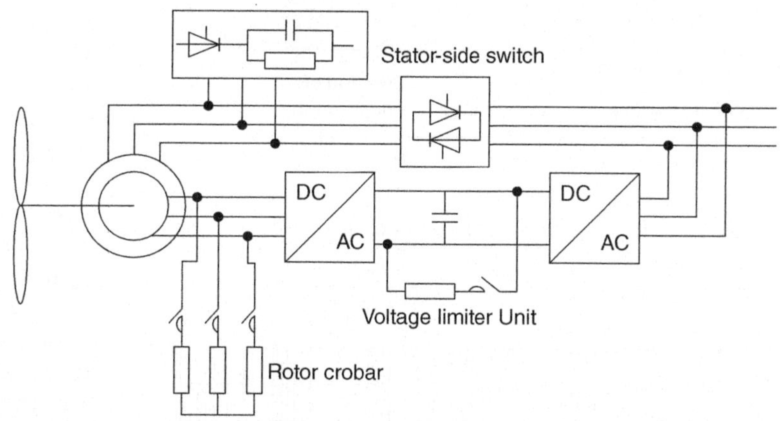

Figure 10.24 LVRT topology based on a STI scheme.

power grid reactive power, but the power grid connection in this control strategy should be done after grid disconnection and the course usually lasts for 200 ms to 300 ms. In this period, there is no exchange of reactive power or energy at all between the DFAG and the power grid so that the DFAG cannot control the power grid voltage. Thus, especially during the power grid fault, the gird stability will be harmed.

Current capacity of the IGBT power electronic device of a rotor converter is increased to enable the rotor converter to withstand a larger current in a short time. The STI upon a fault (when a power grid voltage drop occurs) is avoided but it is possibly triggered by grid voltage recovery only. This scheme avoids the consequences caused by a Crowbar protection circuit and enables the rotor converter to quickly recover the control over exchange of active power and reactive power between the stator side of a DFAG and a power grid [14].

Table 10.3 describes the comparison of FRT characteristics in the rotor side method and the stator side method. The minimum voltage of LVRT is considered to be 15% U_N. In the stator side method, a power electronic device needs to be added, but it is still obviously advantageous because of the trend of WF construction for weak grids as well and the increasingly stricter requirements of grid operators for WTG system's automatic fault-proof operation and performance in grid faults.

The LVRT realization will cause cost increase of generator systems. LVRT realization is a requirement of grid companies, so the costs hereby caused are not related to generator systems or their power generation capacities but included in the overall cost. Table 10.4 shows the similar costs of several LVRT through rough estimation.

10.4.3 HVRT of WTGSs

10.4.3.1 Requirements of Power Grid Specifications for HVRT

DFAG is still partially connected with a power grid, so it will be directly influenced by power grid disturbances.

Table 10.3 Comparison of FRT characteristics in the rotor-side method and that in the stator-side method.

	Rotor-side method	**Stator-side method**
Relative transient short-circuit current	3…5	<2.0
Maximum torque/rated torque	3…4	<2.0
Support for the power grid voltage by means of a reactive current	>100 ms	<20 ms
Reactive power absorption during a fault	Yes	No
DC component in the line current during a fault	Yes	No
Grid current controllability during a fault	Limited	Yes
Necessary power electronic updates	Self-commutated rotor Crowbar and possible DC circuit breaker	Self-commutated stator switch

As a result, the operation may become difficult and there will be a high voltage because of sudden load shedding, a single-phase ground fault, connection of a big capacitor bank, sudden current interruption, and so on. During grid disturbances or faults, voltage rises are less common than voltage drops. Overvoltages of the wind turbines possibly have different amplitudes and durations, depending on the types and degrees of disturbances. Therefore, power grid specifications have different requirements for HVRT. For example, the power grid specification of Australia specifies that each wind turbine should be capable of withstanding an overvoltage of 1.3 p.u. and with duration of 60 ms (see Figure 10.25) [18, 19].

Table 10.4 Comparison of the costs of several LVRT technologies.

LVRT technology	Cost ($/kW)	Duration of ride through
Adding capacitors	600	0.1 s
Boost converter	100–200	5 s
Battery backup	100–200	5 s, 1 h
Super capacitor	300–400	5 s

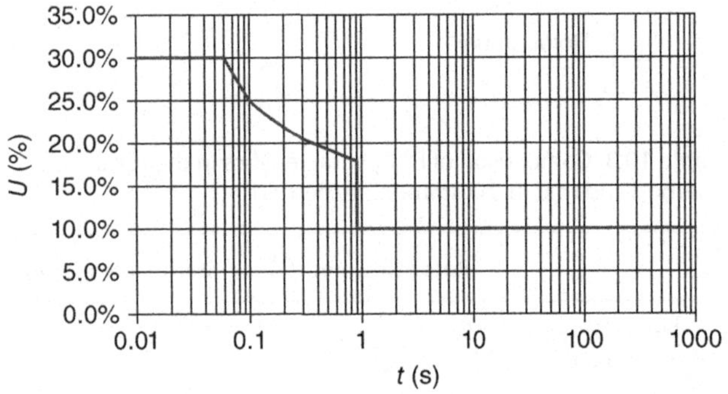

Figure 10.25 Requirements of the Australian power grid specification for HVRT.

An overvoltage possibly makes the power flow in the converter on the line side reverse. Consequently, the current will possibly flow from the power grid to the DC bus of the converter and the DC voltage will therefore increase. To protect the converter, the DC voltage must decrease again to its rated value, otherwise, the converter current must be limited because IGBT is highly sensitive to overcurrent.

As for the requirement of power grid specification for overvoltages, the overvoltages caused by the operation of power producers or users and the overvotlages from grids must be considered. Obviously, grids should avoid the overvoltages caused by system abnormalities. For this purpose, the common method is to set a limit so that a wind turbine can disconnect from a power grid after its voltage reaches the limit. Furthermore, interruption of power generation by transient overvoltages is not expected because grid stability depends on power generation stability. Thus, wind turbines must provide voltage support for grid voltages (stability) during grid faults or even during abnormal operation (see Figure 10.26) to decrease the possibility of disconnecting power producers and users from grids. According to the figure, after the terminal voltage has exceeded the dead zones (between ±10%) of the current operating point, voltage support is necessary. The minimum reactive current or voltage gain should be 2.0 p.u. and, correspondingly, a reactive current of 1.0 p.u. should be provided when the terminal voltage is lower than 50% of the reference voltage. Furthermore, the rise time for the control should be shorter than 20 ms. For controllable voltage support during normal operation, a system may need continuous voltage control as a conventional synchronous generator does. Quick and continuous voltage control also guarantees feed of maximum available reactive currents during faults and smoothes the voltage flickers caused by wind power fluctuations. Usually, big offshore WFs need continuous voltage control. Besides, WFs must contribute to inhibiting the electromechanical oscillations of power systems. For this purpose, power system stability must be considered in voltage controller design.

10.4.3.2 HVRT Scheme

HVRT can be realized through different schemes. According to reference [21], FACTS controllers with self-commutated static converters may be used as a

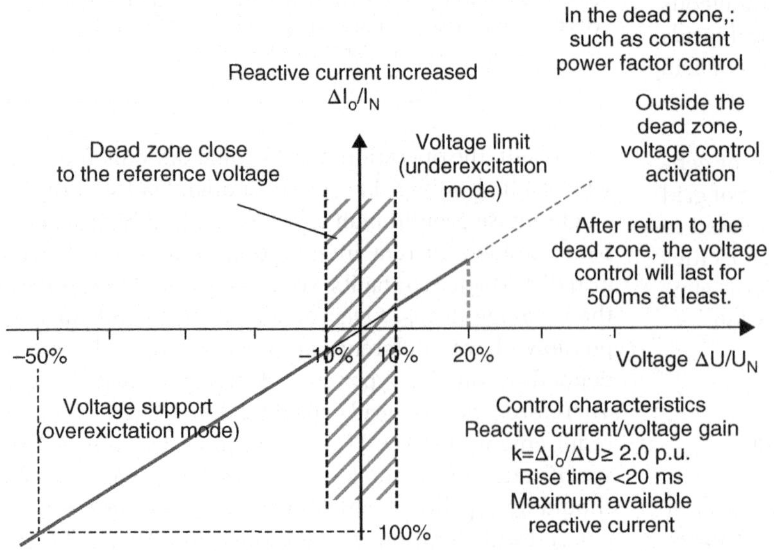

Figure 10.26 Requirements of E.ON Netz (Germany) for reactive power in HVRT.

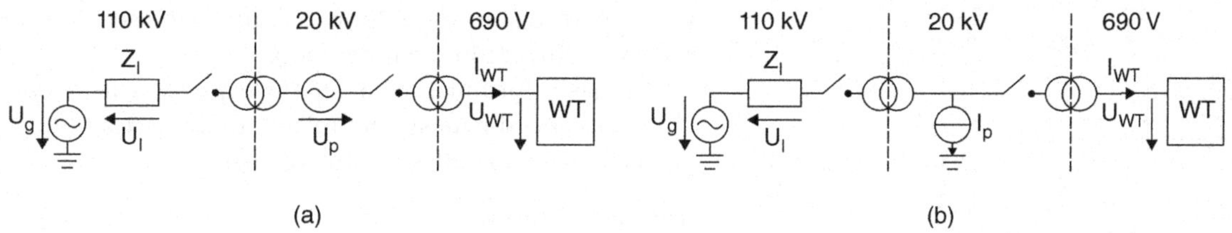

Figure 10.27 Operating principle of voltage rise compensation by FACTS controllers. (a) Serial compensation; (b) Parallel compensation.

controllable voltage source to decrease voltage rise because they can be controlled quickly and exchange active power and reactive power with power systems independently. Parts a and b of Figure 10.27 show FACTS controllers in serial connection and parallel connection, respectively.

In the figure above, it is assumed that a FACTS compensation device is connected between two transformers in connection with a WF (wind turbine) and a power grid. Serial controllers are in serial connection with the line and they can influence the line voltage of the WF (wind turbine) directly. For example, a dynamic voltage restorer (DVR) for serial compensation can dynamically provide a controllable reverse fault voltage to keep a constant voltage for the WF. Parallel controllers are in parallel connection with the line and it can control the line voltage indirectlyby generating an additional voltage drop on a line impedor Z_l to decrease the wind turbine voltage. A STATCOM influences the power transmission system by means of parallel current compensation (see Figure 10.27b). It can be used to reduce flickers or harmonics, depending on the specific control strategy. By means of a shunt current, it generates an additional voltage drop for the system impedor Z_l (including a power transmission line and a transformer impedor) to control voltage of the power transmission line.

Reference [22] describes the schemes and control strategies for HVRT. Figure 10.28 shows the typical structure of a DFAG with HVRT. Short circuit protection Crowbar of the rotor is used for protecting the rotor side converter from overcurrents and DC capacitor overvoltages during grid faults. However, if Crowbar is triggered, it will mean that the motor side converter (MSC) has lost control over the DFAG because it is locked as a result of rotor shortcircuit by the Crowbar resistor. In this period, the DFAG serves as a CAG and absorbs reactive power from a power grid, which is not expected in LVRT. If Crowbar is triggered during HVRT, there will be a large generator torque and uncontrollable active and reactive power output. To avoid triggering Crowbar upon most of faults, a DC chopper is used for limiting the DC voltage and shorting the DC circuit with help of a chopper resistor. For improving the quality of electric energy, a line inductor L and an AC filter are used on the grid side of the converter.

10.4.3.3 Main Control Links

1) **Protection of line side converter (LSC)**

   Figure 10.29 shows a LSC circuit on grid side. In a DFAG system, the LSC function is for keeping a DC voltage, sharing reactive power for the MSC and the LSC and providing the best support for the reactive current. During a power grid fault, additional reactive power will be temporarily provided to support the power grid. Particularly, when the rotor is shorted by a Crowbar resistor, the DFAG will absorb the reactive power and the reactive power absorbed must be compensated by the LSC.

   Figure 10.30 shows the control circuit of a LSC. Part a of the figure describes outer loop control of the LSC. DC voltage control and reactive power control are realized by controlling active current and reactive current of the LSC. Performance of the voltage controller can be strengthened through feedforward control over active current of the MSC, and active current of the MSC can be calculated based on active power and line voltage of the MSC. The set currents should be limited in compliance with rated values of the converters. For proper DC voltage control, the active current is ensured first. If the voltage drops at

Figure 10.28 Structure of a DFAG system with HVRT.

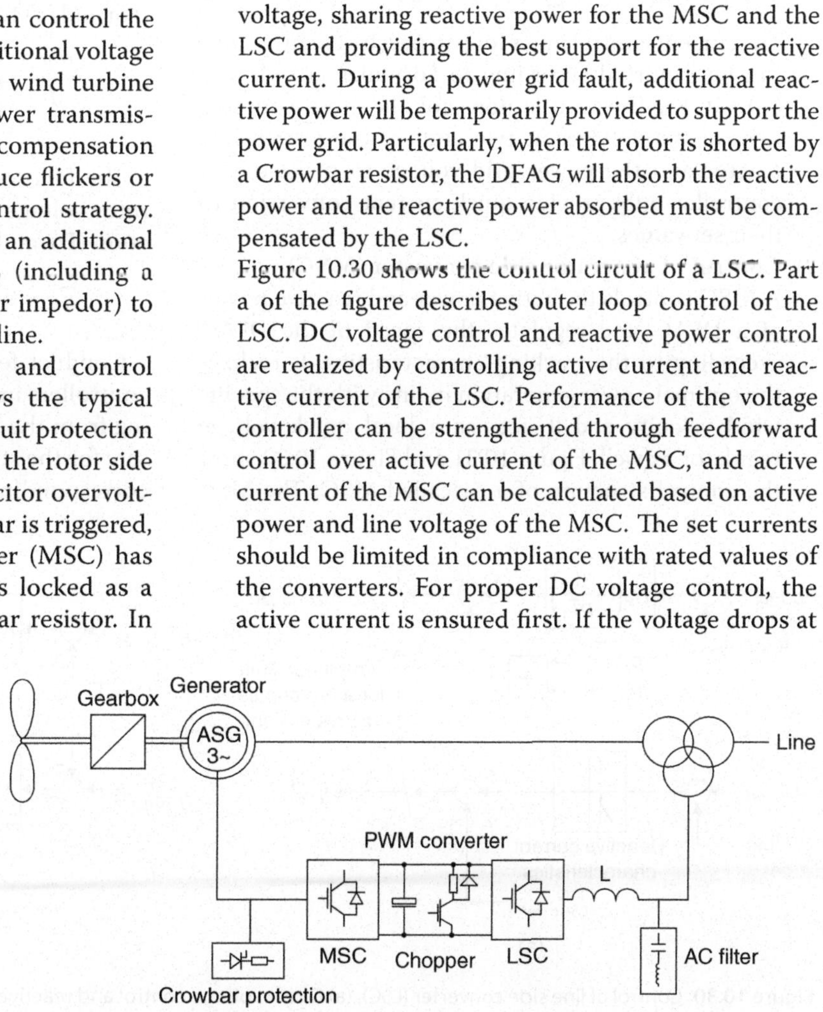

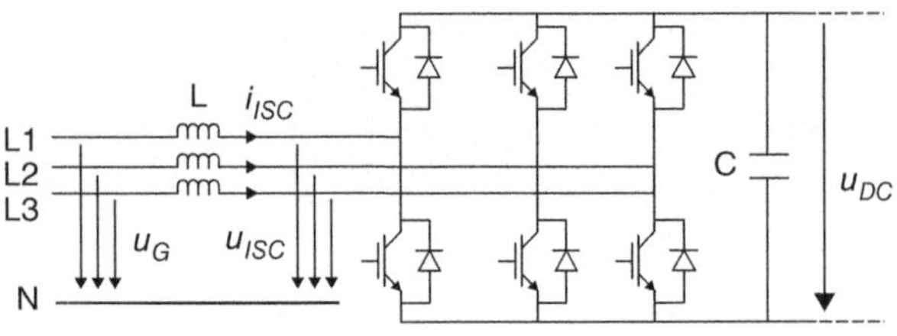

Figure 10.29 LSC circuit.

two ends of the power grid reactor are considered, the converter voltage generated will be (Formula (10.13)):

$$\overline{u}_{LSC}^{\angle ug} = \overline{u}_G^{\angle ug} - l\frac{d\overline{i}_{LSC}^{\angle ug}}{dt} - j\omega_0 l \cdot \overline{i}_{LSC}^{\angle ug} \tag{10.13}$$

In a voltage oriented reference coordinates ($u_{G,q}^{\angle ug} = 0$, $u_{G,d}^{\angle ug} = |\overline{u}_{G,d}^{\angle ug}|$), components d and q of the LSC voltage are as below (Formulas (10.14) and (10.15)):

$$u_{LSC,d}^{\angle ug} = u_{G,d}^{\angle ug} - l\frac{di_{LSC,d}^{\angle ug}}{dt} - j\omega_0 l \cdot i_{LSC,q}^{\angle ug} \tag{10.14}$$

$$u_{LSC,q}^{\angle ug} = -l\frac{di_{LSC,q}^{\angle ug}}{dt} - j\omega_0 l \cdot i_{LSC,d}^{\angle ug} \tag{10.15}$$

The current control inner loop in part b of Figure 10.30 can be exported according to Formulas (10.14) and (10.15). Feedforward is done for the cross-coupling item of the voltages at two ends of the power grid reactor and the power grid voltage to make the PI controller only provide quick current conversion to their set values.

2) **Control of generator side converter (MSC)**
A MSC controls the active power and reactive power of a DFAG and regulates the speed of the DFAG according to the tracking characteristics. It realizes the optimal power generation in line with the specific wind velocity and it acts as a local quick voltage regulator as well. Figure 10.31 and Figure 10.32 show the control structure of a cascaded MSC. The MSC power controls the outer loop to regulate the set value of the rotor current in the inner loop.

The basic formulas of a DFAG expressed in a synchronous coordinate system related to the stator voltage are as below (Formulas (10.16–10.20)):

Voltage formulas:

$$\overline{u}_s = r_s\overline{i}_s + \frac{d\overline{\psi}_s}{dt} + j\omega_s\overline{\psi}_s \tag{10.16}$$

$$\overline{u}_R = r_R\overline{i}_R + \frac{d\overline{\psi}_R}{dt} + j(\omega_s - \omega_R)\overline{\psi}_R \tag{10.17}$$

Flux linkage formulas:

$$\overline{\psi}_s = l_s\overline{i}_s + l_h\overline{i}_R \tag{10.18}$$

$$\overline{\psi}_R = l_h\overline{i}_s + l_R\overline{i}_R \tag{10.19}$$

Motion formula:

$$\frac{d\omega_R}{dt} = \frac{1}{\theta_m}(\psi_{sd}i_{sq} - \psi_{sq}i_{sd} + t_m) \tag{10.20}$$

The following formulas about feedforward current control can be exported when the steady-state operation is considered and the stator resistance is ignored (Formulas (10.21) and (10.22)).

$$\frac{d\overline{\psi}_s}{dt} = \frac{d\overline{\psi}_R}{dt} = 0 \tag{10.21}$$

$$r_s = 0 \tag{10.22}$$

The following complex state formula about the static rotor voltage can be obtained after some algebraic operation (Formula (10.23)).

$$\overline{u}_R = s\frac{x_h}{x_s}\overline{u}_s - js\sigma x_R\overline{i}_R \tag{10.23}$$

Conduct feedforward for the voltage. The PI current controllers in parallel connection need to realize switching from the rotor current to the set value and compensate for the stator resistance only. Figure 10.32 shows the corresponding current control loop.

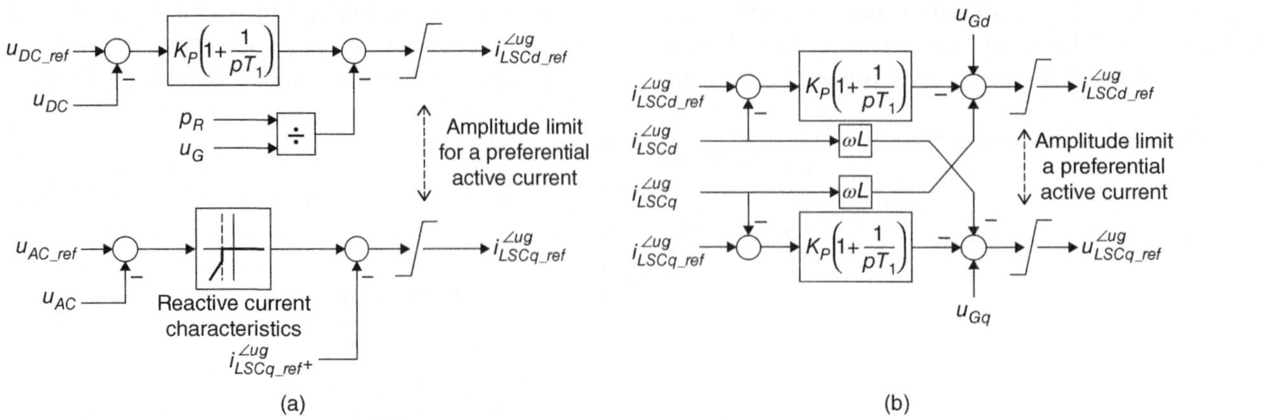

Figure 10.30 Control of line side converter (LSC). (a) Direct voltage control and reactive power control of LSC; (b) Current control of LSC.

Figure 10.31 Structure of active power control and reactive power control of a DFAG.

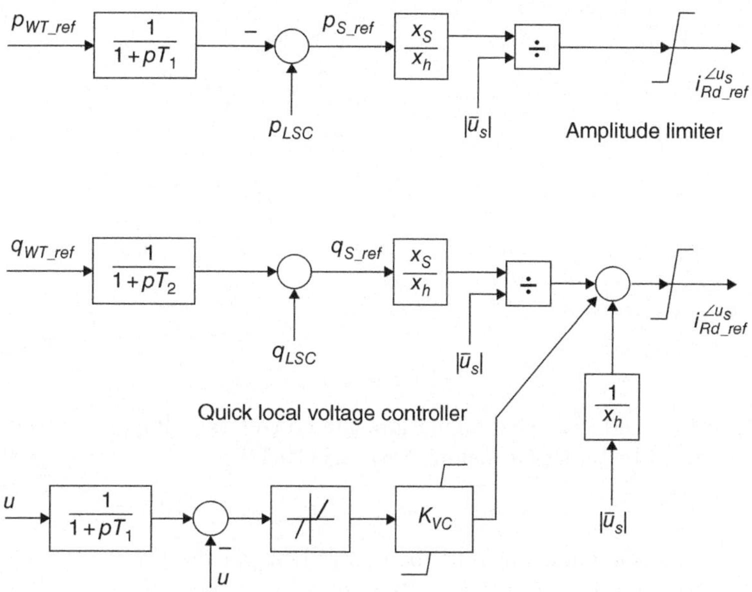

Figure 10.32 Current control structure of a MSC.

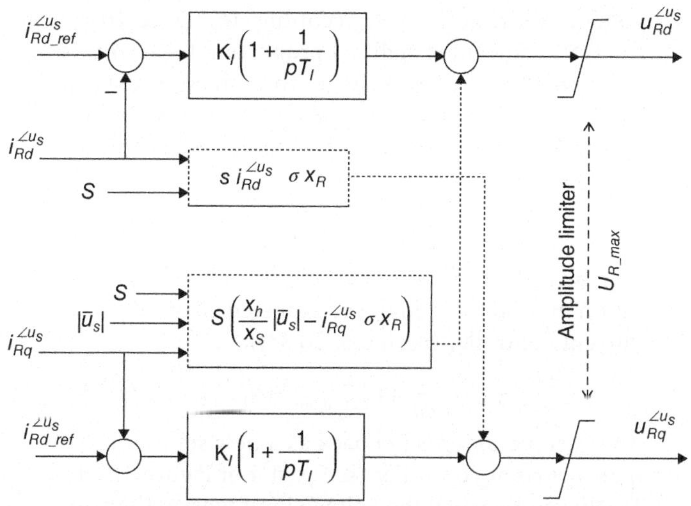

Power formulas (Formulas (10.24–27)):

$$p_{WT} = p_s + p_{LSC} \tag{10.24}$$

$$q_{WT} = q_s + q_{LSC} \tag{10.25}$$

$$p_s = \mathrm{Re}\{\overline{u}_s \overline{i}_s^*\} \tag{10.26}$$

$$q_s = \mathrm{Im}\{\overline{u}_s \overline{i}_s^*\} \tag{10.27}$$

The following formulas can be obtained from Formulas (10.16), (10.18), and (10.21) (Formulas (10.28–30)).

$$\overline{i}_R = i_{Rd} + j i_{Rq} \tag{10.28}$$

$$p_s = -\frac{x_h}{x_s}|\overline{u}_s| i_{Rd} \tag{10.29}$$

$$q_s = \frac{1}{x_s}|\overline{u}_s|^2 + \frac{x_h}{x_s}|\overline{u}_s| i_{Rq} \tag{10.30}$$

Feedforward item of the power control outer loop in Figure 10.33 can be exported according to Formulas (10.29) and (10.30).

10.4.3.4 Impacts of HVRT on Converters

For safe operation of wind turbines during HVRT, the converters must always operate within their permissible voltage ranges. The maximum usable voltage for the converters (obtained by means of DC voltage regulation) is as below (Formula (10.31)):

$$u_{conv,\max} = m_{\max}\frac{\sqrt{3}U_{DC}}{2\sqrt{2}U_{LSCn}} \tag{10.31}$$

During HVRT, there are possibly some problems as a result of voltage rises of the LSC and the MSC. These problems should be properly solved by means of control over the converters (LSC and MSC).

1) **HVRT of LSC**

Figure 10.29 describes the impact of grid overvoltages on a LSC. During HVRT, the power grid voltage is possibly larger than the maximum converter voltage.

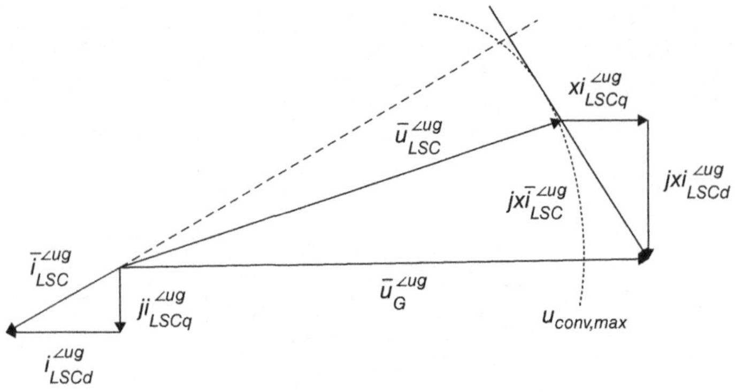

Figure 10.33 Phase diagram of the voltage and current of a LSC for HVRT.

To avoid excessive regulation, the converter voltage must be limited as below (Formula (10.32)):

$$|\bar{u}_{LSC}| = \sqrt{u_{LSC,d}^2 + u_{LSC,q}^2} \leq u_{conv,\max} \quad (10.32)$$

The active current must be controllable all the time to keep the DC voltage. Thus, priority of the active current $i_{LSCd}^{\angle Ng}$ must be considered for limiting output voltage of the converter. According to Figure 10.33, the active current is in direct proportion to the q axis component of the LSC voltage. Thus, in order to keep the q axis component unchanged, the d axis component of the LSC voltage must be limited as below (Formula (10.33)):

$$u_{LSC,d,\lim}^{\angle ug} = \sqrt{u_{conv,\max}^2 - u_{LSC,q}^{2\angle ug}} \quad (10.33)$$

Hereby is the current as a result of LSC voltage limitation in a bid to avoid the unexpected effect of a LSC current controller (Formula (10.34)).

$$i_{LSC,q-ref^*}^{\angle ug} = \frac{1}{\omega L}\left(u_{LSC,d,\lim}^{\angle ug} - u_{G,d}^{\angle ug}\right) \quad (10.34)$$

The current above is fed back as a new set point of a q axis current controller (see part a of Figure 10.31). Furthermore, when the voltage limit works, the controller state variable will be kept.

Thanks to the derived converter voltage limit, HVRT can be protected within a voltage level by means of LSC control, mainly depending on the maximum converter current and the critical current. The critical current possibly results in saturation of the power grid reactor. However, due to the derived LSC voltage limit, a good result is available for symmetric HVRT only and the situation of an asymmetric voltage component has not yet been considered carefully.

To extend to the LSC control over an asymmetric overvoltage, the negative-sequence component $|\bar{u}_{G,2}|$ of a power grid voltage must be considered for the output voltage limit of the LSC. Namely (Formula (10.35)):

$$u_{LSC,d,\lim}^{\angle ug} = \sqrt{u_{conv,\max}^2 - u_{LSC,q}^{2\angle ug}} - |\bar{u}_{G,2}| \quad (10.35)$$

This correction guarantees that negative-sequence component (included as the feedforward item in LSC current control) of a power grid voltage measured will not be decreased by the output limit. In this way, negative-sequence component of the LSC current is minimized, reducing ripple of the DC voltage during asymmetric operation. The negative-sequence voltage can be obtained by means of decomposition of the power grid voltage measured. To realize it, the coordinates should be switched to a reference coordinates rotating around the negative-sequence system and the positive sequence system should be filtered by a low-pass filter. This must be considered for reactive current set point of the LSC.

2) **HVRT of MSC**

Requirements for the HVRT of a MSC are similar to that for the HVRT of a LSC. However, a MSC is not connected with a power grid directly but with the rotor circuit of a DFAG, so its voltage amplitude and frequency during HVRT mainly depend on operating point of the DFAG. To evaluate the impact of an overvoltage on the MSC, DFAG related Formulas (10.16) to (10.30) must be considered.

According to Formula (10.23), the impact of HVRT on the MSC not only depends on amplitude of the stator voltage but also relates to slip(s) of the DFAG, in other words, in the normal speed range of the DFAG, the rotor voltage induced is lower than the maximum converter voltage (even during HVRT). For the MSC, HVRT probably is the sole crucial factor when operating speed of the DFAG is close to its limit.

During operation with the normal voltage, the active current operates first during MSC control to make sure the DFAG can track the active power set point provided in monitoring control. Usually, in this mode, the voltage limit of the MSC does not work. Thus, during ride through (LVRT or HVRT), it is necessary to switch the priority to the reactive current first to make sure the WT meets the requirement of the support voltage specified in the power grid specification. During HVRT, this means that the DFAG operates in the low-excitation mode. According to the power grid specification of Germany, proportional gain of the voltage controller should be 2.0 p.u. at least. However, it is suggested to make the proportional gain higher for better voltage drop.

When the rotor voltage induced is higher than the maximum output voltage of the MSC, the MSC voltage will be limited by an amplitude limiter.

In the case of asymmetry voltage, the MSC may be used for inhibiting the negative-sequence component in the DFAG current too. However, the negative-sequence voltage component induced by the DFAG is amplified by the negative-sequence slip from the stator side to the rotor side (Formula (10.36)).

$$s_{seg} = \frac{\omega_0 + \omega_R}{\omega_0} \tag{10.36}$$

Furthermore, turns ratio of the DFAG from the stator side to the rotor side amplifies the voltage too. As a result, negative-sequence control of the MSC is significantly limited.

In any circumstance, the only reason for the MSC to realize negative-sequence control is to reduce the oscillating torque on the drive chain because the MSC does not participate in DC voltage control. However, the negative-sequence control should adapt to the negative-sequence component between the stator voltage and the power grid voltage. This means that there are no inhabitation actions for voltage unbalance. Without negative-sequence control, the negative-sequence will be shorted as the situation in the damping winding of a synchronous generator. This is expected by a power grid because the negative-sequence current generated contributes to stator voltage balance.

10.5 AGC of a Wind Power Contained Power System

10.5.1 Principle and Function of AGC

Automatic generation control (AGC) is also called active power and frequency automatic control or load frequency control (LFC). In AGC, the active output power of a generator is controlled to track load changes of the related power system to realize power balance for the power system to keep the system frequency at its rated value (such as 50 Hz).

Traditionally, a control area is defined as a part of an interconnected system, and load and power generation in the interconnected system are controlled with the preset rules in a bid to keep the control area within related limits.

Area control error (ACE) refers to deviation of the power generation and net exchange power for frequency recovery from the expected value in a control area. The ACE formula may be changed as below (Formula (10.37)):

$$ACE = -\Delta P_g - \Delta f / R \tag{10.37}$$

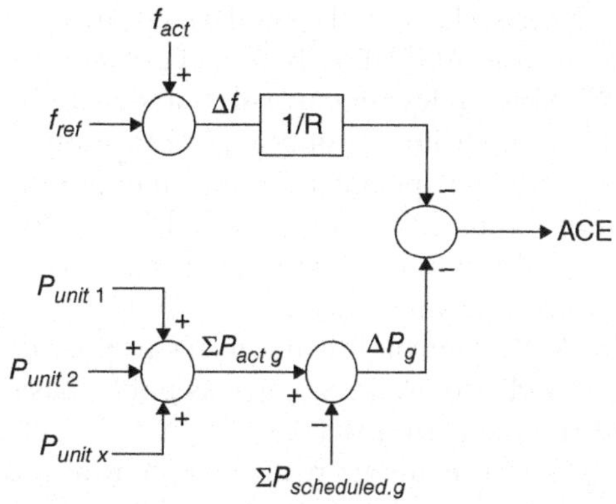

Figure 10.34 ACE in AGC.

In the formula, ΔP_g is deviation (MW) between the actual power generation and expected power generation of a control area; Δf is deviation of the actual frequency from the rated value; $1/R$ is the frequency regulation coefficient (MW/Hz) related to all equipment (generators and loads).

An ideal ACE signal (such as Figure 10.34) should not change greatly. ACE is influenced by random changes directly, so this rule can be statistically handled easily. The so-called standard deviation of ACE should be small.

"Accumulation" is not allowed for ACE. This means that the integral ACE in a proper time should be small. The accumulation of ACE will result in system time errors or so-called "inadvertent interchange errors."

The part under AGC should be the minimum. For example, many ACEs are simple random load fluctuations and need no AGC operation.

In the electric power market system, a control area is understood as a balanced area. In other words, in each balanced area, a company should keep balance. An AGC system should recover the accumulated ACEs to zero within the preset time limit (e.g., it should recover the accumulated ACEs to zero within each 10 minutes by increasing or decreasing set point of the generator).

AGC can be designed into different operation modes (actions), depending on the requirements of systems and the needs of power plants/operators. Usually, the operation modes include: ① **Assist action:** When a large ACE needs a large change of power generation, a generator unit cannot change its output quickly enough even if all actions are taken. In this case, the assist logic changes into driving more generator units to correct the ACE. ② **Filtering of ACE:** The large ACEs in the beginning are possibly random noise and need no tracking by AGC WTGS. Thus, a majority of AGC programs adopt well-designed and self-adaptive non-linear filter schemes in a bid to filter the random noise among the ACEs needing control. ③ **Telemetery failure logic:** It guarantees that AGC does not operate wrongly when using any wrong telemetering value. ④ **Unit control detection:** It detects to find the WTGSs not executing a command for increasing lower pulses and delays the control over them, making AGC distribute its control over other control WTGSs. ⑤ **Ramp control:** It allows AGC to make a WTGS ramp from one output level to another while operating with the specified output change rate. ⑥ **Rate limiting:** In any AGC design, a WTGS should be designed to avoid too quick output changes. ⑦ **Unit control modes**: In a power system, many WTGSs are not in complete AGC control. It is necessary to provide different special modes for WTGSs, such as public basic load, basic load and regulating WTGS.

Usually, modern AGC plans are realized by centralized control centers that can realize telemetering for information about system. Control actions are determined by computer systems and then transmitted to power generator systems via the same telemetering channels. To realize an AGC system, the necessary information includes the output power (MW) of each hydropower plant unit, total output power (MW) of the WF, general output power (MW) generation plan for all the power generator systems controlled in the balanced area, and the system frequency. An output command executed by the AGC program (such as increasing or decreasing orders via the various length pulse) should be transmitted to each power generator system correctly. The local control system should then change reference set point for power generation of the WTGS (increase or decrease in proportion to the pulse length).

10.5.2 AGC Considering Wind Power Generation

On a big WF, variable power generation should be managed. However, total power generation changes of the WF are much flatter than those of each wind turbine on the WF. Therefore, for the objectives of AGC, the wind power generation changes should not be measured on the wind turbines but at the PCCs (such as centralized measurement at the substations). The values obtained in the measurement should be then transmitted to the dispatch center and the control system. The deviation of wind power generation, that is, the deviation of the actual wind power generation from the estimated wind power generation, acts on the ACE. Distribution is then done to the hydropower units or thermal power units under AGC according to the PF. WTGSs are not regarded as units under AGC because they have very limited capabilities to increase their output.

The control modes of an AGC area are expressed in Formulas (10.38) to (10.42) [23]:

$$ACE = -\Delta P - \Delta f/R \tag{10.38}$$

$$\Delta P_g = \Delta P_{h,g} + \Delta P_w \tag{10.39}$$

$$\Delta P_w = \sum P_{w,act} + \sum P_{w,es} \tag{10.40}$$

$$\Delta P_{h,g} = \sum P_{h,act} + \sum P_{h,sch} \tag{10.41}$$

$$\Delta P_{set} = K \times ACE + \frac{1}{T}\int_0^{TI} ACEdt \tag{10.42}$$

In the formula, ΔP_g is power deviation of the power generator system; ΔP_w is power deviation of the WTGS; $\sum P_{w,act}$ is actual power of the WTGS; $\sum P_{w,es}$ is estimated power of the WTGS; $\Delta P_{h,g}$ is power deviation of the hydropower unit; $\Delta P_{h,act}$ is actual generation power of the hydropower unit; $\Delta P_{h,sch}$ is expected generation power of the hydropower unit; ΔP_{set} is set power deviation of the system; K is a constant; ACE is area control error; T is the time needed to drive the ACE to zero; TI is an integral time constant.

If there is wind power, an ACE will be caused by two frequency deviations, including one arising out of hydropower generation and another arising out of wind power generation. When there are several WFs, the power generation of each of them should be telemetered and transmitted to a control center. In this case, the ACE reflects wind power generation as in Figure 10.35.

The controller integrates the ACE within a given time limit and then calculates the average power to be distributed to the units under AGC. The power deviation must be distributed economically with a participation factor (PF) method, and so on.

Additional changes in the power generation of a unit under AGC (Formulas (10.43) and (10.44)):

$$\begin{aligned}
\Delta P_{h1} &= \Delta\lambda / K''_{h1} \\
\Delta P_{h2} &= \Delta\lambda / K''_{h2} \\
&\vdots \\
\Delta P_{hN} &= \Delta\lambda / K''_{hN}
\end{aligned} \tag{10.43}$$

$$\begin{aligned}
\Delta P_g &= \Delta P_{h1} + \Delta P_{h2} + \cdots + \Delta P_{hN} \\
&= \Delta\lambda \sum (1/K''_i)
\end{aligned} \tag{10.44}$$

In the formulas: λ is the marginal cost; K''_{hN} is the second derivative of the cost function. PF of hydropower units (or thermal power units) is below (Formula (10.45)):

$$\Delta P_{hi}/\Delta P_g = (1/K''_{hi}) / \sum (1/K''_i) \tag{10.45}$$

10.6 WF Monitoring and Control

Modern variable-speed WTGs make it possible to control active power and reactive power separately. Reference [24] introduces an overall control method including system monitoring and control for WFs. This control system is based on two control levels as below: A monitoring system controls the active power and reactive power of a whole WF by transmitting set values to all wind turbines while a wind turbine control system makes sure the set values of wind turbine level are realized. The WF monitoring and control proposed in reference [25] focuses on reducing the power output

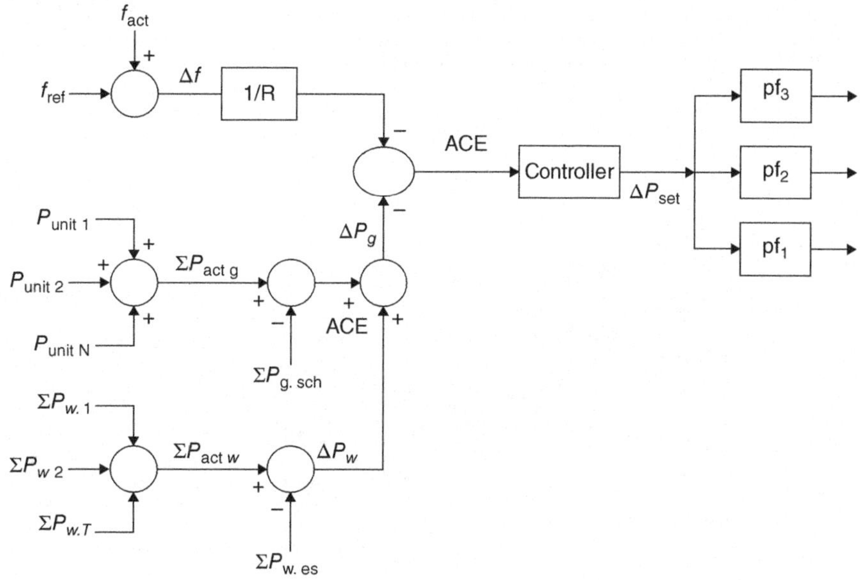

Figure 10.35 AGC when there is wind power.

fluctuations of WFs. It applies to system frequency control and power oscillation damping control as well. The control is realized through the reserve power of independent (external) energy storage equipment or a WF's one or multiple partially loaded wind turbines in the system range. The active power compensation model proposed in reference [26] combines two types of energy storage equipment into a mixed energy storage system to improve the electric energy quality and grid connection stability of distributed power systems. Set value of compensation power of energy storage equipment is calculated through fuzzy control and an energy estimation based dynamic energy optimization algorithm, and a better compensation effect is realized by small-capacity energy storage equipment.

10.6.1 WF Monitoring and Control System

Based on generator control and wind turbine control, a WF monitoring and control system can only be realized on the WFs with variable-speed variable-pitch wind turbines. What will be discussed below is monitoring and control of the WFs with DFAGs.

10.6.1.1 Generator Control

Figure 10.36 shows the structure of a variable-speed variable-pitch wind turbine installed in a WF. The electrical system includes a DFAG. A stator winding is connected with the secondary-side winding of a transformer directly, and a rotor winding is connected with the third winding of a wind turbine transformer with help of a two-way converter made up of back-to-back IGBTs (also called LSC and MSC).

The MSC controls the rotor current (reference coordinate: the stator flux linkage). The d axis current component and the q axis current component decouple the reactive power control and the torque (active power) control. The d axis current of the rotor can control the field current, and the d axis current is for controlling the generator torque to reach the rotating speed needed by the variable speed system. With a controllable current, the LSC transmits rotor power to the power grid while operating at a super-synchronous speed or absorbs rotor power from a power grid while operating at a subsynchronous speed. In vector control, the LSC is also allowed to compensate for reactive power. Through a proper PWM technology, the three-phase choking coil (reactive coil) connected with the LSC and the third winding of transformer in delta connection, the harmonic distortion fed into a power grid can be maintained in corresponding limits.

10.6.1.2 Wind Turbine Control

A WTG control system has two independent control loops, that is, a motor active power control (MAPC) and a motor reactive power control (MRPC).

For an active power control system without commands of grid operators for reference, its control objectives are based on a control strategy for variable-speed variable-pitch wind turbines.

Optimal power when the wind velocity is slower than its rated value: Make the wind turbines operate with the maximum power factor to capture the maximum wind energy. The pitch angle β keeps its optimal value constant, otherwise, it is driven to its best value by means of a change of the tip-speed ratio according to the rotating speed. The reference value of rotating speed is determined by the speed controller (SC) acting on the generator torque.

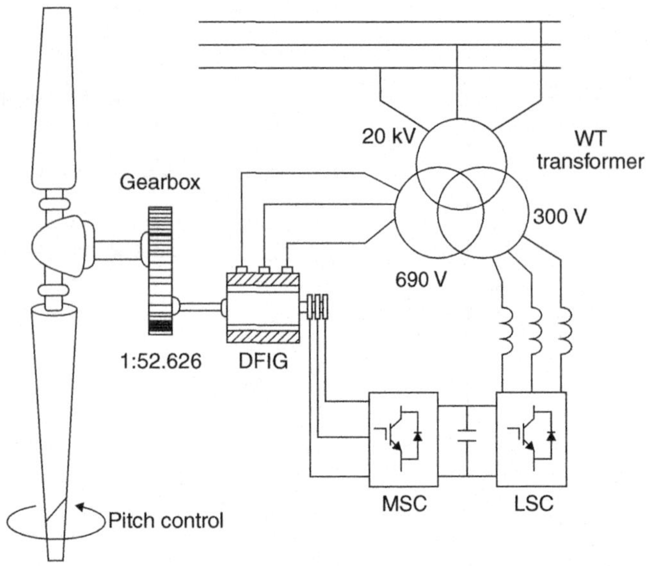

Figure 10.36 Structure of a variable-speed variable-pitch wind turbine.

Operation with the power limit when the wind velocity is faster than its rated value: The power controller must act on the pitch angle to ensure the rated output for the wind turbines. The SC should keep the rating rotating speed. For better connection with a power grid, the wind turbines are not necessary to be negative loads, but some types of power control (such as variable-pitch control) must be available.

Power limit when the wind velocity is slower than its rated value: The power controller must act on the pitch angle to reach the reference power P^*.

Figure 10.37 is the block diagram of a monitoring and control system. The monitoring and control system includes two control loops, that is, a speed control loop acting on the generator torque to control the generator speed and a power control loop acting on the pitch angle to control output power of the wind turbines.

If the reference power is not specified by the monitoring and control system, it should be set as its rated value. If the wind velocity is slower than its rated value, the power error will be negative and the controller output (reference to the pitch angle β^*) will drop until the controller reaches its lower saturated state (the optimal pitch). Thus, the wind turbines will operate with the optimal pitch. The optimal speed as reference value of the SC is obtained from the wind turbine output characteristic curve expressed in an optimal speed function, that is, the so-called maximum power tracking strategy (MPT). The characteristic curve is flattened when the wind velocity is its rated value, so no reference speed is provided in the rated wind velocity range. If the wind velocity is faster than its rated value, the power error will be positive and the pitch angle will be driven to a positive value until the rated power is reached. For the power level, a reference value for the rated speed should be obtained, whereas the SC should make the reference value stable in the operation range. The power controller guarantees the rated output power and the SC guarantees the rated speed, so the generator torque is rated too.

If the reference power is specified by the monitoring and control system, there will be two cases. In the first case, the wind velocity is quick enough. In this case, the power controller will drive the pitch angle to a proper value. However, the given reference value of the SC is not the optimal value (actually lower than the best value), so the tip-speed ratio will be smaller than the optimal value and the power factor will decrease, in other words, the speed control loop will help pitch driving to reduce the power. In the second case, the wind velocity is not quick enough. In this case, the power controller will be saturated with the optimal pitch angle, and the reference value of the optimal speed will be provided for the SC to obtain the maximum power at this wind velocity.

10.6.1.3 Reactive Power Control

A reactive power control system is based on a reactive power controller and a subordinate voltage control loop. The subordinate voltage control loop is used for preventing overvoltage in an attempt to reach the reference reactive power. Principle of reactive power control is as below:

The monitoring and control system determines the reference reactive power Q^*;

The reactive power controller computes the reactive power error and establishes the reference voltage U_G^*;

The motor voltage controller (MVC) computes the voltage error and establishes a reference excitation current I_{RD} for the generator excitation current controller.

In the case of controller saturation (with the maximum or minimum voltage limit), the reference reactive power

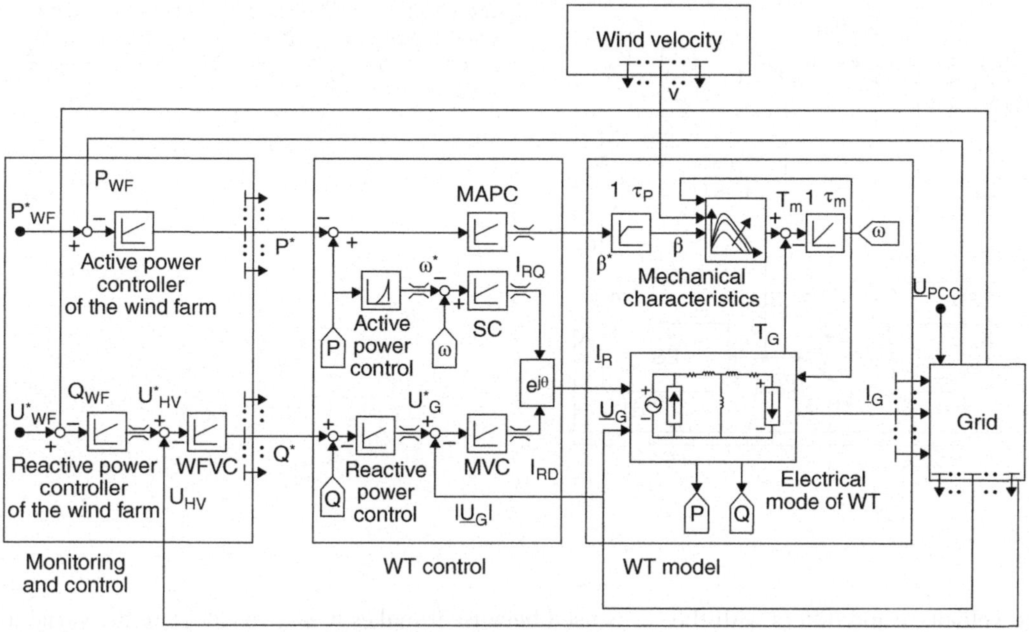

Figure 10.37 Block diagram of a monitoring and control system.

cannot be obtained. In this case, the control system will control the voltage within its limits to ensure the availability of each generator, in other words, the control system will avoid generator disconnection due to the start of voltage protection.

10.6.1.4 Monitoring and Control System

The monitoring and control system is designed to control the WF's active power and reactive power (P_{WF} and Q_{WF}) injecting to the power grid. It is made up of an active power control loop and a reactive power control loop.

The active power control loop is based on a WF active power controller (WFAPC). In grid operation, the WFAPC receives a command from the operator, that is a reference power P^*_{WF}, from a power grid operator, computes the active power error and sets a reference power P^* for the active power controller of each WTG.

If the reference power increases when one or multiple WTGs are operating with the maximum power, the other WTGs will bear the load automatically because the active power controller will increase the computed reference power caused by the power error.

The reactive power control loop is based on a WF reactive power controller (WFRPC) and a subordinate voltage control loop. The subordinate voltage control loop is used for preventing out-of-limit voltage in an attempt to reach the reference reactive power. Principle of reactive power control is as below:

The reference reactive power is usually expected to be set as the unit power factor. The reactive power controller computes the reactive power error and sets a reference voltage U^*_{HV} for the WF substation.

The WF voltage controller (WFVC) computes the voltage error and sets a reference reactive power Q^*_{WF} for the reactive power controller of each WTG.

In the case of controller saturation (with the maximum or minimum voltage limit), the reference reactive power cannot be obtained. In this case, the control system will control the voltage within its limits to ensure the availability of each wind turbine, in other words, the control system will avoid WF disconnection due to the start of voltage protection.

10.6.2 Monitoring and Control Over WFs with External Energy Storage

Figure 10.38 shows the connection of a WF with external energy storage control (ES) to a power system.[25] Through a step-up transformer, the WF may be connected to the high-voltage side (HV) or the medium-voltage side (MV) of a power grid. The energy storage unit is connected with the battery of power system through a power electronic and step-up transformer (not drawn in Figure 10.38). The optimal energy storage capacity should be determined after analyzing data of the WF in a given position and considering the investment cost, the maintenance cost, the energy storage cycle, the energy storage efficiency, the expected control result, and so on. Large-capacity energy storage can possibly handle all possible fluctuations, but it possibly has low economical efficiency.

The power fed by the WF into the power grid ($\sum P_{wt}$) is the difference between power sum of all the wind turbines and grid loss on the WF. The output power of the

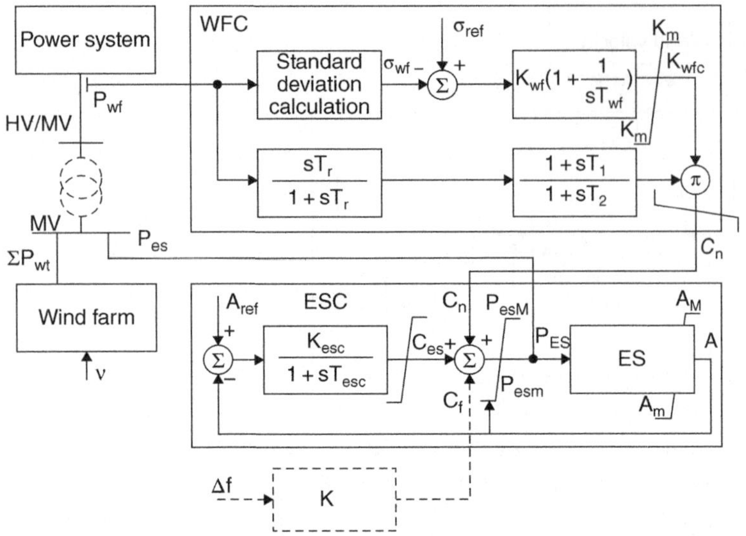

Figure 10.38 Monitoring and control over WFs with external energy storage. (WFC—WF controller; ESC—energy storage controller; ES—energy storage; K—frequency controller).

WF depends on the wind velocity v and differs with the wind velocity, the wind turbines' type, operating points and disturbance type (such as emergency disconnection of the wind turbines). In the steady state, the power fed or consumed of energy storage (P_{es}) is theoretically zero, but it actually differs with the wind velocity.

The total power fed from the WF and the energy storage system into the power system is as below:

$$P_{wf} = \sum P_{wt} + P_{es}$$

The monitoring and control system includes the following two controllers.

WF controller (WFC), which is for minimizing changes of the active output power P_{wf};

Energy storage controller (ESC), which is for maintaining the specified energy storage change level, where $A_{ref}=A_M/2.A_M$refers to the rated energy storage capacity (or the permissible maximum energy storage capacity)

The WF controller includes the following units: a DC block with a time constant (T_r) for filtering the constant (current) component in the output power of a power plant; a compensation unit with two time constants (T_1 and T_2) for compensating the phase shifts caused by objects (the transfer function from control signal c_n to control signal P_{wf}) and the sensor.

For the properly determined compensation time constants T_1 and T_2, efficiency of the WFC depends on gain K_{wfc} only: the larger the gain, the smaller the output power fluctuation of the wind power plant. Usually, the gain may be set and maintained as a constant, but it may be changed during operation too. Controlled by a PI controller, it acts on standard deviation σ_{wf} of output power P_{wf} of the wind power plant. A standard deviation is used because it makes it easy to calculate the variability of the output power. The total harmonic distortion (THD) is related to the standard deviation σ and the mean μ. Their relation is as below (Formula (10.46)):

$$THD = \frac{\sigma}{\sqrt{2}\mu} \tag{10.46}$$

In this way, THD control can be realized through standard deviation control. This is a variable used in electric energy quality related standards. The control scheme proposed can be changed into direct control over the THD easily.

On the equipment considered, the increase of σ_{wf} results in the increase of gain K_{wfc}, resulting in decrease of the power deviation P_{wf} and, by means of feedback, the decrease of σ_{wf}. The control loop must operate quickly for effective control over quick changes of output power of the wind power plant.

The WFC provides a control signal c_n for the energy storage system to change the power fed into the power system or the power P_{es} consumed by the WF.

The control process will possibly result in complete charge or discharge of the energy storage system, which is not expected because it makes proper operation of the WFC impossible. To avoid this problem, the energy storage system should have its own ESC, as ESC is an independent and simple energy storage controller. The ESC opiates relatively slower to avoid efficiency reduction of the WFC, but it should operate quickly enough to avoid complete charge and discharge for the specified disturbance set.

The ESC works according to the energy storage signal c_{es} for charge and discharge of the energy storage system and in return influences power P_{es} and power P_{wf}.

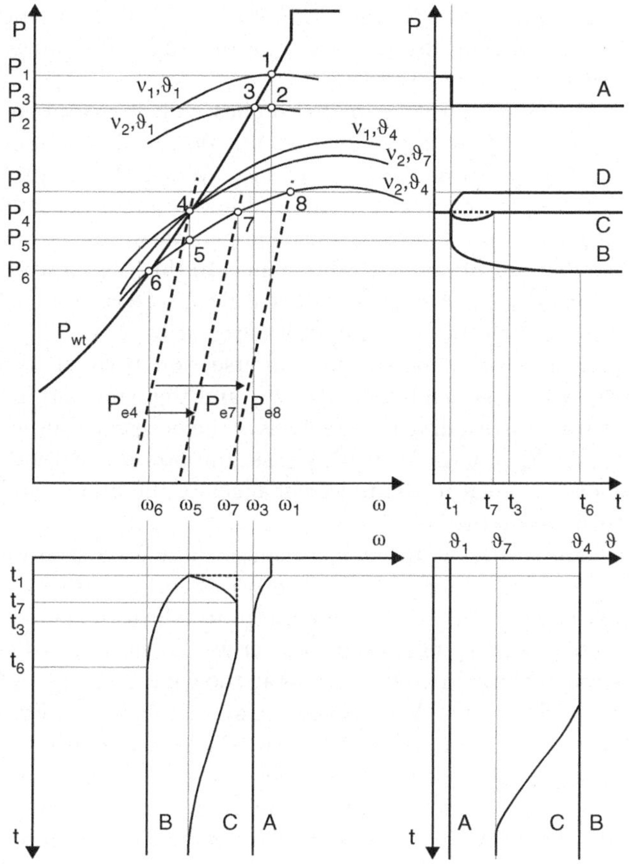

Figure 10.39 Wind turbine characteristic.

The energy A stored is as below (Formula (10.47)):

$$A = \int P_{es} dt \tag{10.47}$$

After the energy stored reaches its limit (A_M or A_m), a proper limit of the energy storage power P_{es}, that is, P_{esM} or P_{esm} (see Figure 10.39) will drop to zero. When the energy stored is within its limit (A_M>A>A_m), the control signal c_n and the energy storage output P_{es} are limited by P_{esM} (=$-P_{esm}$). After P_{esM} is set during simulation, the energy storage controller signal c_{es} is limited to $\pm P_{esM}/2$ and the gain K_{wfc} is limited too (suppose it is between 20 and 50). The standard deviation σ_{wf} can be calculated through Formula (10.48).

$$\sigma_{wf} = \sqrt{\frac{1}{n}\sum_{i=1}^{n}(P_{wf,i} - P_{wf,mean})^2} \tag{10.48}$$

In the formula above, $P_{wf,I}$ is the sampled value; a sampling time step T is 0.01 s; the sampling times n in a time window is 2^{10} times. Thus, the time window width is 10.24 s. The power $P_{wf,mean}$ is a mean value calculated based on the sampling power $P_{wf,i}$.

In the simulation, it is assumed that the power electronic converter is ideal, making compensation unnecessary (T_1=T_2). In fact, the power electronic converter may be treated as an inertialess component (a gain). However, compensation is necessary for the delay caused by the measuring component and the actual energy storage.

The energy storage considered may be used for other kinds of control such as frequency control (e.g., the frequency control box and signal c_f in the short-line box in Figure 10.39). However, the probability to realize frequency control through the energy storage system is limited. The factor limited is the energy to be stored. Actually, the energy storage can minimize frequency changes in a short time, but it cannot influence steady-state value of the frequency. The realization of long-time frequency control needs other energy sources such as fuel cell, gas turbine, or diesel generators that can supply electricity to the power system directly or with help of the energy storage system. Before complete charge or discharge of the energy storage system, these power sources should be started and reach the power generation level needed.

10.6.3 Monitoring and Control over WFs with Power Reserve

Concept of control over the energy storage external batteries is simple. The energy storage system will increase energy only when the output power of a wind power plant (WTGS) decreases or absorbs (consumes) the remaining energy when the output power of a wind power plant (WTGS) increases. Characteristics of the WTGS are not important for the operation type. It is more complicated to use the reserve power of a part-loaded wind power plant because characteristics of the wind turbines of the wind power plant influence characteristics of the whole system.

Figure 10.39 shows utilization of the power reserve concept [25]. In the figure, the normal operation (complete utilization of all available wind power) and part-loaded operation of a WF are compared. A standard wind turbine control mode is used.

The operating point of a wind turbine depends on the intersection between the wind turbine's dynamic power characteristics (depending on the wind velocity v, the pitch angle θ and the wind turbine speed ω) and electrical characteristics (controllable through the rotor current of a DFAG). When the wind turbine has the maximum capture power, the wind turbine characteristic P_{wt} generated is mainly made up of peak of the wind turbine characteristic (such as (v_1, θ_1) and (v_2, θ_2)). Suppose the power plant is operating at non-rated point 1 at the peak of characteristic (v_1, θ_1). θ_1 will be a constant, which is the minimum pitch angle. Drop of the wind velocity from v_1 to v_2 will result in the peak corresponding to (v_2, θ_1) between a steady state and a new operating point (curve A) in two stages. In the first stage, when the rotor speed ω_1 keeps unchanged because of inertia, the power will step to decrease from P_1 to P_2 (from point 1 to point 2). In the second stage, the operating point moves from point 2 to point 3 on the characteristic curve (v_2, θ_1). Thus, when a wind turbine makes the best use of wind energy in this normal operation mode, the output power will change with the wind velocity.

If a wind turbine operates in the power reserve mode and the pitch angle is set higher than its minimum (e.g., $\theta_4 > \theta_1$), the characteristic curve P_{wt} will no longer go through a peak of the wind turbine characteristic curve (v_1, θ_4) but have an intersection with it at a point (such as point 4). Power difference $P_1 - P_4$ is the measure of power reserve. If the wind velocity drops from v_1 to v_4 and the constant pitch angle is θ_4, the transition to a new operating point along curve B including points 4, 5, and 6 when there is no monitoring and control. Point 6 is intersection of the characteristic curve P_{wt} and (v_2, θ_4). The wind turbine characteristic curve P_{wt} is formed by the wind turbine speed regulator (in the box with f (P_g) in Figure 10.40). Thus, the operating point tends to P_{wt}.

The concept of monitoring and control recommended is control over output power changes. If the wind velocity changes from v_1 to v_2, it will be necessary to move the electrical characteristic Pe_4 right to Pe_7 (in DFAG, it is realized by means of rotor current control) to make the characteristic curve (v_2, θ_4) intersect with P_{e7} (point 7) when the power P_{e4} is unchanged. As a result, the wind turbine output power will remain unchanged regardless of drop of the wind velocity.

Meanwhile, the wind turbine speed regulator will operate in line with the wind turbine characteristic f (P_g), thus reducing the pitch angle. The generator controller will move the electrical characteristic curve Pe_7 back to the left. As a result, the operating point will be moved to the left (point 4), making the output power P_4 constant. The whole power plant will respond on curve C.

Actually, the capability to eliminate the power fluctuations due to wind velocity changes mainly depends on the control system and is under different mechanical and electrical constraints. To keep a constant output power, changes of the wind energy must be instantly stored in the wind turbine. The wind turbine rotor is the main but limited storage unit. Therefore, the wind turbine responds along the curve C representing some power transient (change) states.

The operation described complies with the compensation of a wind turbine for the power fluctuations caused by drop of wind velocity. When the control device operates on a WF with multiple wind turbines, its response (power output) must compensate for the power fluctuations of other wind turbines. In this case, a wind velocity drop will make the electrical characteristic curve move to a proper position (not necessarily point 7). The proper position means that intersection between the electrical characteristic curve P_{e8} and the wind turbine characteristic curve (v_2 and θ_4) is at the necessary output power P_8 (curve D). In this case, the power difference must be equal to the power changes of the other wind turbines on the WF.

Figure 10.41 is the block diagram of a WF based on the control strategy above. The WF is connected with a HV grid or a MV grid through step-up transformers. The wind power P_{wf} fed into the power grid by the WF is difference between power sum of all the wind turbines and power loss of the WF's collecting system. Power reserve is provided by one or multiple part-loaded wind turbines. Thus, the unused backup power that can be used (distributed) is equivalent to the energy stored in a conventional energy storage system. P_N is output power of the part-loaded wind turbine(s) and P_{N-1} is output power of the other wind turbine(s).

The monitoring and control system includes two controllers as below:

WFC, which is for minimizing changes of the active output power P_{wf};
PRC, which is for setting and maintaining the reserve power $\Delta P_{ref,N}$ to be fed into the power grid from the selected wind turbine

The WFC in Figure 10.41 has the same structure as in Figure 10.38. The WFC changes output power P_N of the selected wind turbine (with help of signal c_n) to balance output power of the WF. Power P_N is a part of the WF power P_{wf}. The objective is to minimize the output power, so signal c_n can realize quick changes.

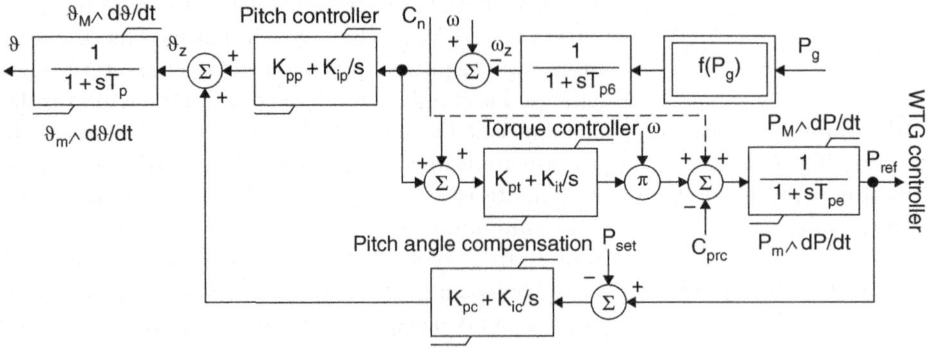

Figure 10.40 Model of wind turbine control system.

Figure 10.41 Monitoring and control over WFs with power reserve, (WFC—WF controller, PRC—power reserve controller, K—frequency controller).

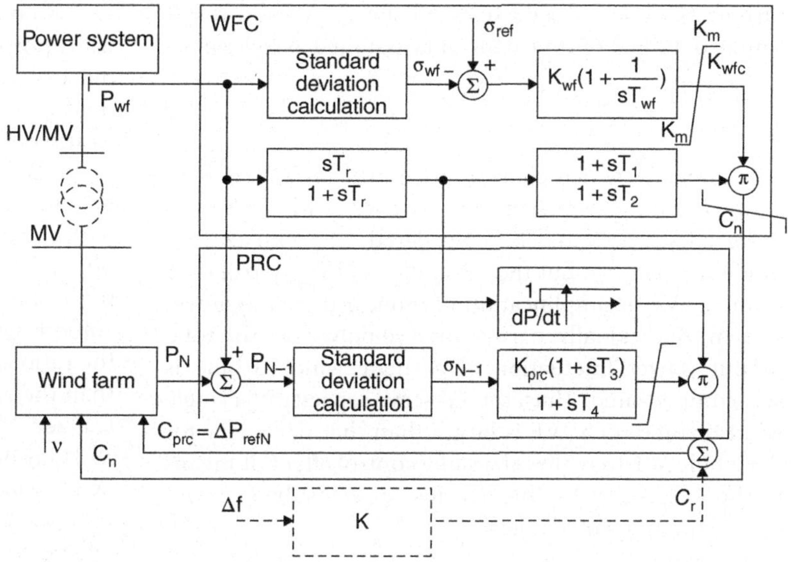

Signal c_n must realize quick changes for output power of the selected wind turbine, so it must exist in the WF's control loop acting on the wind turbine directly.

The power reserve controller (PRC) operates according to the measuring signal P_{N-1}. P_{N-1} is the difference between output power P_{wf} and P_N (i.e., output power of the wind turbine for maintain the backup). In other words, P_{N-1} is output of the wind turbine not controlled by the monitoring and control system. Its standard deviation σ_{N-1} is expressed in ΔP_{refN}. Reserve power of the selected wind turbine can smooth output power of the whole WF. Therefore, output power fluctuation compensation in the positive direction (increase of output power of the selected wind turbine) is covered by reserve power ΔP_{refN}; output power fluctuation compensation in the negative direction (decrease of output power of the selected wind turbine) is restricted by current average output power of the selected wind turbine.

Control signal c_{prc} of the PRC (equal to ΔP_{refN}) must change slowly on the precondition that there is constant active power for backup. Thus, it is introduced in the control loop acting on the wind turbine (pitch angle) directly. According to Figure 10.41, it is introduced following the pitch angle compensation related box.

Signal c_{prc} is a negative forcedly and its upper limit is zero. It operates corresponding to the non-part-loaded wind turbine, in other words, it operates with the maximum power when the set wind velocity is the set value. The lower limit of signal c_{prc} is half of current average output power of the selected wind turbine for maintaining the backup. Upon instantaneous drop of the wind velocity or quick drop of the output power because of stop of the wind turbine, upper limit of c_{prc} will be restricted by the non-linear component too. It is expressed in "dP/dt" in Figure 10.41.

If there are at least two wind turbines are selected for backup, P_N in Figure 10.41 will be the sum of the output power of these wind turbines selected. As a result, output signal ΔP_{refN} of the PRC must be distributed to all the backup wind turbines. For example, it should be in direct proportion to their actual output power.

The arrangement considered applies to frequency control too (see the dotted part in Figure 10.41). Control signal c_f should be added to control signals c_n and/or c_{prc}. If steady-state frequency changes must be obtained, c_f should be added to signal c_{prc}. Standard deviation of the measuring signal should be calculated in the same way above.

Figure 10.40 shows a control system for the wind turbine for power reserve. In this case, the main component compensated by the transfer function should be the torque controller. For effective operation of the WFC, the compensating component parameters could be set as below:

$a_0=0$
$T_1=1/K_{it}$
$T_2=K_{pt}/K_{it}$

Meanwhile, gain K_{prc} for reserve power control is set as "1." A higher gain increases the reserve power but decreases generation power of the wind turbine. The value should be an eclectic one between the WF's capability to minimize output power fluctuations and the WF's efficiency.

In the PRC, the compensating component with time constants T_3 and T_4 should make operation of the PRC dynamically proper. To avoid oscillation of the wind turbine blades, it should respond slowly. A good result can be obtained of time constants assumed as below:

T_3=0 s
T_4=10 s

Technically, which control mode, that is, control over a single wind turbine or control over a whole WF, is more effective should be considered. The two control modes are similar, but the advantage of the control over a whole WF is simplification of monitoring and control system. Additionally, part of reserve power for the total output power fluctuation of a single wind turbine is self-compensated. Thus, total reserve capacity of a single wind turbine on a WF is larger than that necessary for the whole WF to realize the same control effect. It means in the latter case the income loss as a result of power reserve maintenance is less.

10.7 Control of WF Energy Storage System

The output power fluctuation characteristics of a WF may result in problems related to local voltage quality and system frequency quality. If an energy storage system (ESS) is integrated into a WF, it can inhibit output power fluctuations of the WF, control voltage of the WF, and improve value of the wind energy.

10.7.1 Energy Capacity and Rated Converter Power of WF Energy Storage System

The energy capacity of an ESS is specified in its control and energy management scheme. If the ESS needs to provide an energy service in proportion to the WTG power, total energy of the decentralized ESS units equals to energy of the centralized ESS.

The rated power of an ESS is the maximum difference between the actual wind power and the target value. The target value may be the 1 p.u. rated value of the WTG in an extreme condition. For example, if there is no actual wind power, the target value will be 1 p.u. However, in most of cases, the wind power fluctuation is much less than 1 p.u. thus, rated power of the ESS may be r p.u. r is a ratio between rated power of the EE and rated capacity of the WTG; it is equal to or less than 1; its specific value depends on the wind power fluctuation and differs greatly with specific geographical location of the WF (wind turbine). The total power fluctuation of a WF is less than the power fluctuation of a single wind turbine because of the smoothing effect of WTG space distribution. Therefore, the ratio r_{WF} of a WF with a centralized ESS may be less than the ratio r_{WTG} of a single wind turbine with decentralized ESS. Suppose that the rated power of a single WTG is defined as 1 p.u. and that there are m (quantity) WTGs on a WF, the total power capacity of a centralized ESS will be $r_{WF} \times m$ p.u., and that of a decentralized ESS will be $r_{WTG} \times m$ p.u.

The most important two indicators for determining the rated values of a power electronic switch are maximum reverse blocking voltage and maximum forward on-state current. Their product is rated VA of the power electronic switch and sum of rated values of the power electronic switch is rated VA of the converter.

The rated VA of an ESS interface is specified as the power capacity of the ESS, or product of the direct voltage U_{dc} and the maximum average direct current I_{dc}. For each of the two IGBTs in a two-way DC/DC chopper, the maximum blocking voltage is U_{dc} and the maximum current is I_{dc}. The IGBTs have the same rated VA, which is the rated power of the ESS and half of the total rated value of converter. For each IGBT in a typical three-phase two-level PWM AC/DC or DC/AC voltage source converter, the blocking voltage is U_{dc}. If there is no reactive power, the current may be roughly estimated as $I_{dc}/3$. Thus, the rated VA of each switch is $0.577U_{dc}I_{dc}$. The total rated VA of a converter with six door switches is 3.46 times of rated power of the ESS. Table 10.5 introduces the total rated capacities of ESS and converter for different ESS technologies (including energy storage by batteries, superconductors, capacitors, and flywheels

Table 10.5 Estimated rated capacities of ESS and converters based on different ESS technologies and structures.

ESS structure		Decentralized ESS	Centralized ESS
BESS,	Energy capacity of the ESS	Linearly equal to power of the WTG	
SMES,	total rated power of the ESS	$r_{WTG} \cdot m$	$r_{WF} \cdot m$
CAPACITOR$_S$	Topology of the converter	DC/AC	AC/DC/AC
	Total rated VA of the converter	2 $r_{WTG} \cdot m$	3.46 $r_{WF} \cdot m$
FES	Energy capacity of the ESS	Linearly equal to power of the WTG	
	total rated power of the ESS	$r_{WTG} \cdot m$	$r_{WF} \cdot m$
	Topology of the converter	DC/AC	AC/DC/AC
	Total rated VA of the converter	3.46 $r_{WTG} \cdot m$	6.92 $r_{WF} \cdot m$

energy storage) and structures. If reactive power compensation is necessary for AC voltage regulation, a DC/AC converter will need a larger rated value.

If a WF has unapparent smoothing, r_{WF} and r_{WTG} will be the same value and rated values of the ESS will be the same too, but rated capacity of the converter in a centralized ESS will be 1.73 times to two times of that in a decentralized ESS.

10.7.2 Power Fluctuation Inhibition by ESS

When the ESS performance of a WF is evaluated, the universal simulation for the ESS is realized through a power conservation Formula (10.49) and a power energy relationship Formula (10.50) regardless of the specific ESS technology and control method used.

$$P_w = P_{ESS} - P_{w,fil} \tag{10.49}$$

$$E_{ESS}(t) = \int_0^t P_{ESS}(u)du + E_{ESS}(0) \tag{10.50}$$

In the formula, P_w and $P_{w,fil}$ are the wind power before and after filtering, respectively; P_{ESS} and E_{ESS} are the physically limited ESS power and energy, respectively, after the power loss is ignored, that is, the rated values of the ESS.

According to the research, a wind power fluctuation with a medium frequency (0.01 Hz to 1 Hz) will result in a large frequency deviation of a power system (see Figure 10.42). Thus, an ESS is specially designed for inhibiting power fluctuations with a medium frequency. Power of the ESS should be controlled through the following Formula (10.51).

$$P_{ESS} = k(E_{ESS}, P_w)\frac{\tau s}{1+\tau s}P_w \tag{10.51}$$

In the formula above, the energy level of an ESS is monitored and fed back through coefficient k, which is used for avoiding overcharge or overdischarge of the ESS. As in Figure 10.43, coefficient k is the function of the energy level and charging/incharging of the ESS. If the energy level is in the set safe range (between s_1 and s_2 in Figure 10.43), the ESS will operate in the normal operation area and with a coefficient k equaling to 1 and Formula (10.51) will be simplified as a high-pass filter. With a properτ, the power fluctuations between the target frequency range (between 0.01 Hz and 1 Hz) will be absorbed effectively by the ESS, otherwise energy level of the ESS will be in the lower or higher warning area. k differs in charge and discharge of the ESS, in a bid to bring the energy level back to the safe range. However, in these warning areas, performance of the ESS equipment will degrade and be limited to partial inhibition of power fluctuations only.

Effective energy capacity needed by the ESS can be calculated through Formula (10.52). In the formula, c_0 is a coefficient between 0 and 1; $P_{w,\max}$ is the maximum wind power, that is, rated value of the WTG.

$$A_{ESS} = c_0 A_{ESSfull} = c_0 \tau P_{w,\max} \tag{10.52}$$

If c_0 is 1, full capacity of the ESS can be calculated through Formula (10.52). Based on full capacity of the ESS, the normal operation area can cover the whole energy level range (for E_{ESS} between and 0 p.u. and 1 p.u.)

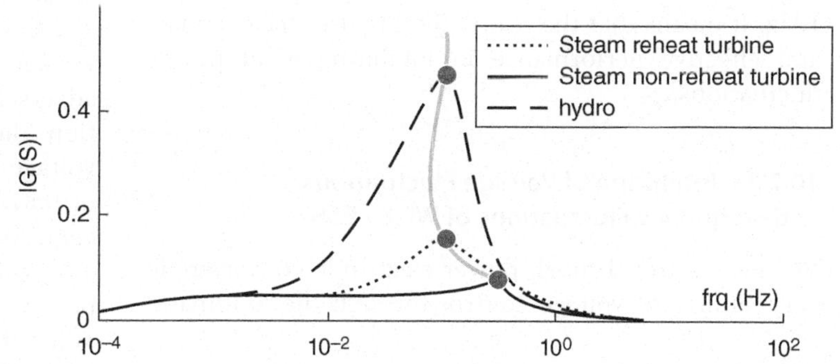

Figure 10.42 Frequency response to power fluctuation G(s)=Δω(s)/ΔPw(s).

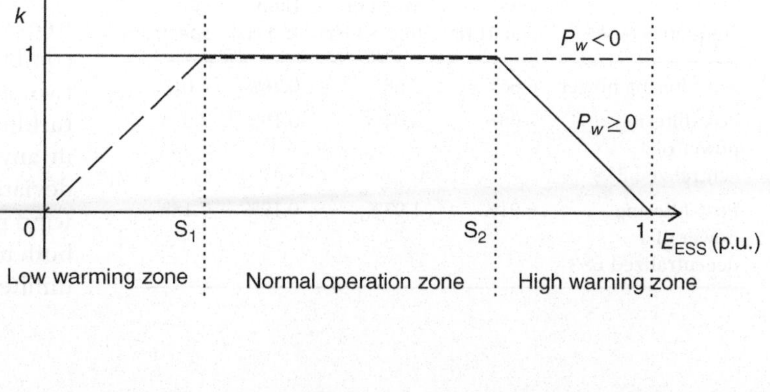

Figure 10.43 Operation area and curve k of ESS (solid line, $Pw \geq 0$; dotted line, $Pw < 0$).

and energy level of the ESS will never be beyond its physical limit even in the worst condition, that is, change of the pre-filtering wind power from zero to full value or in the opposite direction. However, in most of the time, the wind power fluctuation is much less than 1 p.u. (in normal operation), so c_0 may be smaller than 1 and capacity of the ESS may decrease. Upon a major wind power change before filtering, energy level of the ESS will enter the warning area. In this case, the energy level should be properly managed to be kept from its limit and return to the normal operation area. The cost of this is temporary performance degradation. If the c_0 is constant, the total ESS capacity needed by the WF is $c_0\tau\sum P_{w,\max}$, which is the same for both a centralized ESS and a decentralized ESS.

As in Formula (10.53), the effect of wind power fluctuation inhibition is measured with power fluctuation harmonic content (FHC) in the specified frequency range. In the Formula (10.53), F is the frequency range; P_0 is the average power. When F is the whole spectral range, the total FHC can be calculated through the following formula in the same way as THD. A larger FHC suggests a larger power fluctuation in the frequency range.

$$FHC(F) = \frac{\sqrt{\sum_{f_e(F)}(P(f)/\sqrt{2})^2}}{P_0} \tag{10.53}$$

Table 10.6 shows the FHCs. According to comparison result of the two FHCs after ESS power compensation, the two FHCs are both lower than the FHC of the pre-filtering wind power and, in detail, they are approximately 55% or less of the FHC of the pre-filtering wind power when the target frequency range is 0.01 Hz to 1 Hz. It means that the two ESS structures have similar and effective performance in inhibiting wind power fluctuations.

10.7.3 Inhibition of Voltage Fluctuations and Frequency Fluctuations of WF by ESS

Voltage control through power electronic converters is very similar to voltage control through the automatic voltage regulators of synchronous generators, which is based on the theory and operation experiences of STATCOMs. The three-phase AC voltage U_g output by a STATCOM is synchronized with the system voltage U_s through a transformer, and the reactive output power is adjustable through U_g. If U_g is larger than U_s, inductive reactive power will be output; if U_g is smaller than U_s, capacitive reactive power will be output.

The PCC voltage of a WF is adjustable by means of reactive power compensation. On a WF with a centralized ESS, the reactive power can be compensated by the ESS because the ESS power electronic interface (such as a voltage source converter) can provide reactive power and is not related to active power; on a WF with a decentralized ESS, the ESS units are connected with direct buses of the wind turbines. The grid side DC/AC converter of a wind turbine can provide controllable reactive power for voltage regulation. Figure 10.44 shows a simplified diagram of reactive power control structure for voltage regulation. Voltage U_{act} of the bus in connection with the energy storage device is compared with the reference value U_{ref}, and the reference value Q_{ref} (or reference reactive current i_{qref}) necessary for the compensator is obtained through the PI controller. The U/Q droop rate of feedback loop (typical value: 3% to 5%) matches the reactive power shared by wnid trubines. The droop characteristic will result in a small deviation for the steady-state voltage. For a centralized ESS, the reactive power is compensated through an ESS unit, so the droop feedback loop is omissible. Some controllers may include auxiliary signals, such as voltage error of the leading node (hub bus).

According to result of a simulation about a small capacity test system with mixed power generation in reference [19], wind power fluctuations influence local voltage fluctuations and system frequency fluctuations adversely. When there is no reactive power compensation, the wind power not filtered will result in serious voltage fluctuations at the PCC and a minimum voltage of 0.815 p.u; the voltage will drop to 0.847 p.u. (for a centralized ESS) or 0.859 p.u. (for a decentralized ESS). The wind power filtered by an ESS has less fluctuations because voltage fluctuation of the PCC changes slowly. When the reactive power is compensated by a centralized ESS or the WTG converter of a decentralized ESS, voltage of the PCC will be above 0.99 p.u.

Table 10.7 gives the maximum frequency deviation in frequency response of a power system. In the system, combination of three burbines is in the following forms: two steam turbines; one steam turbine and one water turbine; two water turbines. According to Table 10.7, in any of the three cases above, the system frequency deviation due to wind power and that due to filtered wind power by either ESS structure are comparable and both much lower than the frequency deviation caused by unfiltered wind power. In any of the combination forms,

Table 10.6 FHCs of wind power.

Frequency range	Low (<0.01 Hz)	Medium (0.01–1 Hz)	High (>1 Hz)	Full spectrum
Pre-filtering power	6.66%	3.68%	0.26%	7.61%
Post-filtering power of a centralized ESS	7.08%	2.04%	0.19%	7.37%
Post-filtering power of a decentralized ESS	6.91%	1.97%	0.19%	7.18%

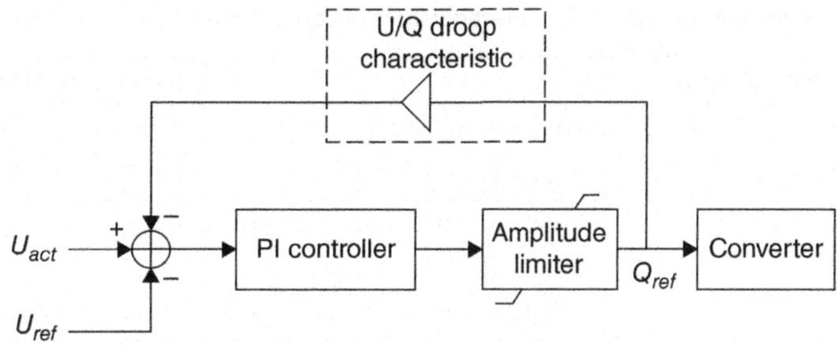

Figure 10.44 Reactive power controller for voltage regulation.

Table 10.7 Maximum system frequency deviation caused by wind power (Hz).

Wind power		Without filtering	Centralized ESS filtering	Decentralized ESS filtering
Combined with thermal power	Δf_{max}	0.06	0.01	0.01
	Δf_{min}	−0.05	−0.03	−0.03
	$\lvert\Delta f\rvert_{max}$	0.06	0.03	0.03
Combined with thermal power and hydro power	Δf_{max}	0.10	0.01	0.01
	Δf_{min}	−0.07	−0.04	−0.03
	$\lvert\Delta f\rvert_{max}$	0.10	0.04	0.03
Combined with hydro power	Δf_{max}	0.21	0.04	0.03
	Δf_{min}	−0.14	−0.07	−0.05
	$\lvert\Delta f\rvert_{max}$	0.21	0.07	0.05

the maximum frequency deviation caused by filtered wind power is less than 50% of that of the frequency deviation caused by unfiltered wind power.

Figure 10.45 specifies the per unit power of a centralized ESS of the system studied and that of a decentralized ESS relative to the WTGs' rated power and the WF's rated power. However, the smoothing effect of the WF changes with the wind direction. According to result of the study on several wind directions in the system, power of the centralized ESS recorded in the two-degree wind direction is 0.25 p.u. and γ_{WTG} is 0.25.

Table 10.8 shows the comparison result of rated power of the converter based on known γ_{WTG} and γ_{WF}. According to the table, in this case, the rated power of a centralized ESS is approximately 30% less than that of a decentralized ESS if smoothing effect of the WF is considered. However, rated capacity of the converter of the centralized ESS is still a large value approximately 1.24 times to 1.43 times of that of the decentralized ESS. Rated value of the converter depends on the ESS technology and structure because a centralized ESS needs to be connected with an AC cable with help of an additional converter.

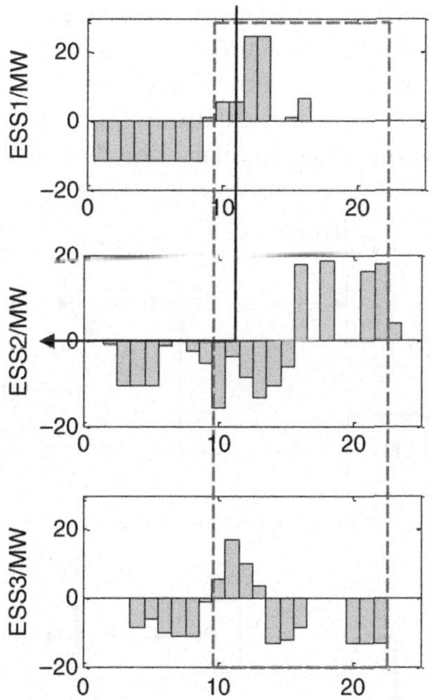

Figure 10.45 Power of decentralized ESS and centralized ESS (p.u.).

10.7.4 Control of Energy Storage System

The control of energy storage equipment is basically realized on two independent levels as below: An inner loop for regulating components d and q of a current; two outer loops for setting current components d and q to realize some performance indicators. Usually, the integral component of current i_q is set for direct regulation of the reactive power or the bus voltage, and current component i_d of the same phase as the system voltage is set according to the fed power or the absorbed power.

Table 10.8 Research about rated values of converter and ESS on a WF in a power system.

ESS structure		Decentralized ESS	Centralized ESS
Total rated power of the ESS		3.5 p.u.	2.5 p.u.
BESS, SMES, CS	Topology of the converter	DC/DC	DC/AC
	Total rated VA of the converter	7 p.u.	8.65 p.u.
FES	Topology of the converter	DC/AC	AC/DC/AC
	Total rated VA of the converter	12.11 p.u.	17.3 p.u.

When a power system is islanded, the energy storage converter may operate as a current source or a voltage source. The operation mode of a power system influences the operation control mode and the power grid connection. The several sections below introduce the different converter control modes in grid-connected operation and islanding operation, respectively [27].

10.7.4.1 Voltage Regulation

Figure 10.46 shows control algorithm for the voltage regulation of a power electronic converter. In the process of realization, direct axis component U_d of the voltage at joint of the energy storage equipment (refer to the d-q axis of the bus voltage) is compared with the reference value $U_{d,ref}$. The difference is fed into the compensator and the compensator will then regulate the reference current $i_{q,ref}$. Usually, the droop characteristic R_v used here is defined as a percentage change of the reference voltage value for rated reactive current (typical droop rate: 3% to 5%). In voltage regulation, some auxiliary signals about contents such as voltage deviation of the control (hub) node may be used.

10.7.4.2 Governor Control

Power electronic converters can be controlled with almost the same method as synchronous generators, with the only difference that the system frequency measured with a phase-locked loop (PLL) is regarded as virtual speed of the power electronic converters. As in Figure 10.47, active power of the power electronic converters is determined in this way by means of combination of the power set point and the governor control signal. The speed droop loop (dotted line) R added can provide power distribution between converters, which applies to the case that there are multiple inverters in the grid parallel connection mode or the islanding mode.

When operation is not related to a power grid, the synchronous mode (without dotted line in Figure 10.47) is needed to avoid a steady-state frequency deviation. When there are multiple converters, this function

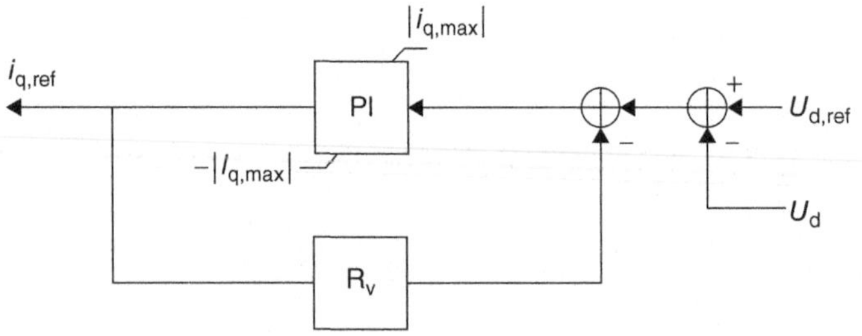

Figure 10.46 Box diagram of voltage control.

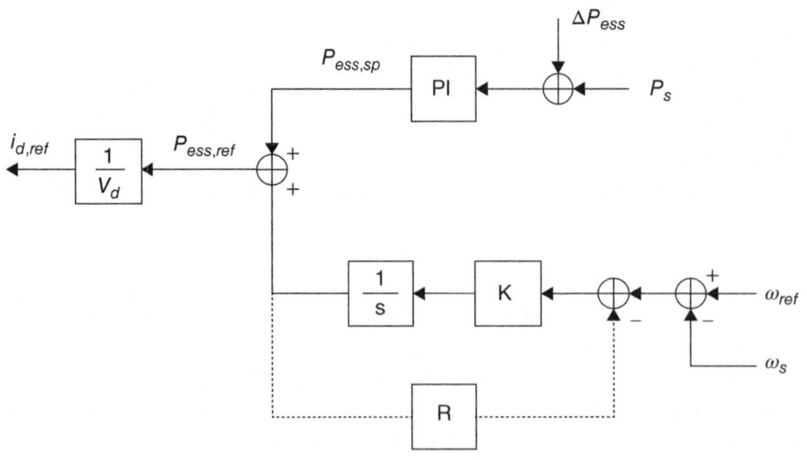

Figure 10.47 Block diagram of synchronous control over ESS converter.

enables an AGC signal to be transmitted between different units for regulation sharing or assigned to a unit that may operate synchronously while the other units are under speed droop control.

When an ESS is connected with a main power grid, the reference power set point is essentially related to the method of energy storage arrangement; during islanding operation, all the reference values will be determined by means of speed regulation control so that the frequency can be maintained. In Figure 10.47, the reference values have been set to make the power flow at the substation zero, which is predictable possible islanding operation; if the system does not need the main grid before the circuit breaker operation, the transient state during islanding operation is usually the minimum. Note that the figure includes an auxiliary signal (ΔP_{ess}) for equipment energy management.

10.7.4.3 Constant Frequency Control

Figure 10.48 is a block diagram about constant frequency control of the converters of an ESS. In the speed regulator control above, the static characteristic of the converter is like that of a current source—governor control and the voltage regulation box work together to determine the reference current ($i_{d,ref}$, $i_{q,ref}$) controlled by an inner loop. Or, the interface may be controlled as a voltage source during islanding operation, particularly when there is a single power source. In this operation mode, the rotational coordinates are not adjusted according to information of the PLL but that of the internal clock. The current control output is completely replaced by the set values of U_d and U_q. If there are multiple inverters, the control will be distributed to a single energy storage unit and the other units will be synchronized with the main unit.

The control mode has a quite simple structure and good performance in islanding mode, particularly when there is only one inverter. This proposes many requirements as below for the main inverter: the information about the circuit breaker state should be known; there should be a sufficient power capacity for the load; a sufficient energy storage capacity should be obtained from the direct bus to ride through the transient state from grid-connected operation to islanding operation.

10.7.4.4 Energy Storage Control

To maximize its benefits, an ESS's output power should be set as the expected objective and its special requirements must be considered in its control algorithm. For a battery ESS, the most effective operation conditions must be known in its control algorithm. A balance must be realized between reference power realization and optimal energy management. For example, the battery system should not operate out of its linear area, which is typically between 20% and 80% of the state-of-charge (SOC). The controller must keep the battery from exhaust or a long overcharge period.

To make the two usually contradictory objectives reasonable, control may have the following two main functions: one function is setting the reference power and correcting the reference value ΔP_{ess} according to need of the system to manage energy state of the energy storage device; the other function is realized with help of modern control means such as the fuzzy inference system, the neural grid or general rules based methods.

The structure may be expressed at high level, and what input there include typical SOC of the energy storage device and other information of the system as well, such as whether islanding operation or non-islanding operation of the energy storage device. In Figure 10.49, output depends on the internally defined specific rules of the structure. When the energy storage component is approaching to its upper (or lower) limit, the usual strategy is to increase (or decrease) the reference power.

In Figure 10.49, some other inputs not related to the ESS are referred, that is, electrical power changes

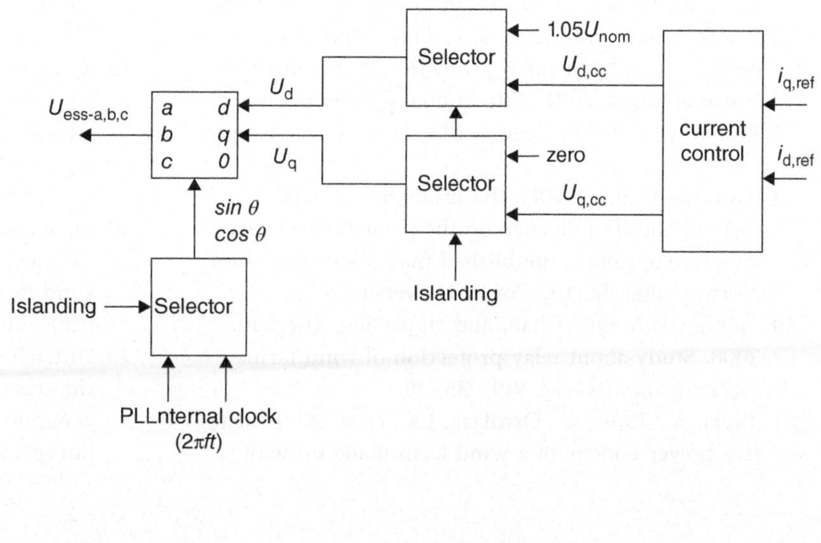

Figure 10.48 Block diagram about constant frequency control of the ESS converter.

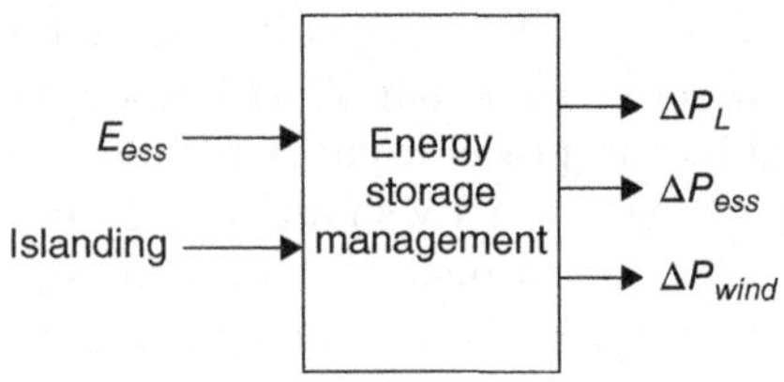

Figure 10.49 General expression of ESS planning method.

(ΔP_{wind}) and load changes (ΔP_L) of the WF. Additionally, there are some other available power sources with controllable power, in a bid to partially reduce the demand for the energy storage device and manage energy state directly. These backup structures are expensive and most of them are not possible to be used except during islanding operation; however, these backup structures are the only choice during islanding operation.

References

1 Daping, Xu, Xinfang, Zhang, and Yibing, Liu. 2005. Overview of relavant control problems of wind turbines. *Electric Power,* **38**(4), 70–74.

2 Iov, F. and Blaabjerg, F. 2009. Power electronics and control for wind power systems. *Power Electronics and Machines in Wind Applications.* IEEE June **24–26**, pp. 1–16.

3 Jingjing, Zeng, Ping, Yang, Chunmei, Xu, and Shiqin, Jiang. 2006. Study on control systems for wind-driven power generation. *Process Automation Instrumentation,* **27**, 32–35.

4 Andrecen, Björn and Johansen, Knud. 2008. Grid code and wind farm control requirements—what to control, why, where and how. 7th International Workshop on Large Scale Integration of Wind Power and on Transmission Networks for Offshore Wind Farms, Spain.

5 Ackermann, Thomas. 2005. *Wind Power in Power Systems.* New York: John Wiley & Sons.

6 Palsson M.P., Toftevaag T., Uhlen K., and Tande, J.O.G. 2003. Control concepts to enable increased wind power penetration. *Power Engineering Society General Meeting,* IEEE, 13-17 July 13–17, pp. 1984–1990.

7 Reichard, M.L., Finney, D., and Garrity, J.T. 2007. Windfarm system protection using eer-to-peer communications. 60th Annual Conference for Protective Relay Engineers. 27-29 March 27–29, pp. 511–521.

8 Hornak, Don and Chau, N. H. "Joe". 2004. Green power – wind generated protection and control considerations. 2004 57th Annual Conference for Protective Relay Engineers. March 30–1 Apr 1, pp. 110–131.

9 Guosheng, Yang. 2008. The protection of wind turbine and its influence on the protection of distribution system. Unpublished master's degree thesis, North China Electric Power University.

10 Yuling, Wen, Qin, Chao, and yibulaying, Tuerxun. 2008. Study about relay protection of wind farm. *Electrical Equipment,* **9**(9), 33–35.

11 Tapia, A., Tapia, G., Ostolaza, J.X., *et al.* 2001. Reactive power control of a wind farm made up with doubly fed induction generators(I). IEEE Porto Tech Conference, September 10–13, Porto Portugal.

12 Tapia, A., Tapia, G., Ostolaza, J.X., *et al.* 2001. Reactive power control of a wind farm made up with doubly fed induction generators(I). IEEE Porto Tech Conference, September 10–13, Porto Portugal.

13 Abbey, C. and Joos, G. 2005. Effect of low voltage ride through (LVRT) characteristic on voltage stability[J]. *IEEE, Transactions on Industry Applications,* **41**(3), 1–7.

14 Xing, Zhang, Xing, Longyun, Zhang, Shuying, Yang, Yong, Yu, and Renxian, Cao. 2008. Low voltage ride-through technologies in wind turbine generation. *Proceedings of the Chinese Society of Universities for Electric Power System and its Automation,* **20**(2), 1–8.

15 Hongliang, Guan, Haixiang, Zhao, Weisheng, Wang, Dai Huizhu, Dai, and Yihan, Yang. 2007. LVRT capability of wind turbine generator and its application. *Transactions of China Electrotechnical Society,* **22**(10), 173–177.

16 Mullane, A., Lightbody, G., and Yacamini, R. 2005. Wind-turbine fault ride-through enhancement[J]. *IEEE Trans on Power Systems,* **20**(4), 1929–1937.

17 Erlich, I., Winter, W., and Dittrich A. 2006. Advanced grid requirements for the integration of wind turbines into the German transmissions system [C]. IEEE Power Engineering Society General Meeting, Montreal, Canada.

18 de Leon. J.A.D., Kehrli. B., and Zalay. A. 2008. How the Lake Bonney wind farm met ESCOSA's, NEMMCO's, and ElectraNet's rigorous interconnecting requirements. Transmission and Distribution Conference and Exposition, T&D. IEEE/PES, pp. 1–5.

19 Li, Wei, and Joós, Géza. 2007. Comparison of energy storage system technologies and configurations in a wind farm. Power Electronics Specialists Conference, IEEE, June 17–21, pp. 1280–1285.

20 Dittrich, A. and Stoev. A. 2005. Comparison of fault ride-through strategies for wind turbines with DFIM generators. Power Electronics and Applications, European Conference.

21 Wessels. C. and Fuchs, F.W. 2009. High voltage ride through with FACTS for DFIG based wind turbines. 13th European Conference on Power Electronics and Applications, pp. 1–10.
22 Feltes, C., Engelhardt, S., Kretschmann, J., Fortmann, J., Koch, F., and Erlich, I. 2008. High voltage ride-through of DFIG-based wind turbines. Power and Energy Society General Meeting - Conversion and Delivery of Electrical Energy in the 21st Century, IEEE pp. 1–8.
23 Gjengedal, Terje. 2003. System control of large scale wind power by use of automatic generation control (AGC). *Quality and Security of Electric Power Delivery Systems*, CIGRE/IEEE PES International Symposium, October **8–10**, pp. 15–21.
24 Luis Rodríguez-Amenedo, José, Arnalte, Santiago, and Burgos, Juan Carlos. 2002. Automatic generation control of a wind farm with variable speed wind turbines. *IEEE Transactions on Energy Conversion*, **17**(2), 279–284.
25 Lubosny, Zbigniew, and Bialek, Janusz W. 2007. Supervisory control of a wind farm. *IEEE Transactions on Power Systems*, **22**(3), 985–994.
26 Wang Hongfu, Wang, Jun, Cao, Jiaju, Qiu, and Xinying, Wei. 2009. An active power compensation model for grid-connected distributed generation system. *Automation of Electric Power Systems*, **33**(8), 94–98.
27 Abbey C., Strunz K., Chahwan J., and Joos, G. 2007. Impact and control of energy storage systems in wind power generation. Power Conversion Conference - Nagoya, April 2–5, 2007 pp. 1201–1206.

11

Operation and Dispatch of a Power System Containing Wind Power

11.1 Overview

Conventional power systems are mainly composed of thermal power plants, hydropower plants, and nuclear power plants, which can maintain the outputs at specified levels, and enable the defined on/off load operation plans to be followed or change the outputs according to the power system requirements. After a large amount of wind power is connected to a power system, the wind power cannot be subject to centralized control completely because it is probably unavailable due to lack of wind when needed by the power system. Therefore, it is important to take the intermittent of wind power generation in various time ranges into consideration and adjust the controllability of wind farms (WFs) according to the operation requirements of power systems.

During the normal operation of a thermal power system, its units are dispatched according to the expected load demand curve, the spinning reserve capacity, and the specified reliability rule, as well as at the lowest operation cost. In unit commitment, output power of the units must be changed to match with the changes of the power system load demand in the dispatch cycles. It includes the capability to follow daily load changes (particularly the load increases in the morning and load decreases at night), the capability of frequency regulation and the capability to provide alternative power sources in case of lack of power generation or other serious incidents. The units for load following are required to comply with normal load fluctuations and unexpected demand changes.

With the rapid development of wind power and other renewable energy sources, countries have put forward different preferential policies and measures, including a requirement that the operators of transmission and distribution systems transmit and distribute the power generated by renewable energy sources while guaranteeing safety and reliability of these systems. The power generated by renewable energy sources should be dispatched first as long as it is allowed by power system operation. The fees related to grid connection and strengthening as well as the loads related to use of transmission systems and distribution systems should be fair and reasonable. Therefore, if massive wind power is connected to an existing power system, different power plants of the power system will be mixed and some conventional thermal power plants will be substituted, thus influencing the task of power system load frequency control. In this case, it is necessary to handle load levels, ramp rates and spinning reserve capacity of the generators again.

In addition, conventional units usually have a forced outage rate of 4% to 5%, which mainly depends on their operation years, technologies, operator experience, and so on. The daily unit commitment is carried out according to expectations rather than possibility. Therefore, unexpected outage of conventional units may happen, making reserve capacity of different types necessary. Generally, this outage cannot be forecasted. As a result, output power of the units will probably drop quickly from full load to zero. Sudden loss of wind power feed-in (it may be a result of internal problems of WFs or external problems of grids) is likely to cause problems similar to the loss of thermal power units, that is, frequency drop. In this case, the other synchronous generators will increase their output power to restore the balance between the power and the demands. However, a WF has multiple types of wind turbines most of which have independent operation, which improve the availability of the WF significantly. Therefore, compared with the loss of a thermal power unit in a similar capacity, the loss of a WF is less important. Even if the biggest WF in a power system is out of service, the short-term reserve capacity objective of the power system will not be influenced.

For the operation of a power system, wind power generation and conventional power generation are different in the aspects below:

1) The wind resources for wind power generation are changeable and instable while the supply of fuels including coal, oil, gas, water, and nucleus in conventional power generation is stable.
2) A single WTGS has a small capacity and its active power output and reactive power output are difficult

Integration of Large Scale Wind Energy with Electrical Power Systems in China, First Edition. Zongxiang Lu and Shuangxi Zhou.

to regulate and control. In particular, due to the limitation of wind conditions, it is impossible to increase the output power. The active output power and reactive output power of conventional power plants can be adjusted flexibly in large ranges.

3) Wind power can be forecasted, but wind power forecast is hard and inaccurate. Thus, making optimal allocation of the output power of units in power systems is complicated.
4) The reserve capacity of grids increases with increasing instable wind power in power systems. When there is no wind power in grids, the load consumption power is borne by conventional units and the reserve capacity of power systems only needs to handle load fluctuations; when there is wind power in grids, the load consumption power is borne by both conventional power units and WTGSs and the reserve capacity of power systems needs to handle both load fluctuations and wind power fluctuations (resulting in accumulation or offset of these two types of fluctuations).

Due to the natural intermittent of wind, the operators and planners of power systems face the following key problems [1, 2]: ① How does wind power generation influence instability of the net load (subtract the wind power from the load)? ② Which degree do the dispatchable reserve generators have? ③ Is the reliability of power systems endangered by wind power instability? ④ How does wind power instability influence the operation cost of power systems? ⑤ How to determine the capacity credit (CC) of a WF? ⑥ What are the effect and value of wind power forecasts? ⑦ How do the impacts of wind power on power systems change with wind power penetration? ⑧ How does wind power influence operation strategies of power system in practice?

In the development and practice of wind power, people are getting increasingly more understanding for the impacts of wind power on power systems. However, they still have some misunderstandings hindering the development of wind power. For example [1, 2]: ① It is believed that there must be 1 MW conventional dispatchable electric power as a reserve capacity for 1 MW wind power. In fact, massive wind power is added into a power system as an energy source rather than a capacity source. Therefore, the electric power for keeping a balance between load demand and power generation of the power system has already existed as a part of the power system. A little more capacity is necessary for regulation or load following, but most of it is usually provided by the existing generators. If the wind power penetration is low (lower than 10% of the total demand), the added demand for short-time reserve will probably smaller than 4% of the total capacity of WTGSs installed. For example, there was more than 6,000 MW of WTGSs installed in the United States in 2005, but there was no conventional generator installed as a reserve generator for wind power. After adding wind power, the reserve generators in the power system will probably used more frequently; however, it is not necessary to add a reserve capacity. ② It is believed that the output power of a WF probably drops to zero in several seconds frequently. In fact, the WTGSs in a WF withstand wind from different directions in any given time, resulting in spatial filter of output power of the WF. If more WFs are added in different positions, there will be an additional filter. Some statistical data of WFs proves that a WTGS needs several minutes or longer time to stop completely. ③ It is believed that upon sudden changes of wind, a WF will not output any longer and the whole power system will break down, which is closely related to the misunderstanding in ②. This is an unreasonable thought. If the WF fails, for example, it loses its boosting transformer and completely trips. It is the same as the case that a conventional power plant loses its boosting transformer. In other cases, the losing of wind power lasts for some time. Viewed from their impacts on a power system, the loss of wind power is not more serious than the loss of a conventional power plant. Furthermore, modern WFs actually help power systems handle major outage or emergencies in other places. The reactive power control and low-voltage penetration of modern WFs actually improve the stability of power systems. ④ The impact of wind power instability on the operation of other parts of a power systems results in an additional operation cost, which is the same as or larger than the wind power value. According to the studies of numerous cases, the additional operation cost due to wind power instability is about 10% of the selling price of wind power. If the wind power penetration is high (more than 20% of the total demand), the additional operation cost of a power system due to wind power instability and uncertainty will be about 1 to 4€/MWh and be highly various in different countries and regions.

The time frame necessary for the comprehensive study of a power system containing wind power ranges from several seconds to several years. The time frame of the dynamic stability study for a power system is milliseconds to several seconds. Evaluation of the contribution to wind power capacity increase for the increasing load damand lasts for one to ten years or a longer time. Evaluation of the operation cost of a power system lasts for one year at least, or, usually, several years. This process needs operation data of several years. The impacts of wind power instability on the operation cost are obtained by means of the inspections of three different time frames as below.

1) Frequency regulation (not interfered by operators but handled through an AGC and a governor) lasts for several seconds to one minute roughly;

2) The time frame in load following (handling of load and power generation forecast changes through economic dispatch, and the operator as well as management according to the daily unit commitment plan) lasts for several minutes to roughly two hours;
3) The time frame in unit commitment (optimization) (handled by the dispatcher according to the load curve forecast and the available unit characteristics) lasts for future several hours to one or several days.

Wind power generation makes the operation of power systems more uncertain. The changes essentially exist on the whole time frame and can be forecasted with an accuracy lower than the previous load forecast. For large power systems, the geographically distributed wind power fluctuates smoothly and quickly. Changes of the natural diversity of power sources is beneficial. Wind power influences a power system more with the increase of its proportion in the power system. Technically, it is possible to integrate massive wind power with power systems and the constraints are the socially acceptable costs and the economically acceptable costs.

This chapter introduces the general requirements for WF operation from the aspects of power generation and power system operation, requirements for the power balance and reserve capacity of power systems due to wind power, combination, and optimal dispatch of the units of power systems with massive wind power; participation of wind power in the power market, and the technologies making WFs similar to conventional power plants.

11.2 General Requirements for WF Operation

The dynamic state of a power system mainly depends on the generators of the power system, so it is important for the grid operator, the WF producer and the technology supplier (WTGS manufacturer) to understand the interaction between the WF and the grid from the following two aspects: How do the WF and its WTGSs respond to faults and events of the grid (sudden voltage drops caused by grid faults)? How does the WF influence stability of the power system (including the factors tending to result in faults and the factors related to grid transient stability, voltage stability, frequency stability, and critical fault clearing time)?

11.2.1 Impacts of WFs on Power System Operation

The impacts of wind power on the operation of a power system mainly relate to scale and flexibility of the power system and proportion of wind power (wind power penetration) in the power system.

The impacts of wind power on the operation of a power system include short-term impacts and long-term impacts. The short-term impacts mainly include the power system's power balance lasting for minutes to hours because of wind power instability. The long-term impacts mainly relate to the contribution of wind power to adequacy of the power system (wind power's capability to meet highly reliable peak loads).

The impacts of wind power on the operation of a power system include local impacts and systematic impacts on the whole power system. A local impact is that the WF interacts with the grid voltage and influences the electricity quality and the voltage control on or close to the WF, which is like that a conventional power plant does. Wind power can provide voltage control and active power control. The WF can decrease loss of the transmission and the distribution grid. A systematic impact is that the WF influences the voltage and the grid power flow. When the WF locates at the load center and has low wind power penetration, it is particularly beneficial to the power system. The WF must add or strengthen infrastructures of the transmission grid and the distribution grid as any power plant in connection with a grid does. To transmit the power generated by a distant and large resource point of wind power to a local center, it is necessary to erect a new line. The connection of wind power into a transmission grid smoothens fluctuations of the geographically distributed wind power. The regulation control capability of wind power (including secondary control and third control) influences efficiencies of other generators in the power system. Table 11.1 introduces overview and classification of the impacts of wind power on a power system.

Connection of massive wind power into a power system is a new challenge for the power system, particularly for the relevant transmission system operator (TSO). Reliable power supply includes balance between the load demand and power generation. When the wind power penetration is low, wind power can be treated as a negative load (e.g., wind power reduces the total load demand), so its impact on the power system is small; when the wind power penetration is high, operating method of the power system needs to be considered again because wind power cannot be dispatched with the same certainty as conventional power generation.

For power systems, their important wind power fluctuations are mainly caused by the daily climate mode; however, the power fluctuations caused by disturbance or gust have not yet generated any impact on power systems.

Short-term limited wind power fluctuations can be well handled through existing power system capacity. For example, it is not necessary to add a secondary reserve capacity (a short-term reserve capacity available in 5 min) even though there is a connection of wind power to a power system, which has been proved by Spain and Ireland. Wind power forecast is important for

Table 11.1 Impacts of wind power on a power system [3].

	Impact or impact factor	Scope of impact	Duration of impact	Contribution of wind power
Short-term impacts	Voltage management	Local	Several minutes	The WF can provide (dynamic) voltage support (depending on the design).
	Production efficiency of thermal/ hydropower plant	Systematic	1–24 hours	After the wind power connected to the power system reaches a specified quantity, it will make the thermal power or hydropower unit operate below the optimal level (such as start, power-on or partially loaded) to reduce operation efficiency of the power system. In this case, the optimal power distribution among the units will be complicated due to instability of the wind power output. This problem can be solved by means of accurate wind power forecast.
	Efficiencies of the transmission grid and the distribution grid	Local or systematic	1–24 hours	It depends on the wind power penetration. The WF will probably results in an additional investment cost or reduce losses of the distribution grid or the transmission grid. For example, if there is massive wind power connected to the power system, the power system will have a larger loss and the transmission grid will probably have blockage because of frequent interregional power exchanges.
	Reserve regulation	Systematic	Several minutes to several hours	If the wind power connected to the power system is not massive, it will not influence the power system's primary reserve capacity or secondary reserve capacity greatly; impact of the wind power connected to the power system on reserves of the power system increases with its quantity; if the wind power connected to the power system is higher than a specified critical point, capacity cost of the reserves need calculating.
	Wind power abandonment	Systematic	Several hours	When the wind power penetration is at high level, the wind power connected to the power system is probably larger than the safe absorption capacity of the power system. In this case, the wind power needs to be reduced to guarantee adequate and reliable operation for the power system.
Long-term impact	Power system reliability (operation and power transmission adequacy)	Systematic	Several years	In some cases (e.g., the power system has a large load, but multiple units are out of service because of faults), wind power can contribute to adequacy of the power system; however, when the wind power connected to the power system has reached a specified quantity, its CC will tend to be a constant.

the daily operation of a power system with high wind power penetration.

The requirement for a long-term power balance is a result of the inherent wind velocity forecast errors of wind power. Wind power is less forecastable than the conventional power generation which the TSO is accustomed to. Day-ahead wind power forecast is probably quite different from actual wind power. In order to indicate the importance of accurate wind power forecasts, the total power curve (Figure 11.1) for all WTGSs is used in a power system in western Denmark. When the wind velocity is from 5 m/s to 15 m/s, the integrated power curve has a so large slope that there is an error, caused by the only 1-m/s-fault forecast, which is about 320 MW between forecast wind power and the actual value. Most of TSOs in Europe are responsible to balance power systems (usually balance wind power generation) or, in other words, they must handle the imbalance between forecast wind power and actual wind power. Therefore, it is usually necessary to use a long-term reserve capacity frequently to realize higher wind power penetration. At present, the power systems in Europe can provide adequate long-term reserve capacity and therefore need no new power plants for reserve capacity [3].

11.2.2 Operation of Power Systems Containing Wind Power in the Eyes of Power System Operator

As the wind power penetration of a power system increases, the power system operator needs to adopt an innovative operation method for the sake of easy connection of wind power and consider the following

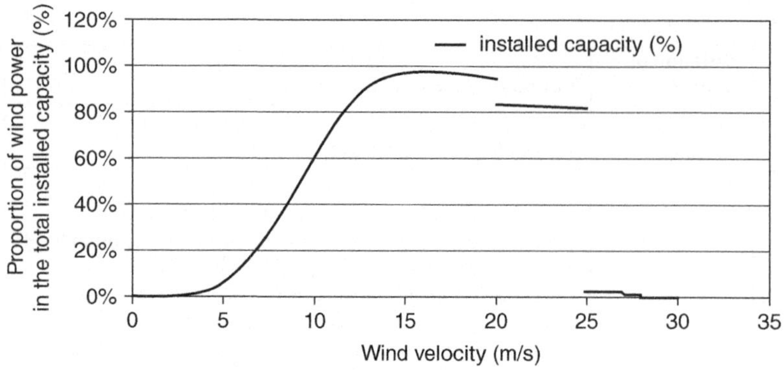

Figure 11.1 Integrated power curve related to installed capacity of wind power in western Denmark.

requirements: To solve the uncertainty of wind power generation, the power system needs an adequate and reliable reserve capacity; the power system needs to meet the emission objectives; the power system operation policy should comply with trading arrangement of the power market; increase of the wind power penetration should not influence dynamic stability of the power system, and so on. To sum up, the operation cost should be minimized based on safety and environmental constraints of the power system [3].

11.2.2.1 TSO

A TSO should make use of the power transmission lines in region of the power system and the tie lines close to the power system in its task to balance and safely operate the power system. The issues to be handled include blockage management, priority of passing, and, in the critical conditions (a low load demand and massive wind power), preferred reduction.

TSOs are used to handling the instability and uncertainty of loads. For a power system with wind power, its instability and uncertainty are quite different from a power system without wind power. It is necessary to consider instability and uncertainty of the load and the wind power output comprehensively and analyze the net load instability (with or without wind power) in different time frames to determine the demand for power balance of the power system.

Voltage control is necessary for a power system (such as a big WF) to adapt to unexpected voltage changes (the voltage probably increases due to instable wind power). If a proper wind power generation technology is adopted, this voltage support can be provided by the WF, otherwise special equipment such as FACTS equipment will be necessary.

Another problem refers to power flow management and possible grid blockage. A specific combination of wind power generation capacity, geographic position, and load demand probably changes amplitude and direction of the power flow in the transmission grid and further power flow in the interconnected system. For this purpose, a TSO needs high-quality tools for wind power forecasts. Moreover, FACTS equipment such as phase-shifting transformer is useful for power flow management.

11.2.2.2 Distribution Grid Operator

WTGSs are usually connected with distribution grids. Distribution grids are characterized in that they do not have positive and effective management as transmission grids do. They must have high capabilities of distributed power generation while not lowering the power supply quality expected by users.

Actually, the "embedded power generation" of wind power is an advantage of grids. In detail, weak distribution grids can be supported by wind power and users on the lines can enjoy a better power supply service. For example, wind power can increase grid voltages and the power electronic equipment of WFs can improve the electricity quality of grids. If consumed in a distribution grid, or, in other words, transmitted to users directly, the wind power generated by a WF will be in the low cost of power transmission. Moreover, if the transmission grid fails, wind power can keep part of a power system operating to avoid power outage of numerous users. Additionally, the increase of wind power has the following problems similar to those in a transmission grid: directions and capacity of the active power flow and the reactive power flow will be changed, which will probably affect the operation of grid control and protection device; there will be more decentralized wind power; traditional design and operation rules of the distribution grid will probably not be applicable any longer; management of the distribution grid will probably must be more positive. This needs proper equipment and design principle development. However, improvement of the distribution grid will benefit the grid operator and the end users indirectly.

11.2.3 Requirements of Power Systems for the Grid-Connected Operation of WFs

Each country (grid company) has a specification for the grid-connected operation of wind power. See Table 11.2

Table 11.2 Basic requirements of a grid specification for a WF [3–5].

Active power control	The WF should control active power positively to guarantee power system frequency stability, avoid line overload, and so on.
Frequency control	Make the power system frequency acceptable to guarantee safe power supply, meet the electricity quality standard, and avoid overload of the electrical equipment.
Frequency/voltage range	The power system can continue to operate even when it has a problem, namely, its voltage or frequency, is not normal.
Voltage control	The WF should realize voltage control and reactive power control and support the grid.
Voltage quality (quick change, flicker, and harmonic)	Comply with related requirements of the applicable specification of the state, and, if necessary, install a dynamic reactive power compensator.
Transformer tap conversion	Some grid specifications make it necessary for a WF to install a transformer with adjustable taps so that the voltage ratio between the WF and a grid can be changed whenever necessary.
WF protection	This is a special requirement for the case that the grid has a fault or disturbance. The relay protection system should operate upon faults, large short-circuit currents after faults, low voltages, or overvoltages. The WF should guarantee normal operation of the grid and support the grid during and after faults. For example, there should be fault ride through (FRT).
WF model and validation	The WF owner or developer should provide a model and system data so that the operator can study interaction between the WF and the power system by means of simulation. There should be monitoring equipment for checking actual characteristics and model of the WF during faults.
Communication and external control	The WF producer should provide important information about lots of parameters (such as voltage, active power, reactive power, operation state, wind velocity, and wind direction) to enable the power system operator to operate the power system correctly. Some grid specifications make it feasible to connect and disconnect WTGSs from the exterior.

for the requirements of a power system for the grid-connected operation of a WF.

More detailed descriptions of some requirements in the table above:

1) **Frequency control:**
   The grid expects the WF to participate in primary control and secondary control as a conventional power plant does. The frequency control includes frequency response capability control, ramp rate control, and active power limit control.
   a) Frequency response: It refers to the capability of a power plant (including WF) to respond to frequency changes of the power system and change active output power.
   b) Ramp rate limit: Limit change rate (ramp rate) of the positive or even negative value output by the active power, so as to avoid large frequency fluctuations due to extreme wind changes or start or stop of a WF.
   c) Frequency range: In a synchronous power system, the normal frequency is the same and usually in a narrow bandwidth. Previously, a grid operator expected the WTGSs to disconnect from the grid upon frequency drops. However, if the WTGS capacity added disconnects upon low frequencies, frequency recovery capability of the power system will be lowered. In order to guarantee safe operation of the power system in normal conditions and avoid the problems during frequency recovery, the grid needs the WTGSs to be connected with the power system in a wide frequency band.
2) **Voltage control:**
   The grid needs the WTGSs to operate continually with the rated voltage and the rated output power, the generator terminal voltage to be constant and the WF's grid-connected point voltage to be in the specified range. When the voltage drop is in the specified range, the WF should be connected with the grid and provide partial reactive power capacity and voltage control.
3) **Fault ride through:**
   The fault ride through or high low voltage ride through capability refers to capability of the WTGSs to keep stable and connect with the grid during grid faults.
4) **Electricity quality:**
   This concept is expressed in a characteristic parameter set for determining impact of the WTGSs on voltage quality of the grid. In principle, this concept can be used for both transmission systems and distribution systems; however, it is more important for distribution systems. The corresponding parameters include active power, reactive power (including the

maximums), voltage fluctuation (flicker), switching times (because of voltage changes), harmonic current, and corresponding quantity.

11.3 Power Balance and System Reserve Capacity

11.3.1 Balance of Electric Power Energy and System Reserve Capacity

During operation of a power system, it is necessary to keep a balance between power generation and power consumption. Power balance is a balance between power source and load of the power system. During planning and operation of the power system, generation capacity of the power system should be determined in compliance with forecast load of the power system. When a certain model is used for the power balance, all power plants are arranged on the typical daily load curve based on given data including load of the power system, installed capacity of the power plants, and hydropower compensation of regulation and by means of the specified rule or optimal objective function, so as to determine operating positions and capacity of the power plants on the load curve and obtain data of each power plant, such as generation capacity, reserve capacity, spare capacity, and surplus water quantity. If random outage of the units is considered, it will be necessary to do reliability analysis and give reliability indexes.

The balance of electric power and energy should be calculated by power plant (such as hydropower plant, coal-fired power plant, nuclear power plant, variable-load power plant, pumped hydroelectric storage (PHS) power plant, and thermal power plant). According to the specific purpose such as daily dispatch of power generation, unit overhaul arrangement, construction progress of a new power plant, the balance of electric power and energy should be calculated each day, week, half a month, month, year, and so on.

Power generation must be adjusted with load changes because of variation and uncertainty of the load and power generation, such as changes of the load and forced outage or overhauls of the units. In order to keep the system frequency in specified range and keep supplying electricity to users, capacity of the generators should, in addition to meeting load demand of the power system, have reserve capacity. The reserve capacity includes a load reserve capacity, an emergency reserve capacity, and an overhaul reserve capacity. In China, the amount of the three reserve capacities above should be 22% of maximum load of the power system, including 3% for the load reserve capacity, 10% for the emergency reserve capacity, and 9% for the overhaul reserve capacity.

The basic parameter for representing energy balance of the power system is frequency. The system frequency will increase when generation output is larger than power consumption or decrease in the opposite situation. A duty of the power system operator is to keep power balance at any time.

The power balance of conventional power systems is realized by means of generation capacity, load management, energy storage, and so on. The balance of power systems is not a new problem, but wind power generation brings lots of new challenges. After wind power reaches a certain level, these new challenges will need to be solved.

There are several operation time ranges of a power system, which is from a second to several days. At the beginning, the primary reserve capacity works automatically according to the frequency fluctuations and prime motors control of the generator quickly responses (usually from 30 s to 60 s). The power imbalance is probably a result of a thermal power unit trip or sudden disconnection from a large load. In order to recover power balance, the prime motors need to respond immediately to make the system frequency stable again. To adapt to this power balance recovery based on almost immediate response, the power system must have a sufficient power generation reserve from the operating units.

Secondary reserve capacity is used for manual or automatic control over active power within 10 minutes to 15 minutes when the frequency deviate from rated value. As shown in Figure 11.2, they support the reserve capacity and operate until replaced by the long-term reserve capacity. These include a spinning reserve capacity (a part-loaded hydropower plant or thermal power plant) and the reserve capacity not in operation (a quick-start gas turbine power plant or load shedding).

The changes of load and power generation resulting in power imbalance are not forecast or preset, the primary control and secondary control maintain to keep the system frequency close to its rated value. The power consumption changes each day, each hour, and even each minute. Thus, in order to keep power balance, each power generation plan must match the load changes. This economic dispatch strategy is made on the scheduled demand trend (the primary control and the secondary control respond to the non-forecast imbalance continuously). For example, a load usually increases in the period from 7 a.m. to noon or after noon, begins to drop upon the daily peak load and reaches its minimum after the midnight.

Some generators need a couple of hours to get started, synchronized, and connected to a grid. It means that the generators available in the peak load period at noon must be started a couple of hours in advance. In many circumstances, units need a long time to go down and probably need a couple of hours for cooling before restart.

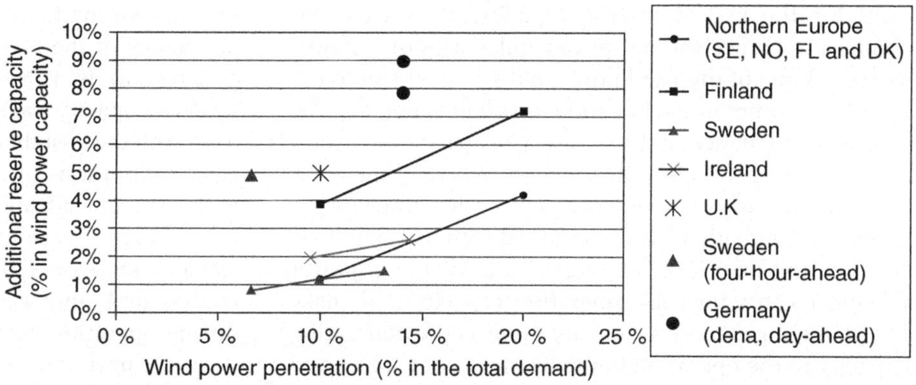

Figure 11.2 Increase of reserve capacity with wind power penetration evaluated in different European studies.

The decision for the units of this type usually includes a time period of a couple of days. The time period is called unit commitment ranging from a couple of hours to a couple of days, depending on the specific generator characteristics and operation practice.

During operation, the task of power balance is usually handed over to power system operators from the power producers. There is a cost benefit because errors between producers and loads are smoothed during system integration. In this case, the control frequency only needs to balance the net imbalance power in the power system. Power system operators have plan information about power generation, power consumption, and use of the interconnected grids. These plans are provided by power system operators, the power market, or other members. They probably make use of online data as well as load forecast and wind power forecast both useful for operation. They operate power systems, arrange the producers with generators and loads to serve as reserve capacity, and encourage the producers to balance the net imbalance of power systems according to their needs.

It must be stressed that the needful work for the power balance is not just dispatching power plants (such as WFs) of specific types but handling all uncertainties of the imbalance between demands and power generation. In addition, the uncertainties managed in the operation of a power system include two types if classified by the composite effect of fluctuations: the fluctuations of demands, and, the fluctuations of conventional power plants and renewable energy power plants. These independent fluctuations are usually not relevant but have a mutual smoothing effect, which is useful for cutting down the operation cost of the power system.

The operation procedures of power system operators change with synchronization systems and countries, and so do terms of the reserve capacity. The reserve capacity includes two types if classified by time range: all the operation reserve capacity within a second or minute time range are primary reserve capacity and all the operation reserve capacity within a ten-minute or hour time range are secondary reserve capacity or third reserve capacity. Primary reserve capacity is also called monetary reserve capacity, frequency response, or automatic reserve capacity. Secondary reserve capacity is also called quick reserve capacity and third reserve capacity is also called long-term reserve capacity (term "load following reserve capacity" are also used for these two types of reserve capacity) [3].

11.3.2 Impacts of Wind Power on the Reserve Capacity of Power Systems

If massive wind power is added to a power system, it will affect the power system in the time ranges of day, hour, minute, or even second as any other large power source does. One of its impacts can be expressed in the additional power reserve capacity needed.

During operation, wind power probably affects primary reserve capacity and secondary reserve capacity. However, the time ranges of related changes of massive wind power are the time ranges (ten-minute/hour) of secondary reserve capacity and the adding of wind power does not affect primary reserve capacity at all. In the range of second or minute time, quick changes output of the total wind power capacity is random as the existing load changes are. When the changes of loads and power generation are considered, the changes due to wind power have small increments. Moreover, the primary reserve capacity configured in a power system depends on outage of the large thermal power plant. Thus, the existing primary reserve capacity in a power system are much larger than the capacity for response to the quick changes of wind power.

Impacts of wind power on secondary reserve capacity or third reserve capacity are related to changeability of the massive wind power in the operation hours (the time in which the power system operator takes control). However, even in an extreme condition, the hourly output change of a decentralized WF in a big region is not larger than ±10% of the rated capacity (a grid covering four northern European countries), for a small region, not larger than ±25% of the rated capacity (a grid in western Denmark).

As for the use of reserve capacity, all unexpected changes influence the net power imbalance of a power system. Some of the wind power output is considered in the plan of other units, but some is still unexpected. The total net imbalance and the reserve capacity necessary for the power system both depend on the imbalance, which is sum of errors of wind power with other power generation and other loads compared with the plan (particularly the load forecast error). In principle, a power producer can adjust its generation capacity and make the planned generation capacity same as the generation capacity in the operation hours by means of trading and replanning its conventional generation capacity. If they do so, the secondary reserve capacity will be influenced by the wind power changes in an hour. Studies show that wind power penetration is larger than 10% of the total power consumption, the impact of wind power on secondary reserve capacity will increase greatly. The greatest impact of wind power lies in the way of planning of conventional units for the sake of load following (the hourly level to the daily level). During load increase in the morning, more generation resources with the load following function must be distributed. If the wind power generation capacity increases and is reliably forecastable, the power system will need a little generation resources with a function of load following. On the contrary, if there is less wind power output during load increase in the morning, more generation resources with a function of load following will be needed. If the output of a WF can be accurately forecast one or two days earlier, it will help the dispatcher to select the units for dispatch. If there is no accurate wind power forecast, the dispatchment of units is uncertain. As a result, a unit is arranged for dispatchment when not needed or not arranged when needed. In this case, degree of mixing of power generation determines the wind power generation capacity: the more flexible units in the power system, the more the acceptable wind power will be.

Results of the previous domestic and foreign studies about the balancing capacity and cost caused by wind power increase were still preliminary. The reasons are that the studies were based on different assumptions; there were no adequate data about uncertainties about wind power and loads; there were no maximum cost benefit choices assumed for system balance.

The time range of wind power forecast is crucial for determination of impacts from which the balancing capacity and cost needed bring to us. In a second or minute time range (primary control), impact of evaluation result of the reserve capacity added is small. It is because that the short-term wind power changes cancel out each other because of their noncorrelation, particularly when they are combined with other quickly changing loads and units. In a time range of 15 minutes to an hour (secondary control), wind power penetration increase strengthens the impact of wind power on net imbalance of the power system. When the wind power penetration is 10% of the total load, the additional reserve capacity caused by wind power are about 2% to 8% of the installed capacity of wind power. In this case, calibrations of errors of the wind power forecasts with different advanced time through reserve capacity have different results. As for determination of reserve capacity increase, day-ahead forecasts are used in some studies and only the degree of wind power changes during operation are observed in some other studies. If there are proper forecast tools and an intra-day dispatch system with the wind power (energy) penetration of 10%, the additional reserve capacity will be 2% to 4% of the installed capacity of wind power [3].

Viewed from power system reliability maintenance, wind power influences secondary reserve capacity most. Due to its random changes, wind power influences primary reserve capacity lightly. When wind power is combined with loads and power generation, its alteration is usually small or can cancel out each other. In countries with massive wind power or solar cell capacity, such as Germany, Denmark, and Spain, their primary reserve capacity necessary for power system reliability maintenance has remained unchanged. According to the comprehensive studies in Germany and the United States, primary reserve capacity needs small increments only. By 2015, wind power will take 14% of the net power consumption of Germany, which can be realized without any additional reserve capacity or additional demand for power plant balancing. Wind power related regulation and reserve capacity can be replaced by conventional generators and realized by means of proper operation modes and without installation or operation of additional power plants [3].

Figure 11.2 introduces the results of several comprehensive studies about additional reserve capacity caused by wind power in Europe. It is difficult to compare these results because the study methods were quite different: Dena in Germany evaluated its reserve capacity demand based on an advanced time of one day; the corresponding evaluations in U.K. and Sweden were based on an advanced time of four hours; the other countries evaluated the impacts of wind power instability on reserve capacity during operation hours. Usually, reserve capacity increase with the increase of wind power penetration.

The additional balanced power of wind power (expressed in percentage of the wind power installed capacity) is a function of wind power capacity penetration.[1] Forecast improvement can reduce the regulating

1 Suppose that mean ratio between a peak load and the mean demand is 1.44 and that the mean wind power capacity factor is 25%, then wind power capacity penetration will be 2.8 times of the wind power penetration.

power. If the wind power penetration is 20% and a continuous method is used for forecasting, 7% of the wind power capacity will need to be balanced. If the forecast is accurate, only 2% of balancing capacity will need to be increased. This value is a relative value and a trend, suggesting the impact of wind power forecast improvement on reserve capacity reduction. The absolute values of reserve capacity depend much on the specific region, power system and power market.

11.3.3 Method and Calculation to Determine the Increment of Wind Power Reserve Capacity

11.3.3.1 Method for Determination of the Reserve Capacity

The reserve capacity of a power system includes load reserve capacity, emergency reserve capacity, and overhaul reserve capacity. Total reserve capacity of the power system is usually about 25% of the maximum load of the power system.

The load reserve capacity is the spinning reserve capacity of the power system. It is used for balancing the transient load fluctuation and the forecast load. Its value is from 2% to 5% of the maximum generation load. The load reserve capacity is rationally distributed between the power plants; coordination of the thermal power plants, the hydropower plant, and the limitation of the line transmission capabilities should be both considered according to the specific seasons. In dry season, the load reserve capacity is usually undertaken by the hydropower plants with a capability of daily regulation at least first.

The emergency reserve capacity is a capacity for avoiding outage of the power system upon an accident of any unit or sudden drop of the expected output of a power plant. It should be put into operation in the specified time (such as 10 min) and at least a part of it (such as 50%) can be put into operation automatically. It should not be less than the generation capacity probably lost in a single accident. Generally speaking, it should be 10% of maximum load of the power system and not be less than capacity of the largest unit in the power system.

The overhaul reserve capacity is necessary for the units scheduled overhauls of the power system. It should be determined in the annual overhaul plans when there is a power balance. According to the preliminary calculation result, it may be 8% to 15% of maximum load of the power system.

The reserve capacity of conventional power systems is easy to be quantized mainly with a deterministic method. For many power systems, the reserve capacity at any time are just enough to make up its maximum lost input (such as the largest unit or power plant) or a percentage of its maximum load. Load forecasts result in some uncertainties, but power system operators know them well and can usually manage them. This deterministic method cannot guarantee safety for a power system in all time, but it assumes that it is rare that any generation capacity loss is larger than the maximum input. Thus, it can be believed that it is not necessary to keep additional reserve capacity in a whole year. If this happens, the power system will have to execute load shedding. This simple method for quantizing reserve capacity demands has been successfully applied in lots of power systems in the world. However, if the wind power penetration increases, we should pay attention to uncertainty of the wind power output. It is because that unexpected wind changes are probably comparable to the trips of large generators.

The reserve capacity of a power system can be determined by means of calculation of the loss of load probability (LOLP). For example, in countries such as the United States, Canada, and Spain, it is required that the LOLP be 0.1 d/a and the corresponding reserve capacity be 28% to 29% of maximum load of the power system; in countries such as Japan and the UK, it is required that the LOLP be 0.1 d/a to 0.4 d/a.

In the studies about wind power reserve capacity, some new methods were proposed. According to reference [6], the reserve capacity margin of a power system in Sweden integrating wind power, hydropower, and thermal power is determined according to forecast errors of both wind power and loads, as well as the ramp rate of the conventional thermal power unit. The determination is based on assumption that correlation of inter-region and various regions wind power forecasts is known, in addition, the assumption that the reserve capacity is related to the probability of an underfrequency due to wind power fluctuations and load fluctuations. According to the study about the impacts of wind power capacity on generator loads, power system reserve capacity availability and generator ramp rate in reference [7], it is believed that power system reserve capacity availability depends on whether the forecast wind power is combined with the unit commitment or not. In reference [8], the technical logic relation of high wind power penetration is quantized by means of primary, secondary reserve capacity, and long-term reserve capacity. In the application of an interconnected power system in Germany, it is believed that it will result in significant changes of a reserve capacity. It is interesting that the authors believe that negative secondary reserve capacity is necessary to avoid the power surplus when a WF has a large unexpected output increase. In reference [9], the impacts of different forecast techniques and methods on the saving of fossil fuels and the spinning reserve capacity of a large power system is evaluated. The evaluation conclusion is that the advantages of the saving fossil fuels and the spinning reserve capacity of a large power system can be strengthened by more accurate forecast techniques than the continuous method. However, the spinning reserve capacity demand increased is calculated as a simple fraction, which is caused by the wind power forecast or its error. Reference [10] discusses the improvement of unit commitment and frequency control when there is massive wind power and tries to determine the constraint of wind power penetration in the worst conditions.

According to reference [11], when the reserve capacity of a power system is quantized, reliability of the power system must be used as an objective in the evaluation of the reserve capacity needed in different conditions. The section below introduces the method to quantize the reserve capacity of a power system based on specified reliability of the power system, the uncertainty of load forecasts and wind power forecasts, and the probability of generation capacity loss.

11.3.3.2 Method to Determine Reserve Capacity When Reliability is Regarded as an Objective

In order to solve any possible insufficient power generation that is unexpected, a power system needs certain reserve capacity. Insufficient power generation is probably a result of generator outage or unexpected load increase or shortage of wind power. Actual changes of wind power loads do not influence the reserve capacity of a power system, but the accuracies of load forecasts and wind power forecasts relate to the reserve capacity of a power system closely because they result in serious uncertainty for the power system.

1) **Power system reliability rule**
   The reserve capacity of power system depends on certain rules for power system reliability. In power system reliability analyses, many different reliability rules are used; most of power system reliability analyzes focus on power generation adequacy calculation based on a consideration of the disconnection probability of load and power generation. This method applies to operation and dispatch of the power system based on a consideration of the disconnection probability of load and power generation. Thus, impacts of the changes of load and power generation can be considered in reserve capacity calculation.
   In this document, the power system reliability rule is defined as the load shedding incidents (LSI) allowed in each year. An LSI is defined as an incident that power generation shortage cannot be solved by adequate reserve capacity. The LSI probably relates to the loss of load expectation (LOLE). LOLE is the product of the LSIs and the mean time of loading shedding. The LSI reliability rule and the LOLE cannot quantize the values of load shedding but the probability of faults.
2) **Generator outage**
   Both the full outage probability (FOP) and partial outage probability (POP) in each hour should be considered. The FOP refers to probability that a unit stop providing its current output completely within an hour. It is assumed that the output of a unit is temporarily unavailable because of a trip. The hourly FOP of a unit relates to the forced outage rate (FOR) and the mean time to repair (MTTR) in the way below (Formula (11.1)):

$$FOP = \frac{FOR}{MTTR} \tag{11.1}$$

   The POP of a unit can be defined in a way similar to FOP. It refers to the probability of transient loss of partial power generation within an hour. The single-state POP method is used in reference [12].
3) **Errors of load forecasts and wind power forecasts**
   The daily load curves are highly repetitive; each load forecast error is not particularly sensitive to the time ranges of load forecasts but usually in proportion to the load of a given hour. The load forecast error in the specific hours h can be simulated as a Gaussian statistical variable and its mean and standard deviation are zero and $\sigma_{load,h}$ respectively. The wind power forecast errors usually increase with the time ranges increase of wind power forecasts. Likewise, the wind power forecast error of a power system with hour h can be simulated as a Gaussian statistical variable and its mean and standard deviation are zero and $\sigma_{wind,h}$, respectively.
   The load forecast errors and the wind power forecast errors are assumed as irrelevant Gaussian statistical variables, so standard deviation of the total forecast error of the power system will be (Formula (11.2)):

$$\sigma_{tital,h} = \sqrt{\sigma^2_{wind,h} + \sigma^2_{load,h}} \tag{11.2}$$

Standard deviation ($\sigma_{wind,h}$) of total wind power forecast error of the power system in hour h may be calculated through Formula (11.3) on the precondition that the correlation of multiple WFs is considered. In the formula, F is the WF quantity; $\sigma_{m,h}$ and $\sigma_{n,h}$ are standard deviations of wind power forecast errors of WFs m and n in hour h, respectively; $\rho_{m,n}$ is coefficient of correlation between the wind power forecast errors of WFs m and n, mainly depending on distances between the WFs.

$$\sigma_{wind,h} = \sqrt{\sum_{m=1}^{F} \sigma^2_{m,h} + 2\sum_{n=1}^{F}\sum_{m=n+1}^{F} \rho_{m,n}\sigma_{m,h}\sigma_{n,h}} \tag{11.3}$$

11.3.3.3 Calculation of Wind Power Reserve Capacity

The hourly reserve capacity of a power system is related with the annual reliability of the power system. It is assumed that the method to configure the annual reserve capacity can keep the same risk for mean hourly load shedding incidents, or, in other words, that one year has 8,760 hours and that the expectation for hourly load shedding incident is LSI. Then, for any hour h, the probability of load shedding (PLS) will be a quotient as below between the annual reliability rule and the annual hours (Formula (11.4)).

$$PLS_h = \frac{LSI}{8760} \tag{11.4}$$

The following three types of LSI should be considered: ① Only total forecast error of the unexpected wind power and loads is larger than reserve capacity of the power system; ② Total forecast error of the unexpected wind power and loads is larger than reserve capacity of the power system while a generator is out of service (completely or partially); ③ Total forecast error of the unexpected wind power and loads is larger than reserve capacity of the power system while a generator is out of service (completely or partially) immediately after the trip of another generator.

In consideration of the small generator outage probability (FOP and POP), it is less likely that three or more generators have outage in a short time. Thus, it makes no great contribution to the annual expected LSI. The annual expected LSI is the sum of the daily LSI probabilities. The contribution to this lies in the following two aspects: the probability of LSI in the normal operation hours (PLSNO), which is shown in area 1 in Figure 11.3; the probability of LSI increased in the time after the outage of a unit, which is shown in area 2 in Figure 11.3 and suggests that the power system is operating with less reserve capacity due to outage of the unit. It is based on an assumption that reliability of the power system recovering its reserve capacity in Hr hour can be rebuilt linearly.

Figure 11.4 indicates the Gaussian distribution of total forecast error of a power system in hour h. The gray area represents the probability that to which degree the total

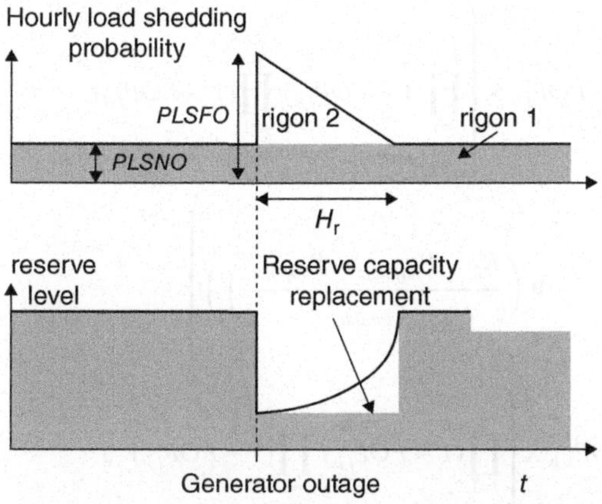

Figure 11.3 Curve of variation of LSI and reserve capacity probability with time (during the complete outage of a generator). *PLSNO*: Probability of LSI in the normal operation hours. *PLSFO*: Immediate PLS following the complete outage of generator *i*.

forecast error of wind power and load is larger than difference between reserve capacity of the power system and the power loss during the complete outage of generator i.

PLSNO has three parts as shown in Formula (11.5). In the formula, $\Phi(X)$ refers to a normalized Gaussian distribution function. ① The probability that total forecast error of the unexpected wind power and loads is larger than reserve capacity of the power system while there is no tripped generator. This corresponds to item 1 in Formula (11.5). ② The probability that total forecast error of the unexpected wind power and loads is larger than reserve capacity of the power system while there is a completely tripped generator. This corresponds to item 2 in Formula (11.5). The gray area in Figure 11.4 corresponds to the probability that total forecast error of the unexpected wind power and loads is larger than difference between reserve capacity of the power system (R_h) and the unavailable power in hour h after the complete trip of generator i. ③ The probability that total forecast error between the unexpected wind power and loads is larger than reserve capacity of the power system while there is a partially tripped generator. This is similar to ② and corresponds to item 3 in Formula (11.5).

$$\begin{aligned} PLSNO_h &= \prod_{i=1}^{NG}(1 - FOP_{i,h})\prod_{i=1}^{NG}(1 - POP_{i,h}) \times \left(1 - \Phi\left(\frac{R_h}{\sigma_{total,h}}\right)\right) \\ &+ \sum_{i=1}^{NG} FOP_{i,h} \times \left(\prod_{\substack{j=1\\ j\neq i}}^{NG}(1 - FOP_{j,h}) \prod_{\substack{j=1\\ j\neq i}}^{NG}(1 - POP_{j,h}) \times \left(1 - \Phi\left(\frac{R_h - P_{nafoi,h}}{\sigma_{total,h}}\right)\right)\right) \\ &+ \sum_{i=1}^{NG} POP_{i,h} \times \left(\prod_{j=1}^{NG}(1 - FOP_{j,h}) \prod_{\substack{j=1\\ j\neq i}}^{NG}(1 - POP_{j,h}) \times \left(1 - \Phi\left(\frac{R_h - P_{nafoi,h}}{\sigma_{total,h}}\right)\right)\right) \end{aligned} \tag{11.5}$$

In the formula above, $PLSNO_h$ is the PLS during normal operation in hour h; $FOP_{i,h}$ is the probability of complete outage, that is, the probability that generator i is completely unavailable in hour h; $POP_{i,h}$ is the POP, that is, the probability that generator i is partially unavailable in hour h; R_h is reserve capacity of the power system in hour h; $\sigma_{total,h}$ is standard deviation of total forecast error of the power system in hour h; $\Phi(X)$ is a standardized (normal) Gaussian distribution function; $P_{nafoi,h}$ is the unavailable power because of the complete outage of generator i in hour h; NG is the generator quantity.

Note that generator outage is a discrete incident and probably happens or does not happen in any specified

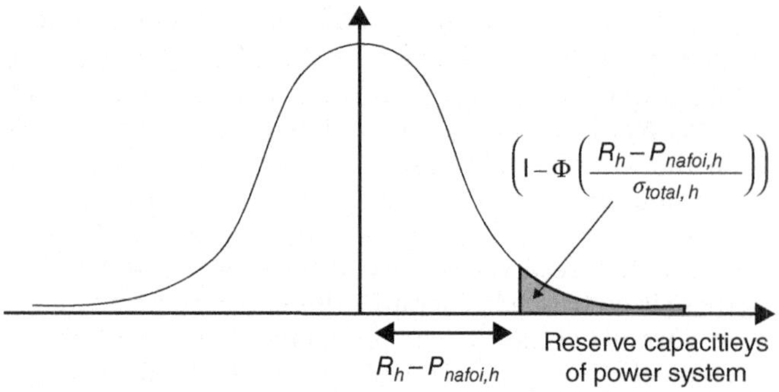

Figure 11.4 Gaussian distribution of total forecast error of a power system in hour *h*.

hour, which contrasts with the continuity of wind power changes and load changes. In this method, it is assumed that all the other reserve capacity is replaced in H_r hours following generator outage. Only the reserve capacity as a compensation for the unexpected changes of wind power and loads is not used. This is obvious in Formula (11.5): $PLSNO_h$ decreases monotonically with the increase of R_h (the reserve capacity reachable for the power system in hour *h*). Reserve capacity of the power system depends on $PLSNO_h$ directly, the maximum risk of load shedding immediately after the outage of a generator (immediate *PLSFO* after complete outage of the generator and immediate *PLSPO* after partial outage of the generator) depends on reserve capacity of the power system before outage. Thus, both *PLSFO* and *PLSPO* depend on $PLSNO_h$ directly.

The immediate PLS in hour *h* after the complete outage of unit *k* ($PLSFO_{k,h}$) can be calculated through Formula (11.6).

$$
\begin{aligned}
&PLSFO_{k,h}\\
&=\prod_{\substack{i=1\\i\neq k}}^{NG}(1-FOP_{i,h})\prod_{\substack{i=1\\i\neq k}}^{NG}(1-POP_{i,h})\\
&\times\left(1-\Phi\left(\frac{R_h-P_{nafok,h}}{\sigma_{total,h}}\right)\right)\\
&+\sum_{\substack{i=1\\i\neq k}}^{NG}FOP_{i,h}\times\left(\prod_{\substack{j=1\\j\neq i\\j\neq k}}^{NG}(1-FOP_{j,h})\prod_{\substack{j=1\\j\neq i\\j\neq k}}^{NG}(1-POP_{j,h})\right.\\
&\left.\times\left(1-\Phi\left(\frac{R_h-P_{nafok,h}-P_{nafoi,h}}{\sigma_{total,h}}\right)\right)\right)\\
&+\sum_{\substack{i=1\\i\neq k}}^{NG}POP_{i,h}\times\left(\prod_{\substack{j=1\\j\neq k}}^{NG}(1-FOP_{j,h})\prod_{\substack{j=1\\j\neq i\\j\neq k}}^{NG}(1-POP_{j,h})\right.\\
&\left.\times\left(1-\Phi\left(\frac{R_h-P_{nafok,h}-P_{napoi,h}}{\sigma_{total,h}}\right)\right)\right)
\end{aligned}
\tag{11.6}
$$

In the formula above, $PLSFO_{k,h}$ is the PLS in hour *h* following the complete outage of unit *k*; $P_{napoi,h}$ is the unavailable power in hour *h* because of the partial outage of generator *i*.

The immediate PLS in hour *h* following the partial outage of unit *k* ($PLSFO_{k,h}$) can be calculated through Formula (11.7). The extra item in the formula is for calculating probability that the unit with residual output (the previously tripped unit) is out of service completely.

$$
\begin{aligned}
&PLSPO_{k,h}\\
&=\prod_{\substack{i=1\\i\neq k}}^{NG}(1-FOP_{i,h})\prod_{\substack{i=1\\i\neq k}}^{NG}(1-POP_{i,h})\\
&\times\left(1-\Phi\left(\frac{R_h-P_{nafok,h}}{\sigma_{total,h}}\right)\right)\\
&+\sum_{\substack{i=1\\i\neq k}}^{NG}FOP_{i,h}\times\left(\prod_{\substack{j=1\\j\neq i\\j\neq k}}^{NG}(1-FOP_{j,h})\prod_{\substack{j=1\\j\neq i\\j\neq k}}^{NG}(1-POP_{j,h})\right.\\
&\left.\times\left(1-\Phi\left(\frac{R_h-P_{nafok,h}-P_{nafoi,h}}{\sigma_{total,h}}\right)\right)\right)\\
&+\sum_{\substack{i=1\\i\neq k}}^{NG}POP_{i,h}\times\left(\prod_{\substack{j=1\\j\neq i\\j\neq k}}^{NG}(1-FOP_{j,h})\prod_{\substack{j=1\\j\neq i\\j\neq k}}^{NG}(1-POP_{j,h})\right.\\
&\left.\times\left(1-\Phi\left(\frac{R_h-P_{nafok,h}-P_{napoi,h}}{\sigma_{total,h}}\right)\right)\right)\\
&+FOP_{k,h}\times\left(\prod_{\substack{j=1\\j\neq k}}^{NG}(1-FOP_{j,h})\prod_{\substack{j=1\\j\neq k}}^{NG}(1-POP_{j,h})\right.\\
&\left.\times\left(1-\Phi\left(\frac{R_h-P_{napok,h}-(P_{nafok,h}-P_{napok,h})}{\sigma_{total,h}}\right)\right)\right)
\end{aligned}
\tag{11.7}
$$

In the formula above, $PLSPO_{i,h}$ is the PLS in hour h following partial outage of generator i.Formula (11.8) indicates the average probability of load shedding in hour h (PLS_h). PLS_h includes two parts, that is, the PLS in normal operation of the power system and the PLS in the outage of a generator. The latter probability refers to some triangular areas (area 2 in Figure 11.4) and is a product of their probabilities in hour h.

$$\begin{aligned} PLS_h = {} & PLSNO_h \\ & + \frac{1}{2}(H_r) \times [FOP_{1,h}, FOP_{2,h}, \cdots, FOP_{NG,h}] \\ & \times \begin{bmatrix} PLSNO_{1,h} - PLSNO_h \\ \vdots \\ PLSNO_{NG,h} - PLSNO_h \end{bmatrix} \\ & + \frac{1}{2}(H_r) \times [POP_{1,h}, POP_{2,h}, \cdots, POP_{NG,h}] \\ & \times \begin{bmatrix} PLSPO_{1,h} - PLSPO_h \\ \vdots \\ PLSPO_{NG,h} - PLSPO_h \end{bmatrix} \end{aligned} \tag{11.8}$$

In the formula above, PLS_h is average probability of load shedding in hour h; H_r is the hours from the outage of a generator to reliability recovery of the power system. PLS_h is probably related to the annual reliability rule mentioned in Formula (11.4) only. R_h, that is, reserve capacity of the power system, cannot be explicitly expressed in any other variable, so it can be solved through the MATLAB optimization toolbox when the $PLSNO_h$ is given. Thus, the $PLSNO_h$ can be changed between lower limit (zero) and upper limit (LSI/8760) of any annual expected LSI with help of solution space search.

11.3.3.4 Main Factors Influencing the Reserve Capacity

In reference [11], the method above is used in the whole power system of Ireland to calculate and analyze some factors, which may influence reserve capacity. All the conventional units in Ireland have total generation capacity of 4,459 MW. The expected installed capacity of wind power by the year of 2010 will be 1,500 MW. Conventional units in the power system have maximum unit capacity of 408 MW. The unit outage probability (FOP and POP) is collected from the historical data of Ireland. It is the quotient between annual times of complete or partial outage of the units and annual generation hours of the units. The FOP ranges from about 0.003 for the most unreliable units to about 0.0006 for the most reliable units. It is assumed that power system reliability recovers in two hours after the generator outage. The wind power forecast error (expressed in standard deviation $\sigma_{w,h}$) is presented in a function of the forecast time range; correlation of wind power forecast errors among different WFs is considered; the total wind power forecast error is calculated through Formula (11.3); the total forecast error is calculated through Formula (11.2). Some calculation results are shown below.

1) **Impact of the wind power penetration power:**
   According to the calculation result about relation between the wind power penetration increment and the reserve capacity based on the forecast time range of three hours and the capacity demand for the different annual LSIs, reserve capacity of the power system must increase with the wind power penetration, otherwise the power system reliability will decline. In the case of wind power installed capacity of 1,500 MW, reserve capacity of the power system must be about 20% more than the situation where there is no wind power.
2) **Impact of the forecast time range:**
   In the case of 1,500 MW wind power and different reliability rules, the forecast time range differs with the reserve capacity needed. With increase of the forecast time range, standard deviation of the total wind power forecast error will increase, so will the reserve capacity needed.
   Usually, a decision for operation of the power system in any specific hour is made earlier (such as hour-ahead or day-ahead). The reserve capacity arranged and committed (in an agreement) for an hour is necessary in operation decision making. When the wind power penetration is high, almost real-time dispatch can employ short-term forecast data. In this case, standard deviation of the wind power forecast error is smaller, so less reserve capacity is needed.
3) **Impact of wind power forecasts:**
   In the case of a reliability rule based on annual three LSIs (LSI=3), relations between the installed capacity of wind power and the reserve capacity of the power system needed when the wind power forecast error is the average, the best and the worst, respectively, is shown below: When the installed capacity of wind power is 1,500 MW, the reserve capacity needed is about 12% more than the situation without wind power when the wind power forecast error is the best, or 44% more than the situation without wind power when the wind power forecast error is the worst. In the case of a reliability rule based on annual three LSIs (LSI=3), a forecast time range of three hours and absence of wind power, the reserve capacity of the power system is 470 MW. In the case of 1,000 MW wind power, the reserve capacity of the power system is 516 MW, which is an increment of 10%. When the wind power forecast errors rather than the load forecast errors are considered, the reserve capacity of the power system is 468 MW, which complies with the wind power forecast errors and the outage of one unit. It suggests that the impacts of wind power uncertainties are similar to that of load uncertainties with the three-hour time range.

Table 11.3 Standard deviation of power system forecast error (wind power capacity: 1,500 MW).

Types of reserve	Time range	σ_{total}(MW)
Primary	5–15 sec	6.0
Secondary	15–90 sec	14.8
Third	90 sec–5 min	27.0
Third 2	5–20 min	54.0
Hourly	20 min–1 h	93.4

4) **Impact of types of reserve capacity:**
A conventional power system has several types of reserve capacity with different time frames. Table 11.3 introduces time frames of the conventional types of reserve capacity and standard deviation of total forecast errors in the time frames when the power system of Ireland has wind power capacity of 1,500 MW.

In a short period, standard deviation of total forecast error of the power system mainly depends on standard deviation of changes of wind power and loads. When the period is shorter than one hour, it is assumed that there is a relation between deviation of total forecast error of the power system in t seconds (σ_t) and standard deviation of total forecast error of the power system in one hour ($\sigma_{1,h}$), which is shown below (Formula (11.9)).

$$\sigma_t = \sqrt{\frac{t}{3600}}\sigma_{1,h} \tag{11.9}$$

According to Table 11.3, the reserve capacity of different types has different time ranges for response. Standard deviation of total forecast error of the power system is small when the time range is 15 s and increases gradually with the time range extending.

An increase of the wind power penetration only has one impact on the reserve capacity with shorter time range owing to the small standard deviation of the wind power forecast error in these shorter time ranges. The reserve capacity working in longer time ranges has the largest increments due to the wind power capacity, so the power system needs some low-cost and quick-start reserve power plants.

11.3.4 Cost of the Additional Wind Power Reserve Capacity

The adding of reserve configurations and applications results in additional costs. In most of cases, the additional wind power reserve capacity can be undertaken by existing conventional power plants. The additional special reserve capacity or additional part-loaded power plants due to wind power result in additional costs. When used as reserve capacity, the fossil fuel plants have higher operation costs while hydropower plants have lower costs. The operation costs of (demand side management) DSM and (energy storage system) ESS are much higher. It is a consensus in most of studies that the additional reserve capacity necessary for high wind power penetration can be provided by the conventional power plants or, in other words, there is no demand for new reserve capacity. For a power system with 10% of wind power penetration (electric energy), the additional cost of following secondary load is estimated at 1€ to 3€ per MWh of wind power. In California, the increment of wind power regulation cost was estimated at 0.1€ per MWh of wind power when the wind power penetration was about 2% (in the year of 2003). Table 11.4 shows the balancing requirements and costs corresponding to different time ranges [3].

Table 11.4 Balance requirements and costs in different time ranges.

Type of reserve capacity	Impacts of WFs on reserve capacity	Influential factors of additional costs
Primary reserve capacity (minute level or less)	Zero	Geographically distributed
Secondary reserve capacity or third reserve capacity (15-minute level or hour level)	Regulating capability Reserve capacity demand: 2% of the rated wind power capacity when the wind power penetration is low or 4% of the rated wind power capacity when the wind power penetration is high Additional cost: 1€ to 3€ per MWh of wind power when the wind power penetration is 10% or 3€ to 4€ per MWh of wind power when the wind power penetration is high	Geographically distributed; wind power forecast level
Load tracking (4- to 12-hour level)	There is an efficiency loss because of wind power fluctuations and forecast errors; there are less fuels used and emitted; most of the expensive units operating at marginal prices (MP) are replaced by wind power units by means of wind power generation capacity forecasts; the forecast errors are brought to the regulationg or balance of power market, or more accurate forecasts are made at time closer to the delivery time.	Forecast error; correlation between wind power and loads; combination of power generation types

When the wind power penetration is 20%, the primary reserve capacity and secondary reserve capacity necessary for the wind power of a power system have a cost smaller than 6$/MWh. This cost differs greatly in different studies, but these differences suggest that additional costs of the reserve capacity because of wind power depend on characteristics of the grids with wind power, adequacy, and characteristics of the existing reserve capacity, and specific requirements of grids for reserve capacity. Europe and the United States have the same study results about this. When the wind power penetration is lower than 20%, the reserve capacity necessary for wind power have moderate costs. In regardless of the different study methods and the different characteristics of the grids evaluated, the study results on this issue in Europe and the United States can be summed up as below:

- The cost of wind power integration is not zero but increases with the wind power compared with conventional power generation or the peak load;
- The additional costs of reserve capacity because of wind power integration are low when the wind power penetration is lower than 20%, but they increase with the wind power penetration;
- The instability and uncertainty of wind power generation, the load changes and the load forecast influence the cost of wind power integration greatly;
- The cost of wind power integration is influenced by the degree of geographical concentration of the WFs too. For big WFs, their space diversity can reduce their output power fluctuations, so their wind power integration costs are smaller.

11.4 Optimal Dispatch of Power Systems Containing Wind Power

Upon increase of the wind power penetration of a power system, the power system operator should adopt an innovative operation mode for easy connection of wind power. The power system operator must consider the following factors on the precondition of efficient and cost-reducing operation: the power system should have adequate and reliable reserve capacity to adapt to the uncertainty of wind power generation; the power system should meet the emission objectives; the power system operation rule should comply with trading arrangement of the power market; safety and dynamic stability of the power system should not be endangered by increase of the wind power penetration, and so on.

In optimal dispatch, it is necessary to evaluate the impacts of wind power generation on power generation dispatch, particularly on the fuel cost, the total emission, and the power system safety. Balance among the fuel cost, the total emission and the power system safety depends on characteristics of the grid, the power generation, and the load. Where there is wind power in the power market, the most important objective will still be safety and minimum operation cost for the power system.

With the increase of the wind power penetration, both the spinning reserve capacity and the power generation cost increase while the total emission decreases correspondingly. Safety of the power system is high when there is a little wind power and low when wind power takes a larger proportion in the power system.

When wind power has a small proportion, economical efficiency, environmental friendliness and safety of the power system can be improved by means of proper dispatch without a necessity to strengthen the power system, and vice versa.

The studies about economic dispatch of the power systems with massive wind power focus on the following aspects [13]: (1) objective functions about different long-term economic dispatch models, including an objective function about the costs of a wind power conversion system (including electricity cost of wind power price, cost of the reserve capacity for wind power compensation, and wind curtailment cost); a double-objective function about combination of the minimum power generation cost and the minimum risk; an objective function about combination of WFs of different types (active power and reactive power have different controllability) in delegated dispatch, an objective function about the total social cost including an environmental cost, and so on. (2) objective functions about different short-term economic dispatch models, including an objective function about statistical safety of the expected cost of preventive safety actions and corrective safety actions against disturbance; an objective function about energy storage in the energy market; an objective function about combination with the spinning reserve capacity; an objective function including a start cost and an outage cost, and so on. (3) Optimal algorithm of the dispatch model is usually realized through a DSM, a particle swarm optimization algorithm, a simulated annealing method, a genetic algorithm, a scene structure method, and so on. (4) Risk dispatch management for the wind power in the power market.

11.4.1 Operation Mode of Power Systems with Massive Wind Power

After connection of massive wind power, a grid needs to regulate its operation mode. For a power system with massive wind power, its operation mode without wind power (during outage of WFs) and with wind power should be both considered. The operation mode with wind power can be further classified into a fuel-saving operation mode and a wind power forecast operation mode. The impacts of different power system operation modes on a power system with massive wind power should be studied. The main contents concerned in the operation of a power system are balance among the power generation, the power demand, and adequate reserve capacity. The reserve capacity of a power system should be regulated to realize the reliability rule in each operation mode. It is assumed that wind power does not provide reserve capacity and that there is a conventional unit commitment objective made up of conventional power generation (MW) for each hour in each operation mode. In order to provide reserve capacity, this demand should be available in the power system [14, 15].

1) **Operation mode without wind power**
This is the basic situation in which there are only conventional power plants working in a power system. For a power system without wind power, its total power generation demand and reserve capacity demand must be satisfied by means of conventional power generation. This operation mode may be a foundation for comparison of the impacts of wind power capacity.
2) **Fuel-saving operation mode**
In this mode, it is assumed that the systematic demands for power generation capacity and reserve capacity are planned when there is no wind power forecast; there are only conventional power plants and any possible wind power contribution is ignored; start and outage of all the units match with the expected demand mode of the power system. However, the WFs are actually running. Thus, wind power contributes to balance between the power generation and the power demand (wind power does not provide any auxiliary service). To be in accordance with the changes of wind power generation, the conventional power plants will reduce their outputs, thus saving fuel. The conventional units can reduce their output until their minimum output; however, there is no conventional power plants that shut down or exit from operation. Therefore, wind power generation is regarded as a negative load, which reduces effective load (net load) of the power system and suggests that the original forecast load demand of the power system is excessive. If one or more unit commitments are removed, more fuel will be saved. However, this action will not be taken because it may result in the pause of wind power. When there is too much wind power to reduce the output of any conventional power plants, any additional wind power will be reduced. This mode is easy and guarantees a safe power system, high reserve capacity, and a better load tracking capacity (because of the reserve capacity of some part-loaded power plants). However, this mode also has noteworthy disadvantages. Sometimes, the conventional power plants have excessive unit commitment and undertake an excessive generation load. As a result, the wind power output must be reduced to comply with the minimum output constraints for conventional generators. This is particularly obvious for a smaller power system because in such a power system the possible wind power generation is ignored when the load demand is low and only a few big (efficient) generators are used. A method that is costly but can provide better load following flexibility is to invoke many small (less-efficient) conventional units. At present, many units are part-loaded, which makes it impossible to completely realize unit thermal efficiency (related to fuel consumption rate) and CO_2 emission reduction because of fuel saving. The units have high efficiency and low emissions during full-load operation but possibly greatly reduced efficiency and low emissions during part-load operation.
For a long period, it is actually assumed in this mode that wind power has a zero CC. Thus, the capital cost for expected increase of the load demand will not reduce any more (conventional power plants). Thus, this mode is simple but probably costly. However, in regardless of its disadvantages above, this conservative unit dispatch plan is accepted by most of operators because it is probably the best mode when the wind power penetration is low (e.g., it is lower than 5% of the annual energy supply capacity). This mode allows ignoring the forecast, reliability of wind power, and the dynamic issue of power systems. It has been used in many wind power researches.
3) **Wind power forecast operation mode**
In this mode, the necessary power generation capacity and reserve capacity of a power system are planned according to a forecast wind power capacity and its error; the necessity to increase reserve capacity of the power system due to wind power forecast uncertainty and the reduction of the demand for conventional power generation because of the forecast wind power output are both considered. In this mode, it is assumed that wind power reduction is not necessary and that no auxiliary service is provided. Thus, the optimal unit commitment is based on net demand forecast. The net demand is the difference between the forecast load demand and the forecast wind power. The most important advantage is that wind power forecasts can reduce needful loads of the conventional power units, so as to realize better benefits and a smaller emission. In this mode, the forecast time range planned is important because the error of an hour-ahead wind power forecast is much smaller than the error of a day-ahead forecast or, in other words, more reserve capacity should be planned for a forecast time range of a couple of hours.

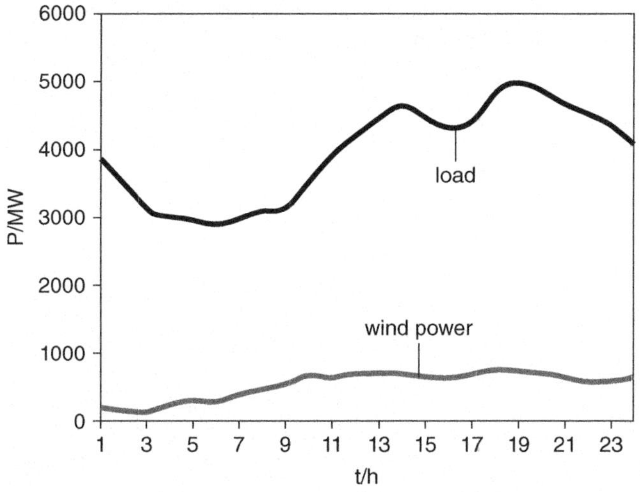

Figure 11.5 Curve about loads and wind power generation in a day.

In reference [14], the power system of Ireland is studied. In the power system, the total installed capacity is 7,000 MW and the installed capacity of wind power is 1,500 MW. Figure 11.5 indicates a curve about the loads and wind power in a day. According to the curve, the mean hourly wind power is 525 MW and the daily wind power capacity factor is 0.35. Table 11.5 lists the mean daily reserve capacity needed in different operation modes.

According to Table 11.5, after wind power connection, the power system witnesses a small increment (4% to 5% of the load power in the operation conditions) of its primary reserve capacity, second reserve capacity, and third reserve capacity. The small increment suggests increase of power system reserve capacity within one hour due to the wind power capacity. The small increment means that most of the reserve capacity are provided for the largest units lost in the power system. With the 1-hour-ahead forecast technique, the hourly reserve capacity increases by 8% (from 455 MW to 491 MW) because of additional changes of the power system in one hour. With the day-ahead forecast technique, the mean hourly reserve capacity increases by 38% (from 455 MW to 626 MW) partially because of hourly changes of the power system. However, with the hour-ahead forecast technique, the power system must have a larger generation capacity to adapt to the possible unexpected changes of the wind power generation capacity happening between the forecast time and the actual time. With forecast time range increasing, there will be much more forecast errors related to the wind power generation capacity, resulting in additional generation capacity necessary for operation. It is assumed that the additional generation capacity is hourly reserve capacity and that the hourly reserve capacity are increased to be 115% of the value (from 455 MW to 978 MW) when there is no wind power through the fuel-saving operation mode. In the fuel-saving operation mode, it is assumed that there is no wind power in a day, so the load of conventional power generation should be reduced for the sake of easy wind power connection. In the fuel-saving operation mode, no conventional power plants should be shut down, so all the generation capacity reduced but still connected with the grid are assumed to be hourly reserve capacity.

Figure 11.6 gives the daily conventional power generation capacity adequate for the power generation capacity and the reserve capacity in different operation modes. Note that the conventional power generation capacity needed in the operation mode without wind power is the same as that needed in the fuel-saving operation mode.

According to Figure 11.6, the conventional power generation capacity needed in the fuel-saving operation mode is much more than that in the day-ahead forecast mode; the conventional power generation capacity needed in the hour-ahead forecast mode is less than that in the day-ahead forecast mode.

It is difficult to quantize all additional costs due to possible impacts on operation of the power system. However, the content above suggests that the additional costs are significantly influenced by operation modes of the power system.

The main influential factors of reserve capacity do not change greatly in a year. The largest unit in the power system influences reserve capacity of the power system greatly. However, the largest unit in the power system of Ireland serves as the basic load power plant, so its output

Table 11.5 Mean daily reserve capacity needed in different operation modes.

Type	Cost	Time range	Reserve capacity (MW)			
			Without wind power	Fuel saving	Hour-ahead forecast	Day-ahead forecast
Primary reserve capacity	€1.73	15 sec	309	322	322	322
Secondary reserve capacity	€1.57	90 sec	315	329	329	329
Third reserve capacity 1	€1.43	5 min	401	417	417	417
reserve capacity 2	€1.43	20 min	420	441	441	441
One hour	€1.08	1 hour	455	978	491	626

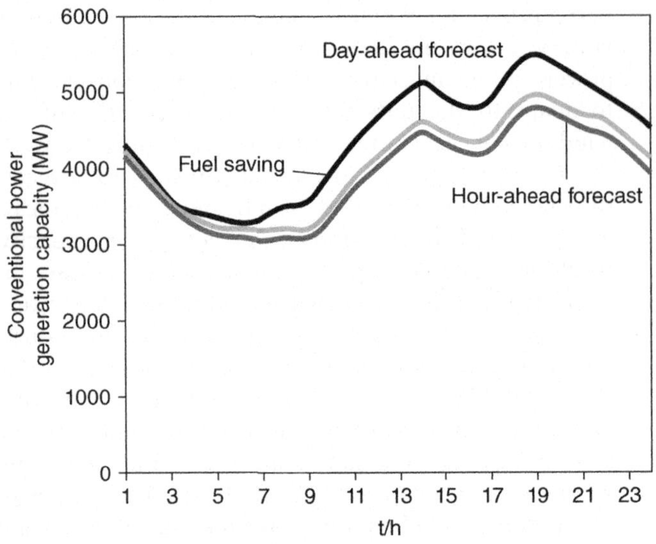

Figure 11.6 Conventional power generation capacity needed in different operation modes.

Table 11.6 Annual cost of reserve capacity.

Mode	Cost of reserve capacity (€m/a)	Additional cost when there is no wind power (%)	Additional cost for each MW when there is 1,500 MW wind power (€/a)
Operation mode without wind power	23.6	-	-
Fuel-saving operation mode	29.4	24.7	3888
Hour-ahead forecast operation mode	24.8	5.1	804
Day-ahead forecast operation mode	26.1	10.5	1656

does not change greatly in a year. Another discovery in the study is that both the load forecast errors and the wind power forecast errors almost do not change in a year. According to the test days and the estimated prices in Table 11.5, the annual costs of power system reserve capacity in Ireland are given in Table 11.6.

According to Table 11.6, for the power system of Ireland with 1,500 MW wind power, the annual cost for reserve capacity is €5.8 million (based on the estimation of 2004) in the fuel-saving operation mode. Operation of the power system is better in the forecast operation modes because the uncertainty of wind power generation in shorter time ranges are lower. In the hour-ahead forecast operation mode, the annual cost of reserve capacity needed by 1,500 MW wind power is €1.2 million. When the power system has an installed wind power capacity of 1,500 MW, the forecast annual cost of reserve capacity needed by a 25 MW WF is about €20,000 in the hour-ahead forecast operation mode or about €100,000 in the fuel-saving operation mode.

11.4.2 Optimal Unit Commitment

11.4.2.1 Concept of Unit Commitment

A power system is a continuously operating dynamic system, so its unit state and active power output should be regulated with its load. In the operation, a continuous dispatch cycle is usually divided into a couple of time periods, each of which is believed to have a constant load. The power system load variation in multiple continuous time periods should be solved by means of unit commitment and dynamic optimal dispatch calculation. In the dispatch cycle, active power output of the units can be calculated only when operation states in the time periods are given by means of dynamic optimal dispatch. If total load of the power system is smaller than sum of minimum active output power of the operating units, some operating units must exit from operation; if total load of the power system is larger than the sum above, some units in outage must be put into operation. The course to determine a start and outage plan for the units in different time periods of a cycle (a day or a week) to minimize the expenses (or costs) of power generation and start in the whole cycle to realize active power optimization while the constraints for load, reserve capacity, units, power system, and environment are all met is called unit commitment (UC) (or economic commitment). The unit commitment plan for operation of the power system is also called the generating plan. It is an operation dispatch function also called pre-dispatch. Unit commitment is expected to determine the units for meeting a certain load demand. The minimum expense (or cost) of unit commitment depends on changes of the loads and the generation capacity (including the wind power capacity). The changes may be caused by natural factors or human factors [16–18].

In the short-term operation dispatch of a power system, there should be an adequate generation capacity in compliance with the forecast load and, additionally, the power system must have reserve capacity because of load forecast uncertainty and possible faults of the units. The reserve capacity with synchronous spinning and ready for load tracking is called a spinning reserve capacity. The spinning reserve capacity can be determined with a deterministic method or a probability method. The deterministic method determines the spinning reserve capacity through a specific or experiential rule. It makes problem solving easy; however, due to different distribution laws of the fault probabilities of all parts in the power system, it is difficult to maintain the same reliability for the power system in its operation through a unit commitment plan, resulting in irrationality between the spinning reserve capacity and the operation cost. Thus, it is a trend to determine the spinning reserve capacity of a power system with the probability method.

Unit commitment and the dispatch determined based on it belong to different categories of optimization. According to the load forecasts with different time ranges and the natures of dispatch tasks, the issues about operation dispatch of a power system can be classified into the following two types: decision-making issues of the resources allocation in a cycle (such as power generation and transmission plans, overhauls, and unit commitment), which can be summed up into unit optimization and task allocation issues with certain resource allocation in a time period (such as economic dispatch and dynamic economic dispatch), which can be summed up into power flow optimization. The main objective of economic dispatch (ED) is to determine economic loads of the units, so as to make the load demand satisfied after unit commitment (UC) is determined.

Unit commitment is an old problem. With the development of generator techniques and optimization algorithm techniques, there are increasingly more new studies about it. Reference [16] proposes a new simulation method for evaluating the impacts of big WFs on the cost, reliability, and environment friendliness of the operation of a power system comprehensively. In the method, a time series is used for forecasting the mean wind velocity in 15 minutes. In unit commitment (UC) economic dispatch (ED), a conventional unit plan is revised through a forecast for the output power of a WF in the next 36 hours. In reference [19], an energy dispatch policy for distributed power generation based on multiple energy sources such as solar energy, wind energy, water energy, combined heat and power, gas turbines, fuel cells and energy storage devices (including cells and super capacitors), and a unit commitment model following a rule of making the best use of renewable energy sources are both studied and a genetic algorithm is used for solution. In reference [20], the optimal unit commitment based on the minimum cost is solved with a hybrid dynamic programming (HDP) method and a simplified dispatch method. The basic idea of HDP is to combine branch and bound algorithm with dynamic programming to obtain more search techniques for unit commitment. On the basis of combination of characteristics of the unit commitment of a power system with the advantages and disadvantages of multiple classical mathematical programming algorithms, reference [21] proposes a new linear mixed integer programming algorithm based on dynamic search. This new algorithm integrates the advantages of algorithms including linear programming algorithm, mixed integer programming algorithm, unit commitment algorithm, and so on. It can solve the unit commitment with safety constraint in a large power system. In combination with the power system reliability probability analysis method, reference [22] proposes the modeling and solution methods for a mathematical model about unit commitment. The operation cost of a power system includes purchasing cost of the spinning reserve capacity, so impacts of random outage rate and fuel micro increasing rate of the units are both considered in solution to enable the power system to have the minimum operation cost in addition to high reliability. In this document, simplex-simulated annealing algorithm (SSA) is used for solving the load economic dispatch subproblem. Reference [23] gives consideration to the issues related to unit commitment of a power system with massive wind power: the wind power uncertainties in unit commitment are considered with a deterministic method rather than a random method; impacts of the wind uncertainty models based on different time ranges are checked; a rolling plan method based on wind power forecasts is proposed. In reference [24], a reliability constraint and an emission constraint are added into the unit commitment formula so that the UC configuration is punished by means of increase of the total UC cost when the UC solution is out of these constraints. In consideration of the necessity of operation for WFs, the WFs in the UC will pay 80% of UC cost of the cheapest conventional unit in the UC. In this way, WFs will be well supported and the free power market will remain unchanged.

Unit commitment is a mixed integer programming algorithm. Its calculation is difficult and massive, which mainly relates to complexity of the mathematical model used.

11.4.2.2 Mathematical Model of Unit Commitment

In fact, optimization of operation or outage of the units of a power system in multiple successive time periods is also caused by the units. Technically speaking, the start and outage of the thermal power units should be realized without start or outage of the boilers or the steam turbines. Frequent start and outage of the boilers and the steam turbines will shorten equipment lives greatly. Thus, operation and outage of the thermal power units should comply with the minimum operation time and the minimum outage time. Economically speaking, start of the thermal power units needs an additional fuel consumption to heat the boilers and the steam turbines until a stable operation state. Thus, in some studies, the optimization of start and outage of units is divided into the following two types: optimal unit commitment and optimal unit start and outage. In the optimization calculation in optimal unit commitment, the energy consumption of start is ignored, simplifying the issue to some extent. Therefore, the usual objective of unit commitment is sum of start consumptions and minimum power generation costs of all units of a power system in a dispatch cycle (Formula (11.10)) [17].

$$\min F = \sum_{t=1}^{T}\sum_{i=1}^{N} P_i^t(1 - u_i^{t-1})ST_i + \sum_{t=1}^{T}\sum_{i=1}^{N} u_i^t f_i(P_i^t) \tag{11.10}$$

In the formula above, i is the serial number of a unit; N is amount of the units participating in unit commitment calculation; t is the serial number of a time period; T is amount of the time periods in the whole dispatch cycle; u^t_i is the state of unit i in time period t (when unit i is operating in time period t, u^t_i is '1'; otherwise, it is '0'); P^t_i is the active power output of unit i in time period t; ST_i is the start consumption of unit i (it is an exponential function of outage time: the longer the outage time of the unit, the larger the start consumption cost needed will be); $f_i(P^t_i)$ is the power generation cost curve of unit i, which is usually expressed in a quadratic function of active output power P^t_i.

Balance between sum of upper and lower limits of active power output of the units in all the time periods and total active load of the power system is the most fundamental constraint considered in unit commitment (Formulas (11.11) and (11.12)). What really makes the unit commitment different from any other issue such as static load distribution or optical power flow is that there are many inter-time-period constraints in unit commitment. There are mainly two types of inter-time-period constraints as below: unit minimum continuous operation time constraint and unit minimum outage time constraint (Formulas (11.13) and (11.14)); unit ramp rate constraint (unit active power maximum increase rate constraint and unit active power maximum decrease rate constraint) (Formula (11.15)). The constraints of the former type are because such equipment of the thermal power units as the boilers and the steam turbines need long time to get heated gradually while switching from outage to operation or from damped-down operation to normal operation. Likewise, it needs a long time to switch from operation to outage too. If these constraints are considered in unit commitment, the calculation results will be much more feasible and practical. Regulation of active power output of the units is realized by means of steam inlet regulation of the steam turbines or water inlet flows of the water turbines, which defines a ramp rate (power increase/decrease rate) constraint of the units (Formulas (11.11)–(11.15)).

$$u_i^t P_i^{\min} \le u_i^t P_i^t \le u_i^t P_i^{\max} \qquad t = 1,\cdots,T, i = 1 \sim N \tag{11.11}$$

$$\sum_{i=1}^{N} u_i^t P_i^t = D^t, t = 1,\cdots,T \tag{11.12}$$

$$(d_{i,on}^t - d_{i,on}^{\min})(u_i^{t-1} - u_i^t) \ge 0 \qquad t = 1,\cdots,T \tag{11.13}$$

$$(d_{i,off}^t - d_{i,off}^{\min})(u_i^t - u_i^{t-1}) \ge 0 \qquad t = 1,\cdots,T \tag{11.14}$$

If both u_i^t and u_i^{t+1} are "1" for unit i, then:

$$-R_i^d \le P_i^{t+1} - P_i^t \le R_i^u \qquad t = 1,\cdots,N, i = 1 \sim N \tag{11.15}$$

In addition, a spinning reserve capacity constraint and a grid transmission limit constraint may be considered for the power system in each time period. Usually, unit commitment studies do not adopt a line transmission power constraint or a simplified constraint with the relatively simpler form based on DC power flow formula. There are studies about dynamic optimal dispatch and unit commitment, in which a line transmission power constraint is expressed in an AC power flow formula constraint.

Compared with issues such as active power optimal dispatch, unit commitment has the following two largest differences: unit commitment is an optimization covering multiple time periods and needs to consider the coupling constraint of multiple time periods; with many continuous variables and integer variables, unit commitment is a high-dimensional, nonlinear and discrete mixed integer programming algorithm. Even if the unit ramp rate and the line transmission constraint are not considered, unit commitment is much more difficult to be solved than issues such as active power optimal dispatch.

11.4.2.3 Calculation Algorithms of Unit Commitment

Unit commitment should be solved with a mathematical optimization algorithm. In this method, the optimizing strategies include analysis, enumeration, randomness, and so on [17]. Generally speaking, the optimization algorithms based on an analysis strategy are more efficient; however, they have poor adaptability to problem changes because they can be only applied when the mathematical models of problems meet specified requirements. The optimization algorithms based on the enumeration strategy have great adaptability but low optimizing efficiency. The optimization algorithms based on the randomness strategy have calculation efficiency lower than that of the optimization algorithms based on the analysis method but usually higher than the optimization algorithms based on the enumeration strategy, and great adaptability to problems. The calculation algorithms of unit commitment are artificial intelligence methods including priority list algorithm, cutting plane algorithm, branch and bound algorithm, unit commitment algorithm, lagrangian relaxation algorithm, genetic algorithm, and so on. These calculation algorithms, based on analysis optimizing, randomness search, or enumeration search, have their own advantages and disadvantages, so it is usually necessary to combine them properly in unit commitment calculation for a better optimization effect and higher calculation efficiency.

11.4.3 Optimal Dispatch Between Wind Power Units and Thermal Power Units

If a power system mainly including thermal power units adopts the fuel-saving operation mode, optimal dispatch between the wind power units and the thermal power units should be solved. Main objectives of the optimal dispatch are to determine the minimum operation cost of the power system (including start and outage cost of the units) and comply with the constraints for the "power system" and the "generators." Usually, fuel cost of the thermal power units is expressed in a quadratic polynomial function. This cost is very small for a wind power company because it needs no fuel in its wind power generation; however, an independent WF owner needs to consider its power generation cost.

11.4.3.1 Mathematical Model

The costs of wind power generation are based on the special contracts about electricity purchase and sale between independent power producers (IPP) and public power companies (i.e., fixed dispatch plans or economic dispatch). Due to its randomness, instability, and low forecast accuracy, wind power generation increases the physical constraints and the economic constraints. Here below are the mathematical models about optimal dispatch between the wind power units and the thermal power units while wind power is considered [20].

1) Objective function (Formula (11.16)):

$$\min F_T = \sum_{t=1}^{T}\sum_{i=1}^{NT} F_i(P_i(t)) \tag{11.16}$$

In the formula above, NT is quantity of the thermal power units in the power system; T is quantity of the time periods (hours); $P_i(t)$ is the generation capacity of thermal power unit i in time period t; $F_i(\bullet)$ is a function about operation cost of thermal power unit i.

2) Power system constraints:
   a) Power balance constraint (Formula (11.17))

$$\sum_{i=1}^{NT} U_i(t)\times P_i(t) + \sum_{j=1}^{NW} P_{WT,j}(t) = P_L(t) \tag{11.17}$$

   In the formula above, $P_{\mathrm{WT,j}}(t)$ is the actual generation capacity of WTGS j in time period t; NW is quantity of the WTGSs in the power system; $U_i(t)$ is the planned state (1: operating; 0: outage) of thermal power unit i in time period t; $P_{\mathrm{L}}(t)$ is the load demand in time period t.

   b) Demand for spinning reserve capacity in operation/outage of the power system (Formula (11.18))

$$\sum_{i=1}^{NT} U_i(t)\times US_i(t) \geq USR_B + ASR_1(P_{WT}(t)) \tag{11.18}$$

   In the inequation above, $US_i(t)$ is reserve capacity contribution to thermal power unit i in time period t during operation; USR_{B} is the spinning reserve capacity in operation of the power system when wind power operation is not considered; ASR_1 is the additional reserve capacity needed when wind power operation is considered (Formula (11.19)).

$$\sum_{i=1}^{NT} U_i(t)\times DS_i(t) \geq ASR_2(P_{WT}(t)) \tag{11.19}$$

   In the inequation above, $DS_i(t)$ is reserve capacity contribution to thermal power unit i in time period t during outage; ASR_2 is the additional reserve capacity needed when wind power outage is considered.

   c) Maximum/minimum output constraints for the thermal power plant (Formulas (11.20) and (11.21)):

$$P_L(t) - P_{WT}(t) \geq ASR_2(P_{WT}(t)) + \sum_{i=1}^{NT} U_i(t)\times P_i^{\min}(t) \tag{11.20}$$

$$\sum_{i=1}^{NT} U_i(t)\times P_i^{\max}(t) + P_{WT}(t) \geq P_L(t) + USR_B + ASR_1(P_{WT}(t)) \tag{11.21}$$

   In the inequations above, $P_i^{\max}(t)$ is the maximum generation capacity of thermal power unit i in time period t; $P_i^{min}(t)$ is the minimum generation capacity of thermal power unit i in time period t.

3) Constraints for the thermal power units
   a) Constraints for reserve capacity during operation/outage of a thermal power unit (Formulas (11.22) and (11.23)):

$$US_i^{max} = d\% \times P_{i,r}^{max} \tag{11.22}$$

$$DS_i^{max} = d\% \times P_{i,r}^{max} \tag{11.23}$$

In the formulas above, US_i^{max} is the maximum reserve capacity contribution to thermal power unit i during operation; DS_i^{max} is the maximum reserve capacity contribution to thermal power unit i during outage; $P_{i,r}^{max}$ is the maximum generation capacity of thermal power unit i; $d\%$ is capacity percentage of the largest unit.

   b) Constraints for spinning reserve capacity increase/decrease of a thermal power unit (Formulas (11.24) and (11.25))

$$US_i(t) = \min[US_i^{max}, P_{i,r}^{max} - P_i(t)] \tag{11.24}$$

$$DS_i(t) = \min[DS_i^{max}, P_i(t) - P_{i,r}^{min}] \tag{11.25}$$

In the formulas above, $P_{i,r}^{max}$ is the maximum generation capacity of thermal power unit i; $P_{i,r}^{min}$ is the minimum generation capacity of thermal power unit i.

   c) Constraints for ramp rate (capacity) upon load increase/decrease of a thermal power unit (Formulas (11.26) and (11.27)):

$$UR_i(t) = \min[UR_i^{max}, P_{i,r}^{max} - P_i(t)] \tag{11.26}$$

$$DR_i(t) = \min[DR_i^{max}, P_i(t) - P_{i,r}^{min}] \tag{11.27}$$

In the formulas above, $UR_i(t)$ is the ramp (load increase) rate of thermal power unit i in hour t; $DR_i(t)$ is the ramp (load decrease) rate of thermal power unit i in time period t; UR_i^{max} is the maximum ramp (load increase) rate of thermal power unit i; DR_i^{max} is the ramp (load decrease) rate of thermal power unit i; $US_i(t)$ is the reserve capacity contribution of thermal power unit i in time period t during operation.

   d) Constraints for the generation capacity of a thermal power unit (Formulas (11.28), (11.29), and (11.30)):

$$P_i^{min}(t) \times U_i(t) \le P_i(t) \le P_i^{max}(t) \times U_i(t) \tag{11.28}$$

$$P_i^{max}(t) = \begin{cases} \min[P_{i,r}^{max}, P_i^{max}(t-1) + UR_i^{max}], \\ \quad \text{if } U_i(t) = U_i(t-1) = 1 \\ \min[P_{i,r}^{max}, P_i(t-1) + SR_i], \\ \quad \text{if } U_i(t) = 1, U_i(t-1) = 0 \end{cases} \tag{11.29}$$

$$P_i^{min}(t) = \begin{cases} \max[P_{i,r}^{min}, P_i(t-1) + DR_i^{max}], \\ \quad \text{if } U_i(t) = U_i(t-1) = 1 \\ P_{i,r}^{min}, \text{ if } U_i(t) = 1, U_i(t-1) = 0 \end{cases} \tag{11.30}$$

   e) Constraint for the minimum operation/outage time (Formula (11.31)):

$$[t_{on,i}(t-1) - T_{on,i}] \times [U_i(t-1) - U_i(t)] \ge 0 \tag{11.31}$$

In the inequation above, $T_{on,i}$ is the minimum operation time of thermal power unit i; $t_{on,i}(t)$ is the time cycles of continuous operation of thermal power unit i until cycle t (Formula (11.32)).

$$[t_{off,i}(t-1) - T_{off,i}] \times [U_i(t) - U_i(t-1)] \ge 0 \tag{11.32}$$

In the inequation above, $t_{off,i}(t)$ is the time cycles of continuous outage of thermal power unit i until cycle t; $T_{off,i}$ is the minimum outage time of thermal power unit i.

4) **Constraints for the WTGSs**
   a) Constraints for wind power generation fluctuations (Formula (11.33)):

$$P_{WT}(t) - P_{WT}(t-1) \le TDR(t), \quad \text{if } P_{WT}(t-1) \le P_{WT}(t) \tag{11.33}$$

In the inequation above, $TDR(t)$ is ramp (load decrease) rate of the power system in time period t (Formula (11.34)).

$$P_{WT}(t-1) - P_{WT}(t) \le TUR(t), \quad \text{if } P_{WT}(t-1) \ge P_{WT}(t) \tag{11.34}$$

In the inequation above, $TUR(t)$ is ramp (load increase) rate of the power system in time period t.

   b) Constraint for the wind power curve (Formula (11.35)):

$$P_{WT,j}^*(t) = \begin{cases} 0 & v(t) \le v_{I,j} \quad \text{or } v(t) > v_{O,j} \\ \varphi_j(v(t)) & v_{I,j} \le v(t) \le v_{R,j} \\ P_{WT,j}^{max} & v_{R,j} \le v(t) \le v_{O,j} \end{cases} \tag{11.35}$$

In the formula above, $v(t)$ is the wind velocity in time period t; $v_{I,j}$ is the cut-in wind velocity of WTGS j; $v_{o,j}$ is the cut-out wind velocity of WTGS j; v_{Rj} is the rated wind velocity of WTGS j; $\varphi_j(\bullet)$ is the wind power curve of WTGS j; $P_{WT,j}^{max}$ is the maximum generation capacity of WTGS j; $P_{WT,j}^*(t)$ is the available generation capacity of WTGS j in time period t.

c) Constraint for the total available wind power (Formula (11.36)):

$$P^*_{WT}(t) = \sum_{i=1}^{NW} P^*_{WT,j}(t) \quad (11.36)$$

$P^*_{WT}(t)$ is the total available wind power in time period t.

d) Constraint for the total actual wind power (Formula (11.37)):

$$0 \leq P_{WT}(t) \leq P^*_{WT}(t) \quad (11.37)$$

$P_{WT}(t)$ is the total actual wind power in time period t.

11.4.3.2 Algorithm for Coordinative Dispatch Between Wind Power Units and Thermal Power Units

When wind power is considered, power generation dispatch is more complicated. The HDP algorithm is used for solving commitment between the wind power units and the thermal power units. In this algorithm, a time range is divided into smaller time periods each of which is usually one hour; in each time period (each hour), the corresponding state indicates integration of the commitment state (operating/outage) of different units in a specified time. The constraints mentioned in constraints (11.17) to (11.37) will be handled in different ways. Inequations (11.18) to (11.21) about operation constraints are evaluated in the effective operation state by means of plan searching and selection. All the plans comply with the requirements for load demand and on/off reserve capacity of the power system. Formula (11.17) and Formulas or Inequations (11.22) to (11.30) all about operation constraints will be handled in an economic dispatch model about constraints for the reserve capacity in each state; Inequations (11.31) to (11.34) about time constraints will be considered during state transition through the HDP algorithm. The total available wind power can be calculated based on the wind velocity with help of the wind power curve (constraint Formulas (11.35) and (11.36)). The total WTG output can be any expected value through pitch control (constraint Formula (11.37)). Here below is the detailed introduction.

1) **Unit commitment model for wind power and thermal power**
The main problem of conventional dynamic programming (CDP) is "disaster of dimensionality." The common practice for finding a method with an acceptable computation time and guaranteeing the optimal solution is to turncate the solution space and only leaves some possible states for the computation in stage 1. When the operation constraints (such as the constraints for the minimum operation or outage time of units, the ramp rates of units, and the wind power fluctuations) are considered, it will be a challenge for the conventional heuristic algorithm because of the complicated states. In addition, these constraints can be used for the solution space to be searched. In order to handle the operation constraints and build a proper search range, it is necessary to adopt two improved structure algorithms. One is for integrating the branch and bound, another is the unit commitment algorithm. In each stage, the branch and bound algorithm can be used for determining the feasible unit states to reduce the states for checking. During search, unit commitment is used for monitoring the possible states saved for each stage. When there are no longer branches in the computation with the lower bound, most of the previously saved costly states will be removed at once. The rough estimation of lower bound is obtained by making the undetermined units operate, limiting their lower generation capacity to zero and ignoring their start costs. The HDP algorithm is mainly expected to properly handle the operation constraints and limit the search range to save the computation time.

2) **Determination of the maximum wind power generation capacity in a state**
The increments and decrements of the wind power output can not be forecast accurately because of large forecast errors, so the power system operator needs an additional emergency reserve capacity to maintain enough frequency control. In fact, the additional spinning reserve capacity increases with increase of the total actual wind power. The models below give consideration to the demand for the additional spinning reserve capacity for operation (or outage) of the wind power units and the total actual wind power.

a) Linear model (Formulas (11.38) and (11.39)):

$$ASR_1(P_{WT}(t)) = r\% \times P_{WT}(t) \quad (11.38)$$

$$ASR_2(P_{WT}(t)) = r\% \times P_{WT}(t) \quad (11.39)$$

In the formula above, $r\%$ is a demand coefficient for the additional reserve capacity increased (or decreased) (linear model); ASR_1 is the demand for additional reserve capacity when wind power operation is considered; ASR_2 is the demand for additional reserve capacity when wind power outage is considered.

b) second-order model (Formulas (11.40) and (11.41)):

$$ASR_1(P_{WT}(t)) = \alpha\% \times P_{WT}(t) + \beta\% \times P^2_{WT}(t) \quad (11.40)$$

$$ASR_2(P_{WT}(t)) = \alpha\% \times P_{WT}(t) + \beta\% \times P^2_{WT}(t) \quad (11.41)$$

In the formulas above, $\alpha\%$ and $\beta\%$ are two demand coefficients for the additional reserve capacity for operation (or outage) (second-order model).

In the determination of the maximum wind power in a state, several constraint techniques are used. The permissible wind power penetration depends on the wind velocity and is restricted by maximum spinning reserve capacity of the power system, minimum generation capacity of the thermal power plant, and maximum ramp (load increase/decrease) rate of the thermal power units of the power system. For simplification, a linear model is used for determining the demand for the additional spinning reserve capacity. In this linear model, the maximum wind power penetration can be calculated through Formula (11.42) as long as the ramp rate and the wind power fluctuation are comparable.

$$P_{WT}(t) = \min[P^{*}_{WT}(t), P_{WT1}(t), P_{WT2}(t), P_{WT3}(t)] \tag{11.42}$$

In the formula above (Formulas (11.43), (11.44), and (11.45)):

$$P_{WT1}(t) = \frac{\sum_{i=1}^{NT} US_i(t) - USR_B}{r\%} \tag{11.43}$$

$$P_{WT2}(t) = \frac{P_L - \sum_{i=1}^{NT} U_i(t) \times P_i^{\min}(t)}{1 + r\%} \tag{11.44}$$

$$P_{WT3}(t) = \frac{\sum_{i=1}^{NT} U_i(t) \times DS_i(t)}{r\%} \tag{11.45}$$

However, if the WTGS output power is larger than the ramp (load decrease) rate of the power system, wind power of the power system will be cut down, namely (Formula (11.46)):

$$P_{WT}(t) = P_{WT}(t-1) + TDR(t) \tag{11.46}$$

In the formula above, $TDR(t)$ is the ramp (load decrease) rate in hour t.

There are no other actions to increase the WTGS output power, so the infeasible states will be removed when the ramp (load increase) rate cannot compensate for decrease of the WTGS output power.

3) **Execution of the economic dispatch model for the reserve capacity constraints in a state**
Optimal dispatch for the power generation of a power system containing both wind power and thermal power includes the generation distribution between the WF and the thermal power plant. It is expected to minimize the total production cost under different constraints. It is assumed that the WTGSs are owned by a power company. In this case, the basic requirement is maximum fuel saving and reliability power supply. Connection of the wind power capacity to the power system needs to change operation mode of the power system, thus increasing the operation cost. However, it is found that saving production cost depends on the wind power penetration. Connection of the wind power capacity to the power system needs more planned additional reserve capacity (it results in much higher operation cost), but the total operation cost is cut down due to increase of the available wind power generation capacity. Percent of decrement of the operation cost caused by the WTGs is larger than percent of increment of the additional costs because of the additional reserve capacity. Therefore, in this case, the proper value of wind power is set as $P_{\mathrm{WT}}(t)$ and the wind power generation capacity is not regulated. This is because high wind power penetration usually minimizes the production cost. Economic dispatch for the reserve capacity modified load constraints because of wind power can be calculated with the optimal algorithm.

4) **Performance improvement through dispatch model simplification**
Computation time of the economic dispatch model is mainly spent by fuel cost evaluation and the economic dispatch for computation of the reserve capacity constraints in the lower bound state. If a problem has too many linear constraints, there will be a long time of solution through a conventional method with oscillation. For quicker solution, a direct search method (DSM) [25, 26] is used for evaluating many possible states of the units. In the case of beginning with the original feasible solution, the optimal solution should be searched along the feasible solution route in all the time. Many of the operation constraints of generators can be properly handled with the DSM, so the computation cost is small. In each step of sequential solution, the solution will be improved with the DSM. In each convergent step following the first one, the step size S_Δ will be made more accurate through formula $S_\Delta = S_\Delta / K_\Delta$ (simplification factor $K_\Delta > 1$) until the step size obtained is smaller than the preset resolution ε. The multi-level convergence is mainly characterized in that it reduces the step size gradually to improve the solution with the DSM. This method is therefore highly applicable to unit commitment cost estimation. Thus, in each time period in the process of HDP, the fuel cost of the massive unit commitment in a time period is computed by means of simplified dispatch with rough convergence (e.g., ε^* =50 MW (ε^* is a preset resolution in the DSM algorithm)) only. After the possible states are obtained, an accurate computation step size (e.g., ε=0.01 MW) should be selected to estimate accumulated cost of each saved state in the next hour accurately. This approximate economic dispatch is for further cutting down the total computation time without lowering quality of the solution.

5) **Steps of HDP solution**
Steps of solution for dispatch between wind power and thermal power with the HDP algorithm:
   a) Read the power system data and set the initial conditions;
   b) Determine the next search range by means of the branch and bound technique;
   c) Determine the maximum wind power generation capacity in a state $[P_{\mathrm{WT}}(\mathrm{t})]$;
   d) Compute the demand for a state of spinning reserve capacity increase or decrease;
   e) Distribute the load provided by the thermal power units, that is, P_L (t) to $P_{\mathrm{WT}}(t)$, by means of simplified dispatch;
   f) Go to step (c) for the next step, otherwise go to step (g);
   g) Does the branch have any more previously saved states? If it does, go to step (b), otherwise go to step (h);
   h) Compute accumulated cost of the saved state again by means of accurate economic dispatch;
   i) Is it the last time period? If it is, go to step (j), otherwise go to step (b);
   j) Back to the result output.

11.4.4 Optimal Dispatch Between Wind Power and Hydropower

Hydropower is a flexible and controllable power source in power systems. For a power system with a hydropower station, particularly a pumped storage hydropower plant, an operation strategy for wind-hydro (W-H) power plants can be adopted. There were many studies about this both at home and abroad [27–34] and most of them focused on short-term economic dispatch. The objective of short-term economic dispatch is to store wind energy by means of water storage in the valley load periods. Upon peak loads, pumped storage plants will implement power generation. The strategy for determining optimal daily operation of WTGSs and PHS units by means of the hourly discrete optimal algorithm is based on the availability of wind power forecasts. W-H operation has the following two purposes: improve the daily economic benefit of operation of WFs; smooth the fluctuations of natural wind power to make the output power of a power plant between its upper limit and lower limit. In W-H operation, the energy storage capacity of a pumped storage hydropower plant has the following applications: store wind power when the electricity price is low and sell upon a high electricity price in each day; when the wind velocity is quick, store energy in reservoirs to make up the gap of wind power and serve as a supplement for wind power to perform the contracts signed with markets or grids [30].

In W-H optimal dispatch, the statistical characteristics of wind power are expressed in a time series of mean and standard deviation of a wind power forecast in a time range of 48 hours. The time series scenes of wind power are determined by means of Monte Carlo simulation. For each scene, the daily optimal operation strategy is determined by means of solution of hourly discrete optimization. The active power constraint and other operation constraints of wind power are considered. Finally, the optimal operation strategy of the W-H power plant is realized based on a small water storage capacity and the hourly wind power generation capacity and the hourly hydro power generation capacity are determined. In addition, the hourly water storage level and the corresponding power consumption are both calculated.

11.4.4.1 Operation Strategy

Reference [31] builds the general structure (Figure 11.7) of a power system containing a pumped storage hydropower plant and a WF and designs six operation strategies. In the figure, there is a WF (WP), a hydropower plant (HP), a conventional power plant (CPS), burning fossil fuels, an upper reservoir (UR), a lower reservoir (LR), a pump station (PS), a load system (LS), and a control system (CS). The six operation strategies determined for the power system in Figure 11.7 are all based on centralized control of all system parts (except for the LS) by a dispatcher.

1) **Strategy 1**
In this strategy, there are two independent electrical systems as below:
   a) The WP is isolated from a conventional grid (in Figure 11.7, switches a and b are off while switch c is on). The energy obtained from wind power generation ($E_{\mathrm{WP}}(\mathrm{t})$) is exclusively provided for the power consumption by the pumps ($D_{\mathrm{PS}}(\mathrm{t})$). Formula (11.47) suggests that the pumps are connected to or disconnected from the power system according to output power changes of the WF. In this strategy, we can always reach the following formula.

$$D_{\mathrm{PS}}(t) = E_{\mathrm{WP}}(t) \tag{11.47}$$

   b) The HP and the CPS are connected to the same grid in time t according to their priorities as below and provide a load demand $D_{\mathrm{LS}}(t)$ as shown in Formula (11.48). Firstly, the CS attempts to provide a demand through energy of the hydropower plant ($E_{\mathrm{HP}}(\mathrm{t})$). If water volume of the UR ($V(\mathrm{t})$) is larger than the technically minimum (V_{min}) or, in other words, $V(\mathrm{t})$ is equal to or larger than V_{min}, the water turbines will be driven for power generation. Secondly, the CS controls energy of the CPS ($E_{\mathrm{CPS}}(\mathrm{t})$) to supply load demand. If the HP

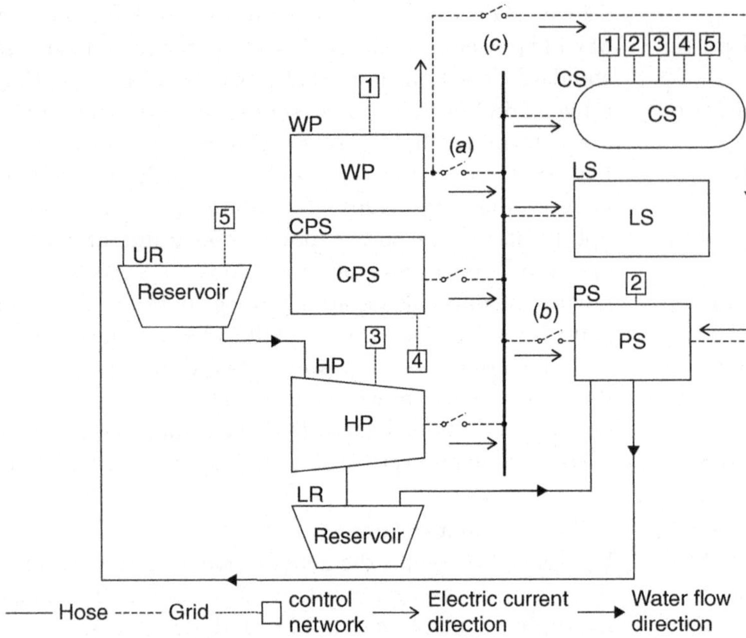

Figure 11.7 General structure of a power system with wind power.

cannot supply the total demand, the CPS will get connected to the power system. In this strategy, we can always reach the following formula.

$$D_{LS}(t) = [1 - \beta]E_{HP}(t) + \beta E_{CPS}(t) \quad (11.48)$$

In the formula above, β is a parameter between 0 and 1. This formula is restricted by Formula or inequation (11.49).

$$\begin{cases} \beta = 0 & E_{HP}(t) \geq D_{LS}(t) \\ 0 < \beta < 1 & 0 < E_{HP}(t) < D_{LS}(t) \\ \beta = 1 & E_{HP}(t) = 0 \end{cases} \quad (11.49)$$

2) **Strategy 2**
This strategy differs from strategy 1 in stage i in initial connection of the HP ($t_{c,i}$) only. The HP implements strategy 2 when volume of the UR (V(t)) equals to its maximum ($V(t)=V_{max}$) only. There is a control volume V_{co}. If it is determined, the subsystem HP will meet the maximum demand (or a percent δ of the maximum demand) in a given period tco. As long as V(t) equals to or smaller than V_{co}, the HP will be disconnected from the power system. According to Figure 11.7, when the UR has an inflow, there is no much energy from the HP fed into the grid in period (a) and in this period, the load demand is exclusively met by the CPS. When water in the UR is exhausted in period (b), the load demand is exclusively met by the HP. In this strategy, it is possible to forecast connection cycle ($t_{c,i}$) and disconnection cycle ($t_{d,i}$) of the HP. Thus, it makes operation and maintenance easy. The connect and disconnect between HP and the CPS are less. Thus, in this strategy, Formulas (11.47) and (11.48) are effective, but Formula (11.48) is not restricted by Formula (11.49) but Formulas (11.50) and (11.51).

$$\beta = 0 \quad \text{When } [t_{c,i} \leq t < t_{d,i}] \rightarrow E_{HP}(t) > D_{LS}(t) \quad (11.50)$$

$$\begin{aligned} &\beta = 1 \quad \text{When } [t_{d,i-1} \leq t < t_{c,i}] \rightarrow E_{HP}(t) = 0 \\ &t_{c,i} \rightarrow V\ (t = t_{c,i}) = V_{max}; \\ &t_{d,i} \rightarrow V\ (t = t_{d,i}) < V_{co}; \\ &t_{d,i-1} \rightarrow V\ (t = t_{d,i-1}) < V_{co} \end{aligned} \quad (11.51)$$

3) **Strategy 3**
In this strategy, all the power systems are connected to the same grid. The objective is to make the load demand (D_{LS}) provided by the renewable energy sources (wind power and hydropower) as much as possible. This is because that the direct connection of wind power to a small- or medium-sized power system probably has serious disadvantages. When the wind power penetration is high, fluctuations of the WP actual active power probably results in imbalance between the power generation and the power demand and further in possible dangerous operation because of frequency fluctuations and voltage fluctuations. In order to avoid possible impacts on safety and stability of the power system, it is therefore necessary to limit the wind power connected to the grid to be a percent α of capacity of the conventional synchronous generators in the power system, which depends on load of the power system. To determine the percent, it is necessary to carry out dynamic studies about the wind velocity fluctuations and other operation conditions (such as size of the power system). α is defined as the maximum percent of wind power in

order to avoid influencing safety and stability of the power system but meet the power load. In this model, α is an input parameter between 0% and 100%.

In this strategy, the operation problems in a grid with high wind power penetration (a large α) can be solved by an HP system, so as to increase the percent of power generation by renewable energy sources in the power system. In this strategy, the CS controls the subsystems with the steps below.

a) Minimum percent (100-α) of the electric energy needed by the LS is provided by the HP and/or the CPS following the sequence below. See Formula (11.52) (Inequations (11.53) and (11.55) are the preconditions of Formula (11.52)). Firstly, if volume of the UR (V(t)) is adequate, the minimum percent (100-α) will be provided by the HP; secondly, if volume of the UR (V(t)) is inadequate, the minimum percent (100-α) will be provided by the HP and the CPS; thirdly, if volume of the UR (V(t)) is smaller than the technical minimum Vmin necessary for water turbine driving, the minimum percent (100-α) will be provided by the CPS.
b) If there is adequate wind, the maximum percent (α) of the power load demand will be provided by the WP. The CS will attempt to connect the pumps continually to realize power balance (see Formula (11.54)). If there is adequate wind and the wind power provided is larger than $\alpha D_{LS}(t)/100$, then (Formulas (11.52), (11.53), (11.54), and (11.55)):

$$D_{LS}(t) = [1-\beta]E_{HP}(t) + \beta E_{CPS}(t) + E_{WP}(t) \tag{11.52}$$

$$\begin{aligned} &[1-\beta]E_{HP}(t) + \beta E_{CPS}(t) \\ &\quad \geq (100-\alpha)D_{LS}(t)/100 \\ &E_{WP} \leq \alpha D_{LS}(t)/100 \end{aligned} \tag{11.53}$$

$$\begin{aligned} &D_{PS}(t) = E_{WP}(t) - \alpha D_{LS}(t)/100, \\ &\quad \text{if } E_{WP}(t) > \alpha D_{LS}(t)/100 \\ &D_{PS}(t) = 0, \quad \text{if } E_{WP}(t) \leq \alpha D_{LS}(t)/100 \end{aligned} \tag{11.54}$$

$$\begin{aligned} &\beta = 0, \quad \text{if } [E_{HP}(t) + E_{WP}(t) \geq D_{LS}(t)]; \\ &[E_{WP}(t) \leq \alpha D_{LS}(t)/100] \\ &0 < \beta < 1, \\ &\text{if } [0 < E_{HP}(t) + E_{WP}(t) < D_{LS}(t)]; \\ &[E_{WP}(t) \leq \alpha D_{LS}(t)/100] \\ &\beta = 1, \quad \text{if } E_{HP}(t) = 0 \end{aligned} \tag{11.55}$$

4) **Strategy 4**

As Formula (11.52), the uncontrollable load demands are supplied by different power generation subsystems, depending on whether the UR has reached its maximum volume (V_{max}). As in Inequation (11.53), if the UR has not reached its maximum volume (β=1 in Formula (11.50)), minimum percent (100-α) of the load demand will be provided by the CPS while maximum percent (α) by the WP, as long as there is adequate wind power. As in Formula (11.54), if wind power is larger than $\alpha D_{LS}(t)/100$, the pumps will be connected continually to pump water from the LR to the UR for the electrical systems balance. If the UR has reached its maximum volume ($t = t_{c,i}$), percent of the load demand provided by the CPS will be provided by the HP (β=0 in Formula (11.50)). Compared with strategy 3, this strategy makes it more possible to forecast the connection-disconnection cycles of the CPS and the HP, maintenance easier and the CPS and the HP connect and disconnect less.

As in Figure 11.7, in period (a) when the UR has an inflow, water energy is not fed into the grid, so the demands will be met by the CPS and the HP; in period (b) when the UR is releasing water, the demands will be met by the HP and the WP.

5) **Strategy 5**

This strategy is an exception of strategy 4. The only difference is that when the UR has an inflow, all the uncontrollable load demands are supplied by the CPS and all the wind power is used for pumping (see Formula (11.56)).

If $t_{d,i-1} \leq t < t_{c,i}$, then:

$$D_{PS}(t) = E_{CPS}(t), \quad \alpha = 0\%; \quad D_{PS}(t) = E_{WP}(t) \tag{11.56}$$

When the UR is releasing water, the load demands are met in the same way as in strategy 4.

To sum up, in the period (a) in Figure 11.7 (when the UR has an inflow), the HP does not feed energy to the grid and the load demands are exclusively met by the CPS; in the period (b) in Figure 11.7 (when the UR is releasing), the load demands are supplied by the HP and the WP.

6) **Strategy 6**

This strategy is an exception of strategy 4. The only difference is that when the UR is releasing water, all the load demands are met by the HP and all the wind power is used for pumping; see Formula (11.57).

If $t_{c,i} \leq t < t_{d,i}$, then

$$D_{LS}(t) = E_{HP}(t); \quad \alpha = 0\%; \quad D_{PS}(t) = E_{WP}(t) \tag{11.57}$$

When the UR has an inflow, the load demands are met in the same way as in strategy 4.

To sum up, in the period (a) in Figure 11.7 (when the UR has an inflow), the HP does not feed energy to the grid and the load demands are met by the CPS and the WP; in the period (b) in Figure 11.7 (when the UR is releasing), the load demands are exclusively met by the HP.

In all the strategies above, volume of the UR in time t ($V(t)$) can be calculated through a function below including the existing capacity ($V(t\text{-}\Delta t)$), the pumping capacity ($V_p(t)$) and the water volume consumed by the hydraulic turbines (HT) ($V_t(t)$) (Formula (11.58)).

$$V(t) \leq V(t - \Delta t) + V_p(t) - V_t(t);$$
$$V(t) \geq V_{\min} \vee V(t) \geq V_{C0} \tag{11.58}$$

11.4.4.2 Optimal Operation of W-H Power Plants

In reference [30], the long-term economic feasibility of coordinative operation of one WF and two reservoirs (including a small hydropower station and a pump station) in Portugal is studied. A pumped storage hydropower power plant (it is probably a small one with a size related to size and characteristics of the WF) is built near the WF to implement operation strategy for the W-H power plant. When there is wind power forecast, the optimal daily operation strategy for the WTGSs and the PHS units is determined with the hourly discrete optimal algorithm.

In operation analysis of the W-H power plant, it is assumed that part of power of the WF is under control so that the minimum wind power for connection to the grid can be guaranteed in regardless of changes of the wind velocity and the WF can, at the beginning of a day, guarantee energy availability for the next 24 hours.

In order to evaluate optimal operation in range of the study, the following data in addition to physical features of the W-H power plant should be input to the model: wind power forecast or wind velocity forecast (the forecast wind velocity and the forecast wind power are random variables), estimate price of the active power, and the output limit curve needed (given by the grid operator according to the operation constraints).

The model should maximize 24-hour operation profit of the W-H power plant (or economic benefit of the on-grid electric energy) and give consideration to main operation constraints and 24-hour discretization of the W-H power plant. It is assumed that wind power is a random variable expressed in hourly values of two sequence, that is, the mean wind power and its standard deviation. Randomly, S (quantity) wind-power series samples can be obtained and each of them stands for a wind power state. Sample vector P_v of wind power can be obtained too. In each case, the following optimization problems should be solved (Formulas (11.59–11.71)).

$$\max \sum_{i=1}^{n} (c_i P_i - c_p P_{pi}) + n c_\alpha \alpha \tag{11.59}$$

$$s.i. \quad P_i = P_{wi} + P_{hi} \tag{11.60}$$

$$P_{Vi} = P_{wi} + P_{pi} + P_{DLi} \tag{11.61}$$

$$E_{i+1} = E_i + t\left(\eta_p P_{pi} - \frac{Ph_i}{\eta_h}\right) \tag{11.62}$$

$$E_1 = E_1^{esp} \tag{11.63}$$

$$E_{n+1} = E_{n+1}^{esp} \tag{11.64}$$

$$\alpha P_i^L \leq P_i \leq P_i^U \tag{11.65}$$

$$P_g^L \leq (P_{wi} + P_{pi}) \leq P_g^U \tag{11.66}$$

$$P_h^L \leq P_{hi} \leq \min\left(P_h^U, \eta_h \frac{E_i}{t}\right) \tag{11.67}$$

$$P_p^L \leq P_{pi} \leq P_p^U \tag{11.68}$$

$$0 \leq E_i \leq E^U \tag{11.69}$$

$$0 \leq \alpha \leq 1.0 \tag{11.70}$$

$$P_{DLi} > 0 \quad i = 1, \ldots, n \tag{11.71}$$

In the formulas and inequations above, the variables are vectors; P is hourly on-grid active power of the W-H power plant; P_w is hourly on-grid active power of the WTGSs; P_h is hourly on-grid active power of the hydropower units; P_p is mean hourly active power during operation of the pumps; P_{DL} is the hourly reserve capacity, that is, the energy on the wind power curve not used for power generation (namely the wind energy decreased if technically possible); E is hourly storage level of the reservoirs; α is a variable, that is, decrease coefficient of the minimum output; P^L and P^U are constraint vectors for the hourly minimum output power and the hourly maximum output power respectively, which are related to the market demand the grid constraints; P_v is vector of the hourly available wind power samples when it is considered; c is vector of the hourly active power price (the time-of-use price may be priced according to the market pricing); c_p is the pumping cost; c_α is a fine when the power generated is lower than the minimum output; E^U is the reservoir storage; η_p is efficiency of the pumping system; η_h is efficiency of the hydropower plant; E_1^{esp} and E_{n+1}^{esp} are initial storage level and final storage level of the reservoirs respectively; P_g^L and P_g^U are minimum and maximum generation capacity of the WF respectively; P_h^L and P_h^U are minimum and maximum generation capacity of the water turbines respectively; P_p^L and P_p^U are actual minimum and maximum power of the pumps respectively; t is the duration of a time period (such as one hour); n is quantity of the discrete time periods.

According to objective Formula (11.59), two objectives can be set: to consider cost of the pumps in the pump station to maximize the hourly on-grid active power benefit of the W-H power plant; to meet the minimum on-grid power. When the dispatch plan cannot be realized (sum of the wind power and the hydropower is smaller than the minimum output), make α smaller than 1.0 to reduce the minimum output; when the inadequate wind power can be compensated by the hydropower, make α 1.0. The minimum power should be guaranteed in any of the n time period, so expression ($c_\alpha \alpha$) should be multiplied by quantity of the discrete time periods. Formula (11.60) suggests that hourly active power output of the W-H power plant is contributed by the hydropower and the directly grid-connected wind power. Formula (11.61) suggests that in the one-hour time periods considered, some of the wind power can be connected to the grid directly while the other used for pumping water are be stored for future use. In some special cases, some available wind power can be discarded. Formula (11.62) indicates energy balance of the reservoirs. At beginning of time period (t+1), energy of the reservoirs is a difference after sum of the initial energy in time period I and energy of the water pumped into the reservoirs is deducted by the on-grid hydropower in the same time period. In Formulas (11.63) and (11.64), initial energy and final energy of the reservoirs should be specified. The initial energy is known as the final energy of the last day. However, the optimal final energy of the day is not known and it depends on the expected operation strategy defined for the next day. To obtain the optimal storage of the reservoirs, the original time range of study, that is, 24 hours, is extended to a time range between 24 hours to 48 hours. The W-H power plant operates in each day, so only the first 24 hours is adopted. Inequation (11.65) suggests that the output power should be in the given range. The output constraints include: operation constraints for the grid which are usually related to thermal limits or stability limit of important lines of the grid in some time periods of operation; constraints for the contract (a daily market or a bilateral market) between the WF and the market. Inequations (11.66) to (11.69) give operation constraints about the WTGSs, the water turbines, the pumping units and the storage capacity. As in Formula (11.67), the maximum hourly hydropower depends on constraints for the related units and available energy of the reservoirs in the time period.

In the Formula and inequations (11.59) to (11.71) about the power system studied, there are 289 variables, 145 formulas, and 530 inequations. This is linear optimization that can be solved with any linear optimization algorithm.

Solutions of the optimization in Formulas and inequations (11.59) to (11.71) are the hourly active hydropower and the hourly active wind power in each 24-hour cycle. If the wind power in each time period is a constant, storage level and pump operation strategy in the time period will be determined. A shorter time step of wind power forecast can make solution distribution more precise. In actual online operation, that is, at the right beginning of service of the pumping units, total output power of the W-H power plant is properly smaller than the minimum output power specified for all the units. In this case, both the pumping units and the water turbines will cover restricted operation at the same time.

After S (quantity) times of simulation (such as 150 times of Monte Carlo simulation), means, maximums and minimums of related variables and envelope curve about output power of the W-H power plant can be obtained. These values can be used for recommending that a small storage be used in the next several hours to realize the optimal operation strategy of the W-H power plant, determining the hourly wind power generation capacity and the hourly hydropower generation capacity, hourly power consumption and hourly storage level of the pumping units and evaluating performance of the solution.

By means of slight modification of the recommended algorithm, the minimum output power of the W-H power plant in the preset wind power forecast scene can be determined. If α has no upper limit (a large value) in Inequation (11.70), namely, $c_\alpha \gg c$ and $c_\alpha \gg c_p$, the optimal algorithm will be expected to determine a strategy for increasing minimum output of the units. The hourly spinning reserve capacity in the next day set by the power system operator is related to the 24-hour minimum output guaranteed for the WF, so increase of the minimum output of the W-H power plant can reduce both the reserve capacity demand and the corresponding auxiliary service cost. In the model above, only the connection of active power is considered; however, a reactive power objective can be added easily and the operation benefit of participation in the auxiliary service can be considered too.

By means of an optimal daily operation strategy, the W-H power plant can improve economic benefit of the WF with its storage capacity because wind power can be connected to the grid first in the non-valley period (the time periods with higher electricity prices). In these time periods, the available wind power may be compensated by hydropower when possible. In the time periods with lower electricity prices or grid blockage, storage of the reservoirs can be increased by the pump station. Compared with the operation of wind power only (there is no hydropower facilities), expected annual mean economic benefit in the W-H operation strategy is significant.

PHS can probably solve any constraints of the grid on installed capacity of the WF. In addition, it improves controllability of output power of the WF and participation of wind power in the power market because it reduces the intermittency of wind power connection to the grid and the conventional reserve capacity managed by the power system operator. PHS can bring about the following incomes: an income because of the hourly electricity price difference; an income by means of storing energy at lower electricity prices when output power of the W-H power plant is its maximum and selling this energy at higher electricity prices in the subsequent time periods.

Based on reference [31], reference [34] protocols reveal a plan about a unite operation of a W-H power system, establishes an economic operation computation and analysis model for a small simulated grid. The model is expected to make the best use of wind power to supply load demand of the power system while making sure the grid operates safely, reliably, and stably. It is now assumed as below: the permissible maximum wind power proportion of the grid is 10% of the maximum shoulder load; the unite operation plan for the power system is that some of the wind power, that is, 10% of the maximum shoulder load, is connected to the grid directly; the PHS unit can be started for phase modulation operation when necessary; the conventional power plant supplies power with a constant capacity, which is a sum of the total base load and 90% of the maximum shoulder load; the peak load is guaranteed by the units connected or disconnected from the W-H power plant; pumped storage hydropower units of the W-H power plant pump water with help of residual energy in the valley periods and store the residual wind power during random fluctuations by means of connection and disconnection with the W-H power plant at proper time. In this way, all the wind power can be input into the grid directly or indirectly and the conventional power plant can operate with a constant output power. The mathematical model is as below:

There is a demand for power balance of the power system, in other words, there should be balance between the total generation capacity $P_{total}(t)$ and the total load $D_{total}(t)$ (Formula (11.72)).

$$P_{total}(t) = D_{total}(t) \tag{11.72}$$

In the formula above (Formulas (11.73) and (11.74)):

$$P_{total}(t) = P_{WF}(t) + P_{HP}(t) + P_{CPS}(t) \tag{11.73}$$

$$D_{total}(t) = D_{PS}(t) + D_{LS}(t) = D_{PS}(t) + D_{LS,base}(t) + D_{LS,mid}(t) + D_{LS,peak}(t) \tag{11.74}$$

In the formula above, $P_{WF}(t)$ is output power of the WF at time t; $P_{HP}(t)$ is output power of the water turbines at time t; $P_{CPS}(t)$ is output power of the conventional power plant at time t; $D_{PS}(t)$ is pumping power of the pumps at time t; $D_{LS}(t)$ is load of the power system at time t, including base load $D_{LS,base}(t)$, shoulder load $D_{LS,mid}(t)$ and peak load $D_{LS,peak}(t)$.

In order to make the best use of wind power while guaranteeing high reliability or high-power supply quality for the power system, the W-H power plant plan above should be based on the following formulas (Formulas (11.75), (11.76), and (11.77)).

$$D_{PS}(t) = \rho g k_{pc}(t) Q_p(t) H_b(t) / \eta_{ps}(t) = P_{WF}(t) - D_{LS,mid\ max}(t) \times 0.1 \tag{11.75}$$

$$D_{LS,base}(t) + D_{LS,mid,max} \times 0.9 = P_{CPS} \tag{11.76}$$

$$D_{LS,peak}(t) = P_{HP}(t) = \rho g k_{tc}(t) Q_h(t) H_n(t) \eta_{hp}(t) \tag{11.77}$$

In the formula above, ρ is the density of water;g is the acceleration of gravity; $Q_p(t)$, $H_b(t)$ and $\eta_{ps}(t)$ are respectively flow, lift, and efficiency of a single pump at time t; $Q_h(t)$, $H_n(t)$ and $\eta_{hp}(t)$ are respectively flow, head, and efficiency of a single water turbine at time t; $k_{pc}(t)$ and $k_{tc}(t)$ are respectively quantities of the operating pumps and the operating water turbines at time t; $D_{LS,\ mid,\ max}(t)$ is maximum shoulder load of the power system. In Formula (11.73), output power of the WF is (Formula (11.78)):

$$P_{WF}(t) = k_w(t) P_{WT}(t) \xi \varphi \tag{11.78}$$

In the formula above, $k_w(t)$ is quantity of the normally operating WTGSs of the WF at time t; ξ and φ are transmission efficiency and generation efficiency of the WTGSs, respectively; $P_{WT}(t)$ is the power of each WTGS at time t, which can be obtained according to the wind velocity v (t) at time t with the interpolation of WTGS power-wind velocity curve provided by the related manufacturer.

If it is assumed that LR of the pumped storage hydropower plant has a relatively large volume (its water level changes can be ignored), heads or lifts of units in the pumped storage hydropower plant can be obtained according to the storage (storage capacity)–water level curve of the UR. Storage $v(t)$ of the UR at time t can be obtained by means of iteration by time step as below, according to the storage $V(t-\Delta t)$ of preceding time step calculation and the present step including pumping capacity $V_p(t)$ of the pumps and the discharge capacity $V_t(t)$ of the water turbines (Formula (11.79)).

$$V(t) = V(t - \Delta t) + V_p(t) - V_t(t); \qquad V(t) \geq V_{min} \tag{11.79}$$

In the formula above, Δt is time step in the iteration; V_{min} is storage of the UR during daily regulation when the generating water level is the lowest. Contents such as wind power generation energy, peaking generation energy of the pumped storage hydropower plant, and pumping power consumption based on the W-H power plant plan can be calculated based on the known data and Formulas (11.72) to (11.79). By means of Formula (11.78), rationality of installed capacity of the pumped storage hydropower plant can be verified according to the fact whether storage capacity of the UR can realize daily regulation cycles. Supply cost of unit electric energy of the power system can be calculated through the formula below (Formula (11.80)).

$$C_{OEL} = \{(C_{I0} - S_0) r / [1 - (1 + r)^{-L}] + C_{O\&M} + C_{EXT}\} \Big/ \left[\int_{0}^{8760} D_{LS\ peak}(t)\, dt \right] \tag{11.80}$$

In the formula above, C_{I0} is the investment cost; S_0 is subsidy of the government for wind power; r is the discount rate; L is the payback period; $C_{O\&M}$ is the annual cost of operation and maintenance; C_{EXT} is other additional costs.

If annual maximum daily load curve, wind velocity daily change curve, WTGS power curve, and pumped storage hydropower plant installed capacity of the power system are given. It is assumed that the minimum head efficiency and the minimum lift efficiency are respectively η_{hp} and η_{ps}, and that the transmission efficiency and the generation efficiency of the WTGSs are respectively ξ and φ, contents such as a daily charge curve about total output of the WF, a daily change curve about power supply capacity of the power system, a curve about operation capacity changes of the pumps and the water turbines at all time, and a storage curve of the UR can be obtained through the W-H power plant model above. The total power supply capacity of the power system meets total load demand. In addition, this model enables comparison of economic and social benefits based on the electricity prices specified for peak periods and valley periods of the grid when there is a pumped storage hydropower plant or not.

11.4.5 Economic Dispatch of Power Systems with Wind Power in the Power Market

Under the existing power market mechanism, some models about the economic dispatch of power systems do not consider such external costs of power generation as environmental pollution and resource consumption or factors such as energy efficiency. The electric energy values of new energy sources are not reasonably realized, which is represented by wind power, making wind power prices less competitive than conventional energy sources such as thermal power and hydropower. Therefore, it is practically significant to develop a model about economic dispatch of power systems without damaging the existing power market mechanism but reasonably reflects the environmental cost and the wind power value. Some domestic and abroad researches have reached a new environmental economic dispatch model for the power systems with wind power based on a lot of relevant studies.[35–37] In the studies, the introduction is considered, which is about an environmental punishment cost and the impacts of wind power on dispatchable capacity and wind power reserve capacity of a power system and ramp rates of the units of the power system. This section introduces only one economic dispatch model for the power systems with wind power. Some issues about the relation between wind power and the power market will be introduced in Section 11.5.

11.4.5.1 Environmental Economic Dispatch Model for Power Systems with Wind Power

In the competitive power market, usually the following two objective functions are adopted in economic dispatch: to guarantee the minimum operation costs for power systems; to guarantee the minimum costs of power purchases from the trading center. Whichever objective function is selected, the output power of different energy sources and the external costs of different units are considered. As a result, the thermal power unit using fossil fuels and taking environment-friendly technical upgrades, which are expected to be preferred in economic dispatch, lose their advantage in competition because of higher electricity prices as a result of environmental protection investments in the same conditions; similarly, the environment-friendly wind power is now difficult to be preferred in dispatch because of a higher investment cost. In consideration of such characteristics of wind power as randomness and limited dispatch, a power system with wind power will have an increasingly increased reserve capacity cost with increase of its wind power penetration. Thus, in the power market, in the economic dispatch of a power system with wind power contained, factors such as the environmental protection cost and the increased cost of the wind power reserve capacity should be considered so that the wind power value can be reflected reasonably.

1) **Nominal environmental compensation cost of thermal power units**
Because of the pollutant emission from thermal power units, the environmental protection cost of thermal power units can be added to the purchase costs of thermal power units in a conventional economic dispatch model. The additional cost is not represented in the electricity purchase costs paid by consumers but a dispatch priority of the related units. In this way, wind power development can be realized without damage to the existing power market mechanism and the existing power system economic dispatch mechanism. The additional cost is defined as a "nominal environmental compensation cost of thermal power units."
The waste gases emitted by thermal power units are pollutants including NO_X, CO_2, SO_X, and so on. Viewed from environmental protection, the pollutant emission characteristic of each thermal power unit can be expressed in a NO_X weight converted from the pollutants emission in a unit time. Formula (11.81) is a pollutant emission characteristic formula [35, 37].

$$f_{e.i} = \partial_i + \beta_i \times P_{Gi} + \gamma_i \times P_{G.i}^2 + \zeta_i \times \exp(\lambda_i \times P_{G.i}) \tag{11.81}$$

In the formula above, ∂_i, β_i, γ_i, ζ_i, and λ_i are respectively the pollutant emission characteristic coefficients of thermal power unit i, which can be obtained by measurement. Thus, the nominal environmental compensation cost of thermal power units can be obtained through Formula (11.82).

$$F_{NEP} = \sum_{i=1}^{n} C_{NEP,i} \times f_{e,i} \tag{11.82}$$

In the formula above, $C_{NEP.i}$ is a function coefficient of the nominal environmental compensation cost (it is set to be a constant for the sake of simplification); n is quantity of the thermal power units in the power system.

2) **Compensation cost of the additional wind power reserve capacity**
The compensation cost of the additional wind power reserve capacity is introduced in Formula (11.83) [35].

$$F_{WRC} = \sum_{j}^{m} C_{WRG,j} \times (P_{W,j} - P_{W,j,av}) \qquad (11.83)$$

In the formula above, $P_{W,j}$ is the expected generation output of WTGS j; $P_{W,j,av}$ is the actual output power of WTGS j; m is quantity of the WTGSs in the power system; $C_{WRC,j}$ is a coefficient of the compensation cost of the additional wind power reserve capacity, which is zero when $P_{W,j}$ is equal to or smaller than $P_{W,j,av}$, or depends on consumption and emission characteristic of the thermal power unit supplying the additional reserve capacity when $P_{W,j}$ is equal to or larger than $P_{W,j,av}$.

3) **Objective function**
According to the analysis above, the total cost of a power system with wind power includes a total electricity purchase cost $\sum_{i=1}^{n} F_{G,i}$ of the thermal power units in the power market, a nominal environmental compensation cost F_{NEP}, a total electricity purchase cost $\sum_{j=1}^{m} F_{W,j}$ of the WTGSs in the power market and a compensation cost of the additional wind power reserve capacity(F_{WRC}). In other words, the objective function for the economic dispatch of a power system with wind power can be expressed in Formula (11.84).

$$\min\left(\sum_{i=1}^{n} F_{G.i} + F_{NEP} + \sum_{j=1}^{m} F_{W.j} + F_{WRC}\right) \qquad (11.84)$$

In the function above, electricity purchase cost of the thermal power units can be expressed in Formula (11.85).

$$\sum_{i=1}^{n} F_{G,i} = \sum_{i=1}^{n} a_i P_{G.i}^2 + b_i P_{G.i} + c_i \qquad (11.85)$$

In the formula above, a_i, b_i, and c_i are the thermal power units' electricity purchase cost characteristic coefficients related to consumption characteristic of the thermal power units, which can be determined according to bidding curve of the thermal power units.
There is no fuel consumption in wind power operation, so the mean electricity purchase cost of wind power in its whole life cycle can be considered to have a rough linear relation with the generation capacity as shown in Formula (11.86) when both investment cost as well as operation and maintenance cost of wind power are considered.

$$\sum_{j=1}^{m} F_{W.j} = \sum_{j=1}^{m} C_{W.j} P_{W.j} \qquad (11.86)$$

In the formula above, $C_{W.j}$ is the electricity purchase cost coefficient of wind power, which can be determined according to bidding curve of the WTGSs; P_{Wj} can be determined according to wind velocities at the WTGSs.

4) **Constraint**
If the grid loss is ignored, the equality constraint will be (Formula (11.87)):

$$\sum_{i=1}^{n} P_{G.i} + \sum_{j=1}^{m} P_{W.j} = \sum P_D \qquad (11.87)$$

In the formula above, $\sum P_D$ is load of the power system.
The inequality constraints will be (Formulas (11.88) and (11.89)):

$$P_{G.i}^{\min} \leq P_{G.i} \leq P_{G.i}^{\max} \qquad (11.88)$$

$$0 \leq P_{W.j} \leq P_{W.j.r} \qquad (11.89)$$

In the inequations above, $P_{G.i}^{\min}$ and $P_{G.i}^{\max}$ are respectively the maximum power output and minimum power output of thermal power unit i; $P_{W.j.r}$ is the rated power output of WTGS j.

11.4.5.2 Solution and Result of the Model

Optimal distribution of loads of the units can be obtained by means of optimization algorithm, solving the Formula (11.84).

If the nominal environmental compensation cost of the units is not considered in the conventional price mechanism, the units influencing the environment more will be more preferred in dispatch. Renewable energy sources such as wind power can be dispatched only when their costs are low enough.

The nominal environmental compensation cost changes the thermal power units' dispatch priority, which drops with increase of the nominal environmental compensation cost, namely, the thermal power units with lower environmental compensation cost coefficients will be dispatched earlier. Because of nominal environmental compensation cost of the thermal power units, environment-friendly renewable energy sources such as wind power will have priority in dispatch, which is beneficial to the development of wind power.

If the compensation cost of additional wind power reserve capacity is considered, and the additional wind power reserve capacity and their compensation cost coefficient are determined reasonably, the priority of wind power in dispatch will be maximized. A dispatch can be done with an output power of wind power closest to the expected value.

According to the result, the economic dispatch model considers both the nominal environmental compensation cost because of pollutant emission characteristic of thermal power units as well as the reserve capacity cost of thermal power units because of wind power. compared with the conventional economic dispatch model based on the consumption characteristic of thermal power units, the model analyzes the economic and social benefits of the electric energy generated by different energy sources better. It provides a method to effectively improve the dispatch priority of environment-friendly units or power sources in the existing power market price mechanism.

11.5 Wind Power Generation and Power Market

With the development of modern wind power generation technologies, the electricity quality of WFs is now much better and wind power now has much higher controllability and reliability. It is less dependent on the auxiliary services of power systems and, in particular, wind power can be still connected with grids and provide reactive power (or voltage) support and active power (or frequency) support for main power systems in grid faults. It enables WFs to participate in the centralized dispatch and control for the whole grids and helps WFs to enter the power market and work as a part.

At present, the proportion of wind power in a power market differs greatly in different countries. In some countries, wind power does not participate in their power markets; the grids absorb all available wind power and wind power is paid at the agreed on-grid electricity prices; in some other countries, wind power participates in the day-ahead or the hour-ahead power markets, but there is no punishment for imbalance (imbalance is a difference between the expected generation capacity and the actual generation capacity). In some power markets, WTGSs are not allowed for bidding and the available wind power is transmitted to power systems. Wind power is paid by the power systems according to sums of the actual market electricity prices and the constant additional costs (bonus). For example, the power markets in northern Europe use the spot power market mode. The wind power in western Denmark participates in the power markets of northern Europe. The electricity price should be calculated as below: it is 0.33 DKK/(kW·h) for a WF with a grid connection period shorter than ten years; the others are subject to the real-time electricity price for a WF with a grid connection period longer than ten years. In addition, a subsidy of 0.27 DKK is given for each kW·h in the specified electric energy quota; or a subsidy of 0.10 DKK is given for each kW·h out of the specified electric energy quota. Each WTGS with a grid connection period longer than ten years and accumulated generation energy larger than the related quota will be removed from the list for preferred electricity purchase. As a European country with the best wind power sources, Ireland has large installed capacity of wind power. Ireland adopts a bilateral power market mode in which power suppliers and power plants trade on a bilateral market. The imbalance between the power generated and the loads because of forecast errors is handled by the designated power plant, that is, ESB. The additional costs because of wind power generation are borne by all power consumers [38, 39].

The analyses in the former sections of this chapter are mainly based on an assumption that WF works under the circumstance of vertical monopoly. This section mainly analyzes the situation that wind power is a part of the competitive power market. In the power market, wind power supplies power in a way different from dispatchable generators. Wind power depends on natural wind and generates no power when there is no wind, so wind power cannot compete with conventional generators. Wind power is more competitive than conventional power generators when government subsidies, tax-free periods, and emission reduction benefits are considered only. This section mainly introduces the impacts of wind power on operation and electricity prices of the power market, the capability of wind power to participate in the power market, the wind power bidding strategy in the power market, and so on.

11.5.1 Impacts of Wind Power on Operation of the Power Market

Wind power generation has the double characteristics of both power generation and loads. WFs sell power directly or indirectly, which is the same as conventional power plants; wind power is basically uncontrollable or controllable in a limited range (in regardless of its reducibility), particularly in a long-time range (a couple of hours or days), which is similar to loads.

The impact of wind power connected to a power system on the grid operation cost is mainly caused by the instability of wind power. The cost of repetitive dispatch of units because of inaccurate wind power forecast increases with wind power forecast errors. Wind power forecast errors result in increase of the whole imbalance costs under pre-contracts.

The participation of wind power in the power market faces many technical challenges. ① Grid constraints, some WFs are connected to medium or high-voltage grids. Long radial overhead lines or cables are necessary for the transmission of power from distant WFs, so the power transmission lines need to be extended to avoid blockage. ② Wind power forecast errors. The impact of wind power instability on the operation costs of power systems cannot be ignored because it is possibly significant in some cases or extremely small in some other cases. It is therefore necessary to estimate the costs caused by wind power forecast error in the power market. For less forecast errors, effective wind power forecast tools are necessary. ③ Difficult operation of power systems. A little wind power cannot influence the operation or control of power systems greatly, but the impact of high wind power penetration on the operation and operation costs of power system is a big challenge. It depends on the WF locations, power system sizes, grid structures, and so on [15].

Each power market needs management. With the connection of massive wind power with unique characteristics to power systems, there is a risk that the operators are not able to be sure about the problems caused by wind power and make corresponding decisions. If the management system is too beneficial to wind power, the other parts of the power systems will probably need to pay for it, which is destructive; if the management system is too strict for wind power, the wind power investments will have loss and the society will not benefit from emission reduction and technically potential power supply safety. Thus, the management needs to encourage innovations and help power companies find the best effective method to integrate wind power into power systems [15].

The participation of wind power in a power market has impacts on such contents of the power market as balance, capacity, and auxiliary services [15, 40].

1) **Wind power results in a larger demand for power market balance.**
   Power systems always have some centralized market forms for balance, which relates to the mechanisms for energy trading in power markets. A balanced power market is necessary. The reason is that the load demands in future cannot be forecast accurately. Making it necessary to balance the power generated and the demands in a real-time way needs to consider the technical constraints, which is difficult to be handled in bilateral contracts and impossible to be handled in centralized constracts. Both power plants using fossil fuels and hydropower plants are controllable. Usually, their output power can be set to be the necessary expected values. Furthermore, wind power has poor controllability. All of these make it necessary to balance power systems.
   A balanced power market is similar to an associated market, but it still has a lot of differences. The balanced market supplies power in a short time, which is sometimes thought to be a spot market. The energy trading in a balanced power market is a little, but its price is particularly important for the whole balanced power market because it reflects the short-term supply-demand balance. The electricity price is low when there is a low load demand and high-power availability or high when there is a high load demand and/or low-power availability. A lasting high electricity price suggests a demand for a generation capacity investment.
2) **Payment for wind power capacity.**
   In some power markets, there is a definite capacity payment based on the generation capacity. A power market without a capacity payment is called a single energy market. The impact of capacity payments is an encouragement for long-term investments because capacity payments guarantee security incomes for investors to some extent. This is different from single energy markets with incomes based on energy yields only. If there are capacity payments, power plants can accept moderate electricity prices. Particularly, if there are no definite capacity payments in energy shortage, the capacity values will be represented in electricity prices, power pool, or bilateral trading. If there is no definite capacity payment mechanism and the related power market is a competitive one, there will probably be unrestricted electricity prices or, in other words, the power market will have to accept the electricity prices no matter they are high or low. Thus, in power shortage, the electricity price is high and includes a capacity payment. If there is a definite capacity payment, the electricity price should be restricted by the administrator to avoid twice capacity payments by the participants. Capacity payments are mainly based on CC. The CC of generators is an additional load for service provision during maintaining the power system reliability. Wind power also has CC, which decreases with increase of the wind power penetration and should be paid in a certain form under the capacity mechanism.
3) **Wind power results in a larger demand for auxiliary services.**
   The reliable operation of a power system is based on auxiliary services including reactive power, blockage management, black start capability, and so on. The auxiliary services provided by a market mechanism keep increasing and wind power probably gets a return from the auxiliary services. Auxiliary services can be provided by centralized markets or through long-term contracts based on competition. For example, for the node prices, coordinative optimal energy, and reserve capacity market in blockage management, reserve capacity, that is, the so-called "spinning wind" can be provided by part-loaded (not in operation with the maximum output power) WTGSs with available wind power. This has an

economic significance. Modern WTGSs or WFs can control reactive output power and participate in voltage control. It is impossible for WTGSs to participate in black start because black start relates to wind availability.

Energy markets, balance, capacity mechanisms, and reserve capacity are closely relevant. If there is an actually real-time effect for bilateral trading and/or a pool power market, a balance mechanism will not be necessary. Reserves and capacity are closely relevant, so there is a reserve contract about the TSO in a long time, the capacity mechanism will be an effective one.

4) **Impacts on market electricity prices.**
   A result of wind power instability is accuracy of generation power forecast and its impact on the cost. The cost of repeated dispatch of units because of inaccurate wind power forecasts increases with the wind power forecast errors. Wind power forecast errors make it more likely for the imbalance costs to increase in pre-contracts. Wind power forecast errors influence the electricity prices of nearly spot market. Inaccurate power forecasts probably result in underestimated or overestimated electricity prices, thus resulting in or saving the costs of users or increasing the incomes of generator manufacturers. The electricity prices paid by users are lower than the actual electricity prices sometimes or higher at other times, which depends on the wind power generation modes and the forecast methods.
5) **Impacts on the market force.**
   Market force is a capability to keep market electricity prices from competitive level to gain profits. Viewed from regulation, market force is a capability to keep the market electricity prices higher than the competitive level for a long time for sellers.
   Market force is harmful for competition, so it is necessary to recognize its disadvantages and take actions to weaken market force. The most significant impacts of market force are transfer of wealth from the users in power markets to suppliers and unfairness for the suppliers exercising market force. In a power system with massive wind power, the market operator can make use of wind power outputs to weaken the market force exercised by the dispatchable generators. Wind power is random and the wind power output is therefore not exactly known by other dispatchable generators, so their market force will be reduced. The WFs with pumped storage hydropower plants or hydropower stations influence market electricity price regulations better.

11.5.2 Power Market with Wind Power

Power markets are special markets because electricity trade has lots of characteristics different from the trade of other markets. In addition, wind power is an instable special power source. For the survival and development of wind power generation in power markets, a proper power market mechanism must be designed to benefit wind power. And wind power needs to strengthen its capability to participate in power markets according to its characteristics to make contributions in different power markets [15, 41, 42].

11.5.2.1 Power Markets with Wind Power

The main products sold in power markets are electricity. There are different mechanisms for electricity sale and their selection depends on the specific market structures and the local environmental conditions. The mechanism selected is probably pool power markets realizing the centralized sale and purchase of electricity, quotation by both electricity sellers and electricity buyers, market clearing, and determination of both prices and quantities. Prices and quantities are used in market plans. Quotation begins before actual power generation and power consumption and ends at a time before the actual power generation and power consumption. The ending time is called closing time. For example, the bidding every half an hour in a day-ahead forecast is available 12 hours in advance.

The mechanism used in market clearing is likely to change according to the centralized unit commitment market, which needs lots of detailed quotation information about costs and technical constraints. A self-commitment type market needs price/quantity (P/Q) pairs for commitment decisions of generators. Broadly speaking, market electricity prices are based on MP, that is, the additional MWh costs in the specified time ranges (usually 30 minutes). Market electricity prices are paid to all generators, whereas all power suppliers provide their energy. The changes of this principle seek the central unit commitment (CUC) market (the prices are based on MPs such as those defined in a 24-hour time frame) in which both start costs and no-load costs are considered. In a market deemed to consider technical constraints, market electricity prices change too. For example, upon the blockage of a power transmission line, the MPs of 1 MWh in specific positions will be locational marginal prices (LMP), which will be different. The CUC market should allow market operators to use cost information (related to technical constraints) and dispatch units with effective methods. In a self-commitment type market, it is necessary to select one-way quotation to reflect basic costs and design related methods properly to make the dispatched electric quantity technically feasible (deliverable).

Wind power generation has small operation costs, which makes it possible to offer a zero-quotation price to the power markets. In another circumstance, WTGSs are probably a price receiver, namely, they provide electricity without cost information and are willing to accept the prices offered by other bidders. Price receivers do not always offer zero bidding prices because there are negative prices in some cases. If a WTGS decides to offer a bidding price higher than zero, it will possibly not be dispatched and gains no income from market clearing.

Extensive use of wind power influences the electricity price. If there is no negative electricity price or technical constraint needing generation capacity reduction when there is wind (wind power generation) and there is a zero-bidding price for wind power, wind power will probably be accepted by the power markets. In this case, wind power will replace other power generation forms and the marginal units will be in lower value positions, thus lower the MP. When there is no wind, the marginal units will be in higher value positions, thus increase the MP.

Therefore, wind power brings about price instability for the energy markets. It means that the expected incomes of wind power from power markets will be a little lower than the expected prices. Electric quantity and prices have negative correlation. Another correlation influences the incomes too. Typically, there is more wind in daytime than night-time and, likewise, the load demands in daytime are larger than those in nighttime, resulting in higher electricity prices in daytime. This positive correlation tends to increase the wind power incomes. Seasons are another influential factor: in some regions, there are more wind and larger load demands in winter, forming positive correlation; in some other regions, there are larger load demands but less wind in summer, forming negative correlation.

Another choice is the bilateral trade in which electricity is sold to the other party such as power suppliers. This bilateral trade can be decided several years earlier. When all bilateral trade information is reported to a power system operator, the bilateral trading can continue until the closing time. However, bilateral trading needs match between the outputs of generators and the demands of power suppliers. These relate to the agglomeration of user demands and change with time including days, weeks, seasons, and so on. It is difficult for wind power generation to meet bilateral trade because wind power is instable. Thus, WTGSs need a balanced power market to purchase electricity to fulfill their commitments for bilateral trading.

11.5.2.2 Markets with Wind Power

In any time, power companies need to provide power plant capacity to meet the expected user demands and maintain certain reserve capacity margin. Even if wind power forecasts have uncertainty, wind power can participate in the trade in the following markets [42].

1) **Retail markets**
   Wind power suppliers can sell electricity to local power companies directly. In addition, retail wheeling markets allow wind power suppliers to supply electricity for the users not in the operation regions of local power companies. Additionally, WFs can be connected to local transmission systems and local distribution systems to supply electricity for the end users in the retail service regions of other power companies.
2) **Electricity energy markets**
   Power markets are very similar to stock markets because they have market participants determining electricity prices by means of the match between power supplied and load demands. Power markets use LMP. If the electricity with the lowest electricity price can be transmitted to all locations, the electricity prices of a single grid will be the same. In the case of power transmission line blockage, electricity will not be transmitted to specific locations automatically. In this case, the electricity price will be increased to meet the load demand and the LMP will be higher in the blocked locations.
   A day-ahead market is a future market (a prepaid market) with electricity supply in each hour in the next day. The hourly LMPs in the next day are calculated on the basis of the bidding prices of supply bidders, the bidding prices of in-demand bidders and the planned bilateral trading. The hourly LMPs are expected values because the mean absolute error (MAE) of day-ahead wind power forecasts is 10% to 15% of the rated capacity of WFs. Wind power suppliers can participate in these markets because wind power forecasts can be more accurate.
   In a spot market, the electricity price is determined in one hour or a shorter time before the day of actual power transmission and the balanced power error is between the day-ahead dispatch plan and the actual load. The LMP in this time period can be calculated according to the actual grid conditions, and the expected MAE is about 5%. It would be best for wind power suppliers to participate in these markets because day-ahead wind power forecasts can be more accurate.
3) **CC markets**
   CC markets are designed to allow wind power suppliers to sell their capacity out of unit commitment and their additionally distributed electricity. The participants of these markets provide their bidding prices for electricity purchase or their maximum electric quantity for selling and bidding prices for electricity sale. The market operators handle bids and settle bidding prices through auction markets according to the capacity clearing prices. After that, the participants will produce or consume energy of the specified quantity in a given time. Wind power producers can participate in these markets because they can provide electricity when necessary.

4) **Auxiliary service markets**
Auxiliary services are a support for the reliable operation of power systems. In normal operation conditions, services including regulation and load tracking are necessary for continuous balance between generation capacity and loads. Basically, sum of the generation capacity, the loads and the positive or negative loss difference stands for frequency increase or decrease of a normal power system. Therefore, regulation refers to the transient regulation of some generators or loads with an AGC.
In an interconnected system, regulation can help maintain its frequency, manage the differences between actual power flows and expected power flows of the control regions and match generation capacity and loads in the control regions. Load following traces hourly load fluctuations with help of generators or loads. Modern WTGSs can provide transient reserve capacity, so they can participate in the dispatch markets as long as there is wind. Most of WFs can participate in the load tracking markets and the additional reserve capacity markets, of which the precondition is improvement of WTGSs' dynamic performance.

11.5.3 Impacts of Wind Power on the Electricity Prices of Power Markets

Structure and operation of the power markets with massive wind power are influenced by operation characteristics of power systems. Wind power generation has characteristics different from many conventional power generation technologies and, in particular, it has poor instability and low forecast accuracy. The instability of wind power refers to changes of the expected output in a time period. Errors of the expected outputs influence the power market electricity prices. In order to gain the maximum benefit from the wind power facilities installed, it is necessary to change the existing operation method. This change should be embodied in the corresponding market structure. For example, the conventional unit commitment is represented in certainty optimization assumed that forecasts of both loads and available units usually accurate. If there is wind power, it is proper for unit commitment to be presented in statistical optimization because of the high probability of wind power forecast errors [15].

The operation of power systems and the arrangement of power markets should be closely combined. For the sake of optimization of power companies, it is necessary to coordinate the technical and economic rules. Some usual practices about operation and markets go against wind power. For example, a long closing time (such as 12 hours) is harmful for wind power because the wind power forecast error in this period is probably big (up to 30%). As a result, wind power operators purchase or sell massive electricity under the balance mechanism. Usually, the electricity price under the balance mechanism is higher when the closing time is shorter or lower when the closing time is longer. A shorter closing time reduces the electric quantity probably influenced by the balance mechanism because that the basic problem is the difficulty of accurate wind power forecasts. For wind power, long-term bilateral contracts depend on the balance mechanism or the generators entering short-term bilateral contracts a couple of hours earlier than the actual operation. The best choice (balance mechanism or bilateral arrangement) for wind power depends on the detailed specifications of markets.

In reference [43], the probability method is used for determining the energy costs caused by forecast errors. It is assumed that the hourly electric energy errors are compensated by auxiliary reserve energy, so WTGSs need to purchase or sell the energy errors due to forecast errors. A forecast error can be simulated through a probability density function about accuracy of the forecast model. On this basis, the hourly forecast energy error and its related trade cost can be calculated. According to a study about three practical wind power cases in Spain, the forecast error cost may be up to 10% of the total sales income of wind power. Wind power integration in big regions and narrowing the time ranges of forecasts can make wind power forecasts close to the situations of spot markets. Improvement of the accuracies of forecast models can cut down the forecast error costs. In reference [44], the possible impacts of the control strategies, installation sites, and wind power penetration of WFs on the power market electricity prices and the total power generation costs are studied.

11.5.3.1 Optimization Formulas

Usually, there are the following two basic power market structures: unilateral auction markets and bilateral auction markets. In unilateral auction markets, generators offer bidding prices and the market electricity prices are determined according to the total demands. The standard objective for settlement in unilateral auction markets is to maximize the social welfares. In unilateral auction markets, maximization of the social welfare is equivalent to maximization of the total power generation cost. The objective of optimization is shown as below (Formula (11.90)) [43]:

$$\min J = \sum_{i=1}^{N} C_i(P_i) \tag{11.90}$$

In the formula above, $C_i(P_i)$ is a cost function of the generators, which is usually a quadratic function of the generation power, that is, $C(P_i) = aP_i^2 + bP_i + c$; J is the total power generation cost; P_i is the output power of generator i; N is quantity of the generators; minimization of the function is based on the following constraints (Formula (11.91)).

1) Power flow formula

$$P_{Gi} - P_{Di} = \sum_j |U_i||U_j|Y_{ij}\cos(\theta_{ij} + \delta_{ij} - \delta_i)$$
$$Q_{Gi} - Q_{Di} = \sum_j |U_i||U_j|Y_{ij}\sin(\theta_{ij} + \delta_{ij} - \delta_i) \qquad (11.91)$$

In the formula above, U_i and U_j are respectively the voltages of nodes i and j; δ_i is the power angle; $Y_{i,j}$ is an element of the node admittance matrix; θ is an angle related to $Y_{i,j}$; P_G and Q_G are active power and reactive power of the generators; P_D and Q_D are respectively active power and reactive power of the loads (Formula (11.92)).

2) Generation constraints

$$P_{Gi}^{\min} \leq P_{Gi} \leq P_{Gi}^{\max}$$
$$Q_{Gi}^{\min} \leq Q_{Gi} \leq Q_{Gi}^{\max} \qquad (11.92)$$

In the inequations above, $P_{Gi}^{\min}$, $P_{Gi}^{\max}$, $Q_{Gi}^{\min}$ and $Q_{Gi}^{\max}$ are respectively the minimum active power, maximum active power, minimum reactive power and maximum reactive power of generator i.

3) Voltage limitation

Voltages of the load nodes should be in acceptable ranges; voltages of the generator nodes are fixed values (Formulas (11.93) and (11.94)).

$$U_{Di}^{\min} \leq U_{Di} \leq U_{Di}^{\max} \quad \forall i \in 1, \cdots, N_L \qquad (11.93)$$
$$|U_i| = constant \quad \forall i \in 1, \cdots, N_G \qquad (11.94)$$

In the inequations above, $U^{\min}$ and $U^{\max}$ are respectively minimum and maximum of the node voltages; N_L is quantity of the load nodes; N_G is quantity of the generator nodes.

4) WF operation constraints

Due to instability of the wind velocities at the WTGSs, the instability is introduced into the related formula through the variable – daily mean wind power (per unit) curve. Location of the curve can be found on the record on the WF. The WF node voltage constraints are treated as generator nodes when the generators are operating in the voltage regulation mode or as load nodes in another operation mode.

11.5.3.2 Expression of a Uniform Market Electricity Price

Power market electricity prices are determined under different pricing mechanisms, namely, pricing is uniform and based on nodes and regions. Under a unified market electricity price mechanism, the model above is solved to obtain the maximum node (micro) incremental cost and set it to be the market electricity price. Thus (Formula (11.95)):

$$\rho \geq \lambda_i \quad \forall i \in 1, \cdots, N \qquad (11.95)$$

In the inequation above, ρ is the uniform power market electricity price; λ_i is the MP of node i; N is amount of the nodes.

11.5.3.3 Wind Power Forecasts

The power P and power Q of a WF are represented in a function of the wind velocity. In order to study the impacts of wind power forecast accuracies on power market electricity prices, the following forecast models, based on two time series, are used. The first forecast model is a persistence model suggesting that the future forecast value ($\hat{X}(i+1)$) is the current observed value ($X(i)$). See the formula below (Formula (11.96)).

$$\hat{X}(i+1) = X(i) \qquad (11.96)$$

The second forecast model is a gray forecast (GM) model in which the data series are accumulated through Formula (11.97) first.

$$X^{(1)}(k) = \sum_{i=1}^{k} X^{(0)}(i) \quad \forall k = 1, \cdots, n \qquad (11.97)$$

In the formula above, $X^{(0)}$ is the original time series; $X^{(1)}$ is the accumulated time series; n is the sampled data length; k and i are respectively steps numbers of the accumulated time series and the original time series.

The value of the future accumulated time series can be forecasted through the GM (1, 1) model expressed in Formula (11.98).

$$\hat{X}^{(1)}(i+1) = \left(X^{(0)}(1) - \frac{b'}{a'}\right)e^{-ai} + \frac{b'}{a'} \qquad (11.98)$$

In the formula above, $\hat{X}^{(1)}$ is forecast value of the accumulated time series; $X^{(0)}$ is initial data of the original time series; a′ and b′ are two parameters of the GM(1, 1) model, which can be estimated as below with the least square method (Formula (11.99)).

$$A = \begin{bmatrix} a' \\ b' \end{bmatrix} = [\beta^T \cdot \beta]^{-1} \cdot \beta^T Y \qquad (11.99)$$

In the formula above (Formulas (11.100), (11.101), and (11.102)):

$$\beta = \begin{bmatrix} -Z^{(1)}(2) & 1 \\ -Z^{(1)}(3) & 1 \\ \vdots & \vdots \\ -Z^{(1)}(n) & 1 \end{bmatrix} \qquad (11.100)$$

$$Y = \begin{bmatrix} X^{(0)}(2) \\ X^{(0)}(3) \\ \vdots \\ X^{(0)}(n) \end{bmatrix} \qquad (11.101)$$

$$Z^{(1)}(i) = \frac{X^{(1)}(i-1) + X^{(1)}(i)}{2} \tag{11.102}$$

Finally, inverse transformation for the forecast accumulated time series is done through Formulas (11.103) and (11.104). The forecast data set is returned to the original time series.

$$\hat{X}^{(0)}(1) = \hat{X}^{(1)}(1) \tag{11.103}$$

$$\hat{X}^{(0)}(i+1) = \hat{X}^{(1)}(i+1) - \hat{X}^{(1)}(i) \quad \forall i = 1,2,3,\cdots \tag{11.104}$$

11.5.3.4 Impacts of WF Factors on Real-Time Power Market Electricity Prices and Total Power Generation Costs

The control strategies, installation positions, and wind power penetration of WFs all influence power market electricity price and total power generation cost. According to the study for the nine-bus grid including a six-bus transmission system and a three-bus distribution system, if WF is connected to bus 8 of the distribution system, the results will be as below (Figure 11.8) [44]:

1) **Impacts of the WF control strategies**
   Impact of control strategies of the WF is mainly impact of reactive power inflow from the WF on the power market electricity price. The modern WF allows to provide reactive power for the grid or absorb reactive power from the grid and the reactive power provided or absorbed by the WF changes the grid power loss, thus generating an impact on the power market electricity price. In this case, the following control strategies can be adopted.
   —Unit power factor (UPF) control mode: WF does not allow exchange of reactive power with the grid, so the reactive power provided or absorbed by the WF (Q_{farm}) is set to be zero, and voltage of grid-connected point (bus 8) of the WF changes according to the optimal power flow (OPF) operation constraint.
   —Voltage regulation (UR) control mode: Voltage of grid-connected point (bus 8) of the WF is regulated to be and kept at 1.0. The reactive power provided or absorbed by the WF changes with the OPF operation constraint as long as apparent power of the WF is within the specified range.
   —power factor (PF) control mode: The reactive power provided or absorbed by the WF is controlled to regulate power factor of grid-connected point of the WF to be 0.8 (leading) and voltage of grid-connected point of the WF changes according to the OPF operation constraint.
   —Local power factor (LPF) control mode: Reactive power of the WF is regulated to make sure the PF of bus 8 is 1.0. It is realized by setting the reactive power provided by the WF to be the reactive power demand of bus 8. This is believed to be a special case of PF regulation. Voltage of grid-connected point of the WF changes according to the OPF operation constraint.

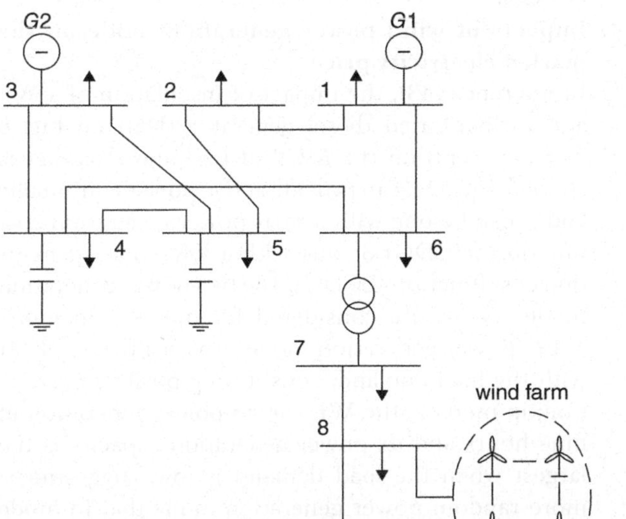

Figure 11.8 Structure of a nine-bus grid.

In the UPF control mode, the grid current increases with the load demand, which results in a larger drop of the voltage of bus 8. If it is allowed to provide the reactive power for the grid through bus 8, the feeder current will decrease, resulting in decrease of both the grid loss and the bus voltage drop. Thus, bus 8 has the lowest voltage in the UPF control mode.

The MCPs in different control modes and the power market electricity price differences (a sort of saving viewed from the demand side) between the UPF control mode and the other three control modes above are calculated with reference to the situation in the UPF control mode. According to the result, in the case of a low load (smaller than 70% of the maximum daily load), the control modes almost have no impact on the market electricity price at all. However, in the case of a high load (particularly the peak load), the market electricity price difference increases to 7.5$/MWh. The power market electricity price is the highest in the UPF control mode and lowest in the UR control mode.

Control mode changes influence the total power generation cost a little (the power generation cost difference is 0.8% in the case of the peak load). Compared with the UPF control mode (reference mode), the biggest economization is realized in the UR control mode.

Compared with the UPF control mode, the UR control mode, the PF control mode and the LPF control mode have a daily mean electric power cost difference (profit) of 950.4$/hr, 895.9$/hr and 741.7$/hr, respectively. These positive daily means suggest that users have smaller electricity bills or that generators have lower incomes. Correspondingly, the daily mean power generation cost differences (the power generation costs saved) are 58.7$/hr, 58.6$/hr and 45.8$/hr, respectively.

To sum up, the uniform market electricity price is influenced by the control strategy changes a little when there are low load demands or greatly when there are high load demands. In addition, the lowest market electricity price is realized in the UR control mode. Due to the control mode changes, the total power generation cost of the WF decreases a little (saved by about 0.8% of the maximum power generation cost).

2) **Impacts of wind power penetration on the market electricity price**

In the study about the impact of wind power penetration of the WF on market clearing price (MCP) of a power market, it is assumed that the WF operates in the UR control mode and that the WF has three rated capacity, that is, 130 MW, 195 MW, and 260 MW. The calculation result suggests as below: ① With increase of rated capacity of the WF, bus voltage regulation needs a smaller reactive power because that bus voltage of the WF increases with the active power provided. ② Total power of the power system decreases with increase of the wind power. ③ The market electricity price drops with increase of the wind power penetration, particularly when there is a high load demand but low wind power. The drop is the smallest when the load demand is the lowest. Compared with the situation when the rated capacity of the WF is 130 MW, the daily mean electric power cost difference (saving) is 1232.5$/hr when the rated capacity is 195 MW or 1859.1$/hr when the rated capacity is 260 MW. ④ The total power generation cost decreases with increase of the wind power penetration. If the wind power capacity is doubled, the net power generation cost will almost be doubled. The daily mean power generation cost differences (saving) corresponding to rated capacity of 195 MW and 260 MW are 876.4$/hr and 1674.1$/hr, respectively.

To sum up, the market electricity price decreases with increase of the wind power penetration (WF capacity). However, this decrease is small when the load demand is low. Furthermore, the total power generation cost decreases with increase of the wind power penetration.

3) **Impacts of installation position of the WF**

Three locations (in Figure 11.8, buses 7, 8, and 9 have different distances from the transmission system: bus 7 is the farthest one and bus 9 is the nearest one) are selected to connect wind power into the distribution system. The WF has a constant capacity of 130 MW and operates in the UR control mode. Changes of the reactive power of bus 8, the MCP and the total power generation cost are checked in the case of different installation sites of the WF.

When the WF is close to the transmission system, more reactive power is needed to regulate bus voltage of the WF to be 1.0 p.u. As a result, the MCP of the power market will increase and the total power generation cost will be influenced a little (the difference is about 1.2% of the maximum power generation cost). According to the calculation result of electric power costs and total power generation costs of the other locations with reference to bus 7, the daily mean electric power cost differences of bus 8 and bus 9 are 1405.7$/hr and 1724.9$/hr, respectively, and the corresponding daily mean power generation cost differences are 94.4$/hr and 81.4$/hr, respectively. When another candidate installation site of the WF is close to the transmission system, more reactive power and a higher MCP of the power market are needed to regulate bus voltage of the WF to be 1.0. Installation site changes of the WF influence the total power generation cost a little (the difference is about 1.2% of the maximum power generation cost).

Thus, control strategy, installation site and wind power penetration of the WF have significant impacts on the nearly real-time power market electricity price and generation income. The impact constitutes a load demand function. These conclusions help power system operators to plan for the installation and integration (to public grids) of new WFs. Furthermore, these conclusions are important for the utilities benefiting from the installation of large WTGSs.

4) **Impacts of wind power generation mode on the market electricity price**

In reference [43], the impact of installation of a WF and a distributed diesel generator (DG) on bus 8 in Figure 11.8 on the MCP of the power market is studied. The DG can provide active power continually and it can be one with a high-power generation cost function (HCDG) or one with a low-power generation cost function (LCDG). The two power generation modes below are considered for the WF: in mode 1, the power generation mode is basically compliant with the load demand, constituting positive correlation; in mode 2, the WF has no power generation in nine hours and its power generation capacity is the largest when the load demand is low, suggesting a more random power generation mode that in mode 1 and constituting negative correlation. Figure 11.9 shows different levels of active power generation of bus 8. Active power of the DG decreases with increase of the DG cost.

Figure 11.9 Active power generation of bus 8.

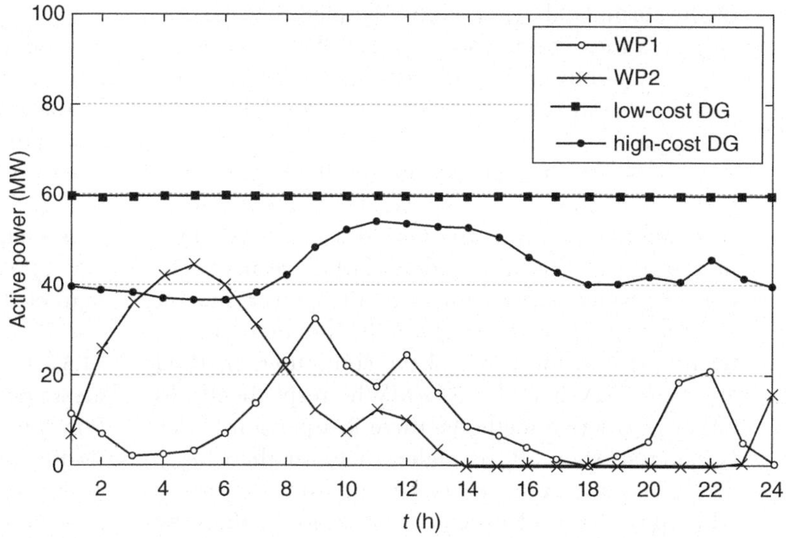

Figure 11.10 MCP.

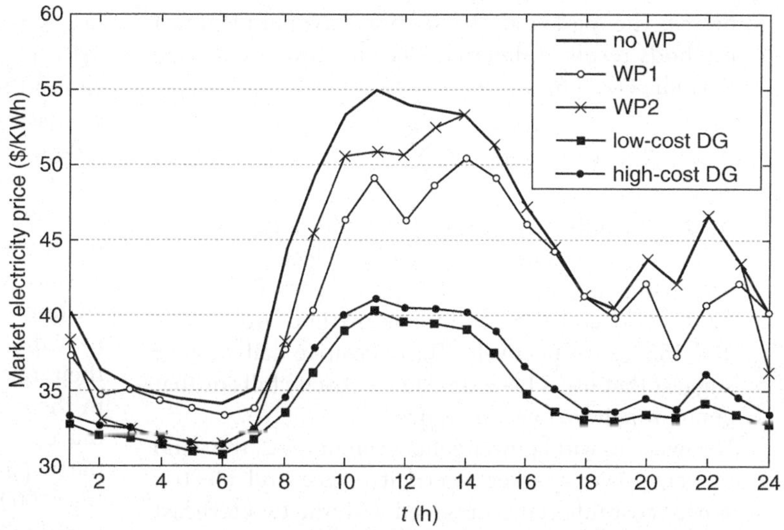

As shown in Figure 11.10, the MCP of the main power system is influenced by the existing DG and the WF. Usually, there will be a lower power market electricity price no matter there is a DG or WF. According to a market electricity price comparison between the case with a WF and the case with a DG, load of the power system probably faces a market electricity price peak, particularly when the wind power generation capacity is low. When there is wind power, large market electricity price fluctuations are caused by the instability of wind power generation; the market electricity price is still lower than the situation where the bus 8 has no wind power generation.

5) **Impacts of wind power forecasts on the nearly real-time market electricity price**

The instability of wind power forecasts influences the nearly real-time market electricity price. Inaccurate power forecasts probably result in an underestimated or overestimated market electricity price, which results in more user costs or economization and increases the incomes of power plants. The electricity prices paid by users are lower than the actual electricity prices sometimes or higher at other times, much depending on the wind power generation modes and the forecast methods. Both wind power generation modes 1 and 2 are considered too. The forecast time range is selected as one hour, so the power market of an hour before is simulated. Two wind power forecast methods, that is, gray method (GM) (1, 1) and continuous method, are used.

The forecast errors of wind power output (including the GM (1, 1) and the continuous method) and market electricity price are expressed in MAE as below (Formula (11.105)):

$$MAE = \frac{\sum |SP_{ac} - SP_{pre}|}{L} \tag{11.105}$$

In the formula above, L is the sampled data length; SP_{actual} is the expected spot market electricity price, which should be calculated based on the actual wind power generation capacity; SP_{pre} is spot market electricity price.

According to the calculation result, MAEs in a power generation mode calculated with the two forecast methods (such as the GM and the continuous method) are 5.34 MW and 5.62 MW, respectively. Forecast value errors influence the power market electricity price, so the MAEs of the spot market electricity price in the GM and the continuous method are 1.25$/MWh and 1.37$/MWh, respectively. In either of the two methods, there is a perid in a day (24 hours) when SP_{ac} is higher or lower than SP_{pre}. Thus, in some cases, users benefit from the errors when purchasing electricity at prices lower than the actual electricity prices. In some other cases, users pay much more money than the actual electricity prices. The market electricity price errors in the two methods result in different electric power costs; see Formula (11.106).

$$C_d = (SP_{ac} - SP_{pre}) \sum_{i=1}^{N_G+N_L} P_i \tag{11.106}$$

In the formula above, C_d is the additional electric power cost because of a market electricity price error. The daily mean electric power cost differences in the GM and the continuous method are 310$/h and 155$/h, respectively. These positive daily means suggest that users have smaller electricity bills or that generators have lower incomes.

Likewise, in wind power generation mode 2, the wind power, power market electricity price and electric power cost difference obtained with the two forecast methods above suggest that the MAEs of forecast power in the GM and the continuous method are 3.28 MW and 4.74 MW, respectively. The MAEs of the corresponding market electricity price in the GM and the continuous method are 0.48$/MW h and 0.81$/MW h, respectively. The annual money saved and lost are expected to be $376,680 and $394,200, respectively.

11.5.4 Quotation Strategies for the Wind Power in Power Markets

Some of the existing wind power in power markets is priced through day-ahead energy markets and hour-ahead energy markets and has no imbalance punishment (imbalance is a difference between the expected wind power generation capacity and the actual wind power generation capacity). In some power markets, WTGSs are not allowed for bidding. The available wind power is transmitted to power systems and paid by the power systems according to sums of the actual market electricity prices and the constant additional costs (bonus). At present, the wind power in China does not participate in the trades of power markets and it is paid by grid companies at the approved on-grid electricity price of wind power. The wind power prices in China are very high and the exceeding parts are shared by all grid users. This section discusses the possible quotation strategies and other related issues of the wind power in power markets [15, 39, 45].

11.5.4.1 Day-Ahead Bidding Mechanism and Market Clearing Process

Power markets have two trading modes, that is, unilateral trading and bilateral trading. In unilateral markets, only power suppliers participate in bidding; in bilateral markets, both power suppliers and power consumers participate in bidding.

The wind power in power markets can offer bidding prices to power markets or not. Usually, WTGSs are price receivers. Wind power producers are usually not allowed to bid with multiple price/quantity (P/Q) pairs as fossil fuel units do. Viewed from this, the optimal bidding strategy of wind power handles the uncertainty of electric quantity more while the optimal bidding strategy of conventional generators focuses on electricity price uncertainty and other information of bidders.

Market participants are expected to submit their 24-hour P/Q pair data at a specified time (such as 12:00) in a day. In some markets, they can submit or revise their related data one day earlier or every four hours in the current day. At a specified time (such as 3 p.m.) one day earlier, there is market clearing and each market participant determines its price and contract level. Since the 0:00 h of the delivery day, the market participants begin to submit their scheduled generation capacity. In the delivery day, the generation imbalance of each power producer is transmitted to the balanced power market and the balanced power market calculates every five minutes or a longer time.

Each generator offers a bidding price to sell its available output power and each company (or another load service entity) offers a bidding price to purchase electricity. After the MCP is determined, all the bidding prices for electricity sale, which are equal to or lower than the MCP, and all the bidding prices for electricity purchase, which are equal to or higher than the MCP, will be accepted while all the bidding prices for electricity sale, which are higher than the MCP, and all the bidding prices for electricity purchase, which are lower than the MCP, will be refused.

There are two types of market clearing as below: uniform market clearing and market clearing with payments in line with bidding prices. In the unified MCP mechanism, each supply bidder, even if its bidding price is lower than the MCP, receives the MCP corresponding to its electric quantity and each in-demand bidder, even if its bidding price is higher than the MCP, pays the MCP. In the market clearing of the second type, payments are not made according to the MCP but the bidding prices of bidders. All the total electricity purchase costs are mean and each in-demand bidder needs to pay the mean bidding price. In consideration of the specificity of electric power, a uniform MCP is a good choice. In most of power markets, unified pricing is adopted and each accepted bidder does not make payments according to its bidding price but the MP.

11.5.4.2 Determination of MCP

The MCP is the lowest price. With it, the accepted supply bidders provide adequate electricity to meet the demands of all accepted in-demand bidders. MCP is the price that the total selling price is equalevant with the buying price. There are two bidding mechanisms: unilateral bidding mechanism in which only power suppliers offer bidding prices; bilateral bidding mechanism in which both power suppliers and users are allowed to offer bidding prices to power markets. Bidders can quote a price for their outputs or demands in the blocked or linear form, as shown in Figure 11.11. In a power market, the bidding by supply bidders and by in-demand bidders are of the same type, that is, block bids or linear bids.

11.5.4.3 Linear Bid Market

1) **Unilateral bidding:**
   In the linear bid market model, the power supply curve about the function of market electricity price (p) of bidder i can be expressed as below (Formula (11.107)):

$$q_i(p) = \frac{p}{m_{si}} \tag{11.107}$$

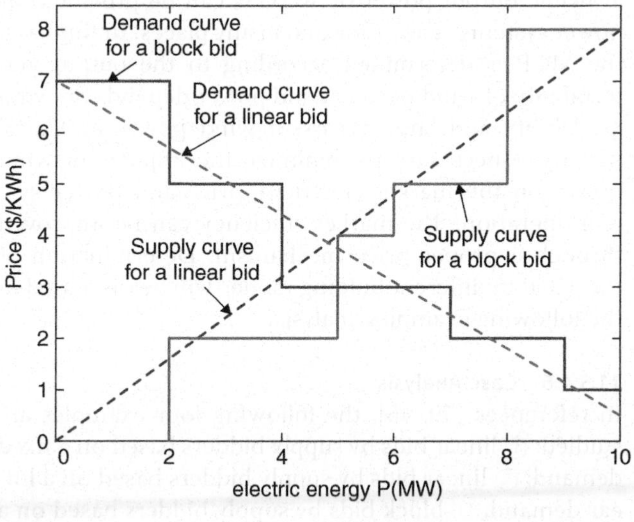

Figure 11.11 Curve of bidding by supply bidders and in-demand bidders.

In the formula above, m_{si} is slope of the power supply curve in Figure 11.11. If quantity of the bidding power suppliers is Ng, the combined power supply curve will be (Formula (11.108)):

$$q(p) = p \sum_{i=1}^{Ng} \frac{1}{m_{si}} \tag{11.108}$$

If demand D if fixed, MCP(p^*) can be calculated through Formula (11.109).

$$p^* \sum_{i=1}^{Ng} \frac{1}{m_{si}} = D \tag{11.109}$$

In the formula above, it is assumed that the bidder has an adequate capacity, so the capacity constraint for any power supplier can be included. If the output power of generator i has a minimum ($q_{\min i}$) and a maximum ($q_{\max i}$) limitation, the power supply curve defined in Formula (11.107) will be changed as below (Formula (11.110)):

$$q_i(p) = \frac{p}{m_{si}}[u(q_i, q_0^{\min}) - u(q_i, q_0^{\max})] \tag{11.110}$$

In the formula above, $u(q, q_0)$ is a function of the unit, which is defined as below (Formula (11.111)):

$$u(q, q_0) = \begin{cases} 1 & \text{when } q \geq q_0 \\ 0 & \text{others} \end{cases} \tag{11.111}$$

Then, the combined power supply curve will be as below (Formula (11.112)):

$$q(p) = \sum_{i}^{Ng} \frac{p}{m_{si}}[u(q_i, q_0^{\min}) - u(q_i, q_0^{\max})] \tag{11.112}$$

Thus, the MCP can be determined by making Formula (11.112) equal to the total demand.

2) **Bilateral bidding market:**
   In unilateral bidding, the power demands are supposed to be fixed, in regardless of the market electricity price. In addition, unilateral bidding is a special case in which in-demand bidders are price receivers for the market electricity price. In the linear bid model, any separate demand curve (it is assumed that rational users have larger power consumptions upon electricity price decrease) can be expressed as below (Formula (11.113)):

$$d_i(p) = \frac{p_{i0} - p}{m_{di}} \tag{11.113}$$

In the formula above, m_{di} is slope of the demand curve; p_{i0} is the intercept on the price axis; see Figure 11.11. If the quantity of user bidders is Nd, the total demand curve will be as below (Formula (11.114)):

$$d(p) = \sum_{i=1}^{Nd} \frac{p_{i0}}{m_{di}} - p \sum_{i=1}^{Nd} \frac{1}{m_{di}} \tag{11.114}$$

The MCP (p^*) can be calculated through Formulas (11.115) and (11.116).

$$p^* \sum_{i=1}^{Ng} \frac{1}{m_{si}} = \sum_{i=1}^{Nd} \frac{p_{i0}}{m_{di}} - p^* \sum_{i=1}^{Nd} \frac{1}{m_{di}} \quad (11.115)$$

$$p^* = \frac{\sum_{i=1}^{Nd} \frac{p_{i0}}{m_{di}}}{\sum_{i=1}^{Ng} \frac{1}{m_{si}} + \sum_{i=1}^{Nd} \frac{1}{m_{di}}} \quad (11.116)$$

Impacts of the generator output and the consumption constraint can be properly considered.

11.5.4.4 Block Bid Market

1) **Unilateral bidding:**
In a block bid market, the power supply function of any generator i can be defined as below (Formula (11.117)):

$$q_i(p) = \sum_{j=o}^{NB} q_{ij}[u(p, p_{i,j}) - u(p_i, p_{i,j+1})] \quad \forall p_{i,j} \in p_i \quad (11.117)$$

In the formula above, q_{ij} is the output power of generator i in block j; $p_{i,j}$ ($p_{i,j} < p_{i,j+1}$) is the price of generator i in block j; NB is the bidding blocks; $u(p, p_{i,j})$ is a function of the unit, which is defined as below (Formula (11.118)):

$$u(p, p_{i,j}) = \begin{cases} 1 & \text{when } p \geq p_{i,j} \\ 0 & \text{others} \end{cases} \quad (11.118)$$

Variable p is a price vector with an ascending order. Thus, there is p_i=[0 2 4 5 8] for Figure 11.11. Total power supply curve of all the generators (quantity: Ng) can be expressed as below (Formula (11.119)):

$$q(p) = \sum_{i=1}^{Ng} \sum_{j=0}^{NB} q_{ij}[u(p, p_{i,j}) - u(p_i, p_{i,j+1})] \quad (11.119)$$

If demand D is fixed, the MCP p^* can be calculated through Formula (11.120).

$$\sum_{i=1}^{Ng} \sum_{j=0}^{NB} q^*_{i,j}[u(p^*, p_{i,j}) - u(p^*, p_{i,j+1})] = D \quad (11.120)$$

$p^* q^*_i$ is the payment for power supplier i; q^*_i is total output of the dispatched power supplier i.

2) **Bilateral bidding:**
The demand curve of any rational users can be expressed as below (Formula (11.121)):

$$d_i(p) = \sum_{j=0}^{N} d_{ij}[u(p, p_{i,j}) - u(p_i, p_{i,j+1})] \quad \forall p_{i,j} \in p_i \quad (11.121)$$

In the formula above, d_{ij} is the power consumption of user i in block j; $p_{i,j}$ ($p_{i,j} > p_{i,j+1}$) is the purchasing price of user i in block j; N is the bidding blocks. Total demand curve of all the users (quantity: Nd) can be expressed as below (Formula (11.122)):

$$d(p) = \sum_{i=1}^{Nd} \sum_{j=0}^{N} d_{ij}[u(p, p_{i,j}) - u(p_i, p_{i,j+1})] \quad (11.122)$$

If the power supply curve is given (a block bid), the MCPp^* can be calculated through Formula (11.123).

$$\sum_{i=1}^{Ng} \sum_{j=0}^{NB} q^*_{ij}[u(p^*, p_{i,j}) - u(p^*, p_{i,j+1})] = \sum_{i=1}^{Nd} \sum_{j=0}^{N} d^*_{ij}[u(p^*, p_{i,j}) - u(p^*, p_{i,j+1})] \quad (11.123)$$

11.5.4.5 MCP Based on WTGS Participation

The MCP, when there is wind power, is different from the MCP when there is no wind power. In a competitive power market, the MCP can be considered in the following two modes: In mode 1, bidding by WTGSs is allowed; wind power is paid at a sum of the MCP and a subsidy; WTGSs do not need to be punished for output power instability because other dispatchable generators bear the instability. In a pay-as-bid market (a pay-as-you-go bidding market), it is risky to be dispatchable. Because of the commitments of governments for green energy sources, the pay-as-bid mode is not suitable. WTGSs are not competitive without government subsidies. Governments will probably cancel the subsidies in future.

In mode 2, which is seemingly more attractive, the available output power of WTGSs can be provided for power systems at any time and in any places. In this case, the MCP is determined according to the output and instability of wind power. Wind power depends on wind availability and large errors of wind power forecasts, making it necessary to minimize the impacts of wind power on the market electricity price and power system operation. The market efficiency can be improved through a proper price mechanism. Determination of the MCP in different bidding modes will be discussed in the following examples analysis.

11.5.4.6 Case Analysis

In references [39, 45], the following four examples are studied: ① linear bids by supply bidders based on a fixed demand; ② linear bids by supply bidders based on a linear demand; ③ block bids by supply bidders based on a fixed demand; ④ block bids by supply bidders based on block demands.

Table 11.7 Parameters of linear bidders.

Bidder	m_{si}(\$/MW)	q_g^{max}(MW)	q_g^{min}(MW)
Bidder 1	0.1	100	20
Bidder 2	0.25	50	10
Bidder 3	0.20	100	10

1) **Linear bids**
Table 11.7 introduces the linear bid parameters of three supply bidders at any unit time (as defined in Formula (11.107)), including the minimums and maximums of the output power of generators. Figure 11.12 shows the total power supply curve obtained through Formula (11.108).
Case 1: Through the intersection between the total bidding curve of the three generators obtained through Formula (11.108) and the fixed demand curve, the MCP and the actual output power of the three generators, that is, 52.63 MW, 21.05 MW, and 26.32 MW, are obtained.
If the load demand is 100 MW and the wind power is 5 MW, the MCP will be (Formula (11.124)):

$$p^* = \frac{D - P_w}{\sum_{i=1}^{Ng} 1/m_{si}} \tag{11.124}$$

In this case: $p^* = \frac{100-5}{\frac{1}{0.1}+\frac{1}{0.25}+\frac{1}{0.2}} = 5.0$ (\$/MWh)
In the formula above, D is a fixed load demand; *Ng* is quantity of the bidding generators (excluding WTGSs).
When there is no wind power, the MCP is \$5.26/MWh; when there is wind power, the bidding generators will have smaller power output and the decrements (Δq_i) can be calculated through Formula (11.125).

$$\Delta q_i = \frac{P_w}{m_{si} \sum_{j=1}^{Ng} 1/m_{sj}} \tag{11.125}$$

In this case: $\Delta q_i = \frac{5}{m_{si}\left(\frac{1}{0.1}+\frac{1}{0.25}+\frac{1}{0.2}\right)}$
According to the calculation result, the three bidding generators have an output power decrement of 2.63 MW, 1.05 MW, and 1.32 MW, respectively.
In order to study the impacts of bidding of WTGSs, WTGSs are allowed to bid on power markets. Figure 11.13 shows wind power output changes and the MCPs based on different bidding success ratios. According to the figure, if wind power (5 MW) is completely dispatched, its bidding price can be any value lower than \$5.0/MWh, which is the MCP as the difference between the total demand and the wind power output. If the bidding price of wind power is zero, wind power can be completely dispatched and the MCP will be \$5.00/MWh. In this case, the total power supply curve is lower than the parallel line of the total power supply curve without wind power. When the bidding success ratio is 1.0, the MCP based on infinite wind power (with an output of massive wind power) is equal to the MCP based on a limited wind power (5 MW in this case). The bidding success ratio m_s of wind power corresponding to the maximum MCPp is infinite; bidding success ratio m_s of wind power corresponding to the minimum MCP is zero. It can be obtained through Formula (11.126).

$$\frac{\partial p}{\partial m_s} = \frac{\partial}{\partial m_s}\left(\frac{D}{\sum_{i=1}^{Ng} 1/m_{si}}\right) = 0 \tag{11.126}$$

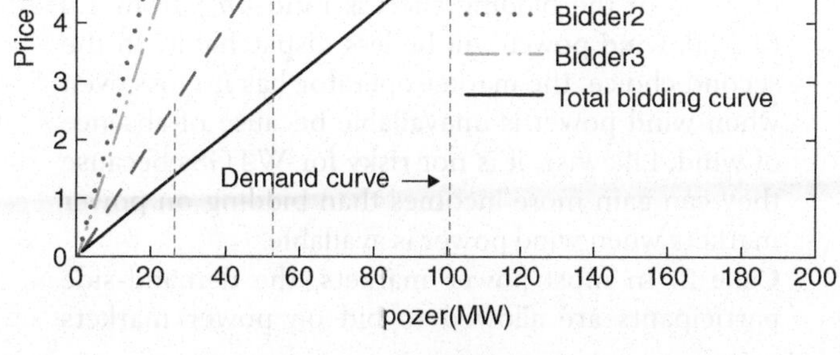

Figure 11.12 Linear bids of power suppliers with fixed demand.

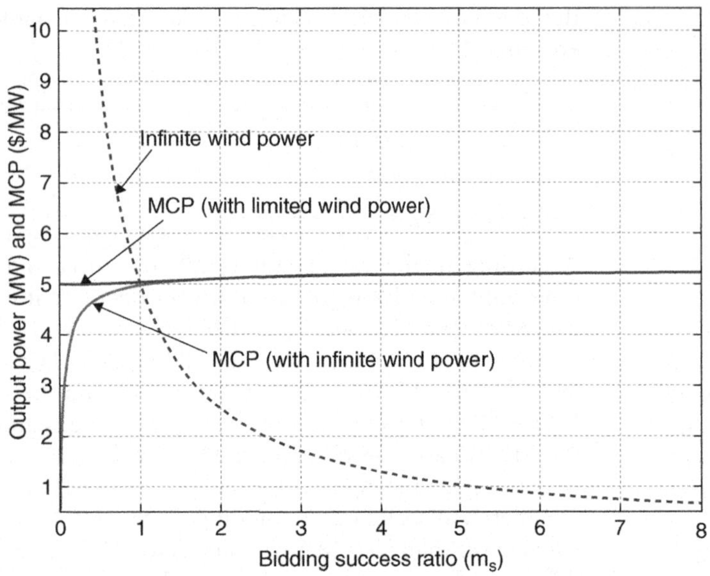

Figure 11.13 MCPs with different bidding success ratios (m_s) of wind power.

Table 11.8 Output power and payment (Case 1).

	Output (MW)		Payment		Payment when II is selected ($)
	m_s>10	m_s<1	m_s>10	m_s<1	
Bidder 1	52.63	50.0	277.01	250.00	263.16
Bidder 2	21.05	20.0	110.80	100.00	105.26
Bidder 3	26.32	25.0	138.51	125.00	138.58
Wind power	0.00	5.0	0.00	25.00	26.32
Total	100.0	100.0	526.32	500.00	526.32

The market operator can make two choices for the MCP: in the first choice, the MCP can be calculated based on a load decrement (the difference between the load demand and the available wind power in the period) and it is $5/MW in this case. The load demand is decreased because of the available wind power, and the MCP will be reduced. In the second choice, wind power is excluded in the total demand in the MCP calculation and output power of the bidding generators is regulated through Formula (11.125), so as to enable the WTGSs and other units to gain benefits. In order to study the financial issues of wind power, the wind power capacity is supposed to be 5 MW. According to Table 11.8, the income from the power market is $526.32 (= MCP×demand). Upon increase of the bidding success ratio (m_s) from 1.0 to 10.0, wind power will be less dispatchable. In the second choice, the market operator has no loss even when wind power is unavailable because of absence of wind. Likewise, it is not risky for WTGSs because they can gain more incomes than bidding on power markets when wind power is available.

Case 2: In most power markets, the demand-side participants are allowed to bid on power markets according to their demand elasticity. Two different bidders are considered here. Table 11.9 shows their bidding data. Figure 11.14 shows the total demand curve of the two in-demand bidders and the total power supply curve of three supply bidders.

According to Figure 11.14, the MCP (the intersection between the total power supply curve and the total demand curve) is $4.713/MWh. The two in-demand bidders have a load demand of 45.73 MW and 43.82 MW, respectively. If a WTGS offers a zero-bidding price (the first choice), the MCP will be the intersection between the parallel new power supply curve with downward trend (P_w MW on the x axis) and the total power supply curve of the three supply bidders. It is similar to the intersection between the parallel line below the demand curve and

Table 11.9 Parameters of in-demand quotation.

	m_{di} ($/MW)	p_{io} ($/MW)
Bidder 1	0.050	7.0
Bidder 2	0.075	8.0

Figure 11.14 Linear bid curve on supply side and demand side.

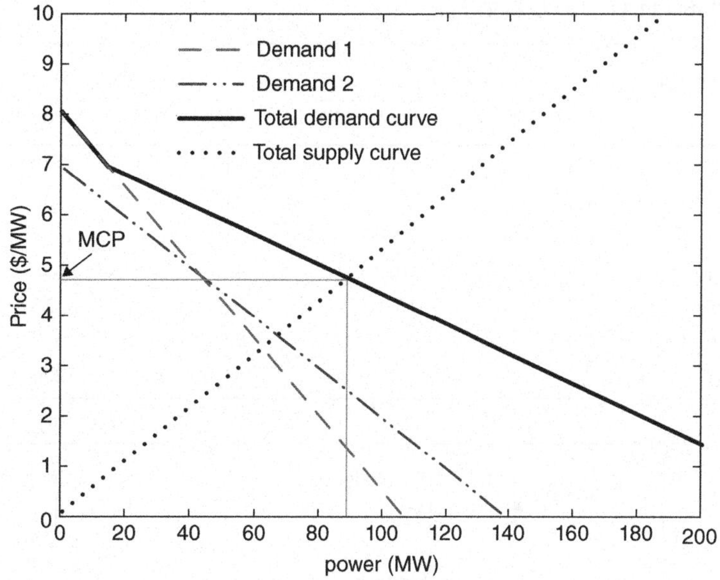

the total power supply curve when there is no wind power. Mathematically, it can be calculated through Formula (11.127) too.

$$p^* = \frac{\sum_{i=1}^{Nd} \frac{p_{io}}{m_{di}} - P_w}{\sum_{j=1}^{Ng} \frac{1}{m_{sj}} + \sum_{i=1}^{Nd} \frac{1}{m_{di}}} \qquad (11.127)$$

In the formula above, Nd is quantity of the in-demand bidders.

When there is 5 MW of wind power, the new MCP is $4.618/MWh. Due to elasticity of the in demand bidders, the MCP on the spot market has dropped, resulting in that power consumption of increases to 47.64 MW and 45.09 MW, respectively, for the two in-demand bidders.

Table 11.10 suggests the payments or incomes of different bidders when there are different MCPs. Each negative stands for a payment by an in-demand bidder. According to the table, if the MCP is calculated with a zero-bidding price of wind power, the WTGSs gain a lot of payment. If it is specified by a power market to fine for load shedding imbalance, the generators (except for the WTGSs) can pay to WTGSs offering a zero-bidding price according to the actual MCP ($4.618/MWh), but WTGSs can be paid according to the maximum MCP. The payments for the in-demand bidders will be made according to the maximum MCP (in the sixth column). When there is no wind power, the other power suppliers need to increase their outputs and are paid by the WTGSs.

2) **Block bids**

Table 11.11 shows the output power (price/power pair) of three bidding generators (bidder 1, bidder 2, and bidder 3) in the block and in the unit time according to the specified price. Figure 11.15 shows the bidding curve and the total power supply curve of these bidding generators.

Case 3: If the fixed demand is 100 MW and there is no wind power, the MCP will be $5.0/MW (Figure 11.15 and Table 11.11) and bidder 1, bidder 2, and bidder 3 will have an output power of 40 MW, 35 MW, and 25 MW, respectively (Figure 11.15). If the bidding

Table 11.10 Output power and payment.

	Without wind power		With wind power		
	Power(MW)	Payment $4.713/MWh	Power(MW)	Payment $4.618/MWh	Payment $4.713/MWh
Bidder 1	47.13	222.14	46.18	213.26	217.64
Bidder 2	18.85	88.86	18.47	85.32	87.05
Bidder 3	23.57	111.10	23.08	106.63	108.81
Wind power	0.00	0.00	5.00	23.09	23.56
Demand 1	45.73	−215.56	47.64	−220.00	−224.53
Demand 2	43.82	−206.54	45.09	−208.20	−212.53

Table 11.11 Data of block bids.

Block	Bidder 1		Bidder 2		Bidder 3	
	Price ($/MW)	Power (MW)	Price ($/MW)	Power (MW)	Price ($/MW)	Power (MW)
1	0.0	7.5	1.0	15	0.0	5
2	1.50	5	2.0	10	2.0	5
3	2.5	7.5	3.0	5	3.0	5
4	4.0	10	4.5	5	4.5	10
5	5.0	15	5.5	15	5.5	10
6	7.0	5	6.5	5	8.0	10
7	---	---	8.0	2.5	---	---

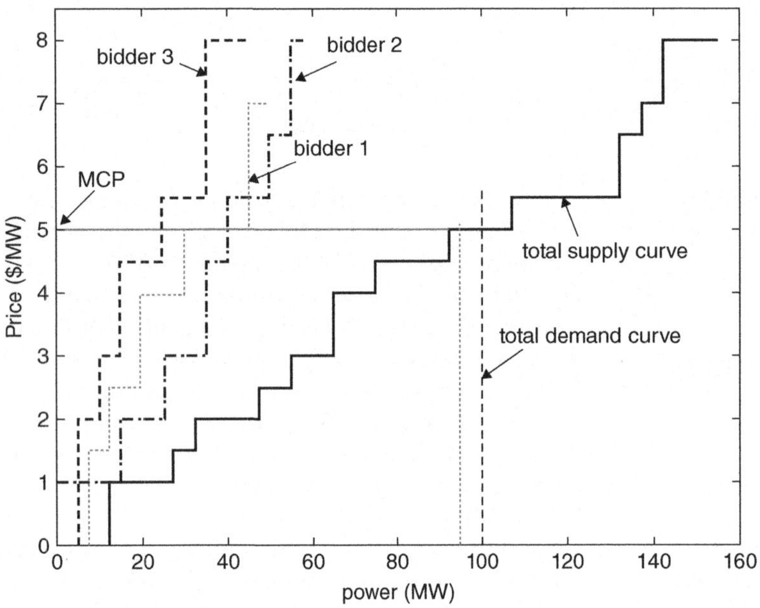

Figure 11.15 Power supply curve of block bids based on a fixed demand.

price of wind power (P_w) is zero, the wind power capacity is 7.5 MW and the MCP remains unchanged. Bidder 1, bidder 2, and bidder 3 will have an output power of (40 $-P_w$) MW, 35 MW, and 25 MW, respectively. If the wind power is larger than 7.5 MW, the MCP will drop to $4.5/MW. Another possibility is that the demand curve and the same as the vertical parts power of supply curve. The MCP can be any value between the left price and the right price. However, this possibility is very low in practical power systems. Usually, proportion of the wind power is not more than 10% to 20% of total capacity of the power system. Therefore, it is proper to pay all the power suppliers at the MCP calculated when the wind power capacity is not considered. When the wind power is available, the power suppliers need to reduce their output power because they have obtained prices higher than the actual MCP and the WTGSs, therefore do not need to pay the imbalance fine. The generators not willing to reduce their output power will surely be paid at a lower MCP.

Case 4: The demand side participants can bid on power markets as supply side participants do. It is assumed that there are two in-demand bidders. Table 11.12 shows their bidding characteristics and Figure 11.16 shows their total demand curve and power supply curve.

According to Figure 11.16, the MCP is 4.2$/MWh. The quantity electric purchased by two in-demand bidders are 41.875 MW and 33.125 MW, respectively. The total power supplied by the generator bidders is 75 MW. If the available wind power is 5 MW and its bidding price is zero, the power supply curve will move by toward right along the x axis. As a result, the MCP will remain unchanged, that is, $4.2/MWh, but the bidders will have larger power consumptions. If there is more available wind power, the MCP can drop more. However, the in-demand bidders must make payments at the MCP calculated without consideration of wind power, so as to avoid any financial risks because of wind power not used.

Table 11.12 Block bids on demand side.

Demand 1	Price ($/MW)	8.0	7.0	6	5	4.2	3
	Load (MW)	10	10	5	10	10	15
Demand 2	Price ($/MW)	7	5.5	4.5	4.2	3.5	3
	Load (MW)	4	10	15	6	5	10

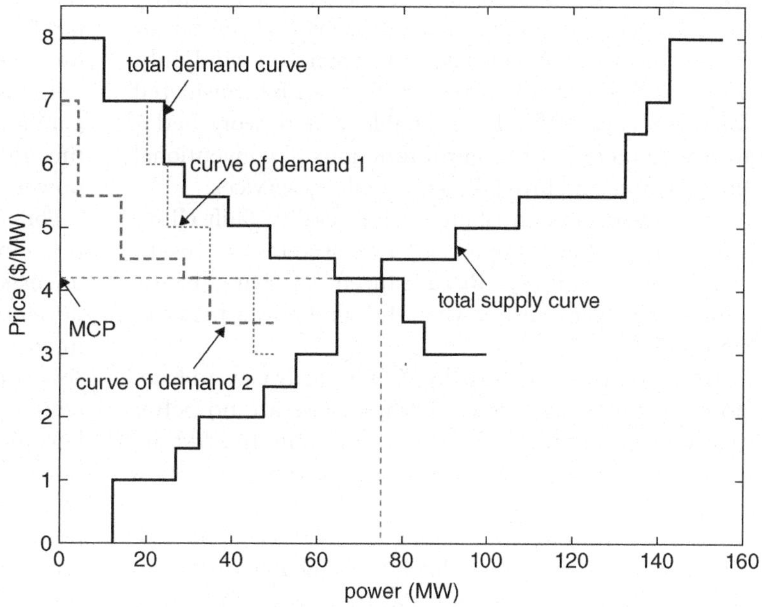

Figure 11.16 Curve of block bids on supply side and demand side.

According to results of the four modes of linear bids and block bids in a competitive power market with wind power, the MCP calculated without consideration of wind power is technically and economically proper. The WTGSs, if with available output power, apply to the same MCP too, but this MCP is probably higher than the MCP calculated with consideration of wind power; the other supply bidders need to reduce their output power according to their bidding success ratios. Wind power will recover its cost and can exist in the power market continually. When wind power is unavailable, the power market has no loss and the supply bidders must prepare to provide all of their loads. Wind power is also important in market power weakening and auxiliary services; however, its cost must be compensated for the sake of successful application. For the sake of free and fair electricity trading, the trading with wind power must be transparent.

The auxiliary services necessary for the quality, safety, and reliability of electricity, such as reactive power support, spinning reserve capacity, load frequency control (AGC), black start capability, and energy balance can be obtained through competitive markets similar to energy markets and based on auction. The output instability of wind power relates to the uncertainty of wind. Thus, on the one hand, if WTGSs are allowed for bidding, a high market cost will probably be caused by more serious power imbalance; on the other hand, if wind power is managed by market operators, active power can be provided in auxiliary services or primary energy markets with help of wind power, thus cutting down the power generation cost.

11.6 Main Technology of WFs to be Conventional Power Plants

11.6.1 Overview

Modern power systems are mainly powered by conventional (main) power plants such as thermal power plants, hydropower plants, and nuclear power plants. Wind power is the clean power source with the most mature technologies, the largest developable capacity and the quickest development in the recent years. However, it is often called an "intermittent power source," an "instable power source" or an "unreliable power source" because of the randomness of wind and the instability of output power of WFs. Can WFs become power sources like conventional power plants?

In the past 25 years, the concept of grid-connected WTGSs has gradually developed to the concept of MW level WTGSs with an advanced control system.

At present, many old WTGSs are subject to moderate control only, but many new WTGSs are probably subject to active control including an active pitch control system, WTGS speed control based on power electronics, output power increase. or decrease (ramp) with controllable methods in permissible conditions of wind, and so on. With the quick development of WTGS technologies, controllable WTGSs are now a majority of the newly installed WTGSs. Active control means that the power output modes of WTGSs do not depend on wind only. The load demands of power systems can be considered for them. New technologies enable WFs to work better to realize most of the control functions of conventional power plants and provide some auxiliary services.

Thus, WFs can contribute to the stability, fault elimination, and voltage support of power systems as conventional power plants do. This advanced WF characteristic is increasingly important when the wind power penetration is high.

The active power control by WFs is costly viewed from power systems, so it is used when there are no better cost-benefit solutions in power systems only. In addition, it is not necessary when the wind power penetration is low.

Because of different spatial characteristics of the WTGSs on a WF (or, more likely, wind changes between the WTGSs), it is very rare that there is an incredible incident [3], namely, all the wind power in a power system is lost because of sudden simultaneous loss of wind for the WTGSs on the WF.

With the worldwide quick development of wind power, the numerous problems of wind power and its power systems have become a hotspot in studies. For the development of wind power, it is significant to find ways to make WFs meet grid operation specifications and easy to be controlled and regulated as conventional power plants do. Here below is a wind power characteristics and studies based summary about the main technologies necessary for wind power to work as conventional power plants, including: ① Start from WTGSs and WFs to seek the methods to improve the controllability of wind power: the related technologies include wind power forecast technologies, WTGS and WF control technologies, WTGS fault ride through technologies, and so on. ② Start from the planning and operation of power systems to seek the technical actions for maximizing the potential of wind power, including improving flexibility and diversity of the units in power systems, strengthening grid structures, providing energy storage devices, using the DSM, and so on.

11.6.2 Wind Power Forecast Technologies

Due to the randomness and instability of wind, a WF has instable output power, thus influencing the power flow of the related grid. With increase of the wind power capacity connected to the grid, the wind power fluctuations will influence reserve capacity of the power system greatly. Unexpected output changes of the WF probably increase the reserve capacity of the power system or make power supply unreliable. In addition, the connection of massive wind power results in an additional loss for the conventional units because the power system has lower operation efficiency when the thermal power units or the hydropower units are not operating at their optimal operation points (during their start, stop, or part-loaded states). Optimal power distribution between the units is more complicated because of the output power instability of the WF.

For a power system, wind power forecast is one of the most direct and valuable methods to lower the uncertainty of its wind power generation plans. Both development of wind power forecast technologies and improvement of wind power forecast accuracy can help solve the problem above. According to the content above, wind power is different from conventional units mainly because of their randomness and instability. Significant improvement of the wind power forecast accuracy will strengthen the predictability and controllability of wind power greatly and enhance the benefits of power systems with wind power. Wind power forecasts can cut down the costs related to the uncertainty of day-ahead dispatch plans (trade in power markets) greatly. Output power forecasts of WFs are a key step to increase the wind power penetration of power systems during the operation of power markets and power systems.

The wind power forecasts are different from load demand forecasts and the forecasts of other power generation forms. Conventional power plants are mainly influenced by physical faults. Based on rich experience, load demand forecasts are more feasible than wind power forecasts. Compared with conventional power plants, wind, a natural phenomenon, is subject to reliable statistical treatment and physical forecasts more.

A wind power forecast should be based on the technologies including precise and complicated numerical weather prediction (NWP) technology, a WF power generation model, a statistical method, the generation capacity forecast every five minutes to one hour during the 48 hours to 72 hours ahead, the quarterly and the annual generation capacity. Based on related studies, preliminary theories and support systems with big errors have been established overseas for wind power forecasts. At present, the forecast error of a WF (root mean square error: RMSE) for a forecast period of 36 hours is between 10% to 20% of installed capacity of the WF, in regardless of which kinds of forecast methods and tools are used. If the grid-connected wind power in the whole region is considered, the forecast error will be below 10%; the forecast error will be about 5% for a forecast period of one to four hours. The bigger the region, the higher the forecast accuracy will be. The longer the forecast period, the lower the forecast accuracy will be, which is similar to load demand forecsts [3, 46].

Wind power forecast technologies are improved continually. The most important factor for wind power forecast improvement is the weather forecast model. The integration between NWP and short-term forecast models can be improved more. For each WF, its available historical data, quality, and quantity of the data acquired from the electronic data exchanged between the WF and the dispatcher influence the wind power forecast accuracy greatly. Another potential factor for improvement is overall forecasts in which the data of multiploe NWPs are aquired or multiple forecast models are properly used. According to relevant studies, the integration of multiple wind power forecast algorithms can improve the wind power forecast accuracy greatly [3].

11.6.3 WTGS/WF Control Technologies

With increase of the grid-connected wind power, grids now have stricter requirements for the connection of wind power to grids. The most important requirements of grid specifications for power system operation include: ① The characteristics and control technologies of active power and frequency response of WTGSs should comply with grid specifications and WTGSs should provide auxiliary services for frequency control. ② The reactive power and voltage regulation characteristics of WFs and WTGSs should comply with grid specifications because of their significant impacts on the voltage stability of power systems and WTGSs should provide auxiliary services for reactive power and voltage control.

Due to their connection with grids through power electronic converters, double-fed WTGSs and direct-drive synchronous generators have good capability of reactive power and voltage control, so they can contribute to the voltages and stability of power systems if operated with proper control strategies. Cage type asynchronous WTGSs have no excitation systems and are excited by generator-terminal shunt capacitors and power systems, so they cannot control the reactive power and voltage of power systems.

For a big WF, reactive power compensation can be added when necessary according to related power system study and analysis; both fixed capacitance compensation with dynamic compensation can be used when the parallel capacitor (reactor) compensation fails to meet the dynamic requirement of the power system.

As for active power control, some countries have proposed some power control requirements and worked out related technical actions for WTGSs on the basis of grid demands, such as control over the total output power of WFs and overall application of balance control, to make WTGSs increase or decrease their output power at specific rates in line with the related requirements of power systems. In some other countries, there are power system requirements for the regulation speeds of WTGSs active power, the permissible maximum active power changes and the permissible minimum active power decrements. Significant improvement of the active power control of WTGSs will reduce the adverse impacts of WTGSs on power systems greatly and give full play to the potential benefits of wind power.

According to Figure 11.17, the WFs working as virtual power plants (VPP) can achieve more power system tasks in addition to pure power generation in their

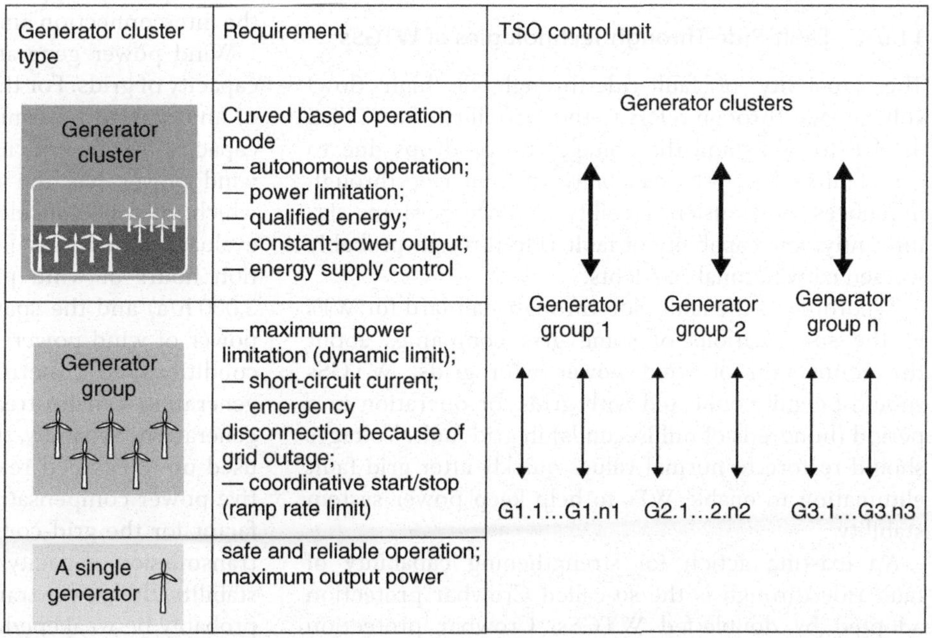

Figure 11.17 WF cluster management system.

operation, or, in other words, they can provide auxiliary services. A cluster including several big (offshore) WFs constitutes a GW-level group, which guarantees a new feasible choice for the transmiting variable output to supply power systems. The new concept of cluster management includes the integration of some geographically decentralized WFs according to specific rules, which needs optimal grid management and optimal (conventional) power generation dispatch. WF clusters will operate and be managed as big conventional power plants do.

It is impossible to forecast wind power completely accurately, making it necessary to minimize differences between the forecast values and the actual values by means of a control strategy about WF cluster management for power generation dispatch plans. In this case, output power is controlled by the dispatch plans determined in line with short-term forecasts. The control strategy has a significant impact on WF operation, so the forecast values with a minute-level forecast period should comply with the actual operation. The plans should be implemented within certain permissible bandwidths (determinated by the forecast errors). For the sake of the best interaction among WFs and the best WF cluster management, the time-variable set points should be generated and updated continually. For this management, it is assumed that WFs and WF cluster regions keep correcting short-term forecasts; control strategies including "output power constraint," "energy control," "capacity control," and "minimum ramp rate" are adopted; the uncontrollable WFs can be supported by the controllable WFs in specific WF clusters. Based on this control strategy, the mixed WF clusters can meet other requirements.

11.6.4 Fault-Ride-Through Technologies of WTGSs

The capability of fault-ride-through or high (low) voltage-ride-through refers to the capability of a WTGS or WF to withstand the voltage rise or drops due to grid faults. The WFs capability of fault-ride-through influences the transient stability of power systems significantly; low capability of fault-ride-through probably worsen conventional accidents.

According to the fault-ride-through standard for WFs in the specifications of some grid companies about the connection of wind power with grids, WTGSs should be still connected with grids for operation in a period (hundreds of milliseconds) in grid fault. Voltages should restore to normal values quickly after grid fault elimination to enable WFs to help keep power system stability.

An existing action for strengthening capability of fault-ride-through is the so-called Crowbar protection adopted by double-fed WTGSs. Crowbar protection itself can avoid damage to converter IGBT switches because of current increase of generator rotor circuits during the short-circuit faults of grids. However, there are different Crowbar operation modes: upon a grid fault, a whole WTGS can be disconnected from the related grid; the related converter disconnects from the rotor while the WTGS is still connected with the grid, making the generators realize grid-connected operation as asynchronous generators with large rotor resistances; the related converter is connected to the generator rotor while the WTGS is still connected with the grid, making it possible to recover normal operation immediately after grid fault elimination. When the voltage drop has lasted for hundreds of milliseconds, the WTGS can even support the grid in the course.

According to the specifications about wind power connection to grids, for a grid in which the installed capacity of wind power has a large proportion (such as 5%) in the total capacity of other power sources, the WFs operating in its region should have low capability of voltage-ride-through. Both double-fed WTGSs and direct-drive WTGSs can have a function of fault-ride-through.

11.6.5 Grid Structure Strengthening

When massive wind power is connected to a grid, it is necessary to strengthen the grid structure. Transmission systems keep developing because they are constructed at the same time with the installation of power facilities. The existing transmission systems in China are designed and integrated by province or region. The grids in different provinces or regions are interconnected for the sake of extensive power balance and exchange, higher reliability, and frequency support in emergencies, which develops the interconnection among the provinces or regions.

Wind power generation influences the transmission capacity of grids. For the increase of a new power source of any type to be connected with a grid, transmission capacity of the grid needs to be increased. However, wind power has the following special characteristics which must be considered in the transmission capacity evaluations of grids [3, 5]: Firstly, the annual utilization hours of wind power is small (usually 2,000 to 3,000 h/a) and the spatial smoothing effect and output power of wind power constitute a function of ambient conditions. In a thermal limit evaluation, wind power generation can be treated as any conventional power generation. Secondly, the cage asynchronous generators used on WFs need reactive power. If there is no reactive power compensation, there will be a lagged power factor for the grid-connected points of the WFs. If the transmission capacity limitation depends on voltage stability, the power transmission from WFs to grids will probably be weakened. However, if continuous reactive power compensation is provided by AC/DC/AC converters, wind power will not influence the maximum transmission power when the transmission capacity limits depend on voltage stability. Furthermore, if the grid-connected points of WFs provide leading power factors, the maximum transmission power of the lines considered will increase, particularly when higher voltages of the grid-connected points of WFs are acceptable. Thirdly, both during and after power system faults, the characteristics of WTGSs are different from those of conventional power plants. Conventional power plants mainly use synchronous generators that can operate continually during the serious voltage transients because of transmission system faults. In order to protect power electronic converters, variable-speed WTGSs need to be disconnected from grids during faults. Upon the trip of many WTGSs because of faults, the faults will have a larger impact on the load. Conversely, it will probably influence the transmission capacity in the regions with massive wind power. The reason is that not a single accident but many accidents need to be considered in a safety evaluation. During faults, the fixed-speed WTGSs probably absorb massive reactive power from grids, resulting in much slower recovery from power system faults and probably influencing the transmission capacity. Fourthly, WFs must be constructed close to wind sources. The best installation sites for WFs are usually distant, broad, and low-population regions. It is more urgent and expensive to strengthen grids in these regions than the places close to industrial loads (conventional power plants are usually in these places). In these regions, transmission capacity of the transmission systems of the WFs do not match with the big WFs increased. In any extreme condition, such as a low regional load demand or a large regional wind power output, the transmission power of a transmission system is limited by thermal limit, voltage and transient stability of the wires.

Usually, the blockage of a grid is not caused by specific power generation technologies because many power sources and grid operators share the same grid and the electric current depends on certain physical laws. In a region with massive and centralized wind power, the blockage of a grid usually happens in gales in two circumstances, that is, high loads and low loads. During low load, blockage of the grid is triggered by wind power only and needs urgent grid strengthening such as upgrading the existing overloaded line or building a new line. If grid strengthening is delayed, the wind power in the region will be further blocked or the grid can operate with help of power generation management only. During high load, blockage of the grid is caused by integration of the high wind power generation and the high conventional power generation. In this case, the conventional power plants provide the part beyond the wind power, and then blockage of the grid can be solved by means of regulation of output power of the conventional power plants. If the existing conventional power plants and the new WF(s) in the same region transmit power to a load center (such as power transmission from Inner Mongolia to Beijing), the transmission bottleneck in the existing transmission system will probably get worse.

The power transmission capabilities of power transmission lines mainly depend on the thermal limits of conductors, voltage stability, and transient power angle stability. The power transmission capabilities of shorter power transmission lines (<100 km) mainly depend on the thermal limits of wires. The current-carrying capability calculation of power transmission lines does not relate to power system structures but to specific power systems because the transmission of maximum active power is influenced by load power factors. If power transmission is realized on a long-distance power transmission line, there will be a voltage stability problem before the thermal limit is reached. The voltage stability and the transient stability limit depend on structure and load of the power system. Thus, different grid strengthening actions should be taken for different influential factors of the transmission limits of grids.

Strengthening the grids of different voltage levels should adapt to the ever-increasing wind power and other power generation technologies as well. Obviously, the demands for grid infrastructure investments are not caused by wind power increase only but related to many market participants. Grid expansion, grid strengthening, and reserve capacity increase do not benefit WFs only but also all power system users. Thus, it is impossible to have a single project or technology to bear the cost of grid infrastructure strengthening. It is meaningless to handle the future infrastructure challenge through a single technology. Wind power is actually the driving force for grid infrastructure strengthening, but it is not or should not be the only technology driving the whole grid infrastructure and the power system to improve their operation for benefits. Therefore, future decision making needs comprehensive methods considering the characteristics of both wind power technologies and other technologies.

Grid strengthening is complicated and needs both short-term actions and long-term actions to realize smooth wind power integration. Short-term actions mainly include optimization of the existing infrastructure and the so-called "soft" actions such as proper management methods. Long-term actions mainly include building bigger grids to absorb massive offshore wind power or the wind power of big wind power bases, making the best use of the smoothening effect of extensive land wind power and improving operation of the existing power systems. The possible types of grids to be strengthened include the high-voltage power transmission lines between regional grids (for the sake of better connection of WFs to big power systems), the long-distance offshore power transmission lines connecting different offshore WFs and load centers and the long-distance power transmission lines between load centers and wind power bases.

In the transmission capacity increase of grids, it is necessary to consider the costs and weigh the additional incomes and costs of transmission capacity increase. Usually, elimination of transmission bottleneck is not believed as the best method. In addition, transmission capacity increase is time consuming. Therefore, when blockage and other related problems are handled and massive wind power is allowed to be connected to a grid, it is necessary to consider other actions including reducing the output power of conventional units (conventional units allow quick regulation) and/or removing the surplus wind power upon the blockage of transmission systems.

11.6.6 Combination of Energy Storage Technologies and Wind Power Technologies

Each grid operator must match power supply and demands in long cycles. Each task probably relates to the specific forms of power generation, energy storage and load control. Table 11.13 sums up the time characteristics necessary for energy storage.

An important characteristic of power generation is that power cannot realize large-scale energy storage (large-scale energy storage is now realizable for pumped storage hydropower plants only). Large-scale energy storage technologies and energy storage devices are the difficulties that attract attention but is unsolved in the power industry and the electrical appliance industry.

The large-scale energy storage of electricity can solve the difficulty of peak-valley differences in power generation; the power supply reliability improvement of power systems can avoid the troubles and losses because of unexpected power interruption; energy storage devices can improve power system stability and avoid power system instability upon serious power system disturbances; energy storage devices are necessary for such instable power generation equipment with renewable energy sources as WTGSs, solar photovoltaic units, or thermal power units to supply power to users stably.

Table 11.13 Time characteristics necessary for energy storage.

Task	Cycle
Smooth output	Discontinuous or second
Load level	12 hours to a day; 40–60 hours each week; quarter (3 months)
Peak load clipping	180–10,000s
Spinning reserve	30–300 s
Voltage stability and frequency regulation	0.5–120 s
Actions against supply interruption	0.12–0.2 s
Stability improvement	0.02–0.2 s

Energy storage technologies have the following two functions in wind power generation. They stabilize power systems: WTGSs provide mean loads and energy storage devices provide short-term peak loads to stabilize wind power outputs. They play a transitional role when WTGSs cannot operate normally.

At present, the energy storage technologies used and studied for WFs mainly include:

① **PHS:**
Pumped storage hydropower plants are the most common and mature energy storage technology. It is currently the only technology for solving the peak-valley differences of power systems extensively. If condition permits, wind power and hydropower can be combined.

② **Flywheel energy storage:**
Upon a valley load, the surplus power is transmitted to a motor to drive a flywheel to accelerate, which needs a couple of hours, such as from the midnight to the early morning. The flywheel then keeps operating quickly until a peak load. After that, the flywheel drives the motor to operate as a generator to convert kinetic energy of the flywheel into electric energy to be transmitted to a grid. The flywheel slows down in this process until about half of its maximum speed. When the kinetic energy is converted into electric energy, the variable-speed and constant-frequency power electronic technology is adopted and the electric energy output frequency may remain unchanged.
Flywheels may be of unit type for the sake of integration into a device of larger power and installation close to a load when necessary, thus realizing expansion when needed and avoiding a transmission loss. Both magnetic suspension bearings and flywheels are sealed in vacuum containers, which reduce the operation loss; the use of a high-strength material enables flywheels to operate more quickly and store more energy; the development of power electronic devices and technology make the technologies of electric energy conversion and frequency control meet the related requirements of power systems completely. Thus, flywheel energy storage is becoming increasingly feasible.

③ **Rechargeable cell energy storage:**
Another energy storage method used for small-scale energy storage is electrochemical energy storage, that is, secondary (or rechargeable) cell energy storage. It is an energy storage technology with a long history. With progress of the cell technologies in the

recent years, the large-scale energy storage based on rechargeable cells has been used in practical production and cell energy storage equipment has become a kind of basic equipment of independent WFs or solar power plants.

④ **Electromagnetic field energy storage:**
It is the true electric energy storage without the limitation by intermediate links. It includes electric field energy storage and magnetic field energy storage. In magnetic field energy storage, superconducting coil energy storage is now promising. With high energy storage densities, superconducting coils can charge and discharge electric energy (active power and reactive power) very quickly. Superconducting energy storage devices have been available commercially in countries such as the United States.
In electric field energy storage, electric energy storage is realized with help of the charge storage capability of capacitors. In this storage method, supercapacitors are promising because they can be made into very small capacitors with a unit of farad because of the extremely high dielectric constants with their dielectrics. The capacitance of a supercapacitor is larger than an ordinary capacitance by several orders of magnitude. Capacitors have the same quick charge and discharge speed of electric energy as superconducting coils, or quicker.

⑤ **Compressed air energy storage and hydrogen energy storage:**
Compressed air energy storage has the same principle as PHS basically. The most important task is to find proper places for the storage of compressed air, such as sealed caves, abandoned mines, and so on.

Hydrogen energy storage is mainly based on the successful development of fuel cells. Its principle is as below: the surplus power of power grids is used for hydrogen generation and the hydrogen obtained is used for the power generation by fuel cells when necessary.

Usually, there are large energy losses during the application of energy storage decisions, namely, the electric energy absorbed is more than the electric energy regenerated. The electric energy for energy storage usually has an input/output efficiency of about 80%.

11.6.7 Improvement of Flexibility and Diversity of the Units in Power Systems

Both wind power and loads are not flexible. The impacts of wind power on a power system much relate to flexibility of the other units in the power system. For the best connection of wind power to a power system, it is necessary to make use of characteristics and flexibility of all units of the power system in the optimal way for the power system. The range of units for selection can be enlarged for higher flexibility of the power system. For example:

- Increase the transmission power between the power systems in different regions;
- Combine multiple instable power sources. For example, combine wind power and solar power. Studies suggest that the combination of multiple instable power sources is more valuable than plenty of a single instable power source. The combination of wind power and a power plant with a limited energy output (such as hydropower plant, a biomass energy power plant or another similar power plant, which cannot operate with its maximum output power continually because of its limited energy availability) can reach a high economic benefit too. In the power systems with massive hydropower, wind power fluctuations can be smoothened by hydropower.
- Distributed power sources such as gas turbines can improve flexibility of the reserve capacity of power systems;
- DSM: Power imbalance can be affected by means of increasing or decreasing load demand. Some load demands (such as the heating or cooling of houses or swimming pools) can be delayed or cut in or out according to signal price signals, realizing new balance between power generation and power consumption without regulation;
- In power system scheduling, increase the units for peaking and decrease the units bearing the base load;
- Build controllable microgrids friendly for the grid, and so on.

11.6.8 Improvement of Dispatching Operation of Grids

Usually, in a grid dispatch, a corresponding power generation plan is made in line with the load forecast for a time period in future. Meanwhile, the grid must maintain certain controllable reserve capacity for possible load fluctuations (differences between the power generation plan and the load forecast).

After the connection of massive wind power to a grid, the traditional dispatch mode has the following problems: uncertainty of output power of a WF increases the net load fluctuation and the error between a day-ahead plan and the actual situation; due to a too large time scale between the day-ahead dispatch and the AGC, errors of the day-ahead plan, loads and random fluctuations of wind power are all borne by the AGC unit, resulting in shortage of regulable capacity of the AGC unit usually; in order to keep active power balance for the power system, the dispatcher has to regulate set power and load of each conventional unit frequently according to power unbalance of the power system, which is a lot of work; such external costs of power generation as environmental pollution and resource consumption are not considered, so are factors such as energy efficiency, failing to reasonably reflect the electric energy values of new energy sources represented by wind power.

The traditional dispatch mode does not apply to the power systems with massive wind power, making a dispatch mode with a smaller time range and a higher accuracy necessary. For example, a multi-time-scale coordinated active power dispatch mode can be used for coordinating a day-ahead dispatch plan, a rolling plan, real-time dispatch and AGC. To be optimal dispatch plans, the day-ahead dispatch plan has enough time for optimal computation and the power generation plan should target at high economic efficiency and be based on safety. The rolling plan keeps correcting the day-ahead dispatch plan in line with the latest information to guarantee a startup cycle of 30 minutes to 60 minutes and comprehensive coordination. On the one hand, the rolling plan corrects the residual power generation plan in a real-time way in line with the latest information to reduce uncertainty of the day-ahead dispatch plan gradually; on the other hand, the rolling plan, by means of multi-level reserve capacity coordination and optimization, reserves adequate correction reserve capacity for the course of real-time dispatch to make the functions of real-time dispatch come true. The real-time dispatch plan follows up the set values in the rolling plan to make further checks and corrections. It adopts a startup cycle of 5 minutes to 15 minutes, use the units with good performance and good implementation plan as buffer units. The real-time dispatch plan regulates output power of the units in line with ultra-short-term load forecasts and wind power forecasts to eliminate uncertainties such as absence of effective follow-up plans for units and an inadequate capacity of the AGC unit. The real-time dispatch corrects errors of the previous level, leaving the remaining errors to be corrected in the next level. The real-time dispatch reserves a regulation space for the AGC unit (second-level) for the sake of normal operation of the AGC unit. The AGC unit (second-level) makes real-time compensation for the error between power generation and power supply and slightly regulates the active power on the basis of real-time dispatch, so as to make the frequency and the tie line power comply with the CPS evaluation objective. Real-time dispatch is for distribution of the load power with high regularity and high amplitude. AGC is for handling quick random fluctuations of the loads with low regularity and low amplitude as well as quick random fluctuations of wind power. By means of control over all other units except for the AGC unit, real-time dispatch is expected to regulate active power balance of the power system to control the AGC unit indirectly, reserve a regulation space for the AGC unit to meet the demand for its error correction.

The dispatch plans should comply with the power source structure features and active power control capabilities of different power systems. These plans should be based on reasonable objective functions and constraints determined following the rules of safety, high economical efficiency, fairness, and so on.

Additionally, adequate controllable reserve capacity should be arranged for dispatch. The grid-connected operation of wind power is another reason for power fluctuations, so both load fluctuations and wind power output fluctuations (differences between wind power forecasts and the actual output power of WFs) should be considered in configuration of controllable reserve capacity of the grid. When the wind power penetration is high, reserve capacity of the grid should be increased correspondingly. When the wind power penetration (energy) is 10%, the additional reserve capacity should be 2% to 4% of the installed capacity of wind power.

To sum up, grids need to build dispatch modes in line with different WF operation characteristics; WFs need to build modes for the active power distribution and control between WTGSs to make the active power of WFs dispatchable. Wind power forecasts (day-ahead ones, short-term ones, and ultra-short-term ones) should be included in generation dispatch plans or reflected as negative loads in load forecasts

11.6.9 Strengthening DSM

Supply side management (SSM) is for meeting the load demands of grids at any time. For higher wind power penetration and better use of wind power, it is necessary to make use of DSM. DSM refers to the power consumption activities for the sake of low-cost power services. The generator terminals' power consumption efficiencies are improved and their power consumption modes are optimized, and the power consumptions and the load demands are reduced while the functions of power use are guaranteed, so as to save energy sources and protect the environment. DSM changes the conventional thought pattern that additional load demands should be met by additional energy supply only. It also establishes a new concept that the energy sources saved on the demand side can be used as replaceable energy sources for suppliers, or, in other words, DSM is a source for the demand side. The optimal configuration and development of power supply sources and DSM sources is a course to bring the electric energy saved by DSM into power scheduling as a source. This course is to compare the plans about power generation, power supply, power consumption, and power saving, work out resource configuration and use plans with the minimum costs by means of combination and selection of the plans to benefit the society, power companies, and users. Thus, DSM can be technically regarded as a power plant for balancing the errors in energy contracts and avoiding costs of the spot markets of reserve power. Finally, DSM can reach the following results: power consumers have smaller costs and higher benefits by means of improve the power consumption efficiency of their equipment; power companies gain better benefits than other investments as a result of lower load demands and less or delayed construction of power plants and power transmission lines; the whole society is benefited because of energy saving and environmental protection.

Due to the specificity of power generation and consumption, DSM must focus on management. For example, an important content of DSM is changing the power consumption modes of users, that is, load shaping management technologies, including peak clipping, valley filling, and peak load shifting. According to power systems' rule that daily loads have peaks and valleys, market-oriented means are taken to cut down or transfer the power demands of users during the peak load periods of grids. The method can increase power consumptions during the valley load periods of grids, so as to change the distribution of power demands on time series, reduce the daily or quarterly peak loads and maximum loads of grids. Furthermore, DSM increase the load rates of grids and utilize existing grid resources effectively to enable grids to reach balance between power supply and power demands with help of a few additional installed capacity. With this new technology, regulator can delay or even avoid new power plant construction and transmission grid expansion to cut down the expected power supply costs. This load shaping management technology can be realized by end users with help of cold and heat storage technologies, energy substitution based operation technologies, operating procedure changes, stagger holiday system regulation, and so on.

Electrical equipment may be time-dependent or non-time-dependent. Time-dependent electrical equipment, such as TV sets and electrical cookware, must operate in the time specified by users. For non-time-dependent electrical equipment represented by heating (steam or water) equipment and cooling equipment, users do not care about their accurate timing or operation outage but pay attention to their ultimate objectives under reasonable constraints.

The management of interruptible loads is an important action for power load management. Upon power supply shortage, interruptible loads are the main method of peak clipping. In this method, a power system dispatcher transmits a request signal to a user in a peak load period of a grid according to a contract between the power company and the user. Part of the power load is interrupted upon response of the user. An interruptible load is equivalent to increase of the reserve capacity of a power system and be capable of stabilizing load of the power system. Interruptible load improves economic efficiency and reliability of operation of the power system, weakening impact of the market power and depress the peak price. Interruptible loads can be extensively used in industrial users in fields of metallurgy, cement, paper-making, textile making, and so on. Interruptible loads can serve as reserve resources for the operation of power systems. Price discounts are for exchanging with users' commitments for load demand decrease when it is necessary for power companies. Quick reserves can be provided in emergencies with help of triggering by low-frequency sensitive relays. This plan benefits power companies because it reduces the instability of load demands.

Interruptible loads are now a main resource for reserve capacity for operation. Heating systems, air-conditioning systems and swimming pool heating or cooling systems have their fixed energy storage capacity, namely, their power can be interrupted without any inherent and apparent impacts. Similar energy storage potentials exist in compressed air supply systems, municipal pumping systems, water desalination and purification systems, aluminium melting, and so on. Typically, these loads can provide transient and complete response without help of decrease warnings. Due to the necessity of valve opening, it probably needs some time for some processes. As a result, heating or cooling loads can be "charged" during the valley load periods of power systems or "discharged" during the peak load periods of power systems. In a power system with wind power, the systems above can cut in during high wind power penetration or cut out during low wind power penetration, thus reducing the impact of high wind power penetration on grids.

In a power system with wind power, the energy of power demands may be transferred to the time periods with high wind power to reduce the impacts of the power demands in peak load periods. This is crucial for maximizing the wind power generation efficiency, improving both financial and environmental benefits and reducing the dependence on the power generation with fossil fuels. For example, if wind power is abundant at 2 a.m. and expected to be less at 6 a.m., the storage of heating water (including the additional heat loss calculated in the nighttime) at 2 a.m. will probably be more effective than water heating at 6 a.m. (wind power is unavailable at this time). The flexibility of DSM is mainly provided by non-time-dependent electrical equipment. This non-time-dependent electrical equipment may have power consumption plans compliant with wind power forecasts or be temporarily powered off to balance out the peak load demands caused by time-dependent electrical equipment.

The common difficulties of load management plans are centralized in control and operation coordination and switching. Traditionally, it is much earlier to monitor a few resources (generators) than numerous small resources (loads) with communication and control technologies. Wireless remote control switches have been used in water heating systems for years. For example, a household hot water tank of a proper size can provide sufficient hot water in a whole day. It heats water with electricity in the seven-hour valley load period in the last night. With new progress of wireless communication technologies, it is now possible to realize real-time bilateral control for the loads of air-conditioning systems. Through a radio pager system, a heater can receive reduction instructions and send back a signal about acceptance or refusing and a system state report as well. Many modern office buildings and other buildings are integrated with energy management control systems. These control systems can control electrical equipment of systems such as heating systems, air-conditioning systems, and lighting systems and usually provide non-local monitoring.

With development of smart grid construction, loads will be more flexible and controllable. DSM will play an increasingly more important role in power systems, particularly those with many distributed power sources.

References

1 DeMeo, Edgar A., Grant, William, Milligan, Michael R., and Schuerger, Matthew J. 2005. Wind plants integration. *Power & Energy Magazine, IEEE,* **3**(6), 38–46.

2 Holttinen, H., Meibom, P., and Ensslin, C., et al. 2008. Impacts of large amount of wind power on design and operation of power systems. 7th international workshop on large scale integration of wind power and transmission network offshore wind farms, May 26–27, Madrid Spain.

3 EWEA report. 2005. Large scale integration of wind energy in the European power supply: Analysis, issues and recommendations. http://www.ewea.org/fileadmin/ewea_documents/documents/publications/grid/051215_Grid_report.pdf

4 State Power Grid Corp. 2006. Technical regulations for wind farm access to power system (for Trial Implementation).

5 Ackermann, Thomas. 2005. *Wind Power in Power Systems.* New York: John Wiley & Sons.

6 Söder, L. 1993. Reserve margin planning in a wind-hydro-thermal power system. *IEEE Trans. on Power Syst.,* **8**(2), 564–571.

7 Persaud, S., Fox, B., Flynn, D. 2003. Effects of large scale wind power on total system variability and operation: Case study of Northern Ireland. *Wind Eng.,* **27**(1), 3–20.

8 Dany, G. 2001. Power reserve in interconnected systems with high wind power production. *Proc. IEEE Porto Power Tech Conf.,* **4**(6). September 10–13, 2001, Porto, Portugal.

9 Watson, S. J., Lanberg, L., and Halliday, J. A. 1994. Application of wind speed forecasting to the integration of wind energy into large scale power system. *Proc. Inst. Elect. Eng., Gen., Transm., Distrib.,* **141**(4), 357–362.

10 Schlueter, R. A., Park, G. L., Lotfalian, M., Shayanfar, H., and Dorsey, J. 1983. Modification of power system operation for significant wind generation penetration. *IEEE Trans. Power App. Syst.,* **PAS-102**(1), 153–161.

11 Doherty, Ronan and O'Malley, Mark. 2005. A new approach to quantify reserve demand in systems with significant installed wind capacity. *IEEE Transactions on Power Systems,* **20**(2), 587.

12 Billington, R. and Allan, R. N. 1994. *Reliability Evaluation in Power Systems,* 2nd ed. New York: Plenum.

13 Boqiang, Ren and Chuanwen, Jiang. 2009. A review on the economic dispatch and risk management considering wind power in the power market. *Renewable and Sustainable Energy Reviews,* **13**, 2169–2174.

14 Doherty, Ronan, Denny, Eleanor, and O'Malley, Mark. 2004. System operation with a significant wind power penetration. Power Engineering Society General Meeting. June 6–10, pp. 1002–1007.

15 Brendan Fox, Brendan, Damian Flynn, Damian, and Leslie Bryans, Leslie, et al. 2007. *Wind Power Integration[M].* London: The Institution of Engineering and Technology.

16 Ummels, Bart C., Gibescu, Madeleine, Pelgrum, Engbert, Kling, Wil L., and Brand, Arno J. 2007. Impacts of wind power on thermal generation unit commitment and dispatch. *IEEE Transactions on Energy Conversion,* **22**(1), 44–51.

17 Shanjun, Yi, Yu, Ma, and Jiannan, Zhang. 2007. Overview on model and algorithm for unit commitment. *Northeast Electric Power Technology,* 7, 40–42.

18 Erkeng, Yu. 1986. Modern power system economic dispatch. The injection of water. Beijing: Conservancy and Electric Power Press.

19 Ming, Ding, Min, Bao, Hongbin, Wu, Wanling, Ma, and Meiqin, Mao. 2008. Unit commitment problem in distributed generation system with multiple energy sources. *Automation of Electric Power Systems,* **32**(6), 46–50.

20 Chun-Lung Chen, Chun-Lung. 2008. Optimal wind–thermal generating unit commitment. IEEE Transactions on Energy Conversion, **23**(1), 273.

21 Xiaolei, Li, Jingyang, Zhou, Erkeng, Yu, and Yi, Pan. 2008. Linear mixed integer programming algorithm for unit commitment based on dynamic search. *Automation of Electric Power Systems,* **32**(21), 18–21, 76.

22 Wen, Sun and Dazhong, Fang. 2008. Unit commitment considering system reliability and economy. *Power System Technology,* **32**(6), 47–51.

23 Tuohy, Aidan, Meibom, Peter, Denny, Eleanor, and O'Malley, Mark. 2009. Unit commitment for systems with significant wind penetration. *IEEE Transactions on Power Systems,* **24**(2), 592–601.

24 Hosseini, S. H. and Abbasi, E. 2007. Probabilistic unit commitment with wind farms considerations. IEEE Large Engineering Systems Conference on Power Engineering, pp. 73–77, October 10–12, 2007, Montreal, Canada.

25 Chen, C. L. and Chen, N. 2001. Direct search method for solving economic dispatch problem considering transmission capacity constraints. *IEEE Trans. Power Syst.,* **16**(4), 764–769.

26 Chen, C. L., Lee, T. Y., and Jan, R. M. 2006. Optimal wind–thermal coordination dispatch in isolated power systems with large integration of wind capacity. *Energy Convers. Manage.,* **47**(18–19), 3456–3472.

27 Furong, Li and Kuri, Bless. 2005. Generation scheduling in a system with wind power. IEEE/PES Transmission and Distribution Conference & Exhibition: Asia and Pacific, Dalian, China.

28 Lee, Tsung-Ying. 2008. Short-term hydroelectric power system scheduling with wind turbine generators using the multi-pass iteration particle swarm optimization approach. *Energy Conversion and Management,* **49**, 751–760.

29 Matevosyan, J. and L. Soder, L. 2006. Optimal daily planning for hydro power system coordinated with wind power in areas with limited export capability, 9th International Conference on Probabilistic Methods Applied to Power Systems, KTH, Stockholm, Sweden, June 11–15.

30 Castronuovo, Edgardo D. and Peças Lopes, J. A. 2004. On the optimization of the daily operation of a wind-hydro power plant. *IEEE Transactions on Power Systems,* **19**(3), 1599.

31 Bueno, C. and Carta, J. A. 2005. Technical-economic analysis of wind-powered pumped hydro storage systems. Part I: Model development [J]. *Solar Energy,* **78**(3), 382–395.

32 Bueno, C. and Carta, J. A. 2005. Technical-economic analysis of wind-powered pumped hydro storage systems. Part II: Model application to the island of El Hierro [J]. *Solar Energy,* **78**(3), 396–405.

33 Pan, Wenxia, Fan, Yongwei, and He, Nanying. 2007. The optimal dispatching research of the wind-hydro hybrid system. Power Engineering Conference International, December 3–6, pp. 977–981.

34 Deyou, Liu, Zhizhong, Tan, and Feng, Wang. 2006. Simulation study on the combined operation system of wind power pumped storage system. *Water Resources and Power,* **24**(6), 39–42.

35 Hetzer, J., Yu, D. C., and Bhattarai, K. 2008. An economic dispatch model incorporating wind power [J]. *IEEE Trans on Energy Conversion,* **23**(2), 603–611.

36 Agrawal, S., Panigrahi, B. K., and Tiwar, M. K. 2008. Multi-objective particle swarm algorithm with fuzzy clustering for electrical power dispatch [J]. *IEEE Trans. on Evolutionary Computation,* **12**(5), 529–541.

37 Yuan Tiejiang, Yuan, Chao Qin, Chao, Tursun, Yibulayin, and Fei, Tong. 2009. An environmental/economic dispatch model for power grid containing wind power generation units and its simulation in electricity market environment. *Power System Technology,* **33**(6), 67–71.

38 Lei Yazhou, Lei and Lightbody, Gordon. 2005. Wind energy and electricity market. *Automation of Electric Power Systems,* **29**(10) 1–5.

39 S. N. Singh, S. N. and Erlich, I. 2008. Strategies for wind power trading in competitive electricity markets. IEEE Transactions on Energy Conversion, **23**(1), 249–256.

40 Zeineldin, H.H., El-Fouly, T.H.M., El-Saadany, E.F., and Salama, M.M.A. 2009. Impact of wind farm integration on electricity market prices. *Renewable Power Generation,* **3**(1), 84– 95.

41 Baldick, R., Helman, U., Hobbs, B.F., and O'Neill, R.P. 2005. Design of efficient generation markets. *Proceedings of the IEEE,* **93**(11), 1998–2012.

42 Abreu, L.V.L. and Shahidehpour, M. 2006. Wind energy and power system inertia. *Power Engineering Society General Meeting,* June 18–22, 2006, Montreal, Canada.

43 Fabbri A., Roma, N.T.G.S., Abbad J.R., and Quezada V.H.M. 2005. Assessment of the cost associated with wind generation prediction errors in a liberalized electricity market. *IEEE Trans. Power Syst.,* **20**(3), 1440–1446.

44 El-Fouly, T.H.M., Zeineldin, H.H., El-Saadany, E.F., and Salama, M.M.A. 2008. Impact of wind generation control strategies, penetration level and installation location on electricity market prices. *Renewable Power Generation,* **2**(3), 162–169.

45 Singh, S.N. and Erlich, I. 2006. Wind power trading options in competitive electricity market. Power Engineering Society General Meeting June 18–22, 2006, Montreal, Canada. Ernst, Bernhard, Oakleaf, Brett, Ahlstrom, Mark L., Lange, Matthias, Oehrlen, Corinna, Lange, Bernhard, Focken, Ulrich, and Rohrig, Kurt. 2007. Predicting the wind. *IEEE Power & Energy Magazine*, **November/December**, 78–89.
46 Rudion, K. 2005. Wind farms with DFIG as virtual power plants. IEEE PES General Meeting, Glasgow.

12

The Evaluation Technology for An Wind Power Integrated Power System

This chapter introduces some evaluation technologies for wind power integrated power systems, such as reliability evaluation, capacity credit (CC) evaluation, wind power value analysis, and, in its end, overall analysis for the maximum wind power penetrating capacities of power systems. These evaluation technologies are significant for the planning and operation evaluations of wind power integrated power systems.

12.1 Reliability Evaluation of Wind Power Integrated Power System

12.1.1 Overview

Due to the graphical decentralization and random output power fluctuation of wind farms (WFs), wind power sources are quite different from the conventional power sources, and the evaluations for power generation system adequacy impacted by the wind energy conversion systems (WECS) are also different from those of the conventional power plants. This section compares the performance between wind power generation systems and conventional power generation systems according to the adequacy, to explain the impacts of wind power on the reliability of power generation systems.

Based on probability sampling for the operation states of all parts in a power system with WFs, the reliability indices of the power system in the i year, such as interruption duration, energy not supplied during interruption, and the interruption frequency, can be calculated. Then the averages of the performance indices of the N (a quantity) samples (year), are the reliability indices of the power system expressed in expectations. The common adequacy indices of a power system are the loss of load expectation (LOLE) and the expected energy not supplied (EENS).

LOLE: Expectation for the hours or days when the system fails to meet the load demand within a given time period (Formula (12.1)):

$$LOLE = \sum_{i=1}^{K} p_i T \tag{12.1}$$

In the formula above, LOLE is the loss of load expectation, which is the hours or days (h/a or d/a) when there is loss of load within a given time period (usually one year); p_i is the probability for the power system to be in state i; K is a set of power system states in which the system fails to meet the load demand in the given time period; T is the given time period.

EENS (MW·h/ a): The expectation for the decreased load demand energy of the power system in the given time period as a result of generation capacity shortage, unexpected equipment shutdown or grid constraints (Formula (12.2)):

$$EENS = \frac{1}{N}\sum_{i=1}^{N} E_i \tag{12.2}$$

$$E_i = 8760 \times \sum_{j=1}^{K} p_j LC_j$$

In the formula above, E_i is the annual energy not supplied (MW·h); p_j is the probability for the power system to be in state j; LC_j is the short energy or the total load to be decreased (MW) when the power system is in state j; K is the annual state number of the power system.

The reliability of a power generation system can be measured by comparing the adequacy change before and after removing different conventional power generators of power system and replaceing by WTGSs. A significant factor to be considered in the adequacy evaluation of a wind power integrated power system is comparing the contributions of a WTGS and conventional power generators to reliability. A method to realize this is to remove different conventional power generators from the wind power integrated power system and then determine the number of the WTGSs (capacity) needed to maintain the specified reliability.

Integration of Large Scale Wind Energy with Electrical Power Systems in China, First Edition. Zongxiang Lu and Shuangxi Zhou.

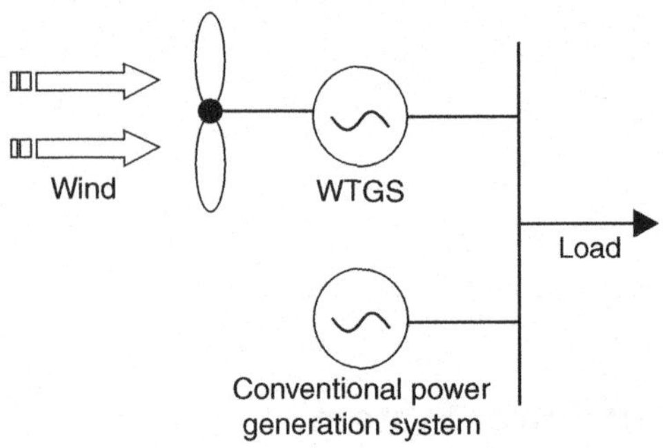

Figure 12.1 Structure of an wind power integrated generation system.

As in Figure 12.1, the wind power integrated generation system containing both wind power generation and conventional power generation supplies electricity for loads, and its adequacy can be evaluated with the sequential Monte Carlo simulation (MCS) method [1].

12.1.2 Reliability Evaluation Model

For the power generation adequacy evaluation, models such as a wind velocity model, a WTGS/WF output model and an outage model are necessary.

12.1.2.1 Wind Velocity Model

In the reliability evaluation of a wind power integrated power system, it is important to simulate the hourly wind velocity accurately. Wind velocity varies with time and space. The wind velocity of a specific hour is related to the wind velocities in the previous several hours. The wind velocity of a WF can be simulated by many methods such as time series analysis method and Weibull parameter analysis method.

1) **Time series wind velocity model**
The time series of an auto-regressive moving average (ARMA) model is used as a wind velocity model, usually expressed as below (Formula (12.3)) [1]:

$$y_t = \varphi_1 y_{t-1} + \varphi_2 y_{t-2} + \ldots + \varphi_n y_{t-n} + \alpha_t - \alpha_{t-1}\theta_1 - \alpha_{t-2}\theta_2 - \ldots - \alpha_{t-m}\theta_m \quad (12.3)$$

In the formula above, y_t is the time series of time t; $\varphi_i (i = 1, 2, 3 \ldots, n)$ and $\theta_j (j = 1, 2, 3 \ldots n)$ are auto-regressive parameters and moving parameters of the wind velocity model, respectively. $\{\alpha_t\}$ is the normal white noise process with a mean of zero and a variance of σ_a^2 $(\alpha_t \in (0, \sigma_a^2))$. Sequence $[n,m]$ and ARMA parameters can be obtained from the wind velocity data acquired in previous years with the help of the least square method.
The hourly wind velocity SW_t in time t can be calculated through Formula (12.4) based on the time series y_t, average wind velocity μ_t and its standard deviation σ_t.

$$SW_t = \mu_t + \sigma_t y_t \quad (12.4)$$

2) **Weibull distribution wind velocity simulation**
According to the analysis for massive measured data, the wind velocity variations of a region approximately obey Weibull distribution. The distribution function is as below (Formula (12.5)) [2]:

$$F_w(v) = P(V \le v) = 1 - \exp[-(v/c)^k] \quad (12.5)$$

In the formula above, k is the shape parameter of Weibull distribution, indicating the degree of skewness of Weibull distribution and with a set range of 1.8 to 2.3; c is the scale parameter, indicating the average wind velocity, and, when the accuracy requirement is not so strict, being the average wind velocity directly. According to the theory of the inverse transformation method, if the random variable $\{X_i\}$ is a uniformly distributed on [0,1]. Its continuously accumulated probability distribution is $F_1(x)$, and the current $\{y_i\}$ distribution function is $F_2(y)$, just make $x = F_2(y)$, that is, $y = F_2^{-1}(y)$. Suppose (Formula (12.6)),

$$x = F_w(v) = 1 - exp[-(v/c)^k] \quad (12.6)$$

Then (Formula (12.7)),

$$v = c[-\ln(1 - x)]^{1/k} \quad (12.7)$$

Both x and $(1-x)$ are random variables uniformly distributed on [0,1], so x can be replaced by $(1 - x)$ and Formula (12.7) can be changed into (Formula (12.8)):

$$v = c[-\ln X_i]^{1/k} \quad (12.8)$$

Thus, according to the probability distribution of wind velocity, a Weibull distribution random number generator $v_i = c[-\ln(X_i)]^{1/k}$ (X_i is a random variable uniformly distributed on [0,1]) generates the sampled hourly wind velocity. Judge whether it is in the normal wind velocity range $(v_{ci} \le v \le v_{co})$ or not and calculate the output power of the WTGS.

3) **Wake effect model** [2–4]
The WTGSs on a WF are in different positions. Due to the influence of the wake, a WTGS in the downwind direction has a wind velocity slower than a WTGS in the upwind direction, which is usually called wake effects (see Figure 12.2). X is the distance between the two WTGSs; R and R_W are the WTGS blade radius and the wake radius, respectively; v_0, v_T, and v_x are the natural wind velocity, the wind velocity flowing through the rotor, and the wind velocity influenced by the wake, respectively.

The main physical factors for the wake effect determination include distance, power and thrust characteristics of the WTGSs, and the turbulence intensity of wind as well. The turbulence intensity of wind influenced by the wake is as below (Formula (12.9)):

$$\sigma/v = (\sigma_G + \sigma_0)/v \quad (12.9)$$

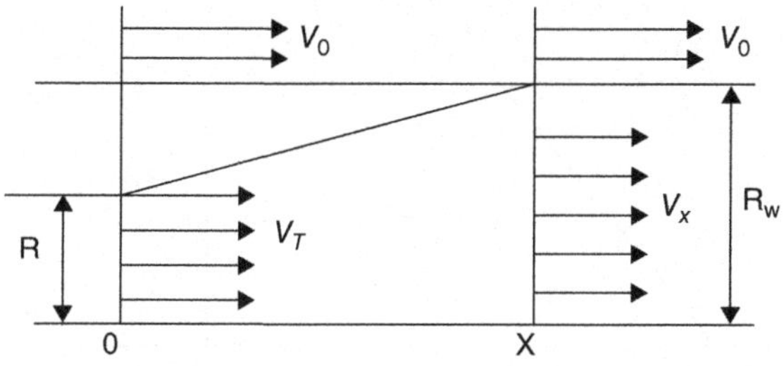

Figure 12.2 Wake effect model (Jensen model) for a flat terrain.

In the formula above, σ_G and σ_0 are the standard variances of the turbulence generated by WTGS and the natural turbulence, respectively; v is the average wind velocity. Usually, σ_G equals to $0.08v$ and σ_0 equals to $0.12v$.

The common WF wind velocity wake models include the *Jensen* model and the *Lissaman* model. The *Jensen* model simulates the wake condition of a flat terrain well and the *Lissaman* model simulates a nonuniform wind velocity field well based on a terrain consideration. The two models can work together to simulate the wind velocity wake loss of a WF roughly.

a) **Jensen model**

As in Figure 12.2, this model was proposed by *N.O. Jensen* in the *Riso* laboratory in Denmark. Suppose a WF is a uniform field; X is the distance between two WTGSs; R and R_W are the WTGS blade radius and the wake radius, respectively; v_0, v_T, and v_x are the natural wind velocity, the wind velocity flowing through the rotor, and the wind velocity influenced by the wake, respectively. Then according to the momentum theory, we can reach the formulas below (Formula (12.10)).

$$\begin{cases} \rho\pi R_W^2 v_X = \rho\pi R^2 v_T + \rho\pi(R_W^2 - R^2)v_0 \\ \dfrac{dR_W}{dt} = k_W(\sigma_G + \sigma_0) \\ \dfrac{dR_W}{dx} = \dfrac{dR_W}{dt}\dfrac{dt}{dx} = k_W(\sigma_G + \sigma_0)/v \end{cases} \tag{12.10}$$

In the formula above, k_W is a constant and can be calculated through the empirical Formula (12.11).

$$k_w = \frac{0.5}{\ln(h/z)} \tag{12.11}$$

In the formula above, h is the hub height; z is the ground roughness coefficient (usually 0.002). v_0 and v_T have the following relation with the thrust coefficient C_T of WTGSs (Formula (12.12)).

$$v_T = v_0\sqrt{1 - C_T} \tag{12.12}$$

Suppose $K = k_W(\sigma_G + \sigma_0)/v$ (K: wake drop coefficient). Then (Formula (12.13)),

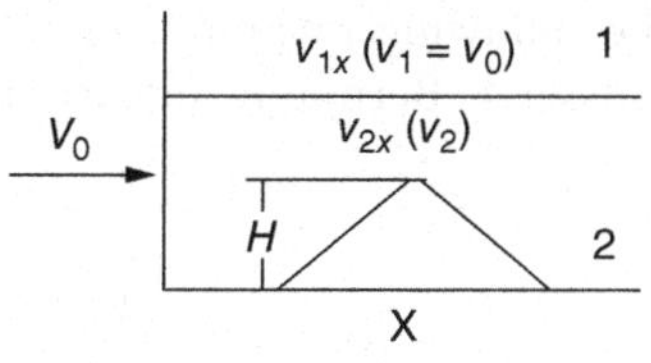

Figure 12.3 Diagram of the Lissaman model.

$$\begin{aligned} v_x &= v_0\left[1 - \left(1 - \sqrt{1 - C_T}\right)\left(\frac{R}{R + KX}\right)^2\right] \\ &= v_0\left[1 - d\right] \end{aligned} \tag{12.13}$$

In the formula above, d is the wind velocity drop coefficient. v_x is the function of C_T, so the wake effect is related to air dynamic characteristics of the WTGSs. Usually, a stalled WTGS has a thrust coefficient larger than that of a variable-pitch WTGS. That's why WTGSs with the same capacity and dimensions have a large wake effect.

b) **Lissaman model**

This model was proposed by *Lissaman* in 1986. Its main objects are the WTGSs with different heights. It can approximate the nonuniform wind velocity fields with a loss better (Figure 12.3).

Suppose there are two neighboring fields including one with a flat terrain (curve 1) and the other one with a more complicated terrain (different height and earth's surface, etc.; curve 2). Wind flows through the two fields at the same time; the edges of the two fields have the same wind velocity v_0. The two WTGSs with the same model are on the two fields, respectively, and their coordinate positions X along the wind velocity direction are both *zero*.

Suppose the wind velocity is v_1 equal to v_0 when X has a flat terrain. It is v_2 not equal to v_0 when X has a more complicated terrain, and it can be converted into a terrain elevation through the formula $v_2 = \left(\frac{h+H}{h}\right)^{\alpha}$, then, after WTGS installed, the wind velocities influenced by the wake in the two circumstances above will be (Formula (12.14)):

$$\begin{cases} v_{1x} = v_1(1 - d_1) \\ v_{2x} = v_2(1 - d_2) \end{cases} \tag{12.14}$$

In the formula above, v_1 and v_2 are the wind velocities when X has a flat terrain or a more complicated terrain, respectively; d_1 and d_2 are the corresponding wind velocity drop coefficients. According to the Jensen model (Formula (12.15)):

$$d_1 = 1 - \left(1 - \sqrt{1 - C_T}\right)\left(\frac{R}{R + KX}\right)^2 \tag{12.15}$$

Suppose the WTGSs have the same loss on the two terrains above, then the pressures in the wakes will be the same and the following formula can be reached through the lossless Bernoulli equation (Formula (12.16)).

$$d_2 = d_1\left(\frac{v_1}{v_2}\right)^2 \tag{12.16}$$

If the Jensen model and the Lissaman model are both used, an accurate overall WF model will be obtained and the wind velocities and output power of the WTGSs in different positions can be therefore calculated. The output power of the whole WF will be the sum of output power of the WTGSs on the WF.

12.1.2.2 Output Power Simulation for WTGS

WTGS output power p_t and wind velocity v have a nonlinear relation. The curve indicating the relation between WTGS output power and wind velocity is called a power characteristic curve which has different shapes and representations.

Figure 12.4 shows a typical WTGS power characteristic curve. The output power can be calculated through Formula (12.17). This relation can be described through the operation parameters of a WTGS. Usually, the operation parameters used include the cut-in wind velocity, the rated wind velocity and the cut-out wind velocity of the WTGS. The output power can be simulated as below based on the hourly wind velocity SW_t [1].

$$P_t = \begin{cases} 0 & 0 \le SW_t \le v_{ci} \\ (A + B\times SW_t + C\times SW_t^2)\,P_r & v_{ci} \le SW_t \le v_r \\ P_r & v_r \le SW_t \le v_{co} \\ 0 & v_{c0} \le SW_t \end{cases} \tag{12.17}$$

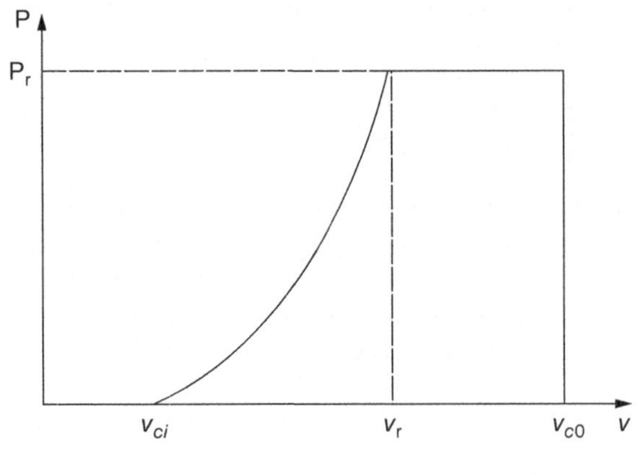

Figure 12.4 Typical output power characteristic curve of a WTGS.

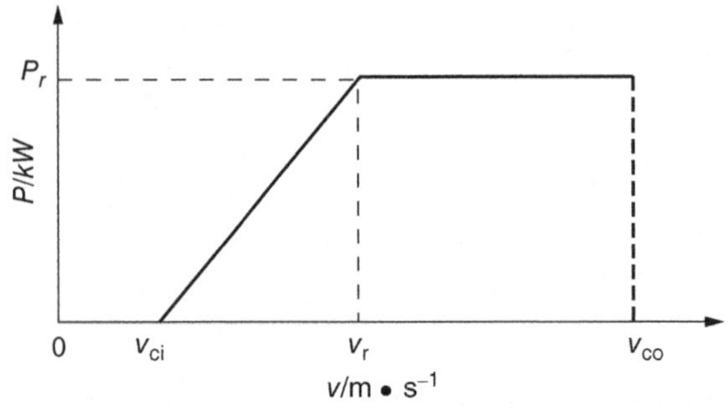

Figure 12.5 Output power linear characteristic curve of a WTGS.

In the formula above, V_{ci}, V_r, V_{co}, and P_r are the cut-in wind velocity, the rated wind velocity, the cut-out wind velocity, and the rated power of the WTGS, respectively; constants A, B, and C are a function about the cut-in wind velocity and the rated wind velocity of the WTGS (Figure 12.5).

$$A = \frac{1}{(v_{ci} - v_r)^2}\left[v_{ci}(v_{ci} + v_r) - 4(v_{ci}v_r)\left(\frac{v_{ci} + v_r}{2v_r}\right)^3\right]$$

$$B = \frac{1}{(v_{ci} - v_r)^2}\left[4(v_{ci} + v_r)\left(\frac{v_{ci} + v_r}{2v_r}\right)^3 - (3v_{ci} + v_r)\right]$$

$$C = \frac{1}{(v_{ci} - v_r)^2}\left[2 - 4\left(\frac{v_{ci} + v_r}{2v_r}\right)^3\right]$$

In a simulation with approximate linear characteristics, the output power characteristic piecewise function expression of the WTGS is as below (Formula (12.18)):

$$P_t = \begin{cases} 0 & v \le v_{ci} \text{ or } v \ge v_{co} \\ P_r\frac{v - v_{ci}}{v_r - v_{ci}} & v_{ci} \le v \le v_r \\ P_r & v_r \le v \le v_{co} \end{cases} \tag{12.18}$$

In the formula above, v is the wind velocity at the hub height of the WTGS; P_r is the rated output power of the WTGS.

12.1.2.3 WTGS Outage Model

The WTGS outage model is a two-state model described with normal operation duration τ_1 and repair time τ_2. The maintenance of land WTGSs lasts for short time and can be done during low wind velocities period, so it may be ignored in the reliability analysis.

It may be believed that the normal operation duration τ_1 and the repair time τ_2 of a WTGS obey exponential distribution, that is, the failure rate λ and the repair rate μ are constants. Thus, the normal operation duration τ_1 and repair time τ_2 can be calculated through the formulas below (Formulas (12.19) and (12.20)) [2].

$$\tau_1 = -\frac{1}{\lambda}\ln\gamma_1 = -MTTF\ln\gamma_1 \tag{12.19}$$

$$\tau_2 = -\frac{1}{\mu}\ln\gamma_2 = -MTTR\ln\gamma_2 \quad (12.20)$$

In the formula above, γ_1 and γ_2 are the uniformly distributed random numbers; MTTF is the mean time to failure; MTTR is the mean time to repair.

If the random outage model of a WTGS in the early period of its life needs to be analyzed, the failure rate curve in this period is usually expressed by means of Weibull distribution. The failure rate function expression is as below (Formula (12.21)):

$$\lambda(t) = \frac{k}{c}\left(\frac{t}{c}\right)^{k-1} \quad (12.21)$$

In the formula above, k is the shape parameter ($k>1$); c is the scale parameter; t is the operating time. It is easy to obtain the sampled value of the normal operating time through the failure function.

According to the statistical result, when the repair rate μ is not a constant, its probability density obeys Weibull distribution and its expression is as below (Formula (12.22)):

$$\mu(t) = \frac{a}{b}t^{a-1}e^{-at/b} \quad (12.22)$$

When the repair time is long, simulation may be done based on exponential distribution and a equals to 1; when the repair time is short, a is between 0 and 1 and its specific value should be determined after the analysis of the actual statistical data.

12.1.2.4 The Outage Model and Load Model of a Conventional Power System

1) **Outage model of a conventional power generation system**

   The outage model of a conventional power generation system is a two-state model. The two states include a normal operation state and a forced outage state. Usually, it may be believed that the normal operation duration τ_1 and the repair time τ_2 obey exponential distribution, or that the failure rate λ and the repair rate μ are constants. Thus, the expressions of normal operation duration and repair time are as Formulas (12.19) and (12.20).

2) **Time series load model**

   The models used frequently in the probability reliability analysis include a peak load model, a hierarchical load model, a clustering model, and a time series load model. A time series load model is used here. For example, an hourly model for 8,760 hours in a year or a load model based on the time series on a chronology can be obtained according to the maximum weekly maximum load in a year, the daily maximum load in a week (seven days) and the hourly maximum load in typical days in four seasons provided on the site. In the load model, loads are given for all buses and the system for 8,760 hours in a year according to the hour sequence. For a grid with one hundred load buses, the number of its load data is the product of 8,760 and 100.

12.1.2.5 Operation Constraints

In a wind power integrated power system, the output power of its conventional power generation systems and WTGSs is balanced with the active power of its load when the grid loss is not considered, namely (Formula (12.23)):

$$\sum \mathrm{P}_{\mathrm{WGi}} + \sum \mathrm{P}_{\mathrm{CGi}} \le \mathrm{P}_{\mathrm{LD}} \quad (12.23)$$

In the formula above, $\mathrm{P}_{\mathrm{WGi}}$, $\sum\mathrm{P}_{\mathrm{CGi}}$, and P_{LD} are output power of the i (an ordinal number) WTGS, the output power of the i (an ordinal number) conventional power generation system, and the active power of the load, respectively. When the wind power penetration limit of the wind power integrated power system is considered, then (Formula (12.24)):

$$\sum \mathrm{P}_{\mathrm{WGi}} \le \delta \times \mathrm{P}_{\mathrm{LD}} \quad (12.24)$$

In the formula above, δ is the wind power penetration limit, which is a ratio between the WF maximum capacity acceptable for the wind power integrated power system and the maximum load of the wind power integrated power system. Thus, the following restrictive relation between the wind power system and the conventional power system can be obtained (Formula (12.25)).

$$\sum P_{WGi} \le \frac{\delta}{1-\delta}\sum P_{CGi} \quad (12.25)$$

12.1.2.6 Monte Carlo Simulation

In the Monte Carlo simulation method, all random states of a wind power integrated power system are simulated on a computer with specific steps or, in other words, a practical process is simulated with the numerical calculation method, and the reliability indices of the wind power integrated power system are then obtained from numerous simulation test results. In the wind power reliability analysis with the Monte Carlo sequential simulation method, we can simulate the actual wind velocity by means of probability sampling according to the characteristic that wind velocity obeys Weibull distribution, and carrying out probability sampling for parts of the wind power integrated power system. Use a two-state climate model in the random sampling for WF outage in consideration of impact of weather on the WF failure rate, and judge the state of the wind power integrated power system by means of combining the WTGSs and the load model. After numerous sample system state tests, count the failure sample number and the test sample number and then calculate the risk index to realize the reliability evaluation for the wind power integrated power system.

The main steps of the adequacy evaluation for wind power integrated power system with the Monte Carlo simulation method include [1, 5]:

1) Create a generation capacity model for the conventional power generation equipments by means of simulation in the chronological sequence;
2) Create a generation capacity model for the WTGSs according to a wind velocity time series ARMA model described in Formula (12.3) and a WTGS output power model described in Formula (12.17);
3) Create an overall generation capacity model of the wind power integrated power system according to the generation capacity models obtained in steps 1 and 2;
4) Estimate a time series load model in the chronological sequence by means of statistics and analysis;
5) Judge the wind power integrated power system's operation state, total loss of load, and total energy not supplied according to the overall generation capacity model, the load model, and the wind power integrated power system's constraints.

In order to meet the simulation accuracy requirement, the simulation should be a multiyear one. On this basis, statistics about reliability indices such as LOLE and EENS should be done to realize reliability evaluation for the wind power integrated power system. The simulation should end upon the specified confidence level. The rule for simulation ending may be as below (Formula (12.26)):

$$\frac{\sigma(X)}{\sqrt{N} \times E(X)} \leq 0.05 \tag{12.26}$$

In the formula above, X is LOLE; N is the sampling years; $E(X)$ is the mean of the function; $\sigma(X)$ is the standard deviation of the function.

12.1.3 Reliability Evaluation of Multiple WFs

Compared with the situation of a WF, the power generation adequacy evaluation of multiple WFs is different because the output power of the WFs' correlation related to wind velocity, wind direction, air temperature needs to be considered in the WF output power model [3, 5].

12.1.3.1 The Joint Probability Distribution of Wind Velocity, Wind Direction, and Air Temperature

For different WFs on the same wind zone, their wind velocities, wind directions, and air temperatures have certain correlation. Their relations should be described by means of joint probability distribution so that their impacts on output power of the WFs can be explored. Suppose there are S (a quantity) WFs and number them with $s(s=1, 2, \ldots, S)$. If the time series of the wind velocities, wind directions, and air temperatures are known, the joint probability distribution P_w (v, D, T) of these physical quantities can be calculated. v $(v^1, v^2, \ldots, v^s)$, D $(D^1, D^2, \ldots, D^s)$ and T $(T^1, T^2, \ldots, T^s)$ are the random wind velocity vector, random wind direction vector, and random temperature vector of the WFs; v^s, D^s, and T^s are wind velocity, wind direction, and air temperature variable of the s (an ordinal number) WF. For easier calculation, the air temperatures may be discretized. For example, "T = 1" and "T = 0" are the operating air temperature and the outage air temperature of a WTGS. Each WF has a main wind direction, so wind directions can be properly simplified according to the probability of wind velocity distribution in different wind directions.

12.1.3.2 The Joint Probability Distribution of WF Output Power

The randomness of WF output power is mainly because of wind velocity variations, wind direct changes, WTGS outage because of the failures of equipments, air temperature changes, and so on. In order to obtain the probability distribution of WF output power, the WF output power in certain wind velocity, wind direction, and air temperature should be considered first. At this time, the randomness of WF output power is completely a result of WTGS outage because of the failures of equipments, so the conditional probability distribution of the WF output power in these conditions can be obtained. After that, the conditional probability distribution of WF output power in all conditions of wind velocity, wind direction, and air temperature and the joint probability distribution of wind velocity, wind direction, and air temperature should be considered, and the probability distribution of WF output power can be obtained with help of the multiplication theorem of probability. Similarly, the joint probability distribution of output power of multiple WFs can be obtained on the basis of the joint probability distribution of wind velocities, wind directions, and air temperatures of these WFs [3].

The joint probability distribution of output power: Suppose the s (an ordinal number) WF have WTGSs of J_S (a quantity) models and there are N_j^S $(j=1, 2, \ldots, J_S)$ (a quantity) WTGSs of the j (an ordinal number) model. In order to describe the operation states of the WTGSs, the column sector formed by the available WTGSs of different models on a WF is called the available WTGS vector expressed in N^s as the formula below (Formula (12.27)).

$$N^S = \left[n_1^S, n_2^S, \cdots, n_{JS}^S\right]^T \tag{12.27}$$

In the formula above, $n_j^S(0, 1, 2, \ldots, N_j^S)$ is the number of the available WTGSs of the j (an ordinal number) model on the s (an ordinal number) WF. According to the definition of the equivalent power characteristic of WTGS, the output power characteristic of the WTGSs of the s (an ordinal number) WF is expressed in matrix $X^S(d)$ as below (Formula (12.28)).

$$X^S(d) = \begin{bmatrix} x_{11}^S & x_{12}^S & \cdots & x_{1Js}^S \\ \cdots & \cdots & \cdots & \cdots \\ x_{I1}^S & x_{I2}^S & \cdots & x_{IJs}^S \end{bmatrix} \tag{12.28}$$

In the formula above, $x_{ij}^S = x_j(i, d)$ refers to the equivalent output power of the WTGSs of the j (an ordinal number) model when the wind velocity is v_i and the wind direction is d. Based on this matrix, random vector $Y^S(d)$ of the output power of the WF when the wind direction is d can be calculated through the formula below (Formula (12.29)):

$$Y^S(d) = \left[y_1^S, y_2^S, \cdots, y_I^S\right] = X^S(d)\,N^S \tag{12.29}$$

In the formula above, y_i^S is the output power of the s (an ordinal number) WF when the wind velocity is v_i and the wind direction is d. In order to calculate the joint probability distribution of the output power of multiple WFs, the output power vector of WFs is defined as below (Formula (12.30)):

$$y = \left[y^1, y^2, \cdots, y^S\right] \tag{12.30}$$

In the formula above, y^s is the output power variable of the s (an ordinal number) WF; the formula is used for calculating the value of y when the wind velocity, the wind direction, and the air temperature are v, d, and t, respectively.

Suppose the outage of WTGSs is mutually independently from the wind velocities, the wind directions, and the air temperatures, then the output power changes of the WFs are also independent when wind velocity, wind direction, and air temperature of the s (an ordinal number) WF are v, d, and t, respectively. Thus, the joint probability distribution $P(y|v,d,t)$ of output power of the WFs is as below (Formula (12.31)):

$$P(y|v,d,t) = \prod_{j=1}^{Is} P(y|v^S, d^S, t^S) \tag{12.31}$$

In the formula above, $P(y|v^S, d^S, t^S)$ is the conditional probability of output power of the s (an ordinal number) WF; see Formula (12.32).

$$P(y^S|v^S, d^S, t^S) = \prod_{j=1}^{Js} P_{gj}^S\left(n_j^S\right) \tag{12.32}$$

In the formula above, $P_{gj}^S(n_j^S)$ is probability of number of the available WTGSs of the j (an ordinal number) model on the s (an ordinal number) WF. For N (a quantity) WTGSs of the same model, their outage is independent s. Therefore, probability of n available $(0 \le n \le N)$ (a quantity) WTGSs is as below (Formula (12.33)):

$$P_g(n) = \binom{N}{n}(1-r)^n r^{(N-n)} \tag{12.33}$$

In the formula above, $\binom{N}{n} = N!/[n!(N-n)!]$. r is the rate of forced outage.

According to the multiplication theorem of probability, the joint probability distribution of the output power of the WFs is as below (Formula (12.34)):

$$P(\gamma) = \sum_{\gamma=y} P_w(v, D, T)\,P(y|v, D, T) \tag{12.34}$$

The calculation of the joint probability distribution of output power may be done with a convolution method, because it will be complicated if done with the formula above directly. If the wind velocity, the wind direction and the air temperature are all constant, and considering the outage of wind power generators, the WTGSs of each model on each WF can be equivalent to a multi-state power generation system, of the output power is a discrete random variable and the equivalent power generation systems have mutually independent output power.

12.1.4 Factors Influencing Power System Adequacy

According to some studies [1, 6, 7], wind power contributes to the reliability of a power system after connected to a grid. Both LOLE and EENS will decrease, suggesting that the adequacy of the power system has been improved by wind power. However, different WFs influence the reliability of power systems quite differently, depending on the wind power capacities, the wind velocity characteristics, and the independence of the WFs, and so on.

12.1.4.1 Impact of Wind Power Capacity

The impact of wind power capacity on LOLE is changing. At the beginning, LOLE changes greatly, wind power contributes significantly to the reliability of a power system, and the risk of the power system reduces quickly. However, with the increase of the wind power capacity, the WTGSs added contribute less to the reliability of the power system and LOLE tends to get "saturated," which is because all the WTGSs on the WFs are limited by the availability of wind on the specific WFs. This suggests the impacts of the randomness and fluctuation of wind power on the reliability of power systems. Thus, for a specific power system, the penetration of wind power is limited by the reliability of the power system. During the increase of the wind power capacity, EENS and LOLE change similarly.

12.1.4.2 Impact of Average Wind Velocity

A WF has different wind power resources when built in different positions and its result, that is, the average wind velocity, is different too. If a WTGS is in place with a wind velocity faster than the average wind velocity, its output power will increase. The larger the average wind velocity is, the larger the corresponding output power will be and the more contribution the WF will make to the reliability of the power system. With the increase of the wind velocity in its whole range, LOLE and EENS will decrease slowly at the beginning and quickly in the middle and tend to increase in the end. This can be explained by the unique nonlinear characteristic of WTGSs. When the wind velocity is slower than the cut-in wind velocity, WTGSs do not operate; and when the wind velocity is quicker than the cut-out wind velocity, WTGSs will trip off for safety. The output power of WTGSs is zero in both of these two cases. The output power of WTGSs increases with the increase of the wind velocity between the cut-in wind velocity and the rated wind velocity and maintains as the rated value after that. However, after the "saturation" of LOLE and EENS, the reliability of the power system is impossible to be improved by means of WTGS adding, even if the average wind velocity increases. Thus, from the point view of energy, adding massive wind power is possibly helpful; however, a considerable conventional generation capacity needs to be added too to keep an acceptable adequacy for the power system.

12.1.4.3 Impact of WF Correlation

It is a clear conclusion that the contribution of wind power to the reliability of a power system mainly depends on the wind conditions of WFs. If there is insufficient wind on the WFs, WTGSs will have no output power and make no contribution to the power system. However, it is unlikely that there is no wind on two independent WFs with a long distance at the same time. Thus, the WTGSs on independent WFs can improve the reliability of power systems. Compared with a single WF, the multiple independent or slightly related WFs with the same total capacity contribute more to the reliability of a power system.

12.1.4.4 Impact of Energy Storage Increase

The ESS of a WF can smooth the output power fluctuations of the WF, improving the continuity of power supply by the wind power system, and enhancing the controllability and capacity value of wind power and the adequacy of the power system.

For the combination of a WF and an ESS, different operation strategies can be adopted. The operation strategies guaranteeing a better reliability benefit for ESS and larger wind power storage capacities for ESSs should be selected.

The charge and discharge characteristics of an ESS influence its reliability benefit greatly. They limit energy storage and have a significant impact on the wind power, which can be stored.

12.1.4.5 Differences of the Contributions of WFs and Conventional Power Plants to the Reliability of Power Systems

The WTGS capacity needed to maintain the specified reliability of a power system is possibly much larger than a related conventional power generation system. Sometimes, it is impossible to maintain the specified reliability of a power system by means of increasing the wind power capacity of a single WF.

In a comparison about the contribution to the reliability of a power system done between a WF and a conventional power plant, the capability of wind power generation to replace conventional power generation is measured through capacity credit (CC). CC is a ratio between a WF's capacity for replacing a conventional power generation system and the rated capacity of the WF while the reliability of the power system is maintained (note that "CC" has different definitions; see section 12.2.1), namely (Formula (12.35)):

$$CC = \left.\frac{\Delta P_{CC}}{P_{WTC}}\right|_{LOLE = cons} \tag{12.35}$$

In the formula above, ΔP_{CC} is a WF's capacity for replacing a conventional power generation system while reliability of the power system is maintained; P_{WTC} is rated capacity of the WF; CC suggests capacity of the conventional power generation system saved after constructing the WF.

12.2 Wind Power Generation CC

12.2.1 Overview

Wind power generation has randomness, so the electric power operators originally regarded it as a completely unreliable power generation, which can provide energy only but not replace some generation capacity. But the wind power producers believe that there is no completely reliable power generation form and wind power generation and that the conventional power generation have no essential difference other than a difference in their equipment availabilities (quantities).

Wind power generation CC is also called available capacity, which is the creditable wind power capacity. It suggests the capability of the WF (plant) to improve power system reliability. If the reliability remains unchanged, a WF added can additionally meet the load undertaken by a conventional power generation system with a certain capacity. CC of the WF is a ratio between the load of the conventional power generation system and installed capacity of the WF. If an existing power system has an insufficient generation capacity, its load must be interrupted partially. If any new power plant is added into the power system, it will be possible that the users of the power system do not need to disconnect from the power system as usual, suggesting reliability improvement of the power system. Figure 12.6 shows changes of the total loss of load of a power system when the power system has wind power or not. Obviously,

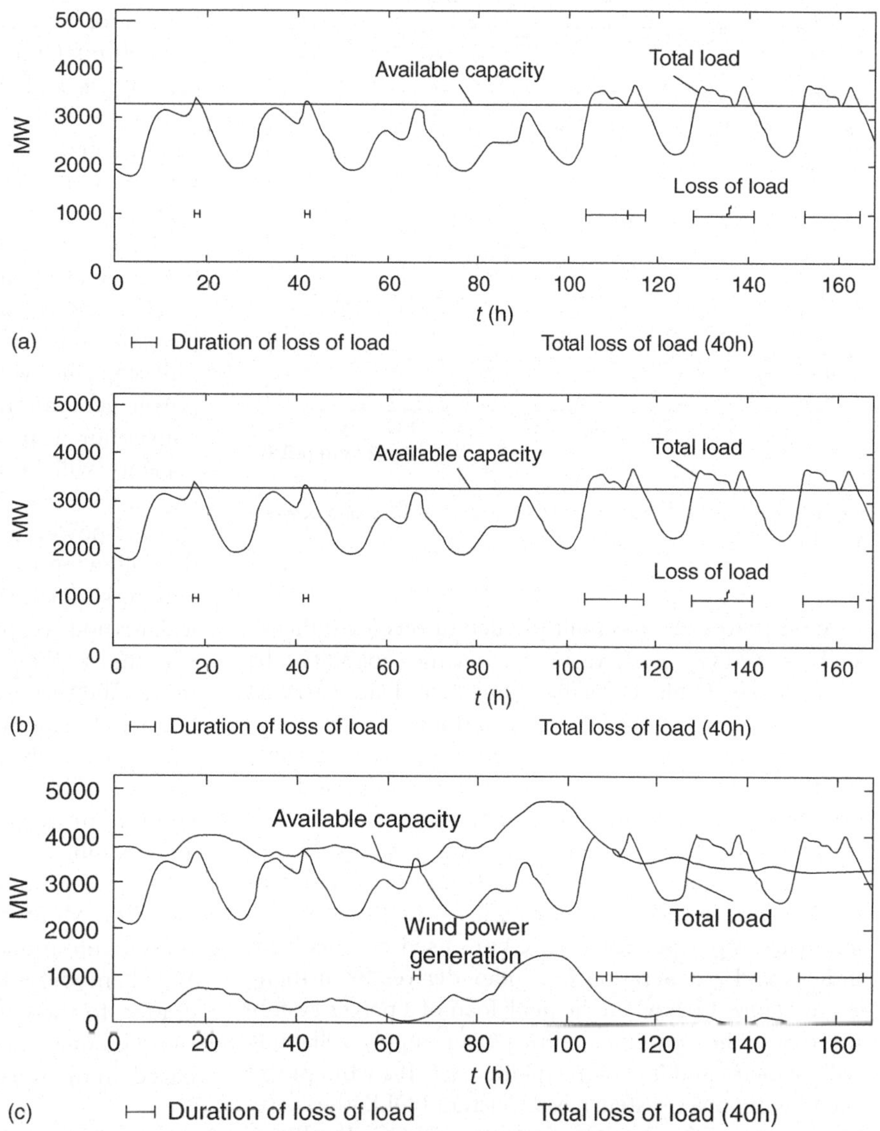

Figure 12.6 Example of CC. (a) CC when there is no wind power; (b) CC when there is wind power; (c) CC when there is wind power (the load is 300 MW larger than that in a).

when there is wind power, the power system has less total loss of load (Figure 12.6b) or, in other words, the power supply reliability is higher or, the power system has a larger total load while its reliability remains unchanged (Figure 12.6c) [8].

The studies about wind power CC are expected to answer the following questions: Can wind power generation replace other conventional power generation forms in power systems? How about the degree of replacement if it can? Can a power system containing wind power meet higher (peak) load demands? These questions relate to the long-time reserve or spinning reserve of power systems [9].

Wind power provides additional load capacities to meet the demand the increase of a power system planned. There is a capacity factor (a ratio P_m/C_{IC} between the average generation power and installed capacity of WTGSs of a WF in a period, or a quotient between the actual power generation capacity and possible maximum power generation capacity in a period), so CC is essentially related to the nature of the wind condition in each place. Wind power can contribute by at most 40% of its installed capacity (based on low wind power penetration and a high capacity factor in the peak load period). When there is a high wind power penetration or a low capacity factor in the peak load period, or when the local wind power output distribution and the system load distribution have negative correlation, it is possible that the wind power contributes to less than 5%. The wind power integration in large geographic regions results in a higher CC.

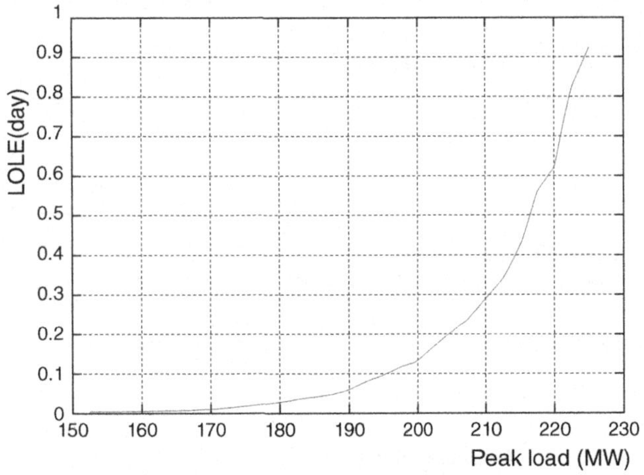

Figure 12.7 Curve of changes of LOLE with peak load of a power system.

Wind power CC has multiple definitions (or indices), such as WTGS effective load carrying capability, the equivalent reliable capacity, equivalent load carrying capability, the equivalent conventional power plant capacity, the guaranteed capacity, and safe capacity [9, 10]. However, most of the definitions (or indices) are based on power system reliability such as the loss of load probability (LOLP) of power systems. Power generation reliability evaluation integrates a power generation model and evaluation indices. LOLE and peak load carrying capability (PLCC) are used here as basic indices and a generation CC index derives from them. As in Figure 12.7, when the peak load of a power system increases, the power system's power supply reliability will decrease and its LOLE will increase. If a wind power integrated power system has a constant LOLE, the corresponding peak load will be defined as PLCC. The PLCC will increase to some extent after a power generation system with a certain capacity is added to the existing power system.

The CC of a WF can be compared in either method below: The comparison between the WF and a conventional power generation system or the comparison between the WF and an ideal power generation system (a completely reliable one). The former comparison method is used here. If a power system has a constant reliability index, the load capacity related proportional relation (ECR) between a WTGS and a conventional power generation system is used as the index for generation CC measurement. The definition of ECR is as below (Formula (12.36)):

$$ECR = \frac{PLCC_{wind} - PLCC_{org}}{PLCC_{conv} - PLCC_{org}} \tag{12.36}$$

In the formula above, $PLCC_{org}$, $PLCC_{wind}$ and $PLCC_{conv}$ are a power system's initial PLCC, PLCC after a WTGS with a certain capacity is added, and PLCC after a conventional power generation system with a certain capacity is added, respectively. According to the definition, a WTGS can replace a conventional power generation system proportionally and the power generation reliability of the power system after the replacement remains unchanged. The reliability provides a risk-based equivalent ratio between a WTGS and a conventional power generation system. If the reliability is 0.2 and a WTGS and a conventional power generation system have the same peak load capacity, the WTGS of one unit capacity will be equivalent to the conventional power generation system of 0.2 unit capacities, or the conventional power generation system of one unit capacity will be equivalent to the WTGS of five unit capacities.

Wind power generation CC can be discussed through the following two methods [11]: the analytical method and the simulation method. The complexity of the analytical method is exponentially related to the scales of power systems, so the simulation method is used increasingly more. The three-state WTGS model for WTGS failures and wind velocity variations can be used; wind velocity variations can be simulated with an auto-regressive moving model. These models give no consideration to the correlation between different WFs, so they have a limited application.

In the wind power generation CC evaluation method based on Monte Carlo simulation, and the wind energy production models includes a wind power model, a WTGS model, and a WF model, and we can give consideration to the correlation between different WFs. With these models, the generation CC can be then calculated based on the reliability index.

12.2.2 Wind Power Generation CC Evaluation

12.2.2.1 Wind Power Model

The motive power (wind power) of WTGS is uncontrollable and whether a WTGS is generating power or not and its output power both depend on the wind velocity, so a wind power model is the basis of generation CC calculation. The method to simulate the correlation between different WFs is a difficulty which is solved by a matrix transformation technology here [11, 12].

A given column vector $z\ (z_1, z_2, \ldots, z_n)^T$ can be changed into a new one as below by means of the matrix transformation (Formula (12.37)).

$$y = Lz + \mu \tag{12.37}$$

In the formula above, L and μ are randomly appointed triangular matrix and column vector. The mathematic expectation $E(y)$ and the incidence matrix Ω_y of vector

y have the following natures (Formulas (12.38) and (12.39)).

$$E(y) = LE(z) + \mu \tag{12.38}$$

$$\Omega_y = L\Omega_z L^T \tag{12.39}$$

Suppose a column vector z $(z_1, z_2, \ldots, z_n)^T$ and its mean μ_z $(0, 0, \ldots, 0)^T$ and covariance matrix are all known.

$$\Omega_z = \begin{bmatrix} 1 & 0 & \cdots & 0 \\ 0 & 1 & \cdots & 0 \\ \vdots & \vdots & \ddots & 0 \\ 0 & 0 & \cdots & 1 \end{bmatrix}$$

Then it will be easy to get a new column vector with the specific mathematic expectation and the correlation characteristic by means of a certain matrix change.

Steps of wind power simulation:

1) Obtain a wind velocity column vector without a correlation characteristic by means of Monte Carlo simulation in compliance with the wind power parameters (average wind velocity and variance) of different WFs;
2) Obtain column vector z (an intermediate variable) by deducting the mean wind velocity from the column vector and then dividing the difference by the wind velocity variance;
3) Decompose the known incidence matrix between different WFs and make Ω_y equal to LL^T to obtain transformation matrix L;
4) Obtain a wind velocity vector ($y = Lz + \mu$) with a correlation characteristic by means of matrix transformation in line with Formula (12.37).

12.2.2.2 WTGS and WF Model

The output power changes of a WF are caused by wind velocity variations, wind direction changes, WTGS outage because of failures, and so on. Suppose the WTGSs on a single WF have almost the same wind velocity and wind direction, it can be believed that all the WTGSs on the WF have the same wind velocity and wind direction. On this basis, the output power of each WTGS can be calculated according to the power characteristic curve of the WTGSs and the output power of the whole WF will be sum of the output power of all the WTGSs. On a big WF, the WTGSs with almost the same wind velocity and wind direction can be divided into a group and the output power of each group of WTGSs can be calculated. In this case, the output power of the whole WF will be sum of the output power of all groups of WTGSs.

Whether a WTGS is generating power or not and its output power both depend on the wind velocity. The power characteristic curve of WTGSs is described through Formula (12.17) here.

WTGS outage because of failures is simulated as below through a WTGS outage model: Each WTGS has a certain failure rate. When it is maintained, its output power is zero; when it operates, its output power depends on wind velocity of the WF.

12.2.2.3 Reliability Index Calculation

For any power system, its reliability index and peak load have the following relation (Formula (12.40)):

$$LOLE = f\,(PLCC) \tag{12.40}$$

However, there is no accurate expression for f, so massive Monte Carlo simulation will need to be done to calculate the LOLE based on a known PLCC. Calculation of the PLCC based on a known LOLE will be much more complicated. To solve this problem, a simplified Newton method is used here for iterative solution of the PLCC based on a known LOLE. Basic steps of the iterative solution are as below [11]:

1) **Determine upper and lower limits of the *PLCC***
   Regard initial $PLCC$ of the power system $PLCC_{\text{down}}$ and make $PLCC_{\text{up}}$ equal to the sum of $PLCC$ and P_{add} (P_{add} is the generation capacity added into the test system).
2) **Calculate the slope**
   Calculate $LOLE_{\text{up}}$ and $LOLE_{\text{down}}$ by means of Monte Carlo simulation based on the known $PLCC_{\text{up}}$ and $PLCC_{\text{down}}$. Slope K can be calculated through the formula below (Formula (12.41))

$$K = \frac{LOLE_{up} - LOLE_{down}}{PLCC_{up} - PLCC_{down}} \tag{12.41}$$

3) **Carry out iterative computation**
   Obtain $PLCC$ by means of iterative computation based on Formula (12.42).

$$PLCC_{next} = PLCC_{ini} - (LOLE_{ini} - LOLE_{down})/K \tag{12.42}$$

   In the formula above, $PLCC_{\text{ini}}$ and $PLCC_{\text{next}}$ are load capacity of the power system before and after the iterative computation, respectively.
4) **Check the iteration completion identifier**
   If the difference between $LOLE_{\text{next}}$ and $LOLE_{\text{ini}}$ is equal to or less than ε (completion identifier ε is set as 0.01 d), the iteration will end, otherwise, the iteration in step 3 will continue.

12.2.2.4 Procedure of Wind Power Generation CC Evaluation

The model and evaluation indices for power generation are introduced above. The basic procedure of wind power generation CC evaluation is introduced below.

1) Calculate $PLCC_{\text{orig}}$ and corresponding $LOLE_{\text{orig}}$ of the power system by means of Monte Carlo simulation.
2) Calculate the increased $PLCC$ based on a known reliability index after a WTGS of a certain capacity is added into the power system.

3) Calculate the increased *PLCC* based on a known reliability index after a conventional power generation system of the same capacity as the WTGS are added into the power system.
4) Calculate the generation CC index ECR based on the results obtained in steps 2 and 3 and Formula (12.36).

12.2.3 Example and Analysis

12.2.3.1 Example Overview

Take the IEEE Reliability Basic Test System (RBTS) [13] as a example. The system has 11 conventional power generation systems with a unit capacity of 5 MW to 40 MW, a total installed capacity of 240 MW, and a maximum load of 185 MW.

The system has eight WFs (100 WTGSs with a total capacity of 22.5 MW). Of each WTGS, the unit capacity is 225 kW and the forced outage rate is 0.05. Table 12.1 and Table 12.2 introduce wind velocity characteristics and incidence matrix of the WF, respectively [11].

For CC calculation, the power generation reliability change of the system after a 10 MW conventional power generation system is added into the system and that after a 12.5 MW conventional power generation system is added into the system are both analyzed.

12.2.3.2 Calculation Result

Figure 12.8 shows change trend of the reliability (LOLE) after 100 WTGSs are added into the standard test system in sequence. According to the figure, power generation reliability of the standard test system has been improved significantly after wind power is added.

Table 12.1 Average wind velocity of WF.

WF	1	2	3	4	5	6	7	8
Average wind velocity (m/s)	7.5	7.5	7.5	7.5	6	6	7.5	6

Table 12.2 Incidence matrix of WF.

WF	1	2	3	4	5	6	7	8
1	1.0	0.2	0.2	0.0	0.0	0.0	0.0	0.0
2	0.2	1.0	0.8	0.2	0.0	0.0	0.0	0.0
3	0.2	0.8	1.0	0.0	0.0	0.0	0.0	0.0
4	0.0	0.2	0.0	1.0	0.0	0.0	0.9	0.0
5	0.0	0.0	0.0	0.0	1.0	0.0	0.0	0.0
6	0.0	0.0	0.0	0.0	0.0	1.0	0.0	0.0
7	0.0	0.0	0.0	0.9	0.0	0.0	1.0	0.0
8	0.0	0.0	0.0	0.0	0.0	0.0	0.0	1.0

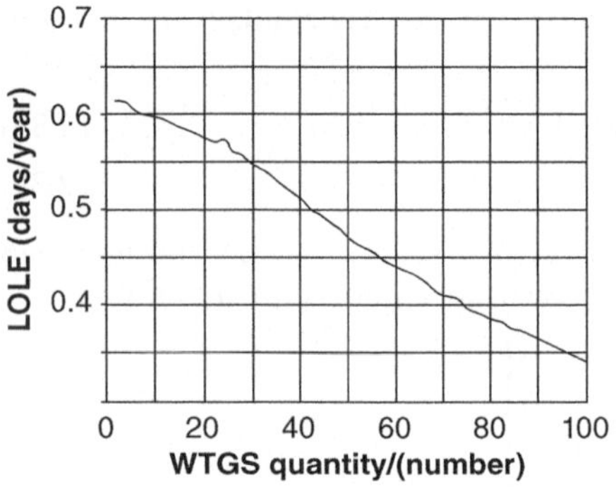

Figure 12.8 Curve of changes of LOLE with WTGS quantity.

Table 12.3 Generation CC.

25 MW wind power PLCC increment (MW)	25 MW conventional power generation system PLCC increment (MW)	CCECR (%)
16.15	6.21	38

Table 12.3 gives the wind power CC corresponding to a LOLE of 0.1275 d/a of the RBTS. According to the table, a 3.8 MW conventional power generation system and a 10 MW WTGS make the same contribution to power generation reliability improvement of the standard test system.

Note: Different regions have diverse grid structures and wind power sources, so the calculation results about the wind power generation CC in different regions may be different.

12.2.3.3 Other Research Results

Figure 12.9 introduces some research results about wind power CC investigated and summed up in reference [9]. According to the figure, the research results are decentralized. A reason of this phenomenon lies in wind conditions and WTGS distances of the WFs studied, which is an explanation of the low wind power CC in Germany (curve 1). For wind power penetration close to zero, all CCs are within the capacity factor ranges of the WFs evaluated. The correlation between wind and loads is useful. One example of this is that the CC of offshore WFs in New York is 40% (curve 9). The wind power CC evaluation results are quite different in different countries and different wind power CC evaluation methods.

In documents such as references [14, 15], some valuable conclusions are reached after a comparison of different wind power CC evaluation methods.

Selection of the CC definition significantly influences the CC obtained. In a single power system, the maximum CC may higher than the minimum CC by 20% at most

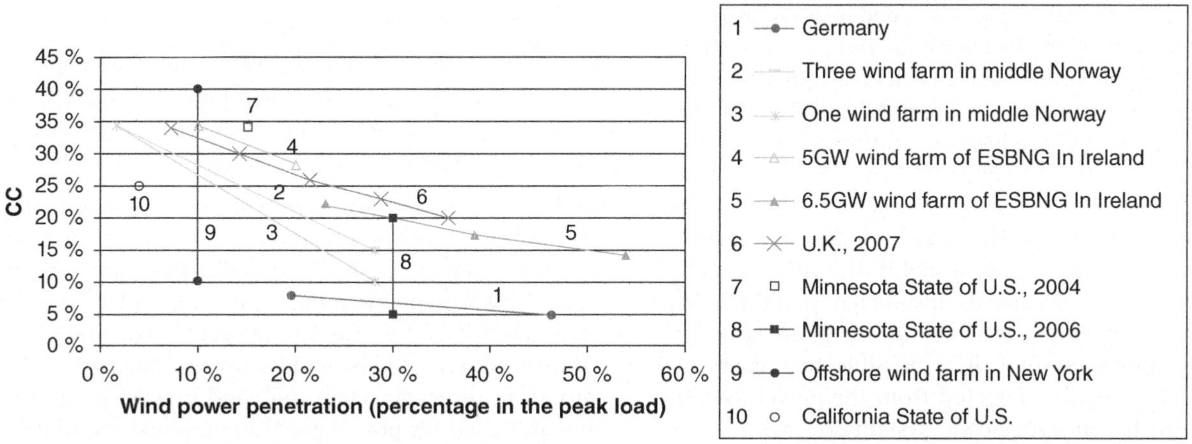

Figure 12.9 Wind power CC.

for the conventional power generation systems or by 30% at most for the WTGSs. However, it can be said that the three methods based on loss of load probability (LOLP) (i.e., equivalent reliable capacity, equivalent load carrying capability, and equivalent conventional capacity) are very consistent. Equivalent reliable capacity and equivalent load carrying capability have no large differences and the only difference is the latter results in larger results. The three definition methods are all based on the way the power generation system finally added influences the whole power generation adequacy (expressed in LOLP of the power system).

The CC definition based on LOLP indicates obvious correlation between LOLP and CC of a power system. If a power generation system is added into a power system with a high LOLP, CC of the power system will tend to be larger (contribute more to the power generation adequacy), but it is not necessary that a power system with a high LOLP has a larger CC than a power system with a low LOLP.

In a power system, the wind power penetration factor has high correlation with CC too, but it is less influential than LOLP. The wind power penetration factor is a ratio between installed capacity $\hat{G}_g$ of the power generation system added and sum of installed capacity $\hat{G}_{tot}$ of the existing power generation system and installed capacity of the power generation system added, namely (Formula (12.43)):

$$\gamma = \frac{\hat{G}_g}{\hat{G}_{tot} + \hat{G}_g} \tag{12.43}$$

In either conventional power generation or wind power generation, CC is related to the penetration factor. On each LOLP level, the CC corresponding to a lower penetration factor is larger than the CC corresponding to a higher penetration factor.

In either a conventional power generation system or a WTGS, CC is lower than the available generation capacity expectation; the available generation capacity expectation in a WTGS is lower than that in a conventional power generation system. However, CC is surely larger than zero. When a WTGS and a conventional power generation system with almost the same annual energy output are compared, CC of the WTGS is usually lower than that of the conventional power generation system.

In a research about wind power CC, the CC definition (or index) selected must be proper. For example, it is proper to be equivalent reliable capacity or equivalent load carrying capability.

12.3 Wind Power Value Analysis

The power plant in a power system aims at providing users with electricity in an economical, reliable, and environment-friendly way. Different power plants can realize this objective through different methods. This section will introduce different requirements for power plants and the way to meet them through wind power generation.

Wind power generation has the following values in power systems: operation cost value, CC value, control value, grid loss reduction value, grid investment value, pollutant emission reduction value, and so on [16].

12.3.1 Values of Conventional Power Plants

From the point view of their control in power systems, different power plants have different characteristics. Usually, the values of a new power plant are believed as marginal values of the other power plants in the existing power system. New power plants have the following values:

Operation cost value: It is the capability of a new power plant to reduce the operation cost of an existing power system. If a new power plant is added into a power system, it will provide electricity for the power system, suggesting generation capacity decrease of the other power plants in the power system. As a result, the other power plants will have lower operation costs.

CC value: CC is the capability of a new power plant to improve the reliability of a power system. The existing power system always has a risk of capacity insufficiency, that is, the so-called LOLP. The risk is usually low. In the case of capacity insufficiency, some loads must be interrupted. If a new power plant is added into the power system, it is possible that users do not need to be disconnected from the power system frequently because the power system now has a larger installed capacity. It suggests that the new power plant has improved reliability of the power system.

Control value: It relates to the capability of a new power plant to track load demands. Continuous control is necessary in a power system because the total power generation capacity must be equal to the system load (including its loss). The ever-changing load makes it necessary for power generation to keep changing. Different power plants have different control capabilities and control values. If a new power plant results in a larger demand for system control, its control value will possibly be negative.

Grid loss reduction value: It relates to the capability of a new power plant to reduce the loss of a grid. Power transmission surely results in a grid loss. Long-distance or low-voltage power transmission results in a high grid loss. Compared with an existing power plant, a new power plant can reduce loss of the power system if it is closer to users because there is less power transmitted in the grid. It means that the new power plant has a special value related to a capability of grid loss reduction. If the grid loss reduction value is negative, it means that the new power plant has resulted in a larger loss of the power system.

Grid investment value: It relates to the capability of a new power plant to reduce the grid investment of a power system. If the new power plant is close to users with increasingly larger demands, it will possibly decrease the demand for new grid investments, showing a special value related to the power plant. If the new power plant has increased the demand for new grid investments, the grid investment value will be negative.

Emission reduction value: A WTGS added can replace conventional power generation systems emitting pollutants. The emissions of CO_2 and SO_2 depend on the contents of carbon and sulphur in fuels and quantities of the fuels burned. Thus, less thermal power units will result in less emissions of CO_2 and SO_2.

12.3.2 Values of WFs

The analysis above about the values of power plants applies to the value analysis of wind power too. Suppose a system is an ideal economic system, its value lies in reduction of the total cost.

Wind power operation cost value: It relates to the capability of wind power to reduce the operation costs of other electric sources. Actually, this value means that 1 kWh generated by wind can replace 1 kWh generated by other electric sources. Hydropower is almost not replaced by wind power because the water stored in reservoir when the wind power is high can be later used for power generation. However, there is an exception that hydropower must have a full output during high flows to avoid overflow. When there are a low load and a gale, overflow is possibly necessary. The surplus electric power in this circumstance, which may be rare can be sold to other neighboring systems.

In a thermal power plant, the power generation systems with lower operation costs are used earlier than those with higher operation costs. It means that WTGSs usually replace the thermal power units with higher operation costs because an objective of power systems is to operate with the minimum costs.

Usually, wind power generation can replace the thermal power generation based on fossil fuels. The replacement can be direct: wind power generation can reduce the power generation capacities of thermal power plants. The replacement can be indirect too in a good wind condition: water is stored by wind power and then used for replacing the power generation based on fossil fuels.

Wind power CC value: It relates to the capability of wind power generation to improve the reliability of power systems. CC decreases with increase of the wind power penetration. Different wind power systems have different CCs. Usually, the wind power systems with lower wind power penetration have higher wind power CCs and those with higher wind power penetration have lower wind power CCs. Thus, it is believed that the capacity benefit of wind power generation is saving of the necessary costs of a necessary additional conventional power plant, including investment cost, operation cost, maintenance cost, and so on.

Wind power control value: It relates to the capability of wind power to take part in balancing the power generation and consumption of power systems. Electric energy cannot be stored, so it must be generated accurately according to the electric energy to be consumed. The handling way of power balance depends on the time frame.

1) **Primary control**
   Primary control aims at keeping a balance between power generation and power consumption. The time frame is about 1 minute. In primary control, it may be supposed as below: most of generators in a power system are normally synchronous and have strong coupling between their shaft speeds and their electrical frequencies, so increase of their shaft speeds will result in increase of their electrical frequencies. It means that excessively less electric power of the power system will result in a lower electric frequency because extra power needs to be obtained from rotational energy of the synchronous motors. After the rotational energy is used up, both the shaft speeds and the electrical frequencies will decrease. In some power plants, their power generation capacities will be increased after their electric frequencies are found decreasing in measurement till their electric frequencies become stable. This method is for short-time power balance for power systems.
   A WF with cage asynchronous generators and extra-control double-fed asynchronous generators or direct-drive synchronous generators provides certain inertial response for the frequency changes of a power system or, in other words, it contributes to primary control. However, the power system therefore has wind power fluctuations, possibly resulting in additional fluctuations of its frequency. For example, if the proportion of wind power is large, a primary control reserve margin will possibly need to be added for wind power fluctuation. Of course, this reserve margin is small, but the reserve margin for load fluctuations is always large.
   The fluctuations of another kind refer to sudden decrease of wind power generation due to sudden trip off of WTGSs caused by factors such as storms. In this case, many WTGSs of the power system possibly have wind velocities quicker than the cut-out wind velocity (such as 25 m/s). However, the power system is unlikely to suffer disconnection of 1,000 MW wind power, so the reserve margin does not need to be increased at all or it needs to be increased slightly. Because 1,000 MW wind power is actually generated by 500 WTGSs (each WTGS generates 2 MW wind power) and it is extremely unlikely that all the 500 WTGSs suffer a storm in one minute.
   Thus, the primary control value of wind power generation is probably a negative one, which is very small.
2) **Secondary control**
   Secondary control refers to the capability of a power system to keep a balance between its power generation and consumption within one hour following the primary control. Because of the primary control, frequency of the power system is probably no longer its rated value (50 Hz), so a reserve margin is needed. In the case of frequency decrease, this is managed by increasing the generation capacities of some power generation systems. It means that some power generation systems must be used as secondary backup power plants and started for power generation upon an excessively low frequency.
   Usually, wind power makes no contribution to secondary control, and an additional secondary backup margin is needed. The reason is that wind power will probably results in an additional load shift lasting for 1 hour. It means that the secondary control value of wind power is negative because the secondary reserve capacity needed to be increased for wind power is much larger than the reserve margin needed in primary control. The increase has the following two reasons: the wind velocity varies more in 1 hour than in 10 minutes; the wind velocity variations of some WFs in different positions on a single wind zone have high correlation.
3) **Daily control and weekly control**
   The daily control has a stricter requirement because the power consumption in daytime is larger than that nighttime. If the wind velocity of the location of a WF is quicker in daytime than nighttime, wind power will be in positive correlation with the load and contribute to necessary daily control of the power system; if the wind velocity of the location of a WF is quicker in nighttime than daytime, wind power will be in negative correlation with the load and daily control value of the WF will be negative.
   With increase of the daily (weekly) load fluctuations of the power system, the other power generation systems must be capable of controlling in a larger range for a purpose of efficiency reduction. The daily (weekly) control value should be discussed with high load, low wind velocity and high hydropower, or high load, high wind velocity, low hydropower, and compared with the situation when there is no wind power. According to the result, the daily (weekly) control value of wind power is a negative one, which is small, which is true at least when there is a little wind power.
4) **Seasonal control**
   A power system with a reservoir should be subject to quarterly control, namely the reservoir may

store water from one season to another season. It is important to plan and dispatch ecumenically in the power systems of this kind. In the process, the uncertainty of water must be considered.

In the four seasons of a year, the load variation mode and the wind power fluctuation mode may be the same or opposite. If they are the same, wind power will have a positive control value; if they are opposite, wind power will have a negative control value.

5) **Multi-year control**

In a power system in which hydropower has the largest proportion, multi-year control is necessary because the hydropower plant has different inflows in different years. It suggests the necessity to store water in different years. However, it is important to determine the quantity of water to be stored according to knowledge about water uncertainty because the inflows in future are uncertain. Particularly, it is possible that the load demands cannot be met in dry years. In this case, the thing concerned is that whether the WF can generate electric power in dry years. The important thing in this case is to consider correlation between the river inflows and the WF's wind power generation capacity. If there is correlation between the river inflows and the WF, it means that the wind power in dry years is slightly lower than the multi-year mean. Compared with thermal power plants, WFs show adverse impacts in their multi-year control requirements (because thermal power plants have the same possibilities in all years). However, a new WF is better than a new hydropower plant because the correlation between a new hydropower plant and an old one is much higher than the correlation between a new WF and a new hydropower plant. It is because that the new hydropower plant is close to the old one, so their inflows have high correlation; the new WF is not necessary to be in the same area as the hydropower plants, and the correlation between the wind in an area and the inflows in another area is limited.

Grid investment value: It relates to the capability of wind power generation to reduce the grid investment of a power system. The necessary things before this value is explored are to figure out the method to select a grid when there is no wind power and to the way to change grid selection when there is wind power generation. It is very difficult to draw a general conclusion from it. An important thing is that grid construction needs an investment, which is included in the WF investment. For a medium-sized offshore WF, cost of the power transmission line between it and the related power transmission system is included in total cost of the WF. If there are no other grid investments, this grid investment will probably be small.

Emission reduction value: Wind power is a clean energy source and emits neither CO_2 nor SO_2 as thermal power plants do. Thus, its emission reduction value is important in wind power value analysis.

12.3.3 Estimation of Grid Loss Reduction by Wind Power

In the analysis of the grid loss reduced by wind power, the issue concerned is the capability of wind power generation to reduce the grid loss of a power system. WFs are usually located close to the distribution grids, which are far from transmission grids and have smaller voltages. This is different from large power plants, which are usually close to transmission grids. If the load close to a WF has the same scale as the WF, the distance of power transmission will not be long and the grid loss will be reduced by means of wind power generation. In this case, the grid loss reduced by wind power generation is positive. However, if there is no load close to a WF, the WF will have to transmit power to a load center and there will be a large grid loss in the course. In this case, the rid loss reduced by wind power generation is negative.

Maximum loss variations are involved here. The power loss on the line impedance R+jX is approximately (Formulas (12.44) and (12.45)):

$$P_L = \frac{P^2+Q^2}{U^2}R = \frac{1+\tan\varphi}{U^2}P^2R \qquad (12.44)$$

$$Q_L = \frac{P^2+Q^2}{U^2}X = \frac{1+\tan\varphi}{U^2}P^2X \qquad (12.45)$$

In the formulas above, P_L and Q_L are active power and reactive power of the line impedor, respectively; P and Q are active power and reactive power of the power transmission line; U is the line voltage; φ is the phase difference between the voltage and the current, that is, power factor angle of the transmission power.

Set $k = R\left(\frac{1+\tan\varphi}{U^2}\right)$ and suppose U and φ are constants, the active power loss will be in direct proportion to square of the transmission power. Suppose the initial transmission power of the line is P_0, the power loss of the line P_{LO} will be kP_0^2 and the line loss percentage will be (Formula (12.46)):

$$P_{LO}(\%) = 100\frac{P_{LO}}{P_0} = 100kP_0 \qquad (12.46)$$

If the transmission power and the grid loss decrease to P_1 and P_{L1}, respectively, because of the local wind power, the grid variation will be small. Thus, the line loss is approximately (Formula (12.47)):

$$P_{L1} \approx P_{LO} - \frac{dP_L}{dP}(P_0-P_1) = P_{LO}-2kP_0(P_0-P_1) \qquad (12.47)$$

Suppose the local power generation ΔP is the difference of P_0 and P_1 and the grid loss decrement ΔP_L is the difference of P_{L0} and P_{L1}, the grid loss percentage will be (Formula (12.48)):

$$\Delta P_L\% = 100\frac{\Delta P_L}{\Delta P} = 100\frac{P_{L0} - P_{L1}}{P_0 - P_1} \approx 100 \times 2kP_{L0} = 2P_{L0}(\%) \tag{12.48}$$

It means that the maximum loss doubles the mean loss.

Increases or decreases the loss of a power system by wind power generation depends on position of the WF, position of the load and position of the power sourced replaced by the WF. Wind power can increase or decrease a grid loss. The important thing is to consider the overall grid loss rather than partial grid loss of the power system.

12.3.4 Value Analysis for Grid-Connected Operation of Wind Power

In reference [17], wind power generation is regarded as a negative load; the operation cost variations of power systems as a result of connection of wind power to grids are analyzed with the reliability evaluation method. On this basis, a formula for wind power operation value calculation is proposed and the impacts of factors such as wind power penetration, wind energy forecast deviation, and WF capacity factor on wind power operation values are discussed through a example.

The actual load and forecast load of an actually operating power system have a deviation. For safe operation reliable power supply of the power system, the power operator needs to adjust power generation plan of the power system in time according to result of the super-short-time load forecast and provide a sufficient spinning reserve capacity and a great load tracking capability to make up sudden load variations and unexpected power unbalance. On the power market, grid companies purchase backup capacities as auxiliary services from power generating companies and pay capacity prices and on-grid prices for them [18].

At present, the power dispatching authorities do not interfere in wind power when compiling power generation plans and, as a result, the generation power of WTGSs depends on the current wind velocities completely. Based on this dispatching rule, combination of the random fluctuations of wind power and the fluctuations of loads results in more uncertain factors, increasing the auxiliary service costs of power systems. A part of the electricity on-grid prices saved because of grid connection of wind power is used for paying the increased auxiliary service costs. If there are no preferential policies for wind power generation, the increased auxiliary service costs will probably be higher than the electricity on-grid prices saved because of grid connection of wind power [19].

Thus, wind power is a kind of special power with randomness and uncertainty. To make the best use of wind power generation equipment and maintain the safe and reliable operation of power systems, the operation modes of power systems need to be properly adjusted. It results in additional expenses. When wind power operation values are calculated, the electricity on-grid price saved and the auxiliary service cost caused both because of grid connection of wind power should be considered.

12.3.4.1 Wind Power Operation Values

1) **Income of wind power generation**

   At present, power dispatching authorities make it necessary for grids to purchase wind power unconditionally and implement protective on-grid prices for wind power when compiling power generation plans. Based on this dispatching rule, the power generation income from the grid connection of wind power in a period t is (Formula (12.49)):

$$f_1(C) = C \times r \times P_{vend} \times t \tag{12.49}$$

   In the formula above, C is the installed capacity of WTGSs; r is capacity factor of the WF; P_{vend} is on-grid price for the wind power.

2) **Power generation cost**

   The on-grid prices of wind power are higher than conventional power generation systems, but the growth of wind power generation is curbed by its high cost prices. A cost price is the minimum on-grid price necessary for the operation of a WF, including construction cost, operation cost, and loan repayment cost of the WF. The power generation cost of grid connection of wind power in a period t is (Formula (12.50)):

$$f_2(C) = C \times r \times P_{\cos t} \times t \tag{12.50}$$

   In the formula above, $P_{\cos t}$ is the cost price of wind power generation; the other items are the same as those in Formula (12.48).

3) **Additional expenses of grid connection of wind power**

   The difference between a power system's power capacity and generation load is called reserve capacity of the power system. For safe, reliable and continuous power supply by grids, power operators make it necessary to reserve backup capacities when compiling power generation plans. A grid company needs to pay a capacity price and an on-grid price for an auxiliary service purchased by it, so the additional expenses caused by grid connection of wind power include a capacity price and an on-grid price too.

Capacity price of auxiliary service: Medium- or short-term backup capacities include ones for load operation and ones for accidents. The backup capacities for load operation refer to the backup capacities set to regulate medium- or short-time load fluctuations of power systems to stabilize the cycles of power systems and bear the unexpectedly increased loads. They are usually 2% to 5% of the maximum generation loads. The lower backup capacities for load operation apply to larger systems and the higher backup capacities for load operation apply to smaller systems. The backup capacities for accidents are necessary for keeping power users from the impacts of unexpected accidents and maintaining the normal power supply of power systems. They are usually about 10% of the maximum generation loads and not smaller than capacities of the largest power generation systems of power systems.

In the case that the wind power of a WF is completely connected to a grid, generation power of the WTGSs on the WF depends on the current wind velocity completely. In actual operation, WTGSs are regarded by power dispatching authorities as negative loads rather than power generation equipment with controllable power. Thus, according to the classification and definitions of backup capacities, wind power generation does not change the backup capacities for accidents but increase the backup capacities for load operation.

According to the statistical data in document [20], the probability that a load fluctuates randomly close to its forecast value has normal distribution and the variance relates to the accuracy of load forecast. According to the probability density function about the fluctuations of a load close to its forecast value in Figure 12.10, the *LOLP* caused by the load fluctuations is (Formula (12.51)):

$$LOLP = \int_{P_1}^{\infty} \frac{1}{\sqrt{2\pi}\sigma_{load}} e^{-\frac{P^2}{2\sigma_{load}^2}} dP \quad (12.51)$$

In the formula above, σ_{load} is standard deviation of the fluctuations of the load close to its forecast value; P_1 is the reserve capacity for load operation of the power system, which is product of total load of the power system and a fixed percentage.

After wind power is added, the standard deviation of total load of the power system close to its forecast value is (Formula (12.52)):

$$\sigma_{new} = \sqrt{\sigma_{load}^2 + \sigma_{wind}^2} \quad (12.52)$$

In the formula above, σ_{wind} is the forecast standard deviation of wind power generation capacity.

To correspond to the new standard deviation of load forecast and maintain the given LOLP, the corresponding reserve capacity for load operation can be calculated through the normal distribution table (Formula (12.53)).

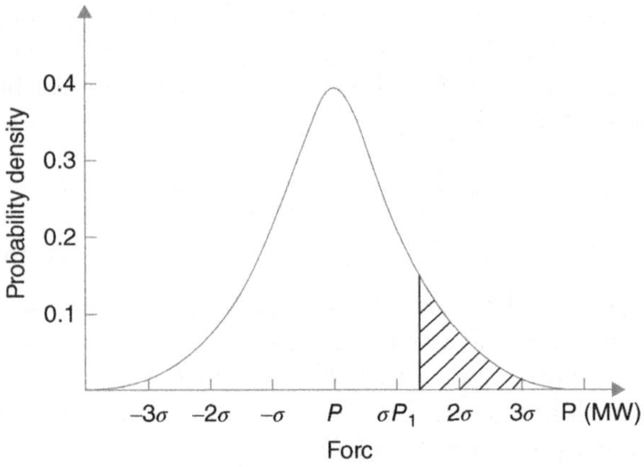

Figure 12.10 Probability density function about load fluctuation.

$$P_2 = \sigma_{new}\Phi^{-1}(1 - LOLP) \quad (12.53)$$

In the formula above, Φ^{-1} is an inverse function about standard normal distribution.

According to the formula above, if the reserve capacity price of a power system is P_c, the reserve capacity price increased in a period of t because of grid connection of wind power will be (Formula (12.54)):

$$f_3 = (P_2 - P_1)P_c t \quad (12.54)$$

On-grid price of auxiliary service: It should be calculated based on the probability after a reserve capacity is called. This probability depends on the accuracy of load forecast. The standard deviation percentage σ is basically in inverse proportion to the power system capacity. For example, σ is about 3% for a power system with a capacity of 2,000 MW, or about 1% for a power system with a capacity of 2,0000 MW [20].

Figure 12.11 shows the density function of a load forecast deviation. According to the figure, the probability p_i for a load to be between the sum of L_{forc} and L_i and the sum of L_{forc}, L_i and ΔL_i is as below (Formula (12.55)):

$$p_i = \int_{L_{forc}+L_i}^{\infty} \frac{1}{\sqrt{2\pi}\sigma_{load}} e^{-\frac{L^2}{2\sigma_{load}^2}} dL - \int_{L_{forc}+L_{i}+\Delta L_i}^{\infty} \frac{1}{\sqrt{2\pi}\sigma_{load}} e^{-\frac{L^2}{2\sigma_{load}^2}} dL \quad (12.55)$$

In the formula above, L_{forc} is the forecast load; σ_{load} is the standard deviation of load forecast.

Thus, when a reserve capacity price is known, the on-grid price for the load increment not in

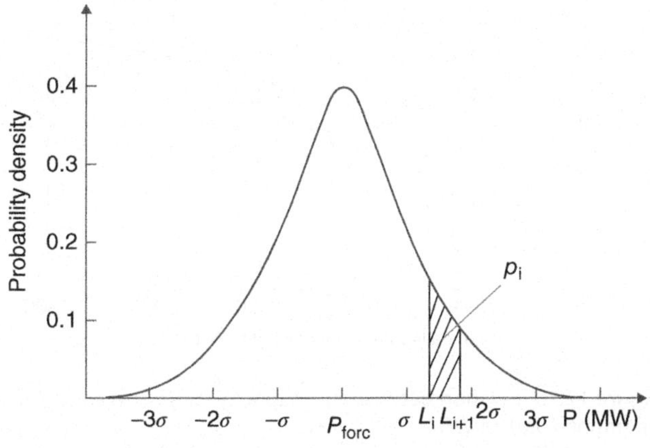

Figure 12.11 Probability density function about load forecast deviation.

the forecast section i in Figure 12.11 is as below (Formula (12.56)):

$$E_i = \beta(\Delta L_i)p_i\Delta L_i \tag{12.56}$$

In the formula above, $\beta(\Delta L_i)$ is the reserve capacity price and relates to the reserve capacity purchased. Usually, the larger the reserve capacity purchased, the larger $\beta(\Delta L_i)$ will be.

Suppose the reserve capacity is P, a grid company will need to pay the following on-grid price in a bid to maintain an electric quantity balance for a power system (Formula (12.57)).

$$E(P) = \sum_{i=1}^{m} \beta(\Delta L_i)p_i\Delta L_i \tag{12.57}$$

In the formula above, m is segments of the reserve capacity and P is the sum of ΔL_1, ΔL_2, and so on till ΔL_m.

Suppose the backup capacities of a power system before and after grid connection are both known, the reserve capacity price increased in a period of t because of grid connection of wind power will be (Formula (12.58)):

$$f_4 = (E(P_2) - E(P_1))t \tag{12.58}$$

In the formula above, P_1 is original reserve capacity of the power system; P_2 is the new reserve capacity of the power system after grid connection of wind power.

4) **Environmental benefit of grid connection of wind power**
The environmental benefit of grid connection of wind power is a fruit of replacement of conventional energy sources by wind power. Conventional energy sources such as coal and petroleum emit pollutants during power generation and countries all over the world are strengthening their control over pollution emission. However, the environmental cost evaluation of pollutant emission is complicated and definite conclusions have not yet reached for many aspects of it. At present, the governments of countries all over the world measure this environmental cost with some simple economic indices only. For example, the Danish government offers a carbon tax subsidy of 0.1 DKK/kWh (or 0.1 yuan/kWh) to WF operators directly. A similar method is used here: Suppose the average waste gas emission of a conventional power source power plant replaced is V (kg/kWh) and the cost of emission of 1 kg waste gas is P_{pol} yuan, the emission cost (power generation cost) saved by a grid-connected WF replacing a fossil fuel power plant in a period of t will be (Formula (12.59)):

$$f_5(C) = C \times r \times V \times P_{pol} \times t \tag{12.59}$$

In the formula above, C is installed capacity of the WF; r is capacity factor of the WF.

5) **Wind power operation value**
To sum up, wind power is a special power. The analysis of its overall value is complicated and overall and should be based on factors such as power generation income, power generation cost, auxiliary service cost, and environmental benefit.
Therefore, wind power value V_p can include the following parts (Formula (12.60)):

$$V_p = f_1 - f_2 - f_3 - f_4 + f_5 \tag{12.60}$$

In the formula above, f_1 is power generation income of grid connection of wind power; f_2 is power generation cost of grid connection of wind power; f_3 and f_4 are capacity price and on-grid price of the auxiliary service; f_5 is the environmental benefit of wind power generation.

12.3.4.2 Example and Discussion

Suppose there is a power system with a capacity of 20 GW; the standard deviation σ of its load forecast is 200 MW [21]; the reserve capacity for load operation is 4% of the maximum load; the average electric quantity price β for the reserve capacity is 280 yuan/MWh; capacity price P_c of the reserve capacity is 112 yuan/MW; the on-grid price P_{vend} and cost price P_{cost} of wind power are 700 yuan/MWh and 500 yuan/MWh, respectively; the average waste gas emission of the conventional energy source power plant replaced is 0.98 kg/kWh; the emission cost of 1 kg waste gas is 0.04 yuan. Additionally, suppose the power manufacturer plans to install some WTGSs to save power generation cost of the conventional energy source power plant and reduce the pollutant emission. The impacts of different factors on the wind power operation values are explored as below [17].

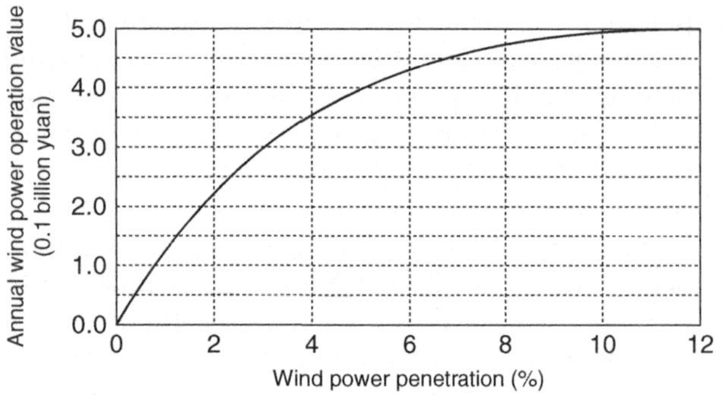

Figure 12.12 Relation between wind power penetration and annual operation value.

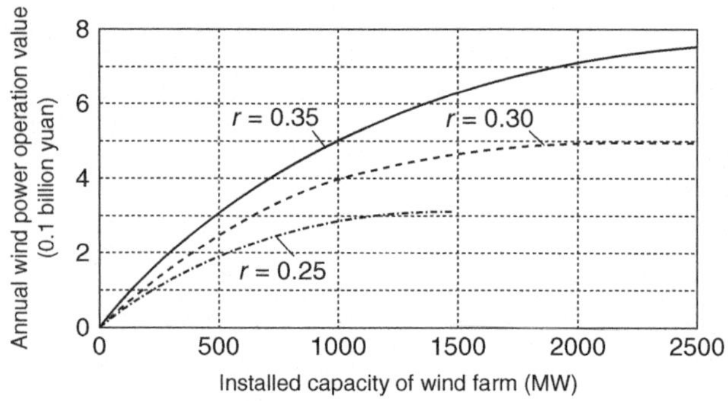

Figure 12.13 Relation between the capacity factor and annual operation value of WF.

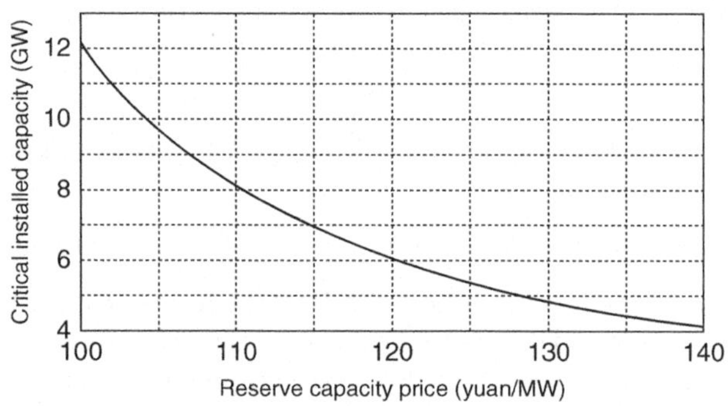

Figure 12.14 Relation between reserve capacity price and critical installed capacity.

1) **Impact of wind power penetration on wind power operation values**
   Wind power penetration is a ratio between the installed capacity of a WF and the total load of a power system. When the power system has a fixed total load, a larger wind power penetration corresponds to a higher auxiliary service cost.
   Figure 12.12 shows the relation between wind power penetration and a WF's annual operation values when a power system has a fixed total load and the standard deviations of capacity factor and wind power forecast of the WF are 0.3 and 0.2, respectively. Obviously, when the WF has a small installed capacity, the wind power operation values and the wind power penetration are in direct proportion; when the WF has an installed capacity of a set value, the marginal operation value of wind power will drop to zero or even a negative and the current installed capacity of the WF has been saturated if viewed economically.
2) **Impact of WF capacity factor on wind power operation values**
   The capacity factor (r) of a WF is closely related to the WF's wind power parameters, WTGS parameters, landform, and so on. It directly indicates operation state and equipment utilization rate of the WF. Figure 12.13 shows the relation between installed capacity and annual operation value of the WF when the WF capacity factor is 0.25, 0.3, and 0.35, respectively. According to the figure, the larger the capacity factor, the higher the wind power operation value will be. Therefore, it is necessary in the planning of a WF to analyze address selection, WTGS selection, and WTGS layout of the WF carefully to get a high capacity factor and a higher return on investment for the WF.
3) **Impact of reserve capacity price on wind power operation values**
   When the WF of a power system has an installed capacity of a set value, the marginal operation value of wind power will drop to zero or even a negative and the installed capacity of the WF at this time is called the critical installed capacity. Figure 12.14 shows change of the critical installed capacity when the reserve capacity price has increased from 100 yuan/MW to 140 yuan/MW. At present, the economical efficiency of wind power generation needs to be improved and the competent government authorities should offer preferential policies for the auxiliary service cost of grid connection of wind power.
4) **Impact of pollutant emission cost on wind power operation values**
   Figure 12.15 describes the impact of the pollutant emission cost saved by means of replacement of a conventional energy source power plant by a WF on the critical installed capacity of the WF. Thus, pollutant emission costs are important in wind power operation value evaluations.
5) **Impact of wind power forecast deviation on wind power operation values**
   The accuracy and advanced duration of wind power forecast are in inverse proportion. Figure 12.16 shows the relation between the variance and advanced duration of wind power forecast of a 10 MW WF. Figure 12.17 shows the relation between the installed capacity and annual operation value of a WF when the wind power generation capacity forecast

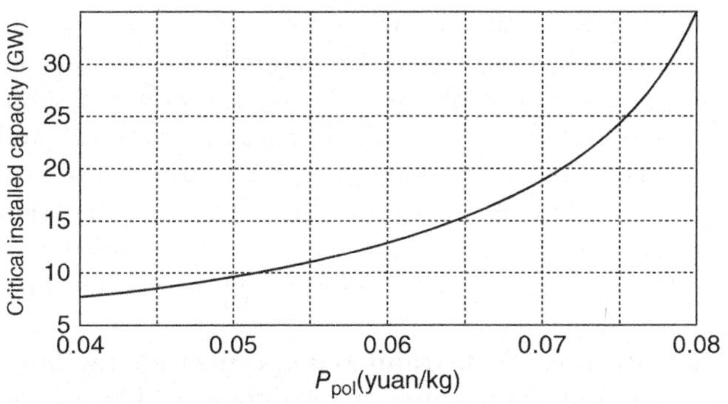

Figure 12.15 Relation between waste gas emission cost and critical installed capacity.

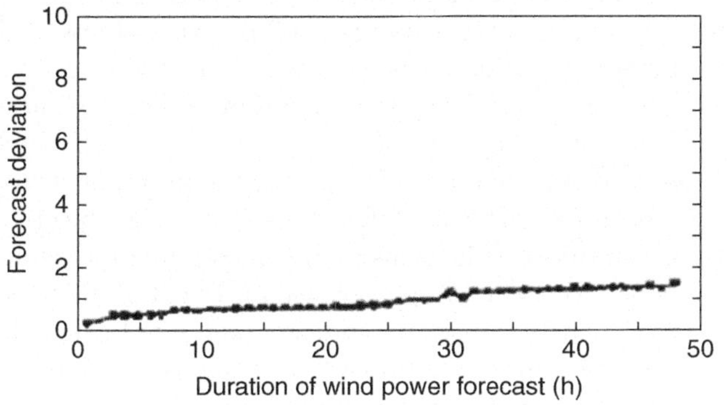

Figure 12.16 Relation between the deviation and advanced duration of wind power forecast.

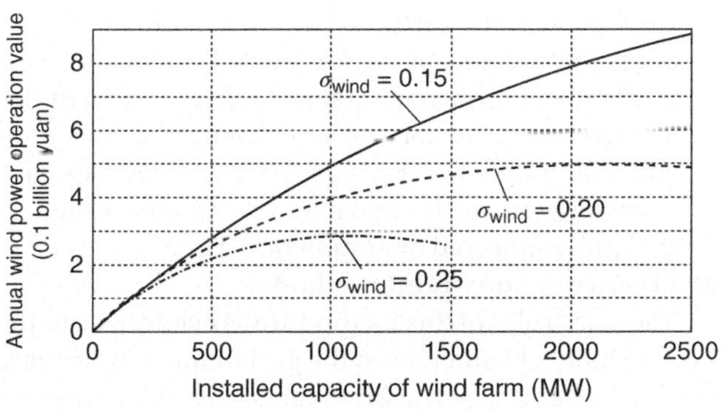

Figure 12.17 Relation between annual operation value and advanced duration of forecast of wind power.

variance is 0.15, 0.2, and 0.25, respectively. Obviously, a higher wind power forecast accuracy can improve the economical efficiency of a power system greatly.

The overall value of wind power generation decides the competitiveness of wind power generation compared with conventional energy sources and influences the growth of wind power generation too. The way to evaluate the values of wind power generation accurately so that a scientific basis can be provided for the compilation of related policies by the electric power planning authorities and other competent government authorities is a question needing to be answered in the growth of wind power. The values of wind power include power generation income, power generation cost, auxiliary service cost, environmental benefit, and so on. Factors such as wind power penetration, wind energy forecast deviation, pollutant emission cost, and WF capacity factor influence the wind power operation value significantly. Therefore, electric power planning authorities and wind power investors should consider impacts of all the factors above carefully when compiling making related decisions and take proper actions for higher wind power operation values and returns on investments.

12.4 Analysis of Maximum Penetrating Capacity of Wind Power

In Chapter 3, the maximum penetrating capacity of a grid-connected WF is discussed and the importance is attached to the relation between a WF and a local grid. The WF's short circuit capacity ratio, that is, a ratio between rated capacity P_{wind} of the WF and short circuit capacity S_{sc} of the point of common coupling (PCC) between the WF and a power system is an important index. The reason is that the short circuit capacity suggests strength of the grid structure: a large short circuit capacity suggests close connection between the node and the power system's power source point because of a small electrical distance between them. The short circuit capacity also suggests degree of sensitivity of voltage of the node to changes of the input power of wind power: a small short circuit capacity ratio (or large short circuit capacity) suggests a great capability of the power system to withstand wind power disturbance.

When the capability of a large power system to receive wind power is measured, it is necessary to consider relation between a WF and the whole power system, including constraints for the power system's frequency, operation stability, and so on. In consideration of this, the concept of wind power penetration limit is usually used. Wind power penetration refers to proportion of the installed capacity of a WF in the total load of a power system. The wind power penetration limit is defined as a ratio between the maximum wind power installed capacity connected to a power system and the maximum load of the power system while the designated technical indices are met.

The wind power penetration limit is for indicating the maximum wind power capacity which can be withstood by a grid of the specified scale. It aims at figuring out the impact of a WF on the frequency of a power system.

To determine this index, the randomness and limited control of wind power should be considered first. In both connection and disconnection of the wind power from the grid, the regulable capacity of the power system should keep the grid frequency within its permissible range.

The maximum penetrating capacity of wind power is limited by the factors above, so related conclusions should be reached based on systematic studies about specific grids in China according to the actual conditions of WFs of power systems. The main methods for analyzing the maximum penetrating capacity of a WF include time-domain simulation method, steady-state power flow simulation method, static safety constraint and optimization method, steady-state frequency constraint method, and so on. These methods have different ranges of application because they attach importance to different influential factors.

12.4.1 Method to Determine the Wind Power Penetration Limit

1) **Time-domain simulation method**
   In this method, suppose the output power level of one or several WFs, select several typical system operation modes, use the digital simulation method to check whether a power system becomes unsafe and unstable at the wind power level to determine the permissible WF connection capacity, and finally change parameters and operation mode of the power system to analyze the factors influencing the maximum penetrating capacity of wind power.
   This method is mainly used for checking the dynamic performance of a power system and probing the interaction between a WF and a power system. In particular, whether a power system can maintain its dynamic stability or not after wind power is connected into a weak grid is an important factor for determining the maximum wind power capacity, which the power system can withstand. This method has a clear physical concept and can be used for checking the impacts and interactions of multiple factors correctly as long as there is a model. However, this is a verifying calculation method, but determination of the maximum wind power, which a specific power system can withstand needs massive simulation calculation.
2) **Steady-state power flow simulation method**
   After connected to a medium- or low-voltage grid, a WF will change the power flow distributions and voltages of a power system partially. This impact was not considered in the original grid planning and design. When a grid connection plan for a WF is determined in the planning period, it is therefore necessary to probe partial node voltage changes of the power system as a result of the grid connection of a WF carefully to ensure normal operation for the WF and the power system. It is necessary to avoid line power overload and out-of-limit node voltage both because of the grid connection of wind power and prevent generation capacity losses because of abnormal shutdown of WTGSs caused by excessively high or low terminal voltages.
3) **Static safety constraint and optimization method**
   The grid connection of wind power changes the power structures and power flow distributions of power systems and influences the static safety of power systems. In addition, the instability of wind power results in a stricter requirement for the spinning reserve capacities of power systems and higher operation costs of conventional power generation systems. Thus, determination of the maximum penetrating capacity of wind power can be changed into determination of the maximum input wind power under specific safety constraints. The objective of optimization is to ensure the maximum input wind power for a WF in a power system. To realize this, the static safety constraint indices used should include power balance, line transmission power limit, upper and lower limits for node voltages, output power limits for the conventional power generation system, and so on.
   In addition to determination of the maximum input wind power, this method can provide the best operation dispatching plans for the other conventional power generation system(s) in a power system. It has a reference value for the grid-connected operation and planning of WFs. However, it does not give consideration to the dynamic constraints related to the grid-connected operation of WFs.
4) **Frequency constraint method**
   The basic rules of this method are as below: Due to the instability of wind power, the grid frequency decrease should be in a permissible range (usual permissible frequency deviation: ± 0.2 Hz) following primary frequency regulation and secondary frequency regulation by the conventional power generation system after interruption of wind power output. In addition, safety and economical efficiency of the whole grid should be guaranteed. In consideration of this, spinning reserve capacity of the conventional power generation system should not be too large but in the range (3% to 5%) set in the specification. This steady-state frequency analysis method is not related to the structures of grids but the loads of grids and the natures and structures of power sources as well. This method is necessary for the grids with smaller proportions of wind power.

12.4.2 Calculation of Maximum Penetrating Capacity of Wind Power Based on Power Balance

The power balance and peak regulation in valley time as a result of wind power randomness are two main factors curbing the grid connection of wind power. In China, there were several times of WTGS outage because of difficult peak regulation in valley time. Peak regulation is more difficult for the power systems with a reverse peak regulation characteristic for their wind power.

In wind power planning, in-depth studies about the capability of grids to connect with wind power should be done to provide a reference in the courses of taking necessary technical actions in grid operation, helping governments compile wind power development plans and reasonably determine the development time sequence.

In the analysis of the maximum penetrating capacity of wind power based on the wind power received in valley time, factors such as grid structure, power system peak regulation capability, grid peak-valley difference, and valley load should be considered. With change of the operation mode, both wind power receiving capability of the grid changes each day, so the wind power receiving capability should be reflected through a daily curve.

The current peak-valley differences are caused by load structures, power consumption modes, working systems, industrial policies, and so on. In some regions in China, WTGSs have forced outage for good frequency quality when the lowest level has been reached in the unconventional peak regulations of thermal power units with valley loads. Calculation of the maximum penetrating capacity of wind power based on power balance is a course to determine the permissible maximum wind power of a power system based on the existing grid structure and conventional power generation conditions.

12.4.2.1 Calculation Method

It is sure that the period with the maximum peak-valley difference in a year is possibly a day in summer or winter. In the peak time, the startup mode should make the spinning reserve capacity of the power source not too large but sufficient; in the valley time, all the hydropower units should shut down (except for the hydropower full-capacity periods) and the output power of all the thermal power units should be made minimum by means of unconventional peak regulation. In this way, the receivable wind power capacity corresponding to each operation mode can be calculated.

The intra-grid power and the interconnecting tie power in the peak time should be converted into grid-supplied power (forecast grid-supplied power) and compared with the forecast load. Their difference should meet the deviation range. After considering the self-supplied power and simultaneity factor, the receiavaible wind power in the valley time can be calculated by the minus of the minimum output level of other power sources from the forecast grid-supplied power [22].

12.4.2.2 Maximum Grid-Supplied Power in Peak Time

The peak power includes output power of the hydropower unit, output power of the thermal power unit, and received power of the interconnecting tie. During grid-supplied power conversion, the auxiliary power ratio and the grid loss should be deducted from output power of the units. The maximum grid-supplied power in the peak time can be calculated according to conditions of the intra-grid power and the interconnecting tie with help of the following formula (Formula (12.61)).

$$P_{\max} = (P_{the} + P_{hyd})(1 - \eta_{pla} - \eta_{net}) + P_{tie}(1 - \eta_{net}) \tag{12.61}$$

In the formula above, $P_{\max}$ is the maximum grid-supplied power; P_{the} is generation power of the thermal power unit and the nuclear power plant; P_{hyd} is generation power of the hydropower plant; η_{pla} is auxiliary power ratio of the power plants; η_{net} is loss of the transmission grid; P_{tie} is the maximum power received by the interconnecting ties.

12.4.2.3 Grid-Supplied Power in Valley Time

The minimum grid-supplied power in the valley time can be calculated through the formula below according to conditions of the intra-grid power and the interconnecting tie (including peak regulation capability of the interconnecting tie) (Formula (12.62)).

$$P_{\min} = (P_{the} + P_{hyd})(1 - \eta_{pla} - \eta_{net}) + P_{tie}\varepsilon(1 - \eta_{net}) \tag{12.62}$$

In the formula above, $P_{\min}$ is the minimum grid-supplied power in the valley time; ε is peak regulation percentage of the interconnecting tie; the other parameters are the same as those in Formula (12.59).

12.4.2.4 Receivable Wind Power Capacity

The receivable wind power capacity of a grid is obtained by deducting the intra-grid power and the interconnecting tie's exchange power from the forecast valley load (Formula (12.63)).

$$P_{wid} = (P_{for} - P_{\min})/(1 - \eta_{wpla}) \tag{12.63}$$

In the formula above, P_{wid} is the receivable wind power in the valley time; P_{for} is the forecast grid-supplied power in the valley time; η_{wpla} is the auxiliary power ratio of the WF.

The receivable wind power capacity can be calculated through the following formula according to the specific simultaneity factor (Formula (12.64)).

$$W_{cap} = P_{wid}/\lambda \tag{12.64}$$

In the formula above, W_{cap} is the receivable wind power capacity of the grid; λ is the intra-grid WF simultaneity factor.

The grid parameters in the calculation above such as thermal power unit auxiliary power ratio, grid loss, interconnecting tie grid loss, WF auxiliary power ratio, WF simultaneity factor, and interconnecting tie peak regulation capability should be determined after counting and checking. They may be different in different grids.

The annual, monthly, and daily maximum receivable wind power capacities of a grid can be obtained through the calculation and analysis above, thus providing a basis for decision making for system planning, wind power development, and operation dispatching.

12.4.3 Calculation of Maximum Penetrating Capacity of Wind Power Based on a Frequency Constraint

12.4.3.1 Mathematic Description of the Frequency Constraint Method

Suppose the installed capacity of a WF is P_{wind}; the wind power load rate is m_i (0~1); output power of the WF is m_iP_{wind}; the installed capacity of a conventional power source is P; the startup capacity of the conventional power source is P_K; proportion of the spinning reserve capacity is m_c; output power of the conventional power source is $(1 - m_c)P_K$. Then, total output of the grid will be (Formula (12.65)):

$$P_0 = m_iP_{wind} + (1 - m_c)P_K \tag{12.65}$$

The conditions allowing the maximum power shortage after disconnection of wind power: After spinning reserve capacity of the conventional power source is completely called through primary frequency regulation and secondary frequency regulation, the grid frequency decrement should not be larger than the permissible frequency deviation Δf and the corresponding power shortage should be the minus of the online capacity of the conventional power source from total output power of the grid, namely (Formula (12.66)),

$$\begin{aligned}\Delta P &= m_iP_{wind} + (1 - m_c)P_K - P_K \\ &= m_iP_{wind} - m_cP_K\end{aligned} \tag{12.66}$$

According to static frequency regulation characteristic of the load, the following formula can be reached (Formula (12.67)).

$$k = \frac{\Delta P/P_0}{\Delta f/f_0} \tag{12.67}$$

In the formula above, k is the load frequency regulation effect coefficient; P_0 is the initial electrical load (the maximum electrical load); f_0 is the initial grid frequency; ΔP is the power shortage; Δf is the permissible frequency deviation.

According to formula $\Delta P = P_0 - P_K$ and Formula (12.65), the following formula can be reached (Formula (12.68)).

$$\frac{P_K}{P_0} = 1 - k\frac{\Delta f}{f_0} \tag{12.68}$$

According to the rule that "After spinning reserve capacity of the conventional power source is completely called through primary frequency regulation and secondary frequency regulation, the grid frequency decrement should not be larger than the permissible frequency deviation Δf" and Formulas (12.65) and (12.66), the following equation can be reached (Formula (12.69)).

$$\frac{m_iP_{wind} - m_cP_K}{P_0} \le k\frac{\Delta f}{f_0} \tag{12.69}$$

According to Formulas (12.65), (12.68), and (12.69), proportion of the installed capacity of the WF in the maximum load should comply with the following equation (Formula (12.70)):

$$\frac{P_{wind}}{P_0} \le \frac{1}{m_i}\left[m_c + (1 - m_c)k\frac{\Delta f}{f_0}\right] \tag{12.70}$$

Similarly, proportion of the installed capacity of the WF in the startup capacity of the conventional power source should comply with the following equation (Formula (12.71)):

$$\frac{P_{wind}}{P_K} \le \frac{1}{m_i}\left[m_c + \frac{k\Delta f}{f_0 - k\Delta f}\right] \tag{12.71}$$

According to Formulas (12.70) and (12.71), the permissible proportion of wind power is related to the grid's steady-state operation frequency, the permissible frequency deviation, the load frequency regulation effect coefficient, the wind power load rate, the permissible spinning reserve capacity of the conventional power source, and so on.

In reference [23], an independent grid in Xinjiang is taken as an example for the calculation and analysis of WF penetration limit through a frequency constraint method. The calculation conditions are as below: power system rated frequency f_0: 50 Hz; permissible: frequency deviation Δf; load frequency regulation effect coefficient K: 3; spinning reserve capacity proportion m_c: 4%; wind power load rate m_i: 1 (for calculating the optimal proportion) or 0.5 (for calculating the maximum proportion). The calculation results are as below: optimal proportion of wind power in the maximum startup capacity of the conventional power source: $P_{wind}/P_K \le 5.21\%$; maximum proportion of wind power in the maximum startup capacity of the conventional power source: $P_{wind}/P_K \le 10.42\%$; optimal proportion of wind power in the maximum load when the spinning reserve capacity is included: $P_{wind}/P_0 \le 5.152\%$; maximum proportion of wind power in the maximum load when the spinning reserve capacity is included: $P_{wind}/P_0 \le 10.304\%$.

According to the calculation results above, we can reach the following conclusions: ① The maximum proportion and optimal proportion of wind power do not increase with increase of the capacity of a conventional power source; ② The proportions of wind power are related to factors such as the grid's steady-state operation frequency, the permissible frequency deviation, the load frequency regulation effect coefficient, the wind power load rate, and the permissible spinning reserve capacity of the conventional power source of which the last three are critical and the load frequency regulation effect coefficient is difficult to be determined because it changes with the specific load type of the power system; ③ In the frequency constraint method, the steady-state frequency is constrained, so it does not relate to the grid structure but the grid load and the power source nature.

12.4.4 Calculation of Maximum Penetrating Capacity of Wind Power Based on Dependent Programming Theory

WTGSs are special power generation equipment. For environmental protection and renewable energy source utilization, it is expected to develop wind power generation as much as possible. However, such characteristics of wind power as changeability and instability make the grid-connected operation of wind power a threat to the safe and stable operation of grids. This threat to a power system is not obvious when the installed capacity of a WF is small. However, with increase of the installed capacity of the WF, this threat is more serious because it adversely affects safe and stable operation of the power system and results in an economic loss. Thus, it is necessary to determine a maximum penetrating capacity of wind power for a specific grid in WF planning and design and practical operation of the power system.

To determine the reasonable scale and maximum capacity of a WF, calculation and systematic analysis methods in line with actual conditions of the related power system should be adopted. Reasonable scale and maximum capacity of the WF depend on operation characteristics of the WF, regulation capabilities of other power generation equipment in the power system, structure of the grid into which wind power is connected, and so on. The determination of maximum penetrating capacity of wind power can be changed into mathematic optimization based on specific safety constraints and mathematic methods such as linear programming [24], chance constrained programming [25], and approximately linear programming [26].

This section sums up the key constraints for grid connection of wind power. On this basis, it changes the determination of the maximum penetrating capacity of wind power into multi-objective optimization based on specific safety constraints and, based on dependent-chance objective programming theory, establishes a calculation method for the maximum penetrating capacity of wind power [27].

12.4.4.1 Key Constraints for Grid Connection of Wind Power

The grid connection of massive wind power will result in some new problems for a power system. The grid connection of wind power has the following main constraints [11, 28, 29]:

1) **Power flow constraint**

The grid structure of a power system is planned and designed according to locations of the power sources and the loads. It should enable the power system to operate normally and, essentially, safely. After a WF is connected to the grid, power source distribution of the power system will be changed. Even if the loads remain unchanged, power flow mode of the power system will change. This change is not considered in the planning stage. Thus, it is significant to probe the impact of grid connection of wind power on power flows.

In order to figure out the impacts of wind power generation on the operation of a power system, massive power flow calculation is needed in planning, design and operation mode studies. The common power flow calculation method needs massive calculation and is difficult to present all actual conditions of the power system. In the probability power flow calculation method, the uncertain factors are analyzed through a probability theory, making it possible to find weaknesses of the power system and probability distributions of operating voltage and branch power flows of the power system.

After WTGSs of a certain capacity are connected to the power system, the probability of out-of-limit power flow can be expressed below (Formula (12.72)):

$$\Pr\{h_1(x,y,\xi) \le 0\} = c_1 \tag{12.72}$$

In the formula above, x is the wind power connected to the power system; y is the power system state vector; ξ is a random wind velocity; $\Pr\{h_1(x,y,\xi) \le 0\}$ is the probability of no out-of-limit power flow of the power system after wind power is connected to it; c_1 is result of the probability power flow calculation (see Section 7.4.2 and reference [30] for the specific calculation).

2) **Voltage stability constraint**

The instability due to grid connection of wind power is mainly voltage instability. Grid connection of wind power has the following main impacts on the voltage stability of a power system: ① The reactive power compensation modes of cage type asynchronous

WTGSs are capacitor compensation and the degrees of compensation are in direct proportion to squares of the voltages of the connection points: If voltage of the power system has dropped greatly, the degrees of reactive power compensation will drop greatly too, but the reactive demand of the WF for the grid will increase. As a result, voltage of the power system will drop continually or even collapse and result in WTGS outage. ② For variable-speed constant-frequency WTGSs, their reactive power can be controlled through a power electronic technology, but their reactive power on the stators can be regulated in a limited range only and is affected by the rotor speeds and capacities of the power electronic converters on the rotor sides. ③ When the power system is still stable after failures and control operation, some WTGSs will trip off because of their low voltage protection and active output power of the WF will decrease correspondingly. ④ If any failure is not eliminated in time, there will be transient voltage instability. ⑤ The WF is usually far from a load center or connected to a weak power system.

With increase of the grid-connected capacity of wind power, voltage instability is increasingly worse. In some regions, it is now a bottleneck curbing the development of wind power. In reference [28], the following formula is given for calculating the voltage stability margins of WF power systems (Formula (12.73)).

$$\Pr\{h_2(x,y,\xi) \le 0\} = c_2 \qquad (12.73)$$

In the formula above, x is the wind power connected to the power system; y is state variable of the power system; ξ is a random wind velocity; $\Pr\{h_2(x,y,\xi) \le 0\}$ is the probability for the power system to be in the voltage stability safe area after wind power is connected to it; c_2 is calculation result of the voltage stability margin.

3) **Economic constraint**

In spite of many preferential policies, wind power is still curbed to some extent because of its high-power generation cost and auxiliary service cost. As described in section 12.3, then the installed capacity of a WF is small, the overall value and installed capacity of the WF are in direct proportion; when the WF has an installed capacity of a set value, the overall value of wind power will not increase any more or even begin to decrease greatly and the current installed capacity of the WF has been saturated if viewed economically.

In the calculation of overall value of wind power through Formula (12.60), factors such as the power generation income, power generation cost, auxiliary service cost, and environmental benefit of wind power should be considered. The overall value of wind power decides the competitiveness of wind power compared with conventional energy sources and the development speed of wind power generation as well.

12.4.4.2 Calculation of Maximum Wind Power Capacity Based on Dependent-Chance Objective Programming

Calculation of the maximum penetrating capacity of a WF depends on many factors. To determine the reasonable scale and maximum capacity of a WF, calculation and systematic analysis methods in line with actual conditions of the related power system should be adopted. Reasonable scale and maximum capacity of the WF depend on operation characteristics of the WF, regulation capabilities of other power generation equipment in the power system, structure of the grid into which wind power is connected, and so on.

According to the analysis above, the maximum connection capacity of a WF has the following main constraints: ① The grid connection of massive wind power will make some node voltages and line power flows out of limit; ② After grid connection of wind power, the power system will have a smaller voltage stability margin; ③ The auxiliary service cost increased because of grid connection of wind power will reduce economical efficiency of the power system.

The determination of maximum penetrating capacity of wind power can be changed into mathematic optimization of the three constraints above. The mathematic programming method for the determination should be characterized as below: ① Constraints are uncertain; ② The optimization objectives are the probabilities for some events to be established; ③ There should be multiple objectives. In reference [31], a dependent–chance programming (DCP) model is established for maximize the chance functions of some events in uncertain environmental constraints. This model complies with the characteristics completely.

According to the dependent–chance programming theory, if the wind power connected to the power system and other manually regulable or changeable physical quantities are regarded as control variables, the physical quantities obtained based on the control variables, such as node voltage amplitude and the phase angle, are regarded as state variables, and the formulas such as the power flow formula of the power system are regarded as constraint formula, the maximum penetrating capacity of the WF can be calculated through the following mathematic formula (Formula (12.74)).

$$\begin{cases} \min\left(P_1 d_1^- + P_2 d_2^- - P_3 d_3^+\right) \\ s.t. \\ \Pr\left\{h_1\left(x,y,\xi\right) \le 0\right\} + d_1^- - d_1^+ = b_1 \\ \Pr\left\{h_2\left(x,y,\xi\right) \le 0\right\} + d_2^- - d_2^+ = b_2 \\ h_3\left(x,y,\xi\right) + d_3^- - d_3^+ = b_3 \\ g\left(x,y,\xi\right) \le 0 \\ d_i^-, d_i^+ \ge 0, i = 1,2,.3 \end{cases}, \qquad (12.74)$$

In the formula above, P_1, P_2 and P_3 are priority factors suggesting relative importance of the objectives (they are $P_1 \gg P_2 \gg P_3$ here); x is the wind power connected to the power system; y is state variable of the power system; ξ is a random wind velocity; $\Pr\{h_1(x, y, \xi) \leq 0\}$ is the overall value of wind power; d_i^+ is the positive deviation of objective i from the target value; d_i^- is the negative deviation of objective i from the target value; $g(x, y, \xi) \leq 0$ is the power flow constraint; b_1 is the expected probability of no out-of-limit power flow after grid connection of wind power; b_2 is the expected probability for the power system to be in the voltage stability safe area after wind power is connected to it; b_3 is the expected overall value of wind power.

Procedure of calculation of the maximum penetrating capacity of wind power based on the dependent–chance programming theory:

1) Determine main constraints and optimization objectives of a power system with a WF.
2) Change the standard model about dependent-chance objective programming into an easier chance function with help of the dependent programming concept in document [31].
3) Input basic parameters of the genetic algorithm.
4) Carry out basic operation of the genetic algorithm based on heredity and variation till the optimal solution or the maximum iteration number is obtained.

12.4.4.3 Example Analysis

In reference [27], the algorithm proposed is checked through a example that is an IEEE 39-node system with 10 generators, 39 nodes, 12 transformers, 34 lines, and total installed capacity of 6,400 MW. In order to check the impacts of wind power generation on the power system comprehensively and determine maximum installed capacity of the power system, some WTGSs are installed at nodes 30, 34, 36, and 37.

According to the analysis above, the determination of maximum penetrating capacity of wind power can be changed into mathematic optimization of the following three objective functions (ranked by importance).

1) The probability of out-of-limit power flow of the power system after the grid connection of wind power should not be higher than 3%.
2) The probability of a voltage stability margin smaller than 0.15 after the grid connection of wind power should not be higher than 10%.
3) The higher the overall value of wind power, the better.

In the first solution obtained after initialization of the genetic algorithm, the installed capacities of nodes 30, 34, 36, and 37 are 952 MW, 443 MW, 14 MW, and 375 MW, respectively; the probability of no out-of-limit power flow is 0.66; the probability for the power system to be in the voltage stability safe area is low, failing to meet the requirements for grid connection of wind power.

Table 12.4 Impact of average wind velocity of a WF on the maximum penetrating capacity of the WF.

Average wind velocity of a WF (m/s)	Proportion of maximum penetrating capacity of the WF in maximum load of the power system (%)
7.4	10.4
7.5	10.7
7.6	11.1

Most of the first-generation samples do not comply with the planning of grid connection of wind power, but the samples have been stabilized at the optimal solution after evolution of 50 generations based on crossover and mutation for a parent population with excellent samples selected through a roulette way. The current installed capacity of wind power accounts for 10.7% of the maximum load of the power system.

Table 12.4 describes the impact of the average wind velocity of a WF on the maximum penetrating capacity of the WF. Obviously, the larger the average wind velocity of the region the WF, the more installed capacity of the WF the power system can receive. It is because that a larger wind velocity corresponds to a better economical efficiency when the installed capacity is constant, thus increasing the installed capacity of the WF.

Table 12.5 describes the impact of the on-grid price of wind power on the maximum penetrating capacity of a WF. According to the table, the economical efficiency of wind power generation needs to be improved.

In the analysis above, the determination of maximum penetrating capacity of wind power is first changed into the multi-objective optimization of constraints such as voltage stability, probability power flow, and economical efficiency and then calculated by means of optimization based on the dependent–chance programming theory. In a practical power system, the key constraints for grid connection of wind power may be different and

Table 12.5 Impact of the on-grid price of wind power on the maximum penetrating capacity of a WF.

On-grid price of wind power (yuan/MWh)	Proportion of maximum penetrating capacity of the WF in maximum load of the power system (%)
690	9.4
695	10.1
700	10.7
705	11.5

should be determined through analysis. According to the example analysis above, the dependent–chance objective programming method makes it easy to consider the constraints for grid connection of wind power in uncertain environmental conditions comprehensively and obtains the maximum penetrating capacity of wind power directly based on random changes of factors such as the wind velocity. Thus, this method is highly applicable.

12.4.5 Actions for Improving Wind Power Penetration

According to domestic and foreign development of wind power generation, wind power penetration is restricted by many factors. From the point view of wind power systems, wind power penetration is affected by wind power forecast, controllability of WTGSs and WFs, correlation between wind power and loads, layouts of WFs in power systems, and so on. From the point view of power systems, wind power penetration is affected by grid strengths, flexibility, and controllability of conventional power plants, reserve capacities of power systems, load characteristics, and so on. Wind is a changeable resource, but it has been found in previous wind power operation and detailed studies about the grid connection of wind powers that numerous reliable and strict technical limitations for wind power can be regulated through grids. In some countries, massive wind power has been received. In Denmark, about 20% of the electric power it needs (43% of the peak load) is wind power. The wind power penetration in Germany is 70% (30% of the peak load) and 11% (30% of the peak load) in both Spain and Portugal. In Ireland, there is an independent system with 9% of wind power penetration (11% of the peak load). The increase of wind power penetration is not technically limited but possibly economically. In the eyes of some people, the cost of increase of wind power penetration is higher than value of the wind power connected to power systems. However, according to the operation experience and detailed grid connection studies of some power systems containing massive wind power in the world, the cost increase as a result of wind power regulation is proper and the additional benefit of more wind energy does not decrease greatly with the increase of wind power. To make wind power penetration 20% to 30%, the main factors to be considered should be economic conditions rather than physical conditions. A main problem needing to be solved is that the wind power integrated power systems have higher changeability and uncertainty [32–34].

The key and first step to confirm and solve the problem is to understand and quantize the impacts of WFs on power systems. In the design and operation of WFs and power systems and the market rule making, different steps can be followed according to the specific factors curbing wind power penetration to improve the capabilities of power systems to receive wind power capacities. These steps include:

1) **Improve the WTGS and WF model**
   Good output power model, performance model, and output power forecast of WFs are necessary for correct understanding of the impacts (on grids) and interactions of WFs and a precondition for problem confirming and solving as well.
2) **Improve the operation characteristics of WFs**
   The existing large and medium modern WFs are usually connected with grids and expected to have control capabilities similar to those of conventional power plants. For example, they are expected to maintain the operation of WTGSs during grid failures (provide fault ride through); their WTGSs should operate in the set frequency ranges; they should be capable of controlling active power during frequency changes; they should be capable of limiting power increase rates (power ramp rates); they should transmit or absorb reactive power (reactive power control) upon request of power systems; they should adjust reactive power according to grid measurements to support grid voltages (voltage control), and so on. These WFs must provide the SCADA information necessary for grid operators. In addition, the WTGSs with power electronic control should regulator response and inertia response. According to the stability studies based on a mature double-fed asynchronous generator model, modern WFs can realize power oscillation damping, support voltage recovery after failures, and improve power system performance. Their capabilities to provide voltage control for the weak nodes of power systems and inertia response for the power systems of which the reliable operation possibly mainly depending on stability constraints are important.
3) **Improve stability of the balanced power of power systems**
   If an extra wind power capacity is added, a conventional power generation system will need better regulation, load tracking and quick start, and the optimal combination of different power generation forms in the power system will change with increase of the wind power capacity in the power system.
4) **Evaluate the impacts of grid connection of wind power carefully**
   In the interested time ranges (such as s, min, and h), it is necessary to understand degree and frequency of load variation of a power system before and after wind power generation, so as to determine the extra power generation combination of balanced power.

5) **Integrate wind power forecasts to grid control center**
Find the operation impact with the highest cost in the combined time frame. At present, the output power forecasts of WFs are important chances to reduce the uncertainty related expenses and risks in the current time frame.
6) **Make better use of available transmission capacities (compared with contracts)**
According to the hourly line load analysis, the annual heavy-load duration of a line is limited hours only. A strong and flexible power transmission system enables the remaining capacity in the low-load operation of a line to be used by other transactions. This is realizable through small changes for the current habits.
7) **Upgrade and extend power transmission systems**
The best wind sources of a country are possibly in remote regions. In this case, new power transmission lines and corridors are needed to transmit the wind sources to markets. Thus, it is significant to extend power transmission systems to make better use of the existing transmission grids to adapt to the increasingly more wind power.
8) **Develop and extend pre-hour and pre-hour open markets with a good development function**
According to the worldwide operation experience, mature, transparent, pre-hour and pre-hour markets are the most economical method to provide the balanced energy necessary for wind power output changeability. It is important for WFs to participate in the pre-hour market to integrate and balance the outputs of WFs in broad geographical regions to minimize the costs of power systems.
9) **Enlarge the balanced areas to larger utilities or resource bases by means of dynamic dispatching or ACE sharing**
According to numerous statistical data, it is useful to integrate power generation forms and loads in broad geographical regions because the diversity of loads reduces amplitudes of the peak loads related to the installed generators and the diversity of wind power reduces the degree and frequency of changeability. It decreases operation hours of the generators with higher dispatching costs and lowers the requirements for operation reserves.
10) **Use energy storage technologies**
In consideration of randomness and, particularly, reverse peak regulation, of wind power, establish peak regulation power plants (such as pumped storage power stations), or big distributed ESSs to change valley electricity into peak electricity to smooth the power curves of grid connection of wind power.
11) **Strengthen management of the demand sides**
Participation and cooperation by the demand sides, such as regulations of the durations and modes of power consumption, are useful for the effective operation of wind power.
In a word, there are many choices for handling the problems as a result of wind power penetration increase. We can know more about the ways to adapt to future wind power penetration from the previous and existing studies and the increasingly more operation experience. The continuous improvement of the capability of WFs to operate as conventional power plants and the intelligentization of power systems will bring us more flexible power systems. With the increase of our understanding for wind power penetration increase and the accumulation of our practical operation experience, the impacts of wind power on power systems and the impacts of changeability and uncertainty of wind power can be solved by means of cost reduction.

References

1 Billinton, Roy and Bai, Guang. 2004. Generating capacity adequacy associated with wind energy. *IEEE Transactions on Energy Conversion*, **19**(3), 641.

2 Shuo, Zhang, Gengyin, Li, Ming, Zhou, and Wei, Liu. 2009. Reliability modeling of large-scale WFs. *Power System Technology*, **33**(13), 37–41, 58.

3 Shuyong, Chen, Huizhu, Dai, Xiaomin, Bia, and Xiaoxin, Zhou. 2000. Reliability model of wind power plants and its application. *Proceedings of the CSEE*, **20**(3), 26–29.

4 Ruimin, Zheng, Jianhua, Li, Zuohong, Li, and Yihua, Liu. 2008. Modeling of large-scale wind farms in the probabilistic power flow analysis considering wake effects. *Journal of Xi'an Jiaotong University*, **42**(12), 1515–1520.

5 Yichun, Wu and Ming, Ding. 2004. Reliability assessment of wind farm. *Electric Power*, **37**(5), 81–84.

6 Hu, P., Karki, R., and Billinton, R. 2009. Reliability evaluation of generating systems containing wind power and energy storage. *IET Gener. Transm. Distrib.*, **3**(8), 783–791.

7 Tande, John Olave G. 2007. Impact of large scale wind power on system adequacy in a regional power system with weak interconnections. Power Engineering Society General Meeting, 2007, IEEE, June 24–28, 2007, Tampa, Florida.

8 Soder, L. and Amelin, M. 2008. A review of different methodologies used for calculation of wind power capacity credit. Power and Energy Society General Meeting - Conversion and Delivery of Electrical Energy in the 21st Century, July 20–24, pp. 1–5.

9 Ensslin, Cornel, Milligan, Michael, Holttinen, Hannele, O'Malley, Mark, and Keane, Andrew. 2008. Current methods to calculate capacity credit of wind power, IEA Collaboration, last accessed November 20, 2017, http://www.ieawind.org.

10 Amelin, Mikael. 2009. Comparison of capacity credit calculation methods for conventional power plants and wind power. *IEEE Transactions on Power Systems*, **24**(2), 685.

11 HaiChao, Wang, Zongxiang, Lu, and Shuangxi, Zhou. 2005. Research on the capacity credit of wind energy resources. *Proceedings of the CSEE*, **25**(10), 103–106.

12 Feijoo, A.E, Cidras, J, and Dornelas, J.L.G. 1999. Wind speed simulation in wind farms for steady-state security assessment of electrical system. *IEEE Trans. Energy Conversion*, **14**(4), 1582–1558.

13 Billinton, R., Kumar, S., and Chowhury, N. 1990. A reliability test system for educational use-basic results. *IEEE Transactions on Power System*, **5**(1), 319–325.

14 Amelin, Mikael. Comparison of capacity credit calculation methods for conventional power plants and wind power. *IEEE Transactions on Power Systems*, **24**(2), 685.

15 Karki, Rajesh. 2004. wind power in power generation planning. CCECE, Niagara Falls.

16 Ackermann, Thomas. 2005. *Wind Power in Power Systems*. New York: John Wiley Sons, Ltd.

17 Shuangxi, Zhou, Haichao, Wang, and Shousun, Chen. 2006. Analysis on operation value of wind power resources. *Power System Technology*, **30**(14), 98–102.

18 Harry, S. and Alex, P. 1999. Competitive procurement of ancillary services by an independent system operator[J]. *IEEE Transactions on Power System*, **14**(2), 498–504.

19 Nabe, Christian A. 2002. Approach to the evaluation of the use of ancillary services by wind generators—the example of Germany[C].The world wind energy conference and exhibitions, Berlin.

20 Lunnong, Tan and Baohui, Zhang. 2003. Reserve capacity for load variation in power market environment. *Electric Power Automation Equipment*, **23**(6), 10–12.

21 Lunnong, Tan. 2003. Problem for security and economy of power system in power market environment[J]. *Journal of University of Hydraulic and Electric Engineering/yichang*, **25**(4), 334–337.

22 Zhiming, Wang, Anlong, Su, and Shun, Lu. 2010. Analysis on capacity of wind power integrated into liaoning power grid based on power balance. *Automation of Electric Power Systems*, **34**(10), 86–90.

23 Lianqing, Song, Jinwu, He, Guang, Yan, Feng, Zhang, and Qin, Chao. 2009. Discussion on the penetration limit of the wind farm in grid. *Renewable Energy Resources*, **27**(3).

24 Yazhou, Lei, Weisheng, Wang, and Yonghua, Yin, et al. 2001. An optimization method for determining wind power penetration limit in power system under static security constraints. *Proceedings of the CSEE*, **21**(6), 25–28.

25 Yazhou, Lei, Weisheng, Wang, and Yonghua, Yin, et al. 2002. Wind power penetration limit calculation based on chance constrained programming. *Proceedings of the CSEE*, **22**(5), 32–35.

26 Guoqiang, Zheng, Hai, Bao, and Shuyong, Chen. 2004. Amending algorithm for wind farm penetration optimization based on approximate linear programming method. *Proceedings of the CSEE*, **24**(10), 68–71.

27 Haichao, Wang, Shuangxi, Zhou, and Zongxiang, Lu. 2009. Calculation of wind power penetration level based on dependent-chance goal programming. *Modern Electric Power*, **26**(3).

28 Haichao, Wang, Jianfeng, Dai, and Shuangxi, Zhou, et al. 2006. Voltage stability margin model for power systems containing wind energy resources. *Journal of Tsinghua University(Science and Technology)*, **46**(7), 1185–1188.

29 Haichao, Wang, Shuangxi, Zhou, Zongxiang, Lu, and Junling, Wu. 2005. A joint iteration method for load flow calculation of power system containing unified wind farm and its application. *Power System Technology*, **29**(18), 59–62.

30 Ming, Ding, Shenghu, Li, and Kai, Huang. 2001. Probabilistic load flow analysis based on monte-carlo simulation. *Power System Technology*, **25**(11), 10–14, 22.

31 Baoding, Liu, Ruiqing, Zhao, and Gang, Wang. 2003. *Uncertainty Planning and Application*. Beijing: Tsinghua University Press.

32 Smith, J. Charles, Milligan, Michael R., DeMeo, Edgar A., and Parsons, Brian. 2007. Utility wind integration and operating impact state of the art. IEEE Transactions on Power Systems, **22**(3), 900–908.

33 Singh, Bharat and Singh, S. N. 2009. Wind power interconnection into the power system: A review of grid code requirements. **22**(5), 1040–6190.

34 Milligan, Michael, Porter, Kevin, DeMeo, Edgar, Denholm, Paul, Holttinen, Hannele, Kirby, Brendan, Miller, Nicholas, Mills, Andrew, O'Malley, Mark, Schuerger, Matthew, and Soder, Lennart. 2009. Wind power myths debunked. *IEEE Power & Energy Magazine*, **7**(6), 89–99.

Index

Integration of Large Scale Wind Energy with Electrical Power Systems in China, First Edition. Zongxiang Lu and Shuangxi Zhou.

Wiley The manufacturer's authorized representative according to the EU General Product Safety Regulation is Wiley-VCH GmbH, Boschstr. 12, 69469 Weinheim, Germany, e-mail: Product_Safety@wiley.com.

Printed and bound by CPI Group (UK) Ltd, Croydon, CR0 4YY
15/09/2026
02217277-0001